CONTENTS

CONTENTS

CONTENTS

CONTENTS

CONTENTS

INTRODUCTION

Welcome to the 2000 edition of the *Annual Abstract of Statistics.* This compendium draws together statistics from a wide range of official and other authoritative sources. Their help is gratefully acknowledged.

Regional information, supplementary to the national figures in this *Annual Abstract,* is published in *Regional Trends,* published by The Stationery Office (TSO).

The five Parts of the *Annual Abstract,* illustrated in the contents, align with their equivalents in the *Britain 2000: The Official Yearbook of the United Kingdom,* so that the two books can be consulted as complementary volumes, painting a picture of the UK in words and figures.

Current data for many of the series appearing in this *Annual Abstract* are contained in other ONS publications, such as *Economic Trends, Monthly Digest of Statistics, Population Trends, Health Statistics Quarterly* and *Financial Statistics.* All are published by TSO.

The name and telephone number of the organisation providing the statistics are shown under each table. In addition, a list of Sources is given at the back of the book, which sets out the official publications or other sources to which further reference can be made.

All the data series published in the *Annual Abstract* are contained on a database, and almost all are stored with a four letter identification code (e.g. LUJS). These codes appear at the start of most columns or rows of data and can be quoted if you contact us requiring any further information about the data.

Definitions and Classification

Time Series. So far as possible annual totals are given throughout, but quarterly or monthly figures are given where these are more suitable to the type of series.

Explanatory notes. Most sections of the are preceded by explanatory notes which should be read in conjunction with the tables. Definitions and explanatory notes for many of the terms occurring in the *Annual Abstract* are also given in the *Annual Supplement* to the *Monthly Digest of Statistics,* published annually in the January editions. Detailed notes on items which appear in both the *Abstract* and *Financial Statistics* are given in an annual supplement to the latter entitled *Financial Statistics: Explanatory Handbook.* The original sources listed in the Sources may also be consulted.

Standard Industrial Classification. A Standard Industrial Classification (SIC) was first introduced into the United Kingdom in 1948 for use in classifying business establishments and other statistical units by the type of economic activity in which they are engaged. The classification provides a framework for the collection, tabulation, presentation and analysis of data about economic activities. Its use promotes uniformity of data collected by various government departments and agencies.

Since 1948 the classification has been revised in 1958, 1968, 1980 and 1992. One of the principal objectives of the 1980 revision was to eliminate differences from the activity classification issued by the Statistical Office of the European Communities (Eurostat) and entitled 'Nomenclature générale des activités économiques dans les Communautés Européennes', usually abbreviated to NACE. In 1990, the European Communities introduced a new statistical classification of economic activities (NACE Rev 1) by regulation. The regulation made it obligatory for the UK to introduce a new Standard Industrial Classification SIC(92), based on NACE Rev 1.

UK SIC(92) is based exactly on NACE Rev 1 but, where it is thought necessary or helpful, a fifth digit has been added to form subclasses of the NACE 1 four digit system. There are 17 sections, 16 subsections, 60 divisions, 222 groups, 503 classes and 253 subclasses. Full details are available from *UK Standard Industrial Classification of Economic Activities 1992* (TSO 1997, price £25.00) and *Indexes to the UK Standard Industrial Classification of Economic Activities 1992* (TSO 1997, price £22.50).

Regional classification. Some tables have been reclassified using the Government Office Regions. This has changed from the Standard Statistical Regions. For further advice please contact the Office for National Statistics Geographic Support Service (01329 813536).

Revisions to contents

Revisions. Some of the figures, particularly for the latest year, are provisional and may be revised in a subsequent issue of the *Annual Abstract.*

This edition of the *Abstract* has again seen a number of changes to the content. The changes are numerous, with most tables being amended at least slightly. The more significant changes are outlined below:

Chapter 14 - Transport and communications. The table titled *International seaborne trade of the United Kingdom* has been removed on advice that the data are not reliable.

Chapter 20 - Agriculture, fisheries and food. The tables titled *Average weekly earnings and hours of agricultural workers* and *Distribution of regular whole-time hired men in agriculture by earnings band* have been removed because the data are no longer available.

Chapter 22 - The tables titled *UK banks: liabilities and assets outstanding, Discount houses, Collecting societies* and *Friendly societies* have been removed because the data are no longer available.

INTRODUCTION

Symbols and conventions used

Change of basis. Where consecutive figures have been compiled on different bases and are not strictly comparable, a footnote is added indicating the nature of the difference.

Units of measurement. The various units of measurement used are listed opposite the inside back cover.

Rounding of figures. In tables where figures have been rounded to the nearest final digit, the constituent items may not add up exactly to the total.

Symbols. The following symbols have been used throughout:

.. = not available (also information suppressed to avoid disclosure).

- = nil or less than half the final digit shown.

† = revised figure: the period marked is the earliest in table.

Contact point

The Editor welcomes any feedback on the content of the *Annual Abstract*, including comments on the format of the data and the selection of topics. Comments and requests for general information should be addressed to:

Daniel Wisniewski
Social and Regional Division
Room D.136
Office for National Statistics
Government Buildings
Cardiff Road
Newport
South Wales NP10 8XG

or

E-mail: annual.abstract@ons.gov.uk

January 2000

Area

1

Area

The United Kingdom comprises Great Britain and Northern Ireland. Great Britain comprises England, Wales and Scotland only.

Physical Features

The United Kingdom (UK) constitutes the greater part of the British Isles. The largest of the islands is Great Britain. The next largest comprises Northern Ireland and the Irish Republic. Western Scotland is fringed by the large island chain known as the Hebrides, and to the north east of the Scottish mainland are the Orkney and Shetland Islands. All these, along with the Isle of Wight, Anglesey and the Isles of Scilly, have administrative ties with the mainland, but the Isle of Man in the Irish Sea and the Channel Islands between Great Britain and France are largely self-governing, and are not part of the United Kingdom. The UK is one of the 15 member states of the European Union (EU).

With an area of about 244 000 sq km (94 000 sq miles), the United Kingdom is just under 1 000 km (about 600 miles) from the south coast to the extreme north of Scotland and just under 500 km (around 300 miles) across at the widest point.

- Highest mountain: Ben Nevis, in the highlands of Scotland, at 1 343 m (4 406 ft)
- Longest river: the Severn, 354 km (220 miles) long, which rises in central Wales and flows through Shrewsbury, Worcester and Gloucester in England to the Bristol Channel
- Largest lake: Lough Neagh, Northern Ireland, at 396 sq km (153 sq miles)
- Deepest lake: Loch Morar in the Highlands of Scotland, 310 m (1 017 ft) deep
- Highest waterfall: Eas a'Chual Aluinn, from Glas Bheinn, in the highlands of Scotland, with a drop of 200 m (660 ft)
- Deepest cave: Ogof Ffynnon Ddu, Wales, at 308 m (1 010 ft) deep
- Most northerly point on the British mainland: Dunnet Head, north-east Scotland
- Most southerly point on the British mainland: Lizard Point, Cornwall
- Closest point to mainland continental Europe: Dover, Kent. The Channel Tunnel, which links England and France, is a little over 50 km (31 miles) long, of which nearly 38 km (24 miles) are actually under the Channel.

1.1 Area of the United Kingdom[1]

At 31st March 1981

	Metric measure					
	Thousand hectares			Square kilometres		
	Total	Land	Inland water[2]	Total	Land	Inland water[2]
United Kingdom	24 410	24 088	322	244 101	240 883	3 218
Great Britain	22 996	22 738	258	229 957	227 377	2 580
England and Wales	15 117	15 028	89	151 168	150 280	888
England	13 041	12 965	76	130 410	129 652	758
Wales	2 076	2 063	13	20 758	20 628	130
Scotland	7 879	7 710	169	78 789	77 097	1 692
Northern Ireland[3]	1 416	1 353	63	14 160	13 532	628

	Imperial measure					
	Thousand acres			Square miles		
	Total	Land	Inland water[1]	Total	Land	Inland water[1]
United Kingdom	60 319	59 523	796	94 248	93 006	1 242
Great Britain	56 824	56 186	638	88 787	87 791	996
England and Wales	37 355	37 136	219	58 366	58 023	343
England	32 225	32 037	188	50 351	50 058	293
Wales	5 129	5 097	32	8 015	7 965	50
Scotland	19 469	19 051	418	30 420	29 767	653
Northern Ireland[3]	3 493	3 339	156	5 467	5 225	249

	Standard regions											
	United Kingdom	North	Yorkshire & Humberside	East Midlands	East Anglia	South East	South West	West Midlands	North West	Wales	Scotland	Northern Ireland
Thousand Square Kilometres	244.1	15.4	15.4	15.6	12.6	27.2	23.8	13.0	7.3	20.8	78.8	14.0

1 Based on 1:50 000 digital information for 1984 Boundaries Commission.
2 Excluding tidal water.
3 Excluding certain tidal waters that are parts of statutory areas in Northern Ireland.

Sources: Ordnance Survey of Northern Ireland; Office for National Statistics; Ordnance Survey

Parliamentary elections

2

2.1 Parliamentary elections[1]

Thousands

	25 Oct 1951	26 May 1955	8 Oct 1959	15 Oct 1964	31 Mar 1966	18 June 1970[1]	28 Feb 1974	10 Oct 1974	3 May 1979	9 June 1983	11 June 1987	9 April 1992	1 May 1997[3]
United Kingdom													
Number of electors	34 919	34 852	35 397	35 894	35 957	39 615	40 256	40 256	41 573	42 704	43 666	43 719	43 830
Average-electors per seat	55.9	55.3	56.2	57.0	57.1	62.9	63.4	63.4	65.5	66.7	67.2	67.2	67.1
Number of valid votes counted	28 597	26 760	27 863	27 657	27 265	28 345	31 340	29 189	31 221	30 671	32 530	33 551	31 286
As percentage of electorate	*81.9*	*76.8*	*78.7*	*77.1*	*75.8*	*71.5*	*77.9*	*72.5*	*75.1*	*71.8*	*74.5*	*76.7*	*71.4*
England and Wales													
Number of electors	30 626	30 591	31 109	31 610	31 695	34 931	35 509	35 509	36 695	37 708	38 568	38 648	38 701
Average-electors per seat	56.5	55.9	56.9	57.8	57.9	63.9	64.3	64.3	66.5	67.2	68.8	68.8	68.6
Number of valid votes counted	25 356	23 570	24 619	24 384	24 116	24 877	27 735	25 729	27 609	27 082	28 832	29 813	27 679
As percentage of electorate	*82.8*	*77.0*	*79.1*	*77.1*	*76.1*	*71.2*	*78.1*	*72.5*	*75.2*	*71.8*	*74.8*	*77.5*	*71.5*
Scotland													
Number of electors	3 421	3 388	3 414	3 393	3 360	3 659	3 705	3 705	3 837	3 934	3 995	3 929	3 951
Average-electors per seat	48.2	47.7	48.1	47.8	47.3	51.5	52.2	52.2	54.0	54.6	55.5	54.6	54.9
Number of valid votes counted	2 778	2 543	2 668	2 635	2 553	2 688	2 887	2 758	2 917	2 825	2 968	2 913	2 817
As percentage of electorate	*81.2*	*75.1*	*78.1*	*77.6*	*76.0*	*73.5*	*77.9*	*74.5*	*76.0*	*71.8*	*74.3*	*74.2*	*71.3*
Northern Ireland													
Number of electors	872	873	875	891	902	1 025	1 027	1 037	1 028	1 050	1 090	1 141	1 178
Average-electors per seat	72.5	72.8	72.9	74.2	75.2	85.4	85.6	86.4	85.6	61.8	64.1	67.1	65.5
Number of valid votes counted	463	647	576	638	596	779	718	702	696	765	730	785	791
As percentage of electorate	*53.1*	*74.1*	*65.8*	*71.7*	*66.1*	*76.0*	*69.9*	*67.7*	*67.7*	*72.9*	*67.0*	*68.8*	*67.1*
Number of Members of Parliament elected:	625	630	630	630	630	630	635	635	635	650	650	651	659
Conservative	320	344	364	303	253	330	296	276	339	396	375	336	165
Labour	295	277	258	317	363	287	301	319	268	209	229	271	418
Liberal Democrat	6	6	6	9	12	6	14	13	11	17	17	20	46
Social Democratic Party	-	-	-	-	-	-	-	-	-	6	5	-	-
Scottish National Party	-	-	-	-	-	1	7	11	2	2	3	3	6
Plaid Cymru	-	-	-	-	-	-	2	3	2	2	3	4	4
Other[2]	4	3	2	1	2	6	15	13	13	18	18	17	20

1 The Representation of the People Act 1969 lowered the minimum voting age from 21 to 18 years with effect from 16 February 1970.
2 Including the Speaker.
3 Provisional.

Source: Home Office: 020 7273 3453

2.2 Parliamentary by-elections

	May 1979 - June 1983	Previous[1] General Election May 1979	June 1983 - June 1987	Previous[1] General Election June 1983	June 1987 - April 1992	Previous[1] General Election June 1987	April 1992 - August 1995	Previous[1] General Election April 1992	May 1997 - November 1999	Previous[1] General Election May 1997
Numbers of by-elections	20		31		24		16		9	
Votes recorded										
By party (percentages)										
Conservative	*23.8*	*33.7*	*16.0*	*23.3*	*23.8*	*33.7*	*18.6*	*40.1*	*32.3*	*32.0*
Labour	*25.7*	*35.2*	*14.9*	*14.8*	*38.9*	*41.1*	*39.5*	*36.2*	*31.5*	*43.3*
Liberal Democrat[2]	*9.0*	*8.0*	*15.0*	*10.9*	*19.1*	*18.6*	*26.2*	*16.4*	*24.8*	*16.8*
Social Democratic Party[2]	*14.2*	-	*5.6*	*3.0*	*3.2*	-	-	-	-	-
Scottish National Party	*1.7*	*1.4*	-	-	*4.8*	*1.8*	*5.8*	*3.1*	*5.8*	*3.7*
Plaid Cymru	*0.5*	*0.4*	*0.3*	*0.3*	*2.3*	*0.6*	*0.5*	*0.2*	-	-
Other	*25.1*[3]	*21.2*[3]	*48.2*[4]	*47.6*[4]	*7.9*	*4.2*	*9.4*	*4.1*	*5.5*	*4.2*
Total votes recorded (percentages)	*100.0*	*100.0*	*100.0*	*100.0*	*100.0*	*100.0*	*100.0*	*100.0*	*100.0*	*100.0*
(thousands)	715	852	1 235	1 410	878	1 130	560	808	244	396

1 Votes recorded in the same seats in the previous General Election. Does not apply to June 1983 in Northern Ireland.
2 The Social Democratic Party was launched on 26 March 1981. An SDP candidate contested a parliamentary seat for the first time at a by-election at Warrington on 16 July 1981. In the 1987 General Election the Liberals and the SDP contested the election jointly as the Liberal/SDP Alliance. Subsequently the Alliance split into its constituent parties and the Liberals are now known as the Liberal Democrats. The SDP effectively ceased to exist when its two remaining MPs lost their seats at the 1992 General Election.
3 The proportions of votes recorded for 'Other' is high because three of the 20 by-elections were in Northern Ireland (two were in the same constituency, Fermanagh and South Tyrone), and as votes recorded for the 1979 General Election have been included twice in the previous General Election Column.
4 The proportion of votes recorded for 'Other' is high because 15 of the 31 by-elections were in Northern Ireland.

Source: Home Office: 020 7273 3453

Overseas aid

3

Overseas aid

Overseas development assistance (Tables 3.1 and 3.2)

The UK development assistance programme is administered by the Department for International Development (DFID) to promote the economic development of recipient countries. It is managed within financial years, the money being voted annually by Parliament. Since 1992 the statistics relating to the programme are also published on a financial year basis and on a calendar year basis for both international aid comparisons and for national purposes such as the balance of payments.

Aid flows can be measured before (gross) or after (net) deduction of repayments of principal on past loans. These tables show only the gross figures.

Assistance is provided in two main ways: *bilateral*, that is directly to governments of recipient countries or to institutions in the United Kingdom for work on behalf of such countries, or *multilateral*, that is to international institutions for their economic development programmes. Table 3.1 shows the main groups of multilateral agencies, the International Development Association being the largest in the World Bank Group.

Bilateral assistance takes various forms:

Project aid is finance for investment schemes primarily designed to increase the physical capital of the recipient country, including contributions for local and recurrent costs.

Sector-wide programmes (typically in education, health or agriculture) comprise a combination of assistance including direct budgetary aid and technical co-operation.

Programme aid is financial assistance to fund imports, sector support programmes or budgetary expenditure, usually as part of a World Bank/IMF co-ordinated structural adjustment programme.

Technical co-operation is the provision of know-how in the form of personnel, training, research and associated costs.

Grants and other aid in kind are used to provide equipment and supplies, and support to the development work of UK and international voluntary organisations.

Humanitarian assistance comprises food aid, disaster relief, short-term refugee relief and disaster preparedness.

Aid and Trade Provision is a special allocation to soften the terms of credit to developing countries by mixing aid funds with private export credits.

The *Commonwealth Development Corporation* invests in productive public or private sector projects in developing countries.

Other Government Departments' expenditure covers debt relief, drug-related assistance and support to voluntary organisations.

Most of the expenditure not allocable by region in Table 3.2, is for assistance provided through organisations in the United Kingdom.

Fuller statistics of the UK's development assistance effort are published annually in *Statistics on International Development* (obtainable from the Library, DFID, Abercrombie House, East Kilbride, Glasgow G75 8EA). International comparisons are available in the OECD Development Assistance Committee's annual report. The latest is *1999 Report: Development Co-operation* (available from The Stationery Office).

3.1 UK Gross public expenditure on aid (GPEX)

£ Thousands

		1990 /91	1991 /92	1992 /93	1993 /94	1994 /95	1995 /96	1996 /97	1997 /98	1998 /99
Bilateral Assistance										
DFID										
Project or Sector Aid	LUJS	173 909	140 925	122 834	119 832	121 812	122 718	106 780	95 851	117 594
Programme Aid	LUJW	151 921	118 126	136 771	121 027	113 510	82 259	94 694	84 462	134 306
Technical Co-operation Projects	LUOS	360 502	375 832	398 637	401 560	430 447	453 149	467 682	496 003	503 878
Grants & Other Aid in Kind	LUOT	133 336	139 006	155 889	171 688	149 556	179 618	181 632	167 568	186 287
Humanitarian Assistance	LUOU	60 159	138 433	142 813	179 601	205 368	141 185	121 842	93 627	116 429
Debt Relief	LUOV	637	489	20 099	23 698	28 144	27 337	23 529	23 054	24 043
Aid and Trade Provision	LUOW	94 427	101 157	93 071	84 700	59 987	72 811	62 283	61 092	57 149
CDC Investments	LUOX	158 819	124 210	223 301	183 354	215 295	279 986	189 129	268 725	159 250
Other Government Departments	LUOY	50 728	38 666	60 475	22 796	33 034	61 606	34 224	170 171	87 147
Total Bilateral Assistance	LUOZ	1 184 438	1 176 844	1 353 890	1 308 256	1 357 153	1 420 669	1 281 795	1 460 553	1 386 084
Multilateral Assistance										
European Community	LUPA	338 044	459 930	476 065	529 823	612 080	693 690	636 418	557 287	724 508
World Bank Group	LUPB	158 825	228 198	249 682	229 613	227 728	206 877	174 398	189 850	185 254
International Monetary Fund	LUPC	1 500	1 500	10 000	20 000	90 000	30 000	20 000	20 000	18 000
UN Agencies	LUPD	94 297	112 875	122 366	142 527	127 880	116 967	130 885	141 279	131 263
Regional Development Banks	LUPE	32 346	69 122	91 933	76 413	71 212	69 513	58 833	60 411	66 295
Other	LUPF	20 762	20 735	21 360	17 829	27 904	27 010	30 648	30 806	30 672
Total Multilateral Assistance	LUPG	645 774	892 360	971 406	1 016 205	1 156 804	1 144 057	1 051 182	999 634	1 155 992
Administrative Costs	LUPH	57 904	63 155	58 823	65 329	70 314	72 975	80 318	89 708	99 571
Total Gross Public Expenditure on Aid	LUPI	1 888 116	2 132 359	2 384 119	2 389 790	2 584 270	2 637 700	2 413 295	2 549 894	2 641 647

Source: Department for International Development: 01355 843612

3.2 Total bilateral gross public expenditure on aid (GPEX) by main recipient countries and regions

£ Thousands

		1990 /91	1991 /92	1992 /93	1993 /94	1994 /95	1995 /96	1996 /97	1997 /98	1998 /99
Top twenty recipients[1]										
India	LUPJ	100 751	135 983	116 108	102 367	98 622	155 138	112 140	117 223	113 865
Tanzania	LUPK	41 596	32 430	62 457	32 499	31 633	29 989	56 432	53 291	78 924
Bangladesh	LUPM	51 600	59 217	64 934	56 068	57 441	55 392	42 405	39 520	69 918
Uganda	LUPN	32 211	27 836	34 707	40 551	46 615	40 837	49 095	59 262	64 307
Ghana	LUPL	29 219	44 659	41 736	26 190	23 530	26 365	28 339	30 382	63 844
Malawi	LUPP	43 272	26 093	27 378	27 015	41 174	35 779	44 877	30 829	52 395
Kenya	LUPR	49 957	38 351	37 514	31 503	30 691	29 379	26 491	30 078	43 011
China	LUPS	19 515	21 243	34 914	33 630	28 344	32 788	31 701	39 176	39 541
South Africa	LUPT	8 093	9 325	11 558	14 504	15 706	17 529	22 830	27 704	34 144
Zambia	LUPO	49 868	34 444	47 978	47 752	56 629	59 376	40 629	46 904	33 134
Montserrat	LUPU	4 218	2 996	3 271	5 689	5 815	6 565	14 449	38 315	32 068
Russian Federation	LUPW	1	26 686	11 795	20 749	32 753	30 727	32 258	33 513	31 234
Zimbabwe	LUPX	33 627	47 690	44 268	41 390	35 394	20 181	21 103	13 807	29 391
Mozambique	LUPV	26 593	18 001	34 177	34 744	25 068	22 139	26 447	48 510	29 158
Pakistan	LUPY	44 824	37 236	35 063	48 197	63 487	60 241	59 367	32 817	28 608
Indonesia	LUPZ	25 158	34 422	32 679	34 885	51 404	57 258	38 471	48 667	25 614
Sudan	LUQB	24 613	21 840	9 956	15 529	12 760	5 997	4 534	5 677	24 706
Guyana	LUPQ	21 104	8 031	4 215	7 571	8 168	14 675	7 036	92 488	22 893
Nepal	LUQC	19 050	16 513	13 646	13 220	15 475	16 485	15 373	17 252	17 209
Rwanda	LUQA	389	214	312	1 761	45 284	11 958	9 676	6 198	13 776
Total Top Twenty	LUQD	625 659	643 210	668 666	635 814	725 993	728 798	683 651	811 612	847 741
Total Other Countries	LUQE	558 779	533 635	685 223	672 443	631 159	691 871	598 144	648 941	538 343
Regional Totals										
Africa	LUQF	505 941	440 598	526 833	457 991	475 626	438 025	445 340	452 689	584 766
America	LUQG	125 330	92 905	118 685	114 208	132 407	192 396	107 357	278 910	153 532
Asia	LUQH	348 367	403 782	424 574	410 387	415 812	467 516	400 458	402 675	358 004
Europe	LUQI	19 604	71 340	99 362	134 713	137 995	128 798	138 874	114 299	86 823
Pacific	LUQJ	33 751	22 462	28 715	14 935	16 291	13 989	12 458	27 845	20 285
World Unallocated[2]	LUQK	151 445	145 757	155 720	176 022	179 023	179 945	177 308	184 136	182 673
Total Bilateral GPEX	LUQL	1 184 438	1 176 845	1 353 889	1 308 257	1 357 153	1 420 669	1 281 795	1 460 553	1 386 084

1 Top twenty recipients in 1998/99.
2 World Unallocated comprises block grants to the British Council, Natural Resources Institute, VSO, NGOs, Research Institutions and Commonwealth Organisations based in the UK, and some ATP Technical Co-operation.

Source: Department for International Development: 01355 843612

Defence

4

Defence

This section includes figures on Defence expenditure, on the size and role of the Armed Forces and on related support activities.

Much of the material in this section can be found in *UK Defence Statistics 1999 (TSO).*

Formation of the armed forces ***(Table 4.1)***
This table shows the number of units which comprise the "teeth" elements of the Armed Forces and excludes supporting units.

Service personnel ***(Tables 4.3 to 4.6 and 4.10)***
The Regular Forces consist entirely of volunteer members serving on a whole-time basis.

Locally Entered Personnel are recruited outside the United Kingdom for whole-time service in special formations with special conditions of service and normally restricted locations. The Brigade of Gurkhas is an example.

The Regular Forces are supported by Reserves and Auxiliary Forces. There are both regular and volunteer Reserves. Regular Reserves consist of former Service personnel with a Reserve liability. Volunteer Reserves are open to both former Service personnel and civilians. The call out liabilities of the various reserve forces differ in accordance with their roles.

All three Services run cadet forces for young people and the Combined Cadet Force, which is found in certain schools where education is continued to the age of 17 or above, may operate sections for any or all of the Services.

Deployment of Service personnel ***(Table 4.6)***
The source from which the individual national totals are compiled is different from that used to obtain the total UK strength, and consequently the figures for England, Wales, Scotland and Northern Ireland do not add up to the UK figure. Royal Navy and Royal Marines personnel on board ship are included in the UK figure if the ship was in home waters on the situation date or otherwise against the appropriate overseas area. The overseas figures include personnel who are on loan to countries in the areas shown.

Service married accommodation and Defence land holdings ***(Table 4.7)***
Accommodation is provided for Service families in the United Kingdom and abroad, partly by building to approved standards and partly by renting accommodation. Permanent holdings in the United Kingdom include a small number of unfurnished hirings taken on from local authorities which are not recorded separately. Multiple hirings relate to accommodation built by private developers and leased by the Federal German Authorities and the Government of Gibraltar on behalf of British Forces.

The table also presents statistics of land and foreshore in the United Kingdom owned by the Ministry of Defence or over which it has limited rights under grants or licences. Land declared as surplus to Defence requirements is also included.

Civilian personnel ***(Table 4.8)***
This table gives an analysis of the number of civilians employed in the various management areas. The attribution of civilian staff can change according to circumstances so that figures for successive years may not always be comparable.

Some civilians from the United Kingdom serve tours of duty overseas. There were 4 400 such civilians serving abroad on 1 April 1998. Other civilian staff are engaged overseas to work locally as circumstances demand.

Health ***(Tables 4.9 - 4.11)***
The Services operate a number of hospitals in this country and in areas abroad where there is a significant British military presence. These hospitals take as patients, members of all three Services and their dependants; in addition, the hospitals in the United Kingdom take civilian patients under arrangements agreed with the National Health Service. Medical support is also supplied by Service medical staff at individual units, ships and stations.

Sickness, medical discharges and deaths of UK Service personnel ***(Table 4.11)***
It should be noted that, whereas the Royal Air Force content is for all episodes of bedded sickness lasting two days or more terminating in the year, that for the Army covers only those cases admitted to medical units (including hospitals) for two days or more, and the Royal Navy episodes exclude less than seven days sickness on shore.

Search and rescue ***(Table 4.12)***
This table covers incidents in which Rescue Coordinating Centres (RCCs) in the United Kingdom coordinated search and rescue (SAR) action in which elements of the Armed Forces were involved. The table also includes urgent medical incidents in which the Forces SAR facilities gave assistance (eg inter-hospital transfers).

Defence services and the civilian community ***(Table 4.13)***
The Ministry of Defence helps the civil community in a variety of ways, for example by providing assistance in time of natural disasters or other emergencies and by undertaking community projects which are of training value to the Services. In some cases facilities established primarily for defence purposes also provide benefits to the general public. Overall figures are not available, but the table presents information on a number of these activities and facilities.

Service assistance may be provided during an industrial dispute at the request of the civil ministries in order to maintain services essential to the life of the community (eg maintenance of emergency fire services in 1977/78).

The Royal Navy Fishery Protection squadron operates within the UK fishery limits on behalf of Fishery departments, who pay some of the running costs of the squadron.

The Hydrographer Office is the national authority responsible for hydrographic and oceanographic surveys and nautical charting. Many of the maps and charts are now available electronically, as well as in traditional paper format.

4.1 Formation of the UK armed forces

As at 1 April

Number

		Front Line Units[1]	1989	1990	1991	1992	1993	1994	1995	1996	1997	1998	1999
Royal Navy[2]													
Submarines	KCGA	Vessels	26	24	24	20	18	18	16	15	15	15	15
Carriers and assault ships	KCGB	Vessels	4	4	3	3	3	4	3	5	5	5	6
Cruisers, Destroyers and Frigates	KCGC	Vessels	41	43	41	43	38	32	30	36	35	35	35
Mine counter-measure	KCGE	Vessels	36	37	35	31	30	18	18	18	18	19	20
Patrol ships and craft	KCGF	Vessels	33	32	31	26	24	30	33	32	34	28	24
Fixed wing aircraft	KCGG	Squadrons	3	3	3	3	3	3	3	3	3	3	3
Helicopters[3]	KCGH	Squadrons	14	13	13	13	13	11	11	15	15	12	12
Royal Marines	KCGI	Commandos	3	3	3	3	3	3	3	3	3	3	3
Army													
Royal Armoured Corps	KCGJ	Regiments	19	19	19	19	14	12[4]	11	11	11	11	11
Royal Artillery	KCGK	Regiments	22	22	22	21	18	16[4]	16	16	16	15	15
Royal Engineers	KCGL	Regiments	13	13	13	13	13	12	10	10	10	10[5]	10
Infantry	KCGM	Battalions	55	55	55	55	48	45	41	41	40	40	40
Special Air Service	KCGN	Regiments	1	1	1	1	1	1	1	1	1	1	1
Army Air Corps	KCGO	Regiments	4	4	4	5	6	6	5	5	5	5	5
Royal Air Force													
Strike/attack	KCGP	Squadrons	11	11	11	9	8	6	6	6	6	6	5
Offensive support	KCGQ	Squadrons	5	5	5	5	5	5	5	5	5	5	5
Air defence	KCGR	Squadrons	9	9	11	9	7	6	6	6	6	6	5
Maritime patrol	KCGS	Squadrons	4	4	4	4	3	3	3	3	3	3	3
Reconnaissance	KCGT	Squadrons	3	3	5	5	5	5	5	5	5	5	5
Airborne early warning	KCGU	Squadrons	1	1	1	1	1	1	1	2	2	2	2
Air transport and tankers and helicopters[6]	KCGV	Squadrons	15	15	15	16	16	15	14	14	13	14	14
Search and rescue	KCGX	Squadrons	2	2	2	2	2	2	2	2	2	2	2
Surface to air missiles	KCGY	Squadrons	8	7	7	6	5	5	5	6	6	6	4
Ground defence[7]	KCGZ	Squadrons	6	6	5	5	5	5	5	5	5	5	5

1 Units which comprise the 'teeth' elements of the Armed Forces.
2 Excludes vessels undergoing construction, major refit, conversion, or on stand-by, etc.
3 Excludes training, search and rescue, fleet training and support squadrons.
4 Includes one training regiment.
5 Does not include the Military Works force.
6 Includes UK and Germany-based helicopter Squadrons and the Royal Squadron.
7 Includes the Queen's Colour Squadron, which since 1994 has a war role.

Source: Ministry of Defence/DASA (Tri-Service): 020 7218 4535

4.2 UK Defence expenditure

£ millions

		1987 /88	1988 /89	1989 /90	1990 /91	1991 /92	1992 /93	1993 /94	1994 /95	1995 /96	1996 /97	1997 /98
Total expenditure at outturn prices	KDAA	18 856	19 072	20 755	22 298	24 562	23 762	23 424	22 519	21 517	22 041	20 910
of which:												
Expenditure on personnel	KDAB	7 212	7 572	8 099	8 811	9 934	10 504	9 835	9 313	8 524	8 633	8 259
of the Armed Forces	KDAC	4 032	4 300	4 539	4 811	5 499	5 634	6 907	6 490	6 150	6 223	5 938
of the retired Armed Forces[1]	KDAD	1 079	1 060	1 197	1 406	1 613	1 950	..	..	..	..	..
of civilian staff	KDAE	2 101	2 212	2 363	2 594	2 822	2 920	2 928	2 823	2 374	2 409	2 321
Expenditure on equipment	KDAF	8 270	8 038	8 536	8 838	9 758	8 711	9 207	8 819	8 537	9 100	9 003
Sea systems	KDAG	2 797	2 633	2 890	2 955	3 142	2 891	2 589	2 441	2 110	2 190	2 142
Land systems	KDAH	1 700	1 554	1 738	1 927	2 157	1 846	1 806	1 642	1 576	1 806	1 658
Air systems	KDAI	3 230	3 085	3 102	3 197	3 574	3 152	3 246	3 184	3 356	3 507	3 843
Other	KDAJ	543	766	806	759	885	822	1 566	1 552	1 495	1 597	1 360
Other expenditure	KDAK	3 374	3 462	4 120	4 649	4 870	4 547	4 382	4 387	4 456	4 308	3 648
Works, buildings and land	KDAL	1 453	1 411	1 900	2 067	2 090	1 780	2 073	2 402	2 065	1 904	1 253
Miscellaneous stores and services	KDAM	1 921	2 051	2 220	2 582	2 780	2 767	2 309	1 985	2 391	2 404	2 395
Adjusted defence budget at 1996/97 prices[2]	KZBG	28 571	27 202	27 486	27 202	28 139	25 316	24 437	23 867	22 148	22 345	21 042

1 From 1990/91 to 1994/95, includes financial assistance to pre-1973 war widows.
2 The Adjusted Defence Budget takes account of major definitional changes in defence spending and of major transfers of responsibility to and from Governmental Departments. It therefore provides a more consistent and reliable guide to trends. This is the basic upon which defence spending figures are presented in the MoD Department Return, the Statement on the Defence Estimates and certain Treasury publications. It excludes the element of receipts from the sale of the Married Quarters Estate. 1996/97 prices have been calculated using the GDP deflator.

Source: Ministry of Defence/DASA (Tri-Service): 020 7218 4535

4.3 UK Defence: service manpower strengths[1]

As at 1 April

Thousands

		1989	1990	1991	1992	1993	1994	1995	1996	1997	1998	1999
UK service personnel												
All services:												
Male	KCEB	295.2	288.2	279.2	273.8	255.9	236.7	216.7	206.1	196.0	194.4	192.5
Female	KCEC	16.5	17.5	18.8	19.6	18.9	17.8	16.6	15.7	14.8	15.7	16.1
Total	KCEA	311.6	305.7	298.1	293.4	274.8	254.5	233.3	221.8	210.8	210.1	208.6
Naval service:												
Male	KCEE	61.2	59.6	57.9	57.5	54.9	51.5	47.0	44.7	41.9	41.2	40.4
Female	KCEF	3.5	3.6	4.2	4.6	4.5	4.3	3.9	3.6	3.3†	3.3	3.3
Total	KCED	64.6	63.2	62.1	62.1	59.4	55.8	50.9	48.3	45.2†	44.5	43.7
Army:												
Male	KCEJ	148.9	145.8	140.3	137.6	12.7	116.0	105.1	102.4	102.1	102.4	102.1
Female	KCEK	6.7	7.0	7.3	7.8	7.6	7.1	6.6	6.5	6.7	7.4	7.6
Total	KCEI	155.6	152.8	147.6	145.4	134.6	123.0	111.7	108.8	108.8	109.8	109.7
Royal Air Force:												
Male	KCEM	85.1	82.9	81.1	78.7	74.1	69.2	64.6	59.1	52.0	50.8	50.0
Female	KCEN	6.4	6.8	7.3	7.3	6.8	6.5	6.1	5.6	4.9	5.0	5.2
Total	KCEL	91.4	89.7	88.4	86.0	80.9	75.7	70.8	64.7	56.9	55.8	55.2
UDR/Royal Irish(HS):	KIUX	6.3	6.2	6.1	6.0	5.6	5.4	5.3	5.0	4.8	4.6	4.4
Service personnel locally entered/ engaged:												
Total	KCEO	9.1	9.0	9.1	9.0	8.2	7.2	5.2	4.8	4.3†	4.0	4.0
Regular Reserves:												
Royal Naval Services	KCEP	27.6	28.4	28.5	27.5	22.0	23.3	23.3	22.7	24.1	24.8	24.7
Army	KCES	175.3	183.4	187.7	188.6	190.1	192.5	195.3	195.5	190.1	186.0	180.4
Royal Air Force	KCET	37.5	40.1	42.5	44.8	46.1	46.4	45.5	45.2	45.3†	43.9	42.4
Total	KCER	240.5	252.0	258.7	260.8	258.3	262.2	264.1	264.6	259.4†	254.7	247.5
Volunteer Reserves and Auxiliary Forces:												
Naval Service[2]	KCEV	6.9	7.0	7.0	5.8	5.6	4.6	3.7	3.5	3.6	3.7	3.8
Territorial Army	KCEX	72.5	72.5	73.3	71.3	68.5	65.0	59.9	57.3	57.6	57.0	51.8
Home Services Force[3]	KCEZ	3.0	3.2	3.3	2.9	0.2	..	..	..	..	..	..
Royal Air Force	KCFA	1.6	1.8	1.8	1.9	1.7	1.8	1.3	1.2	1.4	1.6	1.7
Total	KCEU	90.4	90.6	91.6	87.9	76.1	71.3	64.9	62.0	62.5†	62.4	57.4
Cadet Forces:[4]												
Naval Service	KCFC	26.5	26.2	27.0	28.8	22.1	21.7	20.9	21.2	25.0	20.6	19.9
Army	KCFD	65.9	65.7	64.6	64.8	65.1	65.2	52.5	65.6	65.1	64.9†	65.7
Royal Air Force[5]	KCFE	46.7	44.2	43.1	43.6	43.7	43.8	52.3	42.4	41.8	41.2	42.7
Total	KCFB	139.2	136.0	134.7	137.2	135.8	130.7	137.1	129.0	131.8†	126.7	128.3
University Forces:[6]												
Naval Service	LQZI	..	..	..	..	..	..	..	..	0.7	0.7	0.7
Army	LQZJ	..	..	..	..	..	..	..	..	4.0	4.8	4.7
Royal Air Force	LQZK	..	..	..	..	..	..	..	..	0.8	0.8	0.9
Total	LQZL	..	..	..	..	..	..	..	..	5.5	6.4	6.3

1 This table does not include Gurkhas.
2 The Royal Naval Auxiliary Service is not included in this table. They were disbanded on 31 March 1994.
3 By 1 April 1994, the Home Service Forces had been disbanded.
4 Includes the appropriate component of the combined Cadet Force but excludes officers and training and administrative staff.
5 The Air Cadet component of the 1999 figure is at 1 December 1998, not 1 April 1999.
6 Data for University Forces was unavailable before 1997.

Source: Ministry of Defence/DASA (Tri-Service): 020 7218 4535

4.4 Intake of UK regular forces from civilian life: by Service

Number

		1988 /89	1989 /90	1990 /91	1991 /92	1992 /93	1993 /94	1994 /95	1995 /96	1996 /97	1997 /98	1998 /99
All Services:												
Male	KCJB	30 821	32 297	26 782	24 357	13 067	10 616	11 154	15 495	19 121	20 169	22 545
Female	KCJC	3 042	4 115	4 428	3 550	1 750	1 333	1 853	2 175	3 038	3 336	3 431
Total	KCJA	33 863	36 412	31 210	27 907	14 817	11 949	13 007	17 670	22 165	23 505	25 976
Naval service:												
Male	KCJE	5 535	5 759	5 712	5 570	1 888	1 272	961	2 007	3 397	3 966	4 106
Female	KCJF	700	860	1 199	1 013	384	250	340	353	562	634	661
Total	KCJD	6 235	6 619	6 911	6 583	2 272	1 532	1 301	2 360	3 959	4 600	4 767
Army:												
Male	KCJJ	19 898	20 378	15 955	15 544	10 289	8 764	9 491	11 510	13 480	13 374	14 988
Female	KCJK	1 450	1 706	1 547	1 594	1 098	812	1 193	1 376	2 042	2 005	1 975
Total	KCJI	21 348	22 084	17 502	17 138	11 387	9 576	10 684	12 886	15 522	15 379	16 963
Royal Air Force:												
Male	KCJM	5 395	6 166	5 122	3 243	890	580	702	1 978	2 250	2 829	3 451
Female	KCJN	885	1 543	1 675	943	268	261	320	446	434	697	795
Total	KCJL	6 280	7 709	6 797	4 186	1 158	841	1 022	2 424	2 684	3 526	4 246

Source: Ministry of Defence/DASA (Tri-Service): 020 7218 4535

4.5 Outflow of UK regular forces: by service[1]

Number

		1988 /89	1989 /90	1990 /91	1991 /92	1992 /93	1993 /94	1994 /95	1995 /96	1996 /97	1997 /98	1998 /99
All Services:												
Male	KDNA	36 536	39 461	36 017	30 057	31 138	29 699	31 046	25 746	29 325	21 775†	24 459
Female	KDNB	2 742	3 106	3 062	2 748	2 517	2 429	2 994	3 118	3 675	2 483†	2 976
Total	KDNC	39 278	42 567	39 079	32 805	33 655	32 128	34 040	28 864	33 000	24 258†	27 435
Naval Service:												
Male	KDND	6 474	7 377	7 428	6 105	4 655	4 613	5 504	4 314	6 191	4 650	4 925
Female	KDNE	591	657	680	662	455	493	681	631	940	621	606
Total	KDNF	7 065	8 034	8 108	6 767	5 110	5 106	6 185	4 945	7 131	5 271	5 531
Army:												
Male	KDNI	22 823	23 760	21 582	18 304	20 993	19 633	20 234	13 935	13 758	13 103†	15 282
Female	KDNJ	1 233	1 301	1 242	1 147	1 336	1 288	1 651	1 508	1 596	1 275†	1 733
Total	KDNK	24 056	25 061	22 824	19 451	22 329	20 921	21 885	15 443	15 354	14 378†	17 015
Royal Air Force:												
Male	KDNL	7 239	8 324	7 007	5 648	5 490	5 453	5 308	7 497	9 376	4 022	4 252
Female	KDNM	918	1 148	1 140	939	726	648	662	979	1 139	587	637
Total	KDNN	8 157	9 472	8 147	6 587	6 216	6 101	5 970	8 476	10 515	4 609	4 889

1 Comprises all those who left the Regular Forces and includes deaths.

Source: Ministry of Defence/DASA (Tri-Service): 020 7218 4535

4.6 Deployment of UK service personnel

As at 1 July

Thousands

		1988	1989	1990	1991	1992	1993	1994	1995	1996	1997	1998
UK Service personnel, Regular Forces:												
In United Kingdom[8]	KDOB	226.2	220.2	215.9	207.9	210.7	204.3	192.1	182.2	177.4	171.6	173.4
England	KDOC	189.8	183.3	179.6	173.3	174.2	167.1	156.5	149.4	146.6	142.6	144.6
Wales	KDOD	5.8	5.4	5.3	5.2	5.2	5.1	4.8	5.2	4.3	3.3	3.2
Scotland	KDOE	19.3	20.0	19.3	18.2	18.8	19.4	18.4	16.9	15.5	13.9	14.2
Northern Ireland[2]	KDOF	11.2	11.2	11.5	11.2	12.6	12.5	12.3	9.9	10.5	11.5	11.0
Overseas	KDOG	90.4	88.4	88.9	89.3	78.9	65.5	56.2	50.0	48.5	42.6	43.1
Germany[3,4,6]	KDOH	65.6	63.9	63.2	69.7	60.0	45.6	37.2	33.4	20.8	21.2	20.3
Elsewhere in Continental Europe[4,6]	KDOI	6.9	7.0	6.8	–	–	–	–	–	11.7†	6.2	6.9
Gibraltar	KDOJ	1.8	1.6	1.8	1.0	1.0	1.3	1.5	0.6	0.6	0.5	0.5
Cyprus	KDOL	4.5	4.8	4.8	5.2	4.8	4.7	4.1	4.4	4.0	3.9	3.6
Mediterranean, Near East and Gulf	KDOM	3.4	1.8	1.3	1.1	2.2	2.8	2.4	0.6	0.5	0.3	1.2
Hong Kong	KDON	2.3	2.1	2.1	2.3	2.1	1.9	1.7	0.9	0.9	–	–
Elsewhere in the Far East	KDOP	0.3	0.6	0.5	0.5	2.1	0.3	0.3	0.3	0.3	1.5	0.3
Other locations[1,5]	KDOQ	5.6	6.6	8.4	9.5	6.7	8.8	8.9	9.4	9.8	9.0†	10.4
Total[9]	KDOA	315.8	307.8	303.1	297.2	289.6	270.9	248.3	232.2	225.9†	214.2	216.5
Locally entered service personnel:												
United Kingdom	KDOS	1.3	1.2	1.4	1.4	1.3	1.4	1.5	1.8	1.7	2.1†	2.1
Gibraltar	KDOT	0.1	0.1	0.1	0.1	0.2	0.2	0.2	0.2	0.2	0.2	0.4
Hong Kong	KDOV	5.8	5.5	5.5	5.5	5.2	4.4	3.7	1.8	1.6	0.7†	–
Brunei	KDOW	0.8	0.9	0.9	0.8	0.8	0.8	0.7	0.9	0.7	1.0†	0.9
India/Nepal	KDOX	1.4	1.4	1.1	1.3	1.3	1.2	0.2	0.5	0.7	0.9†	0.5
Total	KDOK	9.4	9.0	8.9	9.1	8.8	7.9	7.1	5.1	4.9	4.7†	4.0

1 From 1984 to 1992 the England, Wales and Scotland national figures include personnel who were UK based but temporarily deployed in the South Atlantic.These have been included also in the Overseas numbers against 'Other locations'.
2 The figures for Northern Ireland include all personnel who are serving on emergency tours of duty but exclude the former Ulster Defence Regiment, now the Home Services element of the Royal Irish Regiment.
3 Prior to 1994, personnel serving in Northern Ireland on emergency tours of duty but remaining under the command of the Commander-in-Chief, British Army of the Rhine, are included in these numbers. Personnel serving on emergency tours of duty in other overseas areas are included in the numbers for that area.
4 These figures include personnel stationed in Berlin and Sardinia.
5 These figures include Defence Attaches/Advisers and their staffs.
6 From 1991-1995, figures for the Federal Republic of Germany and elsewhere in Continental Europe were combined.
7 Figures may not equate to totals because of rounding.
8 The figures for Service personnel in each country are obtained from a different source from that used to obtain the United Kingdom total. Consequently, the sum of the national figures can differ from the United Kingdom total.
9 The figures include service personnel who are on loan to countries in the areas shown. The Royal Navy and Royal Marines figures include personnel who are at sea at the situation date. All Defence Attaches and Advisers and their staffs are included under "Other Locations" and not identified within specific areas.

Source: Ministry of Defence/DASA (Tri-Service): 020 7218 4535

4.7 Service married accommodation and Defence land holdings

As at 1 April

			1989	1990	1991	1992	1993	1994	1995	1996	1997	1998	1999
Married accommodation		Thousands											
United Kingdom: total	KDPA	"	77.2	75.0	73.6	72.3	71.8	71.1	69.7	72.1†	68.6	67.3	65.5
Permanent holdings[1]	KDPB	"	77.2	75.0	73.6	72.3	71.8	71.1	69.7	72.1†	68.6	67.3	65.5
Hirings	KDPC	"	–	–	–	–	–	–	–	–	–	–	–
Overseas: total[2]	KDPD	"	45.2	46.7	48.9	45.1	42.4	36.9	28.8	26.3	..	..	..
Permanent holdings	KDPE	"	20.9	21.3	24.2	22.1	21.7	21.2	16.8	15.5	..	..	..
Hirings	KDPK	"	24.3	25.4	24.7	23.0	20.7	15.7	12.0	10.8	..	..	..
Land holdings		Thousand hectares											
United Kingdom													
Land[3]	KDPF	"	223.3	222.6	224.8	225.4	224.1	223.5	221.0	222.6	221.0	220.0	220.2
Foreshore[3]	KDPH	"	18.0	18.0	18.0	18.0	18.4	18.5	18.4	18.5	18.6	18.6	18.6
Rights held[4]	KDPJ	"	101.1	103.4	121.3	120.7	100.2	100.4	122.3	124.3	124.5	124.5	124.8
Defence land													
Used for agricultural purposes	KDPL	"	109.6	110.8	109.9	109.3	107.1	107.6	107.9	107.4	96.2	103.5	114.5
Used for grazing only	KDPM	"	61.4	61.2	63.9	62.6	62.4	61.3	61.8	60.7	51.9	59.6	65.5
Full agricultural use	KDPN	"	48.2	49.6	46.0	46.7	44.6	46.3	46.1	46.7	44.3	43.9	49.0

1 Including a few unfurnished hirings taken from local authorities and from the Scottish Special Housing Association.
2 This information is no longer available.
3 Freehold and leasehold.
4 Land and foreshore not owned by, or leased to MOD, over which the Department has limited rights under grants or licenses.

Source: Ministry of Defence/DASA (Tri-Service): 020 7218 4535

4.8 UK Defence civilian manpower strengths

As at 1 April

Thousands

		1989	1990	1991	1992	1993	1994	1995	1996[8]	1997	1998	1999
Ministry of Defence civilians												
Centre[5,7]												
Non-industrial	KDQE	17.2	17.7	21.8	22.9	23.7	18.9	17.3	20.0	21.2	20.6	20.0
Industrial	KDQF	0.3	0.4	0.7	0.8	0.8	0.8	0.7	0.9	1.1	1.1	1.0
Trading funds[4]												
Non-industrial	KYCU	..	..	..	..	..	..	..	12.1	13.4	12.3	11.9
Industrial	KYCV	..	..	..	..	..	..	..	1.9	1.7	1.3	1.1
Naval Service[3]												
Non-industrial	KYCW	14.8	15.1	14.9	14.9	15.5	15.8	15.2	13.7	12.6	12.2	11.6
Industrial	KYCX	17.1	16.4	15.8	15.2	14.1	13.0	12.0	9.7	9.0	8.0	6.8
Army												
Non-industrial	KDQK	21.9	22.1	21.4	21.2	22.0	22.5	22.1	21.4	20.7	20.5†	20.3
Industrial	KDQL	17.8	17.7	17.0	17.0	15.1	14.9	14.5	13.7	13.2	11.8	10.3
Royal Air Force												
Non-industrial	KDQM	12.2	12.4	10.2	10.7	10.3	10.3	10.1	9.9	10.0	10.1	10.6
Industrial	KDQN	7.8	8.0	7.7	7.7	7.2	7.0	6.7	6.5	6.4	6.1	7.1
Procurement Executive[6,7]												
Non-industrial	KDQO	22.6	22.6	22.0	21.0	16.4	15.3	15.1	–	–	–	–
Industrial	KDQP	9.5	9.1	8.6	8.1	4.1	3.1	2.5	–	–	–	–
Total UK based[2]	KDQB	141.3	141.4	140.2	139.5	129.2	121.6	116.1	109.9	109.2	104.0†	100.9
Non-industrial	KDQC	88.7	89.9	90.3	90.7	87.9	82.8	79.8	77.1	77.9	75.7†	74.5
Industrial	KDQD	52.6	51.5	49.9	48.8	41.3	38.8	36.3	32.7	31.3	28.3	26.4
Locally engaged overseas	KDQA	31.2	30.9	28.9	27.0	26.6	20.7	16.9	16.9	15.7	15.0	14.8
Non-industrial	KDQT	10.2	10.1	9.9	9.7	9.6	7.3	6.4	7.1	7.0	6.7	6.7
Industrial	KDQU	21.1	20.8	19.1	17.3	17.0	13.4	10.5	9.8	8.7	8.2	8.0
Total[1]	KFHT	172.5	172.3	169.1	166.5	155.8	142.3	133.0	126.8	124.9	119.0†	115.7

1 Prior to April 1995 all part-timers were counted as half of full time. From that date they are counted as the number of hours worked as a proportion of normal conditioned hours. The average part timer works about 60% of full time.
2 Including UK-based civilians serving overseas.
3 The Devonport and Rosyth Dockyards were contractorised with effect from 6 April 1987.
4 Prior to becoming Trading Funds the Defence Evaluation Agency and the Hydrographic Office were included in Procurement Executive. From 1991 to 1995 Metrological Office was included in Central.
5 From 1989 the 'Centre' figure includes all MOD Police. From 1991 the 'Centre' figure includes staff in the Meteorological Office (transferred from 'Air Force'), staff in the Chemical Defence Establishment (transferred from 'Procurement Executive'), and all personnel serving overseas in support of the services (other than in BAOR or RAF Germany). Communications staff are excluded after 1993.
6 Atomic Weapons Establishment was contractorised with effect from 1 April 1992.
7 From 1996, those staff in the procurement executive are included in "Centre" figures.

Source: Ministry of Defence/DASA (Tri-Service): 020 7218 4535

4.9 Service hospitals

		1985	1986	1987	1988[4]	1989	1990	1991	1992	1993[5]	1994	1995
Average number of beds[1]												
United Kingdom	KDLA	1 986	1 959	1 846	1 765	1 661	1 441	1 264	1 375	1 348	1 358	1 209
Overseas	KDLB	1 184	1 117	1 029	985	985	825	815	723	581	445	345
Average number of admission by percentage[2]												
United Kingdom:	KDLE	62.0	59.2	58.2	59.6	59.7	60.0	56.2	57.5	53.0	50.4	43.3
Service personnel	KDLF	26.5	25.1	24.2	23.3	22.5	24.1	26.5	23.3	21.3	21.5	18.8
Service dependants	KDLG	7.8	6.5	6.9	6.0	5.6	5.4	4.9	4.8	3.5	2.5	1.5
NHS/Other	KDLH	27.7	27.7	27.0	30.3	31.6	30.4	24.8	29.3	28.2	26.4	23.0
Overseas:	KDLI	50.6	48.5	47.9	47.0	43.5	41.9	40.1	45.7	39.9	46.1	44.1
UK Service personnel	KDLJ	17.7	16.6	16.3	15.1	14.2	15.1	14.2	17.1	16.2	21.8	22.0
Service dependants	KDLK	20.6	20.6	21.9	22.1	21.3	23.9	23.1	25.4	20.9	21.3	19.9
Others	KDLL	12.4	11.4	9.7	9.6	8.0	2.9	2.8	3.1	2.8	3.0	2.2
Average number of admissions per bed[2]												
United Kingdom	KDLM	37.8	36.8	38.3	41.3	42.8	43.7	40.9	39.9	37.3	38.4	29.2
Overseas	KDLN	35.3	36.4	38.1	39.1	36.0	38.4	33.6	37.8	36.0	35.6	35.9
Average days in hospital per patient												
United Kingdom	KDLO	6.0	5.9	5.6	5.4	5.1	5.0	5.0	5.2	5.1	4.8	4.9
Overseas	KDLP	5.2	4.9	4.6	4.5	4.4	4.0	4.1	4.2	4.1	4.3	4.5
Outpatient attendances (thousands)[3]												
United Kingdom:	KDLQ	373.7	373.8	359.6	374.7	371.5	362.1	281.3	306.7	291.1	239.8	199.9
Service personnel	KDLR	119.3	117.2	116.4	119.8	111.4	109.0	93.3	103.9	102.8	96.5	82.2
Service dependants	KDLS	48.7	46.2	41.1	39.4	38.4	37.1	30.7	31.4	27.0	20.0	13.0
NHS/Others	KDLT	205.7	210.3	202.1	215.5	221.7	216.1	157.4	171.5	161.3	123.3	104.7
Overseas:	KDLU	229.6	212.9	202.8	179.6	171.2	138.7	134.8	138.7	123.9	81.5	64.2
UK Service personnel	KDLV	64.4	59.6	56.5	50.2	50.8	51.4	49.8	49.4	46.1	31.6	25.9
Service dependants	KDLW	102.9	96.5	94.0	84.1	81.8	75.8	75.2	78.9	67.5	42.2	33.1
Others	KDLX	62.3	56.8	52.2	45.3	38.5	11.4	9.9	10.8	10.2	7.7	5.2

1 These relate to the numbers of available beds.
2 Based on the number of beds as defined above.
3 In addition 3 300 patients were seen at outlying clinics of the hospitals and some 19 000 day patients were treated in the UK and 3 000 overseas in 1992.
4 Changes to the Statistical package used in certain hospitals to produce figures from 1988 onwards, mean that these figures are not fully comparable with those for earlier years.
5 Provisional.

Source: Ministry of Defence/DASA (Information Services): 01225 472963

4.10 Strength of uniformed UK medical staff[1]

At 1 April

Number

		1989	1990	1991	1992	1993	1994	1995	1996	1997	1998	1999
Qualified doctors:												
Naval Service	KDMA	279	275	267	263	270	268	262	238	211	209†	209
Army	KDMB	547	501	649	530	522	490	473	430†	433	431	437
Royal Air Force	KDMC	350	374	371	375	368	359	334	273	223	207	197
All Services	KDMD	1 176	1 150	1 287	1 168	1 160	1 117	1 069	941†	867	847	843
Qualified dentists:												
Naval Services	KDME	81	83	77	75	76	75	73	62	65	64†	65
Army	KDMF	200	197	194	183	178	168	154	145	139	140	141
Royal Air Force	KDMG	117	120	120	121	119	115	110	104	87	80	85
All Services	KDMH	398	400	391	379	373	358	337	311	291	284†	291
Support staff:[2]												
Naval Services	KDMI	1 244	1 249	1 254	1 308	1 361	1 401	1 352	1 294	1 025	946†	935
Army	KDMJ	4 696	4 706	4 941	4 297	4 288	3 761	3 381	3 284†	3 017	3 094	3 002
Royal Air Force	KDMK	2 032	1 964	1 974	1 948	2 038	1 837	1 795	1 404	1 213	1 194	1 205
All Services	KDML	7 972	7 919	8 169	7 553	7 687	6 999	6 245	5 982†	5 255	5 234	5 142

1 Includes staff employed at units (including ships) and in hospitals.
2 Includes all members of the Nursing Services/Nursing Corps.

Source: Ministry of Defence/DASA (Tri- Service): 020 7218 4535

4.11 Health indicators for UK regular forces[1]

		1975	1980	1985	1990	1992	1993	1994	1995	1996	1997[3]	1998
Deaths[2]												
Total Number	LUIC	422	370	319	320	267	241	219	207	146	164	165
Male	LUID	414	367	314	314	263	230	212	198	144	155	157
Female	LUIE	8	3	5	6	4	11	7	9	2	9	8
Rates per thousand[3]												
Tri-service	LUIF	1.24	1.11	0.96	1.03	0.90	0.87	0.86	0.88	0.65	0.76	0.77
Navy	LUIG	0.81	1.07	0.81	0.80	0.78	0.75	0.68	0.62	0.57	0.85	0.59
Army[4]	LUIH	1.45	1.16	0.90	1.10	1.03	0.96	1.03	0.99	0.75	0.82	0.86
RAF	LUII	1.21	1.05	1.18	1.06	0.76	0.81	0.70	0.89	0.53	0.58	0.75
By Diagnosis												
Diseases	LUIJ	0.53	0.36	0.33	0.33	0.34	0.26	0.27	0.26	0.18	0.27	0.26
Injuries	LUIK	0.72	0.75	0.63	0.70	0.56	0.61	0.59	0.62	0.47	0.49	0.51
Others[5]	LUIL	–	–	–	–	–	–	–	–	–	–	–
Medical Discharges[6,7]												
Total Number	LUIM	1 822	1 010	1 398	1 419	1 251	1 220	1 345	1 626	1 727	1 636	1 598
Male	LUIN	1 746	962	1 365	1 367	1 175	1 134	1 190	1 406	1 416	1 407	1 327
Female	LUIO	76	48	33	52	76	86	155	220	311	229	271
Rates per thousand[3]												
Tri-service	LUIP	5.38	3.03	4.19	4.56	4.21	4.41	5.29	6.92	7.70	7.61	7.50
Navy	LUIQ	4.90	2.16	3.54	4.26	3.27	4.05	4.89	6.29	7.18	7.89	9.58
Army[4]	LUIR	6.35	3.66	5.97	6.32	6.08	6.03	7.62	9.60	10.47	9.38	9.01
RAF	LUIS	3.99	2.56	1.41	1.63	1.56	1.85	1.63	2.96	3.14	3.88	2.77
By Diagnosis												
Diseases	LUIT	4.53	2.53	3.16	3.95	3.64	3.75	4.09	4.60	4.77	4.93	4.91
Injuries	LUIU	0.84	0.50	1.03	0.61	0.57	0.66	1.20	2.31	2.93	2.67	2.40
Others[5]	LUIV	–	–	–	–	–	–	–	–	–	–	0.19
Causes of medical discharge[8]												
Musclo-skeletal disease	LUIW	..	0.66	1.42	2.39	2.31	2.22	2.55	2.67	3.04	2.71	2.77
Training & exercise injuries	LUIX	..	0.05	0.20	0.11	0.19	0.23	0.40	0.80	1.19	1.03	0.86
Mental Diseases	LUIY	..	0.49	0.37	0.30	0.30	0.38	0.41	0.54	0.42	0.56	0.64
Fall & jump injuries	LUIZ	..	0.07	0.17	0.09	0.04	0.05	0.11	0.38	0.51	0.51	0.49
All other causes	LUJA	..	1.77	2.03	1.67	1.37	1.52	1.82	2.52	2.54	2.80	2.74
Reported sickness events (provisional)[9,10]												
Total Number	LUJB	61 900	54 727	49 412	46 504	43 718	42 788	36 822	36 546	26 795	25 104	22 398
Male	LUJC	56 800	49 753	45 227	42 312	38 818	37 920	32 562	32 151	23 477	21 783	19 243
Female	LUJD	5 100	4 974	4 185	4 192	4 900	4 868	4 260	4 395	3 318	3 321	3 155
Rates per thousand[3]												
Tri-Service	LUJE	182	164	148	149	147	155	145	155	119	117	105
Navy	LUJF	..	157	140	108	105	110	99	99	87	90	76
Army[4]	LUJG	..	159	147	150	135	134	129	132	105	91	81
RAF	LUJH	..	180	156	177	199	223	206	234	169	188	177
By Diagnosis												
Diseases	LUJI	141	127	116	115	116	126	115	123	90	91	82
Injuries	LUJJ	34	31	26	27	23	21	23	23	23	20	17
Others[5]	LUJK	8	7	6	7	8	8	7	9	7	6	6
Average Strength[11]												
Tri-Service (thousands)	LUJL	339.2	332.8	333.7	311.1	296.9	276.5	254.2	235.0	224.3	215.1	213.2
Navy	LUJM	76.1	72.1	69.5	62.7	61.5	58.4	54.4	50.2	47.2	44.9	44.2
Army	LUJN	169.2	170.1	170.9	159.1	150.9	138.6	125.7	114.8	113.5	113.1	113.3
RAF	LUJO	93.9	90.6	93.3	89.3	84.5	79.4	74.1	70.0	63.6	57.1	55.7

1 These data have been revised since the previous edition as a result of updated information.
2 Includes deaths occuring on or off duty.
3 Rates per thousand have been calculated by taking the number of episodes in each year and dividing by the average strength for the year.
4 Army figures include Gurkhas, except those shown for 1975.
5 Used where no classifactory diagnosis was reported or where the person was not sick, e.g. admissions for investigation, preventive measures or elective surgery.
6 Those medically discharged out of the Services before the completion of their engagement.
7 Medical discharges by service in 1975 are for males only.
8 Main cause of medical discharge based on common causes in 1998.
9 Reported sickness events are based on the following events reported to DASA; all hospital events; all medical centre events over 48 hours and all sick at home events lasting more than 6 days. Deaths are also included but out-patient treatments are not. Readers should be aware that these data should be regarded as indicators of health activity only, they do not cover all aspects of health care since primary care and outpatients activity are not included. In addition DASA (Medical Statistics) are aware that not all sickness events are reported. Efforts are being made to quantify and improve this under reporting of sickness events.
10 For 1975 reported sickness events have been based on two days or more, whether occuring on or off duty, and terminating during the year. Episodes of Army personnel sickness not requiring admission into medical units and of Navy personnel sick on shore for less than 7 days are excluded. The number of reported sickness events for 1975 is rounded to the nearest hundred.
11 Average strength is calculated by taking th average strength at 1 Jan, 1 Apr, 1 Jul, 1 Oct of the year in question and 1 Jan of the following year.
12 Rates may not add up due to roundings.
13 Information is compiled from the recent information collected by DASA (Medical Statistics) as at 10 May 1999.

Source: Ministry of Defence/DASA (Medical Statistics): 01225 472963

4.12 Search and rescue operations at home

Number

		1988	1989	1990	1991	1992	1993	1994	1995	1996	1997	1998
Call outs[1]												
of helicopters[2]	KCME	1 734	1 736	1 851	1 779	1 997	1 858	1 766	2 059	1 904	1 753	1 720
of other aircraft[2]	KCMF	81	84	86	87	69	77	82	85	70	82	72
of HM ships and Auxilliary Vessels[2]	KCMG	1	7	1	5	5	8	6	8	1	3	2
of mountain rescue teams	KCMH	80	76	83	92	125	149	140	125	162	88	82
Persons rescued[3]:total	KCMI	1 234	1 275	1 866	1 458	1 379	1 568	1 336	1 532	1 489	1 145	1 164
by helicopters	KCMJ	1 194	1 241	1 418	1 423	1 358	1 475	1 294	1 488	1 391	1 132	1 153
by HM ships and Auxilliary Vessels	KCMK	–	–	–	–	–	–	–	–	–	–	–
by mountain rescue teams[4]	KCML	40	34	448	35	21	93	42	40	98	13	11
Incidents: total	KCMM	1 599	1 675	1 851	1 783	2 015	1 879	1 791	2 035	1 896	1 739	1 687

1 More than one element of the Search and Rescue (SAR) services called to the same incident. Consequently, the number of callouts is likely to be greater than the number of incidents.
2 Not permanently on stand-by.
3 Civilian or Military personnel who were removed from a hazardous environment or were transported from the scene by SAR Units in order to receive urgent medical attention.
4 People assisted by Mountain Rescue Teams, but subsequently transported from the scene by helicopter are credited as having been rescued by the helicopter unit concerned. The total incidents figure also includes any HMCG incidents under the control of ARCC Kinloss. :

Source: Ministry of Defence/DASA (Tri-Service): 020 7218 4535

4.13 UK Defence services and the civilian community

		1988	1989	1990	1991	1992	1993	1994	1995	1996	1997	1998
Military aid to civil ministries during industrial disputes												
(Manweeks of Service personnel deployed)	KCMN	14 762	11 550	–	–	–	–	–	1 827	256	389	1 394
Fishery protection[1]												
Vessels boarded	KCMO	2 202	2 284	1 992	2 284	2 465	2 540	2 080	1 878	2 224	1 758	1 829
Hydrographic services												
New charts produced	KCMP	212	148	240	96	94	–	–	–	–	–	–
New editions of charts	KCMQ	323	231	190	276	258	–	–	–	–	–	–
Charts printed(millions)	KCMR	2.6	2.5	2.9	3.1	3.1	–	–	–	–	–	–

1 This item relates to activities of the RN Fishery Squadron operating within British fishing limits under contract to MAFF. Boardings carried out by vessels of the Scottish Office Agriculture and Fisheries Department and the Department of Agriculture for Northern Ireland are not included.

Source: Ministry of Defence/DASA (Tri-Service): 020 7218 4535

Population and vital statistics

5

Population and vital statistics

This section begins with a summary of population figures for the United Kingdom and constituent countries for 1851 to 2031 and for Great Britain from 1801 (Table 5.1). Table 5.2 analyses the elements of population change. Table 5.3 gives details of the national sex and age structures for years up to the present date, with projected figures up to the year 2016. Legal marital condition of the population is shown in Table 5.4. The distribution of population at regional and local levels is summarised in Table 5.5.

In the main, historical series relate to census information, while mid-year estimates, which make allowance for under-enumeration in the census, are given for the recent past and the present (from 1961 onwards).

Population *(Tables 5.1 - 5.3)*

Figures shown in these tables relate to the population enumerated at successive censuses, (up to 1951), mid-year estimates (from 1961 to 1996) and population projections (up to 2011 or 2031).

Definition of resident population
The estimated population of an area includes all those usually resident in the area, whatever their nationality. HM Forces serving abroad are excluded from, but non-UK Armed Forces stationed here are included within the estimates of resident population. Students are taken to be resident at their term-time addresses.

The current series of estimates are updated annually. Starting with estimates derived from results of the 1991 Census of Population, allowance is made for subsequent births, deaths and migration and for estimated underenumeration in the 1991 Census.

Table 5.4 shows the population estimates by marital status. The 1991 figures for England and Wales have been rebased using 1991 Census results. Rebased population estimates by marital status for England and Wales for 1991-1995 were published in an ONS First Release in October 1997 and are available from the Population Estimates Unit.

Projected resident population of the United Kingdom and constituent countries *(Tables 5.1 - 5.3)*

These projections are prepared by the Government Actuary, in consultation with the Registrars General, as a common framework for use in national planning in a number of different fields. New projections are made at least every second year on assumptions regarding future fertility, mortality and migration which seem most appropriate on the basis of the statistical evidence available at the time. The population projections in Tables 5.1 - 5.3 are based on the estimates of the population of the United Kingdom at mid-1998 made by the Registrars General.

Geographical distribution of the population *(Table 5.5)*

The population enumerated in the censuses for 1911-1951, the mid-year population estimates for later years, and areas (1991 in km^2), are provided for standard regions of the United Kingdom, for metropolitan areas, for broad groupings of local authority districts by type within England and Wales, and for some of the larger cities. Projections of future sub-national population levels are prepared from time to time by the Registrar General, but are not shown in this publication.

Migration into and out of the United Kingdom *(Tables 5.7 and 5.8)*

A migrant into the United Kingdom is defined as a person who has resided abroad for a year or more and on entering has declared the intention to reside here for a year or more; and vice versa for a migrant from the United Kingdom. The estimates shown are derived from the International Passenger Survey (IPS), a sample survey covering the principal air and sea routes between the United Kingdom and overseas but excluding routes to and from the Irish Republic. Migration between the Channel Islands and the Isle of Man and the rest of the world has been excluded from these tables with effect from 1988. It is also highly likely that the data exclude asylum seekers and persons admitted as short-term visitors who are subsequently granted an extension of stay for a year or more for other reasons, for example as students or on the basis of marriage. After taking account of persons leaving the United Kingdom for a short-term period who stay overseas for periods longer than originally intended, adjustment is needed to net migration ranges from about 10 thousand in 1981 to 32 thousand in 1997, an average of approximately 30 thousand. The overall net inflow for 1997, including migration with the Irish Republic, is about 92 thousand.

Acceptances for settlement in the United Kingdom *(Table 5.9)*

This table presents in geographic regions, the statistics of individual nationalities, arranged alphabetically within each region. The figures are on a different basis from those derived from IPS (Tables 5.6 and 5.7) and relate only to people subject to immigration control. Persons accepted for settlement are allowed to stay indefinitely in the United Kingdom. They exclude temporary migrants such as students and generally relate only to non-EEA nationals. Settlement can occur several years after entry to the country.

Applications received for asylum in the United Kingdom excluding dependants *(Table 5.10)*

This table shows statistics of applications for asylum in the United Kingdom. Figures are shown of the main applicant nationalities by geographic region. The basis of assessing asylum applications, and hence of deciding whether to grant asylum in the United Kingdom, is the 1951 United Nations Convention on Refugees.

Divorces *(Tables 5.12 and 5.13)*

Table 5.12 gives figures for dissolutions and annulments by duration of marriage and age of wife. Scottish figures prior to 1979 relate only to marriages which took place in Scotland. Data in Table 5.13 give petitions filed for divorce, divorces by fact proven and separations.

Births *(Tables 5.14 and 5.15)*

For Scotland and Northern Ireland the number of births relate to those registered during the year. For England and Wales the figures up to and including 1930-32 are for those registered while later figures relate to births occurring in each year.

Deaths *(Tables 5.18 and 5.19)*

The figures relate to the number of deaths registered during each calendar year. However, from 1993 onwards, the figures for England and Wales represent occurrences. This change has little effect on annual totals.

Changes in coding practices for England and Wales, particularly coding of underlying cause of death, from January 1993 have led to some differences in the pattern of cause of death (eg. pneumonia) as compared with previous years. These changes result in a return to the rules in place prior to 1984 and bring England and Wales back into line with Scotland, Northern Ireland, and other countries. For further details, see the Introduction to *Mortality Statistics: cause 1993*, Series DH2, no. 20, HMSO (1993).

Life tables *(Table 5.22)*

The interim life tables are constructed from the estimated populations in 1996-98 and the deaths occurring in those years for England and Wales and registered in those years for Scotland and Northern Ireland.

5.1 Population summary

Thousands

	United Kingdom			England and Wales			Wales	Scotland			Northern Ireland		
	Persons	Males	Females	Persons	Males	Females	Persons	Persons	Males	Females	Persons	Males	Females
Enumerated population: census figures													
1801	..	..	..	8 893	4 255	4 638	587	1 608	739	869	..	..	..
1851	22 259	10 855	11 404	17 928	8 781	9 146	1 163	2 889	1 376	1 513	1 442	698	745
1901	38 237	18 492	19 745	32 528	15 729	16 799	2 013	4 472	2 174	2 298	1 237	590	647
1911	42 082	20 357	21 725	36 070	17 446	18 625	2 421	4 761	2 309	2 452	1 251	603	648
1921[1]	44 027	21 033	22 994	37 887	18 075	19 811	2 656	4 882	2 348	2 535	1 258	610	648
1931[1]	46 038	22 060	23 978	39 952	19 133	20 819	2 593	4 843	2 326	2 517	1 243	601	642
1951	50 225	24 118	26 107	43 758	21 016	22 742	2 599	5 096	2 434	2 662	1 371	668	703
1961	52 709	25 481	27 228	46 105	22 304	23 801	2 644	5 179	2 483	2 697	1 425	694	731
Resident population: mid-year estimates													
	DYAY	BBAB	BBAC	BBAD	BBAE	BBAF	KGJM	BBAG	BBAH	BBAI	BBAJ	BBAK	BBAL
1963	53 625	25 992	27 633	46 973	22 787	24 186	2 664	5 205	2 500	2 705	1 447	705	741
1964	53 991	26 191	27 800	47 324	22 979	24 346	2 677	5 209	2 501	2 707	1 458	711	747
1965	54 350	26 368	27 982	47 671	23 151	24 521	2 693	5 210	2 501	2 709	1 468	716	752
1966	54 643	26 511	28 132	47 967	23 296	24 671	2 702	5 201	2 496	2 704	1 476	719	757
1967	54 959	26 673	28 286	48 272	23 451	24 821	2 710	5 198	2 496	2 702	1 489	726	763
1968	55 214	26 784	28 429	48 511	23 554	24 957	2 715	5 200	2 498	2 702	1 503	733	770
1969	55 461	26 908	28 553	48 738	23 666	25 072	2 722	5 209	2 503	2 706	1 514	739	776
1970	55 632	26 992	28 641	48 891	23 738	25 153	2 729	5 214	2 507	2 707	1 527	747	781
1971	55 928	27 167	28 761	49 152	23 897	25 255	2 740	5 236	2 516	2 720	1 540	755	786
1972	56 097	27 259	28 837	49 327	23 989	25 339	2 755	5 231	2 513	2 717	1 539	758	782
1973	56 223	27 332	28 891	49 459	24 061	25 399	2 773	5 234	2 515	2 719	1 530	756	774
1974	56 236	27 349	28 887	49 468	24 075	25 393	2 785	5 241	2 519	2 722	1 527	755	772
1975	56 226	27 361	28 865	49 470	24 091	25 378	2 795	5 232	2 516	2 716	1 524	753	770
1976	56 216	27 360	28 856	49 459	24 089	25 370	2 799	5 233	2 517	2 716	1 524	754	770
1977	56 190	27 345	28 845	49 440	24 076	25 364	2 801	5 226	2 515	2 711	1 523	754	769
1978	56 178	27 330	28 849	49 443	24 067	25 375	2 804	5 212	2 509	2 704	1 523	754	770
1979	56 240	27 373	28 867	49 508	24 113	25 395	2 810	5 204	2 505	2 699	1 528	755	773
1980	56 330	27 411	28 919	49 603	24 156	25 448	2 816	5 194	2 501	2 693	1 533	755	778
1981	56 352	27 409	28 943	49 634	24 160	25 474	2 813	5 180	2 495	2 685	1 538	754	783
1982	56 318	27 391	28 927	49 613	24 148	25 466	2 806	5 167	2 490	2 677	1 538	754	784
1983	56 377	27 429	28 948	49 681	24 190	25 491	2 807	5 153	2 484	2 669	1 543	756	788
1984	56 506	27 511	28 995	49 810	24 270	25 540	2 806	5 146	2 482	2 664	1 550	760	791
1985	56 685	27 611	29 074	49 990	24 369	25 621	2 810	5 137	2 479	2 658	1 558	763	795
1986	56 852	27 698	29 153	50 162	24 456	25 706	2 820	5 123	2 474	2 649	1 567	768	798
1987	57 009	27 789	29 220	50 321	24 546	25 775	2 833	5 113	2 470	2 643	1 575	773	802
1988	57 158	27 876	29 282	50 487	24 641	25 846	2 854	5 093	2 461	2 632	1 578	774	804
1989	57 358	27 989	29 368	50 678	24 750	25 928	2 869	5 097	2 463	2 634	1 583	777	806
1990	57 561	28 118	29 443	50 869	24 872	25 998	2 878	5 102	2 466	2 636	1 589	780	809
1991	57 808	28 246	29 562	51 100	24 995	26 104	2 891	5 107	2 470	2 637	1 601	781	820
1992	58 006	28 362	29 645	51 277	25 099	26 178	2 899	5 111	2 473	2 638	1 618	791	828
1993	58 191	28 474	29 718	51 439	25 198	26 241	2 906	5 120	2 479	2 642	1 632	797	835
1994	58 395	28 592	29 803	51 621	25 304	26 317	2 913	5 132	2 486	2 646	1 642	802	840
1995	58 606	28 727	29 878	51 820	25 433	26 387	2 917	5 137	2 489	2 647	1 649	805	844
1996	58 801	28 856	29 946	52 010	25 557	26 453	2 921	5 128	2 486	2 642	1 663	812	851
1997	59 009	28 990	30 019	52 211	25 684	26 527	2 927	5 123	2 484	2 638	1 675	821	854
1998	59 237	29 128	30 108	52 428	25 817	26 611	2 933	5 120	2 484	2 636	1 689	827	861
Resident population: projections (mid-year)[2]													
2001	59 954	29 581	30 372	53 137	26 258	26 879	2 950	5 109	2 485	2 624	1 708	839	869
2006	60 860	30 147	30 714	54 021	26 799	27 222	2 969	5 098	2 489	2 609	1 742	858	883
2011	61 773	30 696	31 077	54 915	27 328	27 587	2 993	5 087	2 492	2 595	1 771	876	895
2021	63 642	31 717	31 925	56 763	28 325	28 437	3 047	5 058	2 487	2 570	1 821	904	917

1 Figures for Northern Ireland are estimated. The population at the Census of 1926 was 1 257 thousand (608 thousand males and 649 thousand females).
2 These projections are 1998-based. More detail given in Table 5.3.

Sources: Office for National Statistics: 01329 813233;
General Register Office (Northern Ireland);
General Register Office (Scotland);
Government Actuary's Department: 020 7211 2622

5.2 Population changes

Thousands

		Average annual change				
	Population[1] at start of period	Overall annual change	Births	Deaths[2]	Excess of births over deaths	Net migration and other adjustments[3]
United Kingdom						
1901 - 1911	38 237	385	1 091	624	467	-82
1911 - 1921	42 082	195	975	689	286	-92
1921 - 1931	44 027	201	824	555	268	-67
1931 - 1951	46 038	213	793	603	190	22
1951 - 1961	50 225	258	839	593	246	12
1961 - 1971	52 807	310	962	638	324	-14
1971 - 1981	55 928	42	736	666	69	-27
1981 - 1991	56 352	146	757	655	103	43
1991 - 1998	57 808	204	748	637	112	93
1998 - 2001[4]	59 237	236	714	630	84	155
2001 - 2011	59 954	178	701	614	87	95
2011 - 2021	61 773	185	712	620	92	95
England and Wales						
1901 - 1911	32 528	354	929	525	404	-50
1911 - 1921	36 070	182	828	584	244	-62
1921 - 1931	37 887	207	693	469	224	-17
1931 - 1951	39 952	193	673	518	155	38
1951 - 1961	43 758	244	714	516	197	47
1961 - 1971	46 196	296	832	560	272	23
1971 - 1981	49 152	48	638	585	53	-5
1981 - 1991	49 634	147	664	576	89	58
1991 - 1998	51 100	190	662	561	101	89
1998 - 2001[4]	52 428	236	634	556	78	159
2001 - 2011	53 137	178	623	542	81	97
2011 - 2021	54 915	185	635	547	88	97
Scotland						
1901 - 1911	4 472	29	131	76	54	-25
1911 - 1921	4 761	12	118	82	36	-24
1921 - 1931	4 882	-4	100	65	35	-39
1931 - 1951	4 843	13	92	67	25	
1951 - 1961	5 096	9	95	62	34	-25
1961 - 1971	5 184	3	97	63	34	-30
1971 - 1981	5 236	-6	70	64	6	-11
1981 - 1991	5 180	-7	66	63	3	-10
1991 - 1998	5 107	2	62	61	1	1
1998 - 2001[4]	5 120	-4	57	59	-2	-2
2001 - 2011	5 109	-2	55	56	-1	-1
2011 - 2021	5 087	-3	55	57	-2	-1
Northern Ireland						
1901 - 1911	1 237	1	31	23	8	-6
1911 - 1921	1 251	1	29	22	7	-6
1921 - 1931	1 258	-2	30	21	9	-11
1931 - 1951	1 243	6	28	18	10	-4
1951 - 1961	1 371	6	30	15	15	-9
1961 - 1971	1 427	11	33	16	17	-6
1971 - 1981	1 540	0	28	17	11	-11
1981 - 1991	1 538	6	27	16	12	-5
1991 - 1998	1 601	12	25	15	10	3
1998 - 2001[4]	1 689	-4	23	15	8	-2
2001 - 2011	1 708	-2	22	15	7	-1
2011 - 2021	1 771	-3	21	16	5	-1

1 Census enumerated population up to 1951; mid-year estimates of resident population from 1961 to 1998 and mid-1998-based projections of resident population thereafter.

2 Including deaths of non-civilians and merchant seamen who died outside the country. These numbered 577 000 in 1911-1921 and 240 000 in 1931-1951 for England and Wales; 74 000 in 1911-1921 and 34 000 in 1931-1951 for Scotland; and 10 000 in 1911-1926 for Northern Ireland.

3 Changes in census visitor balance, in Armed Forces, asylum seekers, etc.

4 1998-based national population projections.

Sources: Government Actuary's Department: 020 7211 2622; Office for National Statistics: 01329 813233; General Register Office (Northern Ireland); General Register Office (Scotland)

5.3 Age distribution of the resident population

Thousands

		United Kingdom												
		Population enumerated in Census			Estimated mid-year resident population					Projected mid-year resident population[1]				
		1901	1931	1951	1961	1971	1981	1991	1998	2001	2006	2011	2016	2021
Persons: All ages	KGUA	38 237	46 038	50 225	52 807	55 928	56 352	57 808	59 237	59 954	60 860	61 773	62 729	63 642
Under 1	KGUK	938	712	773	4 274[3]	899	730	794	715	712	696	699	711	713
1 - 4	KABA	3 443	2 818	3 553	..	3 654	2 725	3 092	2 956	2 875	2 813	2 781	2 822	2 857
5 - 9	KGUN	4 106	3 897	3 689	3 819	4 684	3 677	3 674	3 914	3 781	3 591	3 513	3 485	3 538
10 - 14	KGUO	3 934	3 746	3 310	4 267	4 232	4 470	3 501	3 795	3 926	3 790	3 601	3 523	3 495
15 - 19	KGUP	3 826	3 989	3 175	3 748	3 862	4 735	3 739	3 670	3 752	3 977	3 842	3 653	3 576
20 - 29	KABB	6 982	7 865	7 154	6 570	7 968	8 113	9 298	7 910	7 606	7 651	8 094	8 184	7 862
30 - 44	KABC	7 493	9 717	11 125	10 529	9 797	10 956	12 221	13 347	13 854	13 538	12 477	11 845	12 118
45 - 59	KABD	4 639	7 979	9 558	10 605	10 202	9 540	9 500	10 820	11 247	11 993	12 734	13 579	13 273
60 - 64	KGUY	1 067	1 897	2 422	2 788	3 222	2 935	2 888	2 818	2 859	3 244	3 824	3 532	4 000
65 - 74	KBCP	1 278	2 461	3 689	3 977	4 764	5 195	5 067	4 965	4 908	4 988	5 476	6 388	6 642
75 - 84	KBCU	470	844	1 555	1 885	2 159	2 675	3 136	3 205	3 264	3 344	3 382	3 561	4 000
85 Plus	KGVD	61	113	224	346	485	602	896	1 122	1 170	1 235	1 349	1 446	1 569
School ages (5-15)	KBWU	..	..	7 649	..	9 704	9 086	7 855	8 439	8 470	8 186	7 842	7 726	7 735
Under 18	KGUD	..	..	13 248	..	15 798	14 470	13 184	13 582	13 555	13 275	12 843	12 702	12 726
Pensionable ages[2]	KFIA	2 387	4 421	6 828	7 747	9 123	10 031	10 597	10 730	10 800	11 219	11 915	12 040	12 210
Males: All ages	KGWA	18 492	22 060	24 118	25 528	27 167	27 409	28 246	29 128	29 581	30 147	30 696	31 234	13 717
Under 1	KGWK	471	361	397	2 194[3]	461	374	407	366	364	356	358	364	365
1 - 4	KBCV	1 719	1 423	1 818	..	1 874	1 399	1 588	1 516	1 473	1 440	1 424	1 445	1 463
5 - 9	KGWN	2 052	1 967	1 885	1 956	2 401	1 889	1 888	2 005	1 939	1 841	1 800	1 786	1 813
10 - 14	KGWO	1 972	1 892	1 681	2 185	2 175	2 295	1 800	1 948	2 013	1 945	1 847	1 807	1 792
15 - 19	KGWP	1 898	1 987	1 564	1 897	1 976	2 424	1 925	1 884	1 923	2 034	1 966	1 868	1 828
20 - 29	KBCW	3 293	3 818	3 509	3 288	4 024	4 103	4 738	4 055	3 901	3 904	4 115	4 158	3 994
30 - 44	KBCX	3 597	4 495	5 461	5 237	4 938	5 513	6 141	6 763	7 053	6 925	6 383	6 043	6 157
45 - 59	KBUU	2 215	3 753	4 493	5 137	4 970	4 711	4 732	5 387	5 602	5 989	6 407	6 875	6 753
60 - 64	KGWY	490	894	1 061	1 250	1 507	1 376	1 390	1 380	1 401	1 592	1 883	1 744	1 989
65 - 74	KBWL	565	1 099	1 560	1 605	1 999	2 264	2 272	2 290	2 295	2 373	2 622	3 074	3 206
75 - 84	KBWM	196	335	617	675	716	921	1 151	1 237	1 297	1 383	1 458	1 574	1 787
85 plus	KGXD	23	36	70	105	126	141	214	297	319	363	433	497	569
School ages (5-15)	KBWV	..	..	3 895	..	4 982	4 666	4 038	4 328	4 345	4 198	4 021	3 961	3 965
Under 18	KGWD	..	..	6 753	..	8 108	7 429	6 776	6 966	6 950	6 804	6 583	6 508	6 521
Pensionable ages[2]	KFIB	785	1 471	2 247	2 385	2 841	3 325	3 637	3 824	3 911	4 119	4 513	5 145	5 562
Females: All ages	KGYA	19 745	23 978	26 107	27 279	28 761	28 943	29 562	30 108	30 372	30 714	31 077	31 495	31 925
Under 1	KGYK	466	351	376	2 079[3]	437	356	387	349	347	340	341	347	348
1 - 4	KBWN	1 724	1 397	1 735	..	1 779	1 326	1 505	1 440	1 403	1 373	1 357	1 377	1 394
5 - 9	KGYN	2 054	1 930	1 804	1 863	2 283	1 788	1 786	1 909	1 841	1 750	1 713	1 699	1 725
10 - 14	KGYO	1 962	1 854	1 629	2 083	2 057	2 175	1 701	1 848	1 913	1 845	1 754	1 717	1 703
15 - 19	KGYP	1 928	2 002	1 611	1 851	1 887	2 311	1 814	1 785	1 828	1 943	1 876	1 785	1 747
20 - 29	KBWO	3 690	4 047	3 644	3 282	3 945	4 009	4 560	3 855	3 704	3 747	3 979	4 026	3 869
30 - 44	KBWP	3 895	5 222	5 663	5 292	4 859	5 443	6 080	6 583	6 801	6 613	6 094	5 802	5 961
45 - 59	KBWR	2 424	4 226	5 065	5 467	5 231	4 829	4 769	5 433	5 645	6 004	6 328	6 703	6 520
60 - 64	KGYY	577	1 003	1 361	1 539	1 715	1 559	1 498	1 438	1 458	1 652	1 941	1 788	2 011
65 - 74	KBWS	713	1 361	2 127	2 372	2 765	2 931	2 795	2 674	2 613	2 615	2 854	3 314	3 435
75 - 84	KBWT	274	509	937	1 210	1 443	1 755	1 986	1 968	1 968	1 961	1 924	1 987	2 213
85 Plus	KGZD	38	77	154	241	359	461	682	825	851	872	916	949	1 000
School ages (5-15)	KBWW	..	..	3 753	..	4 722	4 421	3 817	4 111	4 125	3 988	3 820	3 765	3 770
Under 18	KGYD	..	..	6 495	..	7 690	7 041	6 408	6 616	6 605	6 471	6 260	6 194	6 205
Pensionable ages[2]	KFIC	1 601	2 950	4 580	5 362	6 282	6 706	6 961	6 906	6 889	7 099	7 401	6 895	6 648

1 1998-based projections are made as described in the introductory note on page 24.
2 The pensionable age population is that over state retirement age. The 2011, 2016 and 2021 figures take account of planned changes in retirement age from 65 for men and 60 for women at present to 65 for both sexes. This change will be phased in between April 2010 and March 2020.
3 This is for ages 0 - 4.

Sources: Government Actuary's Department: 020 7211 2622; Office for National Statistics: 01329 813233

5.3 Age distribution of the resident population

continued

Thousands

		England						Wales				
		Estimated mid-year resident population			Projected population[1]			Estimated mid-year resident population			Projected population[1]	
		1981	1991	1998	2001	2011		1981	1991	1998	2001	2011
Persons: All ages	KCCI	46 821	48 208	49 495	50 187	51 922	KERY	2 813	2 891	2 933	2 950	2 993
Under 1	KCCJ	598	663	599	600	589	KFAC	36	39	34	33	33
1 - 4	KCCK	2 235	2 574	2 475	2 414	2 345	KFBX	136	154	140	136	131
5 - 9	KCCL	3 011	3 035	3 264	3 161	2 960	KFCA	185	187	192	182	165
10 - 14	KCCM	3 666	2 880	3 141	3 268	3 024	KFCB	222	178	195	198	174
15 - 19	KCCN	3 897	3 083	3 033	3 104	3 215	KFCC	233	188	186	193	188
20 - 29	KCEG	6 734	7 790	6 606	6 370	6 792	KFCD	381	422	362	348	394
30 - 44	KCEH	9 175	10 231	11 197	11 654	10 548	KFCE	536	587	612	627	552
45 - 59	KCEQ	7 948	7 920	9 060	9 423	10 686	KFCF	485	486	553	571	607
60 - 64	KCEW	2 449	2 399	2 340	2 377	3 211	KFCG	158	154	150	152	201
65 - 74	KCGD	4 347	4 222	4 127	4 081	4 580	KFCH	272	284	273	266	293
75 - 84	KCJG	2 249	2 645	2 694	2 739	2 829	KFCI	139	165	177	182	182
85 Plus	KCKJ	511	763	957	997	1 142	KFCK	29	47	59	62	75
School ages (5-15)	KCWX	7 451	6 473	7 009	7 060	6 594	KFCL	453	399	425	420	374
Under 18	KCWY	11 871	10 899	11 297	11 314	10 804	KFCM	721	665	674	666	611
Pensionable ages[2]	KEAA	8 403	8 870	8 968	9 025	9 983	KFEB	525	576	585	587	638
Males: All ages	KEAB	22 795	23 588	24 378	24 807	25 845	KFEI	1 365	1 407	1 439	1 451	1 482
Under 1	KEAC	306	340	307	307	302	KFEJ	18	20	17	17	17
1 - 4	KEAD	1 147	1 322	1 270	1 236	1 201	KFEK	70	79	72	70	67
5 - 9	KEAE	1 547	1 561	1 673	1 622	1 517	KFEL	95	96	99	93	84
10 - 14	KEAF	1 883	1 482	1 613	1 676	1 551	KFFA	113	92	100	102	89
15 - 19	KECA	1 996	1 588	1 558	1 593	1 646	KFFN	119	97	95	99	96
20 - 29	KECB	3 404	3 974	3 385	3 265	3 449	KFHA	193	214	189	181	201
30 - 44	KECC	4 623	5 148	5 691	5 954	5 406	KFHB	270	294	308	317	283
45 - 59	KECD	3 938	3 957	4 519	4 701	5 394	KFHW	240	242	275	284	303
60 - 64	KECE	1 154	1 159	1 151	1 169	1 584	KFQO	73	74	73	75	99
65 - 74	KECF	1 902	1 900	1 913	1 918	2 201	KFQV	118	128	126	125	141
75 - 84	KECG	777	975	1 045	1 094	1 227	KFUK	48	60	68	72	78
85 plus	KECH	119	183	254	273	369	KFUL	7	11	16	17	24
School ages (5-15)	KECI	3 827	3 330	3 595	3 622	3 381	KFUV	232	206	218	215	192
Under 18	KECJ	6 096	5 604	5 796	5 802	5 537	KFVE	370	342	345	341	313
Pensionable ages[2]	KECK	2 798	3 058	3 212	3 285	3 797	KFVF	173	199	210	214	242
Females: All ages	KEJV	24 026	24 620	25 117	25 380	26 077	KFVL	1 448	1 484	1 495	1 499	1 510
Under 1	KEJW	292	324	292	293	288	KFYW	18	19	16	16	16
1 - 4	KEJX	1 088	1 253	1 205	1 178	1 145	KFZJ	66	75	68	66	64
5 - 9	KEKP	1 464	1 474	1 592	1 539	1 444	KGCK	90	91	94	89	80
10 - 14	KEKQ	1 783	1 399	1 528	1 593	1 473	KGCM	109	86	95	97	85
15 - 19	KEKR	1 901	1 495	1 475	1 511	1 569	KGCN	114	91	91	95	92
20 - 29	KEKS	3 330	3 816	3 221	3 105	3 344	KGCO	189	208	173	166	192
30 - 44	KENR	4 553	5 083	5 506	5 701	5 141	KGCP	265	294	304	310	269
45 - 59	KEOQ	4 009	3 964	4 540	4 721	5 291	KGGZ	246	244	278	287	304
60 - 64	KEOZ	1 295	1 239	1 189	1 208	1 627	KGIY	85	80	76	77	102
65 - 74	KEQJ	2 445	2 323	2 214	2 163	2 380	KGKR	154	156	147	141	152
75 - 84	KEQK	1 472	1 670	1 649	1 645	1 601	KGTQ	91	105	109	110	103
85 Plus	KEQL	392	580	704	724	773	KGTZ	22	36	44	45	51
School ages (5-15)	KEQM	3 625	3 143	3 413	3 438	3 214	KGVG	221	194	207	205	182
Under 18	KEQN	5 775	5 295	5 501	5 513	5 267	KGVH	351	323	329	324	298
Pensionable ages[2]	KEQO	5 605	5 812	5 756	5 740	6 186	KGVK	352	377	375	373	396

1 1998-based projections are made as described in the introductory note on page 24.

2 The pensionable age population is that over state retirement age. The 2011 figures take account of planned changes in retirement age from 65 for men and 60 for women at present to 65 for both sexes. This change will be phased in between April 2010 and March 2020.

Sources: Government Actuary's Department: 020 7211 2622; Office for National Statistics: 01329 813233

5.3 Age distribution of the resident population

continued

Thousands

		Scotland						Northern Ireland				
		Estimated mid-year resident population			Projected population[1]			Estimated mid-year resident population			Projected population[1]	
		1981	1991	1998	2001	2011		1981	1991	1998	2001	2011
Persons: All ages	KGVP	5 180	5 107	5 120	5 109	5 087	KIOY	1 538	1 601	1 689	1 708	1 771
Under 1	KHAQ	69	66	58	56	55	KIOZ	27	26	24	23	22
1 - 4	KHCT	249	259	243	230	218	KIPA	104	104	98	95	87
5 - 9	KHDN	348	321	325	313	275	KIPN	133	131	132	125	113
10 - 14	KHDQ	433	313	324	325	284	KIPP	148	129	135	135	120
15 - 19	KHDT	459	340	324	325	315	KIPQ	146	128	127	130	124
20 - 29	KHDU	771	834	696	651	662	KIPR	227	253	247	237	245
30 - 44	KHDV	971	1 088	1 176	1 195	1 009	KIPS	273	315	362	378	368
45 - 59	KHFK	880	853	932	966	1 100	KIPT	227	241	275	288	341
60 - 64	KHOZ	260	265	257	256	320	KIPU	68	70	71	74	93
65 - 74	KHTU	460	441	442	439	460	KIPV	116	119	122	122	143
75 - 84	KHUO	232	259	260	267	288	KIPW	55	67	74	77	84
85 Plus	KHUQ	49	69	82	86	101	KIPX	13	17	24	25	32
School ages (5-15)	KHVV	871	698	713	703	616	KIPY	311	285	292	287	257
Under 18	KIMT	1 377	1 153	1 145	1 117	1 012	KIQL	502	466	466	458	416
Pensionable ages[2]	KIMU	882	910	920	926	993	KIQM	221	241	257	262	300
Males: All ages	KIMV	2 495	2 470	2 484	2 485	2 492	KIQN	754	781	827	839	876
Under 1	KIMW	35	34	30	29	28	KIQO	14	13	12	12	11
1 - 4	KIMX	128	133	124	118	112	KIQP	53	54	50	49	45
5 - 9	KIMY	178	164	166	160	141	KIQQ	69	67	67	64	58
10 - 14	KIMZ	222	161	166	167	146	KIQR	76	66	69	69	61
15 - 19	KINA	234	174	165	165	161	KIQS	75	66	65	67	64
20 - 29	KINB	390	423	354	332	337	KIQT	117	127	128	123	128
30 - 44	KINC	483	543	586	596	507	KIQU	137	156	179	186	186
45 - 59	KIND	424	415	457	475	542	KIQV	109	118	135	142	167
60 - 64	KINE	118	124	122	122	155	KIQW	32	32	34	35	45
65 - 74	KINR	194	192	197	197	213	KIRJ	50	52	54	55	68
75 - 84	KINS	77	91	96	101	118	KIRK	20	24	28	30	35
85 plus	KINT	11	16	21	22	31	KIRL	4	4	6	7	10
School ages (5-15)	KINU	446	358	365	360	317	KIRM	160	146	150	147	132
Under 18	KINV	706	591	586	572	520	KIRN	258	239	239	235	213
Pensionable ages[2]	KINW	282	299	314	321	362	KIRO	73	81	88	91	112
Females: All ages	KINX	2 685	2 637	2 636	2 624	2 595	KIRP	783	820	861	869	895
Under 1	KINY	33	32	28	28	27	KIRQ	13	13	12	11	11
1 - 4	KINZ	121	126	118	112	106	KIRR	51	51	48	47	43
5 - 9	KIOA	170	157	159	153	133	KIRS	65	64	64	61	55
10 - 14	KIOB	211	153	158	158	138	KIRT	72	63	65	65	59
15 - 19	KIOC	225	166	158	159	154	KIRU	70	62	62	63	60
20 - 29	KIOO	381	411	342	319	325	KISH	110	125	119	114	118
30 - 44	KIOP	488	545	590	599	502	KISI	137	159	183	191	182
45 - 59	KIOQ	456	437	475	491	558	KISJ	118	123	139	146	175
60 - 64	KIOR	142	141	135	135	165	KISK	37	38	37	38	47
65 - 74	KIOS	265	249	245	241	247	KISL	66	67	68	67	76
75 - 84	KIOT	155	168	164	166	170	KISM	36	43	46	47	49
85 Plus	KIOU	38	53	61	63	70	KISN	9	13	17	18	22
School ages (5-15)	KIOV	424	340	348	343	299	KISO	151	139	142	140	125
Under 18	KIOW	671	562	559	545	492	KISP	244	227	228	223	203
Pensionable ages[2]	KIOX	600	611	606	605	631	KISQ	148	160	168	171	188

1 1998-based projections are made as described in the introductory note on page 24.

2 The pensionable age population is that over state retirement age. The 2011 figures take account of planned changes in retirement age from 65 for men and 60 for women at present to 65 for both sexes. This change will be phased in between April 2010 and March 2020.

Sources: Office for National Statistics: 01329 813233; General Register Office (Scotland); General Register Office (Northern Ireland); Government Actuary's Department: 020 7211 2622

5.4 Marital condition (de jure) : estimated population

Thousands

		United Kingdom[2]								England and Wales						
		Males				Females				Males				Females		
		1971	1981	1991[1]		1971	1981	1991[1]		1971	1981	1991		1971	1981	1991
All ages:																
Single	KQCA	12 120	12 168	12 874	KQDP	11 131	10 860	11 231	KRPL	10 507	10 614	11 320	KUBS	9 584	9 424	9 829
Married	KQCB	14 067	13 791	13 237	KQDQ	14 130	13 856	13 364	KRPM	12 522	12 238	11 745	KVCC	12 566	12 284	11 838
Widowed	KQCC	779	793	828	KQDR	3 178	3 331	3 370	KRPN	682	698	731	KVCD	2 810	2 939	2 978
Divorced	KQCD	201	657	1 306	KQDS	321	897	1 598	KRPO	187	611	1 200	KVCE	296	828	1 459
Age groups:																
0 - 14: Single	KQCE	6 912	5 956	5 683	KQDT	6 557	5 646	5 378	KRPP	5 984	5 181	4 991	KVCF	5 672	4 910	4 720
15 - 19: Single	KQCF	1 936	2 400	1 916	KQDU	1 726	2 203	1 778	KRPQ	1 677	2 095	1 677	KVCG	1 491	1 923	1 554
Married	KQCG	39	24	9	KQDV	160	107	36	KRPR	34	20	8	KVCH	142	93	32
Widowed	KQCH	–	–	–	KQDW	–	–	–	KRPS	–	–	–	KVCI	–	–	–
Divorced	KQCI	–	–	–	KQDX	–	–	–	KRPT	–	–	–	KVCJ	–	–	–
20 - 24: Single	KQCJ	1 376	1 618	2 004	KQDY	857	1 150	1 616	KRPU	1 211	1 420	1 764	KVCK	745	1 007	1 421
Married	KQCK	781	542	285	KQDZ	1 251	930	558	KRPV	689	466	249	KVCL	1 113	811	490
Widowed	KQCL	–	1	–	KQYZ	2	2	1	KRPW	–	1	–	KVCM	2	2	1
Divorced	KQCM	3	11	13	KQZA	10	30	32	KRPX	3	10	12	KVCN	9	27	29
25 - 34: Single	KQCN	722	1 020	1 925	KQZB	379	563	1 280	KRPY	637	906	1 718	KVCO	326	496	1 135
Married	KQCO	2 762	2 847	2 393	KQZC	2 980	3 160	2 831	KRPZ	2 450	2 508	2 100	KVCP	2 635	2 791	2 488
Widowed	KQCP	4	5	3	KQZD	14	16	9	KRQA	4	4	2	KVCQ	12	13	8
Divorced	KQCQ	41	163	270	KQZE	68	236	345	KRQB	38	151	245	KVCR	63	218	312
35 - 44: Single	KQCR	366	359	548	KQZF	242	198	322	KRQC	317	316	482	KVEH	201	170	280
Married	KQCS	2 838	2 845	3 002	KQZG	2 871	2 878	3 118	KRQD	2 513	2 519	2 658	KVEI	2 529	2 540	2 760
Widowed	KQCT	15	14	14	KQZH	57	49	41	KRQE	13	12	12	KVEJ	48	41	34
Divorced	KQCU	52	191	423	KQZI	72	240	487	KRQF	48	178	388	KVEK	66	222	444
45 - 54: Single	KQCV	326	293	287	KQZJ	298	203	168	KRQG	279	254	251	KVEL	248	169	144
Married	KQCW	2 924	2 640	2 648	KQZK	2 891	2 598	2 629	KUAR	2 605	2 338	2 347	KVEM	2 570	2 292	2 322
Widowed	KQCX	55	44	37	KQZL	218	176	138	KUBA	47	38	31	KVEN	187	150	118
Divorced	KQCY	50	144	315	KQZM	75	171	361	KUBB	46	134	290	KVEO	69	158	332
55 - 59: Single	KQCZ	140	149	116	KQZN	177	131	82	KUBC	118	128	101	KVEP	148	108	69
Married	KQDA	1 399	1 336	1 186	KQZO	1 293	1 267	1 114	KUBD	1 250	1 192	1 050	KVEQ	1 154	1 130	982
Widowed	KQDB	57	53	40	KQZP	247	216	159	KUBE	49	46	34	KVER	213	186	136
Divorced	KQDC	21	53	103	KQZQ	33	68	117	KUBF	20	49	95	KVES	31	63	107
60 - 64: Single	KQDD	125	115	120	KQZR	208	131	97	KUBG	105	98	104	KVET	174	109	80
Married	KQDE	1 276	1 149	1 121	KQZS	1 098	1 046	1 023	KUBH	1 140	1 029	997	KVEU	985	936	908
Widowed	KQDF	89	74	73	KQZT	382	328	289	KUBI	77	64	63	KVEV	332	284	250
Divorced	KQDG	17	38	76	KQZU	27	55	89	KUBJ	16	35	70	KVEW	25	51	82
65 - 74: Single	KQDH	159	178	177	KQZV	396	315	214	KUBK	132	149	150	KMGN	332	263	176
Married	KQDI	1 567	1 773	1 754	KQZW	1 254	1 431	1 464	KUBL	1 407	1 594	1 574	KMGO	1 133	1 291	1 317
Widowed	KQDJ	258	267	261	KQZX	1 087	1 112	1 002	KUBM	226	234	229	KMGP	959	978	879
Divorced	KQDK	15	46	80	KQZY	28	72	116	KUBN	14	43	74	KMGQ	26	68	107
75 and over: Single	KQDL	58	81	99	KQZZ	291	320	295	KUBO	47	67	81	KMGR	249	270	250
Married	KQDM	480	633	840	KRPI	332	440	592	KUBP	434	573	763	KMGS	304	401	541
Widowed	KQDN	301	336	401	KRPJ	1 172	1 433	1 731	KUBQ	266	300	360	KMGT	1 056	1 285	1 554
Divorced	KQDO	3	11	26	KRPK	7	24	50	KUBR	3	11	25	KMGU	6	23	46

1 Mid-1991 population estimates by marital status are still provisional.
2 Figures for Northern Ireland are more approximate than those for other countries: at each age/sex the proportions enumerated in the census in the various marital status groups have been applied to the estimated total (all statuses) population.

Sources: Office for National Statistics: 01329 813233; General Register Office (Northern Ireland); General Register Office (Scotland)

5.4 Marital condition (de jure) : estimated population

continued

Thousands

		Scotland								Northern Ireland[2]							
		Males				Females				Males				Females			
		1971	1981	1991		1971	1981	1991		1971	1981	1991[1]		1971	1981	1991[1]	
All ages:																	
Single	KJPS	1 197	1 149	1 135	KJVG	1 155	1 066	1 029	KJWV	417	406	411	KJYK	392	370	373	
Married	KJPT	1 227	1 227	1 159	KJVH	1 243	1 240	1 181	KJWW	318	327	339	KJYL	321	332	350	
Widowed	KJPU	79	77	78	KJVI	297	317	306	KJWX	19	18	19	KJYM	71	75	80	
Divorced	KJPV	13	42	97	KJVJ	24	62	122	KJWY	1	4	11	KJYN	2	6	18	
Age groups:																	
0 - 14: Single	KJPW	695	563	492	KJRI	660	535	468	KJWZ	232	212	200	KJYO	224	201	191	
15 - 19: Single	KJPX	195	230	173	KJVL	178	214	163	KJXA	64	75	66	KJYP	58	67	61	
Married	KJPY	5	4	1	KJVM	15	11	2	KJXB	1	1	–	KJYQ	3	3	1	
Widowed	KJPZ	–	–	–	KJVN	–	–	–	KJXC	–	–	–	KJYR	–	–	–	
Divorced	KJQQ	–	–	–	KJVO	–	–	–	KJXD	–	–	–	KJYS	–	–	–	
20 - 24: Single	KJQR	120	149	184	KJVP	81	109	151	KJXE	45	48	57	KJYT	31	34	46	
Married	KJQS	75	60	26	KJVQ	111	93	49	KJXF	18	16	9	KJYU	27	26	16	
Widowed	KJQT	–	–	–	KJVR	–	–	–	KJXG	–	–	–	KJYV	–	–	–	
Divorced	KJQU	–	1	1	KJVS	1	3	2	KJXH	–	–	–	KJYW	–	–	–	
25 - 34: Single	KJQV	61	87	160	KJVT	37	51	113	KJXI	25	27	43	KJYX	16	16	30	
Married	KJQW	243	266	224	KJVU	267	290	259	KJXJ	70	73	73	KJYY	77	80	86	
Widowed	KJQX	–	1	–	KJVV	2	2	1	KJXK	–	–	–	KJYZ	–	1	1	
Divorced	KJQY	3	11	23	KJVW	5	17	29	KJXL	–	1	2	KJZA	–	2	5	
35 - 44: Single	KJQZ	35	32	49	KJVX	30	21	32	KJXM	13	11	14	KJZB	11	8	9	
Married	KJRW	259	252	265	KJVY	271	262	276	KJXN	67	74	81	KJZC	70	76	83	
Widowed	KJRX	2	2	2	KJVZ	7	6	5	KJXO	–	–	–	KJZD	2	2	2	
Divorced	KJRY	3	12	31	KJWA	6	17	38	KJXP	–	1	4	KJZE	–	2	6	
45 - 54: Single	KJRZ	34	28	26	KJWB	37	25	18	KJXQ	13	10	10	KJZF	13	9	7	
Married	KJTD	254	241	232	KJWC	255	244	237	KJXR	64	61	69	KJZG	65	63	70	
Widowed	KJTE	6	5	4	KJWD	25	21	15	KJXS	1	1	1	KJZH	6	6	5	
Divorced	KJTF	3	9	22	KJWE	6	12	26	KJXT	–	1	3	KJZI	1	1	4	
55 - 59: Single	KJTG	16	15	11	KJWF	22	17	10	KJXU	6	5	4	KJZJ	7	6	4	
Married	KJTH	120	116	107	KJWG	112	110	105	KJXV	29	28	28	KJZK	27	27	28	
Widowed	KJTI	6	6	5	KJWH	27	23	18	KJXW	1	1	1	KJZL	6	6	5	
Divorced	KJTJ	1	3	8	KJWI	2	5	9	KJXX	–	–	1	KJZM	–	–	1	
60 - 64: Single	KJTK	15	12	11	KJWJ	26	16	12	KJXY	6	5	5	KJZN	8	6	5	
Married	KJTL	109	96	99	KJWK	92	88	92	KJXZ	27	24	25	KJZO	21	22	24	
Widowed	KJTM	10	8	8	KJWL	40	35	30	KJYA	2	2	2	KJZP	9	9	8	
Divorced	KJTN	1	2	5	KJWM	2	4	7	KJYB	–	–	1	KJZQ	–	–	1	
65 - 74: Single	KJUY	20	21	18	KJWN	50	39	27	KJYC	8	8	8	KJZR	14	13	10	
Married	KJUZ	129	144	143	KJWO	98	113	119	KJYD	31	35	38	KJZS	23	27	31	
Widowed	KJVA	27	27	26	KJWP	104	109	95	KJYE	6	6	6	KJZT	24	26	26	
Divorced	KJVB	1	2	5	KJWQ	2	4	8	KJYF	–	–	1	KJZU	–	–	1	
75 and over: Single	KJVC	7	10	11	KJWR	34	39	34	KJYG	4	4	4	KJZV	9	10	11	
Married	KJVD	36	48	62	KJWS	22	31	43	KJYH	11	13	16	KJZW	6	8	10	
Widowed	KJVE	28	29	33	KJWT	93	121	141	KJYI	7	7	8	KJZX	23	26	34	
Divorced	KJVF	–	1	1	KJWU	–	1	3	KJYJ	–	–	–	KJZY	–	–	–	

1 Mid-1991 population estimates by marital status are still provisional.

2 Figures for Northern Ireland are more approximate than those for other countries: at each age/sex the proportions enumerated in the census in the census in the various marital status groups have been applied to the estimated total (all statuses) population.

Sources: Office for National Statistics: 01329 813233; General Register Office (Northern Ireland); General Register Office (Scotland)

5.5 Geographical distribution of the population

Thousands

		Area[1] Square Km	Population enumerated in Census			Mid-year population estimates				
			1911	1931	1951	1961	1971	1981	1991	1998
United Kingdom	KIUR	241 752	42 190	46 074	50 287	52 807	55 928	56 352	57 808	59 237
Great Britain	KISR	228 269	40 831	44 795	48 854	51 380	54 388	54 815	56 207	57 548
England	KKOJ	130 423	33 650	37 359	41 159	43 561	46 412	46 821	48 208	49 495
Standard Regions										
North	KKNA	15 415	2 729	2 938	3 009	3 246	3 152	3 117	3 092	3 082
Yorkshire and Humberside	KKNB	15 411	3 896	4 319	4 567	4 630	4 902	4 918	4 983	5 043
East Midlands	KKNC	15 627	2 467	2 732	3 118	3 108	3 652	3 853	4 035	4 169
East Anglia	KKND	12 570	1 191	1 231	1 381	1 489	1 688	1 894	2 082	2 181
South East	KKNE	27 224	11 613	13 349	14 877	16 346	17 125	17 011	17 637	18 387
South West	KKNF	23 829	2 818	2 984	3 479	3 436	4 112	4 381	4 718	4 901
West Midlands	KKNG	13 004	3 277	3 743	4 423	4 762	5 146	5 187	5 265	5 333
North West	KKNH	7 342	5 659	6 062	6 305	6 545	6 634	6 459	6 396	6 398
Government Office Regions	JZBT	130 412	..	..	..	43 561	46 412	46 821	48 208	49 495
North East	JZBU	8 594	..	..	..	..	2 679	2 636	2 603	2 590
North West	JZBV	13 510	..	..	..	..	5 446	5 418	5 436	5 481
Merseyside	JZBW	655	..	..	..	..	1 662	1 522	1 450	1 409
Yorkshire and The Humber	JZBX	15 400	..	..	..	..	4 902	4 918	4 983	5 043
East Midlands	JZBY	15 627	..	..	..	..	3 652	3 853	4 035	4 169
West Midlands	JZBZ	13 004	..	..	..	..	5 146	5 187	5 265	5 333
South West	JZCA	23 825	..	..	..	..	4 112	4 381	4 718	4 901
Eastern	JZCB	19 120	..	..	..	..	4 454	4 854	5 150	5 377
Greater London	JZCC	1 578	..	..	..	..	7 529	6 806	6 890	7 187
South East	JZCD	19 098	..	..	..	..	6 830	7 245	7 679	8 004
Wales	KKNI	20 766	2 421	2 593	2 599	2 635	2 740	2 813	2 891	2 933
Scotland	KGJB	77 179	4 761	4 843	5 096	5 184	5 236	5 180	5 107	5 120
Northern Ireland	KGJC	13 570	1 251	1 280[3]	1 371	1 427	1 540	1 538	1 601	1 689
Greater London	KKNJ	1 578	7 161	8 110	8 197	7 977	7 529	6 806	6 890	7 187
Inner London[2]	KISS	321	4 998	4 893	3 679	3 481	3 060	2 550	2 627	2 761
Outer London[2]	KITF	1 258	2 162	3 217	4 518	4 496	4 470	4 255	4 263	4 427
Metropolitan areas of England & Wales	KITG	6 974	9 716	10 770	11 365	11 686	11 862	11 353	11 166	11 148
Tyne and Wear	KGJN	540	1 105	1 201	1 201	1 241	1 218	1 155	1 130	1 116
West Yorkshire	KGJP	2 034	1 852	1 939	1 985	2 002	2 090	2 067	2 085	2 113
South Yorkshire	KGJO	1 559	963	1 173	1 253	1 298	1 331	1 317	1 302	1 304
West Midlands	KGJQ	899	1 780	2 143	2 547	2 724	2 811	2 673	2 629	2 628
Greater Manchester	KGJR	1 286	2 638	2 727	2 716	2 710	2 750	2 619	2 570	2 577
Merseyside	KGJS	655	1 378	1 587	1 663	1 711	1 662	1 522	1 450	1 409
Principal Metropolitan Cities[2]	KITH	1 535	3 154	3 906	3 915	4 204	3 910	3 550	3 451	3 439
Newcastle	KGJT	112	267	286	292	336	312	284	278	276
Leeds	KGJX	562	446	483	505	710	749	718	717	727
Sheffield	KGJV	367	455	512	513	581	579	548	529	531
Birmingham	KGKF	265	526	1 003	1 113	1 179	1 107	1 021	1 007	1 013
Manchester	KGKJ	116	714	766	703	657	554	463	439	430
Liverpool	KGKM	113	746	856	789	741	610	517	481	461
Other metropolitan districts[2]	KITI	5 438	6 562	6 864	7 450	7 482	7 952	7 803	7 716	7 709
Non-metropolitan districts of England & Wales	KITJ	142 637	19 194	21 072	24 196	26 533	29 761	31 475	33 043	34 092
Non-metropolitan cities[2,4]	KITK	2 094	..	..	..	4 670	4 715	4 617	4 676	..
Incl. Kingston-upon-Hull	KKNZ	71	278	314	299	302	288	274	267	262
Leicester	KKOA	73	227	239	285	286	285	283	285	294
Nottingham	KKNX	75	260	269	308	311	302	278	281	287
Bristol	KKNV	110	357	397	443	436	433	401	397	402
Plymouth	KITL	80	207	215	225	240	249	253	254	253
Stoke-on-Trent	KKOD	93	235	277	275	276	265	252	253	252
Cardiff	KKOB	120	182	224	244	289	291	281	294	321
Industrial districts[2,4]	KITM	17 132	..	..	..	6 004	6 486	6 713	6 852	..
New Towns[2,4]	KITN	5 054	..	..	..	1 552	1 895	2 194	2 382	..
Resort, port and retirement districts[2,4]	KITO	10 512	..	..	..	2 828	3 184	3 368	3 626	..
Urban and mixed urban/rural districts[2,4]	KITP	38 683	..	..	..	7 240	8 821	9 446	9 964	..
Remoter, mainly rural districts[2,4]	KITQ	69 161	..	..	..	4 239	4 661	5 137	5 544	..
City of Edinburgh local government district	KGKU	135	320	439	467	..	478	446	440	..
City of Glasgow local government district	KGKT	157	784	1 088	1 090	..	983	774	689	..
Belfast	KGKV	73	387	438[3]	444	..	..	315	294	..

1 Areas as constituted in 1991. Population figures for earlier years may relate to different areas where boundaries have changed.

2 Details of the classification by broad area type are given in recent issues of the ONS annual reference volume "Key Population and Vital Statistics; local and health authority areas" (Series VS). The ten broad area types include all local authorities in England and Wales.

3 Figures for Northern Ireland and the City of Belfast relate to the 1937 Census.

4 The breakdown of non-metropolitan districts by area type has not been provided for the mid-1996 population. This is because the effect of boundary changes due to the major local government reorganisation on 1 April 1995 and 1 April 1996 (particularly in Wales) which make the comparison of 1991 and 1996 data with data for earlier years invalid.

Sources: Office for National Statistics: 01329 813233; General Register Office (Northern Ireland); General Register Office (Scotland)

5.6 Population by age and ethnic group, Great Britain[1]

Average over the period Summer 1998 to Spring 1999

Percentages and Thousands

	Age-group											All ages	
	0 to 4	5 to 9	10 to 14	15 to 19	20 to 24	25 to 29	30 to 34	35 to 44	45 to 59	60 to 74	75 and over	%	000's
Ethnic group													
Black - Caribbean	8	7	7	6	5	7	12	18	15	13	*3*	*100*	485
Black - African	13	10	8	8	6	10	13	19	9	*3*	*0.4*	*100*	365
Black - Other (non-mixed)	*14*	*14*	*13*	*10*	*6*	*8*	15	15	*2*	*2*	*1*	*100*	108
Black - Mixed	24	21	15	11	*7*	*6*	*6*	*7*	*2*	*0*	*0.3*	*100*	192
Indian	7	7	8	8	8	8	9	18	16	9	*2*	*100*	942
Pakistani	12	10	11	11	9	10	8	12	10	6	*1*	*100*	591
Bangladeshi	14	14	13	*10*	*8*	*8*	*8*	*12*	*7*	*5*	*0.3*	*100*	257
Chinese	*3*	*4*	*6*	*9*	14	*11*	*10*	22	14	*6*	*2*	*100*	159
Other - Asian (non-mixed)	*8*	*7*	*6*	*5*	*8*	*10*	*10*	24	*17*	*5*	*1*	*100*	191
Other - Other (non-mixed)	*9*	*9*	*9*	*8*	*8*	*7*	11	19	15	*4*	*1*	*100*	165
Other - Mixed	20	17	15	9	8	7	7	8	*6*	*2*	*1*	*100*	248
All ethnic minority groups	11	10	10	8	8	8	10	16	12	6	*1*	*100*	3 702
White	6	6	6	6	6	7	8	14	19	14	*7*	*100*	53 090
All ethnic groups[2]	7	7	6	6	6	7	8	15	18	13	*7*	*100*	56 807

1 Estimates whose relative standard errors exceed 20% appear in italics and should not be used in any analysis - they are provided as an approximate indication only

2 Includes ethnic group not stated.

Source: Office for National Statistics, Labour Force Survey

5.7 Migration into and out of the United Kingdom, estimates from International Passenger Survey*

Analysis by usual occupation[1] and sex

Thousands

	Total			Professional and managerial			Manual and clerical			Not gainfully employed[2]		
	Persons	Males	Females	Persons	Males	Females	Persons	Males	Females	Persons	Males	Females
Inflow												
	KGOA	KGOB	KGOC	KGOD	KGOE	KGOF	KGOG	KGOH	KGOI	KGOJ	KGOK	KGOL
1987	212	105	107	63	45	18	49	22	27	100	38	62
1988	216	109	107	67	44	23	44	26	18	105	40	66
1989	250	110	140	76	47	28	49	22	26	125	40	86
1990	267	135	132	93	61	32	53	26	27	121	48	72
1991	267	122	144	80	51	29	56	23	33	131	48	83
1992	216	99	117	62	39	23	44	17	27	111	43	68
1993	213	101	112	66	41	25	43	21	22	105	39	66
1994	253	126	127	82	49	34	56	31	25	115	46	68
1995	245	130	115	86	57	29	46	20	26	113	53	60
1996	272	130	143	89	53	36	57	25	33	125	52	74
1997	285	143	142	93	59	34	44	24	20	147	60	87
Outflow												
	KGPA	KGPB	KGPC	KGPD	KGPE	KGPF	KGPG	KGPH	KGPI	KGPJ	KGPK	KGPL
1987	210	108	102	64	38	26	57	29	28	89	41	48
1988	237	125	113	68	44	23	52	25	27	118	55	63
1989	205	108	97	71	44	27	49	24	26	85	41	44
1990	231	113	118	75	47	27	56	25	31	100	41	59
1991	239	120	119	82	48	34	50	28	21	108	44	63
1992	227	113	114	82	50	33	47	23	24	98	40	58
1993	216	113	103	70	38	32	45	24	20	101	51	50
1994	191	92	98	55	32	23	48	24	25	87	37	50
1995	192	102	90	62	42	20	42	23	19	88	38	50
1996	216	105	111	84	53	31	46	23	24	86	30	56
1997	225	121	103	86	57	29	48	23	25	91	42	50
Balance												
	KGRA	KGRB	KGRC	KGRD	KGRE	KGRF	KGRG	KGRH	KGRI	KGRJ	KGRK	KGRL
1987	2	–3	5	–1	7	–8	–8	–7	–1	11	–2	14
1988	–21	–15	–6	–1	–1	–	–8	1	–9	–13	–16	3
1989	44	1	43	5	4	1	–1	–1	1	40	–1	41
1990	36	22	14	19	14	5	–3	1	–4	21	7	13
1991	28	2	26	–2	3	–5	6	–5	11	24	4	19
1992	–11	–14	3	–21	–11	–10	–4	–6	3	13	3	10
1993	–2	–12	10	–4	3	–8	–2	–4	2	3	–12	15
1994	62	34	28	27	17	11	8	8	–	27	9	18
1995	54	28	26	24	15	9	5	–3	7	25	16	9
1996	56	24	32	5	1	5	11	2	9	39	22	18
1997	60	22	38	8	2	5	–4	1	–5	56	18	38

*Data are from the International Passenger Survey and exclude migration with the Irish Republic and other categories. See Explanatory notes, at the beginning of the Population and Vital Statistics chapter.

Data also excludes the Channel Islands and Isle of Man from 1988.

1 Refers to regular occupation before migration.
2 Includes housewives, students, children and retired persons.

Source: Office for National Statistics: 01329 813255

5.8 Migration into and out of the United Kingdom, estimates from International Passenger Survey*

Analysis by citizenship and country of last or next residence

Thousands

	All migrants	British citizens						European Union citizens[1] (excl British)			
			Country of last/next residence						Country of last/next residence		
	Total	Total	European[1] Union	Old[2] Common-wealth	New[3] Common-wealth	United States of America	Other countries	Total	European[1] Union	Other Europe	Other countries
Inflow											
	KEZR	KGLA	KGLB	KGLC	KGLD	KGLE	KGLF	KGLG	KGLH	KGLI	KGLJ
1987	212	99	33	18	18	12	17	26	22	1	2
1988	216	89	27	20	18	8	16	27	26	–	2
1989	250	104	24	31	18	12	19	31	29	–	2
1990	267	106	30	29	22	10	15	37	33	–	4
1991	267	117	39	30	23	10	15	32	29	–	3
1992	216	99	47	22	13	8	10	24	23	–	2
1993	213	92	31	24	19	10	9	25	23	–	2
1994	253	118	45	22	21	16	14	31	27	1	3
1995	245	91	32	16	18	13	13	41	36	–	5
1996	272	104	33	21	22	14	13	54	49	1	4
1997	285	97	36	23	12	8	19	61	56	–	5
Outflow											
	KEZS	KGMA	KGMB	KGMC	KGMD	KGME	KGMF	KGMG	KGMH	KGMI	KGMJ
1987	210	130	41	42	12	18	17	21	17	1	3
1988	237	143	41	52	17	15	19	24	17	3	4
1989	205	122	27	47	18	13	17	22	18	–	4
1990	231	135	34	49	15	22	15	30	25	2	3
1991	239	137	46	42	19	13	18	32	23	2	6
1992	227	133	45	34	15	16	23	17	14	–	3
1993	216	127	45	34	16	16	16	23	21	–	3
1994	191	108	34	28	13	15	18	23	19	–	4
1995	192	118	38	34	13	17	16	20	16	–	4
1996	216	139	52	37	18	16	17	24	18	–	5
1997	225	131	40	37	12	15	26	32	27	1	4
Balance											
	KEZT	KGNA	KGNB	KGNC	KGND	KGNE	KGNF	KGNG	KGNH	KGNI	KGNJ
1987	2	-32	-8	-25	7	-6	–	4	5	–	-1
1988	-21	-54	-14	-32	1	-7	-2	4	8	-2	-2
1989	44	-18	-3	-16	–	–	2	9	11	–	-2
1990	36	-30	-4	-20	7	-12	-1	7	8	-1	1
1991	28	-20	-7	-12	4	-3	-2	–	5	-2	-3
1992	-11	-34	2	-12	-3	-8	-13	8	9	–	-1
1993	-2	-35	-13	-10	2	-7	-7	2	2	–	–
1994	62	10	11	-6	8	1	-4	8	9	1	-2
1995	54	-27	-6	-17	5	-5	-4	21	21	–	1
1996	56	-36	-19	-16	4	-1	-3	30	31	1	-1
1997	60	-34	-5	-14	-1	-8	-7	29	29	-1	1

*Data are from the International Passenger Survey and exclude migration with the Irish Republic and other categories. See Explanatory notes, at the beginning of the Population and Vital Statistics chapter.

Data also excludes the Channel Islands and Isle of Man from 1988.

1 Figures for the European Union have been revised for all the years in this table to show the Union as it was constituted on 1 January 1995.
2 Figures for all years include South Africa in the Old Commonwealth.
3 Figures for all years include Pakistan in the New Commonwealth.

Source: Office for National Statistics: 01329 813255

5.8 Migration into and out of the United Kingdom, estimates from International Passenger Survey*

continued

Analysis by citizenship and country of last or next residence

Thousands

	Commonwealth citizens: Country of last/next residence										Other foreign citizens: Country of last/next residence				
	Total	Australia	Canada	New Zealand	South Africa	Bangladesh, India, Sri Lanka	Pakistan	Other African Commonwealth	Caribbean Commonwealth	Other countries	Total	European[1] Union	Other Europe	United States of America	Other countries
Inflow															
	KGLK	KGLL	KGLM	KGLN	KTDK	KGLO	KGLP	KGLQ	KGLR	KGLT	KGLU	KGLV	KGLW	KGLX	KGLY
1987	55	7	3	8	2	12	6	6	1	9	33	1	2	14	16
1988	54	11	2	7	2	9	5	5	1	10	45	1	6	14	24
1989	71	10	3	10	4	12	6	8	1	13	43	2	4	18	19
1990	71	17	3	10	1	11	4	9	2	10	53	6	7	17	22
1991	68	12	4	7	1	10	7	9	2	14	50	6	7	14	23
1992	52	9	2	6	–	7	5	6	2	14	40	2	7	9	21
1993	53	11	3	5	2	9	4	5	–	10	44	2	12	12	18
1994	52	9	2	6	1	7	4	8	–	12	53	6	14	11	22
1995	63	12	5	7	2	8	3	4	–	19	50	2	10	11	26
1996	64	15	3	7	4	8	7	6	1	10	50	–	7	16	27
1997	80	15	5	7	5	16	5	7	1	19	47	–	7	12	28
Outflow															
	KGMK	KGML	KGMM	KGMN	KTDL	KGMO	KGMP	KGMQ	KGMR	KGMT	KGMU	KGMV	KGMW	KGMX	KGMY
1987	31	9	5	4	–	2	1	2	1	4	28	2	2	13	11
1988	36	6	2	3	1	3	2	6	2	8	34	–	5	14	15
1989	27	7	2	2	1	3	2	3	1	5	34	2	3	16	12
1990	31	8	3	5	2	1	1	4	1	5	34	–	4	18	12
1991	34	7	4	5	–	2	2	3	–	10	36	1	3	16	16
1992	29	6	2	5	1	2	1	2	1	7	48	1	11	19	17
1993	32	8	2	4	1	3	1	3	1	7	34	2	3	16	14
1994	29	5	2	4	1	2	3	2	1	7	31	1	9	8	13
1995	27	6	1	4	1	1	1	2	1	6	27	–	5	8	14
1996	29	7	2	3	2	2	–	2	–	8	24	1	6	5	11
1997	34	6	1	5	4	2	2	2	1	12	28	2	5	8	12
Balance															
	KGNK	KGNL	KGNM	KGNN	KTDM	KGNO	KGNP	KGNQ	KGNR	KGNT	KGNU	KGNV	KGNW	KGNX	KGNY
1987	25	–2	–2	4	2	10	5	4	–	5	5	–	–	1	4
1988	18	5	–	4	–	6	3	–	–1	2	11	–	2	–1	9
1989	44	4	2	8	3	9	4	4	–	8	9	–1	1	2	7
1990	40	9	–	5	–1	10	3	4	2	5	19	6	4	–1	10
1991	34	6	–	2	1	7	5	6	2	4	14	4	4	–2	7
1992	23	4	–	1	–1	5	4	3	1	7	–8	1	–4	–10	4
1993	21	3	1	1	1	6	3	2	–	2	10	–	9	–4	4
1994	23	4	–	2	–	5	1	6	–1	5	22	4	5	4	9
1995	36	6	4	3	1	7	3	2	–1	13	23	2	6	3	12
1996	35	8	1	5	2	6	7	4	1	1	27	–1	1	11	16
1997	46	8	4	3	1	14	4	4	–	8	20	–2	2	4	16

*Data are from the International Passenger Survey and exclude migration with the Irish Republic and other categories. See Explanatory notes, at beginning of Population and Vital Statistics chapter.

Data also excludes the Channel Islands and Isle of Man from 1988.

1 Figures for the European Union have been revised for all the years in this table to show the Union as it was constituted on 1 January 1995. This includes Austria, Finland and Sweden.

Source: Office for National Statistics: 01329 813255

5.9 Acceptances for settlement by nationality
United Kingdom

Number of persons

Geographical region and nationality		1995	1996	1997	1998
		All acceptances for settlement			
All nationalities	KGFA	55 480	61 730	58 720	69 790
Europe					
European Economic Area					
Austria	KGFM	10	–	–	–
Belgium	KGFB	10	–	–	–
Denmark	KGFC	10	10	–	10
Finland	KGFO	10	–	–	–
France	KGFD	30	30	10	30
Germany	KGFE	40	10	10	30
Greece	KGFF	10	10	10	20
Iceland	KOSQ	–	–	–	–
Italy	KGFG	20	20	10	20
Luxembourg	KGFH	–	–	–	–
Netherlands	KGFI	20	10	20	30
Norway	KGFQ	10	10	10	–
Portugal	KGFJ	20	20	20	110
Spain	KGFK	–	–	10	20
Sweden	KGFR	20	10	10	–
European Economic Area[1]	KGFL	220	120	110	270
Remainder of Europe					
Bulgaria	KGFW	150	200	160	180
Cyprus	KGFN	220	240	240	280
Former Czechoslovakia	KGFX	190	290	280	400
Of which,					
Czech Republic[2]	LQLS	..	..	..	240
Slovakia	LQLT	..	..	..	160
Hungary	KGFZ	130	200	180	180
Malta	KGFP	80	80	60	70
Poland	KGGA	580	640	570	580
Romania	KGGB	190	270	220	240
Switzerland	KGFS	150	210	200	250
Turkey	KGFT	1 170	3 720	4 230	2 360
Former USSR	KGGC	610	820	870	1 180
Of which,					
Estonia	LQLU	..	..	..	20
Latvia	LQLV	..	..	..	40
Lithuania	LQLW	..	..	..	50
Russia[3]	LQLX	..	..	..	860
Ukraine	LQLY	..	..	..	140
Other former USSR	LQLZ	..	..	..	80
Former Yugoslavia	KGFU	550	680	590	1 500
Of which,					
Croatia	LQMA	..	..	..	180
Slovenia	LQMB	..	..	..	10
Yugoslavia[4]	LQMC	..	..	..	870
Other former Yugoslavia	LQMD	..	..	..	440
Other Europe	KOSO	20	20	30	60
Remainder of Europe	KOSP	4 030	7 370	7 640	7 300
Europe	KGGE	4 250	7 500	7 740	7 570
Americas					
Argentina	KGGF	60	50	50	60
Barbados	KGGG	70	80	60	100
Brazil	KGGH	330	390	330	350
Canada	KGGI	940	970	980	1 050
Chile	KGGJ	40	60	70	50
Colombia	KGGK	280	270	240	370
Guyana	KGGM	190	200	170	180
Jamaica	KGGN	1 400	1 420	1 030	1 120
Mexico	KGGO	90	110	130	120
Peru	KGGP	100	80	110	100
Trinidad and Tobago	KGGQ	360	340	280	320
USA	KGGR	3 960	4 030	3 900	3 940
Venezuela	KGGT	40	50	50	60
Other Americas	KOSR	310	430	390	2 950
Americas	KGGU	8 180	8 470	7 790	10 780
Africa					
Algeria	KGGV	440	400	370	710
Angola	KOSS	40	30	60	90
Congo (Dem. Rep.)[5]	KOST	120	90	90	120

Geographical region and nationality		1995	1996	1997	1998
		All acceptances for settlement			
Africa(continued)					
Egypt	KGGW	270	330	330	360
Ethiopia	KGGX	170	140	210	190
Ghana	KGGY	1 820	1 970	1 290	1 550
Kenya	KGHA	530	590	500	530
Libya	KGHB	70	110	120	160
Mauritius	KGHC	390	460	380	450
Morocco	KGHD	430	460	410	430
Nigeria	KGHE	3 260	3 220	2 540	2 950
Sierra Leone	KGHF	440	570	470	540
Somalia	KGHG	760	680	990	2 950
South Africa	KGHH	1 300	1 040	1 290	2 260
Sudan	KGHI	150	360	2 180	470
Tanzania	KGHJ	250	220	150	220
Tunisia	KGHK	80	90	70	110
Uganda	KGHL	440	1 040	690	500
Zambia	KGHM	190	220	180	210
Zimbabwe	KGHN	330	360	340	410
Other Africa	KOSU	490	600	550	870
Africa	KGHO	12 000	12 970	13 200	16 090
Asia					
Indian sub-continent					
Bangladesh	KGHP	3 280	2 720	2 870	3 630
India	KGHQ	4 860	4 620	4 650	5 430
Pakistan	KGHR	6 310	6 250	5 560	7 350
Indian sub-continent	KGHS	14 450	13 590	13 080	16 420
Middle East					
Iran	KGHT	1 120	1 720	1 060	930
Iraq	KGHU	540	1 580	1 610	1 650
Israel	KGHV	290	290	280	300
Jordan	KGHW	150	120	150	130
Kuwait	KGHX	30	20	40	60
Lebanon	KGHY	400	660	640	590
Saudi Arabia	KGHZ	40	40	30	50
Syria	KGIA	100	110	120	110
Yemen	KOSV	160	180	150	230
Other Middle East	KOSW	60	70	90	120
Middle East	KGIB	2 880	4 790	4 160	4 180
Remainder of Asia					
China	KGIC	1 130	1 180	1 230	1 550
Hong Kong[6]	KOSX	1 310	1 240	900	810
Indonesia	KGID	100	90	90	120
Japan	KGIE	1 870	1 780	1 760	1 880
Malaysia	KGIF	660	610	500	550
Philippines	KGIG	1 090	1 030	890	950
Singapore	KGIH	170	150	160	120
South Korea	KOTE	260	270	220	270
Sri Lanka	KGII	1 370	2 180	1 620	2 100
Taiwan	KOSY	40	60	80	100
Thailand	KGIJ	520	550	500	540
Other Asia	KOSZ	270	350	420	530
Remainder of Asia	KGIL	8 790	9 500	8 370	9 520
Asia	KGIM	26 120	27 880	25 610	30 120
Oceania					
Australia	KGIN	2 020	2 120	1 920	2 200
New Zealand	KGIO	1 390	1 360	1 150	1 440
Other Oceania	KOTA	40	40	30	50
Oceania	KGIP	3 450	3 520	3 100	3 690
British Overseas citizens	KGIQ	690	620	540	960
Stateless	KGIS	780	780	740	580
All nationalities	KGFA	55 480	61 730	58 720	69 790

1 Includes Liechtenstein.
2 Includes Czechoslovakian passport holders.
3 Includes Soviet Union passport holders.
4 Includes holders of passports of the former Yugoslavia.
5 Democratic Republic of the Congo, formerly known as Zaire.
6 Includes Hong Kong stateless persons.

Source: Home Office: 020 8760 8289

5.10 Applications[1] received for asylum, excluding dependants, by nationality
United Kingdom

Number of principal applicants

Nationality		1990[2]	1991[2]	1992[2]	1993[2]	1994[2]	1995[2]	1996[2,3]	1997[2]	1998[2]
Europe										
Albania	LQME	–	15	100	70	75	110	105	445	560
Czech Republic	LQMF	–	–	–	5	5	15	55	240	515
Poland	DMLX	20	20	90	155	360	1 210	900	565	1 585
Romania	KEAV	305	555	305	370	355	770	455	605	1 015
Slovak Republic	LQMG	5	5	10	10	30	50	55	290	835
Turkey	KEAW	1 590	2 110	1 865	1 480	2 045	1 820	1 495	1 445	2 015
Former USSR	KEAX	100	245	270	385	595	795	1 400	2 015	2 820
Former Yugoslavia	KIAL	15	320	5 635	1 830	1 385	1 565	1 030	2 260	7 980
Other	KEAY	165†	420	160	235	510	720	980	1 285	425
Total	KEAZ	2 200	3 685	8 435	4 535	5 360	7 050	6 475	9 145	17 745
Americas										
Colombia	KEBZ	175	140	280	380	405	525	1 005	1 330	425
Ecuador	KYDB	–	5	15	60	105	250	435	1 205	280
Other	KECS	70	65	170	305	380	565	330	295	270
Total	KECT	245	210	465	745	890	1 340	1 765	2 825	975
Africa										
Algeria	KOTB	25	45	150	275	995	1 865	715	715	1 260
Angola	KECU	1 685	5 780	245	320	605	555	385	195	150
Congo (Dem. Rep.)	KEEH	2 590	7 010	880	635	775	935	680	690	660
Ethiopia	KECW	2 340	1 685	680	615	730	585	205	145	345
Gambia	DMMA	5	20	10	25	140	1 170	245	125	45
Ghana	KECX	1 330	2 405	1 600	1 785	2 035	1 915	780	350	225
Ivory Coast	DMLZ	135	1 415	310	330	705	245	125	70	95
Kenya	KOTC	50	70	110	630	1 130	1 395	1 170	605	885
Liberia	DMMB	120	145	100	90	140	390	330	205	70
Nigeria	KECY	135	335	615	1 665	4 340	5 825	2 900	1 480	1 380
Sierra Leone	KOTD	20	75	325	1 050	1 810	855	395	815	565
Somalia	KECZ	2 250	1 995	1 575	1 465	1 840	3 465	1 780	2 730	4 685
Sudan	KEEE	340	1 150	560	300	330	345	280	230	250
Tanzania	DMMC	10	25	30	110	205	1 535	225	90	80
Uganda	KEEG	2 125	1 450	295	595	360	365	215	220	210
Other	KEEI	655	3 400	120	360	765	1 030	810	815	1 475
Total	KEEJ	13 870	27 500	7 630	10 295	16 960	22 545	11 290	9 515	12 380
Middle East										
Iran	KEEK	455	530	405	365	520	615	585	585	745
Iraq	KEEL	985	915	700	495	550	930	965	1 075	1 295
Lebanon	KEGW	1 110	755	380	285	215	150	145	160	155
Other	KEGX	100	340	495	370	695	600	455	515	595
Total	KEGY	2 650	2 540	1 980	1 520	1 985	2 295	2 150	2 335	2 785
Asia										
Afghanistan	DMLY	175	210	270	315	325	580	675	1 085	2 395
China	KEGZ	240	525	330	215	425	790	820	1 945	1 925
India	KEIL	1 530	2 075	1 450	1 275	2 030	3 255	2 220	1 285	1 030
Pakistan	KEIM	1 475	3 245	1 700	1 125	1 810	2 915	1 915	1 615	1 975
Sri Lanka	KEIN	3 330	3 765	2 085	1 965	2 350	2 070	1 340	1 830	3 505
Other	KEIO	135	675	270	285	575	1 075	920	810	1 110
Total	KEJO	6 890	10 495	6 100	5 175	7 515	10 685	7 885	8 570	11 940
Other, and nationality not known[4]	KEJP	350	410	–	100	125	50	80	105	190
Grand Total	KEJQ	26 205	44 840	24 605	22 370	32 830	43 965	29 640	32 500	46 015

1 Figures do not include overseas applications.
2 Figures rounded to the nearest 5.
3 Includes revisions to the number of in-country applications lodged between September and December 1996.
4 Where the nationality was not known between 1991 and 1994 the most likely nationality was recorded.

Source: Home Office: 020 8760 8297

5.11 Marriages

Numbers

		1987	1988	1989	1990	1991	1992	1993	1994	1995	1996	1997[2,3]
United Kingdom												
Marriages	KKAA	397 937	394 049	392 042	375 410	349 739	356 013	341 608	331 232	322 251	317 514	310 218
Persons marrying per 1 000 resident population	KKAB	*14.0*	*13.8*	*13.7*	*13.1*	*12.1*	*12.3*	*11.7*	*11.3*	*11.0*	*10.8*	*10.5*
Previous marital status												
Bachelors	KKAC	296 290	289 493	288 478	276 512	256 538	258 567	245 996	236 619	227 717	221 826	216 237
Divorced men	KKAD	89 814	92 755	92 033	88 199	83 069	87 419	85 824	85 261	85 743	87 113	85 625
Widowers	KKAE	11 833	11 801	11 531	10 699	10 132	10 027	9 788	9 352	8 791	8 575	8 356
Spinsters	KKAF	301 073	293 551	291 516	279 442	259 084	260 252	248 063	237 241	228 462	221 697	216 776
Divorced women	KKAG	85 238	89 066	89 234	85 608	81 224	86 361	84 268	85 220	85 396	87 618	85 648
Widows	KKAH	11 626	11 612	11 294	10 360	9 431	9 400	9 277	8 771	8 393	8 199	7 794
First marriage for both partners	KMGH	260 459	253 150	251 572	240 729	222 369	222 142	210 567	200 910	192 078	185 293	181 135
First marriage for one partner	KMGI	76 445	76 744	76 850	74 496	70 884	74 535	72 925	72 040	72 023	72 937	70 743
Remarriage for both partners	KMGJ	61 033	64 155	63 620	60 185	56 486	59 336	58 116	58 282	58 150	59 284	58 340
Males												
Under 21 years	KKAI	24 269	20 608	19 070	15 930	13 271	11 031	8 767	7 091	6 302	5 497	5 126
21-24	KKAJ	118 355	109 482	102 977	92 270	79 877	74 458	65 129	56 877	48 432	42 488	36 875
25-29	KKAK	119 808	120 939	123 491	122 800	115 637	118 255	114 101	111 108	105 218	101 647	97 345
30-34	KKAL	51 389	53 865	56 442	56 966	56 970	62 470	63 848	65 490	68 245	69 867	70 904
35-44	KKAM	48 598	51 329	51 411	49 984	48 147	51 125	50 553	51 310	53 350	56 513	58 292
45-54	KKAN	19 788	21 544	22 329	21 996	20 915	23 290	23 841	24 136	24 786	26 252	26 472
55 and over	KKAO	15 730	16 282	16 322	15 464	14 922	15 384	15 369	15 220	14 918	15 250	15 204
Females												
Under 21 years	KKAP	68 629	59 284	54 256	45 626	38 305	32 618	26 839	22 903	20 643	18 485	17 254
21-24	KKAQ	140 509	134 122	128 411	119 037	105 505	102 494	93 125	84 171	75 071	66 191	59 549
25-29	KKAR	90 911	95 338	100 531	103 209	99 851	105 223	104 517	102 803	100 644	99 651	97 932
30-34	KKAS	36 643	39 680	41 989	42 794	43 617	48 514	49 546	52 359	54 819	57 752	58 589
35-44	KKAT	36 978	39 534	40 290	38 983	37 582	40 075	40 090	41 213	43 115	45 969	47 267
45-54	KKAU	15 001	16 570	17 172	16 825	16 473	18 504	18 800	19 280	19 720	21 025	21 038
55 and over	KKAV	9 260	9 521	9 393	8 936	8 406	8 585	8 691	8 503	8 239	8 441	8 589
England and Wales												
Marriages	KKBA	351 761	348 492	346 697	331 150	306 756	311 564	299 197	291 069	283 012	278 975	272 536
Persons marrying per 1 000 resident population	KKBB	*14.0*	*13.8*	*13.7*	*13.1*	*12.0*	*12.2*	*11.6*	*11.3*	*10.9*	*10.7*	*10.4*
Previous marital status												
Bachelors	KKBC	258 750	252 780	252 230	241 274	222 823	224 152	213 476	206 077	198 208	193 306	188 268
Divorced men	KKBD	82 315	84 991	84 035	80 282	74 860	78 473	76 986	76 633	76 967	78 003	76 839
Widowers	KKBE	10 696	10 721	10 432	9 594	9 073	8 939	8 735	8 359	7 837	7 666	7 429
Spinsters	KKBF	262 958	256 221	254 763	243 825	224 812	225 608	214 987	206 332	198 603	192 707	188 457
Divorced women	KKBG	78 219	81 691	81 702	77 994	73 408	77 542	75 904	76 857	76 869	78 939	77 098
Widows	KKBH	10 584	10 580	10 232	9 331	8 536	8 414	8 306	7 880	7 540	7 329	6 981
First marriage for both partners	KMGK	226 308	219 791	218 904	209 043	192 238	191 732	181 956	174 200	166 418	160 680	156 907
First marriage for one partner	KMGL	69 092	69 419	69 185	67 013	63 159	66 296	64 551	64 009	63 975	64 653	62 911
Remarriage for both partners	KMGM	56 361	59 282	58 608	55 094	51 359	53 536	52 690	52 860	52 619	53 642	52 718
Males[1]												
Under 21 years	KKBI	20 541	17 578	16 312	13 772	11 416	9 471	7 540	6 173	5 520	4 877	4 574
21-24	KKBJ	102 907	95 029	89 263	79 818	68 547	63 932	55 963	49 072	42 711	36 713	31 907
25-29	KKBK	105 458	106 349	108 834	107 784	100 891	102 942	99 314	96 848	91 607	88 338	84 644
30-34	KKBL	46 063	48 112	50 409	50 600	50 403	55 012	56 129	57 854	60 014	61 582	62 265
35-44	KKBM	44 237	46 688	46 549	45 038	43 013	45 364	44 863	45 518	47 330	50 038	51 654
45-54	KKBN	18 055	19 754	20 365	19 991	18 929	20 925	21 440	21 794	22 349	23 661	23 688
55 and over	KKBO	14 500	14 982	14 965	14 147	13 557	13 918	13 948	13 803	13 481	13 766	13 804
Females[1]												
Under 21 years	KKBP	59 705	51 717	47 529	40 022	33 428	28 541	23 469	20 250	18 343	16 510	15 439
21-24	KKBQ	122 999	117 239	112 048	103 653	91 338	88 553	80 470	72 937	65 126	57 296	51 766
25-29	KKBR	80 062	88 833	88 662	90 629	87 244	91 735	91 134	89 941	87 680	86 838	85 352
30-34	KKBS	32 932	35 609	37 452	38 032	38 425	42 675	43 559	46 119	48 216	50 799	51 405
35-44	KKBT	33 818	36 135	36 678	35 315	33 755	35 660	35 662	36 651	38 367	40 889	41 838
45-54	KKBU	13 690	15 173	15 661	15 290	14 891	16 619	16 969	17 409	17 791	18 992	18 938
55 and over	KKBV	8 555	8 786	8 667	8 209	7 675	7 781	7 934	7 762	7 489	7 651	7 798

1 The figures for England and Wales include an assumed distribution of 'Age not stated'.
2 Provisional.
3 Later figures not available.

Source: Office for National Statistics: 01329 813772

5.11 Marriages

continued

Numbers

		1988	1989	1990	1991	1992	1993	1994	1995	1996	1997	1998
Scotland												
Marriages	KKCA	35 599	35 326	34 672	33 762	35 057	33 366	31 480	30 663	30 242	29 611	29 668
Persons marrying per 1 000 population	KKCB	*14.0*	*13.9*	*13.6*	*13.2*	*13.7*	*13.0*	*12.3*	*11.9*	*11.8*	*11.6*	*11.6*
Previous marital status												
Bachelors	KKCC	27 622	27 186	26 636	25 549	26 106	24 609	23 004	22 126	21 454	20 994	20 987
Divorced men	KKCD	7 062	7 224	7 108	7 344	8 027	7 879	7 654	7 741	8 048	7 845	7 934
Widowers	KKCE	915	916	928	869	924	878	822	796	740	772	747
Spinsters	KKCF	28 150	27 606	26 940	25 979	26 274	25 103	23 248	22 410	21 799	21 303	21 241
Divorced women	KKCG	6 574	6 832	6 869	7 048	7 935	7 469	7 487	7 542	7 718	7 621	7 754
Widows	KKCH	875	888	863	735	848	794	745	711	725	687	673
First marriage for both partners	KEZV	24 628	24 032	23 529	22 401	22 588	21 214	19 644	18 822	18 071	17 751	17 677
First marriage for one partner	KEZW	6 516	6 728	6 518	6 726	7 204	7 284	6 964	6 892	7 111	6 795	6 874
Remarriage for both partners	KEZX	4 455	4 566	4 625	4 635	5 265	4 868	4 872	4 949	5 060	5 065	5 117
Males												
Under 21 years	KKCI	2 259	2 042	1 626	1 378	1 172	902	680	577	452	406	421
21-24	KKCJ	10 956	10 295	9 454	8 572	7 958	6 870	5 693	4 915	4 191	3 494	3 147
25-29	KKCK	11 140	11 122	11 430	11 244	11 587	11 266	10 812	10 209	10 056	9 495	9 439
30-34	KKCL	4 609	4 831	5 061	5 290	6 022	6 214	6 176	6 574	6 574	6 911	6 988
35-44	KKCM	3 953	4 135	4 210	4 383	4 950	4 803	4 864	5 021	5 412	5 649	5 945
45-54	KKCN	1 558	1 721	1 760	1 733	2 083	2 093	2 050	2 124	2 288	2 459	2 412
55 and over	KKCO	1 124	1 180	1 131	1 162	1 285	1 218	1 205	1 243	1 269	1 197	1 316
Females												
Under 21 years	KKCP	5 692	4 951	4 161	3 589	3 060	2 461	1 959	1 728	1 423	1 302	1 289
21-24	KKCQ	12 801	12 386	11 666	10 735	10 503	9 506	8 291	7 264	6 474	5 568	5 248
25-29	KKCR	8 959	9 163	9 755	9 771	10 457	10 368	9 895	9 904	9 818	9 574	9 764
30-34	KKCS	3 311	3 728	3 925	4 331	4 822	5 002	5 145	5 401	5 675	5 927	6 036
35-44	KKCT	2 964	3 120	3 194	3 333	3 847	3 774	3 917	4 025	4 378	4 722	4 726
45-54	KKCU	1 233	1 345	1 351	1 386	1 673	1 606	1 648	1 689	1 794	1 844	1 900
55 and over	KKCV	639	633	620	617	695	649	625	652	680	674	705
Northern Ireland												
Marriages	KKDA	9 960	10 019	9 588	9 221	9 392	9 045	8 683	8 576	8 297	8 071	7 826
Persons marrying per 1 000 population	KKDB	*12.6*	*12.7*	*12.1*	*11.7*	*11.5*	*11.0*	*10.6*	*10.4*	*10.0*	*9.6*	*9.3*
Previous marital status												
Bachelors	KKDC	9 093	9 062	8 602	8 166	8 309	7 911	7 538	7 383	7 066	6 975	6 689
Divorced men	KKDD	702	774	809	865	919	959	974	1 035	1 062	941	1 014
Widowers	KKDE	165	183	177	190	164	175	171	158	169	155	123
Spinsters	KKDF	9 182	9 147	8 677	8 293	8 370	7 973	7 661	7 449	7 191	7 016	6 767
Divorced women	KKDG	621	700	745	768	884	895	876	985	961	929	932
Widows	KKDH	157	172	166	160	138	177	146	142	145	126	127
First marriage for both partners	KEZY	8 733	8 636	8 157	7 730	7 822	7 397	7 066	6 838	6 542	6 477	6 188
First marriage for one partner	KEZZ	809	937	965	999	1 035	1 090	1 067	1 156	1 173	1 037	1 080
Remarriage for both partners	KFBI	418	446	466	492	535	558	550	582	582	557	558
Males												
Under 21 years	KKDI	771	716	532	477	388	325	236	205	168	146	144
21-24	KKDJ	3 497	3 419	2 998	2 758	2 568	2 296	2 111	806	1 584	1 474	1 187
25-29	KKDK	3 452	3 535	3 586	3 502	3 726	3 521	3 434	3 402	3 253	3 206	3 122
30-34	KKDL	1 144	1 202	1 305	1 277	1 436	1 505	1 466	1 657	1 711	1 728	1 785
35-44	KKDM	688	727	736	751	811	887	932	999	1 063	989	1 081
45-54	KKDN	232	243	245	253	282	308	292	313	303	325	321
55 and over	KKDO	176	177	186	203	181	203	212	194	215	203	186
Females												
Under 21 years	KKDP	1 875	1 776	1 443	1 288	1 017	909	694	572	552	513	439
21-24	KKDQ	4 082	3 977	3 718	3 432	3 438	3 149	2 943	2 681	2 421	2 215	1 951
25-29	KKDR	2 547	2 706	2 825	2 836	3 031	3 015	2 967	3 060	2 995	3 006	3 018
30-34	KKDS	761	809	837	861	1 017	985	1 095	1 202	1 278	1 257	1 331
35-44	KKDT	435	492	474	494	568	654	645	723	702	707	750
45-54	KKDU	164	166	184	196	212	225	223	240	239	256	233
55 and over	KKDV	96	93	107	114	109	108	116	98	110	117	104

Sources: Office for National Statistics: 01329 813772; General Register Office (Scotland); General Register Office (Northern Ireland)

5.12 Divorce
England and Wales, Scotland

Number

		1987	1988	1989	1990	1991	1992	1993	1994	1995	1996	1997	1998
England and Wales													
Decrees absolute, granted:													
Number[1]	KKEA	151 007	152 633	150 872	153 386	158 745	160 385	165 018	158 175	155 499	157 107	146 689	..
Rate per 1 000 married couples	KKEB	*12.7*	*12.8*	*12.7*	*13.0*	*13.5*	*13.7*	*13.9*	*13.4*	*13.1*	*13.8*	*13.0*	..
Duration of marriage:													
0-4 years	KKEC	35 423	35 582	35 719	36 299	37 779	36 898	37 252	35 695	34 507	34 924	31 767	..
5-9 years	KKED	43 150	42 617	42 108	42 061	42 735	43 745	46 536	44 769	44 304	44 609	41 260	..
10-14 years	KKEE	26 194	26 545	26 281	27 310	28 791	29 285	30 156	28 073	27 365	27 332	26 215	..
15-19 years	KKEF	19 576	20 132	19 418	19 819	20 127	20 160	20 233	19 200	18 943	19 321	18 027	..
20 years and over	KKEG	26 664	27 747	27 327	27 881	29 294	30 290	30 836	30 431	30 370	30 912	29 408	..
Not stated	KKEH	–	10	19	16	19	7	5	7	10	9	12	..
Age of wife at marriage:													
16-19 years	KKEI	46 097	44 693	42 612	41 116	40 594	39 731	38 810	34 068	31 319	29 927	25 579	..
20-24 years	KKEJ	68 345	69 489	69 424	71 489	74 050	74 698	76 580	73 287	71 355	71 123	66 167	..
25-29 years	KKEK	19 049	20 267	20 369	21 701	24 025	25 172	27 177	28 358	29 439	31 396	31 022	..
30-34 years	KKEL	7 983	8 441	8 590	8 909	9 608	9 939	10 593	11 007	11 585	12 335	12 094	..
35-39 years	KKEM	4 403	4 501	4 643	4 880	5 024	5 200	5 673	5 615	5 800	6 051	5 767	..
40-44 years	KKEN	2 416	2 530	2 541	2 598	2 727	2 872	3 091	3 064	3 121	3 254	3 156	..
45 years and over	KKEO	2 714	2 712	2 693	2 693	2 717	2 766	2 819	2 769	2 870	3 021	2 904	..
Age of wife at divorce:													
16-24 years	KKEP	19 066	17 693	16 628	15 454	14 960	13 482	12 924	10 956	9 783	8 615	6 871	..
25-29 years	KKEQ	34 209	34 504	34 483	35 121	35 582	34 853	35 362	32 608	30 563	30 075	26 435	..
30-34 years	KKER	28 995	29 406	29 757	31 295	33 195	34 901	36 300	35 848	35 538	36 274	33 967	..
35-39 years	KKES	24 934	24 685	24 170	24 421	25 661	26 577	28 162	27 195	27 550	28 727	27 715	..
40-44 years	KKET	19 403	20 873	20 647	21 263	21 979	21 783	21 891	20 765	20 739	20 774	20 125	..
45 years and over	KKEU	24 400	25 462	25 168	25 816	27 349	28 782	30 374	30 796	31 316	32 633	31 564	..
Not stated	KKEV	–	10	19	16	19	7	5	7	10	9	12	..
Divorces in which there were:													
No children[2]	KKEW	46 770	47 049	46 910	47 119	48 115	46 979	47 652	48 286	48 560	48 800	45 556	..
1 or more children[2]	KKEX	104 237	105 584	103 962	106 267	110 630	113 406	117 366	109 889	106 939	108 307	101 133	..
Scotland													
Decrees absolute, granted:													
Number[1]	KKFA	12 133	11 472	11 659	12 272	12 399	12 479	12 787	13 133	12 249	12 308[4]	12 222	12 384
Rate per 1 000 married[3] couples	KKFB	*10.2*	*9.8*	*10.0*	*10.5*	*10.6*	*10.8*	*11.1*	*11.5*	*10.8*	*10.9*	*11.0*	*11.2*
Duration of marriage:													
0-4 years	KKFC	2 173	1 986	2 013	2 208	2 142	2 085	2 092	2 095	1 908	1 914	1 793	1 766
5-9 years	KKFD	3 544	3 353	3 420	3 546	3 508	3 610	3 722	3 790	3 399	3 432	3 224	3 360
10-14 years	KKFE	2 351	2 227	2 245	2 361	2 484	2 454	2 539	2 592	2 407	2 310	2 385	2 456
15-19 years	KKFF	1 670	1 592	1 633	1 617	1 718	1 675	1 745	1 786	1 698	1 709	1 804	1 729
20 years and over	KKFG	2 395	2 314	2 348	2 540	2 547	2 655	2 689	2 870	2 837	2 934	3 016	3 073
Age of wife at marriage:													
16-20 years	KKFH	6 134	5 660	5 565	5 600	5 592	5 378	5 406	5 306	4 600	4 420	4 142	3 984
21-24 years	KKFI	3 778	3 577	3 708	4 185	4 147	4 198	4 252	4 532	4 336	4 341	4 321	4 414
25-29 years	KKFJ	1 274	1 236	1 363	1 377	1 545	1 685	1 812	1 926	1 887	1 933	2 151	2 314
30-34 years	KKFK	423	466	479	497	514	575	612	628	654	697	791	824
35-39 years	KKFL	224	219	235	275	249	301	312	329	338	393	360	382
40-44 years	KKFM	142	124	131	139	148	138	152	163	196	234	199	198
45 years and over	KKFN	158	153	143	159	142	153	164	166	166	198	173	185
Age not stated	KKFO	–	37	35	40	62	51	77	83	72	92	85	83
Age of wife at divorce:													
16-24 years	KKFP	1 504	1 285	1 139	1 199	1 038	963	844	767	622	583	426	377
25-29 years	KKFQ	2 920	2 676	2 818	2 938	2 932	2 807	2 775	2 750	2 353	2 269	2 021	1 957
30-34 years	KKFR	2 443	2 378	2 442	2 611	2 741	2 785	3 037	3 045	2 747	2 708	2 736	2 767
35-39 years	KKFS	1 942	1 917	1 927	1 891	2 037	2 092	2 212	2 390	2 290	2 307	2 469	2 562
40-44 years	KKFT	1 482	1 381	1 436	1 614	1 665	1 685	1 771	1 788	1 734	1 761	1 819	1 951
45 years and over	KKFU	1 842	1 798	1 862	1 979	1 924	2 096	2 071	2 310	2 431	2 587	2 667	2 687
Age not stated	KKFV	–	37	35	40	62	51	77	83	72	93	84	83
Divorces in which there were:													
No children aged under 16	KKFW	5 989	5 887	6 091	6 555	6 521	6 927	6 951	7 390	7 515	8 167	9 761	9 231
1 or more children under 16	KKFX	6 144	5 585	5 568	5 717	5 878	5 552	5 836	5 743	4 734	4 141	2 461	3 153

1 Includes decrees of divorce and of nullity.
2 Children of the family as defined by the Matrimonial Causes Act 1973.
3 Rates are calculated using the average of the estimated married male and female populations.
4 Includes 9 cases where the duration of marriage was not recorded.

Sources: Office for National Statistics: 01329 813772; General Register Office (Scotland); Scottish Courts Administration

5.13 Divorce proceedings

Number

		England and Wales										
		1988	1989	1990	1991	1992	1993	1994	1995	1996	1997	1998
Dissolution of marriage[1]												
Petitions filed[2]	KKGA	182 804	184 610	191 615	179 103	189 329†	184 471	175 510	173 966	178 005	163 787	165 602
On grounds of:[3]												
Adultery	KKGB	50 250	51 650	..	..	..	..	..	..	..	..	..
Behaviour	KKGC	88 260	89 040	..	..	..	..	..	..	..	..	..
Desertion	KKGD	2 180	2 040	..	..	..	..	..	..	..	..	..
Separation (2 years and consent)	KKGE	30 860	30 610	..	..	..	..	..	..	..	..	..
Separation(5 years)	KKGF	9 830	10 100	..	..	..	..	..	..	..	..	..
Adultery and behaviour	KKGG	1 000	820	..	..	..	..	..	..	..	..	..
Adultery and desertion	KKGH	120	100	..	..	..	..	..	..	..	..	..
Behaviour and desertion	KKGI	70	160	..	..	..	..	..	..	..	..	..
Other	KKGJ	240	70	..	..	..	..	..	..	..	..	..
By husbands[3]	KKGK	49 130	49 150	..	..	..	..	..	..	..	..	..
By wives[3]	KKGL	133 670	135 090	..	..	..	..	..	..	..	..	..
Decrees nisi granted	KKGM	154 788	151 309	157 344	153 258	149 126†	160 625	154 241	155 739	157 585	148 326	143 879
Decrees absolute granted	KKGN	152 139	150 477	155 239	155 927	156 787	162 579	154 873	153 337	155 310	145 886	141 345
Nullity of marriage[4]												
Petitions filed[2]	KKGO	604	478	665	619	535	634	822	881	702	485	747
By husbands[3]	KKGP	260	220	..	..	..	..	..	..	..	..	..
By wives[3]	KKGQ	350	260	..	..	..	..	..	..	..	..	..
Decrees nisi granted	KKGR	389	365	430	508	369†	365	705	425	332	248	474
Decrees absolute granted	KKGS	494	395	467	417	435	410	1 017	516	669	298	267
Judicial separation												
Petitions filed[2]	KKGT	2 925	2 741	2 874†	2 588	2 434	2 252	4 358	3 349	2 795	1 078	1 374
By husbands[3]	KKGU	270	270	..	..	..	..	..	..	..	..	..
By wives[3]	KKGV	2 660	2 470	..	..	..	..	..	..	..	..	..
Decrees granted	KKGW	1 917	1 678	1 794	1 747	1 452†	1 413	1 350	1 543	1 099	589	518

1 Excluding petitions in which divorce is asked for as an alternative to nullity.
2 The breakdown of petitions filed is based on actual figures for the Principal Registry of the Family Division and the county courts (from 1990).
3 This data is no longer available.
4 Including cases in which dissolution is asked for in the alternative.

Sources: Office for National Statistics: 01329 813772; The Court Service

5.13 Divorce proceedings

continued

Number

		Scotland										
		1988	1989	1990	1991	1992	1993	1994	1995	1996	1997	1998
Divorce												
Actions in which final judgment given	KKHA	11 472	11 659	12 272	12 399	12 479	12 776	13 123	12 249	12 308	12 222	12 384
On grounds of[1]:												
Adultery	KKHG	1 309	1 291	1 232	1 198	1 136	1 092	1 099	956	946†	909	833
Desertion	KKHH	83	100	68	82	58	56	103	72	61	33	28
Behaviour	KKHI	3 559	3 532	3 847	3 622	3 407	3 757	3 711	3 203	3 184	3 081	3 005
2 years non-cohabitation	KKHJ	4 879	5 104†	5 368	5 274	5 823	5 799	6 076	5 846	5 833	5 773	6 118
5 years non-cohabitation	KKHK	1 626	1 632	1 757	1 919	2 052	2 072	2 134	2 166	2 284	2 426†	2 400
At instances of: Husbands	KKHM	3 195	3 220†	3 410	3 469	3 614	3 593	3 854	3 704	3 745[2]	3 956	4 055
Wives	KKHN	8 277	8 439	8 862	8 930	8 865	9 183	9 269	8 545†	8 563[2]	8 266	8 329
Duration of marriages where divorce or separation granted												
Under 1 year	KKHR	24	27	26	24	33	40	28	49	47	48†	7
1 - 2 years	KKHS	173	168	195†	179	166	165	163	131	179	167	83
2 - 5 years	KKHT	1 789	1 818	1 987	1 939	1 886	1 881	1 898	1 728	1 688	1 637†	1 373
5 - 10 years	KKHU	3 353	3 420	3 546	3 508	3 610	3 720	3 789	3 399	3 432	3 226†	3 363
10 - 20 years	KKHV	3 819	3 878	3 978	4 202	4 129	4 281	4 376	4 105	4 028†	4 147	4 618
20 years and over	KKHW	2 314	2 348	2 540	2 547	2 655	2 689	2 869	2 837	2 934	2 997†	2 940
Actions in which there were children of marriage under 16 years of age	KKHX	5 585	5 568	5 717	5 878	5 552	5 834	5 922	4 734	..	..	..

1 The grounds given show the allegations made-divorce is granted on the grounds of irretrievable breakdown of marriage under the Divorce (Scotland) Act 1976.
2 Estimated from data available.

Sources: Scottish Courts Administration: 0131 221 6814; Office for National Statistics: 01329 813772

		Northern Ireland										
		1988	1989	1990	1991	1992	1993	1994	1995	1996	1997	1998
Petitions filed												
Nullity of marriage	KKHZ	1	6	7	5	1	2	5	5	5	7	2
Divorce	KKIA	2 217	2 385	2 258	2 591	2 597	2 670	2 610	2 875	2 695	2 808	2 760
Judicial separation	KKIB	6	16	16	23	17	44	57	84	63	70	64
Divorces: Decrees nisi granted by facts proved												
High Court												
Adultery	KHYK	132	124	160	161	150	157	133	115	113	118	159
Behaviour	KHYL	159	184	213	214	239	268	303	300	301	311	375
Desertion	KHYM	8	12	8	11	14	9	8	5	7	7	11
Separation (2 years and consent)	KHYN	564	682	608	685	597	566	604	556	484	502	569
Separation (5 years)	KHYO	261	298	321	332	341	288	316	326	326	283	363
Combination of more than one ground	KSPB	–	–	–	–	–	–	–	24	18	27	40
Other grounds[1]	KHYP	17	10	5	16	9	1	2	–	–	–	–
Total	KHYQ	1 141	1 310	1 315	1 419	1 350	1 289	1 366	1 326	1 249	1 248	1 517
County Court												
Adultery	KHYR	45	51	50	62	69	66	71	67	63	67	72
Behaviour	KHYS	63	61	71	115	118	139	145	152	175	218	201
Desertion	KHYT	3	8	14	12	17	10	11	7	6	8	6
Separation (2 years and consent)	KHYU	304	344	331	459	552	552	590	627	586	632	669
Separation (5 years)	KHYV	141	217	187	253	255	326	337	355	332	348	418
Other grounds[1]	KHYW	50	38	22	24	25	2	–	1	8	11	21
Total	KHYX	606	719	675	925	1 036	1 095	1 154	1 209	1 170	1 284	1 387

1 Includes several facts proved.

Sources: Northern Ireland Court Service: 028 9032 8594; Office for National Statistics: 01329 813772

5.14 Births

Annual averages or calendar years

Thousands

	Live births				Rates				
	Total	Male	Female	Sex ratio	Crude birth rate[1]	General fertility rate[2]	TPFR[3]	Still-births[4]	Still-birth rate[4]
United Kingdom									
1900 - 02	1 095	558	537	1 037	28.6	115.1	..	..	..
1910 - 12	1 037	528	508	1 039	24.6	99.4	..	..	..
1920 - 22	1 018	522	496	1 052	23.1	93.0	..	..	..
1930 - 32	750	383	367	1 046	16.3	66.5	..	..	..
1940 - 42	723	372	351	1 062	15.0	..	1.89	26	..
1950 - 52	803	413	390	1 061	16.0	73.7	2.21	18	..
1960 - 62	946	487	459	1 063	17.9	90.3	2.80	18	..
1970 - 72	880	453	427	1 064	15.8	82.5	2.36	12	13
1980 - 82	735	377	358	1 053	13.0	62.5	1.83	5	7
	BBCA	KBCZ	KBCY	KMFW	KBCT	KBCS	KBCR	KBCQ	KMFX
1983	721	371	351	1 058	12.8	60.2	1.77	4	6
1984	730	373	356	1 049	12.9	60.3	1.77	4	6
1985	751	385	366	1 053	13.3	61.4	1.80	4	6
1986	755	387	368	1 053	13.3	61.1	1.78	4	5
1987	776	398	378	1 053	13.6	62.3	1.82	4	5
1988	788	403	384	1 049	13.8	63.2	1.84	4	5
1989	777	398	379	1 051	13.6	62.4	1.81	4	5
1990	799	409	390	1 049	13.9	64.2	1.84	4	5
1991	793	406	386	1 052	13.7	63.6	1.82	4	5
1992	781	400	380	1 052	13.5	63.4	1.80	3	4
1993	762	391	371	1 054	13.1	62.4	1.76	4	6
1994	751	385	365	1 054	12.9	61.6	1.74	4	6
1995	732	375	357	1 052	12.5	60.1	1.71	4	6
1996	733	376	357	1 055	12.5	60.1	1.72	4	6
1997	727†	373†	354	1 051	12.3	59.5	1.72	4	5
1998	717	366	351	1 042	12.1	58.7	1.70	4	5
England and Wales									
1900 - 02	932	475	458	1 037	28.6	114.7	..	..	..
1910 - 12	884	450	433	1 040	24.5	98.6	..	..	..
1920 - 22	862	442	420	1 051	22.8	91.1	..	..	..
1930 - 32	632	323	309	1 047	15.8	64.4	..	27	..
1940 - 42	607	312	295	1 057	15.6	61.3	1.81	22	..
1950 - 52	683	351	332	1 058	15.6	72.1	2.16	16	..
1960 - 62	812	418	394	1 061	17.6	88.9	2.77	16	..
1970 - 72	764	394	371	1 061	15.6	81.4	2.31	10	13
1980 - 82	639	328	311	1 053	12.9	61.8	1.81	4	7
	BBCB	KMFY	KMFZ	KMGA	KMGB	KMGC	KMGD	KMGE	KMGF
1983	629	323	306	1 056	12.7	59.7	1.76	4	6
1984	637	326	311	1 049	12.8	59.8	1.75	4	6
1985	656	337	320	1 054	13.1	61.0	1.78	4	6
1986	661	339	322	1 052	13.2	60.6	1.77	4	5
1987	682	350	332	1 053	13.6	62.0	1.81	3	5
1988	694	355	339	1 048	13.8	63.0	1.82	3	5
1989	688	352	335	1 051	13.6	62.5	1.80	3	5
1990	706	361	345	1 048	13.9	64.3	1.84	3	5
1991	699	358	341	1 052	13.7	63.6	1.82	3	5
1992	690	354	336	1 053	13.4	63.5	1.80	3	4
1993	673	346	328	1 056	13.1	62.6	1.76	4	6
1994	665	341	323	1 055	12.9	61.9	1.75	4	6
1995	648	332	316	1 051	12.5	60.4	1.72	4	6
1996	649	333	316	1 055	12.5	60.5	1.73	4	5
1997	643†	330†	314†	1 051	12.3	59.8†	1.73	3	5
1998	636	326	310	1 051	12.1	59.0	1.72	3	5

See footnotes on the second part of this table.

Source: Office for National Statistics: 020 7533 5113

5.14 Births

Annual averages or calendar years

continued

Thousands

	Live births				Rates				
	Total	Male	Female	Sex ratio	Crude birth rate[1]	General fertility rate[2]	TPFR[3]	Still-births[4]	Still-birth rate4
Scotland									
1900 - 02	132	67	65	1 046	29.5	120.6	..	..	..
1910 - 12	123	63	60	1 044	25.9	107.4	..	..	..
1920 - 22	125	64	61	1 046	25.6	105.9	..	..	..
1930 - 32	93	47	45	1 040	19.1	78.8	..	..	..
1940 - 42	89	46	43	1 051	18.5	73.7	..	4	..
1950 - 52	91	47	44	1 060	17.9	81.4	2.41	2	..
1960 - 62	102	53	50	1 060	19.7	97.8	2.98	2	..
1970 - 72	84	43	41	1 057	16.1	83.3	2.46	1	13
1980 - 82	68	35	33	1 051	13.1	62.2	1.80	-	6
	BBCD	KMEU	KMEV	KMEW	KMEX	KMEY	KMEZ	KMFM	KMFN
1983	65	34	31	1 071	12.6	58.9	1.70	–	6
1984	65	33	32	1 037	12.7	58.4	1.68	–	6
1985	67	34	33	1 048	13.0	59.5	1.71	–	5
1986	66	34	32	1 061	12.9	58.5	1.68	–	6
1987	66	34	32	1 053	13.0	58.8	1.68	–	5
1988	66	34	32	1 059	13.0	59.1	1.68	–	5
1989	63	33	31	1 049	12.5	56.9	1.61	–	5
1990	66	34	32	1 057	12.9	58.8	1.67	–	5
1991	67	34	33	1 056	13.1	59.8	1.70	–	6
1992	66	34	32	1 044	12.9	59.2	1.67	–	5
1993	63	32	31	1 046	12.4	57.4	1.62	–	6
1994	62	31	30	1 038	12.0	55.9	1.58	–	6
1995	60	31	29	1 043	11.7	54.5	1.55	–	7
1996	59	31	29	1 061	11.6	54.0	1.55	–	6
1997	59	31†	29	1 058†	11.6	54.3	1.58†	–	5
1998	57	29	28	1 060	11.2	52.6	1.55	–	6
Northern Ireland									
1900 - 02	..	..	..	..	..	..	..	..	..
1910 - 12	..	..	..	..	..	..	..	..	..
1920 - 22	31	16	15	1 048	24.2	105.9	..	..	..
1930 - 32	26	13	12	1 047	20.5	78.8	..	..	..
1940 - 42	27	14	13	1 078	20.8	73.7	..	..	..
1950 - 52	29	15	14	1 066	20.9	81.4	..	..	..
1960 - 62	31	16	15	1 068	22.5	111.5	3.47	7	23
1970 - 72	31	16	15	1 074	20.4	105.7	3.13	3	14
1980 - 82	28	14	13	1 048	18.0	87.5	2.59	-	8
	BBCE	KMFO	KMFP	KMFQ	KMFR	KMFS	KMFT	KMFU	KMFV
1983	27	14	13	1 068	17.7	84.0	2.56	–	7
1984	28	14	13	1 069	17.9	84.2	2.49	–	6
1985	28	14	13	1 054	17.7	83.1	2.45	–	6
1986	28	15	14	1 075	18.0	83.7	2.46	–	4
1987	28	14	14	1 056	17.7	82.2	2.41	–	6
1988	28	14	14	1 055	17.6	82.0	2.39	–	5
1989	26	13	13	1 061	16.5	76.9	2.23	–	5
1990	26	14	13	1 048	16.7	78.0	2.26	–	4
1991	26	14	13	1 065	16.5	76.0	2.18	–	5
1992	26	13	13	1 039	15.9	72.9	2.09	–	5
1993	25	13	12	1 027	15.3	71.0	2.01	–	5
1994	24	12	12	1 054	14.9	68.6	1.95	–	6
1995	24	12	11	1 079	14.5	66.8	1.91	–	6
1996	25	12	12	1 031	14.8	68.0	1.95	–	6
1997	24	12	12	1 047	14.5	67.5	1.93	–	5
1998	24	12	12	1 037	14.1	65.5	1.89	–	5

1 Rate per 1 000 population.
2 Rate per 1 000 women aged 15 - 44.
3 Total period fertility rate is the average number of children which would be born per woman if women experienced the age-specific fertility rates of the period in question throughout their child-bearing life span. UK figures for the years 1970-72 and earlier are estimates.
4 Figures given are based on stillbirths of 28 completed weeks gestation or more. On 1 October 1992 the legal definition of a stillbirth was altered to include babies born dead between 24 and 27 completed weeks gestation. Between 1 October and 31 December 1992 in the UK there were 258 babies born dead between 24 and 27 completed weeks gestation (216 in England and Wales, 35 in Scotland and 7 in Northern Ireland). If these babies were included in the stillbirth figures given, the stillbirth rates would be 5 for the UK and England and Wales while the Scotland and Northern Ireland stillbirth rate would remain as stated.

Sources: Office for National Statistics: 020 7533 5113; General Register Office (Scotland); General register Office (Northern Ireland)

5.15 Birth occurrence inside and outside marriage by age of mother

Thousands

	Inside marriage						Outside marriage					
	All ages	Under 20	20 - 24	25 - 29	Over 30	Mean age (Years)	All ages	Under 20	20 - 24	25 - 29	Over 30	Mean age (Years)
United Kingdom												
	KKEY	KKEZ	KKFY	KKFZ	KKGX	KKGY	KKGZ	KKIC	KKID	KKIE	KKIF	KKIG
1961	890	55	273	280	282	27.7	54	13	17	10	13	25.5
1971	828	70	301	271	185	26.4	74	24	25	13	12	23.8
1981	640	36	193	231	180	27.3	91	30	33	16	13	23.4
1986	597	21	159	231	185	27.9	158	45	60	31	22	23.7
1987	598	18	153	235	192	28.1	178	48	68	37	26	23.9
1988	590	16	144	235	195	28.2	198	51	76	42	29	24.1
1989	571	14	130	229	198	28.4	207	49	79	46	32	24.3
1990	576	13	121	233	209	28.6	223	51	83	53	37	24.5
1991	556	10	109	224	213	28.9	236	50	87	58	41	24.8
1992	540	9	98	216	218	29.1	241	46	86	62	46	25.1
1993	520	8	87	204	221	29.3	242	44	84	64	50	25.4
1994	511	7	78	195	231	29.6	240	41	80	65	55	25.7
1995	486	6	69	180	232	29.8	246	42	79	66	60	25.9
1996	473	6	61	170	237	30.1	260	45	80	69	66	26.0
1997	460[†]	6	55	159	240	30.3[†]	267	47	79[†]	71[†]	71	26.1
1998	447	6	51	149	243	30.5	270	49	77	70	74	26.2
Great Britain												
	KKIH	KKII	KKIJ	KKIK	KKIL	KKIM	KKIN	KKIO	KKIP	KKIQ	KKIR	KKIS
1961	859	53	264	270	272	27.7	53	13	17	10	13	25.5
1971	797	68	293	261	176	26.4	73	24	25	13	12	23.8
1981	614	34	186	223	171	27.2	89	29	32	16	13	23.3
1986	572	20	153	222	177	27.9	155	44	59	30	22	22.9
1987	574	17	147	227	184	28.0	174	46	66	36	25	23.4
1988	566	16	138	226	186	28.2	194	49	74	42	29	23.6
1989	549	13	125	220	190	28.4	202	48	77	45	32	24.2
1990	554	12	116	225	201	28.6	218	49	81	52	36	24.6
1991	535	10	105	216	205	28.9	231	48	85	57	41	24.8
1992	520	9	94	208	210	29.1	235	45	84	61	46	25.1
1993	500	7	84	196	213	29.3	236	42	82	62	49	25.4
1994	492	7	75	188	222	29.6	235	41	78	63	53	25.7
1995	468	6	66	173	223	29.8	240	40	77	65	59	25.9
1996	455	6	59	163	227	30.1	254	44	78	68	65	26.0
1997	442[†]	6	53	152	231	30.3[†]	261[†]	46	76	69	69	26.2
1998	430	6	49	143	233	30.5	263	48	74	68	73	26.3

Source: Office for National Statistics: 020 7533 5113

5.16 Live births by age of mother

Number

All live births - United Kingdom

Age-group:	Under 20	20 - 24	25 - 29	30 - 34	35 - 39	40 - 44	45 and over	All ages
	KMDV	KMDW	KMDX	KMDY	KMDZ	KMES	KMET	KMBZ
1988	66 961	219 965	276 869	159 821	53 802	9 590	548	787 566
1989	63 173	208 972	274 882	164 215	55 537	9 966	540	777 285
1990	63 007	203 490	286 014	176 826	58 425	10 301	549	798 612
1991	59 722	196 243	282 186	182 739	60 583	10 453	580	792 506
1992	54 935	184 397	277 397	189 118	63 808	10 809	553	781 017
1993	51 467	171 109	267 462	193 643	66 273	11 128	582	761 713
1994	47 878	157 483	259 422	202 954	70 905	11 441	536	750 671
1995	47 652	147 067	246 062	204 664	73 981	12 014	585	732 049
1996	50 797	141 101	238 902	210 585	78 387	12 838	638	733 375
1997	52 853†	133 274†	229 468†	212 235†	84 557†	13 741†	618†	726 812†
1998	54 823	127 245	218 112	212 953	88 776	14 465	640	717 080

Age-specific fertility rates - United Kingdom

Age-group:	Under 20	20 - 24	25 - 29	30 - 34	35 - 39	40 - 44	45 and over	All ages
	KMBR	KMBS	KMBT	KMBU	KMBV	KMBW	KMBX	KMBY
1988	32.4	94.6	124.6	82.3	28.0	4.8	0.3	63.1
1989	31.9	91.2	120.6	82.8	29.3	4.9	0.3	62.3
1990	33.1	90.8	122.9	86.5	31.0	5.0	0.3	64.1
1991	32.9	88.9	119.9	86.5	32.0	5.0	0.3	63.6
1992	31.8	85.5	117.6	87.1	33.2	5.5	0.3	63.4
1993	30.9	81.8	114.2	86.8	33.8	5.8	0.3	62.4
1994	28.8	78.2	112.1	88.5	35.4	6.0	0.3	61.6
1995	28.3	75.7	108.5	87.2	35.9	6.4	0.3	60.1
1996[1]	29.7	76.2	106.7	88.6	36.9	6.8	0.3	60.1
1997	30.3†	75.4†	104.6†	88.9†	38.7†	7.1†	0.3	59.5
1998	30.7	74.1	102.0	89.8	39.5	7.4	0.3	58.7

All live births - England and Wales

Age-group:	Under 20	20 - 24	25 - 29	30 - 34	35 - 39	40 - 44	45 and over	All ages
	KGSA	KGSB	KGSC	KGSD	KGSE	KGSF	KGSG	KGSH
1988	58 741	193 726	243 460	140 974	47 649	8 520	507	693 577
1989	55 543	185 239	242 822	145 320	49 465	8 845	491	687 725
1990	55 541	180 136	252 577	156 264	51 905	9 220	497	706 140
1991	52 396	173 356	248 727	161 259	53 644	9 316	519	699 217
1992	47 861	163 311	244 798	166 839	56 650	9 696	501	689 656
1993	45 121	151 975	235 961	171 061	58 824	9 986	539	673 467
1994	42 026	140 240	229 102	179 568	63 061	10 241	488	664 726
1995	41 938	130 744	217 418	181 202	65 517	10 779	540	648 138
1996	44 667	125 732	211 103	186 377	69 503	11 516	587	649 485
1997	46 372†	118 589†	202 792†	187 528†	74 900†	12 332†	582†	643 095†
1998	48 285	113 537	193 144	188 499	78 881	12 980	575	635 901

Age-specific fertility rates - England and Wales

Age-group:	Under 20	20 - 24	25 - 29	30 - 34	35 - 39	40 - 44	45 and over	All ages
	KGSI	KGSJ	KGSK	KGSL	KGSM	KGSN	KGSO	KGSP
1988	32.5	94.6	124.0	82.4	27.9	4.8	0.4	63.0
1989	32.0	91.7	120.4	83.2	29.4	4.9	0.3	62.5
1990	33.3	91.4	122.6	86.9	31.1	5.0	0.3	64.2
1991	33.0	89.3	119.4	86.7	32.1	5.1	0.3	63.6
1992	31.7	86.2	117.3	87.2	33.4	5.5	0.3	63.5
1993	31.0	82.7	114.1	87.0	34.1	5.9	0.3	62.6
1994	29.0	79.4	112.1	88.7	35.8	6.1	0.3	61.9
1995	28.5	76.8	108.6	87.3	36.2	6.5	0.3	60.4
1996	29.8	77.5	106.9	88.6	37.2	6.9	0.3	60.5
1997	30.2	76.6†	104.8†	88.8†	38.9†	7.3†	0.3	59.8†
1998	30.9	75.5	102.2	89.9	39.8	7.5	0.3	59.0

Sources: Office for National Statistics: 020 7533 5113; General Register Office (Northern Ireland); General Register Office (Scotland)

5.16 Live births by age of mother

continued

Number

	All live births - Scotland							
Age-group:	Under 20	20 - 24	25 - 29	30 - 34	35 - 39	40 - 44	45 and over	All ages
	KGTA	KGTB	KGTC	KGTD	KGTE	KGTF	KGTG	KGTH
1988	6 166	18 836	23 947	12 844	3 863	534	22	66 212
1989	5 729	17 191	22 936	13 131	3 852	619	22	63 480
1990	5 608	16 970	24 227	14 393	4 147	604	24	65 973
1991	5 537	16 753	24 196	15 225	4 595	686	32	67 024
1992	5 215	15 424	23 591	15 985	4 877	674	23	65 789
1993[1]	4 750	13 923	22 758	16 088	5 049	697	23	63 337
1994	4 303	12 637	21 851	16 705	5 346	736	26	61 656
1995	4 280	11 913	20 395	16 803	5 799	811	26	60 051
1996	4 546	11 023	19 515	17 045	6 129	891	32	59 308
1997	4 835	10 607	18 782	17 455	6 740	936	19	59 440
1998	4 802	9 804	17 477	17 207	6 893	1 027	42	57 318

	Age-specific fertility rates - Scotland							
Age-group:	Under 20	20 - 24	25 - 29	30 - 34	35 - 39	40 - 44	45 and over	All ages
	KGTI	KGTJ	KGTK	KGTL	KGTM	KGTN	KGTO	KGTP
1988	31.9	88.6	120.3	72.2	22.9	3.2	0.1	59.1
1989	31.2	82.5	113.7	72.1	22.9	3.6	0.1	56.9
1990	31.9	82.5	117.3	76.8	24.5	3.4	0.2	58.8
1991	33.4	82.4	116.6	78.5	26.8	3.8	0.2	59.8
1992	33.1	77.7	113.7	80.8	27.8	3.9	0.1	59.3
1993	31.2	72.4	109.7	79.7	28.0	4.1	0.1	57.4
1994	28.4	68.2	106.1	81.2	28.9	4.4	0.2	56.0
1995	28.1	66.7	100.8	80.5	30.5	4.8	0.2	54.5
1996	29.6	64.6	97.9	81.6	31.4	5.2	0.2	54.0
1997	30.9	65.5	97.0	83.5	33.9	5.3	0.1	54.3
1998	30.4	62.7	94.2	82.6	34.1	5.7	0.3	52.6

	All live births - Northern Ireland							
Age-group:	Under 20	20 - 24	25 - 29	30 - 34	35 - 39	40 - 44	45 and over	All ages
	KMDF	KMDG	KMDH	KMDI	KMDJ	KMDK	KMDL	KMDM
1988	2 054	7 403	9 462	6 003	2 790	536	19	27 767
1989	1 900	6 542	9 124	5 764	2 220	502	27	26 080
1990	1 858	6 384	9 210	6 169	2 373	477	28	26 499
1991	1 789	6 134	9 263	6 255	2 344	451	29	26 265
1992	1 860	5 664	9 008	6 294	2 278	439	29	25 572
1993	1 596	5 211	8 743	6 494	2 400	445	20	24 909
1994	1 549	4 606	8 469	6 681	2 498	464	22	24 289
1995	1 434	4 410	8 249	6 659	2 665	424	19	23 860
1996	1 584	4 346	8 284	7 163	2 755	431	19	24 582
1997	1 646	4 078	7 894†	7 252	2 917	473	17	24 277†
1998	1 736	3 904	7 491	7 247	3 002	458	23	23 861

	Age-specific fertility rates - Northern Ireland							
Age - group:	Under 20	20 - 24	25 - 29	30 - 34	35 - 39	40 - 44	45 and over	All ages
	KMDN	KMDO	KMDP	KMDQ	KMDR	KMDS	KMDT	KMDU
1988	30.8	115.1	159.3	113.7	48.0	11.2	0.5	81.9
1989	29.2	101.9	151.6	107.1	46.3	10.5	0.6	76.9
1990	29.3	100.1	152.0	111.6	48.7	10.0	0.6	78.0
1991	28.9	97.8	147.7	106.7	46.6	9.1	0.6	75.9
1992	30.5	90.3	141.9	104.5	43.7	8.9	0.6	73.4
1993	27.6	82.9	134.1	104.7	44.5	9.1	0.4	71.0
1994	26.2	74.0	131.0	105.8	44.8	9.4	0.5	68.5
1995	23.7	72.0	129.2	103.7	46.1	8.5	0.4	66.8
1996	25.9	72.2	128.7	109.3	46.1	8.5	0.4	68.0
1997	29.4	72.0	121.9	107.7	47.0	8.9	0.3	67.5
1998	27.9	69.7	118.6	109.4	47.8	8.5	0.5	65.5

1 Live births for Scotland where the age of mother was not distributed, includes 49 cases in 1993, 52 cases in 1994, 24 cases in 1995, 127 cases in 1996, 66 cases in 1997 and 66 cases in 1998.

Sources: Office for National Statistics: 020 7533 5113; General Register Office (Northern Ireland); General Register Office (Scotland)

5.17 Legal abortions

Total by age for residents

Numbers

England and Wales

	All ages	Under 15	15	16 - 19	20 - 24	25 - 29	30 - 34	35 - 39	40 - 44	45 and over	Not stated
1986	147 619	924	2 970	33 819	45 316	28 656	18 005	12 977	4 521	409	22
1987	156 191	907	2 858	35 167	49 256	31 243	18 960	12 639	4 757	390	14
1988	168 298	859	2 709	37 928	54 067	34 584	20 000	12 681	5 047	412	11
1989	170 463	803	2 580	36 182	54 880	36 604	21 284	12 713	5 020	388	9
1990	173 900	873	2 549	35 520	55 281	38 770	22 431	12 956	5 104	404	12
1991	167 376	886	2 272	31 130	52 678	38 611	23 445	13 035	4 901	408	10
1992	160 501	905	2 095	27 589	49 052	38 430	23 870	13 252	4 844	452	12
1993	157 846	964	2 119	25 806	46 846	38 139	24 690	13 885	4 889	494	14
1994	156 539	1 080	2 166	25 223	44 871	38 081	25 507	14 156	5 008	440	7
1995	154 315	946	2 324	24 945	43 394	37 254	25 759	14 352	4 868	457	16
1996	167 916	1 098	2 547	28 790	46 356	39 311	28 228	16 118	5 027	428	13
1997	170 145	1 020	2 414	29 947	44 960	40 159	28 892	16 858	5 413	482	-
1998	177 871	1 103	2 656	33 236	45 766	40 366	30 449	18 174	5 576	511	34

Scotland

	All ages	Under 15	15	16 - 19	20 - 24	25 - 29	30 - 34	35 - 39	40 - 44	45 and over
1986	9 628	74	236	2 529	2 985	1 744	1 081	708	249	22
1987	9 460	70	210	2 417	2 996	1 729	1 082	697	242	17
1988	10 128	65	218	2 529	3 304	1 970	1 107	663	257	15
1989	10 209	53	209	2 561	3 202	1 968	1 229	706	266	15
1990	10 219	54	186	2 539	3 242	2 063	1 161	700	253	21
1991	11 068	77	203	2 571	3 486	2 253	1 445	743	262	28
1992	10 818	73	174	2 377	3 389	2 291	1 444	799	254	17
1993	11 076	92	193	2 300	3 368	2 447	1 492	891	264	29
1994	11 392	78	215	2 312	3 486	2 431	1 648	877	315	30
1995	11 143	79	233	2 169	3 399	2 438	1 609	887	296	33
1996	11 978	87	236	2 362	3 571	2 603	1 801	960	331	27
1997	12 109	85	204	2 431	3 444	2 651	1 854	1 093	322	25
1998[1]	12 424	73	212	2 694	3 404	2 740	1 799	1 143	338	21

1 Provisional.

Sources: Office for National Statistics; Scottish Executive

5.18 Deaths: analysis by age and sex

Annual averages or calendar years

Number

United Kingdom

	All ages[1]	Under 1 year	1-4	5-9	10-14	15-19	20-24	25-34	35-44	45-54	55-64	65-74	75-84	85 and over
Males														
1900 - 02	340 664	87 242	37 834	8 429	4 696	7 047	8 766	19 154	24 739	30 488	37 610	39 765	28 320	6 563
1910 - 12	303 703	63 885	29 452	7 091	4 095	5 873	6 817	16 141	21 813	28 981	37 721	45 140	29 397	7 283
1920 - 22	284 876	48 044	19 008	6 052	3 953	5 906	6 572	13 663	19 702	29 256	40 583	49 398	34 937	7 801
1930 - 32	284 249	28 840	11 276	4 580	2 890	5 076	6 495	12 327	16 326	29 376	47 989	63 804	45 247	10 022
1940 - 42	314 643	24 624	6 949	3 400	2 474	4 653	4 246	11 506	17 296	30 082	57 076	79 652	59 733	12 900
1950 - 52	307 312	14 105	2 585	1 317	919	1 498	2 289	5 862	11 074	27 637	53 691	86 435	79 768	20 131
1960 - 62	318 850	12 234	1 733	971	871	1 718	1 857	3 842	8 753	26 422	63 009	87 542	83 291	26 605
1970 - 72	335 166	9 158	1 485	1 019	802	1 778	2 104	3 590	7 733	24 608	64 898	105 058	82 905	30 027
1980 - 82	330 495	4 829	774	527	652	1 999	1 943	3 736	6 568	19 728	54 159	105 155	98 488	31 936
1990 - 92	312 521	3 315	623	372	396	1 349	2 059	4 334	6 979	15 412	40 424	87 849	106 376	43 032
	KHUA	KHUB	KHUC	KHUD	KHUE	KHUF	KHUG	KHUH	KHUI	KHUJ	KHUK	KHUL	KHUM	KHUN
1978	336 395	5 220	866	726	729	2 067	2 055	3 944	6 912	22 439	57 381	110 384	92 574	31 098
1979	339 568	5 447	748	708	707	2 042	1 969	4 012	6 868	21 828	56 944	110 172	96 026	32 097
1980	332 370	5 174	792	609	659	2 022	1 940	3 786	6 698	20 577	55 176	107 089	96 301	31 547
1981	329 145	4 759	771	517	666	2 008	1 919	3 761	6 544	19 740	53 770	104 950	97 881	31 859
1982	329 971	4 555	760	456	632	1 966	1 971	3 661	6 462	18 867	53 531	103 426	101 281	32 403
1983	328 824	4 230	695	469	609	1 834	1 899	3 601	6 537	18 238	54 493	100 469	103 038	32 712
1984	321 095	3 995	725	423	580	1 708	1 999	3 595	6 425	17 647	53 715	95 420	102 513	32 350
1985	331 562	4 003	728	393	583	1 612	2 031	3 452	6 728	17 316	52 502	97 458	109 241	35 515
1986	327 160	4 219	653	384	444	1 676	2 067	3 668	6 712	16 814	50 352	95 987	108 123	36 061
1987	318 282	4 105	657	377	470	1 612	2 125	3 776	6 793	15 950	47 675	93 348	105 773	35 621
1988	319 119	4 110	680	433	460	1 525	2 160	3 983	6 860	16 016	46 001	91 893	107 082	37 916
1989	320 193	3 799	699	414	398	1 537	2 118	3 968	6 832	15 560	43 693	90 304	109 450	41 421
1990	314 601	3 614	674	376	406	1 487	2 197	4 354	6 991	15 507	41 983	88 458	107 451	41 103
1991	314 427	3 377	636	395	404	1 417	2 049	4 270	7 102	15 493	40 256	88 014	107 416	43 598
1992	308 535	2 954	559	346	377	1 144	1 932	4 379	6 845	15 236	39 033	87 075	104 261	44 394
1993[2]	317 796	2 746	582	325	401	1 072	1 907	4 442	6 672	15 631	38 734	90 160	105 693	49 431
1994	303 333	2 660	497	319	400	1 041	1 829	4 741	6 661	14 983	36 469	86 896	98 982	47 855
1995	310 722	2 595	447	314	388	1 115	1 810	4 748	6 754	15 644	36 068	85 459	103 324	52 056
1996	305 323	2 562	489	267	352	1 104	1 693	4 746	6 789	15 796	35 033	81 333	102 090	53 069
1997	300 414	2 391	456	300	364	1 111	1 712	4 583	6 667	15 689	33 707	77 870	101 365	54 199
1998	300 160	2 327	463	283	343	1 058	1 539	4 684	6 902	15 825	33 778	75 718	101 468	55 772
Females														
1900 - 02	322 058	68 770	36 164	8 757	5 034	6 818	8 264	18 702	21 887	25 679	34 521	42 456	34 907	10 099
1910 - 12	289 608	49 865	27 817	7 113	4 355	5 683	6 531	15 676	19 647	24 481	32 813	46 453	37 353	11 828
1920 - 22	274 772	35 356	17 323	5 808	4 133	5 729	6 753	14 878	18 121	24 347	34 026	48 573	45 521	14 203
1930 - 32	275 336	21 072	9 995	3 990	2 734	4 721	5 931	12 699	15 373	24 695	39 471	59 520	56 250	18 886
1940 - 42	296 646	17 936	5 952	2 743	2 068	4 180	5 028	11 261	14 255	23 629	42 651	70 907	71 377	24 658
1950 - 52	291 597	10 293	2 098	880	625	1 115	1 717	5 018	8 989	18 875	37 075	75 220	92 848	36 844
1960 - 62	304 871	8 887	1 334	627	522	684	811	2 504	6 513	16 720	36 078	73 118	105 956	51 117
1970 - 72	322 968	6 666	1 183	654	459	718	900	2 110	5 345	15 594	36 177	75 599	109 539	68 024
1980 - 82	330 269	3 561	585	355	425	733	772	2 099	4 360	12 206	32 052	72 618	117 760	82 743
1990 - 92	328 218	2 431	485	259	255	520	714	1 989	4 340	9 707	25 105	61 951	115 467	104 994
	KIUA	KIUB	KIUC	KIUD	KIUE	KIUF	KIUG	KIUH	KIUI	KIUJ	KIUK	KIUL	KIUM	KIUN
1978	330 782	3 908	715	480	497	818	823	2 343	4 719	13 914	33 353	75 433	115 730	78 049
1979	336 009	4 026	617	428	462	701	738	2 244	4 544	13 667	33 274	75 610	118 859	80 839
1980	329 149	3 938	596	409	442	771	811	2 157	4 460	12 583	32 349	73 672	116 461	80 500
1981	328 829	3 402	599	352	424	738	737	2 083	4 309	12 275	31 625	72 476	117 458	82 351
1982	332 830	3 342	561	304	410	689	767	2 057	4 312	11 759	32 183	71 705	119 362	85 379
1983	330 277	3 126	568	318	374	719	698	1 914	4 318	11 384	32 197	69 266	118 940	86 455
1984	323 823	3 005	537	304	344	665	722	1 932	4 269	10 947	32 262	66 432	116 649	85 756
1985	339 094	3 027	574	314	355	626	729	1 852	4 397	10 581	32 010	68 505	122 445	93 679
1986	333 575	2 961	561	275	307	635	769	1 882	4 387	10 211	29 954	67 313	120 663	93 657
1987	326 060	2 972	550	265	288	614	733	1 974	4 454	10 177	29 037	65 570	117 266	92 160
1988	330 059	2 951	552	264	251	612	745	1 915	4 615	9 887	28 154	65 020	117 731	97 362
1989	337 540	2 743	551	271	268	598	773	1 955	4 506	9 834	27 324	64 575	120 975	103 167
1990	327 198	2 658	489	249	273	534	700	1 967	4 463	9 718	26 350	62 019	116 357	101 421
1991	331 754	2 448	512	280	264	538	738	2 005	4 295	9 699	24 952	62 200	116 924	106 899
1992	325 703	2 187	455	249	228	489	704	1 994	4 262	9 705	24 013	61 635	113 119	106 663
1993[2]	340 685	2 084	436	239	283	465	659	2 121	4 204	9 973	23 900	63 767	114 905	117 649
1994	324 303	1 989	410	205	232	406	626	2 053	4 285	10 081	22 401	62 069	106 816	112 730
1995	334 771	1 931	370	224	250	449	592	2 140	4 203	10 389	22 093	60 988	110 247	120 895
1996	330 701	1 904	355	214	224	493	589	2 140	4 215	10 301	21 406	57 889	109 578	121 393
1997	329 332	1 862	333	215	239	487	574	1 960	4 323	10 412	20 999	55 687	108 276	123 965
1998	329 012	1 752	347	213	215	486	568	1 971	4 289	10 430	20 874	54 200	107 135	126 532

1 In some years the totals include a small number of persons whose age was not stated.

2 See chapter text.

Source: Office for National Statistics: 020 7533 5249

5.18 Deaths: analysis by age and sex

Annual averages or calendar years

continued

Number

	England and Wales													
	All ages[1]	Under 1 year	1-4	5-9	10-14	15-19	20-24	25-34	35-44	45-54	55-64	65-74	75-84	85 and over
Males														
1900 - 02	288 886	76 095	32 051	7 066	3 818	5 611	7 028	15 869	21 135	26 065	31 600	33 568	23 835	5 144
1910 - 12	257 253	54 678	24 676	5 907	3 348	4 765	5 596	13 603	18 665	24 820	32 217	38 016	24 928	6 036
1920 - 22	240 605	39 796	15 565	5 151	3 314	4 901	5 447	11 551	17 004	25 073	34 639	42 025	29 685	6 455
1930 - 32	243 147	23 331	9 099	3 844	2 435	4 354	5 580	10 600	14 041	25 657	41 581	54 910	39 091	8 624
1940 - 42	268 876	19 393	5 616	2 834	2 051	3 832	3 156	9 484	14 744	25 983	50 058	68 791	51 779	11 158
1950 - 52	266 879	11 498	2 131	1 087	778	1 248	1 947	4 990	9 489	23 815	46 948	75 774	69 496	17 677
1960 - 62	278 369	10 157	1 444	812	742	1 523	1 624	3 278	7 524	22 813	54 908	77 000	73 180	23 364
1970 - 72	293 934	7 818	1 259	860	677	1 524	1 788	3 079	6 637	21 348	56 667	92 389	73 365	26 522
1980 - 82	290 352	4 168	657	452	555	1 716	1 619	3 169	5 590	16 909	47 144	92 485	87 338	28 551
1990 - 92	275 550	2 926	545	325	338	1 157	1 757	3 717	6 057	13 258	34 977	77 063	94 672	38 757
	KHVA	KHVB	KHVC	KHVD	KHVE	KHVF	KHVG	KHVH	KHVI	KHVJ	KHVK	KHVL	KHVM	KHVN
1978	295 505	4 513	739	628	619	1 776	1 761	3 325	5 849	19 307	49 976	97 194	82 067	27 751
1979	297 862	4 731	641	602	605	1 727	1 649	3 426	5 822	18 607	49 586	96 764	85 091	28 611
1980	291 869	4 471	668	517	546	1 745	1 613	3 203	5 710	17 693	48 053	94 188	85 300	28 162
1981	289 022	4 119	651	447	573	1 734	1 576	3 181	5 535	16 889	46 858	92 189	86 774	28 496
1982	290 166	3 914	652	391	546	1 669	1 668	3 122	5 526	16 144	46 521	91 079	89 940	28 994
1983	289 419	3 654	604	391	514	1 580	1 635	3 071	5 581	15 632	47 315	88 622	91 531	29 289
1984	282 357	3 443	610	348	501	1 484	1 728	3 033	5 512	15 113	46 904	83 728	90 983	28 970
1985	292 327	3 510	638	328	503	1 374	1 738	2 953	5 776	14 838	45 704	85 695	97 362	31 908
1986	287 894	3 724	573	325	380	1 429	1 746	3 104	5 767	14 370	43 637	84 437	96 201	32 201
1987	280 177	3 637	578	309	404	1 389	1 811	3 218	5 823	13 678	41 367	82 021	94 060	31 882
1988	280 931	3 649	587	374	402	1 279	1 802	3 367	5 855	13 701	39 791	80 870	95 306	33 948
1989	281 290	3 368	606	371	337	1 325	1 782	3 380	5 947	13 407	37 680	79 012	97 027	37 048
1990	277 336	3 207	593	333	338	1 295	1 889	3 714	6 060	13 342	36 405	77 604	95 539	37 017
1991	277 582	2 966	554	341	354	1 208	1 760	3 687	6 160	13 316	34 853	77 227	95 815	39 341
1992	271 732	2 606	487	302	322	969	1 621	3 751	5 952	13 117	33 674	76 357	92 662	39 912
1993[2]	279 561	2 407	510	276	340	912	1 596	3 813	5 784	13 416	33 347	78 881	93 754	44 525
1994	267 555	2 367	432	278	331	843	1 550	4 065	5 769	12 923	31 320	76 270	88 230	43 177
1995	274 449	2 305	391	269	340	910	1 533	4 043	5 880	13 487	30 973	74 970	92 291	47 057
1996	268 682	2 272	441	236	291	925	1 409	4 064	5 843	13 565	30 066	71 046	90 708	47 816
1997	264 865	2 137	412	267	325	947	1 442	3 940	5 707	13 484	28 907	68 024	90 207	49 066
1998	264 707	2 070	413	240	291	875	1 292	4 013	5 895	13 595	29 052	66 099	90 450	50 422
Females														
1900 - 02	269 432	60 090	30 674	7 278	4 010	5 265	6 497	15 065	18 253	21 474	28 424	35 307	29 118	7 977
1910 - 12	242 079	42 642	23 335	5 883	3 519	4 522	5 256	12 742	16 363	20 611	27 571	38 489	31 363	9 782
1920 - 22	229 908	29 178	14 174	4 928	3 456	4 719	5 533	12 244	15 142	20 580	28 633	41 010	38 439	11 871
1930 - 32	233 915	16 929	8 013	3 338	2 293	3 969	5 039	10 716	13 022	21 190	33 798	50 844	48 531	16 234
1940 - 42	253 702	14 174	4 726	2 265	1 695	3 426	4 198	9 470	12 093	20 413	36 814	60 987	61 891	21 550
1950 - 52	252 176	8 367	1 727	732	520	893	1 365	4 131	7 586	16 161	31 875	65 087	81 154	32 579
1960 - 62	266 849	7 409	1 103	527	444	591	700	2 147	5 576	14 389	31 083	63 543	93 548	45 789
1970 - 72	284 181	5 677	1 020	562	396	620	806	1 814	4 585	13 417	31 222	65 817	96 952	61 293
1980 - 82	290 026	3 064	511	301	365	635	670	1 821	3 740	10 420	27 606	63 023	103 676	74 194
1990 - 92	288 851	2 161	420	227	217	455	625	1 718	3 765	8 347	21 466	53 783	101 752	93 914
	KIVA	KIVB	KIVC	KIVD	KIVE	KIVF	KIVG	KIVH	KIVI	KIVJ	KIVK	KIVL	KIVM	KIVN
1978	290 396	3 368	607	417	421	691	710	2 042	4 004	11 911	28 607	65 411	102 058	70 149
1979	295 157	3 447	527	368	386	622	635	1 938	3 868	11 671	28 624	65 646	104 807	72 618
1980	289 516	3 428	518	349	373	667	696	1 861	3 771	10 757	27 857	64 087	102 728	72 424
1981	288 868	2 902	529	302	368	650	642	1 821	3 742	10 513	27 211	62 762	103 554	73 872
1982	291 695	2 861	485	253	353	588	672	1 781	3 708	9 990	27 751	62 221	104 745	76 287
1983	290 189	2 727	489	269	332	629	597	1 655	3 708	9 786	27 792	59 913	104 844	77 448
1984	284 524	2 594	454	260	302	575	621	1 676	3 658	9 343	27 764	57 813	102 744	76 720
1985	298 407	2 631	497	260	308	544	630	1 583	3 803	9 111	27 664	59 285	108 099	83 992
1986	293 309	2 589	491	248	272	562	674	1 646	3 834	8 761	25 785	58 360	106 463	83 624
1987	286 817	2 635	489	237	246	525	639	1 708	3 897	8 774	25 000	56 858	103 354	82 455
1988	290 477	2 621	498	232	218	542	650	1 670	4 025	8 448	24 104	56 567	103 666	87 236
1989	295 582	2 440	472	241	226	531	650	1 678	3 925	8 406	23 336	55 932	106 000	91 745
1990	287 510	2 357	434	220	230	472	616	1 702	3 875	8 337	22 511	53 770	102 440	90 546
1991	292 462	2 192	439	248	222	462	644	1 729	3 703	8 369	21 303	54 156	103 268	95 727
1992	286 581	1 933	387	214	199	432	615	1 722	3 717	8 336	20 585	53 423	99 548	95 470
1993[2]	299 238	1 835	374	194	246	394	575	1 802	3 625	8 914	20 423	55 245	100 947	104 964
1994	285 639	1 753	364	187	204	357	535	1 771	3 669	8 688	19 039	53 921	94 197	100 954
1995	295 234	1 677	333	196	210	382	502	1 859	3 644	9 001	18 891	52 987	97 162	108 390
1996	291 453	1 687	320	175	196	430	507	1 852	3 658	8 852	18 244	50 195	96 679	108 658
1997	290 416	1 663	297	177	209	426	490	1 718	3 737	9 016	17 949	48 293	95 508	110 933
1998	290 308	1 555	309	177	189	407	480	1 724	3 678	9 066	17 927	46 894	94 713	113 189

1 In some years the totals include a small number of persons whose age was not stated.

2 See chapter text.

Source: Office for National Statistics: 020 7533 5249

5.18 Deaths: analysis by age and sex

continued

Annual averages or calendar years

Number

	Scotland													
	All ages[1]	Under 1 year	1-4	5-9	10-14	15-19	20-24	25-34	35-44	45-54	55-64	65-74	75-84	85 and over
Males														
1900 - 02	40 224	9 189	4 798	1 083	672	1 069	1 292	2 506	2 935	3 591	4 597	4 531	3 117	834
1910 - 12	35 981	7 510	3 935	962	595	826	910	1 969	2 469	3 325	4 356	5 113	3 182	813
1920 - 22	34 649	6 757	2 847	710	489	747	791	1 616	2 128	3 314	4 785	5 624	3 928	911
1930 - 32	32 476	4 426	1 771	610	365	568	706	1 352	1 848	2 979	5 095	6 906	4 839	1 010
1940 - 42	36 384	3 973	1 011	449	321	668	888	1 643	2 090	3 348	5 728	8 556	6 317	1 337
1950 - 52	32 236	1 949	349	175	105	200	265	693	1 267	3 151	5 574	8 544	8 094	1 871
1960 - 62	32 401	1 578	222	121	102	146	185	456	1 013	2 986	6 682	8 505	7 980	2 425
1970 - 72	32 446	944	168	119	93	178	233	396	875	2 617	6 641	10 176	7 383	2 624
1980 - 82	31 723	451	80	56	71	206	233	423	776	2 280	5 601	10 152	8 804	2 591
	KHWA	KHWB	KHWC	KHWD	KHWE	KHWF	KHWG	KHWH	KHWI	KHWJ	KHWK	KHWL	KHWM	KHWN
1978	32 432	479	87	63	78	211	210	479	854	2 508	5 888	10 734	8 299	2 542
1979	32 884	490	79	79	71	219	215	442	805	2 599	5 851	10 804	8 543	2 687
1980	31 669	481	93	65	78	190	223	421	778	2 316	5 628	10 248	8 571	2 577
1981	31 700	435	71	50	66	208	250	439	816	2 330	5 506	10 193	8 788	2 548
1982	31 801	436	77	53	69	220	225	410	733	2 195	5 669	10 015	9 052	2 647
1983	31 196	380	67	53	65	185	178	406	764	2 131	5 769	9 414	9 204	2 580
1984	30 731	389	87	53	65	172	202	429	696	2 017	5 493	9 337	9 222	2 569
1985	31 147	342	57	49	58	174	208	390	759	1 959	5 486	9 339	9 569	2 757
1986	31 111	334	66	44	49	177	238	436	757	1 967	5 354	9 169	9 574	2 946
1987	30 384	331	54	46	47	163	212	415	779	1 870	5 131	9 058	9 383	2 895
1988	30 195	324	64	39	42	181	246	475	808	1 915	4 997	8 763	9 314	3 027
1989	31 025	331	62	24	45	150	246	445	719	1 721	4 889	9 028	9 922	3 443
1990	29 617	297	62	31	50	138	240	502	745	1 734	4 512	8 635	9 499	3 172
1991	29 312	299	59	42	34	150	211	441	757	1 741	4 382	8 657	9 209	3 330
1992	29 334	265	51	28	36	123	238	511	731	1 716	4 313	8 541	9 225	3 556
1993[2]	30 504	240	50	39	37	107	225	490	725	1 817	4 375	9 031	9 470	3 898
1994	28 416	212	42	27	48	133	212	538	715	1 684	4 114	8 575	8 446	3 670
1995	28 791	197	37	30	30	152	195	563	698	1 746	4 144	8 449	8 604	3 946
1996	29 223	206	41	23	46	139	212	556	755	1 845	4 087	8 259	8 926	4 128
1997	28 305	186	32	22	27	114	208	521	788	1 794	3 876	7 909	8 791	4 037
1998	28 132	183	37	34	39	134	200	524	843	1 796	3 828	7 746	8 585	4 183
Females														
1900 - 02	39 891	7 143	4 477	1 162	747	1 058	1 246	2 625	2 732	3 130	4 485	5 273	4 305	1 508
1910 - 12	36 132	5 854	3 674	981	618	836	910	2 149	2 473	2 909	3 960	5 636	4 588	1 552
1920 - 22	34 449	5 029	2 602	687	489	711	889	1 947	2 266	2 828	4 157	5 587	5 443	1 814
1930 - 32	32 377	3 319	1 602	527	339	568	666	1 508	1 812	2 731	4 380	6 630	6 178	2 117
1940 - 42	33 715	2 852	921	373	283	595	656	1 382	1 672	2 528	4 630	7 674	7 613	2 536
1950 - 52	31 525	1 432	284	115	84	185	293	714	1 127	2 188	4 204	8 157	9 310	3 431
1960 - 62	30 559	1 107	170	80	63	72	87	287	762	1 897	4 115	7 752	9 991	4 177
1970 - 72	30 978	694	118	69	46	73	74	231	608	1 769	4 036	7 823	10 112	5 324
1980 - 82	32 326	337	49	37	44	74	73	213	493	1 456	3 565	7 781	11 333	6 871
	KIWA	KIWB	KIWC	KIWD	KIWE	KIWF	KIWG	KIWH	KIWI	KIWJ	KIWK	KIWL	KIWM	KIWN
1978	32 691	351	73	37	61	93	86	236	598	1 651	3 871	8 208	11 028	6 398
1979	32 863	388	69	37	58	59	81	247	551	1 635	3 776	8 045	11 309	6 608
1980	31 630	350	51	41	44	77	90	222	547	1 511	3 587	7 673	10 988	6 449
1981	32 128	345	46	35	43	68	69	213	453	1 414	3 556	7 935	11 144	6 807
1982	33 221	317	50	35	45	78	60	203	479	1 444	3 552	7 735	11 867	7 356
1983	32 258	266	51	33	33	67	76	201	504	1 317	3 568	7 558	11 340	7 244
1984	31 614	283	62	32	37	72	78	205	475	1 320	3 703	6 979	11 134	7 234
1985	32 820	282	55	37	34	63	76	207	481	1 179	3 563	7 449	11 604	7 790
1986	32 356	247	50	16	24	50	77	188	441	1 181	3 372	7 251	11 476	7 983
1987	31 630	232	44	21	34	60	70	195	429	1 160	3 301	7 032	11 262	7 790
1988	31 762	219	33	22	21	54	63	197	470	1 115	3 250	6 879	11 361	8 078
1989	33 992	223	56	23	27	54	104	224	470	1 156	3 279	7 052	12 100	9 224
1990	31 910	213	32	16	34	46	68	204	468	1 099	3 109	6 685	11 233	8 703
1991	31 729	174	54	22	31	57	74	225	463	1 070	2 974	6 542	11 059	8 984
1992	31 603	184	50	23	21	45	73	225	442	1 109	2 816	6 663	10 944	9 008
1993[2]	33 545	172	45	34	27	55	60	258	460	1 089	2 793	6 918	11 330	10 304
1994	30 912	170	29	11	19	33	74	229	495	1 102	2 723	6 617	10 008	9 402
1995	31 709	178	26	16	26	50	70	231	435	1 100	2 601	6 449	10 452	10 075
1996	31 448	159	24	31	21	49	67	218	453	1 172	2 573	6 206	10 256	10 219
1997	31 189	130	23	28	21	43	71	199	496	1 128	2 480	5 985	10 164	10 421
1998	31 032	137	26	28	19	55	68	198	485	1 106	2 416	5 955	9 913	10 626

1 In some years the totals include a small number of persons whose age was not stated.
2 See chapter text.

Sources: Office for National Statistics: 020 7533 5249; General Register Office (Scotland)

5.18 Deaths: analysis by age and sex

Annual averages or calendar years

continued

Number

	Northern Ireland													
	All ages[1]	Under 1 year	1-4	5-9	10-14	15-19	20-24	25-34	35-44	45-54	55-64	65-74	75-84	85 and over
Males														
1900 - 02	11 554	1 958	985	280	206	367	446	779	669	832	1 413	1 666	1 368	585
1910 - 12	10 469	1 697	841	222	152	282	311	569	679	836	1 148	2 011	1 287	434
1920 - 22	9 622	1 491	596	191	150	258	334	496	570	869	1 159	1 749	1 324	435
1930 - 32	8 626	1 083	406	126	90	154	209	375	437	740	1 313	1 988	1 317	388
1940 - 42	9 383	1 258	322	117	102	153	202	379	462	751	1 290	2 305	1 637	405
1950 - 52	8 197	658	105	55	36	50	77	179	318	671	1 169	2 117	2 178	583
1960 - 62	8 080	499	67	38	27	49	48	108	216	623	1 419	2 037	2 131	816
1970 - 72	8 786	396	58	40	32	76	83	115	221	643	1 590	2 493	2 157	881
1980 - 82	8 420	211	37	20	26	77	92	144	202	539	1 414	2 518	2 346	795
	KHXA	KHXB	KHXC	KHXD	KHXE	KHXF	KHXG	KHXH	KHXI	KHXJ	KHXK	KHXL	KHXM	KHXN
1978	8 458	228	40	35	32	80	84	140	209	624	1 517	2 456	2 208	805
1979	8 822	226	28	27	31	96	105	144	241	622	1 507	2 604	2 392	799
1980	8 832	222	31	27	35	87	104	162	210	568	1 495	2 653	2 430	808
1981	8 423	205	49	20	27	66	93	141	193	521	1 406	2 568	2 319	815
1982	8 004	205	31	12	17	77	78	129	203	528	1 341	2 332	2 289	762
1983	8 209	196	24	25	30	69	86	124	192	475	1 409	2 433	2 303	843
1984	8 007	163	28	22	14	52	69	133	217	517	1 318	2 355	2 308	811
1985	8 088	151	33	16	22	64	85	109	193	519	1 312	2 424	2 310	850
1986	8 155	161	14	15	15	70	83	128	188	477	1 361	2 381	2 348	914
1987	7 721	137	25	22	19	60	102	143	191	402	1 177	2 269	2 330	844
1988	7 993	137	29	20	16	65	112	141	197	400	1 213	2 260	2 462	941
1989	7 878	100	31	19	16	62	90	143	166	432	1 124	2 264	2 501	930
1990	7 648	110	19	12	18	54	68	138	186	431	1 066	2 219	2 413	914
1991	7 533	112	23	12	16	59	78	142	185	436	1 021	2 130	2 392	927
1992	7 469	83	21	16	19	52	73	117	162	403	1 046	2 177	2 374	926
1993[2]	7 731	99	22	10	24	53	86	139	163	398	1 012	2 248	2 469	1 008
1994	7 362	81	23	14	21	65	67	138	177	376	1 035	2 051	2 306	1 008
1995	7 482	93	19	15	18	53	82	142	176	411	951	2 040	2 429	1 053
1996	7 418	84	7	8	15	40	72	126	191	386	880	2 028	2 456	1 125
1997	7 244	68	12	11	12	50	62	122	172	411	924	1 937	2 367	1 096
1998	7 321	74	13	9	13	49	47	147	164	434	898	1 873	2 433	1 167
Females														
1900 - 02	12 735	1 537	1 013	317	277	495	521	1 012	902	1 075	1 612	1 876	1 484	614
1910 - 12	11 397	1 369	808	249	218	325	365	785	811	961	1 282	2 328	1 402	494
1920 - 22	10 415	1 149	547	193	188	299	331	687	713	939	1 236	1 976	1 639	518
1930 - 32	9 044	824	380	125	102	184	226	475	539	774	1 293	2 046	1 541	535
1940 - 42	9 229	910	305	105	90	159	174	409	490	688	1 207	2 246	1 873	572
1950 - 52	7 896	494	87	33	21	37	59	173	276	526	996	1 976	2 384	834
1960 - 62	7 463	371	61	20	15	21	24	70	175	434	880	1 823	2 417	1 151
1970 - 72	7 809	295	45	23	17	25	20	65	152	408	919	1 959	2 475	1 407
1980 - 82	7 917	160	26	17	17	23	29	65	127	329	881	1 813	2 752	1 678
	KIXA	KIXB	KIXC	KIXD	KIXE	KIXF	KIXG	KIXH	KIXI	KIXJ	KIXK	KIXL	KIXM	KIXN
1978	7 695	189	35	26	15	34	27	65	117	352	875	1 814	2 644	1 502
1979	7 989	191	21	23	18	20	22	59	125	361	874	1 919	2 743	1 613
1980	8 003	160	27	19	25	27	25	74	142	315	905	1 912	2 745	1 627
1981	7 833	155	24	15	13	20	26	49	114	348	858	1 779	2 760	1 672
1982	7 914	164	26	16	12	23	35	73	125	325	880	1 749	2 750	1 736
1983	7 830	133	28	16	9	23	25	58	106	281	837	1 795	2 756	1 763
1984	7 685	128	21	12	5	18	23	51	136	284	795	1 640	2 770	1 802
1985	7 867	114	22	17	13	19	23	62	113	291	783	1 771	2 742	1 897
1986	7 910	125	20	11	11	23	18	48	112	269	797	1 702	2 724	2 050
1987	7 613	105	17	7	8	29	24	71	128	243	736	1 680	2 650	1 915
1988	7 820	111	21	10	12	16	32	48	120	324	800	1 574	2 704	2 048
1989	7 966	80	23	7	15	13	19	53	111	272	709	1 591	2 875	2 198
1990	7 778	88	23	13	9	16	16	61	120	282	730	1 564	2 684	2 172
1991	7 563	82	19	10	11	19	20	51	129	260	675	1 502	2 597	2 188
1992	7 519	70	18	12	8	12	16	47	103	260	612	1 549	2 627	2 185
1993[2]	7 902	77	17	11	10	16	24	61	119	270	684	1 604	2 628	2 381
1994	7 752	66	17	7	9	16	17	53	121	291	639	1 531	2 611	2 374
1995	7 828	76	11	12	14	17	20	50	124	288	601	1 552	2 633	2 430
1996	7 800	58	11	8	7	14	15	70	104	277	589	1 488	2 643	2 516
1997	7 727	69	13	10	9	18	13	43	90	268	570	1 409	2 604	2 611
1998	7 672	60	12	8	7	24	20	49	126	258	531	1 351	2 509	2 717

1 In some years the totals include a small number of persons whose age was not stated.

2 See chapter text.

Sources: Office for National Statistics: 020 7533 5249; General Register Office (Northern Ireland)

5.19 Deaths: analysed by cause

International Statistical Classification of Diseases, Injuries and Causes of Death

Ninth Revision, 1979

Number

		ICD 9 code	England and Wales 1991[1]	1992[1]	1993[1,5]	1994	1995[4]	1996[4]	1997	1998
Total deaths	KHEA		570 044	558 313	578 799	553 194	569 683	560 135	555 281	555 015
Deaths from natural causes [2]	KHEB		549 706	538 677	559 649	534 354	550 936	541 429	536 422	536 396
Infectious and parasitic diseases[3]	KHEC	001-139	2 406	2 633	3 257	3 318	3 682	3 636	3 496	3 410
Intestinal infectious diseases	KJZZ	001-009	169	240	194	222	277	339	384	418
Tuberculosis of the respiratory system	KHEH	010-012	334	309	423	418	353	310	289	283
Other tuberculosis, including late effects	KHEI	013-018,137	240	263	193	180	161	166	148	170
Whooping cough	KHEK	033	–	1	–	3	2	2	1	4
Meningococcal infection	KHEM	036	170	162	173	149	196	235	242	210
Measles	KHEP	055	1	2	4	1	1	–	3	3
Malaria	KHER	084	11	9	4	11	4	11	12	8
Syphilis	KHES	090-097	15	22	6	11	9	6	9	3
Neoplasms	KHET	140-239	145 355	145 963	142 535	141 747	141 297	139 459	137 618	138 306
Malignant neoplasm of stomach	KHEU	151	8 427	8 285	7 548	7 590	7 077	6 756	6 613	6 442
Malignant neoplasm of trachea, bronchus and lung	KHEV	162	34 190	33 662	32 614	32 143	31 627	30 810	29 976	30 199
Malignant neoplasm of breast	KHEW	174-175	13 869	13 755	13 115	12 918	12 623	12 246	12 047	11 835
Malignant neoplasm of uterus	KWUP	179+182	1 494	1 389	1 308	1 241	1 329	1 303	1 291	1 296
Malignant neoplasm of cervix	KWUQ	180	1 668	1 647	1 483	1 370	1 339	1 315	1 225	1 158
Leukaemia	KHEY	204-208	3 687	3 616	3 559	3 507	3 540	3 464	3 587	3 551
Benign and unspecified neoplasms	KHEZ	210-229,239	1 337	1 368	1 544	1 580	1 784	1 647	1 624	1 605
Endocrine, nutritional and metabolic diseases and immunity disorders	KHFA	240-279	10 538	10 605	7 924	7 430	7 883	7 502	7 383	7 542
Diabetes mellitus	KHFB	250	8 087	8 067	6 266	5 938	6 240	5 994	5 890	5 938
Nutritional deficiencies	KHFC	260-269	109	104	85	66	80	70	65	72
Other metabolic and immunity disorders[3]	KMBO	270-279	1 664	1 747	1 119	1 052	1 138	1 048	1 038	1 169
Diseases of blood and blood-forming organs	KHFD	280-289	2 446	2 417	1 974	1 898	1 929	1 986	2 008	1 937
Anaemias	KHFE	280-285	1 110	1 056	786	660	711	724	681	641
Mental disorders	KHFF	290-319	13 500	12 950	7 780	8 042	9 149	9 296	9 725	10 430
Diseases of nervous system and sense organs	KHFG	320-389	11 889	11 577	9 143	9 010	9 724	9 772	9 772	10 035
Meningitis	KHFH	320-322	233	208	218	170	209	245	224	216
Diseases of the circulatory system	KHFI	390-459	261 834	254 683	257 989	242 213	243 390	237 669	228 446	226 677
Rheumatic heart disease	KHFJ	393-398	2 193	2 136	1 834	1 719	1 714	1 682	1 481	1 629
Hypertensive disease	KHFL	401-405	3 340	3 144	3 052	2 800	2 882	3 026	3 084	3 122
Ischaemic heart disease	KHFM	410-414	150 090	145 904	146 302	135 440	133 861	129 047	122 432	121 037
Diseases of pulmonary circulation and other forms of heart disease	KHFN	415-429	18 820	18 352	21 341	25 795	26 891	26 170	26 609	26 377
Cerebrovascular disease	KHFO	430-438	68 669	66 291	61 172	58 768	59 957	59 723	57 747	57 516
Diseases of the respiratory system	KHFP	460-519	63 273	60 388	90 981	81 485	91 298	88 630	92 517	90 192
Influenza	KHFQ	487	248	262	439	62	251	179	347	129
Pneumonia	KHFR	480-486	28 504	26 257	54 624	48 917	55 318	54 137	56 719	54 631
Bronchitis, emphysema	KHFS	490-492	6 773	6 070	5 715	4 969	4 907	4 249	4 116	3 523
Asthma	KHFT	493	1 884	1 791	1 701	1 516	1 459	1 349	1 439	1 366
Diseases of the digestive system	KHFU	520-579	18 508	18 742	18 399	18 635	19 466	19 946	20 406	21 025
Ulcer of stomach and duodenum	KHFV	531-533	4 304	4 296	4 222	4 111	3 999	4 111	3 959	3 935
Appendicitis	KHFW	540-543	108	130	124	113	101	102	125	144
Hernia of the abdominal cavity and other intestinal obstruction	KHFX	550-553,560	1 967	2 025	1 855	1 814	1 975	1 967	2 106	2 042
Chronic liver disease and cirrhosis	KHFY	571	3 102	3 056	2 979	3 244	3 612	3 789	4 107	4 494
Diseases of the genito-urinary system	KHFZ	580-629	6 464	5 306	6 727	6 812	7 118	6 752	6 757	6 946
Nephritis, nephrotic syndrome and nephrosis	KHGA	580-589	3 234	2 072	3 401	3 246	3 283	3 057	2 930	2 946
Hyperplasia of prostate	KHGB	600	413	343	275	283	273	235	247	207
Complications of pregnancy, childbirth, etc	KHGC	630-676	45	45	36	50	45	41	35	43
Abortion	KHGD	630-639	6	6	3	6	8	4	2	7
Diseases of the skin and subcutaneous tissue	KHGE	680-709	930	907	1 019	1 107	1 088	1 075	1 025	1 070
Diseases of the musculo-skeletal system	KHGF	710-739	5 417	5 376	3 559	3 406	3 646	3 517	3 559	3 566
Congenital anomalies	KHGG	740-759	1 643	1 565	1 338	1 301	1 290	1 227	1 283	1 247
Certain conditions originating in the perinatal period	KHGH	760-779	250	242	259	147	148	149	131	124
Birth trauma, hypoxia, birth asphyxia and other respiratory conditions	KHGI	767-770	78	67	169	133	116	114	99	94
Signs, symptoms and ill-defined conditions	KHGJ	780-799	5 208	5 278	6 729	7 754	9 783	10 772	12 292	13 846
Sudden infant death syndrome	KMBP	798-0	912	456	391	371	315	345	327	236
Deaths from injury and poisoning[2]	KHGK	E800-E999	17 286	16 681	16 354	16 091	16 049	16 061	16 311	16 201
All accidents	KHGL	E800-E929	11 049	10 435	10 396	10 219	10 156	10 479	10 661	10 351
Motor vehicle accidents	KHGM	E810-E825	4 470	4 114	3 437	3 279	3 123	3 184	3 184	2 946
Suicide and self-inflicted injury	KHGN	E950-E959	3 893	3 952	3 719	3 619	3 570	3 445	3 424	3 614
All other external causes	KHGO	(E930-E949)	2 344	2 294	2 239	2 253	2 323	2 137	2 226	2 236

1 On 1 January 1986, a new certificate for deaths within the first 28 days of life was introduced. It is not possible to assign one underlying cause of death from this certificate. The 'cause' figures in this table for 1986 onwards exclude all deaths at ages under 28 days.

2 Within certain main categories only selected causes of death are shown.

3 Deaths assigned to AIDS & AIDS - related diseases are included in ICD 270 279 for England and Wales up to 1992 and Scotland up until 1995. Northern Ireland has always assigned such deaths to the Chapter on Infectious Diseases (001 - 139). England and Wales adopted this practice from 1993 and Scotland from 1996 onwards.

4 Data for 1995 and 1996 has been amended to show the number of deaths that occurred during the calendar year. Data up to 1992 gives the number of deaths that were registered in the calendar year and subsequent years give the number that occurred during the calendar year.

5 See chapter text.

Source: Office for National Statistics: 020 7533 5249

5.19 Deaths: analysed by cause

continued

International Statistical Classification of Diseases, Injuries and Causes of Death

Ninth Revision, 1979

Number

			Scotland							
		ICD 9 code	1991	1992	1993	1994	1995	1996	1997	1998
Total deaths	KHHA		61 041	60 937	64 049	59 328	60 500	60 671	59 494	59 164
Deaths from natural causes[1]	KHHB		58 217	58 098	61 306	56 710	57 864	57 941	56 932	56 579
Infectious and parasitic diseases	KHHC	001-139	310	270	340	306	326	493	431	486
Intestinal infectious diseases	KFBP	001-009	14	18	19	19	17	22	19	34
Tuberculosis of respiratory system	KHHH	010-012	46	40	45	41	31	38	48	36
Other tuberculosis, including late effects	KHHI	013-018,137	38	23	42	24	–	15	23	22
Whooping cough	KHHL	033	–	–	–	1	–	–	1	–
Meningococcal infection	KHHN	036	10	9	13	8	15	12	12	20
Measles	KHHQ	055	1	–	–	2	1	–	–	–
Malaria	KHHS	084	1	–	1	2	–	1	–	1
Syphilis	KHHT	090-097	2	4	1	3	1	2	1	–
Neoplasms	KHHU	140-239	15 031	15 312	15 619	15 394	15 462	15 419	15 054	14 907
Malignant neoplasm of stomach	KHHV	151	833	828	766	754	758	699	710	680
Malignant neoplasm of the trachea, bronchus and lung	KHHW	162	4 209	4 308	4 299	4 237	4 221	4 126	4 106	3 984
Malignant neoplasm of breast	KHHX	174-175	1 282	1 256	1 285	1 278	1 249	1 200	1 161	1 147
Malignant neoplasm of uterus	KWUR	179+182	119	131	115	133	98	105	106	110
Malignant neoplasm of cervix	KWUS	180	175	183	170	154	147	138	144	145
Leukaemia	KHHZ	204-208	268	284	288	290	283	319	291	329
Benign neoplasms	KHIA	210-229,239	112	138	121	122	118	194	126	110
Endocrine, nutritional and metabolic diseases and immunity disorders	KHIB	240-279	776	742	775	768	738	722	727	797
Diabetes mellitus	KHIC	250	530	504	513	503	461	526	510	574
Nutritional deficiencies	KHID	260-269	6	17	16	14	6	16	7	12
Other metabolic and immunity disorders[2]	KMBN	270-279	197	193	215	217	235	140	167	175
Diseases of blood and blood-forming organs	KHIE	280-289	170	183	173	121	129	180	200	222
Anaemias	KHIF	280-285	68	75	71	60	67	76	71	81
Mental disorders	KHIG	290-319	1 110	1 133	1 322	1 306	1 583	1 595	1 611	1 725
Diseases of nervous system and sense organs	KHIH	320-389	947	877	879	853	832	852	900	894
Meningitis	KHII	320-322	26	27	17	24	18	14	17	6
Diseases of the circulatory system	KHIJ	390-459	29 166	28 776	29 909	27 138	27 079	26 728	25 911	25 253
Rheumatic heart disease	KHIL	393-398	180	186	201	175	178	154	183	151
Hypertensive disease	KHIM	401-405	316	323	325	312	324	281	294	277
Ischaemic heart disease	KHIN	410-414	16 866	16 536	16 925	15 234	14 977	14 650	14 013	13 419
Diseases of pulmonary circulation and other forms of heart disease	KHIO	415-429	2 234	2 291	2 470	2 200	2 351	3 063	2 900	2 784
Cerebrovascular disease	KHIP	430-438	7 968	7 861	8 392	7 684	7 748	7 130	6 959	6 900
Diseases of the respiratory system	KHIQ	460-519	7 068	6 999	8 409	6 981	7 668	7 863	7 891	8 011
Influenza	KHIR	487	28	29	127	11	33	45	83	12
Pneumonia	KHIS	480-486	3 785	3 729	4 495	3 757	4 021	4 158	4 028	4 064
Bronchitis, emphysema	KHIT	490-492	378	388	424	357	372	348	320	288
Asthma	KHIU	493	161	115	129	119	120	122	113	120
Diseases of the digestive system	KHIV	520-579	2 059	2 122	2 162	2 192	2 252	2 440	2 428	2 578
Ulcer of stomach and duodenum	KHIW	531-533	368	387	372	380	359	344	306	314
Appendicitis	KHIX	540-543	10	10	8	11	7	9	14	11
Hernia of abdominal cavity	KHIY	550-553, 560	176	197	200	183	173	196	168	211
Chronic liver disease and cirrhosis	KHIZ	571	476	446	494	555	607	723	767	806
Diseases of the genito-urinary system	KHJA	580-629	805	888	857	816	928	839	904	890
Nephritis, nephrotic syndrome and nephrosis	KHJB	580-589	541	574	581	562	606	549	576	570
Hyperplasia of prostate	KHJC	600	12	20	16	11	15	11	9	7
Complications of pregnancy, childbirth, etc	KHJD	630-676	9	7	7	9	6	6	4	5
Abortion	KHJE	630-639	–	–	1	–	–	–	–	–
Diseases of the skin and subcutaneous tissue	KHJF	680-709	82	82	100	101	80	75	87	94
Diseases of the musculo-skeletal system	KHJG	710-739	270	306	269	279	303	252	268	284
Congenital anomalies	KHJH	740-759	193	209	216	170	176	192	182	176
Certain conditions originating in the perinatal period	KHJI	760-779	213	216	174	191	178	181	140	165
Birth trauma, hypoxia, birth asphyxia and other respiratory conditions	KHJJ	767-770	105	107	101	104	90	94	71	82
Signs, symptoms and ill-defined conditions	KHJK	780-779	300	280	344	331	365	337	383	398
Sudden infant death syndrome	KMBM	7980	90	64	58	48	48	43	52	79
Deaths from injury and poisoning[1]	KHJL	E800-E999	2 532	2 535	2 489	2 372	2 395	2 497	2 373	2 379
All accidents	KHJM	E800-E929	1 734	1 580	1 469	1 413	1 439	1 497	1 398	1 388
Motor vehicle accidents	KHJN	E810-E825	513	468	402	354	422	363	384	387
Suicide and self-inflicted injuries	KHJO	E950-E959	525	569	615	624	623	596	599	649
All other external causes	KMBQ	(E930-E949) (E960-E999)	273	386	405	335	333	404	376	342

1 Within certain main categories only selected causes of death are shown.
2 See chapter text.

Sources: Office for National Statistics: 020 7533 5249; General Register Office (Scotland)

5.19 Deaths: analysed by cause

continued

International Statistical Classification of Diseases, Injuries and Causes of Death
Ninth Revision, 1979

Number

		ICD 9 code	Northern Ireland 1991	1992	1993	1994	1995	1996	1997	1998
Total deaths	KHKA		15 096	14 988	15 633	15 114	15 310	15 218	14 971†	14 933
Deaths from natural causes[1]	KHKB		14 256	14 303	14 871	14 325	14 516	14 528	14 276†	14 271
Infectious and parasitic diseases	KHKC	001-139	44	41	55	39	44	54	61	53
Intestinal and infectious diseases	KHKD	001-009	1	–	–	2	1	1	–	–
Tuberculosis of the respiratory system	KHKE	010-012	6	8	9	6	13	–	8	4
Other tuberculosis, including late effects	KHKF	013-018,137	2	5	6	3	1	3	3	2
Whooping cough	KHKG	033	–	–	–	–	–	–	–	–
Meningococcal infection	KHKH	036	6	7	4	4	4	5	4	6
Measles	KHKI	055	–	–	–	1	–	–	–	–
Malaria	KHKJ	084	–	–	1	–	–	–	–	–
Syphilis	KHKK	090-097	–	–	1	–	–	–	–	–
Neoplasms	KHKL	140-239	3 552	3 621	3 705	3 665	3 585	3 715	3 669†	3 769
Malignant neoplasm of stomach	KHKM	151	233	213	194	201	170	200	171	215
Malignant neoplasm of the trachea, bronchus and lung	KHKN	162	788	771	812	768	752	816	773†	775
Malignant neoplasm of breast	KHKO	174-175	342	308	339	338	329	309	267	299
Malignant neoplasm of uterus	KWUT	179+182	28	31	32	37	28	28	35	34
Malignant neoplasm of cervix	KWUU	180	38	33	33	38	19	45	26	33
Leukaemia	KHKQ	204-208	83	66	84	94	83	92	107	93
Benign neoplasms and neoplasms of unspecified nature	KHKR	210-229,239	38	38	55	47	51	60	53	121
Endocrine, nutritional and metabolic diseases	KHKS	240-279	68	60	75	74	86	81	106	85
Diabetes mellitus	KHKT	250	41	23	44	45	43	49	77	55
Nutritional deficiencies	KHKU	260-269	1	–	–	1	2	–	1	1
Other metabolic and immunity disorders[2]	KHKV	270-279	23	26	25	20	38	28	17	26
Diseases of blood and blood-forming organs	KHKW	280-289	31	18	29	29	29	22	20	24
Anaemias	KHKX	280-285	18	11	8	15	10	11	9	14
Mental disorders	KHKY	290-319	68	52	56	91	78	100	138	145
Diseases of nervous system and sense organs	KHKZ	320-389	168	181	189	187	224	236	235†	240
Meningitis	KHLA	320-322	10	7	15	4	1	7	8	4
Diseases of the circulatory system	KHLB	390-459	6 986	7 112	7 137	7 011	6 929	6 633	6 505	6 367
Rheumatic heart disease	KHLC	393-398	56	58	47	30	47	29	38	29
Hypertensive disease	KHLD	401-405	67	69	67	79	94	75	73	72
Ischaemic heart disease	KHLE	410-414	4 223	4 313	4 245	4 168	4 086	3 856	3 764	3 654
Diseases of pulmonary circulation and other forms of heart disease	KHLF	415-429	627	633	678	633	641	655	635	680
Cerebrovascular disease	KHLG	430-438	1 711	1 691	1 722	1 738	1 690	1 653	1 646	1 602
Diseases of the respiratory system	KHLH	460-519	2 493	2 423	2 756	2 398	2 656	2 749	2 664†	2 627
Influenza	KHLI	487	5	6	17	3	9	6	8	2
Pneumonia	KHLJ	480-486	1 683	1 547	1 814	1 595	1 781	1 817	1 775†	1 727
Bronchitis, emphysema	KHLK	490-492	150	143	151	116	104	137	99	108
Asthma	KHLL	493	47	53	58	30	42	31	32	44
Diseases of the digestive system	KHLM	520-579	395	405	445	424	449	483	450†	499
Ulcer of stomach and duodenum	KHLN	531-533	98	97	97	100	87	102	95	80
Appendicitis	KHLO	540-543	2	–	1	3	6	–	4	2
Hernia of the abdominal cavity and other intestinal obstruction	KHLP	550-553,560	45	51	43	50	39	52	42	47
Chronic liver disease and cirrhosis	KHLQ	571	60	69	66	66	71	88	68	104
Diseases of the genito-urinary system	KHLR	580-629	272	238	261	250	251	254	243†	265
Nephritis, nephrotic syndrome and nephrosis	KHLS	580-589	183	168	171	165	162	168	175	164
Hyperplasia of prostate	KHLT	600	8	6	5	3	7	3	4	5
Complications of pregnancy, childbirth, etc	KHLU	630-676	1	–	–	–	–	1	–	1
Abortion	KHLV	630-639	–	–	–	–	–	–	–	–
Diseases of the skin and subcutaneous tissue	KHLW	680-709	36	27	33	27	37	29	25	36
Diseases of the musculo-skeletal system	KHLX	710-739	54	43	35	31	40	35	44	30
Congenital anomalies	KHLY	740-759	90	77	116	87	91	72	59†	70
Certain conditions originating in the perinatal period	KHLZ	760-779	82	63	62	69	93	68	67†	64
Birth trauma, hypoxia, birth asphyxia and other respiratory conditions	KHMA	767-770	31	26	22	29	27	28	24†	21
Signs, symptoms and ill-defined conditions	KHMB	780-799	40	46	40	44	55	88	92	149
Sudden infant death syndrome	KHMC	7980	15	11	6	7	9	15	8	4
Deaths from injury and poisoning[1]	KHMD	E800-E999	719	581	639	688	663	598	593†	569
All accidents	KHME	E800-E929	492	376	391	430	391	402	427	381
Motor vehicle accidents	KHMF	E810-E825	195	165	152	172	140	121	153†	129
Suicide and self-inflicted injuries	KHMG	E950-E959	129	107	129	138	122	124	120	126
All other external causes	KHMH	(E930-E949) (E960-E999)	98	98	119	120	150	72	45†	62

1 Within certain main categories only selected causes of death are shown.
2 See chapter text.

Sources: Office for National Statistics: 020 7533 5249; General Register Office (Northern Ireland)

5.20 Infant and maternal mortality

	Deaths of Infants under 1 year of age per thousand live births												Maternal deaths per thousand live births[2]			
	United Kingdom			England and Wales[1]			Scotland			Northern Ireland						
	Total	Males	Females	Total	Males	Females	Total	Males	Females	Total	Males	Females	United Kingdom	England and Wales	Scotland	Northern Ireland
1900 - 02	142	156	128	146	160	131	124	136	111	113	123	103	4.71	4.67	4.74	6.03
1910 - 12	110	121	98	110	121	98	109	120	97	101	110	92	3.95	3.67	5.65	5.28
1920 - 22	82	92	71	80	90	69	94	106	82	86	95	77	4.37	4.03	6.36	5.62
1930 - 32	67	75	58	64	72	55	84	94	73	75	83	66	4.54	4.24	6.40	5.24
1940 - 42	59	66	51	55	62	48	77	87	66	80	89	70	3.29	2.74	4.50	3.79
1950 - 52	30	34	26	29	33	25	37	42	32	40	45	36	0.88	0.79	1.09	1.09
1960 - 62	22	25	19	22	24	19	26	30	22	27	30	24	0.36	0.36	0.37	0.43
1970 - 72	18	20	16	18	20	15	19	22	17	22	24	20	0.17	0.17	0.17	0.12
1980 - 82	12	13	10	11	13	10	12	13	10	13	15	12	0.09	0.09	0.14	0.06
1990 - 92	7	8	6	7	8	6	7	8	6	7	8	6	0.07	0.07	0.10	-
	KKAW	KKAX	KKAY	KKAZ	KKBW	KKBX	KKBY	KKBZ	KKCW	KKCX	KKCY	KKCZ	KKDW	KKDX	KKDY	KKDZ
1978	13.3	14.8	11.7	13.2	14.7	11.6	12.9	14.5	11.2	15.9	17.3	14.5	0.10	0.11	0.06	–
1979	12.9	14.4	11.3	12.8	14.4	11.1	12.8	13.9	11.8	14.8	15.6	13.9	0.11	0.12	0.10	0.04
1980	12.2	13.4	10.6	12.0	13.3	10.7	12.1	13.6	10.4	13.4	15.1	11.5	0.11	0.11	0.15	0.07
1981	11.2	12.7	9.5	11.1	12.6	9.4	11.3	12.3	10.2	13.2	14.7	11.6	0.09	0.09	0.19	0.04
1982	11.0	12.3	9.5	10.8	12.2	9.4	11.4	12.9	9.8	13.6	14.8	12.4	0.07	0.07	0.09	0.07
1983	10.1	11.3	8.9	10.1	11.3	8.9	9.9	11.3	8.5	12.1	13.9	10.1	0.09	0.09	0.12	0.15
1984	9.6	10.7	8.4	9.5	10.6	8.3	10.3	11.7	8.9	10.5	11.4	9.6	0.09	0.08	0.12	0.11
1985	9.4	10.4	8.3	9.4	10.4	8.2	9.4	10.0	8.7	9.6	10.6	8.5	0.07	0.07	0.13	0.07
1986	9.5	10.9	8.1	9.6	11.0	8.0	8.8	9.9	7.7	10.2	11.0	9.2	0.07	0.07	0.11	–
1987	9.1	10.3	7.9	9.2	10.4	7.9	8.5	9.7	7.2	8.7	9.6	7.7	0.06	0.07	0.03	0.04
1988	9.0	10.2	7.7	9.0	10.3	7.7	8.2	9.5	6.8	8.9	9.6	8.2	0.06	0.06	0.12	0.07
1989	8.4	9.5	7.2	8.4	9.6	7.3	8.7	10.2	7.2	6.9	7.4	6.3	0.08	0.08	0.06	–
1990	7.9	8.8	6.8	7.9	8.9	6.8	7.7	8.8	6.6	7.5	8.1	6.8	0.08	0.08	0.06	–
1991	7.4	8.3	6.3	7.4	8.3	6.4	7.1	8.7	5.3	7.4	8.3	6.4	0.07	0.06	0.13	0.04
1992	6.6	7.4	5.7	6.6	7.4	5.8	6.8	7.9	5.7	6.0	6.4	5.6	0.07	0.07	0.11	–
1993	6.3	7.0	5.6	6.3	7.0	5.6	6.5	7.4	5.6	7.1	7.8	6.3	0.06	0.05	0.11	–
1994	6.2	6.9	5.4	6.2	6.9	5.4	6.2	6.8	5.6	6.1	6.5	5.6	0.08	0.08	0.15	–
1995	6.2	6.9	5.4	6.1	6.9	5.3	6.2	6.4	6.1	7.1	7.5	6.6	0.07	0.07	0.10	–
1996	6.1	6.8	5.4	6.1	6.9	5.4	6.2	6.7	5.5	5.8	6.7	4.8	0.07	0.07	0.10	0.04
1997	5.8	6.4	5.3	5.9	6.5	5.3	5.3	6.1	4.5	5.6	5.5	5.8	0.06	0.06	0.07	–
1998	5.7	6.3	5.0	5.7	6.4	5.0	5.6	6.2	4.9	5.6	6.1	5.1	0.07	0.07	0.09	0.04

1 From 1937 to 1956 death rates are based on the births to which they relate in the current and preceding years.
2 Deaths in pregnancy and childbirth.

Sources: Office for National Statistics: 020 7533 5641; General Register Office (Northern Ireland); General Register Office (Scotland)

5.20 Infant and maternal mortality

continued

Deaths per thousand live births

		Analysis by sex of infant													
		1985	1986	1987	1988	1989	1990	1991	1992	1993	1994	1995	1996	1997	1998
Total															
United Kingdom:															
Stillbirths[2]	KHNQ	5.5	5.3	5.0	4.9	4.7	4.6	4.7	4.3	5.7	5.8	5.6	5.5	5.3	5.4
Perinatal[2]	KHNR	10.1	9.6	9.0	8.8	8.3	8.1	8.1	7.7	9.0	8.9	8.9	8.7	8.3	8.3
Neonatal	KHNS	5.4	5.3	5.0	4.9	4.7	4.5	4.4	4.3	4.2	4.1	4.2	4.1	3.9	3.8
Post neonatal	KHNT	4.0	4.2	4.1	4.1	3.7	3.3	3.0	2.3	2.2	2.1	2.0	2.0	2.0	1.9
England and Wales:															
Stillbirths[2]	KHNU	5.5	5.3	5.0	4.9	4.7	4.6	4.6	4.3	5.7	5.7	5.5	5.4	5.3	5.3
Perinatal[2]	KHNV	10.1	9.6	8.9	8.7	8.3	8.1	8.0	7.6	8.9	8.9	8.7	8.6	8.3	8.2
Neonatal	KHNW	5.4	5.3	5.1	4.9	4.8	4.6	4.4	4.3	4.2	4.1	4.1	4.1	3.9	3.8
Post neonatal	KHNX	4.0	4.3	4.1	4.1	3.7	3.3	3.0	2.3	2.1	2.1	2.0	2.0	2.0	1.9
Scotland:															
Stillbirths[2]	KHNY	5.5	5.8	5.1	5.4	5.0	5.3	5.5	4.9	6.4	6.1	6.6	6.4	5.3	6.1
Perinatal[2]	KHNZ	9.8	10.2	8.9	8.9	8.7	8.7	8.6	8.5	9.6	9.0	9.6	9.2	7.8	8.7
Neonatal	KHOA	5.5	5.2	4.7	4.5	4.7	4.4	4.4	4.6	4.0	4.0	4.0	3.9	3.2	3.6
Post neonatal	KHOB	3.9	3.6	3.8	3.7	4.0	3.3	2.7	2.2	2.5	2.2	2.2	2.2	2.1	2.0
Northern Ireland:															
Stillbirths[2]	KHOC	6.4	4.4	6.1	5.0	5.1	4.4	4.7	4.9	5.2	6.3	6.1	6.3	5.4	5.1
Perinatal[2]	KHOD	11.1	9.5	9.8	9.3	8.2	7.6	8.4	8.2	8.8	9.7	10.4	9.4	8.2	8.1
Neonatal	KHOE	5.6	6.0	4.8	5.4	4.0	4.0	4.6	4.1	4.9	4.2	5.5	3.7	4.2	3.9
Post neonatal	KHOF	4.0	4.2	3.8	3.6	2.9	3.5	2.8	1.9	2.1	1.9	1.6	2.0	1.4	1.7
Males															
United Kingdom:															
Perinatal[2]	KHOG	11.2	10.4	9.7	9.5	9.3	8.8	8.7	8.3	9.7	9.6	9.4	9.1	8.7	8.8
Neonatal	KHOH	5.9	6.0	5.7	5.5	5.4	5.0	4.9	4.8	4.7	4.6	4.6	4.6	4.2	4.2
Infant mortality	KHOI	10.4	10.9	10.3	10.2	9.5	8.8	8.3	7.4	7.0	6.9	6.9	6.8	6.4	6.3
England and Wales:															
Perinatal[2]	KHOK	11.3	10.4	9.8	9.4	9.2	8.8	8.6	8.1	9.5	9.6	9.3	9.0	8.7	8.8
Neonatal	KHOL	6.0	6.0	5.7	5.5	5.4	5.1	4.8	4.7	4.6	4.6	4.6	4.6	4.2	4.3
Infant mortality	KHOM	10.4	11.0	10.4	10.3	9.6	8.9	8.3	7.4	7.0	6.9	6.9	6.9	6.5	6.4
Scotland:															
Perinatal[2]	KHOO	10.1	10.9	8.9	9.7	10.0	9.1	9.8	9.8	10.8	9.6	10.1	10.0	8.1	9.6
Neonatal	KHOP	5.5	5.6	5.3	5.3	5.3	4.7	5.5	5.5	4.5	4.4	4.1	4.3	3.4	4.0
Infant mortality	KHOQ	10.0	9.9	9.7	9.5	10.2	8.8	8.7	7.9	7.4	6.8	6.4	6.7	6.1	6.2
Northern Ireland:															
Perinatal[2]	KHOS	11.3	9.7	10.5	10.9	8.7	8.7	7.9	8.6	10.3	10.4	10.4	10.1	8.5	8.9
Neonatal	KHOT	6.3	6.2	5.3	6.0	4.5	4.8	5.5	4.5	5.9	4.3	5.7	4.3	4.3	4.4
Infant mortality	KHOU	10.6	11.0	9.6	9.6	7.4	8.1	8.3	6.4	7.8	6.5	7.5	6.7	5.5	6.1
Females															
United Kingdom:															
Perinatal[2]	KHOW	9.1	8.8	8.1	8.0	7.4	7.4	7.5	7.0	8.2	8.2	8.3	8.2	7.9	7.7
Neonatal	KHOX	4.8	4.6	4.3	4.3	4.0	4.0	3.8	3.8	3.6	3.6	3.7	3.6	3.5	3.3
Infant mortality	KHOY	8.3	8.1	7.9	7.7	7.2	6.8	6.3	5.7	5.6	5.4	5.4	5.4	5.3	5.0
England and Wales:															
Perinatal[2]	KHPA	8.9	8.7	8.0	8.0	7.4	7.4	7.5	7.0	8.3	8.1	8.1	8.2	7.9	7.7
Neonatal	KHPB	4.8	4.5	4.3	4.3	4.0	4.0	3.9	3.9	3.6	3.6	3.6	3.6	3.6	3.3
Infant mortality	KHPC	8.2	8.0	7.9	7.7	7.3	6.8	6.4	5.8	5.6	5.4	5.3	5.4	5.3	5.0
Scotland:															
Perinatal[2]	KHPE	9.5	9.4	9.0	8.1	7.3	8.2	7.4	7.1	8.3	8.4	9.2	8.4	7.5	7.9
Neonatal	KHPF	5.5	4.8	4.1	3.7	4.1	4.1	3.2	3.7	3.5	3.6	3.9	3.5	2.9	3.2
Infant mortality	KHPG	8.7	7.7	7.2	6.8	7.2	6.6	5.3	5.7	5.6	5.6	6.1	5.5	4.5	4.9
Northern Ireland:															
Perinatal[2]	KHPI	10.9	9.2	9.2	7.6	7.6	6.4	9.0	7.9	7.2	8.9	10.5	8.6	8.0	7.3
Neonatal	KHPJ	4.9	5.8	4.4	4.7	3.5	3.2	3.7	3.6	3.9	4.1	5.2	3.1	4.0	3.4
Infant mortality	KHPK	8.5	9.2	7.7	8.2	6.3	6.8	6.4	5.6	6.3	5.6	6.6	4.8	5.8	5.1

1 Provisional.
2 Deaths per 1 000 live and stillbirths. On 1 October 1992 the legal definition of a stillbirth was altered from baby born dead after 28 completed weeks gestation or more, to one born dead after 24 completed weeks gestation or more. The 258 stillbirths of 24 to 27 weeks gestation which occurred between 1 October and 31 December 1992 are excluded from this table.

Sources: Office for National Statistics: 020 7533 5641; General Register Office (Northern Ireland); General Register Office (Scotland)

5.21 Death rates per 1 000 population[1]

United Kingdom

Analysis by age and sex

	All ages	0-4	5-9	10-14	15-19	20-24	25-34	35-44	45-54	55-64	65-74	75-84	85 and over
Males													
1900 - 02	18.4	57.0	4.1	2.4	3.7	5.0	6.6	11.0	18.6	35.0	69.9	143.6	289.6
1910 - 12	14.9	40.5	3.3	2.0	3.0	3.9	5.0	8.0	14.9	29.8	62.1	133.8	261.5
1920 - 22	13.5	33.4	2.9	1.8	2.9	3.9	4.5	6.9	11.9	25.3	57.8	131.8	259.1
1930 - 32	12.9	22.3	2.3	1.5	2.6	3.3	3.5	5.7	11.3	23.7	57.9	134.2	277.0
1940 - 42	..	..	..	..	..	..	..	..	..	..	..	..	..
1950 - 52	12.6	7.7	0.7	0.5	0.9	1.4	1.6	3.0	8.5	23.2	55.2	127.6	272.0
1960 - 62	12.5	6.4	0.5	0.4	0.9	1.1	1.1	2.5	7.4	22.2	54.4	123.4	251.0
1970 - 72	12.4	4.6	0.4	0.4	0.9	1.0	1.0	2.4	7.3	20.9	52.9	116.3	246.1
1980 - 82	12.1	3.2	0.3	0.3	0.8	0.9	0.9	1.9	6.3	18.2	46.7	107.1	224.9
1990 - 92	11.1	2.0	0.2	0.2	0.7	0.9	0.9	1.8	4.6	14.3	38.7	92.9	195.7
	KHZA	KHZB	KHZC	KHZD	KHZE	KHZF	KHZG	KHZH	KHZJ	KHZK	KHZL	KHZM	KHZN
1977	12.1	3.5	0.3	0.3	0.9	1.0	1.0	2.1	7.0	19.2	49.0	110.7	229.4
1978	12.3	3.4	0.3	0.3	0.9	1.0	1.0	2.1	7.0	19.3	49.3	111.8	232.0
1979	12.4	3.8	0.3	0.3	0.9	1.0	1.0	2.1	6.9	19.2	48.8	112.0	236.0
1980	12.1	3.4	0.3	0.3	0.8	0.9	0.9	2.0	6.5	18.7	47.2	108.6	227.0
1981	12.3	3.2	0.3	0.3	0.9	0.9	1.0	2.0	6.4	18.4	47.5	110.6	239.5
1982	12.0	2.9	0.3	0.3	0.8	0.9	0.9	1.8	6.1	17.9	46.4	106.4	225.0
1983	12.0	2.7	0.3	0.3	0.8	0.8	0.9	1.8	5.9	18.0	46.3	104.6	221.0
1984	11.7	2.6	0.3	0.3	0.7	0.9	0.9	1.7	5.7	17.5	44.9	100.9	212.8
1985	12.0	2.6	0.2	0.3	0.7	0.8	0.9	1.8	5.6	17.5	45.0	104.8	223.4
1986	11.8	2.6	0.2	0.2	0.7	0.9	0.9	1.7	5.5	17.0	43.6	102.0	217.1
1987	11.5	2.5	0.2	0.3	0.7	0.9	0.9	1.7	5.2	16.4	42.0	97.3	194.3
1988	11.5	2.5	0.2	0.3	0.7	0.9	0.9	1.7	5.1	16.0	41.2	96.6	193.9
1989	11.5	2.3	0.2	0.2	0.7	0.9	0.9	1.7	4.9	15.3	40.4	97.0	199.0
1990	11.2	2.2	0.2	0.2	0.7	0.9	1.0	1.8	4.8	14.8	39.5	94.3	187.8
1991	11.1	2.0	0.2	0.2	0.7	0.9	0.9	1.8	4.7	14.2	38.7	93.3	203.5
1992	10.9	1.8	0.2	0.2	0.6	0.9	0.9	1.7	4.4	13.8	37.9	91.2	195.8
1993[2]	11.2	1.7	0.2	0.2	0.6	0.9	1.0	1.7	4.4	12.8	36.8	92.7	213.3
1994	10.6	1.6	0.2	0.2	0.6	0.9	1.0	1.7	4.1	12.8	36.8	90.3	190.5
1995	10.8	1.5	0.2	0.2	0.6	0.9	0.9	1.7	4.2	13.5	38.4	91.3	188.0
1996	10.6	1.6	0.1	0.2	0.6	0.9	1.0	1.7	4.2	12.4	35.4	86.5	195.5
1997	10.4	1.5	0.1	0.2	0.6	0.9	1.0	1.6	4.1	11.8	33.9	83.4	191.9
1998	10.3	1.5	0.1	0.2	0.6	0.9	1.0	1.6	4.1	11.7	33.1	82.0	188.1
Females													
1900 - 02	16.3	47.9	4.3	2.6	3.5	4.3	5.8	9.0	14.4	27.9	59.3	127.0	262.6
1910 - 12	13.3	34.0	3.3	2.1	2.9	3.4	4.4	6.7	11.5	23.1	50.7	113.7	234.0
1920 - 22	11.9	26.9	2.8	1.9	2.8	3.4	4.1	5.6	9.3	19.2	45.6	111.5	232.4
1930 - 32	11.5	17.7	2.1	1.5	2.4	2.9	3.3	4.6	8.3	17.6	43.7	110.1	246.3
1940 - 42	..	..	..	..	..	..	..	..	..	..	..	..	..
1950 - 52	11.2	6.0	0.5	0.4	0.7	1.0	1.4	2.3	5.3	12.9	35.5	98.4	228.8
1960 - 62	11.2	4.9	0.3	0.3	0.4	0.5	0.8	1.8	4.5	11.0	30.8	87.3	218.5
1970 - 72	11.3	3.6	0.3	0.2	0.4	0.4	0.6	1.6	4.5	10.5	27.5	76.7	196.1
1980 - 82	11.4	2.3	0.2	0.2	0.3	0.4	0.5	1.3	3.9	9.9	24.8	67.2	179.5
1990 - 92	11.1	1.6	0.1	0.2	0.3	0.3	0.4	1.1	2.9	8.4	22.2	58.5	154.6
	KHZO	KHZP	KHZQ	KHZR	KHZS	KHZT	KHZU	KHZV	KHZW	KHZX	KHZY	KHZZ	KHZI
1977	11.3	2.7	0.2	0.2	0.4	0.4	0.5	1.5	4.2	10.1	25.5	71.0	183.1
1978	11.5	2.8	0.2	0.2	0.4	0.4	0.6	1.5	4.3	10.1	25.7	70.5	184.5
1979	11.6	2.8	0.2	0.2	0.3	0.4	0.6	1.4	4.3	10.2	25.6	70.8	187.1
1980	11.4	2.7	0.2	0.2	0.3	0.4	0.5	1.3	4.0	10.0	25.0	67.7	181.3
1981	11.6	2.5	0.2	0.2	0.3	0.4	0.5	1.3	3.9	10.0	25.5	70.4	190.6
1982	11.5	2.3	0.2	0.2	0.3	0.4	0.5	1.2	3.8	9.9	24.8	66.8	179.4
1983	11.4	2.1	0.2	0.2	0.3	0.3	0.5	1.2	3.6	9.8	24.7	65.1	177.2
1984	11.2	2.0	0.2	0.2	0.3	0.3	0.5	1.2	3.5	9.7	24.3	62.6	170.2
1985	11.7	2.0	0.2	0.2	0.3	0.3	0.5	1.2	3.4	9.9	24.7	64.8	178.4
1986	11.5	2.0	0.2	0.2	0.3	0.3	0.5	1.1	3.3	9.5	24.0	63.2	172.4
1987	11.2	2.0	0.2	0.2	0.3	0.3	0.5	1.1	3.3	9.4	23.3	60.8	159.9
1988	11.3	1.9	0.2	0.2	0.3	0.3	0.5	1.2	3.2	9.2	23.2	60.4	162.3
1989	11.5	1.8	0.2	0.2	0.3	0.3	0.5	1.1	3.1	9.1	23.2	61.4	164.7
1990	11.1	1.7	0.1	0.2	0.3	0.3	0.5	1.1	3.0	8.8	22.4	58.9	156.7
1991	11.2	1.6	0.2	0.2	0.3	0.3	0.4	1.1	2.9	8.4	22.3	58.9	156.9
1992	11.0	1.4	0.1	0.1	0.3	0.3	0.4	1.1	2.8	8.1	22.0	57.8	150.3
1993[2]	11.5	1.3	0.1	0.2	0.3	0.3	0.5	1.1	2.9	7.5	21.6	58.0	162.3
1994	10.9	1.3	0.1	0.1	0.2	0.3	0.4	1.1	2.8	7.6	21.7	57.5	148.4
1995	11.1	1.2	0.1	0.1	0.3	0.3	0.4	1.1	2.7	8.0	22.6	59.8	150.4
1996	11.1	1.2	0.1	0.1	0.3	0.3	0.5	1.1	2.7	7.3	21.2	56.8	153.7
1997	11.0	1.2	0.1	0.1	0.3	0.3	0.4	1.1	2.7	7.1	20.6	55.2	153.9
1998	10.9	1.2	0.1	0.1	0.3	0.3	0.4	1.0	2.7	7.0	20.3	54.4	153.3

1 The figures 1974 to 1980 incorporate the revised intercensal estimates for England and Wales and Northern Ireland, but the old series for Scotland.

Sources: Office for National Statistics: 020 7533 5249; General Register Office (Northern Ireland); General Register Office (Scotland)

5.22 Life tables

Interim life tables, 1996-98

Age(x)	United Kingdom Males l_x	United Kingdom Males e^0_x	United Kingdom Females l_x	United Kingdom Females e^0_x	England and Wales Males l_x	England and Wales Males e^0_x	England and Wales Females l_x	England and Wales Females e^0_x
0 years	100 000	74.6	100 000	79.6	100 000	74.8	100 000	79.8
5 years	99 226	70.2	99 385	75.1	99 221	70.4	99 382	75.3
10 years	99 156	65.2	99 329	70.2	99 151	65.5	99 330	70.4
15 years	99 064	60.3	99 268	65.2	99 062	60.5	99 268	65.4
20 years	98 770	55.5	99 129	60.3	98 780	55.7	99 132	60.5
25 years	98 336	50.7	98 968	55.4	98 365	50.9	98 975	55.6
30 years	97 884	45.9	98 784	50.5	97 929	46.1	98 794	50.7
35 years	97 380	41.1	98 534	45.6	97 441	41.4	98 547	45.8
40 years	96 753	36.4	98 135	40.8	96 834	36.6	98 154	41.0
45 years	95 782	31.7	97 497	36.0	95 892	31.9	97 631	36.2
50 years	94 295	27.2	96 496	31.4	94 454	27.4	96 559	31.6
55 years	91 866	22.8	94 873	26.9	92 110	23.0	94 976	27.0
60 years	87 810	18.8	92 309	22.6	88 175	18.9	92 479	22.7
65 years	81 356	15.0	88 256	18.5	81 889	15.2	88 534	18.6
70 years	71 426	11.8	81 737	14.7	72 123	11.8	82 153	14.8
75 years	57 325	9.0	71 659	11.4	58 110	9.1	72 190	11.5
80 years	40 327	6.7	57 710	8.5	41 062	6.8	58 321	8.6
85 years	22 945	5.0	39 825	6.2	23 481	5.0	40 395	6.3

Interim life tables, 1996-98

Age(x)	Scotland Males l_x	Scotland Males e^0_x	Scotland Females l_x	Scotland Females e^0_x	Northern Ireland Males l_x	Northern Ireland Males e^0_x	Northern Ireland Females l_x	Northern Ireland Females e^0_x
0 years	10 000	72.4	10 000	77.9	10 000	74.2	10 000	79.5
5 years	9 925	67.9	9 942	73.4	9 931	69.7	9 937	75.0
10 years	9 917	63.0	9 933	68.4	9 924	64.7	9 931	70.0
15 years	9 906	58.1	9 927	63.5	9 914	59.8	9 925	65.1
20 years	9 866	53.3	9 911	58.6	9 878	55.0	9 910	60.2
25 years	9 807	48.6	9 890	53.7	9 831	50.2	9 896	55.3
30 years	9 746	43.9	9 868	48.8	9 785	45.5	9 880	50.3
35 years	9 679	39.2	9 840	43.9	9 733	40.7	9 855	45.5
40 years	9 596	34.5	9 793	39.1	9 673	35.9	9 818	40.6
45 years	9 469	29.9	9 713	34.4	9 577	31.3	9 761	35.9
50 years	9 273	25.5	9 585	29.9	9 415	26.8	9 657	31.2
55 years	8 943	21.3	9 381	25.5	9 167	22.4	9 494	26.7
60 years	8 422	17.5	9 063	21.3	8 752	18.3	9 238	22.4
65 years	7 634	14.0	8 566	17.3	8 055	14.7	8 813	18.3
70 years	6 506	11.0	7 789	13.8	7 001	11.5	8 164	14.5
75 years	5 023	8.5	6 678	10.7	5 561	8.8	7 140	11.3
80 years	3 376	6.3	5 214	7.9	3 837	6.6	5 705	8.4
85 years	1 822	4.7	3 465	5.6	2 132	4.9	3 903	6.1

Note Column l_x shows the number who would survive to exact **age**(x), out of 100 000 or 10 000 born, who were subject throughout their lives to the death rates experienced in the three-year period indicated. Column e^0_x is 'the expectation of life', that is, the average future lifetime which would be lived by a person aged exactly x if likewise subject to the death rates experienced in the three-year period indicated. See introductory notes.

Source: Government Actuary's Department: 020 7211 2622

Education

6

Education

In general, the detailed provision and delivery of education in the United Kingdom is not subject to central adminstration. School quality assessment is undertaken in England by an inspectorate (OFSTED), in Wales by the Office of Her Majesty's Chief Inspector of Schools (OHMCI), in Scotland by Her Majesty's Inspectors of Schools (HMI) and in Northern Ireland by the Education and Training Inspectorate (ETI). Further education quality control is undertaken by the FEFC Inspectorate. In Wales external quality assurance is secured on behalf of the FEFCW through a contracted service provided by OHMCI. Higher Education Institutions (HEIs) are subject to quality assurance procedures of the HE Funding Councils and the Quality Assurance Agency (QAA). In practice, within the quanta of funds distributed by Funding Councils and Education Departments, Institutions are managed by their governing bodies which are able to select their own staff and decide how best to allocate funds to suit local needs. Some central control is however, exercised over the curriculum taught during compulsory schooling. In particular the Education Reform Act 1988 introduced the National Curriculum in England and Wales. However, unlike England Wales and Northern Ireland, the curriculum in Scotland is not statutorily prescribed, and pupils aged 5 - 14 study a broad curriculum based on national guidelines. The 1988 Act also provided for local authorities to delegate the management of school budgets to the schools themselves and for the schools to seek withdrawal from LEA control and become self-governing.

The four home Government Departments dealing with education statistics are:

- Department for Education and Employment:
  - which deals with all sectors of education in England.
- National Assembly for Wales (NAFW):
  - which deals with schools and higher and further education in Wales, excluding matters connected with the remuneration, superannuation and misconduct of teachers. From 1 April 1993, WOED has responsibility for the colleges of the University of Wales.
- Scottish Executive (SE):
  - all sectors of education in Scotland.
- Department of Education, Northern Ireland:
  - schools, institutes of further education, teacher training colleges and universities in Northern Ireland.

Each of the four home Education Departments have overall responsibility for funding the schools sectors in their own country. The FE and HE Funding Councils of England and of Wales and the Scottish HE Funding Council are responsible respectively for the distribution of funds made available by the Secretaries of State to FE colleges in England and Wales, and HEIs in England, Wales and Scotland. The SE is responsible for allocating funds to all but three of the FE colleges in Scotland. The remaining 3 are still funded by the relevant regional authorities. DENI allocates funds to further and higher education establishments in Northern Ireland.

Statistics for the separate systems obtained in England, Wales, Scotland and Northern Ireland are collected and processed separately in accordance with the particular needs of the responsible Departments. Since 1994/95 the data collection for all HEIs in the UK has been undertaken by the Higher Education Statistics Agency (HESA). This includes the former UFC funded UK universities previously collected by the Universities Statistical Record. There are some structural differences in the information collected for schools, further and higher education in each of the four home countries. In some tables the GB/UK data presented are an amalgamation from sources which are not entirely comparable.

Stages of education

There are five stages of education: nursery, primary, secondary, further and higher education. Primary and secondary education is compulsory for all children between the ages of five and sixteen years and the transition is normally at age eleven. The non compulsory fourth stage, further education, covers non-advanced education which can be taken at both further (including tertiary) education colleges, higher education institutions and increasingly in secondary schools. The fifth stage, higher education, is study beyond A levels and their equivalent which, for most full-time students, takes place in higher education institutions.

Nursery education

Many children under 5 attend state nursery schools or nursery classes attached to primary schools. Others may attend playgroups in the voluntary sector or in private run nurseries. In England and Wales many primary schools also operate an early admission policy where they admit children under 5 into what are called reception classes.

Primary education

Primary education normally provides for pupils aged five to eleven. In Northern Ireland children who attain the age of four on or before 1 July are required to commence primary school the following September. In Scotland children generally commence primary school when aged between 4.5 and 5.5 years, transferring to secondary school after seven years of primary schooling. The great majority of public sector primary schools take both boys and girls in mixed classes.

Middle schools

Some local education authorities in England operate a system of middle schools which cater for pupils on either side of the transition age, and these are deemed either primary or secondary according to the age range of the pupils.

Secondary education

Provision of secondary education in an area may include any combination of types of schools. The pattern is a reflection of historical circumstances and of the policy adopted by the local education authority. Comprehensive schools, on a variety of patterns as to forms of organisation and the age of the pupils attending, form the vast majority. In their 'pure' form, comprehensive schools admit pupils without reference to ability and aptitude, and cater for all the children in a neighbourhood. In 1997/98, over 86 percent of secondary school pupils in England attended comprehensive schools while all secondary schools in Wales are comprehensive schools. The majority of education authority secondary schools in Scotland are comprehensive in character and offer six years of secondary education; however in remote areas there are several two-year and four-year secondary schools.

Specialist schools

City technology colleges, technology colleges and language colleges, sports colleges and arts colleges operate in England. The Specialist Schools Programme enables secondary schools to develop a strength in a particular subject area, often in partnership with an employer with an interest in some specialism, while still delivering a broad and balanced education through the National Curriculum. City technology colleges take the form of charitable companies limited by guarantee and are registered as independent schools.

Special schools

Special schools, both day and boarding, provide education for children with special educational needs, though the great majority are educated in ordinary schools. All children attending special schools are offered a curriculum designed to overcome their learning difficulties and to enable them to become self-reliant.

Further education

The term further education may be used in a general sense to cover all non-higher education, and embraces courses up to but not beyond, GCE A level or its equivalent. When undertaken immediately after compulsory schooling, it can be taken by continuing in school or transferring to further education colleges. Increasingly adults return to study at any stage of their lives, and are catered for full-time or part-time by FE colleges, or by adult education centres or by further education courses in some higher education institutions. From 1 April 1993 sixth form colleges are included in the further education sector.

Higher education

Higher education is defined as the academic standard beyond GNVQ/NVQ level 3, GCE A level or equivalent. This includes higher degrees and postgraduate diplomas and certificates, first degree and Edexcel (formerly BTEC) Higher National Certificate/Diploma as well as a wide range of professional qualifications at the higher education level. As a result of the 1992 Further and Higher Education Act former polytechnics and some other higher education institutions were designated as universities in 1992/93. Students normally attend HE courses at higher education institutions, but some attend at further education colleges. Some students also attend institutions which do not receive public grant (such as the University of Buckingham), and these numbers are excluded from the tables.

6.1 Number of schools or departments by type and establishments of further and higher education

Academic years

		1986 /87	1987 /88	1988 /89	1989 /90	1990 /91	1991 /92	1992 /93	1993 /94	1994 /95	1995 /96	1996[1] /97	1997[2] /98
United Kingdom: Public sector mainstream													
Nursery	KBFK	1 271	1 298	1 312	1 337	1 364	1 380	1 406	1 451	1 477	1 486	1 537†	1 685
Primary	KBFA	24 609	24 482	24 344	24 268	24 135	23 958	23 829	23 673	23 516	23 426	23 306	23 213
of which grant maintained	KPGK	..	..	..	..	..	..	75	265	415	453	489†	514
Secondary[3]	KBFF	5 091	5 020	4 894	4 876	4 790	4 731	4 648	4 496	4 479	4 462	4 439†	4 435
of which grant maintained	KPGL	..	..	..	..	50	132	266	564	634	654	665†	680
of which 6th form colleges	KPGM	..	..	..	..	..	116	119	..	..	..	..	..
Non - maintained mainstream	KBFU	2 544	2 546	2 542	2 492	2 508	2 488	2 476	2 478	2 433	2 436	2 523†	2 501
of which City Technology Colleges (CTC's)	KPGN	..	..	..	..	..	13	14	15	15	15	15	15
Special - all	KBFP	1 915	1 900	1 873	1 851	1 830	1 792	1 768	1 742	1 749	1 567	1 529†	1 518
maintained	KPVX	..	..	..	..	..	..	..	1 670	1 638	1 458	1 429†	1 420
of which grant maintained	KSOB	..	..	..	..	..	..	..	..	2	9	18	21
non maintained	KPGO	..	..	..	..	..	..	–	72	111	109	100	98
Pupil referral units	KXEP	..	..	..	..	..	..	..	..	..	287	333	333
Universities (including Open University)[4]	KAHG	48	48	48	48	48	48	48	88	89	89	88	88
All other further and higher education institutions	KJPQ	728	724	727	706	698	694	688	751	739	736	696†	606
Higher education institutions	KPVY	..	..	..	..	..	..	111	69	67	66	63	63
Further education institutions[5]	KSNY	..	..	..	..	..	..	577	682	672	670	633†	543[7]
of which 6th form colleges	KPGP	..	..	..	..	..	..	..	117	110	110	110	110
England: Public sector mainstream													
Nursery	KBAK	558	558	559	566	560	561	561	552	551	547	544	533
Primary	KBAA	19 432	19 319	19 232	19 162	19 047	18 926	18 828	18 683	18 551	18 480	18 392	18 312
of which grant maintained	KPGQ	..	..	..	..	..	13	75	260	410	448	483	508
Secondary[3]	KBAF	4 221	4 153	4 035	3 976	3 897	3 847	3 773	3 629	3 614	3 594	3 569	3 567
of which grant maintained	KPGR	..	..	..	..	50	130	262	554	622	642	652	667
of which 6th form colleges	KPGS	106	104	112	113	114	115	117	..	..	..	..	..
Non - maintained	KBAU	2 276	2 273	2 271	2 283	2 289	2 271	2 263	2 268	2 261	2 266	2 273	2 244
of which City Technology Colleges (CTC's)	KPVZ	..	..	..	..	7	13	14	15	15	15	15	15
Special - all	KBAP	1 470	1 443	1 414	1 398	1 380	1 352	1 327	1 310	1 291	1 263	1 239	1 229
maintained	KPGT	..	..	..	..	..	..	..	1 238	1 218	1 191	1 170	1 164
of which grant maintained	KSOC	..	..	..	..	..	..	..	..	2	9	18	21
non maintained	KPGU	..	..	..	..	..	..	..	72	73	72	69	65
Pupil referral units	KXEQ	..	..	..	..	..	..	..	..	..	287	309	309
Universities (including Open University)[4]	KAHM	37	37	37	37	37	37	37	72	72	72	71	71
All other further and higher education institutions	KJPR	480	475	474	466	460	449	438	511	502	503	501	501
Higher education institutions	KPXA	..	..	..	..	..	..	85	49	49	50	48	48
Further education institutions	KPWC	..	..	..	..	..	..	353	462	453	453	453	453
of which 6th form colleges	KPGV	..	..	..	..	..	..	..	115	110	110	110	110
Wales: Public sector mainstream													
Nursery	KBBK	59	58	56	55	54	52	52	52	52	52	51	51
Primary	KBBA	1 762	1 753	1 743	1 729	1 717	1 697	1 697	1 698	1 691	1 681	1 681	1 681
of which grant maintained	KPGW	..	..	..	..	..	..	..	5	5	5	6	6
Secondary[3]	KBBF	234	233	232	231	230	229	229	227	227	228	229	229
of which grant maintained	KPGX	..	..	..	..	..	2	4	10	11	11	12	12
of which 6th form colleges	KPGY	1	2	2	2	2	2	2	..	..	..	..	..
Non - maintained	KBBU	69	67	66	67	71	68	65	64	62	62	59	59
Special	KBBP	65	65	64	62	61	61	61	59	56	54	51	51
Pupil referral units	KZBF	..	..	..	..	..	..	..	..	..	24	24	24
Universities[4]	KAHS	1	1	1	1	1	1	1	1	2	2	2	2
All other further and higher education institutions	KJQP	41	41	42	38	38	38	34	37	34	32	30	30
Higher education institutions	KSNZ	..	..	..	..	..	..	..	9	7	5	4	4
Further education institutions	KPGZ	..	..	..	..	..	..	25	28	27	27	26	26
of which 6th form colleges	KPHA	..	..	..	..	..	..	..	2	–	–	–	–
Scotland: Public sector mainstream													
Nursery	KBDK	568†	596	612	633	659	680	705	758	783	796	851	1 010
Primary	KBDA	2 417†	2 418	2 384	2 378	2 372	2 364	2 347	2 341	2 336	2 332	2 313	2 300
Secondary	KBDF	440†	438	434	429	424	419	412	408	406	405	403	401
of which grant maintained	KSOA	..	..	..	..	..	..	..	..	1	1	1	1
Non - maintained	KBDU	111†	118	118	123	131	131	130	125	90	87	172[6]	176[6]
Special - all	KBDP	356†	349	345	345	343	333	334	327	355	201	192	191
maintained	KYCZ	356†	349	345	345	343	333	334	327	317	164	161	158
non-maintained	KYDA	..	..	..	..	..	..	..	..	38	37	31	33
Universities[4]	KAHX	8	8	8	8	8	8	8	13	13	13	13	13
All other further and higher education institutions	KJRA	70	69	69	64	64	66	61	55	56	56	55	56
Higher education institutions	KPWE	..	..	..	..	..	..	15	9	9	9	9	9
Further education institutions	KPHB	..	..	..	..	..	..	46	46	47	47	46	47
Evening centres	KAHZ	109	111	114	110	110	115	129	122	128	126	91	..[7]
Northern Ireland: Grant aided mainstream													
Nursery	KBEK	86	86	85	85	85	88	88	89	91	91	91	91
Primary	KBEA	998	992	985	999	999	964	957	951	938	923[8]	920[8]	920[8]
Secondary	KBEF	196	196	193	240	239	236	234	232	232	236	238	238
Non - maintained	KBEU	87	88	87	19	17	18	18	21	20	21	19	22
Special	KBEP	24	46	46	46	46	46	46	46	47	47	47	47
Universities[4]	KIAD	2	2	2	2	2	2	2	2	2	2	2	2
Colleges of education	KIAE	2	2	2	2	2	2	2	2	2	2	2	2
Further education colleges	KIAG	26	26	26	26	24	24	24	24	17	17	17	17

1 Revised to include 1996/97 schools and FE college data for Wales.
2 Provisional. Schools data for Wales and further education colleges data for England are for 1996/97.
3 From 1 April 1993 excludes sixth form colleges in England and Wales which were re-classified as further education colleges.
4 From 1 April 1993 includes former polytechnics and colleges which became universities as a result of the Further and Higher Education Act 1992.
5 Includes Evening Centres for Scotland.
6 Schools in Scotland with more than one department have been counted once for each department e.g. a school with Nursery, Primary and Secondary departments has been counted 3 times.
7 Due to changes in data collection, Scottish Evening Centres are not available.
8 Excludes Preparatory Departments in Grammar Schools.

Source: Education Departments: 01325 392756

6.2 Full-time and part-time pupils in school by age and gender

United Kingdom

All schools at January [1]

		1988	1989	1990	1991	1992[5]	1993	1994[6]	1995	1996	1997	1998[7]
Age at 31 August[8]												
Number (thousands)												
England	KBIA	7 610	7 553	7 557	7 617	7 712	7 842	7 883	8 013	8 110	8 195	8 261
Wales	KBIB	481	480	480	482	487	493	498	504	..	..	501
Scotland	KBIC	840	828	821	821	826	832	841	845	846	848	850
Northern Ireland[4]	KBID	341	341	340	341	343	346	349	351	353	354	352
United Kingdom[4]	KBIE	9 273	9 203	9 199	9 260	9 368	9 513	9 571	9 714	9 813	9 905	9 973
Boys and girls												
2 - 4[9]	KBIF	912	937	979	1 012	1 045	1 082	1 103	1 135	1 145	1 149	1 150
5 - 10	KBIG	4 180	4 252	4 312	4 346	4 356	4 392	4 462	4 517	4 581	4 627	4 666
11	KBIH	668	643	658	714	735	727	713	718	717	742	747
12 - 14	KBII	2 199	2 097	2 011	1 975	2 021	2 112	2 179	2 179	2 158	2 151	2 180
15	KBIK	818	764	726	691	664	638	651	700	721	717	701
16	KBIL	296	297	289	295	308	313	259	259	279	289	289
17	KBIM	171	185	194	197	206	215	180	181	191	206	216
18 and over	KBIN	27	27	29	32	34	35	23	24	22	22	24
Boys												
14	KBIO	398	376	358	344	332	338	367	376	371	365	368
15	KBIP	420	393	373	355	341	327	333	358	368	366	358
16	KBIQ	146	146	141	144	151	154	128	127	136	141	140
17	KBIR	86	93	96	96	100	104	88	88	92	100	104
18 and over	KBIS	15	15	16	17	18	18	12	13	12	12	13
Girls												
14	KBIT	375	356	339	326	314	322	347	358	355	349	350
15	KBIU	397	372	354	336	323	311	318	342	353	351	344
16	KBIV	150	151	147	151	156	159	131	132	142	148	149
17	KBIW	85	93	98	100	106	111	92	93	99	106	112
18 and over	KBIX	12	12	14	15	16	16	11	11	10	10	11

1 In Wales the school census date is September whereas that for the rest of the United Kingdom remains at January. In Scotland, figures are at the previous September (the same academic session).
2 Estimated; these figures include 1984-85 data for Scotland.
3 Estimated; these figures include 1987-88 data for Scotland.
4 From 1989/90 includes Northern Ireland voluntary grammar schools formerly allocated to the independent sector.
5 From 1991/92 figures for independent schools in England include pupils aged less than 2 years.
6 From 1 April 1993 excludes 6th form colleges in England and Wales which were reclassified as further education colleges.
7 Provisional. Includes 1996/97 data for Wales.
8 Age data for Northern Ireland are at 1 July from 1992/93 for grant aided schools. It remains the same January for the independent schools.
9 Includes the so called rising 5s (ie those pupils who become 5 during the Autumn term).

Source: Education Departments: 01325 392756

6.3 Number of pupils and teachers: pupil/teacher ratios by school type

United Kingdom

At January[1]

		1988	1989	1990	1991	1992	1993	1994[2]	1995[2]	1996[2]	1997[2]	1998[3]
All schools or departments[4]												
Total												
Pupils (thousands)												
Full-time and full-time equivalent of part-time	KBCA	9 100.7	9 022.7	9 010.0	9 062.3	9 162.1	9 300.0	9 351.9	9 487.4	9 587.3	9 663.7	9 725.8
Teachers[5] (thousands)	KBCB	531.1	528.0	532.5	531.6	534.9	537.8	533.2	532.6	534.0	533.0	532.8
Pupils per teacher:[6]												
United Kingdom[5]	KBCC	17.1	17.1	16.9	17.0	17.1	17.3	17.5	17.8	18.0	18.1	18.3
England	KBCD	17.2	17.1	17.0	17.2	17.2	17.4	17.7	18.1	18.2	18.3	18.5
Wales	KBCE	18.0	18.1	18.1	18.2	18.2	18.2	18.4	18.7	18.7	18.8	18.8
Scotland	KBCF	15.8	15.7	15.3	15.2	15.3	15.4	15.5	15.4	15.5	15.8	15.8
Northern Ireland	KBCG	18.4	18.3	18.3	18.1	18.3	18.1	17.8	17.3	17.1	16.7	16.7
Public sector mainstream schools or departments												
Nursery												
Pupils (thousands)												
Full-time and full-time equivalent of part-time	KBFM	57.6	58.4	59.4	60.4	60.9	61.6	61.6	62.2	61.8	61.5	61.9
Teachers[5,6] (thousands)	KBFN	2.7	2.7	2.7	2.8	2.8	2.9	2.8	2.8	2.9	2.9	3.0
Pupils per teacher	KBFO	21.4	21.6	21.8	21.5	21.6	21.6	21.6	21.9	21.3	21.3	20.7
Primary												
Pupils (thousands)												
Full-time and full-time equivalent of part-time	KBFB	4 598.9	4 662.8	4 747.7	4 812.3	4 849.5	4 923.0	4 997.7	5 065.4	5 144.9	5 184.9	5 213.3
Teachers[5,6] (thousands)	KBFD	210.1	213.3	219.0	220.6	222.6	224.4	225.5	225.9	227.0	227.0†	225.6
Pupils per teacher	KBFE	21.9	21.9	21.7	21.8	21.8	21.9	22.2	22.4	22.7	22.8	23.1
Secondary												
Pupils (thousands)												
Full-time and full-time equivalent of part-time	KBFG	3 701.5	3 551.7	3 491.6	3 473.3	3 534.5	3 605.8	3 587.5	3 655.6	3 677.4	3 714.1	3 741.1
Teachers[5,6] (thousands)	KBFH	244.9	237.0	236.6	232.3	232.7	233.6	228.4	228.4	228.2	228.9	228.7
Pupils per teacher	KBFI	15.1	15.0	14.8	15.0	15.2	15.4	15.7	16.0	16.1	16.2	16.4
Special schools												
Pupils (thousands)(ft and fte of pt)												
all	KBFQ	120.9	117.6	114.6	112.5	112.6	113.2	114.1	115.4	114.1	114.2	114.5
maintained	KPGE	..	..	..	..	..	..	108.6	108.3	107.7	107.8	108.4
non-maintained[7]	KPGF	..	..	..	..	..	..	5.5	7.1	6.4	6.4	6.0
Teachers[6] (thousands)(ft and fte of pt)												
all	KBFS	19.3	19.2	19.6	19.6	19.6	19.6	19.2	19.2	18.6	18.3	18.1
maintained	KPGG	..	..	..	..	..	..	18.0	17.5	17.1	16.8	16.7
non-maintained[7]	KPGH	..	..	..	..	..	..	1.2	1.7	1.5	1.4	1.4
Pupils per teacher[6]												
all	KBFT	6.3	6.1	5.8	5.7	5.7	5.8	5.9	6.0	6.2	6.2	6.3
maintained	KPGI	..	..	..	..	..	..	6.0	6.2	6.3	6.4	6.5
non-maintained[7]	KPGJ	..	..	..	..	..	..	4.5	4.3	4.4	4.4	4.4

1 In Scotland and in Wales the school census date is September whereas that for the rest of the United Kingdom remains at January.
2 From 1 April 1993 excludes sixth form colleges in England & Wales which were reclassified as further education colleges.
3 Provisional. Includes 1996/97 data for Wales.
4 From 1980 onwards includes non-maintained schools or departments, including independent schools in Scotland.
5 Figures of teachers and of pupil/teacher ratios take account of the full-time equivalent of part-time teachers.
6 Up to 1993/94 includes unqualified teachers for England and Scotland. From 1994/95 qualified teachers only for all countries.
7 England and Scotland only.

Source: Education Departments: 01325 392756

6.4 Full-time and part-time pupils with special educational needs (SEN)[1], 1997/98[2]

United Kingdom

By type of school

Thousands and percentages

	United Kingdom[2]	England	Wales	Scotland	Northern Ireland
All schools					
Total pupils	9 972.8	8 260.7	510.2	849.6	352.4
SEN pupils with statements	282.0	242.3	16.6	15.2	7.9
Incidence (%)[3]	2.8	2.9	3.3	1.8	2.3
Maintained schools[4]					
Nursery					
Total pupils	110.7	48.8	3.1	53.3	5.5
SEN pupils with statements	0.7	0.4	..	0.2	0.1
Incidence (%)[3]	0.6	0.9	0.2	0.4	1.0
Placement (%)[5]	0.2	0.2	..	1.2	0.7
Primary[6]					
Total pupils	5 381.9	4 460.7	292.7	440.6	187.8
SEN pupils without statements[7]	822.4	821.3	..	1.1	..
SEN pupils with statements	79.3	67.0	6.2	3.9	2.2
Pupils with statements - incidence (%)[3]	1.5	1.5	2.1	0.9	1.2
Pupils with statements - placement (%)[5]	28.1	27.7	37.4	25.6	27.7
Secondary					
Total pupils	3 741.1	3 072.8	200.3	314.9	153.1
SEN pupils without statements[7]	481.1	479.7	..	1.5	..
SEN pupils with statements	86.6	74.0	6.5	4.3	1.9
Pupils with statements - incidents (%)[3]	2.3	2.4	3.2	1.4	1.2
Pupils with statements - placement (%)[5]	30.7	30.5	39.2	28.1	23.9
Special[8,9]					
Total pupils	109.9	93.5	3.7	8.1	4.7
SEN pupils with statements	101.7	87.9	3.5	6.4	3.8
Incidence (%)[3]	92.5	94.1	97.1	79.4	80.9
Placement (%)[5]	36.0	36.3	21.4	42.1	47.7
Pupil referral units[8]					
Total pupils	8.2	7.7	0.5	..	..
SEN pupils with statements[7]	1.9	1.8	0.1	..	..
Incidence (%)[3]	23.7	23.2	30.3	..	..
Placement (%)[5]	0.7	0.7	0.9	..	..
Other schools					
Independent					
Total pupils	614.9	572.2	9.9	31.7	1.2
SEN pupils with statements[7]	6.7	6.5	0.2	0.1	..
Incidence (%)[3]	1.1	1.1	1.9	0.2	..
Placemant (%)[5]	2.4	2.7	1.1	0.4	..
Non-maintained special[8]					
Total pupils	6.1	5.0	..	1.1	..
SEN pupils with statements	5.1	4.7	..	0.4	..
Incidents (%)[3]	83.9	95.0	..	34.2	..
Placements (%)[5]	1.8	1.9	..	2.5	..

1 For Scotland, pupils with a Record of Needs.
2 Provisional. UK totals include 1996/97 data for Wales.
3 Incidence of pupils - the number of pupils with statements within each school type expressed as a proportion of the total number of pupils on roll in each school type.
4 Grant-Aided schools in Northern Ireland.
5 Placement of pupils - the number of pupils with statements within each school type expressed as a proportion of the number of pupils with statements in all schools.
6 Includes nursery classes (except for Scotland, where they are included with Nursery schools) and reception classes.
7 UK totals are slight undercounts.
8 England figures exclude dually registered pupils.
9 Including general and hospital special schools.

Source: Education Departments: 01325 392756

6.5 GCE, GCSE, SCE and GNVQ/GSVQ qualifications obtained by pupils and students at a typical age[1,2], other ages and all ages

United Kingdom

	1995/96[3]			1996/97[3]			1997/98[4]		
	Examinations taken at			Examinations taken at			Examinations taken at		
	Typical age[1,2]	Other ages	All ages[5]	Typical age[1,2]	Other ages	All ages[5]	Typical age[1,2]	Other ages	All ages
Persons									
Pupils with GCE A level/SCE H grade passes or equivalent[6]									
2 or more A, 3 or more H[7]	180	6	204	195	5	219	253	7	260
1 A, or 2 H	41	26	78	41	24	77	49[8]	21[8]	70[8]
Pupils with GCSE or equivalent[9]									
5 or more A*-C/1-3 grades[9]	291	6	331	292	6	333	329	7	335
1-4 A*-C/1-3 grades[10,11]	170	182	372	166	175	359	178	158	336
No grades A*-C but at least 1 grade D-G[12,13]	144	83	237	141	91	240	147	87	234
No graded results	12	27	51	12	28	53	22	26	47
Males[6]									
Pupils with GCE A level/SCE H grade passes or equivalent[6]									
2 or more A, 3 or more H[7]	84	3	95	90	3	101	116	4	119
1 A, 1 or 2 H	19	9	34	19	8	33	23[8]	8[8]	30[8]
Pupils with GCSE or equivalent[9]									
5 or more A*-C/1-3 grades[9]	133	3	151	133	3	152	149	3	153
1-4 A*-C/1-3 grades[10,11]	85	80	175	83	77	170	92	71	162
No grades A*-C but at least 1 grade D-G[12,13]	87	38	131	86	43	134	88	41	129
No graded results	7	12	25	7	13	26	12	11	23
Females[6]									
Pupils with GCE A level/SCE H grade passes or equivalent[6]									
2 or more A, 3 or more H[7]	96	3	109	105	3	118	138	4	141
1 A, 1 or 2 H	21	17	44	21	15	43	26[8]	14[8]	40[8]
Pupils with GCSE or equivalent[9]									
5 or more A*-C/1-3 grades[9]	158	3	180	159	3	181	179	3	182
1-4 A*-C/1-3 grades[10,11]	86	102	196	83	98	189	87	87	174
No grade A*-C but at least 1 grade D-G[12,13]	57	45	106	55	48	107	58	46	104
No graded results	5	15	26	5	16	26	10	14	24

1 For GCSE's and equivalent, those in all schools and further education who were 15 at the start of the academic year, i.e. 31 August; pupils in Year S4 in Scotland.
2 For GCE A level and equivalent, pupils in schools and students in further education institutions aged 16-18 at the start of the academic year in England, Wales and Northern Ireland; pupils in Scotland generally sit Highers one year earlier and the figures tend to relate to the result of pupils in Year S5/S6.
3 Schools only for Scotland and Northern Ireland.
4 GCSE data for Northern Ireland relate to schools only and figures, other than 5 or more grades A* - C, refer to 1996/97.
5 Age breakdown is not available for Scotland data, therefore data are included in the 'All Ages' column only.
6 2 AS level count as 1 A level pass. A levels only counted for Northern Ireland.
7 Includes Advanced level GNVQ/GSVQ. An Advanced GNVQ/GSVQ is equivalent to 2 GCE A levels or AS equivalents/3 SCE Higher grades.
8 Data for Northern Ireland relate to FE colleges only.
9 Includes Intermediate and Foundation level GNVQ/GSVQ.
10 GNVQ/GSVQ Intermediate Part 1, Full and language unit are equivalent to 2, 4 and 0.5 GCSE grades A*-C/SCE Standard grades 1-3 respectively. Figures include those with 4.5 GCSE's.
11 Includes pupils with 1 AS level for England and Wales and 1 SCE Higher grade for Scotland.
12 Grades D-G at GCSE and Scottish SCE Standard grades 4-7.
13 GNVQ/GSVQ Foundation Part 1, Full and language unit are equivalent to 2, 4 and 0.5 GCSE grades D-G/SCE Standard grades 4-7 respectively

Source: Education Department: 01325 392756

6.6 Post compulsory education and training: students and starters, 1997/98[4,5,6]

United Kingdom (home and overseas students)

Students in further and higher[1] education by level, mode of study[2], gender and age[3]

Thousands

	Postgraduate level		First degree[7]		Other undergraduate[7]		Total higher education[8]		Further education[9]		Total FE/HE students	
	Full-time	Part-time	Full-time	Part-time	Full-time	Part-time	Full-time	Part-time	Full-time	Part-time	Full-time	Part-time
All												
Age under 16	-	0.1	0.1	-	-	0.1	0.1	0.2	5.9	22.5	6.0	22.7
16	-	-	0.5	-	0.8	0.2	2.1	0.4	247.6	57.4	249.8	57.8
17	0.1	-	10.5	0.1	4.8	0.8	16.5	1.1	210.9	65.2	227.4	66.2
18	0.1	0.1	146.5	0.4	18.3	3.2	168.9	5.5	103.5	59.9	272.4	65.5
19	0.1	0.1	183.4	1.5	22.8	4.4	212.0	9.3	41.2	45.5	253.2	54.8
20	0.8	0.1	182.4	2.5	20.9	5.0	208.6	11.2	23.6	38.7	232.2	49.8
21	11.3	1.0	126.6	3.5	15.2	5.2	156.0	13.1	17.8	37.7	173.8	50.8
22	17.9	3.3	62.5	3.8	11.0	5.9	93.2	16.3	14.5	39.0	107.7	55.3
23	17.6	5.4	33.5	3.8	8.0	6.9	60.6	19.7	12.4	41.2	73.0	60.9
24	14.9	7.7	21.8	3.7	5.9	7.5	43.7	22.6	11.7	43.5	55.5	66.0
25	12.2	9.2	16.2	3.8	4.8	8.6	34.3	25.4	11.2	44.9	45.5	70.3
26	9.6	9.7	13.3	3.9	4.0	9.4	27.8	26.6	10.4	43.9	38.2	70.5
27	7.4	9.2	10.6	3.5	3.2	9.4	22.1	25.7	9.8	43.7	31.9	69.3
28	6.3	9.0	9.3	3.5	2.8	9.4	19.2	25.5	9.5	43.3	28.6	68.8
29	5.2	8.8	8.2	3.5	2.6	9.6	16.7	25.4	9.2	42.7	25.9	68.0
30+	36.9	140.5	73.5	49.5	26.8	223.2	146.8	462.9	119.0	903.0	265.8	1 366.0
Unknown	0.3	2.9	0.6	1.2	0.7	12.3	1.9	17.3	5.3	44.4	7.2	61.6
All ages	140.8	207.0	899.3	88.1	152.7	321.3	1 230.4	708.0	863.6	1 616.4	2 094.0	2 324.4
Males												
Age under 16	-	-	-	-	-	-	0.1	0.1	3.4	11.3	3.4	11.4
16	-	-	0.2	-	0.3	0.1	0.9	0.2	121.7	31.5	122.7	31.7
17	-	-	4.7	-	2.2	0.4	7.4	0.6	102.1	36.7	109.4	37.3
18	0.1	0.1	68.7	0.3	8.5	2.0	79.3	3.6	52.8	32.9	132.0	36.5
19	0.1	0.1	86.9	0.9	10.6	2.7	100.7	6.0	22.3	23.0	123.0	29.0
20	0.4	0.1	87.3	1.5	9.1	2.8	99.3	6.6	12.7	17.8	111.9	24.4
21	5.1	0.4	62.6	2.0	6.1	2.6	75.5	7.0	8.9	15.4	84.5	22.4
22	8.7	1.5	33.0	2.1	4.5	2.7	47.3	8.0	7.1	15.3	54.4	23.3
23	9.1	2.5	18.6	2.1	3.5	3.0	32.0	9.2	6.1	16.2	38.1	25.4
24	7.9	3.7	12.0	1.9	2.5	3.1	23.1	10.4	5.7	16.8	28.8	27.3
25	6.2	4.6	8.9	1.8	2.1	3.7	17.9	11.8	5.3	17.5	23.2	29.4
26	5.1	4.9	7.2	1.8	1.7	3.9	14.5	12.2	4.9	17.4	19.4	29.7
27	4.1	4.7	5.6	1.6	1.3	4.2	11.6	12.1	4.5	17.4	16.1	29.6
28	3.7	4.6	4.8	1.5	1.2	4.2	10.0	12.1	4.3	17.4	14.3	29.5
29	3.0	4.7	4.2	1.5	1.0	4.4	8.6	12.2	4.0	17.0	12.6	29.2
30+	21.4	71.7	30.2	17.2	8.2	93.0	64.6	204.8	47.3	332.3	111.9	537.1
Unknown	0.2	1.3	0.3	0.4	0.3	4.6	1.0	6.7	2.6	18.3	3.5	25.1
All ages	75.1	104.7	435.4	36.7	63.0	137.3	593.7	323.9	415.6	654.3	1 009.3	978.2
Females												
Age under 16	-	-	-	-	-	0.1	-	0.1	2.5	11.2	2.6	11.3
16	-	-	0.3	-	0.5	0.1	1.2	0.2	125.9	25.8	127.1	26.1
17	-	-	5.7	-	2.7	0.4	9.1	0.5	108.9	28.5	118.0	29.0
18	0.1	-	77.8	0.2	9.8	1.2	89.6	1.9	50.7	27.0	140.4	28.9
19	-	-	96.5	0.6	12.2	1.7	111.3	3.3	18.8	22.5	130.2	25.8
20	0.5	0.1	95.2	1.0	11.8	2.3	109.3	4.5	11.0	20.9	120.3	25.5
21	6.2	0.6	64.0	1.5	9.1	2.6	80.4	6.1	8.9	22.3	89.3	28.4
22	9.2	1.8	29.4	1.7	6.5	3.2	46.0	8.3	7.3	23.7	53.3	32.0
23	8.5	2.9	14.9	1.8	4.6	4.0	28.6	10.5	6.4	25.0	34.9	35.5
24	7.0	4.0	9.8	1.8	3.4	4.4	20.6	12.1	6.0	26.6	26.7	38.8
25	5.9	4.7	7.3	2.0	2.7	4.9	16.4	13.6	5.9	27.3	22.3	40.9
26	4.4	4.9	6.1	2.1	2.3	5.5	13.2	14.3	5.6	26.5	18.8	40.8
27	3.3	4.5	5.0	1.9	1.9	5.2	10.6	13.5	5.2	26.2	15.8	39.7
28	2.6	4.4	4.5	2.0	1.6	5.2	9.1	13.4	5.1	25.9	14.3	39.3
29	2.1	4.2	4.0	2.0	1.6	5.3	8.1	13.2	5.2	25.6	13.3	38.8
30+	15.6	68.8	43.3	32.3	18.6	130.2	82.2	258.1	71.7	570.8	153.9	828.8
Unknown	0.1	1.5	0.3	0.8	0.4	7.7	0.9	10.5	2.8	26.0	3.7	36.6
All ages	65.7	102.4	464.0	51.4	89.7	183.9	636.7	384.1	448.0	962.1	1 084.7	1 346.2

1 Includes Open University students.
2 Full-time includes sandwich, and for Scotland, short full-time. Part-time comprises both day and evening, including block release (except Scotland) and open/distance learning.
3 Ages as at 31 August 1997 (1 July for Northern Ireland and 31 December for Scotland).
4 Figures are for students on courses at a particular point during the year. Therefore students starting courses after this date will not be counted.
5 Provisional. Includes 1996/97 further education institution data for England and Wales.
6 Figures are not directly comparable with similar figures in earlier volumes since those refer to enrolments rather than headcounts.
7 Due to a change in reporting practice in 1997/98 most Open University students were recorded as 'Other Undergraduate', whereas in previous years most were recorded as 'First Degree'.
8 Data are not available by level for higher education students in further education institutions in England and are included in total higher education columns only.
9 Excludes approximately 177 000 students in further education institutions in England since the information cannot be broken down in this way. External institutions and specialist designated colleges are also excluded.

Source: Education Departments: 01325 392756

6.7 Students in further and higher[1] education by type of course, mode of study[2], gender and subject group[2]

Home and overseas[3] students — Academic year 1997/98[4,5,6] — Thousands

	Postgraduate level		First degree[7]		Other undergraduate[7]		Total higher education		Further education		Total FE/HE enrolments	
	Full-time	Part-time	Full-time	Part-time	Full-time	Part-time	Full-time	Part-time	Full-time	Part-time	Full-time	Part-time
Persons												
Subject group												
Medicine & Dentistry	4.6	8.2	28.5	0.1	0.2	0.2	33.3	8.4	-	-	33.3	8.4
Allied Medicine	4.2	13.6	45.3	21.1	45.1	29.3	94.6	64.0	5.3	10.6	99.9	74.7
Biological Sciences	9.1	8.2	63.0	2.8	2.5	1.4	74.7	12.3	0.2	0.6	74.8	12.9
Agriculture	2.1	1.7	10.0	0.2	4.5	1.0	16.6	3.0	1.6	2.9	18.3	5.9
Physical Sciences	10.5	7.6	48.7	2.2	1.6	2.0	60.8	11.7	0.8	2.3	61.6	14.0
Mathematical Science	8.9	8.5	59.6	4.5	12.3	10.3	80.8	23.3	3.1	22.1	83.9	45.4
Engineering & Technology	13.6	13.5	80.0	7.5	11.7	12.4	105.3	33.5	8.3	17.1	113.6	50.6
Architecture	4.7	5.6	22.9	5.5	3.4	4.6	31.1	15.6	5.4	6.3	36.5	22.0
Social Sciences	22.1	23.8	104.8	10.8	11.4	16.9	138.2	51.4	7.3	19.9	145.4	71.3
Business & Financial	15.8	46.1	101.4	11.2	30.9	38.3	148.1	95.7	14.4	36.4	162.5	132.0
Documentation	2.4	2.6	13.2	0.7	1.7	1.3	17.3	4.6	1.7	4.4	18.9	9.0
Languages	6.0	6.8	59.4	3.1	2.6	11.8	68.1	21.7	0.9	9.8	68.9	31.5
Humanities	5.4	8.0	33.1	2.4	0.6	10.1	39.1	20.5	0.6	1.3	39.7	21.8
Creative arts	4.9	4.2	71.0	2.6	9.9	4.7	85.9	11.6	10.0	8.7	95.9	20.3
Education[8]	2.7	13.7	6.8	1.2	0.9	5.7	10.4	20.6	3.4	16.8	13.8	37.4
ITT and INSET[8]	21.0	28.3	41.4	2.5	0.2	7.6	62.5	38.4	-	0.1	62.5	38.5
Combined, gen	2.7	6.5	110.2	9.8	13.3	163.7	126.3	180.0	10.8	33.7	137.1	213.7
Unknown[9]	..	..	..	..	..	..	37.6	91.5	789.9	1 423.5	827.5	1 515.0
All subjects	140.8	207.0	899.3	88.1	152.7	321.3	1 230.4	708.0	863.6	1 616.4	2 094.0	2 324.4
Males												
Subject group												
Medicine and Dentistry	2.2	4.3	13.7	-	0.1	0.1	16.0	4.4	-	-	16.0	4.4
Allied Medicine	1.5	4.0	11.3	2.1	6.0	3.0	18.9	9.0	0.4	2.5	19.2	11.5
Biological Sciences	4.1	3.5	24.5	1.0	1.3	0.6	29.9	5.1	0.1	0.2	30.0	5.3
Agriculture	1.2	1.0	4.4	0.1	2.6	0.6	8.2	1.8	1.2	1.8	9.4	3.6
Physical Sciences	7.0	5.1	30.9	1.3	1.0	1.1	38.9	7.5	0.4	1.0	39.3	8.5
Mathematical Science	6.6	6.2	45.3	3.5	9.7	6.7	61.6	16.4	2.1	8.3	63.8	24.7
Engineering & Technology	11.1	11.8	68.3	7.0	10.2	11.5	89.6	30.3	7.7	18.5	97.3	45.8
Architecture	3.1	3.8	17.3	4.5	2.8	3.6	23.2	11.8	5.2	5.9	28.4	17.7
Social Sciences	10.8	10.7	43.9	4.8	3.1	4.2	57.9	19.8	2.0	7.0	59.9	26.8
Business & Financial	9.3	27.3	49.9	5.0	13.6	15.9	72.9	48.3	5.4	10.4	78.2	58.7
Documentation	0.9	1.0	5.2	0.2	0.8	0.5	6.9	1.7	0.9	1.7	7.8	3.4
Languages	2.4	2.7	17.1	1.0	0.7	4.5	20.2	8.1	0.4	3.5	20.5	11.7
Humanities	3.0	4.2	15.3	0.9	0.2	3.6	18.5	8.8	0.2	0.5	18.7	9.2
Creative arts	2.2	1.9	29.8	0.9	4.6	1.6	36.6	4.4	3.8	2.3	40.4	6.7
Education[8]	1.0	5.2	3.2	0.3	0.3	2.0	4.5	7.5	1.5	6.6	6.0	14.1
ITT and INSET[8]	7.1	8.0	8.0	0.5	-	2.3	15.1	10.8	-	-	15.1	10.8
Combined, gen	1.5	3.8	47.4	3.7	5.9	75.7	54.7	83.1	5.3	13.8	60.0	96.9
Unknown[9]	..	..	..	..	..	..	20.2	45.2	379.1	573.2	399.3	618.4
All subjects	75.1	104.7	435.4	36.7	63.0	137.3	593.7	323.9	415.6	654.3	1 009.3	978.2
Females												
Subject group												
Medicine & Dentistry	2.5	3.9	14.8	-	0.1	0.1	17.4	4.0	-	-	17.4	4.0
Allied Medicine	2.6	9.7	33.9	19.0	39.1	26.4	75.7	55.0	4.9	8.1	80.6	63.1
Biological Sciences	5.0	4.6	38.5	1.8	1.3	0.8	44.7	7.2	0.1	0.4	44.8	7.6
Agriculture	0.9	0.7	5.6	0.1	1.8	0.4	8.4	1.2	0.4	1.1	8.9	2.3
Physical Sciences	3.5	2.4	17.9	0.8	0.6	0.9	21.9	4.2	0.4	1.3	22.3	5.5
Mathematical Science	2.3	2.4	14.3	1.0	2.6	3.6	19.1	6.9	1.0	13.8	20.1	20.7
Engineering & Technology	2.5	1.7	11.7	0.6	1.4	0.9	15.7	3.2	0.7	1.5	16.4	4.7
Architecture	1.6	1.8	5.7	1.0	0.6	1.0	7.9	3.8	0.2	0.4	8.0	4.2
Social Sciences	11.2	13.1	60.9	6.0	8.3	12.6	80.3	31.7	5.2	12.9	85.6	44.5
Business & Financial	6.4	18.8	51.4	6.1	17.4	22.4	75.2	47.4	9.0	26.0	84.2	73.4
Documentation	1.5	1.6	8.0	0.5	0.9	0.8	10.4	2.9	0.8	2.7	11.2	5.6
Languages	3.6	4.2	42.3	2.2	2.0	7.3	47.9	13.6	0.5	6.2	48.4	19.8
Humanities	2.4	3.8	17.8	1.5	0.3	6.4	20.6	11.7	0.4	0.8	20.9	12.5
Creative arts	2.7	2.3	41.2	1.8	5.3	3.1	49.2	7.2	6.2	6.4	55.5	13.6
Education[8]	1.7	8.5	3.6	1.0	0.6	3.7	5.9	13.1	1.9	10.2	7.8	23.3
ITT and INSET[8]	13.9	20.3	33.4	2.0	0.1	5.4	47.4	27.7	-	0.1	47.4	27.7
Combined, gen	1.2	2.7	62.9	6.1	7.4	88.0	71.5	96.9	5.5	19.9	77.0	116.8
Unknown[9]	..	..	..	..	..	..	17.4	46.4	410.8	850.3	428.2	896.6
All subjects	65.7	102.4	464.0	51.4	89.7	183.9	636.7	384.1	448.0	962.1	1 084.7	1 346.2

1 Includes Open University courses allocated throughout subject headings.
2 Full-time includes sandwich, part-time comprises both day and evenings.
3 Includes 273 400 students from abroad.
4 Figures are for students on courses at a particular point during the year. Therefore students starting courses after this date will not be counted.
5 Provisional - Includes 1996/97 further education institutions data for England and Wales.
6 Figures are not directly comparable with similar figures in earlier editions since those refer to enrolments rather than headcounts.
7 Due to a change in reporting practice in 1997/98, most Open University students were recorded as 'Other Undergraduate' whereas in previous years most were recorded as 'First Degree'.
8 Students in Scotland and Northern Ireland on in-service training courses are included in Education.
9 Includes further education institutions in England and Wales which are not available by subject, or, for higher education students, by level but excludes approximately 177 000 students in further education in England.

Source: Education Departments:01325 392756

6.8 Students[1] obtaining higher education qualifications[2] by type of course, gender and subject group, 1996/97[3]

Thousands

	Sub-degree[4]	First degree[5]	Postgraduate[5] PHD's & equivalent	Postgraduate[5] Other	Postgraduate[5] Total	Total higher education
Persons						
Subject group						
Medicine & Dentistry	-	5.8	0.9	1.8	2.7	8.5
Allied Medicine	13.1	14.5	0.5	3.6	4.1	31.7
Biological Sciences	1.5	15.4	1.6	1.9	3.5	20.3
Agriculture	1.5	2.8	0.3	0.9	1.2	5.6
Physical Sciences	1.2	14.1	1.8	2.3	4.1	19.4
Mathematical Science	5.5	13.0	0.6	3.6	4.2	22.7
Engineering & Technology	6.2	23.0	1.8	5.8	7.5	36.8
Architecture	2.2	7.5	0.1	3.0	3.1	12.8
Social Sciences	7.9	31.3	0.8	12.9	13.7	52.9
Business & Financial	18.9	29.6	0.3	19.0	19.3	67.8
Documentation	0.8	3.3	-	1.9	2.0	6.1
Languages	1.4	16.5	0.5	2.6	3.1	21.0
Humanities	0.6	10.3	0.5	2.3	2.8	13.7
Creative arts	4.6	18.0	0.1	2.9	3.0	25.6
Education	1.1	2.0	0.3	2.9	3.2	6.2
ITT and INSET	2.7	12.1	-	25.6	25.7	40.5
Combined, gen	7.4	36.3	0.1	6.4	6.5	50.2
Unknown[6,7]	10.5	2.8	-	1.2	1.2	14.5
All subjects	87.3	258.2	10.2	100.5	110.7	456.2
Males						
Subject group						
Medicine and Dentistry	-	2.8	0.5	0.8	1.3	4.1
Allied Medicine	1.3	3.1	0.3	0.9	1.2	5.6
Biological Sciences	0.6	5.9	0.8	0.8	1.6	8.1
Agriculture	0.9	1.4	0.2	0.5	0.7	3.0
Physical Sciences	0.7	8.8	1.4	1.5	2.9	12.4
Mathematical Science	4.0	9.5	0.5	2.7	3.2	16.7
Engineering & Technology	5.5	19.7	1.5	4.8	6.3	31.5
Architecture	1.8	5.9	0.1	1.9	2.0	9.7
Social Sciences	2.1	13.8	0.5	5.9	6.4	22.3
Business & Financial	7.4	14.3	0.2	11.2	11.4	33.2
Documentation	0.4	1.3	-	0.7	0.7	2.4
Languages	0.4	4.7	0.2	0.9	1.2	6.3
Humanities	0.3	4.8	0.3	1.2	1.5	6.5
Creative arts	2.0	7.5	0.1	1.2	1.3	10.8
Education	0.3	0.9	0.1	1.0	1.2	2.4
ITT and INSET	0.8	2.5	-	8.0	8.0	11.3
Combined, gen	3.0	15.9	0.1	4.1	4.2	23.2
Unknown[6,7]	5.2	1.1	-	0.6	0.6	6.9
All subjects	36.8	123.8	6.8	48.9	55.7	216.3
Females						
Subject group						
Medicine & Dentistry	-	3.0	0.4	0.9	1.3	4.4
Allied Medicine	11.8	11.4	0.3	2.7	2.9	26.1
Biological Sciences	0.9	9.5	0.8	1.1	1.9	12.2
Agriculture	0.7	1.4	0.1	0.4	0.5	2.6
Physical Sciences	0.5	5.2	0.5	0.8	1.3	7.0
Mathematical Science	1.5	3.5	0.1	0.9	1.0	6.0
Engineering & Technology	0.7	3.4	0.2	1.0	1.2	5.3
Architecture	0.5	1.5	-	1.1	1.1	3.1
Social Sciences	5.7	17.6	0.3	7.0	7.3	30.6
Business & Financial	11.5	15.3	0.1	7.8	7.9	34.6
Documentation	0.4	2.0	-	1.2	1.2	3.7
Languages	1.0	11.8	0.2	1.7	1.9	14.7
Humanities	0.3	5.5	0.2	1.1	1.3	7.1
Creative arts	2.6	10.5	-	1.6	1.7	14.8
Education	0.7	1.1	0.1	1.9	2.0	3.8
ITT and INSET	1.9	9.6	-	17.6	17.6	29.2
Combined, gen	4.4	20.4	-	2.2	2.3	27.0
Unknown[6,7]	5.3	1.7	-	0.7	0.7	7.7
All subjects	50.4	134.4	3.4	51.7	55.1	239.9

1 Includes students on Open University courses.
2 Excludes qualifications from the private sector.
3 Includes 1995/96 data for higher education institutions in England and further education colleges in Scotland.
4 Excludes students who successfully completed courses for which formal qualifications are not awarded.
5 Excludes some 0.9 thousand postgraduate qualification awards in higher education institutions not broken down further.
6 Further education institutions in England for which a subject breakdown is not available.
7 Includes further education institutions in Wales for which the standard subject breakdown is not available.

Source: Education Departments: 01325 392756

6.9 Teachers and lecturers by type of establishment, gender and graduate status

(i) Full-time

	1989/90[2]		1990/91[2]		1991/92[2]		1992/93[2]		1993/94[2]		1994/95[2]		1995/96[2,3]		1996/97[2,5]	
	All (000's)	% Graduate	All (000's)	% Graduate	All (000's)	% Graduates	All (000's)	% Graduate	All (000's)	% Graduate	All (000's)	% Graduate	All (000's)	% Graduate	All (000's)	% Graduate
United Kingdom																
Males																
Schools																
Public sector mainstream																
Primary[5]	38	43	38	45	37	47	37	49	36	52	36	55	35	57	35	60
Secondary[6,7]	125	69	121	70	118	71	116	72	112	72	110	74	108	75	106	77
Non-maintained mainstream[6,8]	20	85	21	84	21	86	20	86	20	87	20	87	21	85	21	..
All special	6	41	6	43	6	44	6	46	6	47	5	50	5	50	5	53
All schools[9]	190	65	185	65	181	67	179	68	174	69	171	70	170	72	167	72
Establishments of further and higher education[10]	63	47	62	48	60	48	56	49	..	..	..	..	84	84	87	85
of which																
FEIs[11]	..	..	..	..	..	..	..	..	..	..	..	..	34[12]	64[13]	34[12]	64[13]
HEIs[10]	..	..	..	..	..	..	..	..	..	..	47	95	49	97	53	97
All establishments[9]	283	64	276	64	266	66	260	67	..	..	218	76	254	76	254	76
Females																
Schools																
Public sector mainstream																
Primary[5]	171	32	171	34	172	37	173	39	176	42	177	45	176	48	177	51
Secondary[6,7]	115	62	112	62	113	63	113	65	113	65	113	68	114	69	115	71
Non-maintained mainstream[6,8]	23	58	24	58	25	59	25	60	26	62	26	62	27	63	28	..
All special	13	33	13	34	13	35	13	37	13	37	12	39	12	41	12	43
All schools[9]	322	45	320	46	323	48	324	50	327	51	328	54	330	56	331	58
Establishments of further and higher education[10]	27	49	28	49	29	49	28	49	..	..	..	..	44	78	46	79
of which																
FEIs[11]	..	..	..	..	..	..	..	..	..	..	..	..	26[12]	65[13]	26[12]	65[13]
HEIs[10]	..	..	..	..	..	..	..	..	..	..	16	93	18	95	20	95
All establishments[9]	358	45	355	47	358	48	357	50	..	..	343	56	374	59	378	60
England and Wales[2]																
All schools																
Males	168	64	163	65	161	66	158	67	153	68	151	70	148	72	145	72
Females	278	45	276	47	278	48	279	50	282	52	283	55	284	58	286	60
Scotland[2,8]																
All schools																
Males	15	70	15	71	15	71	14	73	14	73	14	74	15	74[3]	15	74[4]
Females	33	37	32	38	32	38	32	41	32	41	32	45	33	45[3]	33	45[4]
Northern Ireland[2,8]																
All schools																
Males	7	68	7	69	6	70	6	72	6	74	6	75	6[3]	75[3]	6[4]	75[4]
Females	12	50	12	52	12	55	12	58	13	60	13	63	13[3]	63[3]	13[4]	63[4]

(ii) Part-time[14]

	1989/90[15]	1990/91[15]	1991/92	1992/93	1993/94	1994/95	1995/96	1996/97[16]
	All (000's)	All (000's)	All (000's)	All (000's)	All (000's)	All (000's)	All (000's)	All (000's)
United Kingdom (Persons)								
Public sector mainstream								
Primary[5]	..	..	..	..	16	17	19	19
Secondary[6,7]	..	..	..	..	15	16	18	17
Non-maintained mainstream[6,8]	..	..	..	..	8	8	9	9
All special	..	..	..	..	2	1	2	1
All schools	29	30	32	33	41	43	47	46

1 Includes estimated data for England and Wales.
2 Includes some estimated data for each of the countries.
3 1995/96 schools data for Northern Ireland and schools graduate data for Scotland are not available. However 1994/95 data have been used for United Kingdom totals.
4 Provisional. 1996/97 schools data are for Northern Ireland and schools graduate data for Scotland are not available therefore data shown are for 1994/95.
5 Includes nursery schools.
6 From 1989-90 voluntary grammar schools in Northern Ireland are recorded in the maintained sector.
7 From 1 April 1993 excludes sixth form colleges which were reclassified as further education colleges.
8 Excludes independent schools in Scotland and Northern Ireland.
9 Totals include a number of teachers classified as miscellaneous.
10 Excludes Open University.
11 Includes all academic staff, both temporary and permanent in Scotland.
12 Excludes Wales.
13 England and Northern Ireland only.
14 Full-time equivalents of part-time teachers.
15 Great Britain only.
16 In addition there are 56,100 part-time lecturers in further and higher education institutions.

Source: Education Departments: 01325 392756

6.10 Full-time academic teaching and research staff at higher education institutions

United Kingdom

Academic years

Number

	Professors	Readers and senior lecturers	Lecturers and assistant lecturers	Other	Total	Percentage annual change
	KBGU	KBGV	KBGW	KBGX	KBGY	KBGZ
1974/75[1]	3 906	7 398	19 883	914	32 101	*2.1*
1975/76[1]	3 989	7 617	19 722	880	32 208	*0.3*
1976/77[1]	4 124	7 877	20 022	712	32 735	*1.6*
1977/78[1]	4 164	8 163	19 980	677	32 984	*0.8*
1978/79[1]	4 225	8 454	20 302	714	33 695	*2.2*
1979/80[1]	4 337	8 734	20 518	661	34 250	*1.6*
1980/81[1]	4 382	8 809	20 460	646	34 297	*0.1*
1981/82[1]	4 351	8 777	20 045	562	33 735	*−1.6*
1982/83[1]	4 017	8 284	18 885	456	31 642	*−6.2*
1983/84[1]	3 893	8 145	18 595	463	31 096	*−1.7*
1984/85[1,2]	3 807	7 942	18 737	557	31 043	*−0.2*
1985/86[1]	3 959	8 025	18 850	578	31 412	*1.2*
1986/87[1]	4 070	8 074	18 711	577	31 432	*0.1*
1987/88[1]	4 160	8 291	18 268	542	31 261	*−0.5*
1988/89[1]	4 093	8 266	17 778	484	30 621	*−2.0*
1989/90[1]	4 261	8 618	17 903	558	31 340	*2.3*
1990/91[1]	4 520	8 842	17 830	669	31 861	*1.7*
1991/92[1]	4 872	9 270	17 824	672	32 638	*2.4*
1992/93[1]	5 226	9 650	17 854	717	33 447	*2.5*
1993/94[1]	5 545	9 890	18 275	787	34 497	*3.1*

	Professors	Senior lecturers & researchers	Lecturers	Researchers	Other	Total	Percentage annual change
	KBGU	KTEE	KTEF	KTEG	KBGX	KBGY	KBGZ
1994/95[3]	6 762	16 949	39 295	4 662	4 842	72 510	..
1995/96[3]	7 947	17 457	41 503	5 412	5 582	77 901	*7.4*
1996/97[3,4]	8 222	17 591	39 820	4 809	5 589	76 031	*−2.4*
1997/98[3,4]	8 441	17 293	39 497	4 999	5 684	75 914	*−0.2*

1 Full-time teaching and research staff in posts wholly financed from general university funds excluding the Open University and the former polytechnics and central institutions who obtained university status in 1992.

2 Includes Ulster Polytechnic which merged with the University of Ulster in October 1984.

3 Full-time academic staff of at least 25% full-time equivalence who are wholly institutionally financed for all publicly-funded higher education institutions.

4 Due to changes in the definition of the HESA standard staff population from 1996/97 onwards, it is not advisible to make direct comparisons with figures for 1994/95 and 1995/96.

Sources: 74/75 to 93/94 Higher Education Funding Council for England: 0117 931 7395;
94/95 to 97/98 Higher Education Statistics Agency: 01242 255577

6.11 Students in higher education institutions
United Kingdom

Academic years — Number

		1987[1,2] /88	1988[1,2] /89	1989[1,2] /90	1990[1,2] /91	1991[1,2] /92	1992[1,2] /93	1993[1,2] /94	1994[4,5] /95	1995[4,5] /96	1996[4,5,6] /97	1997[4,5] /98
Full - time students												
Total full - time students	KBKA	320 920	333 547	350 981	370 254	401 657	435 617	470 565	1 076 630	1 107 841	1 138 595	1 166 127
of which												
Total overseas students	KBKB	38 512	39 756	41 532	43 496	45 462	47 278	49 817	123 620	144 777	160 056	172 316
Total postgraduate students	KBKC	56 117	58 354	60 696	65 282	73 579	78 490	85 105	129 711	135 348	140 916	143 521
Total women students	KBKD	132 674	140 647	151 247	161 895	177 834	197 339	215 500	530 540	533 689	581 419	605 760
Overseas students as a percentage of total	KBKE	*12.0*	*11.9*	*11.8*	*11.7*	*11.3*	*10.9*	*10.6*	*11.5*	*13.1*	*14.1*	*14.8*
Postgraduate students as a percentage of total	KBKF	*17.5*	*17.5*	*17.3*	*17.6*	*18.3*	*18.0*	*18.1*	*12.1*	*12.2*	*12.4*	*12.3*
Women students as a percentage of total	KBKG	*41.3*	*42.2*	*43.1*	*43.7*	*44.3*	*45.3*	*45.8*	*49.3*	*50.0*	*51.1*	*51.9*
At undergraduate level												
Total	KBKH	264 803	275 193	290 285	304 972	328 078	357 127	385 460	946 919	972 493	997 679	1 022 606
Men	KBKI	150 808	154 783	161 126	167 438	178 183	190 712	204 050	472 227	478 818	480 153	483 793
Women	KBKJ	113 995	120 410	129 159	137 534	149 895	166 415	181 410	474 692	493 675	517 526	538 813
From United Kingdom												
Total	KBKK	246 122	255 733	269 590	283 418	305 227	332 986	360 043	864 067	873 240	886 801	902 170
Men	KBKL	139 010	142 722	148 567	154 544	164 715	176 721	189 477	427 827	426 687	422 852	422 839
Women	KBKM	107 112	113 011	121 023	128 874	140 512	156 265	170 566	436 240	446 553	463 949	479 331
From overseas												
Total	KBKN	18 681	19 460	20 695	21 554	22 851	24 141	25 417	82 852	99 253	110 878	120 436
Men	KBKO	11 798	12 061	12 559	12 894	13 468	13 991	14 573	44 400	52 131	57 301	60 954
Women	KBKP	6 883	7 399	8 136	8 660	9 383	10 150	10 844	38 452	47 122	53 577	59 482
At postgraduate level												
Total	KBKQ	56 117	58 354	60 696	65 282	73 579	78 490	85 105	129 711	135 348	140 916	143 521
Men	KBKR	37 438	38 117	38 608	40 921	45 640	47 566	51 015	73 863	75 334	77 023	76 574
Women	KBKS	18 679	20 237	22 088	24 361	27 939	30 924	34 090	55 848	60 014	63 893	66 947
From United Kingdom												
Total	KBKT	38 286	38 058	39 859	43 340	50 968	55 353	60 705	88 943	89 824	91 738	91 641
Men	KBKU	22 124	22 887	23 305	25 171	29 686	31 730	34 448	48 092	47 333	47 685	46 337
Women	KBKV	14 162	15 171	16 554	18 169	21 282	23 623	26 257	40 851	42 491	44 053	45 304
From overseas												
Total	KBKW	19 831	20 296	20 837	21 942	22 611	23 137	24 400	40 768	45 524	49 178	51 880
Men	KBKX	15 314	15 230	15 303	15 750	15 954	15 836	16 567	25 771	28 001	29 338	30 237
Women	KBKY	4 517	5 066	5 534	6 192	6 657	7 301	7 833	14 997	17 523	19 840	21 643
Part - time students												
Total part-time students	KBKZ	46 062	50 097	53 850	58 604	66 438	75 506	83 311	490 683	612 253	617 584	633 937
At undergraduate level												
Total	KBLA	9 927	10 179	10 558	11 692	13 155	14 460	15 923	285 069	377 482	394 928	390 457
Men	KBLB	4 541	4 514	4 640	4 963	5 592	5 861	6 093	131 008	161 615	163 539	159 532
Women	KBLC	5 386	5 665	5 918	6 729	7 563	8 599	9 830	154 061	215 867	231 389	230 925
From United Kingdom												
Total	KBLD	9 405	9 659	10 084	10 983	12 598	13 884	15 502	274 052	362 813	385 471	379 561
Men	KBLE	4 319	4 278	4 428	4 687	5 340	5 642	5 903	125 654	154 842	159 622	154 723
Women	KBLF	5 086	5 381	5 656	6 296	7 258	8 242	9 599	148 398	207 971	225 849	224 838
From overseas												
Total	KAJX	522	520	474	709	557	576	421	11 017	14 669	9 457	10 896
Men	KAJY	222	236	212	276	252	219	190	5 354	6 773	3 917	4 809
Women	KAJZ	300	284	262	433	305	357	231	5 663	7 896	5 540	6 087
At postgraduate level												
Total	KAKA	36 135	39 918	43 292	46 912	53 283	61 046	67 388	205 614	234 771	222 656	243 480
Men	KAKB	23 030	24 494	25 981	27 474	30 606	34 637	37 786	111 106	125 533	113 959	124 827
Women	KAKC	13 105	15 424	17 311	19 438	22 677	26 409	29 602	94 508	109 238	108 697	118 653
From United Kingdom												
Total	KAKD	32 511	35 997	39 129	42 161	47 123	53 064	57 853	176 538	197 871	194 105	213 428
Men	KAKE	20 554	21 854	23 119	24 137	26 380	29 087	31 270	91 324	100 780	96 084	106 696
Women	KAKF	11 957	14 143	16 010	18 024	20 743	23 977	26 583	85 214	97 091	98 021	106 732
From overseas												
Total	KAKG	3 624	3 921	4 163	4 751	6 160	7 982	9 535	29 076	36 900	28 551	30 052
Men	KAKH	2 476	2 640	2 862	3 337	4 226	5 550	6 516	19 782	24 753	17 875	18 131
Women	KAKI	1 148	1 281	1 301	1 414	1 934	2 432	3 019	9 294	12 147	10 676	11 921

1 Excluding the Open University and the former polytechnics and central institutions who obtained university status in 1992.
2 Overseas students defined by fee-paying status. From 1980/81 most European Union students paid home fees and are therefore excluded from the overseas figures and are shown as home students.
3 Includes Ulster Polytechnic which merged with the University of Ulster in October 1984.
4 Students in all publicly funded higher education institutions.
5 Full-time includes sandwich students. Part-time includes other modes.
6 From 1996/97 onwards the HESA standard HE population excludes those students studying for the whole of their programme of study outside of the UK.

Sources: 86/87 to 93/94 Higher Education Funding Council for England: 0117 931 7395;
94/95 to 97/98 Higher Education Statistics Agency: 01242 255577

6.12 Students in H.E.I.s: UK: New admissions[6] taking courses, & numbers of full-time students by country of home residence & university residence

Academic years

Number

			1987[1] /88	1988[1] /89	1989[1] /90	1990[1] /91	1991[1] /92	1992[1] /93	1993[1] /94	1994[9] /95	1995[9] /96	1996[9] /97	1997[9] /98
New students admitted													
(full-time & sandwich)													
Men	KBMA		49 907	52 409	56 397	58 309	62 764	68 667	72 807	236 460	234 666	232 766	240 142
Women	KBMB		38 721	40 758	46 298	48 797	53 870	61 970	64 604	220 808	227 566	237 802	252 718
New students taking courses													
(full and part-time)													
Men	KBMC		215 817	221 908	230 355	240 796	260 021	278 776	298 944	340 574	320 172	318 392	325 982
Women	KBMD		151 165	161 736	174 476	188 062	208 074	232 347	254 932	341 083	348 964	367 247	382 570
Full-time students (all years)													
Total	KBME	Men	188 246	192 900	199 734	208 359	223 823	238 278	255 065	546 090	554 152	557 176	560 367
	KBMF	Women	132 674	140 647	151 247	161 895	177 834	197 339	215 500	530 540	553 689	581 419	605 760
Postgraduate advanced	KBMG	Men	37 438	38 117	38 608	40 921	45 640	47 566	51 015	49 192	49 125	49 551	49 898
(first years only)	KBMH	Women	18 679	20 237	22 088	24 361	27 939	30 924	34 090	41 583	43 986	46 140	48 755
First degree	KBMI	Men	147 707	151 507	157 198	163 003	173 148	184 585	197 068	154 721	155 767	154 964	162 245
(first years only)	KBMJ	Women	110 474	116 522	124 440	132 093	143 240	158 001	171 785	147 575	152 712	158 625	169 828
First diploma	KBMK	Men	194	185	160	184	192	199	184	..	..	..	..
(first years only)	KBML	Women	137	109	89	129	132	100	98	..	..	..	..
Others	KBMM	Men	2 907	3 091	3 768	4 251	4 843	5 982	6 798	..	..	..	..
(first years only)	KBMN	Women	3 384	3 779	4 630	5 312	6 523	8 314	9 527	..	..	..	..
Other undergraduate	KTDR	Men	..	..	..	..	..	..	..	32 547	29 774	28 251	27 999
(first years only)	KTDS	Women	..	..	..	..	..	..	..	31 650	30 868	33 037	34 135
Part-time students (all years)													
Total	KBMO	Men	27 571	29 008	30 621	32 437	36 198	40 498	43 879	242 114	287 148	277 498	284 359
	KBMP	Women	18 491	21 089	23 229	26 167	30 240	35 008	39 432	248 569	325 105	340 086	349 578
Postgraduate advanced	KBMQ	Men	23 030	24 494	25 981	27 474	30 606	34 637	37 786	44 027	33 594	33 499	34 496
(first years only)	KBMR	Women	13 105	15 424	17 311	19 438	22 677	26 409	29 602	42 556	37 933	40 640	43 248
First degree	KBMS	Men	2 883	3 004	3 179	3 378	3 792	4 142	4 471	24 436	13 015	19 077	12 110
(first years only)	KBMT	Women	3 638	3 840	4 005	4 309	4 750	5 580	6 636	27 974	19 554	12 641	18 736
First diploma	KBMU	Men	155	115	108	102	100	62	67	..	..	..	..
(first years only)	KBMV	Women	79	104	97	108	103	89	107	..	..	..	..
Occasional	KBMW	Men	1 503	1 395	1 353	1 483	1 448	1 438	1 555	..	..	..	..
(first years only)	KBMX	Women	1 669	1 721	1 816	2 312	2 405	2 573	3 087	..	..	..	..
Other undergraduate	KTDT	Men	..	..	..	..	..	..	..	35 651	38 897	39 486	39 234
(first years only)	KTDU	Women	..	..	..	..	..	..	..	49 745	63 911	69 728	67 868
Country of home residence:[3]													
(full-time only)													
United Kingdom	KBMY		275 365	284 436	297 483	311 515	338 244	367 276	395 049	953 010	963 064	978 539	993 811
Foreign and Commonwealth[8] countries	KBMZ		45 555	49 111	53 498	58 739	63 413	68 341	75 516	123 620	144 777	160 056	172 316
University residence:[4,5]													
(full-time and part-time)													
Institution maintained property	KBNA	Men	86 442	87 326	87 776	88 632	89 667	97 044	102 225	140 944	148 931	161 181	161 969
(all years)	KBNB	Women	58 023	59 948	63 045	65 662	68 481	77 359	81 685	124 863	136 208	153 451	160 600
Own home[7]	KBNC	Men	67 576	69 992	75 988	81 625	91 945	96 339	102 103	165 896	184 137	222 462	241 222
(all years)	KBND	Women	47 509	51 814	57 439	62 601	71 673	77 004	85 415	177 473	211 858	269 520	300 608
Parental/guardian home	KBNE	Men	23 280	24 693	25 111	26 170	28 583	32 049	35 744	72 120	68 640	69 791	78 200
(all years)	KBNF	Women	19 045	20 878	22 144	24 104	26 437	30 533	34 290	65 996	64 348	70 455	82 289

1 Excluding the Open university and the former polytechnics and central institutions who obtained university status in 1992.
2 Includes Ulster Polytechnic which merged with the University of Ulster in October 1984. Data for some courses in the categories First diploma and Others have been re-coded in 1985/86.
3 Stateless and unknown domicile included in the overseas figure.
4 Including those students on courses 'not of university standard', but not those sandwich students undertaking the industrial part of their training away from the university or college at the time of the student count, which is the end of the autumn term (31 December).
5 Excludes students whose type of residence is other than those listed or unknown.
6 New admission are taken to be first-year students.
7 Accommodation owned or rented by students.
8 Total for other EU and overseas students.
9 Students in all publicly funded higher education institutions.

Sources: 86/87 to 93/94 Higher Education Funding Council for England: 0117 931 7395; 94/95 to 97/98 Higher Education Statistics Agency: 01242 255577

6.13 Higher education institutions: full-time students analysed by subject group of study

United Kingdom

Academic years

Thousands

		1987[1] /88	1988[1] /89	1989[1] /90	1990[1] /91	1991[1] /92	1992[1] /93	1993[1] /94	1994[3] /95	1995[3] /96	1996[3] /97	1997[3] /98
Medicine and dentistry:												
Men	KAKJ	14.8	14.7	14.5	14.2	14.2	14.1	14.4	15.3	15.4	15.6	15.9
Women	KAKK	11.5	11.8	12.0	12.2	12.6	13.3	14.0	15.2	15.7	16.7	17.3
Subjects allied to medicine:												
Men	KAKL	3.1	3.3	3.5	3.8	4.2	4.7	5.4	13.6	14.9	17.2	19.3
Women	KAKM	5.9	6.1	6.7	7.6	8.9	11.1	13.0	50.1	55.6	69.2	76.8
Biological sciences:												
Men	KAKN	10.5	10.8	11.2	11.8	12.7	13.7	14.9	24.1	25.7	28.2	29.9
Women	KAKO	11.3	11.9	13.2	14.4	15.8	18.0	20.2	33.5	36.7	40.7	44.7
Veterinary science												
Men	KTDV	..	..	..	..	..	..	..	1.1	1.1	1.1	1.0
Women	KTDW	..	..	..	..	..	..	..	1.5	1.7	1.8	1.9
Agriculture and related subjects:												
Men	KAKP	3.7	3.7	3.6	3.6	3.6	3.7	3.8	7.3	7.5	7.2	6.8
Women	KAKQ	2.6	2.7	2.7	2.8	3.0	3.2	3.3	5.6	5.9	6.0	6.4
Physical sciences:												
Men	KAKR	20.5	20.6	21.5	22.2	23.7	25.1	26.7	41.1	40.8	40.1	38.9
Women	KAKS	6.7	6.9	7.6	8.4	9.5	10.7	11.8	20.8	21.7	22.1	21.9
Mathematical sciences:												
Men	KAKT	14.7	15.7	16.8	17.9	19.3	20.1	21.5	11.6	11.2	11.2	11.6
Women	KAKU	4.6	5.0	5.5	6.1	6.4	6.6	7.1	6.3	6.4	6.5	6.7
Computer science												
Men	KTDX	..	..	..	..	..	..	..	40.5	42.7	45.1	47.6
Women	KTDY	..	..	..	..	..	..	..	10.3	10.6	11.0	11.9
Engineering and technology:												
Men	KAKV	37.4	37.1	37.9	39.2	41.8	44.7	47.5	94.3	93.8	89.9	86.4
Women	KAKW	4.3	4.7	5.1	5.8	6.5	7.0	7.6	16.1	16.3	15.8	15.4
Architecture, building and planning:												
Men	KAKX	4.0	4.0	4.1	4.5	5.0	5.3	5.5	26.1	25.7	24.4	22.8
Women	KAKY	1.5	1.6	1.8	1.8	1.9	2.1	2.2	7.7	7.8	7.6	7.8
Social studies: (social, economics & political studies)												
Men	KAKZ	24.7	25.7	26.4	28.0	30.0	32.0	34.3	40.4	40.2	39.3	38.8
Women	KALA	21.3	22.7	23.4	25.2	27.3	29.9	32.8	49.8	51.4	52.6	53.8
Law												
Men	KTEA	..	..	..	..	..	..	..	18.9	18.8	18.5	18.8
Women	KTEB	..	..	..	..	..	..	..	21.0	21.6	22.8	24.1
Business and administrative studies:												
Men	KALB	10.1	10.7	11.6	12.1	13.5	14.6	15.7	64.6	66.7	68.3	69.2
Women	KALC	6.5	7.0	7.7	7.9	8.9	10.1	11.2	59.7	63.0	66.1	67.7
Mass communications and documentation: (librarianship & information science)												
Men	KALD	0.4	0.5	0.6	0.7	0.9	0.9	1.0	4.7	5.6	6.1	6.3
Women	KALE	0.6	0.8	0.9	1.0	1.2	1.3	1.5	7.4	8.4	9.4	9.9
Languages and related disciplines:												
Men	KALF	9.8	10.2	10.6	11.4	13.4	13.5	14.7	20.5	20.4	20.4	20.4
Women	KALG	21.5	22.4	23.7	25.2	27.4	30.0	32.2	46.0	47.0	47.5	48.5
Humanities:												
Men	KALH	9.9	10.1	10.7	11.5	12.6	13.6	14.6	18.2	18.3	18.7	18.7
Women	KALI	8.2	8.5	9.2	10.1	11.3	12.5	14.0	19.2	19.8	20.3	20.9
Creative arts: (creative arts & design)												
Men	KALJ	2.0	2.3	2.3	2.3	2.5	2.7	3.0	29.1	31.8	32.8	35.0
Women	KALK	2.9	3.4	3.6	3.5	3.9	4.2	4.6	37.9	41.4	43.4	46.7
Education:												
Men	KALL	4.4	4.4	4.4	4.4	5.3	5.5	6.1	22.0	21.5	20.7	19.6
Women	KALM	6.6	7.3	8.2	8.9	10.0	11.6	12.3	58.1	57.1	54.2	53.4
Multidisciplinary: (combined)												
Men	KALN	18.4	19.2	20.0	20.8	22.3	24.2	26.2	51.3	52.0	52.7	53.1
Women	KALO	16.6	17.9	19.9	21.1	23.1	25.8	27.8	63.1	65.5	67.7	70.0

1 Excludes the Open University and the former polytechnics and central institutions who obtained university status in 1992.

2 A new subject classification was introduced in 1985 when certain subjects were reclassified and combinations of subjects were classified separately for the first time.

3 Students in all publicly funded higher education institutions.

Sources: 86/87 to 93/94 Higher Education Funding Council for England: 0117 931 7395; 94/95 to 97/98 Higher Education Statistics Agency: 01242 255577

6.14 Higher education institutions: degrees and diplomas obtained by full-time students
United Kingdom

Calendar years

Number

		1985[1,2]	1986[1]	1987[1]	1988[1]	1989[1]	1990[1]	1991[1]	1992[1]	1993[1]	1994[4,5]/95	1995[4,5]/96	1996[4,5]/97	1997[4,5]/98
Degrees														
First degrees - honours:														
Men	KBLJ	36 119	35 823	35 845	36 735	37 473	38 324	39 360	41 908	43 642	..	..	..	..
Women	KBLK	26 122	25 988	26 601	27 799	28 578	30 121	31 659	34 423	36 983	..	..	..	..
First degrees - ordinary[3]:														
Men	KBLL	6 057	5 405	5 518	5 520	5 301	4 973	4 879	4 980	4 891	..	..	..	..
Women	KBLM	3 911	3 696	3 853	3 702	3 601	3 745	3 739	3 582	3 591	..	..	..	..
First degree totals														
Men	KTEC	..	..	..	..	..	..	..	..	..	104 977	111 050	111 027	111 152
Women	KTED	..	..	..	..	..	..	..	..	..	106 864	115 576	119 388	122 458
Higher degrees:														
Men	KBLN	16 802	17 354	18 559	19 176	20 424	20 905	22 689	24 929	27 568	19 634[6]	23 743[6]	24 404[6]	26 294[6]
Women	KBLO	6 754	7 219	8 008	8 676	9 794	10 419	11 912	14 034	16 374	20 973[6]	25 283[6]	26 743[6]	29 469[6]
Total:														
Men	KBLH	58 978	58 582	60 059	61 431	63 198	64 202	66 928	71 817	76 101	124 611	134 793	135 431	137 446
Women	KFGW	36 787	36 903	38 475	40 177	41 973	44 285	47 310	52 039	56 948	127 837	140 859	146 131	151 927
Diplomas and certificates:														
Men	KBLP	6 447	5 867	6 800	6 659	6 621	6 712	7 914	8 811	9 470	16 606[7]	15 661	15 201	14 608
Women	KBLQ	5 751	5 706	6 303	6 611	6 599	7 000	8 997	9 881	11 584	17 199[7]	16 975	19 000	18 880

1 Excluding the Open University and the former polytechnics and central institutions who obtained university status in 1992.
2 Includes Ulster Polytechnic which merged with the University of Ulster in October 1984.
3 Includes some degrees where the class is not recorded.
4 Students in all publicly funded higher education institutions.
5 Academic, not calendar year.
6 All postgraduate qualifications.
7 Other undergraduate qualifications.

Sources: 85 to 93 Higher Education Funding Council for England: 0117 931 7395; 94/95 to 97/98 Higher Education Statistics Agency: 01242 255577

Labour market

Labour market

Labour Force Survey *(Tables 7.1 to 7.3, 7.6, 7.9, 7.10, 7.12 and 7.15 to 7.17)*

Background

The LFS is the largest regular household survey in the United Kingdom. LFS interviews are conducted continuously throughout the year. In any 3 month period, a nationally representative sample of approximately 120,000 people aged 16 or over in around 61,000 households are interviewed. Each household is interviewed five times, at 3 monthly intervals. The initial interview is done face-to-face by an interviewer visiting the address. The other interviews are done by telephone wherever possible. The survey asks a series of questions about respondents' personal circumstances and their labour market activity. Most questions refer to activity in the week before the interview.

The concepts and definitions used in the LFS are agreed by the International Labour Organisation (ILO) - an agency of the United Nations. The definitions are used by European Union member countries and members of the Organisation for Economic Co-operation and Development (OECD).

The Labour Force Survey was carried out every two years from 1973 to 1983. The ILO definition was first used in 1984. This was also the first year in which the survey was conducted on an annual basis with results available for every spring quarter (representing an average of the period from March to May). The survey moved to a continuous basis in spring 1992 in Great Britain and in winter 1994/5 in Northern Ireland, with average quarterly results published 4 times a year for seasonal quarters: spring (March to May), summer (June to August), autumn (September to November) and winter (December to February). From April 1998, results are published 12 times a year for the average of 3 consecutive months.

Strengths and limitations of the LFS

The LFS produces coherent labour market information on the basis of internationally standard concepts and definitions. It is a rich source of data on a wide variety of labour market and personal characteristics. It is the most suitable source for making comparisons between countries. The LFS is designed so that households interviewed in each three month period constitute a representative sample of UK households. The survey covers those living in private households and nurses in National Health Service accommodation. Students living in halls of residence have been included since 1992 as information about them is collected at their parents' address.

However the LFS has its limitations. It is a sample survey and is therefore subject to sampling variability. The survey does not include people living in institutions such as hostels or residential homes. 'Proxy' reporting (when members of the household are not present at the interview, another member of the household answers the questions on their behalf) can affect the quality of information on topics such as earnings, hours worked, benefit receipt and qualifications. Around one third of interviews are conducted 'by proxy', usually by a spouse or partner but sometimes by a parent or other near relation.

Sampling Variability

Survey estimates are prone to *sampling variability.* The easiest way to explain this concept is by example. In the September to November 1997 period, ILO unemployment in Great Britain (seasonally adjusted) stood at 1,847,000. If we drew another sample for the same period we could get a different result, perhaps 1,900,000 or 1,820,000.

In theory, we could draw many samples, and each would give a different result. This is because each sample would be made up of different people who would give different answers to the questions. The spread of these results is the sampling variability. Sampling variability is determined by a number of factors including the sample size, the variability of the population from which the sample is drawn and the sample design. Once we know the sampling variability we can calculate a range of values about the sample estimate that represents the expected variation with a given level of assurance. This is called a confidence interval. For a 95% confidence interval we expect that in 95% of the samples (19 times out of 20) the confidence interval will contain the true value that would be obtained by surveying the entire population. For the example given above, we can be 95% confident that the true value was in the range 1,791,000 to 1,903,000.

Unreliable estimates

Very small estimates have relatively wide confidence intervals making them unreliable. For this reason, the ONS does not publish LFS estimates below 10,000.

Non-Response

Non-response can introduce bias to a survey, particularly if the people not responding have characteristics that are different from those who do respond. The LFS has a response rate of around 80 per cent to the first interview, and over 90 per cent of those who are interviewed once go on to complete all five interviews. These are relatively high levels for a household survey. Any bias from non-response is minimised by *weighting* the results.

Weighting (or grossing) converts sample data to represent the full population. In the LFS, the data are weighted separately by age, sex and area of residence to population estimates based on the Census. Weighting also adjusts for people not in the survey and thus minimises non-response bias.

Labour Force Survey Concepts and Definitions

Discouraged workers - a sub-group of the economically inactive population, defined as those neither in employment nor unemployed (on the ILO measure) who said they would like a job and whose main reason for not seeking work was because they believed there were no jobs available.

Economically active - people aged 16 and over who are either in employment or ILO unemployed.

Economic activity rate - the percentage of people aged 16 and over who are economically active.

Economically inactive - people who are neither in employment nor unemployed. This group includes, for example, all those who were looking after a home or retired.

Employment - people aged 16 or over who did at least one hour of paid work in the reference week (whether as an employee or self-employed); those who had a job that they were temporarily away from (on holiday, for example); those on Government-supported training and employment programmes (from spring 1983); and those doing unpaid family work (from spring 1992).

Employees - the division between employees and self-employed is based on survey respondents' own assessment of their employment status.

Full Time - the classification of employees, self-employed and unpaid family workers in their main job as full-time or part-time is on the basis of self-assessment. Up until autumn 1995, people who were on government work-related training programmes are classified as full-time or part-time according to whether their usual hours of work per week were over 30 or 30 and under; from winter 1995/96 onwards, the full-time/part-time classification for this group has been changed to self-assessment, in line with the other groups outlined above. People on Government-supported training and employment programmes who are at college in the survey reference week are classified, by convention, as part-time.

Government-supported training and employment programmes - comprise all people aged 16 and over participating in one of the Government's employment and training programmes (Youth Training, Training for Work and Community Action), together with those on similar programmes administered by Training and Enterprise Councils in England and Wales, or Local Enterprise Companies in Scotland.

Hours worked - respondents to the LFS are asked a series of questions enabling the identification of both their usual hours and their actual hours. Total hours include overtime (paid and unpaid) and exclude lunchbreaks.

ILO unemployment - the International Labour Office (ILO) measure of unemployment used throughout this supplement refers to people without a job who were available to start work in the two weeks following their LFS interview and who had either looked for work in the four weeks prior to interview or were waiting to start a job they had already obtained. This definition of unemployment is in accordance with that adopted by the 13th International Conference of Labour Statisticians, further clarified at the 14th ICLS, and promulgated by the ILO in its publications.

ILO unemployment (rate) - the percentage of economically active people who are unemployed on the ILO measure.

ILO unemployment (duration) - defined as the shorter of the following two periods: (a) duration of active search for work; and (b) length of time since employment.

Part Time - see full-time

Second jobs - jobs which LFS respondents hold in addition to a main full-time or part -time job.

Self-employment - See Employees

Temporary employees - in the LFS these are defined as those employees who say that their main job is non permanent in one of the following ways: fixed period contract; agency temping; casual work; seasonal work; other temporary work.

Unpaid Family Workers - the separate identification from spring 1992 of this group in the LFS is in accordance with international recommendations. The group comprises persons doing unpaid work for a business they own or for a business that a relative owns.

Persons employed in local authorities *(Table 7.8)*

The full-time equivalents for local authorities are derived by applying factors to the numbers of part-time workers in three groups based on average hours worked in each group nationally.

Jobseekers allowance claimant count *(Tables 7.13 and 7.14)*

Jobseekers Allowance claimant count - This is a count of all those people who are claiming unemployment-related benefits at Employment Service local offices and who have declared that they are unemployed, capable of, available for, and actively seeking work during the week in which their claim is made. All people claiming unemployment related benefits on the day of the monthly count are included in the claimant count, irrespective of whether they are actually receiving benefits.

Average earnings index *(Tables 7.21 and 7.22)*

The Average Earnings Index (AEI) is designed to measure changes in the level of earnings i.e. wage inflation in Great Britain. Average earnings are calculated as the total wages and salaries paid by firms, divided by the number of employees paid. Like all indices, changes are measured against a base year, whose index value is set to 100. The current base year is 1995 for Table 7.21 and March 1996 for Table 7.22.

Labour market

The Average Earnings Index was the subject of two reviews at the beginning of 1999. These were, *"Review of Methodology for the Average Earnings Index"* R Chambers and D Holmes, University of Southampton December 1998, and *"Review of the Revisions to the Average Earnings Index"* report submitted by Sir Andrew Turnbull and Mervyn King, The Stationery Office, March 1999. They made a number of recommendations for change in the methodology underpinning the index, and set out a long-term project for development. Work to implement the recommendations is underway and regular updates on progress will be published in the ONS' journal Labour Market Trends.

The AEI is published monthly in the Labour Market Statistics First Release. The main indicator of growth, the headline rate, is based on the annual change in the seasonally adjusted index values for the latest 3 months compared with the same period a year ago. The use of a 3-month average reduces the level of volatility seen in the data on a month-on-month basis.

Strengths of the AEI
The AEI, based on monthly survey data, is a timely indicator of changes in the level of earnings.

Limitations of the AEI
The index is not adjusted for any changes in the composition of the workforce such as changes in the share of full time and part time workers, or in the share of skilled and unskilled workers. Similarly, the index does not account for changes in the number of hours worked, or any temporary factors that affect earnings.

The sample of the Monthly Wages and Salaries Survey on which the AEI is based is not designed to provide information on the level of earnings. The sample is not completely representative of the economy as firms with fewer than 20 employees are excluded, as are the earnings of self employed persons.

The AEI only covers earnings in Great Britain as earnings information is not collected for Northern Ireland and regional data are not available.

Vacancies at jobcentres *(Table 7.27)*

Vacancy - This is a job opportunity notified by an employer to a Jobcentre.

Unfilled vacancy - (also known as "Stock of vacancies") This is the number of vacancies which have not been filled or cancelled on the count date.

Inflow of vacancy - (also known as "Notified vacancies") This is the number of job opportunities notified by employers to Jobcentres in the period between two successive count dates.

Outflow of vacancy - This is a derived statistic which represents the total of vacancies filled plus cancelled between count dates. This concept can also be expressed as "vacancy stock" at the beginning of the period plus notified vacancies (inflows) minus "vacancy stock" at the end of the period.

Placings - This is the number of Jobseekers placed into employment by individual Jobcentres.

Earnings *(Table 7.19 - 7.25)*

The total gross remuneration employees receive before any statutory deductions (tax, national insurance). Income in kind and pension funds are excluded.

7.1 Summary for United Kingdom labour force

Thousands, Spring each year, not seasonally adjusted

	All aged 16+[1]	All aged 16 & over				Economic activity rate 16-59/64 (%)[2]	Employment rate - all aged 16+ (%)[3]	Employment rate 16-59/64 (%)[5]	ILO unemployment rate (%)[4]
		Total economically active	Total in employment	ILO unemployed	Economically inactive				
All									
	BEAJ	BEAM	BEAP	BEAS	BEAV	BEAY	BEBK	BEBN	BEBQ
1989	44 978	28 764	26 689	2 075	16 214	80.0	59.3	74.2	7.2
1990	45 107	28 909	26 935	1 974	16 198	80.2	59.7	74.7	6.8
1991	45 226	28 813	26 400	2 414	16 413	79.8	58.4	73.0	8.4
1992	45 310	28 581	25 812	2 769	16 728	78.8	57.0	71.1	9.7
1993	45 400	28 447	25 511	2 936	16 954	78.4	56.2	70.2	10.3
1994	45 465	28 433	25 697	2 736	17 033	78.2	56.5	70.6	9.6
1995	45 574	28 426	25 973	2 454	17 148	78.0	57.0	71.1	8.6
1996	45 725	28 552	26 219	2 334	17 172	78.1	57.3	71.6	8.2
1997	45 898	28 716	26 682	2 034	17 182	78.2	58.1	72.5	7.1
1998	46 056	28 713	26 947	1 766	17 343	78.0	58.5	73.1	6.1
1999	46 212	28 992	27 251	1 741	17 220	78.4	59.0	73.6	6.0
Men									
	BEAK	BEAN	BEAQ	BEAT	BEAW	BEAZ	BEBL	BEBO	BEBR
1989	21 706	16 434	15 219	1 215	5 272	88.3	70.1	81.8	7.4
1990	21 801	16 483	15 318	1 165	5 318	88.3	70.3	82.1	7.1
1991	21 871	16 401	14 887	1 514	5 470	87.7	68.1	79.6	9.2
1992	21 924	16 187	14 321	1 865	5 737	86.3	65.3	76.3	11.5
1993	21 985	16 021	14 035	1 986	5 964	85.6	63.8	74.8	12.4
1994	22 050	15 996	14 171	1 825	6 053	85.2	64.3	75.4	11.4
1995	22 132	15 981	14 374	1 607	6 151	84.7	64.9	76.1	10.1
1996	22 232	15 992	14 446	1 546	6 240	84.6	65.0	76.3	9.7
1997	22 341	16 023	14 720	1 304	6 317	84.4	65.9	77.4	8.1
1998	22 441	15 997	14 906	1 091	6 444	83.9	66.4	78.1	6.8
1999	22 542	16 120	15 031	1 088	6 422	84.1	66.7	78.4	6.8
Women									
	BEAL	BEAO	BEAR	BEAU	BEAX	BEBJ	BEBM	BEBP	BEBS
1989	23 272	12 330	11 470	860	10 942	70.9	49.3	65.9	7.0
1990	23 307	12 427	11 617	809	10 880	71.3	49.8	66.6	6.5
1991	23 354	12 412	11 512	900	10 942	71.0	49.3	65.8	7.2
1992	23 386	12 395	11 491	904	10 991	70.6	49.1	65.4	7.3
1993	23 415	12 426	11 476	949	10 989	70.6	49.0	65.1	7.6
1994	23 416	12 436	11 526	910	10 979	70.6	49.2	65.3	7.3
1995	23 442	12 445	11 599	846	10 997	70.6	49.5	65.6	6.8
1996	23 493	12 561	11 773	788	10 932	71.1	50.1	66.5	6.3
1997	23 557	12 692	11 962	731	10 865	71.4	50.8	67.2	5.8
1998	23 614	12 716	12 042	674	10 898	71.5	51.0	67.6	5.3
1999	23 671	12 872	12 219	653	10 798	72.1	51.6	68.3	5.1

See chapter text for definitions.
1 Population in private households, and from 1992, student halls of residence.
2 Total economically active as a percentage of all persons of working age.
3 Total employed as a percentage of all persons 16 and over.
4 Total ILO unemployment as a percentage of all economically active.
5 Total employed as a percentage of all persons of working age (men 16-64, women 16-59).

Source: ONS, Labour Force Survey

7.2 Employment status, full-time/part-time, second jobs, temporary employees
United Kingdom

Thousands, Spring each year, not seasonally adjusted

	All in employment[1,5]					Total employment[5]		Employees[5]		Self-employed[5]			
	Total[3]	Employees	Self employed	Unpaid family workers[4]	Government[2] supported training & employment programmes	Full-time	Part-time	Full-time	Part-time	Full-time	Part-time	Workers with second jobs[6]	Temporary employees
All													
	BEAP	BEBT	BEBW	BEBZ	BECD	BECG	BECJ	BECM	BECP	BECS	BECV	BECY	BEDE
1989	26 689	22 656	3 528	–	498	20 840	5 810	17 588	5 065	2 973	554	1 076	1 216
1990	26 935	22 887	3 572	–	471	21 030	5 889	17 773	5 112	3 008	563	1 101	1 195
1991	26 400	22 531	3 416	–	437	20 499	5 885	17 367	5 163	2 896	520	1 106	1 199
1992	25 812	22 018	3 227	181	386	19 753	6 052	16 835	5 180	2 683	544	988	1 220
1993	25 511	21 812	3 184	151	364	19 377	6 128	16 569	5 241	2 605	578	1 057	1 274
1994	25 697	21 907	3 301	146	343	19 406	6 282	16 527	5 374	2 692	608	1 164	1 410
1995	25 973	22 187	3 355	140	291	19 642	6 327	16 734	5 451	2 729	626	1 307	1 547
1996	26 219	22 553	3 286	127	254	19 665	6 551	16 853	5 698	2 642	644	1 307	1 586
1997	26 682	23 003	3 335	118	226	19 981	6 697	17 173	5 829	2 646	687	1 267	1 706
1998	26 947	23 408	3 257	101	181	20 201	6 741	17 529	5 875	2 541	714	1 202	1 661
1999	27 251	23 810	3 176	100	165	20 418	6 827	17 819	5 987	2 488	686	1 295	1 628
Men													
	BEAQ	BEBU	BEBX	BECA	BECE	BECH	BECK	BECN	BECQ	BECT	BECW	BECZ	BEDF
1989	15 219	12 204	2 695	–	315	14 347	846	11 650	553	2 516	178	485	449
1990	15 318	12 298	2 715	–	302	14 387	920	11 693	603	2 529	185	527	452
1991	14 887	12 016	2 599	–	264	13 958	919	11 374	640	2 437	161	519	466
1992	14 321	11 577	2 438	55	251	13 307	1 009	10 905	670	2 260	179	451	517
1993	14 035	11 370	2 384	43	237	12 993	1 039	10 680	690	2 184	200	477	563
1994	14 171	11 417	2 480	49	224	13 052	1 116	10 668	748	2 267	213	515	622
1995	14 374	11 599	2 544	43	188	13 201	1 171	10 782	817	2 313	231	549	720
1996	14 446	11 784	2 461	41	160	13 199	1 246	10 879	904	2 225	236	552	709
1997	14 720	12 069	2 474	37	140	13 389	1 329	11 068	1 000	2 220	252	557	785
1998	14 906	12 368	2 393	28	118	13 563	1 341	11 360	1 006	2 122	270	528	745
1999	15 031	12 531	2 356	35	110	13 645	1 382	11 475	1 053	2 098	256	550	777
Women													
	BEAR	BEBV	BEBY	BECB	BECF	BECI	BECL	BECO	BECR	BECU	BECX	BEDD	BEDG
1989	11 470	10 452	833	–	183	6 493	4 964	5 938	4 512	456	376	591	767
1990	11 617	10 589	857	–	169	6 643	4 968	6 079	4 509	479	379	574	742
1991	11 512	10 515	817	–	174	6 541	4 966	5 993	4 522	459	358	586	733
1992	11 491	10 441	789	126	135	6 445	5 042	5 930	4 510	423	365	537	702
1993	11 476	10 441	800	108	127	6 384	5 089	5 889	4 551	422	378	580	711
1994	11 526	10 490	820	97	119	6 355	5 166	5 859	4 626	425	395	649	788
1995	11 599	10 588	811	97	103	6 441	5 156	5 953	4 634	416	395	758	827
1996	11 773	10 769	824	85	95	6 465	5 306	5 973	4 795	417	408	754	878
1997	11 962	10 935	861	80	86	6 592	5 368	6 105	4 829	426	435	710	921
1998	12 042	11 040	864	74	63	6 638	5 401	6 169	4 869	419	445	673	916
1999	12 219	11 280	820	65	55	6 773	5 444	6 344	4 934	390	430	745	850

1 See chapter text for definitions.
2 Those on employment and training programmes are classified as in employment. Some of those on programmes may consider themselves to be employees or self employed so appear in other categories.
3 Includes those who did not state whether they were employees or self employed in 1989-91.
4 Unpaid family workers have been classified as in employment since Spring 1992. Some may have reported themselves as employees or self employed in 1988-91.
5 People whose main job is full or part time.
6 Second jobs reported in LFS in addition to person's main full or part time job.

Source: ONS, Labour Force Survey

7.3 Employment
United Kingdom

Age group by numbers and rates

Numbers in thousands, rates in percentage, Spring each year, nsa

	All aged 16 & over	16-59/64	16-17	18-24	16-24	25-34	35-49	50-64 (m) 50-59 (f)	65+ (m) 60+ (f)
	Numbers								
All									
	BEAP	BEDH	BEDK	BEDN	BEDQ	BEDT	BEDW	BEDZ	BEEC
1990	26 935	26 149	819	4 577	5 396	6 782	9 156	4 815	787
1991	26 400	25 622	750	4 184	4 935	6 781	9 186	4 720	777
1992	25 812	24 996	648	3 814	4 462	6 720	9 176	4 638	816
1993	25 511	24 738	551	3 579	4 130	6 809	9 218	4 581	773
1994	25 697	24 915	560	3 435	3 995	6 913	9 320	4 686	782
1995	25 973	25 178	584	3 332	3 916	7 002	9 462	4 798	795
1996	26 219	25 450	635	3 282	3 917	7 013	9 624	4 896	769
1997	26 682	25 880	674	3 232	3 906	7 145	9 688	5 141	802
1998	26 947	26 175	668	3 204	3 871	7 099	9 821	5 383	773
1999	27 251	26 437	650	3 207	3 857	7 024	9 986	5 570	814
Men									
	BEAQ	BEDI	BEDL	BEDO	BEDR	BEDU	BEDX	BEEA	BEED
1990	15 318	15 027	427	2 488	2 915	3 966	5 097	3 048	291
1991	14 887	14 603	388	2 227	2 616	3 940	5 074	2 974	284
1992	14 321	14 021	332	1 998	2 329	3 846	4 980	2 866	300
1993	14 035	13 780	276	1 879	2 155	3 860	4 973	2 792	255
1994	14 171	13 907	286	1 823	2 109	3 924	5 038	2 836	264
1995	14 374	14 086	295	1 780	2 074	3 979	5 142	2 891	288
1996	14 446	14 181	323	1 738	2 061	3 970	5 189	2 961	265
1997	14 720	14 451	331	1 736	2 068	4 027	5 240	3 116	269
1998	14 906	14 633	334	1 724	2 058	4 019	5 323	3 233	273
1999	15 031	14 745	322	1 724	2 046	3 949	5 408	3 341	287
Women									
	BEAR	BEDJ	BEDM	BEDP	BEDS	BEDV	BEDY	BEEB	BEEE
1990	11 617	11 122	392	2 089	2 481	2 816	4 059	1 767	496
1991	11 512	11 020	362	1 957	2 319	2 842	4 112	1 747	493
1992	11 491	10 975	316	1 817	2 133	2 874	4 197	1 771	515
1993	11 476	10 958	275	1 701	1 976	2 949	4 244	1 790	518
1994	11 526	11 008	275	1 612	1 886	2 989	4 282	1 851	518
1995	11 599	11 091	289	1 552	1 841	3 023	4 320	1 907	507
1996	11 773	11 269	312	1 543	1 856	3 042	4 435	1 936	504
1997	11 962	11 429	343	1 495	1 838	3 118	4 448	2 025	533
1998	12 042	11 542	334	1 479	1 813	3 080	4 498	2 151	500
1999	12 219	11 693	328	1 483	1 811	3 074	4 578	2 229	527
	Rates[1]								
All									
	BEBK	BEBN	BEEF	BEEI	BEEL	BEEO	BEER	BEEU	BEEX
1990	59.7	74.7	55.0	74.2	70.4	77.3	81.7	65.3	7.8
1991	58.4	73.0	52.8	69.6	66.4	75.7	80.7	64.5	7.7
1992	57.0	71.1	46.9	64.9	61.5	74.0	79.7	63.3	8.0
1993	56.2	70.2	41.6	63.0	58.9	74.1	79.1	62.0	7.6
1994	56.5	70.6	43.0	62.7	58.9	74.7	79.2	62.5	7.7
1995	57.0	71.1	43.3	63.1	59.1	75.5	79.5	63.1	7.8
1996	57.3	71.6	44.5	64.7	60.3	75.8	79.8	63.5	7.5
1997	58.1	72.5	46.0	65.4	61.0	77.8	80.0	64.5	7.8
1998	58.5	73.1	45.8	65.5	61.0	78.4	80.7	65.4	7.5
1999	59.0	73.6	44.9	65.5	60.8	79.4	81.1	66.1	7.9
Men									
	BEBL	BEBO	BEEG	BEEJ	BEEM	BEEP	BEES	BEEV	BEEY
1990	70.3	82.1	55.7	79.0	74.4	89.4	91.0	70.0	8.3
1991	68.1	79.6	53.2	72.5	68.8	86.9	89.2	68.7	8.1
1992	65.3	76.3	46.9	66.6	62.8	83.6	86.6	66.2	8.5
1993	63.8	74.8	40.5	64.8	60.2	83.0	85.4	64.1	7.1
1994	64.3	75.4	42.8	65.0	60.7	83.6	85.6	64.4	7.4
1995	64.9	76.1	42.6	65.9	61.1	84.5	86.3	64.9	7.9
1996	65.0	76.3	44.1	66.9	61.9	84.4	85.9	65.7	7.2
1997	65.9	77.4	44.1	68.6	63.0	86.2	86.3	67.2	7.3
1998	66.4	78.1	44.7	68.8	63.3	87.3	87.1	67.8	7.4
1999	66.7	78.4	43.4	68.8	63.0	87.7	87.4	68.5	7.7
Women									
	BEBM	BEBP	BEEH	BEEK	BEEN	BEEQ	BEET	BEEW	BEEZ
1990	49.8	66.6	54.1	69.2	66.3	64.9	72.4	58.5	7.5
1991	49.3	65.8	52.4	66.5	63.8	64.2	72.2	58.3	7.5
1992	49.1	65.4	47.0	63.1	60.1	64.1	72.9	59.0	7.8
1993	49.0	65.1	42.6	61.1	57.6	65.0	72.8	58.9	7.9
1994	49.2	65.3	43.3	60.2	57.0	65.6	72.8	59.7	7.9
1995	49.5	65.6	44.0	60.3	57.0	66.3	72.6	60.4	7.7
1996	50.1	66.5	45.0	62.4	58.6	66.8	73.6	60.3	7.7
1997	50.8	67.2	47.9	62.1	58.8	69.1	73.7	60.7	8.1
1998	51.0	67.6	46.9	62.0	58.5	69.3	74.1	62.2	7.6
1999	51.6	68.3	46.5	62.0	58.5	70.9	74.7	62.8	8.0

1 Total in employment as a percentage of all persons in the relevant group.

Source: ONS, Labour Force Survey

7.4 Distribution of the workforce

At mid-June each year

Thousands, seasonally adjusted

		1989	1990	1991[7]	1992	1993	1994	1995	1996	1997	1998	1999
United Kingdom												
Claimant unemployed[1]	KAMO	1 786	1 615	2 301	2 734	2 919	2 644	2 308	2 146	1 598	1 361	1 269
Males	KAMP	1 276	1 191	1 744	2 095	2 242	2 024	1 761	1 630	1 223	1 038	972
Females	KAMQ	510	424	557	639	677	620	547	516	375	323	297
Workforce jobs[2]	KAMR	27 294†	27 567	26 670	25 973	25 629	25 776	26 063	26 728	27 239	27 508	27 747
Males	KAMS	15 489†	15 551	14 866	14 281	13 940	14 000	14 163	14 378	14 735	14 847	15 006
Females	KAMT	11 805†	12 015	11 804	11 692	11 689	11 775	11 901	12 350	12 504	12 661	12 742
HM Forces[3]	KAMU	308	303	297	290	271	250	230	221	210	210	208
Males	KAMV	291	286	278	270	252	232	214	206	195	194	192
Females	KAMW	16	18	19	20	19	18	16	16	15	16	16
Self-employment jobs[4]	KAMX	3 866†	3 918	3 756	3 441	3 446	3 548	3 608	3 609	3 598	3 475	3 497
Males	KAMZ	2 908†	2 949	2 828	2 573	2 537	2 630	2 680	2 671	2 617	2 519	2 546
Females	KANA	958†	969	928	868	910	918	928	937	981	956	951
Employees jobs[5]	KANB	22 658†	22 907	22 248	21 902	21 586	21 661	21 985	22 702	23 257	23 699	23 929
Males	KANC	11 999†	12 049	11 536	11 225	10 949	10 939	11 113	11 380	11 815	12 057	12 199
Females	KAND	10 659	10 858	10 712†	10 677	10 636	10 722	10 872	11 322	11 442	11 642	11 731
of whom												
Total, production and construction industries	KANF	6 408	6 285	5 756	5 395	5 082	5 060	5 108	5 227†	5 369	5 464	5 306
Total, all manufacturing industries	KANG	4 851	4 733	4 319	4 096	3 913	3 928	4 026	4 110†	4 166	4 144	3 991
Government-supported trainee[6]	KANH	462	438†	369	340	326	317	240	195	173	124	112
Males	KANI	291	268†	225	213	203	199	156	121	107	77	68
Females	KANJ	171	171†	144	128	124	118	84	74	66	47	43
Great Britain												
Claimant unemployed[1]	KANN	1 682	1 520	2 202	2 629	2 815	2 545	2 221	2 059	1 535	1 303	1 217
Males	KANO	1 199	1 119	1 668	2 014	2 162	1 948	1 693	1 563	1 173	992	932
Females	KANP	483	401	534	615	654	598	528	496	362	310	286
Workforce jobs[2]	KANQ	26 670†	26 923	26 012	25 324	24 983	25 116	25 380	26 048	26 537	26 799	27 034
Males	KANR	15 136†	15 188	14 497	13 922	13 583	13 637	13 790	14 011	14 357	14 464	14 620
Females	KANS	11 533†	11 734	11 515	11 402	11 400	11 479	11 591	12 037	12 180	12 335	12 413
HM Forces[3]	KANT	308	303	297	290	271	250	230	221	210	210	208
Males	KANU	291	286	278	270	252	232	214	206	195	194	192
Females	KANV	16	18	19	20	19	18	16	16	15	16	16
Self-employment jobs[4]	KANW	3 777†	3 824	3 660	3 354	3 362	3 463	3 515	3 524	3 508	3 389	3 411
Males	KANX	2 834†	2 871	2 748	2 501	2 464	2 556	2 603	2 600	2 545	2 447	2 474
Females	KANY	943†	953	912	853	898	907	913	924	963	942	937
Employee jobs[5]	KANZ	22 133	22 370	21 707	21 359	21 039	21 103	21 410	22 123†	22 660	23 091	23 314
Males	KAOA	11 726	11 773	11 260	10 951	10 675	10 660	10 827	11 095†	11 519	11 756	11 894
Females	KAOB	10 406	10 597	10 447	10 408	10 365	10 443	10 583	11 029†	11 141	11 335	11 420
of whom												
Total, production and construction industries	KAOC	6 267	6 142	5 616	5 260	4 950	4 928	4 973	4 094†	5 228	5 322	5 165
Total, all manufacturing industries	KAOD	4 747	4 628	4 215	3 995	3 814	3 827	3 922	4 006†	4 058	4 036	3 885
Government-supported trainees[6]	KAOE	452	426†	348	322	310	301	224	179	159	110	100
Males	KAOF	285	259†	211	201	192	189	145	111	98	68	60
Females	KAOG	167	166†	137	121	118	112	79	68	61	42	39

Note: Because the figures have been rounded independently totals may differ from the sum of the components. Also the totals may include some employees whose industrial classification could not be ascertained.

1 Claimant unemployment: those people who were claiming unemployment-related benefits (Unemployment Benefit, Income Support or National Insurance credits) at Employment service local offices on the day of the monthly count. The seasonally adjusted claimant unemployment series allows for all relevant changes which, unless adjusted for, would distort comparisons over time.

2 The workforce jobs (formerly workforce in employment) comprises employee jobs, self- employment jobs (from the LFS), HM Forces and government supported trainees.

3 HM Forces figures, provided by the Ministry of Defence, represent the total number of UK service personnel, male and female, in HM Regular Forces, wherever serving and including those on release leave.

4 Estimates of self-employed jobs are based on the results of the Labour Force Survey. The Northern Ireland estimates are not seasonally adjusted.

5 Estimates of employee jobs from December 1987 to August 1989 include an allowance based on the Labour Force Survey to compensate for persistent undercounting in the regular sample enquiries (*Employment Gazette*, October 1989 p56). For all dates, individuals with two jobs as employees of different employers are counted twice.

6 Includes all participants on government training and employment programmes who are receiving some work experience on their placement but who do not have a contract of employment (those with a contract are included in the employee jobs series). The numbers are not subject to seasonal adjustment.

7 Estimates from September 1989 are based on data provided by the New Panel of employers (see article on p191 of the April 1992 edition of *Employment Gazette*).

Source: Earnings and Employment Division, ONS: 01928 792566

7.5 Employee jobs
Analysis by industry based on the Standard Industrial Classification 1992
At June in each year

Thousands, not seasonally adjusted

		SIC 1992	United Kingdom 1994	1995	1996	1997	1998	1999		Great Britain 1994	1995	1996	1997	1998	1999
All sections	KAOH	A - Q	21 698†	22 025	22 706	23 253	23 687	23 913	KAPN	21 141	21 452	21 128†	22 657	23 080	23 299
Index of production and construction industries	KAOI	C - F	5 049	5 097	5 216†	5 357	5 451	5 292	KAPO	4 917	4 963	5 083†	5 216	5 310	5 151
Index of production industries	KAOJ	C - E	4 185	4 259	4 334†	4 390	4 357	4 199	KAPP	4 078	4 149	4 224†	4 276	4 244	4 088
of which, manufacturing industries	KAOK	D	*3 923*	*4 021*	*4 106†*	*4 162*	*4 140*	*3 984*	KAPQ	*3 823*	*3 918*	*4 002†*	*4 055*	*4 033*	*3 878*
Service industries	KAOL	G - Q	16 352	16 658	17 213†	17 604	17 938	18 304	KAPR	15 944	16 236	16 785†	17 166	17 488	17 847
Agriculture, hunting and forestry and fishing	KAOM	A/B	296†	270	277	292	298	317	KAPS	280	253	261†	275	282	301
Agriculture hunting and forestry	KPHI	A	290†	264	271	285	291	310	KOVW	274	247	255†	269	275	294
Agriculture hunting & related activities	KPHJ	01	281†	255	259	274	279	299	KOVX	265	238	243	258†	264	283
Fishing	KPHK	B	7	6	6	7†	7	7	KOVY	6	5	6†	6	6	6
Mining and quarrying	KPHL	C	71	67	78†	79	76	72	KOVZ	69	65	76†	77	74	70
Mining and quarrying of energy producing materials	KPHM	CA	42	37	43†	46	44	41	KOWA	42	37	43†	46	43	40
Mining	KAPG	10/12	..	..	..	..	..	..	KOWB	16	11	15†	16	15	15
Extraction of crude petroleum	KPHN	11	..	..	..	..	..	..	KOWC	26	26	28	30†	29	26
Mining and quarrying except of energy producing materials	KPHO	CB(13/14)	28	30	34	33†	32	31	KOWD	27	28	33	31†	30	29
Energy and water supply industries	KAOO	C/E	262	237	228†	227	217	215	KOWE	255	231	222†	221	211	209
Manufacturing	KPHP	D	3 923	4 021	4 106†	4 162	4 140	3 984	KOWF	3 823	3 918	4 002†	4 055	4 033	3 878
Manufacture of food products Beverages and tobacco	KPHQ	DA	447	445	445†	471	484	475	KOWG	427	426	426†	451	464	456
Of food	KPHR	151 to 158	..	..	..	..	..	..	KOWH	362	368	372†	395	405	398
Of beverages and tobacco	KPHS	159/16	..	..	..	..	..	..	KOWI	65	57	54†	56	59	58
Manufacture of textiles and textile products	KPHT	DB	354	343	337†	335	314	275	KOWJ	329	319	313†	311	291	256
Of textiles	KPHU	17	193	186	184	184†	172	157	KOWK	181	175	174†	173	162	148
Of made-up textile articles except apparel	KPHV	174	..	..	..	..	..	..	KOWL	37	33	34†	33	33	31
Of textiles excluding made-up textile	KPHW	Rest of 17	..	..	..	..	..	..	KOWM	145	142	140	139†	129	118
Of wearing apparel,dressing and dyeing of fur	KPHX	18	161	157	153†	151	141	118	KOWN	147	143	139†	138	129	108
Manufacture of leather and leather products including footwear	KPHY	DC	44	39	39	34†	30	27	KOWO	43	38	38	34†	29	26
Of leather and leather goods	KPHZ	191/192	..	..	..	..	..	..	KOWP	16	13	14	12†	11	10
Of footwear	KPIA	193	..	..	..	..	..	..	KOWQ	28	25	24	22†	18	16
Manufacture of wood and wood products	KPIB	DD(20)	91	82	84†	87	85	84	KOWR	88	79	81†	83	82	81
Manufacture of pulp paper and paper products,publishing and printing	KPIC	DE	458	465	474†	473	478	467	KOWS	453	459	467†	467	471	460
Of pulp paper and paper products	KPID	21	125	124	123	117†	114	107	KOWT	123	122	121†	115	112	104
Publishing printing and reproduction of recorded media	KPIE	22	334	341	350†	356	364	360	KOWU	330	337	337	352†	360	356
Manufacture of coke refined petroleum products and nuclear fuel	KPIF	DF(23)	35	30	32†	32	27	26	KOWV	35	30	32†	32	27	26
Of refined petroleum products		232							KOWW	19	17	18†	19	15	15
Manufacture of chemicals,chemical products and man-made fibres	KPIG	DG(24)	246	255	254†	254	253	249	KOWX	243	251	250†	250	249	244
Manufacture of rubber and plastics	KPIH	DH(25)	203	224	230†	241	242	235	KOWY	197	218	224†	235	235	228
Manufacture of other non-metallic mineral products	KPII	DI(26)	154	151	146†	149	147	141	KOWZ	150	146	141†	145	143	136
Manufacture of basic metals and fabricated metal products	KPIJ	DJ	546	550	569†	566	551	529	KOXA	541	545	564†	561	545	523
Of basic metals	KPIK	27	128	132	135	131†	127	118	KOXB	128	131	134	131†	126	117
except machinery	KPIL	28	418	419	435†	435	425	412	KOXC	413	414	430†	430	419	406

Sources: Department of Manpower Services (Northern Ireland); Earnings and Employment Division, ONS: 01928 792566

7.5 Employee jobs

continued

Analysis by industry based on the Standard Industrial Classification 1992

At June in each year

Thousands, not seasonally adjusted

		SIC 1992	United Kingdom 1994	1995	1996	1997	1998	1999		Great Britain 1994	1995	1996	1997	1998	1999
Manufacture of Machinery & Eqpt.Nec	KPIM	DK(29)	383	397	402†	400	395	377	KOXD	377	391	395†	394	389	370
Manufacture of electrical and optical equipment	KPIN	DL	449	488	510†	520	532	509	KOXE	442	479	501	509†	521	498
Of office machinery and computers	KPIO	30	45	52	49†	47	49	47	KOXF	45	51	49	46	47	45
Of electrical machinery and apparatus	KPIP	31	153	166	178†	181	183	170	KOXG	150	163	175†	178	180	168
Of electric motors etc control apparatus and insulated cable	KPIQ	311 to 313	..	..	..	..	..	..	KOXH	88	97	99	101†	105	99
Of accumulators, primary cells, batteries, lamps & electrical eqpt.	KPIR	314 to 316	..	..	..	..	..	..	KOXI	62	66	76†	76	76	69
Radio television & communication eqpt	KPIS	32	113	124	130†	132	132	127	KOXJ	109	119	126	126†	127	122
Of electronic components	KPIT	321	..	..	..	..	..	..	KOXK	50	53	53	51†	48	45
Of radio TV and telephone apparatus, sound and video recorders	KPIU	322/323	..	..	..	..	..	..	KOXL	59	66	73	75†	79	77
Of medical precision & optical eqpt, watches	KPIV	33	139	146	153†	160	167	165	KOXM	138	145	152	159†	166	163
Manufacture of transport equipment	KPIW	DM	339	360	388†	389	396	381	KOXN	328	349	377†	378	384	368
Of motor vehicles and trailers	KPIX	34	173	204	227†	228	232	219	KOXO	170	200	223†	224	228	215
Of other transport equipment	KPIY	35	166	157	162†	162	165	162	KOXP	158	149	154†	154	156	153
Manufacturing NEC Of furniture	KPIZ	DN(36/37)	174	192	195†	210	206	207	KOXQ	171	188	192†	206	203	204
Electricity gas and water supply	KPJA	361	192	170	150†	148	141	143	KOXR	98	111	110†	121	122	121
Electricity gas steam and hot		E							KOXS	186	165	146†	144	137	139
water supply	KPJB	40	..	..	..	..	..	..	KOXT	141	122	102	104†	99	101
Collection purification and distribution of water	KPJC	41	..	..	..	..	..	..	KOXU	45	43	44†	40	38	38
Construction	KPJD	F(45)	864	838	882†	967	1 095	1 092	KOXW	840	814	859†	940	1 066	1 063
Services	KPJE	G - Q	16 352	16 658	17 213†	17 604	17 938	18 304	KOXX	15 944	16 236	16 785†	17 166	17 488	17 847
Wholesale and retail trade; Repair of motor vehicles, motorcycles and personal household goods	KPJF	G (50 - 52)	3 653	3 707	3 816†	3 933	4 025	4 086	KOXY	3 570	3 619	3 728†	3 840	3 928	3 989
Sale maintenance and repair of motor vehicles,retail of automotive fuel	KPJG	50	515	511	563†	557	553	555	KOXZ	504	498	551†	545	539	542
Sale of motor vehicles, motorcycles and parts,motorcycle repair and sale of automotive fuel	KPJH	501/503 - 505	..	..	..	..	..	..	KOYA	328	327	376†	381	374	370
Maintenance and repair of motor vehicles	KPJI	502	..	..	..	..	..	..	KOYB	176	171	175	164†	166	172
Wholesale trade and commission trade except motor vehicles	KPJJ	51	917	956	1 009†	1 080	1 104	1 135	KOYC	898	936	989†	1 059	1 083	1 114
Wholesale on a fee of contract basis	KPJK	511	..	..	..	..	..	..	KOYD	33	38	43†	43	48	58
Wholesale agricultural raw materials and live animals	KPJL	512	..	..	..	..	..	..	KPLD	23	22	21†	22	25	24
Wholesale food beverages & tobacco	KPJM	513	..	..	..	..	..	..	KPLE	185	180	192†	195	193	187
Wholesale household goods	KPJN	514	..	..	..	..	..	..	KPLF	194	200	226†	245	246	251
Wholesale of non-agricultural intermediate products waste & scrap	KPJO	515	..	..	..	..	..	..	KPLG	222	234	232†	231	234	244
Wholesale machinery eqpt. & supplies	KPJP	516	..	..	..	..	..	..	KPLH	182	189	201†	230	244	247
Other wholesale	KPJQ	517	..	..	..	..	..	..	KPLI	60	73	74†	91	93	102
Retail trade except of motor vehicles and motorcycles;repair of personal and household goods	KPJR	52	2 221	2 240	2 244†	2 296	2 368	2 396	KPLJ	2 169	2 185	2 188†	2 237	2 306	2 333
Non-specialised stores selling mainly food beverages & tobacco	KPJS	5211/5221-4,5227	..	..	..	..	..	..	KPLK	813	820	836†	840	884	901
Other non-specialised stores second hand shops & sales not in stores	KPJT	5212/525-526	..	..	..	..	..	..	KPLL	298	316	312†	338	348	359

Sources: Department of Manpower Services (Northern Ireland); Earnings and Employment Division, ONS: 01928 792566

7.5 Employee jobs

Analysis by industry based on the Standard Industrial Classification 1992

continued

At June in each year

Thousands, not seasonally adjusted

		SIC 1992	United Kingdom 1994	1995	1996	1997	1998	1999		Great Britain 1994	1995	1996	1997	1998	1999
Alcoholic & other beverages, tobacco	KPJU	5225 to 5226	..	..	..	..	..	..	KPLM	34	25	43†	56	67	60
Pharmaceutical & medical goods cosmetics & toilet articles	KPJV	523	..	..	..	..	..	..	KPLN	134	129	129	121†	136	143
Clothing footwear & leather goods	KPJW	5242/5243	..	..	..	..	..	..	KPLO	224	235	225†	249	244	237
Textile furniture lighting eqpt. electrical household appliances radio and TV paints glass hardware and household goods nec.	KPJX	5241/5244-46	..	..	..	..	..	..	KPLP	273	277	265†	267	263	260
Books newspapers & stationary, other retail in specialised stores	KPJY	5247/5248	..	..	..	..	..	..	KPLQ	374	365	361†	348	348	354
Repair of personal & household goods	KPJZ	527	..	..	..	..	..	..	KPLR	18	17	16†	17	16	18
Hotels and restaurants	KPKA	H	1 198	1 262	1 274†	1 299	1 309	1 326	KPLS	1 173	1 235	1 245†	1 270	1 276	1 293
Hotels camp sites short-stay accom.	KPKB	551/552	..	..	..	..	..	..	KPLT	317	331	313†	301	290	306
Restaurants	KPKC	553	..	..	..	..	..	..	KPLU	311	333	337†	343	358	358
Bars	KPKD	554	..	..	..	..	..	..	KPLV	362	393	396†	429	429	429
Canteens and catering	KPKE	555	..	..	..	..	..	..	KPLW	183	178	200	197†	200	200
Transport storage & communication	KPKF	I	1 314	1 307	1 325†	1 346	1 399	1 462	KPLX	1 293	1 285	1 302†	1 322	1 375	1 437
Land transport,transport via pipelines	KPKG	60	488	472	470†	466	482	498	KPLY	479	463	460†	455	472	487
Transport via railways	KPKH	601	..	..	..	..	..	..	KPLZ	126	105	75	43†	34	32
Other land transport & via pipelines	KPKI	602/603	..	..	..	..	..	..	KPMA	353	358	385†	412	438	455
Water transport	KPKJ	61	23	27	25†	22	20	23	KPMB	23	26	24†	21	20	22
Air transport	KPKK	62	57	56	61†	71	78	83	KPMC	56	55	60	71†	77	82
Supporting and auxiliary transport activities,activities of travel agents	KPKL	63	316	321	324	340†	358	374	KPMD	312	317	320	336†	353	370
Travel agencies and tour operators	KPKM	633	..	..	..	..	..	..	KPME	75	82	88†	97	105	114
Post and telecommunications	KPKN	64	431	431	445†	447	461	485	KPMF	423	424	438†	439	453	477
National post & courier activities	KPKO	641	..	..	..	..	..	..	KPMG	236	239	256†	249	261	269
National post activities	KPKP	6411	..	..	..	..	..	..	KPMH	195	189	203	212†	213	220
Courier activities	KPKQ	6412	..	..	..	..	..	..	KPMI	41	50	53†	37	49	49
Telecommunications	KPKR	6420	..	..	..	..	..	..	KPMJ	187	185	182†	190	191	207
Financial Intermediation	KPKS	J	980	998	972†	996	1 028	1 036	KPMK	966	984	959†	982	1 014	1 021
Financial intermediation except insurance and pension funding	KPKT	65	578	601	572†	575	588	598	KPML	569	592	564†	566	579	588
Insurance and pension funding except compulsory social security	KPKU	66	223	216	216†	231	233	233	KPMM	220	213	214†	229	231	230
Activities auxiliary to financial intermediation	KPKV	67	180	181	184†	190	207	205	KPMN	178	179	181	187†	204	203
Except insurance and pension funding	KPKW	671	..	..	..	..	..	..	KPMO	38	44	47	63†	67	68
Auxiliary to insurance and pension funding	KPKX	672	..	..	..	..	..	..	KPMP	139	134	135	125†	138	135
Real estate renting & business activities	KPKY	K	2 484	2 635	2 977†	3 174	3 301	3 415	KPMQ	2 455	2 604	2 944†	3 138	3 262	3 373
Real estate activities	KPKZ	70	253	263	271†	288	282	293	KPMR	251	260	269†	285	280	290
Activities with own property, letting of own property	KPLA	701/702	..	..	..	..	..	..	KPMS	141	156	152†	154	154	160
Activities on a fee or contract basis	KPLB	703	..	..	..	..	..	..	KPMT	109	105	117†	131	126	131

Sources: Department of Manpower Services (Northern Ireland); Earnings and Employment Division, ONS: 01928 792566

7.5 Employee jobs
Analysis by industry based on the Standard Industrial Classification 1992

continued

At June in each year

Thousands, not seasonally adjusted

		SIC 1992	United Kingdom 1994	1995	1996	1997	1998	1999		Great Britain 1994	1995	1996	1997	1998	1999
Renting of machinery and eqpt. without operator & of personal & household goods	KPLC	71	123	118	124†	139	131	141	KPMU	121	116	122†	137	129	139
Construction and civil engineering machinery	KOUU	7132	..	..	..	..	..	..	KPMV	40	37	36	37†	41	41
All other goods and equipment	KOUV	Rest of 71	..	..	..	..	..	..	KPMW	81	79	86†	100	88	97
Computer and related equipment	KOUW	72	190	209	293†	340	388	427	KPMX	189	208	291†	338	385	425
Research and development	KOUX	73	92	87	94†	93	92	92	KPMY	90	86	93†	91	91	91
Other business activities	KOUY	74	1 827	1 958	2 195†	2 314	2 408	2 461	KPMZ	1 804	1 934	2 170†	2 286	2 378	2 428
Legal, accounting,book-keeping & auditing activities	KOUZ	741	..	..	..	..	..	..	KPNA	514	542	612†	615	644	663
Legal activities	KOVA	7411	..	..	..	..	..	..	KPNB	190	191	196†	204	223	230
Accounting, book-keeping auditing, tax consultancy	KOVB	7412	..	..	..	..	..	..	KPNC	146	138	160†	138	150	160
Market research business and consultancy activities	KOVC	7413/7414	..	..	..	..	..	..	KPND	143	163	181†	187	189	183
Management activities of holding co.	KOVD	7415	..	..	..	..	..	..	KPNE	34	50	75†	86	82	90
Architectural engineering activities and related technical consultancy, technical testing	KOVE	742/743	..	..	..	..	..	..	KPNF	361	332	338†	322	310	318
Advertising	KOVF	744	..	..	..	..	..	..	KPNG	54	63	81†	83	85	95
Industrial cleaning	KOVG	747	..	..	..	..	..	..	KPNH	392	403	398†	387	401	402
Public admin. & defence, compulsory social security	KOVH	L(75)	1 441	1 404	1 406†	1 358	1 331	1 329	KPNI	1 382	1 345	1 346†	1 299	1 272	1 270
Education	KOVI	M(80)	1 853	1 864	1 883†	1 888	1 886	1 946	KPNJ	1 792	1 802	1 818†	1 824	1 822	1 880
Health and social work	KOVJ	N	2 469	2 511	2 510†	2 537	2 550	2 544	KPNK	2 380	2 419	2 419†	2 444	2 457	2 451
Human health, vetinary activities	KOVK	851/852	..	..	..	..	..	..	KPNL	1 480	1 517	1 465†	1 536	1 551	1 561
Social work activities	KOVL	853	..	..	..	..	..	..	KPNM	900	902	954†	908	907	890
Other community social and personal service activities,private households with employed persons, extra-territorial organisations and bodies	KOVM	O - Q	960	969	1 050†	1 074	1 109	1 161	KPNN	932	943	1 024†	1 047	1 081	1 133
Sewage & refuse disposal; sanitation	KOVN	90	77	75	71†	78	78	77	KPNO	74	73	69†	76	76	74
Activities of membership organisations	KOVO	91	193	189	202	201†	220	221	KPNP	184	180	194	193†	212	212
Recreational cultural and sporting activities	KOVP	92	532	541	559†	573	588	627	KPNQ	520	530	547†	560	575	614
Motion picture video radio TV news agencies & entertainment activities	KOVQ	921 to 924	..	..	..	..	..	..	KPNR	130	133	162†	162	173	192
Library archives museums and other cultural activities	KOVR	925	..	..	..	..	..	..	KPNS	76	78	78†	79	85	85
Sporting activities and other recreational activities	KOVS	926/927	..	..	..	..	..	..	KPNT	314	320	307†	319	318	338
Other service activities,private households with employed persons, extra territorial organisations	KOVT	93/95/99	158	164	217†	222	222	236	KPNU	154	160	214†	218	218	233
Washing dry cleaning of textile and fur products	KOVU	9301	..	..	..	..	..	..	KPNV	42	41	42†	41	39	40
Hairdressing, other beauty treatment physical and well-being activities	KOVV	9302/9304	..	..	..	..	..	..	KPNW	83	74	78†	74	76	73

Sources: Department of Manpower Services (Northern Ireland); Earnings and Employment Division, ONS: 01928 792566

7.6 Weekly hours worked[1]
United Kingdom

Hours, Spring each year, not seasonally adjusted

	All workers weekly hours[2]		Average actual weekly hours of work		
	Total (millions)	Average	Full-time workers[2]	Part-time workers[3]	Second jobs[4]
All					
	BEFA	BEFD	BEFG	BEFJ	BEFM
1989	925	34.7	39.6	15.3	9.0
1990	926	34.4	39.1	15.4	9.2
1991	908	34.4	39.3	15.2	9.6
1992	856	33.3	38.4	14.8	9.3
1993	852	33.5	38.8	14.9	9.2
1994	865	33.8	39.2	15.2	9.1
1995	883	34.1	39.5	15.3	9.1
1996	886	33.9	39.5	15.3	8.8
1997	892	33.5	39.0	15.2	9.3
1998	900	33.4	38.9	15.2	9.0
1999	912	33.5	39.0	15.5	8.9
Men					
	BEFB	BEFE	BEFH	BEFK	BEFN
1989	616	40.6	41.6	14.9	10.5
1990	614	40.1	41.2	15.0	10.0
1991	596	40.1	41.2	14.8	10.6
1992	554	38.9	40.3	14.3	10.3
1993	548	39.3	40.8	14.4	10.2
1994	558	39.6	41.2	14.9	9.8
1995	571	39.9	41.6	14.7	10.0
1996	571	39.7	41.5	14.7	9.7
1997	574	39.1	41.1	14.7	10.6
1998	580	39.0	41.0	14.9	9.6
1999	582	38.9	40.8	14.9	9.5
Women					
	BEFC	BEFF	BEFI	BEFL	BEFO
1989	309	27.0	35.1	15.4	7.8
1990	312	26.9	34.6	15.4	8.5
1991	312	27.1	35.3	15.2	8.6
1992	302	26.4	34.6	14.9	8.5
1993	304	26.5	34.9	15.0	8.5
1994	308	26.8	35.2	15.3	8.5
1995	312	26.9	35.1	15.4	8.5
1996	315	26.8	35.3	15.4	8.2
1997	318	26.6	34.9	15.3	8.3
1998	319	26.5	34.8	15.3	8.5
1999	329	27.0	35.2	15.6	8.5

1 Average hours actually worked in the reference week which includes hours worked in second jobs (see chapter text for definitions).
2 Main and second job.
3 Main job only.
4 Second jobs reported in the LFS in addition to persons main full time job.

Source: ONS, Labour Force Survey

7.7 Civil Service staff[1]

Analysis by ministerial responsibility

At 1 April each year

Full-time equivalents[2] (thousands)

		1987	1988	1989	1990	1991	1992	1993	1994	1995	1996	1997	1998
Agriculture, Fisheries and Food[20]	BCDA	11.0	11.0	11.0	11.0	11.0	11.0	11.0	11.0	10.6	10.8	10.1	10.8
Cabinet Office[3,6,21,24,31,54]	KPQI	2.0	2.0	2.0	2.0	2.0	2.0	6.0	12.0	11.7	12.3	7.8	6.2
Chancellor of the Exchequer's Departments:[3]													
Customs and Excise	BCDC	26.0	26.0	27.0	27.0	27.0	27.0	25.0	25.0	24.1	23.2	23.1	23.4
Inland Revenue	BCDD	68.0	67.0	67.0	67.0	67.0	70.0	68.0	64.0	59.1	56.5	54.0	53.4
Department for National Savings	BCDE	8.0	7.0	7.0	7.0	7.0	6.0	6.0	6.0	5.4	4.7	4.3	4.1
Treasury and others[3,16,27,62]	BCDF	9.0	8.0	8.0	10.0	10.0	10.0	5.0	5.0	4.3	6.0	5.1	5.1
Total	BCDB	111.0	109.0	110.0	111.0	111.0	113.0	105.0	99.0	92.9	90.3	86.4	86.0
Education[6]	BCDG	2.0	3.0	3.0	3.0	3.0	3.0	3.0	2.0	2.5	–	–	–
Employment[3]	BCDH	61.0	59.0	55.0	53.0	50.0	58.0	58.0	55.0	49.6	–	–	–
Education and Employment[21,29]	KTDI	..	..	..	..	..	..	..	..	..	40.8	34.1	33.6
Energy[12]	BCDI	1.0	1.0	1.0	1.0	1.0	1.0	–	–	–	–	–	–
Environment[13,24,30,31,55]	BCDJ	34.0	33.0	31.0	29.0	26.0	23.0	18.0	10.0	9.4	10.9	9.6	–
Environment, Transport & Regions[55,64]	JYXS	..	..	..	..	..	..	..	..	..	..	..	21.2
Foreign and Commonwealth[52]	BCDK	10.0	10.0	10.0	10.0	10.0	10.0	10.0	8.0	7.5	7.1	6.6	5.4
International Development[52]	JYXU	..	..	..	..	..	..	..	..	..	..	..	1.1
Health[9,10,11,27,28]	BAKR	–	–	11.0	8.0	7.0	7.0	7.0	7.0	6.2	4.8	4.7	4.6
Home[53,60,61]	BCDL	40.0	40.0	41.0	43.0	45.0	50.0	52.0	51.0	51.4	50.8	50.4	50.7
Legal Departments[8]	KPQJ	23.0	25.0	28.0	28.0	29.0	29.0	30.0	29.0	28.7	27.8	26.3	25.4
National Heritage[15,53,56]	KPQK	–	–	–	–	–	1.0	1.0	1.0	1.0	1.0	1.0	–
Culture, Media & Sport[56,63]	JYXT	..	..	..	..	..	..	..	..	..	..	..	0.6
Northern Ireland	KPQD	–	–	–	–	–	–	–	–	0.2	0.2	0.2	0.2
Scotland[4]	BCDN	13.0	13.0	12.0	13.0	13.0	13.0	14.0	13.0	12.1	11.7	11.8	12.0
Social Security[9,32]	BAKS	–	–	84.0	82.0	80.0	81.0	86.0	90.0	89.2	91.5	93.1	87.2
Health and Social Security[9]	KAZL	98.0	103.0	–	–	–	–	–	–	–	–	–	–
Trade and Industry[3,5,12,17,19,21,22,23,25,54]	KAZM	15.0	15.0	15.0	14.0	13.0	13.0	13.0	12.0	11.1	11.2	10.3	10.4
Transport[18,26,55]	BCDR	14.0	14.0	14.0	16.0	15.0	15.0	15.0	14.0	12.9	11.3	11.4	–
Welsh Office	BCDS	2.0	2.0	2.0	2.0	2.0	2.0	3.0	2.0	2.2	2.1	2.2	2.1
Total civil departments	BCDU	436.0	439.0	431.0	425.0	418.0	431.0	431.0	418.0	400.8	384.6	366.0	357.6
Defence[7,14,57,58,59]	BCDW	165.0	144.0	142.0	142.0	141.0	140.0	130.0	122.0	116.1	109.9	109.2	104.2
Total all departments	BCDX	601.0	583.0	573.0	567.0	559.0	571.0	560.0	540.0	516.9	494.5	475.2	461.7
of whom													
Non-industrial staff	BCDY	*510.0*	*510.0*	*503.0*	*499.0*	*495.0*	*509.0*	*509.0*	*494.0*	*474.1*	*458.7*	*439.3*	*430.5*
Industrial staff	BCDZ	*91.0*	*73.0*	*70.0*	*68.0*	*64.0*	*61.0*	*52.0*	*46.0*	*42.0*	*35.9*	*36.0*	*32.8*

NOTE: Machinery of Government changes prior to 1 April 1985 are given in the *Annual Supplement of Definitions and Explanatory Notes* published in the January edition of *Monthly Digest*. Figures may not add exactly to the totals shown due to rounding.

1 The figures include non-industrial and industrial staff but exclude casual or seasonal staff (normally recruited for short periods of not more than twelve months) and employees of the Northern Ireland Civil Service.

2 Figures included are measured as 'full-time equivalent' staff. Part-time staff are recorded as a proportion of full-time employees according to the proportion of a full week that they work. This method of calculation was introduced from 1 April 1995 and figures have been recalculated back to April 1987 to enable comparisons over time to be maintained.

3 On 31 July 1989 a new department, the Central Statistical Office, was formed under the responsibility of the Chancellor of the Exchequer. It incorporated staff from the Department of Trade and Industry, the Cabinet Office, and the Department of Employment. On 1 October 1990 the Paymaster General's Office (now The Office of HM Paymaster General) became the responsibility of the Chancellor of the Exchequer. In mid-April 1992 responsibility for the Central Office of Information and HM Stationery Office was transferred from the Chancellor of the Exchequer to the Cabinet Office, and in mid-July 1992 some functions of HM Treasury were also transferred to the Cabinet Office.

4 Departments of the Secretary of State for Scotland and the Lord Advocate.

5 The Office of Electricity Regulation was formed under the responsibility of the Secretary of State for Trade and Industry on 11 September 1989.

6 With effect from mid-April 1992, responsibility for Science was transferred from the Department for Education to the Cabinet Office/OPSS.

7 From 6 April 1987 commercial management was introduced at the Devonport and Rosyth dockyards and around 4 100 non-industrial and 12 500 industrial staff were excluded from the staffing count.

8 On 20 July 1987 a new department, the Serious Fraud Office was formed.

9 With effect from 25 July 1988, the Department of Health and Social Security was split into the Department of Health and the Department of Social Security.

10 On 1 April 1990 around 3 300 staff at the Department of Health were transferred to the NHS and therefore are no longer in the staffing count.

11 Includes Office of Population, Censuses and Surveys, before 1 April 1996.

12 With effect from mid-April 1992, the Department of Trade and Industry took over responsibility for most Energy issues.

13 The Projects Division of PSAS was sold to Tarmac in December 1992 resulting in 700 staff being transferred to the private sector. The sale of Building Management businesses during 1993/94 resulted in 2 730 staff being transferred to the private sector.

14 3 620 non-industrial and 2 700 industrial staff of the MOD Atomic Weapons Establishment (AWE) were excluded from April 1993.

15 During April 1992 a new department, the Department of National Heritage, was formed from parts of six other departments.

16 FORWARD was privatised on 14 February 1994, and 400 staff were excluded from the staffing count from that date.

17 On 21 December 1991 the Insurance Services Group of ECGD, comprising some 460 staff, transferred to the private sector.

18 On 17 December 1993 the Driver and Vehicle and Operator Information Technology comprising some 320 staff were transferred to the private sector.

19 On 31 March 1994 Warren Spring Laboratory was merged with AEA Technology and 80 staff were excluded from the staffing count.

20 On 1 April 1995 the Meat Hygiene Service Executive Agency was formed. It was staffed mainly by personnel transferred from Local Authorities.

21 On 5 July 1995 a new department, the Department for Education and Employment was formed with staff from Employment Main, Employment Service and the Department for Education. The Labour Market Statistics Group of the Employment Department went to ONS; Industrial Relations Division of ED moved to DTI; responsibility for HSE moved from ED to DoE; responsibility for ACAS moved from ED to DTI. The Office of Science and Technology moved from OPSS to DTI. The Competitiveness Division and Deregulation Unit moved from DTI to OPS.

22 On 30 September 1995 the National Physical Laboratory was privatised and 533 staff were excluded from the count.

23 On 1 November 1995 the National Engineering Laboratory was privatised and 217 staff were excluded from the count.

24 On 1 January 1996 responsibility for SAFE and The Buying Agency was transferred from DoE to OPS.

25 On 31 March 1996 the Laboratory of the Government Chemist was privatised and 270 staff were excluded from the count.

26 On 31 March 1996 Transport Research Laboratory was privatised and 450 staff were excluded from the count.

27 On 1 April 1996 OPCS merged with the CSO to form the Office for National Statistics under the responsibility of the Chancellor of the Exchequer.

28 On 1 April 1996 870 staff transferred from Regional Health Authorities to DH.

Footnotes continue on next page

Source: Cabinet Office (OPS): 020 7270 5744

7.7 Civil Service staff[1]

continued

Analysis by ministerial responsibility

At 1 April each year

Full-time equivalents [2] (thousands)

Footnotes continued

29 On 1 April 1996 1 000 staff transferred from the Employment Service to DSS due to the introduction of the Job Seekers Allowance scheme.
30 On 1 April 1996 HM Inspectorate of Pollution moved to the Environment Agency (NDPB) from the DoE and 340 staff were excluded from the count.
31 On 1 April 1996 responsibility for PACE (the successor body to PH) was transferred from DoE to OPS.
32 The Resettlement Agency was abolished with effect from 1 April 1996.
33 On 1 April 1996 the Metropolitan Police Forensic Science Laboratory merged with the Forensic Science Service (FSS) and 361 staff transferred to the FSS.
34 On 1 April 1996 8 permanent staff dealing with pay and grading delegation transferred from HM Treasury to Cabinet Office (OPS).
35 On 1 May 1996 26 permanent staff working in Voluntary and Community Division transferred from the Home Office to the Department of National Heritage.
36 On 1 May 1996 the Natural Resources Institute transferred to the University of Greenwich. 303 permanent staff ceased to be civil servants.
37 On 1 July 1996 47 permanent staff at Companies House transferred to Capita a private sector company.
38 On 1 August 1996 Chessington Computer Centre (CCC) was privatised and 370 permanent staff ceased to be civil servants. CCC was sold to a consortium made up of a Management and Employee Buyout Team, Integris UK and Close Brothers.
39 On 5 September 1996 the Recruitment and Assessment Agency was sold to Capita Group plc. 140 permanent staff ceased to be civil servants.
40 On 30 September 1996 HM Stationery Office was sold to National Publishing Group. 2 580 permanent staff ceased to be civil servants.
41 On 30 September 1996 the Occupational Health Service Agency was sold to BMI Health Services. 100 permanent staff ceased to be civil servants.
42 On 30 September 1996 the Teachers Pension Agency transferred to Capita Management Services Ltd a private sector company. 380 permanent staff ceased to be civil servants.
43 On 16 January 1997 6 permanent staff dealing with IT support services in the Cabinet Office transferred to Digital a private sector company.
44 On 19 March 1997 the Building Research Establishment (BRE) agency was sold to the BRE Management Bid Team.
45 From 20 March 1997 the Crown Prosecution Service and Serious Fraud Office began operating on Next Steps Lines.
46 On 1 April 1997 150 permanent staff in the Department for Education and Employment Information Service Division transferred under TUPE to F1 Group
47 On 1 April 1997 Paymaster was sold to EDS. 560 staff ceased to be civil servants.
48 On 1 April 1997 ADAS was sold to ADAS plc and 1 190 staff ceased to be civil servants.
49 On 1 April 1997 the Farming and Rural Conservation Agency (formerly part of ADAS) was established to take over ADAS functions that remained in the public sector.
50 On 1 April 1997 the Government Car and Despatch Agency (formerly part of SAFE) was established and 220 permanent staff from SAFE moved to the new agency.
51 On 1 April 1997 26 permanent staff at the Crown Estate Office ceased to be counted as civil servants.
52 Following the General Election on 3 May 1997, the Overseas Development Administration was renamed Department for International Development and made independent of the Foreign and Commonwealth Office.
53 Following the General Election on 3 May 1997, responsibility for voluntary organisations and charities (18 staff) transferred from the Department for National Heritage to the Home Office.
54 On 6 May 1997, following the General Election, Competitiveness Division of the Office of Public Service (27 staff) transferred to the Department of Trade and Industry.
55 On 16 June 1997, following the General Election, the Departments of the Environment and Transport merged under one Minister to form the Department of the Environment, Transport and Regions.
56 On 14 July 1997, following the General Election, the Department of National Heritage was renamed the Department for Culture, Media and Sport.
57 On 29 August 1997 Commander-in-Chief of Land Command (MOD) contracted out 83 permanent non-industrial staff to Primary Management Aldershot Ltd (PMAL).
58 On 31 December 1997 Armed Forces Personnel Administration (MOD) contracted out around 650 permanent non-industrial staff to EDS.
59 On 1 March 1998, the Fleet Maintenance and Repair function of the Naval Bases & Supplies Agency (MOD) was transferred to Fleet Support Ltd. and 1 140 staff were excluded from the count.
60 On the 1 April 1998, the Police Information & Technology Organisation (HO) became an Executive NDPB and 325 staff were excluded from the count.
61 On 1 April 1998 the National Criminal Intelligence Service (HO) became a Service Authority and 564 staff were excluded from the count.
62 Debt Management Agency (HMT) was established on 1 April 1998.
63 On 1 April 1998 Historic Royal Palaces Agency (DCMS) became an Executive NDPB.
64 On 1 April 1998 Marine Safety and Coastguard Agencies merged to form the Maritime and Coastguard Agency (DETR).

Source: Cabinet Office (OPS): 020 7270 5744

7.8 Persons employed in local authorities[1]

Thousands (Full-time equivalents)[2]

		1989	1990	1991	1992	1993	1994	1995	1996	1997	1998	1999
Service: England												
Education												
lecturers and teachers[3]	BCHA	485.9	492.0	486.4	476.1	395.8	376.6	376.6	373.2	365.5	362.7†	364.4
others	BCHB	362.0	365.0	363.0	359.7	316.3	316.2	321.8	318.6	326.3	331.2†	342.9
Construction	BCHC	97.2	93.7	85.7	77.1	70.9	67.4	63.5	58.8	55.0	51.9†	52.3
Transport	BCHD	2.6	2.4	1.9	1.5	1.4	1.4	1.4	1.4	1.4	1.4	1.4
Social services	BCHE	232.0	236.0	235.6	232.3	225.8	233.3	234.3	230.6	227.3	221.3†	215.9
Public libraries and museums	BCHF	33.5	33.6	33.3	32.4	31.9	31.8	32.7	31.5	31.4	31.1†	31.5
Recreation, parks and baths	BCHG	78.1	77.2	75.7	72.7	68.6	67.2	65.2	63.6	60.7	59.8†	58.1
Environmental health	BCHH	19.1	19.1	18.8	19.6	19.0	18.5	17.4	17.2	17.4	16.9†	17.2
Refuse collection and disposal	BCHS	33.0	29.9	27.1	23.6	21.8	20.5	25.1	24.5	24.6	24.0†	23.7
Housing	BCHT	61.3	64.5	65.5	65.4	65.5	65.7	68.0	66.8	65.5	64.7†	65.4
Town and country planning	BCHU	21.7	22.9	23.1	23.1	22.5	22.2	25.9	27.1	27.3	26.8†	27.7
Fire service												
regulars	BAIV	34.3	34.5	34.5	34.3	34.2	33.9	33.8	33.6	33.3	33.1	33.8
others	BAIW	5.7	6.0	5.7	5.8	5.7	5.7	5.8	5.7	5.6	5.6	6.2
Other services[4]	BCHM	232.3	243.7	244.7	238.0	226.7	217.0	191.3	183.8	177.9	173.1†	178.1
Total of above	BCHN	1 698.7	1 720.5	1 701.0	1 661.6	1 506.1	1 477.4	1 462.8	1 436.4	1 419.2	1 403.6†	1 418.6
Police Service												
police: all ranks	BCHO	118.9	120.2	120.9	120.8	121.4	121.0	118.8	118.4	118.5	119.2†	118.8
cadets	BAIX	0.3	0.3	0.3	0.3	0.2	0.1	0.1	0.3	0.3	0.1	–
traffic wardens	BAIY	4.5	4.5	4.7	4.9	4.9	4.7	4.5	4.3	4.1	3.5†	3.3
civilians	BAIZ	40.6	42.2	44.0	44.9	46.7	46.9	48.1	49.8	50.0	50.3†	49.6
Magistrates courts	BAJA	9.3	9.4	10.0	10.0	10.0	9.9	10.0	9.9	9.2	8.9†	9.0
Probation staff												
officers	BAJB	6.3	6.5	6.5	6.9	7.1	7.2	7.2	6.9	6.9	6.2†	6.2
others	BAJC	6.6	7.0	7.3	7.5	7.9	8.0	8.5	7.2	7.1	5.4†	5.4
Total law and order	BAJD	186.5	190.1	193.7	195.3	198.2	197.8	197.2	196.8	196.1	193.6†	192.3
Agency staff	BAJE	1.6	1.3	1.5	1.1	1.4	1.6	0.9	0.8	0.8	0.8	–
Total(excluding special employment and training measures)	BCHR	1 886.8	1 911.9	1 896.2	1 858.0	1 705.7	1 676.8	1 660.9	1 634.0	1 616.1	1 598.0†	1 610.9

		1989	1990	1991	1992	1993	1994	1995	1996	1997	1998	1999
Wales												
Education												
lecturers and teachers[3]	BCGA	32.2	32.0	32.3	31.1	28.2	28.2	27.9	28.5	27.5	27.6†	28.3
others	BCGB	22.9	22.3	23.3	22.7	20.0	19.7	20.8	22.1	22.0	22.3	23.3
Construction	BCGC	7.4	7.4	7.0	6.5	5.9	5.8	5.9	6.0	5.5	5.5†	5.7
Transport	BCGD	–	–	–	–	–	–	–	–	–	..	..
Social services	BCGE	15.0	15.0	15.0	16.0	16.0	17.0	18.3	18.5	19.0	18.8†	19.2
Public libraries and museums	BCGF	1.5	1.6	1.7	1.7	1.7	1.7	1.7	1.5	1.6	1.5†	1.6
Recreation, parks and baths	BCGG	5.8	5.8	5.9	5.8	6.0	6.0	6.3	5.8	6.4	6.1†	6.1
Environmental health	BCGH	1.3	1.4	1.4	1.5	1.5	1.5	1.5	1.6	1.5	1.3†	1.4
Refuse collection and disposal	BCGI	1.7	1.6	1.7	1.6	1.6	1.5	2.2	2.9	2.9	2.9	2.8
Housing	BCGJ	2.8	2.8	2.9	3.0	3.1	3.1	3.4	3.3	3.6	3.6	3.8
Town and country planning	BCGK	1.5	1.5	1.6	1.7	1.8	1.8	2.2	1.9	2.0	2.0	2.1
Fire service												
regulars	BAKT	1.8	1.8	1.8	1.8	1.8	1.8	1.8	1.9	1.9	1.9	1.9
others	BAKU	0.4	0.3	0.4	0.4	0.4	0.4	0.3	0.3	0.3	0.3	0.4
Other services	BCGM	18.5	19.0	18.8	18.7	23.5	19.8	15.6	14.2	14.6	14.0†	15.0
Total of above	BCGN	112.9	112.9	114.3	112.5	111.5	108.1	107.9	108.5	108.8	107.8†	111.6
Police Service												
police: all ranks	BCGO	6.6	6.5	6.6	6.6	6.6	6.5	6.4	6.4	6.6	6.6	6.6
cadets	BAKV	–	–	–	–	–	–	–	–	–	..	..
traffic wardens	BAKW	0.2	0.2	0.2	0.2	0.2	0.2	0.2	0.2	0.2	0.2	0.2
civilians	BAKX	1.9	2.0	2.1	2.1	2.1	2.2	2.2	2.5	2.4	2.4†	2.5
Magistrates courts	BAKY	0.6	0.6	0.6	0.6	0.7	0.7	0.7	0.6	0.6	0.6	0.7
Probation staff												
officers	BAKZ	0.4	0.4	0.4	0.4	0.4	0.4	0.4	0.4	0.4	0.3†	0.3
others	BALA	0.3	0.3	0.3	0.4	0.4	0.4	0.4	0.4	0.4	0.3†	0.3
Total law and order	BALB	9.9	10.1	10.2	10.3	10.4	10.4	10.3	10.5	10.6	10.4†	10.6
Agency staff	BALC	–	–	–	–	–	–	–	–	–	..	..
Total(excluding special employment and training measures)	BCGR	122.7	122.9	124.4	122.8	121.9	118.5	118.2	119.0	119.4	118.2†	122.2

1 Figures are based on surveys undertaken on behalf of central and local government by the ONS and the Scottish Executive and the Convention of Scottish Local Authorities (COSLA).

2 Based on the following factors to convert part-time employees to approximate full-time equivalents: for teachers and lecturers in further education, 0.11; teachers in primary and secondary education and all other non-manual employees, 0.53; manual employees, 0.41.

3 An estimated 97 000 FTEs were lost to the Local Authority sector, when most further education institutions became the responsibility of the Further Education Funding Council (FEFC) on 1 April 1993.

4 Including civil service departments (eg engineers and treasurers) and all services not shown separately.

Sources: Department of the Environment, Transport and the Regions; Joint Staffing Watch; National Assembly for Wales

7.8 Persons employed in local authorities[1]

continued

Thousands (Full-time equivalents)[2]

Service		1989	1990	1991	1992	1993	1994	1995	1996	1997	1998	1999
Scotland												
Education												
lecturers and teachers[3,4,5]	BCMA	59.6	59.7	59.4	59.9	54.0	53.2	53.1	52.6	51.8	53.3†	54.0
others[6]	BCMB	42.0	39.6	37.8	37.1	35.8	33.0	24.5†	25.0	21.8	22.2	23.4
Construction	BCMC	15.2	13.3	13.4	13.6	13.1	12.9	21.7†	20.1	15.4	15.9	15.3
Transport	BCMD	0.7	0.7	0.8	0.8	0.8	0.9	0.9	0.9	0.9	0.9	0.9
Social Services	BCME	34.9	36.2	36.2	36.8	37.3	38.3	39.1	38.4	37.9	37.0†	34.2
Public libraries and museums	BCMF	4.3	4.4	4.4	4.5	4.6	4.7	4.7	4.2	4.2	4.2†	4.1
Recreation, leisure and tourism	BCMG	13.7	13.8	13.6	13.3	13.7	13.9	13.9	12.8	11.3	11.3†	8.3
Environmental health	BCMH	2.4	2.3	2.5	2.5	2.7	2.7	2.6	2.5	2.9	3.1†	3.1
Cleansing	BCMI	8.9	8.4	8.2	8.1	8.2	8.1	8.0	7.7	6.7	6.6†	7.2
Housing	BCMJ	6.9	7.3	7.3	7.2	7.4	7.8	7.9	7.9	8.4	8.5†	8.8
Physical planning	BCMK	1.8	2.0	2.0	2.1	2.2	2.2	2.2	2.2	2.6	3.3†	3.1
Fire service	BCML	5.2	5.1	5.2	5.2	5.3	5.3	5.2	5.3	5.5	5.7†	5.7
Other services[7]	BCMM	39.6	44.3	46.8	48.1	48.0	52.1	51.5	51.1	48.1	44.1†	46.1
Total of above	BCMN	235.1	237.1	237.6	239.2	233.1	235.1	235.3	230.7	217.5	216.1†	214.2
Police service												
police (all ranks)	BCMO	13.6	13.7	13.9	13.9	14.2	14.2	14.4	14.3	14.1	14.2†	14.2
others[8]	BCMP	4.8	4.7	4.6	4.8	5.1	5.2	5.2	5.0	5.3	5.4†	5.4
Administration of district courts	BCMQ	0.1	0.2	0.2	0.1	0.2	0.2	0.2	0.2	0.2	0.2	0.2
Total (excluding special employment measures) and training	BCMR	253.6	255.7	256.3	258.0	252.6	254.7	255.1	250.2	237.1	235.9†	234.0

1 See previous page.

2 Based on the following factors to convert part-time employees to approximate full-time equivalents: for lecturers and teachers, 0.40; non-manual staff (excluding teachers), 0.58; manual employees, 0.46.

3 Includes only those part-time staff employed in vocational further education (ie courses of an academic nature or those leading to a qualification).

4 Figures from June 1993 are not directly comparable to those for earlier periods because of the transfer, on 1 April 1993, from local authority control of 9176 full-time equivalent staff employed in further education.

5 Lecturers are not included from 1 April 1993. See note 4 for details.

6 Includes school-crossing patrols.

7 Including central services departments (eg engineers and finance) and all services not shown separately.

8 Includes civilian employees of police forces and traffic wardens.

Sources: Department of the Environment, Transport and the Regions: 0117 987 8675;
Joint Staffing Watch;
National Assembly for Wales;
Scottish Executive;
Scottish Joint Staffing Watch.

7.9 Duration of ILO unemployment
United Kingdom

Thousands, Spring each year, not seasonally adjusted

		Duration of ILO unemployment[1]									
										All 1 year or more	
	All ILO unemployed	Less than 3 months	3 months & less than 6 months	6 months & less than 1 year	1 year & less than 2 years	2 years & less than 3 years	3 years & less than 4 years	4 years & less than 5 years	5 years or more	Number	As % of total
All											
	BEAS	BEFP	BEFS	BEFV	BEFY	BEGB	BEGE	BEGH	BEGK	BEGN	BEGQ
1989	2 075	647	306	333	252	133	103	71	229	788	38.0
1990	1 974	686	324	310	211	107	73	51	210	653	33.1
1991	2 414	834	466	434	276	113	67	41	179	676	28.0
1992	2 769	668	500	607	529	174	75	37	179	993	35.9
1993	2 936	600	474	599	612	287	109	57	196	1 262	43.0
1994	2 736	609	388	488	514	310	166	81	179	1 249	45.7
1995	2 454	568	386	422	404	243	143	102	182	1 074	43.8
1996	2 334	600	381	419	344	189	128	85	185	931	39.9
1997	2 034	599	317	326	288	148	83	72	197	789	38.8
1998	1 766	592	325	263	217	109	68	42	148	584	33.1
1999	1 741	620	326	276	209	87	45	39	138	518	29.7
Men											
	BEAT	BEFQ	BEFT	BEFW	BEFZ	BEGC	BEGF	BEGI	BEGL	BEGO	BEGR
1989	1 215	306	169	180	152	86	86	53	182	559	46.0
1990	1 165	346	170	172	137	72	55	41	171	476	40.9
1991	1 514	455	288	281	189	76	50	31	142	488	32.3
1992	1 865	396	320	409	380	125	55	28	152	740	39.6
1993	1 986	348	313	385	439	212	84	43	160	938	47.2
1994	1 825	345	229	312	362	236	124	63	152	937	51.3
1995	1 607	300	239	267	279	182	115	79	144	799	49.7
1996	1 546	331	236	268	235	143	104	68	160	710	45.9
1997	1 304	323	194	198	195	103	63	58	166	585	44.9
1998	1 091	311	189	170	138	76	47	35	123	419	38.4
1999	1 088	339	200	170	141	61	31	31	113	377	34.6
Women											
	BEAU	BEFR	BEFU	BEFX	BEGA	BEGD	BEGG	BEGJ	BEGM	BEGP	BEGS
1989	860	341	137	153	100	47	16	18	47	229	26.6
1990	809	340	154	138	74	35	18	10	40	177	21.9
1991	900	378	178	153	87	36	17	10	37	188	20.9
1992	904	272	180	198	149	49	20	..	28	254	28.1
1993	949	251	161	214	173	74	25	15	37	323	34.1
1994	910	264	159	176	152	74	42	18	26	312	34.3
1995	846	268	147	154	125	61	28	23	39	275	32.6
1996	788	269	144	151	109	46	24	17	25	222	28.1
1997	731	276	123	128	94	45	20	14	31	203	27.8
1998	674	280	136	93	79	32	21	..	25	165	24.4
1999	653	281	125	106	68	27	13	..	25	141	21.6

1 See chapter text for definitions.
2 ".." sample size too small for reliable estimate.

Source: ONS, Labour Force Survey

7.10 ILO Unemployment
United Kingdom
Age group by numbers and rates

Numbers in thousands, rates in percentage, Spring each year, nsa

	All aged 16 and over	16-59/64	16-17	18-24	16-24	25-34	35-49	50-64 (m) 50-59 (f)	65+ (m) 60+ (f)
	Numbers								
All									
	BEAS	BEGT	BEGW	BEGZ	BEHC	BEHF	BEHI	BEHL	BEHO
1990	1 974	1 939	105	505	610	531	471	326	35
1991	2 414	2 373	131	643	775	656	577	365	41
1992	2 769	2 738	123	694	817	781	714	426	32
1993	2 936	2 902	112	755	867	801	748	487	33
1994	2 736	2 710	122	649	771	765	710	464	25
1995	2 454	2 436	122	584	706	691	652	387	17
1996	2 334	2 314	143	534	677	659	618	360	19
1997	2 034	2 012	145	463	608	535	545	324	22
1998	1 766	1 746	133	409	542	481	448	276	20
1999	1 741	1 722	142	398	540	431	475	275	19
Men									
	BEAT	BEGU	BEGX	BEHA	BEHD	BEHG	BEHJ	BEHM	BEHP
1990	1 165	1 148	60	306	366	299	245	239	16
1991	1 514	1 496	71	414	485	401	338	272	18
1992	1 865	1 850	71	469	541	517	459	333	15
1993	1 986	1 974	62	504	566	530	500	378	12
1994	1 825	1 815	66	433	499	510	454	352	10
1995	1 607	1 599	69	382	451	449	407	293	..
1996	1 546	1 535	87	359	446	415	401	273	11
1997	1 304	1 292	79	302	381	339	340	232	11
1998	1 091	1 082	73	257	330	289	263	200	..
1999	1 088	1 079	89	247	335	252	292	200	..
Women									
	BEAU	BEGV	BEGY	BEHB	BEHE	BEHH	BEHK	BEHN	BEHQ
1990	809	790	45	199	244	232	226	88	19
1991	900	877	61	229	290	255	240	92	23
1992	904	888	52	224	276	264	255	93	16
1993	949	928	49	251	300	271	248	109	21
1994	910	895	56	216	272	255	256	113	15
1995	846	837	53	202	255	242	245	94	..
1996	788	780	56	175	231	244	218	87	..
1997	731	719	65	162	227	196	205	92	11
1998	674	664	60	152	212	192	184	75	10
1999	653	643	53	152	205	179	183	75	10
	Rates[1]								
All									
	BEBQ	BEHR	BEHU	BEHX	BEIA	BEID	BEIG	BEIJ	BEIM
1990	6.8	6.9	11.3	9.9	10.2	7.3	4.9	6.3	4.3
1991	8.4	8.5	14.9	13.3	13.6	8.8	5.9	7.2	5.0
1992	9.7	9.9	16.0	15.4	15.5	10.4	7.2	8.4	3.7
1993	10.3	10.5	16.8	17.4	17.3	10.5	7.5	9.6	4.1
1994	9.6	9.8	17.9	15.9	16.2	10.0	7.1	9.0	3.2
1995	8.6	8.8	17.3	14.9	15.3	9.0	6.4	7.5	2.1
1996	8.2	8.3	18.3	14.0	14.7	8.6	6.0	6.9	2.4
1997	7.1	7.2	17.7	12.5	13.5	7.0	5.3	5.9	2.7
1998	6.1	6.3	16.6	11.3	12.3	6.3	4.4	4.9	2.5
1999	6.0	6.1	17.9	11.0	12.3	5.8	4.5	4.7	2.3
Men									
	BEBR	BEHS	BEHV	BEHY	BEIB	BEIE	BEIH	BEIK	BEIN
1990	7.1	7.1	12.3	11.0	11.2	7.0	4.6	7.3	5.3
1991	9.2	9.3	15.4	15.7	15.6	9.2	6.2	8.4	5.9
1992	11.5	11.7	17.7	19.0	18.8	11.9	8.4	10.4	4.9
1993	12.4	12.5	18.5	21.1	20.8	12.1	9.1	11.9	4.6
1994	11.4	11.5	18.8	19.2	19.1	11.5	8.3	11.0	3.7
1995	10.1	10.2	18.9	17.7	17.9	10.1	7.3	9.2	..
1996	9.7	9.8	21.2	17.1	17.8	9.5	7.2	8.4	4.1
1997	8.1	8.2	19.3	14.8	15.6	7.8	6.1	6.9	4.0
1998	6.8	6.9	18.0	13.0	13.8	6.7	4.7	5.8	..
1999	6.8	6.8	21.6	12.5	14.1	6.0	5.1	5.7	..
Women									
	BEBS	BEHT	BEHW	BEHZ	BEIC	BEIF	BEII	BEIL	BEIO
1990	6.5	6.6	10.3	8.7	9.0	7.6	5.3	4.7	3.7
1991	7.2	7.4	14.3	10.5	11.1	8.2	5.5	5.0	4.4
1992	7.3	7.5	14.0	11.0	11.5	8.4	5.7	5.0	3.1
1993	7.6	7.8	15.1	12.9	13.2	8.4	5.5	5.7	3.9
1994	7.3	7.5	17.0	11.8	12.6	7.9	5.6	5.7	2.9
1995	6.8	7.0	15.6	11.5	12.2	7.4	5.4	4.7	..
1996	6.3	6.5	15.1	10.2	11.1	7.4	4.7	4.3	..
1997	5.8	5.9	16.0	9.7	11.0	5.9	4.4	4.3	2.0
1998	5.3	5.4	15.2	9.3	10.5	5.9	3.9	3.4	2.0
1999	5.1	5.2	14.0	9.3	10.2	5.5	3.9	3.3	1.9

1 Total ILO unemployment as a percentage of all economically active persons in the relevant age group.
2 ".." sample size too small for reliable estimate.

Source: ONS, Labour Force Survey

7.11 Claimant count by age and duration[1]

United Kingdom

Thousands, not seasonally adjusted

			1988	1989	1990	1991	1992	1993	1994	1995	1996	1997	1998
Annual averages													
All ages													
All	GEYP	1	2 425.9	1 841.3	1 651.9	2 237.9	2 749.7	2 946.4	2 680.5	2 356.8	2 167.4	1 653.5	1 381.0
Up to 6 months	GEYQ	2	1 004.0	808.0	835.2	1 210.1	1 302.4	1 294.3	1 148.5	1 037.9	968.4	809.4	762.0
Over 6 and up to 12 months	DPBZ	3	431.0	319.9	282.0	446.3	585.2	587.7	509.9	443.5	413.0	273.6	246.1
All over 12 months	DPBY	4	990.9	713.4	534.7	581.5	862.2	1 064.4	1 022.0	875.5	785.9	570.5	372.9
All over 24 months	GEYA	5	661.6	469.7	331.8	302.2	358.7	514.3	569.7	507.0	448.2	334.2	207.3
Aged 18 to 24													
All	GEYB	6	685.3	528.1	484.4	678.3	805.6	825.4	731.8	635.8	562.1	423.9	348.9
Up to 6 months	GEYC	7	372.0	311.4	312.9	437.7	455.4	437.3	390.1	350.5	316.5	261.7	238.1
Over 6 and up to 12 months	GEYD	8	139.0	102.0	86.9	137.4	176.4	175.2	151.4	131.4	117.8	78.1	62.8
All over 12 months	GEYE	9	174.3	114.7	84.6	103.2	173.9	212.9	190.4	154.0	127.8	84.1	48.0
All over 24 months	GEYF	10	86.0	50.3	32.1	30.6	43.6	71.3	74.6	59.8	46.4	30.5	15.1
Aged 25 to 49													
All	GEYG	11	1 163.0	922.6	854.5	1 193.5	1 501.5	1 634.2	1 503.1	1 336.6	1 241.7	947.1	791.1
Up to 6 months	GEYH	12	453.3	393.7	420.4	620.8	669.9	669.0	596.3	541.7	508.3	426.1	409.0
Over 6 and up to 12 months	GEYI	13	204.0	164.1	151.5	245.5	319.8	317.2	276.2	245.6	231.5	153.8	145.4
All over 12 months	GEYJ	14	505.7	364.8	282.6	327.2	511.8	647.9	630.6	549.3	502.0	367.1	236.7
All over 24 months	DPQX	15	346.7	237.7	170.5	163.7	212.0	325.0	370.4	334.0	299.2	224.0	133.5
Aged 50 and over													
All	DPQY	16	498.9	388.2	311.2	362.0	431.3	469.7	428.6	367.1	346.0	265.3	226.1
Up to 6 months	DPQZ	17	123.8	101.1	100.1	147.7	166.7	173.1	147.3	130.2	127.8	105.6	101.0
Over 6 and up to 12 months	DPRA	18	70.8	53.5	43.5	63.1	88.2	93.4	80.6	64.9	62.3	40.6	37.1
All over 12 months	DPRB	19	304.2	233.7	167.5	151.2	176.4	203.3	200.7	171.9	155.9	119.0	88.1
All over 24 months	DPRC	20	228.9	181.7	129.2	107.9	103.1	118.0	124.6	113.3	102.6	79.7	58.6

Relationship between rows: 1=2+3+4; 6=7+8+9; 11=12+13+14; 16=17+18+19

1 Count of claimants of unemployment-related benefits.

Sources: Office for National Statistics: 020 7533 6094; Benefits Agency

7.12 ILO unemployment rates[1]

Percentages, Spring each year, not seasonally adjusted

		1992	1993	1994	1995	1996	1997	1998	1999
United Kingdom	BEBQ	9.7	10.3	9.6	8.6	8.2	7.1	6.1	6.0
North East	BENU	11.9	12.0	12.5	11.4	10.8	9.8	8.2	10.1
North West (GOR) & Merseyside	BENV	10.1	10.8	10.3	9.0	8.4	6.9	6.6	6.2
North West (GOR)	BENW	9.2	9.8	9.6	8.3	7.3	6.3	5.6	5.5
Merseyside	BENX	14.0	15.3	13.5	11.7	13.3	9.6	10.9	9.6
Yorkshire and Humber	BENY	10.1	10.0	9.9	8.7	8.1	8.1	7.0	6.6
East Midlands	BENZ	8.8	9.1	8.3	7.5	7.4	6.3	4.9	5.2
West Midlands	BEOA	10.7	11.8	10.0	9.0	9.2	6.8	6.3	6.8
Eastern	BEOB	7.7	9.2	8.2	7.5	6.2	5.9	5.0	4.2
London	BEOC	12.0	13.2	13.1	11.5	11.3	9.1	8.1	7.5
South East (GOR)	BEOD	7.8	8.0	7.1	6.4	6.0	5.2	4.3	3.6
South West	BEOE	9.1	9.2	7.5	7.8	6.3	5.2	4.5	4.7
England	BEOF	9.7	10.3	9.5	8.6	8.1	6.9	6.0	5.8
Wales	BEOG	8.9	9.6	9.3	8.8	8.3	8.4	6.7	7.1
Scotland	BEOH	9.5	10.2	10.0	8.3	8.7	8.5	7.4	7.4
Northern Ireland	BEOI	12.3	12.5	11.7	11.0	9.7	7.5	7.3	7.2

1 See chapter text for definitions.

Sources: ONS, Labour Force Survey; Department of Economic Development, Northern Ireland

7.13 Claimant count rates[1,2]
Analysis by Government Office Regions[3]

Percentages, seasonally adjusted

		1988	1989	1990	1991	1992	1993	1994	1995	1996	1997	1998
Annual averages												
United Kingdom	BCJE	8.0	6.2	5.8	8.0	9.7	10.3	9.3	8.0	7.3	5.5	4.7
Great Britain	DPAJ	7.8	6.0	5.6	7.9	9.6	10.2	9.2	7.9	7.2	5.5	4.6
North East	DPDM	13.0	10.9	9.6	11.2	12.1	12.9	12.4	11.3	10.3†	8.4	7.5
North West	IBWC	10.1	8.2	7.4	9.2	10.5	10.6	9.9	8.5	7.8	6.1	5.3
Yorkshire and the Humber	DPBI	9.3	7.3	6.6	8.7	9.9	10.2	9.6	8.6	7.9	6.3†	5.5
East Midlands	DPBJ	7.1	5.4	5.1	7.2	9.0	9.5	8.7	7.4	6.7	4.9†	4.0
West Midlands	DPBN	8.8	6.5	5.7	8.4	10.3	10.8	9.9	8.1	7.2	5.5†	4.7
East	DPDP	4.9	3.4	3.6	6.4	8.7	9.4	8.1	6.6	5.9	4.1†	3.3
London	DPDQ	6.6	5.1	5.0	8.0	10.5	11.6	10.7	9.4	8.6	6.4†	5.3
South East	DPDR	4.1	2.8	3.0	5.8	8.0	8.8†	7.3	5.9	5.1	3.4	2.7
South West	DPBM	6.0	4.4	4.3	6.9	9.2	9.5	8.1	6.8	6.1	4.3	3.5
Wales	DPBP	9.9	7.4	6.7	9.0	10.0	10.3	9.3	8.5	8.0	6.4	5.6
Scotland	DPBQ	11.2	9.4	8.2	8.8	9.4	9.7	9.3	7.9	7.7	6.4	5.7
Northern Ireland	DPBR	15.0	14.0	12.8	12.9	13.8	13.7	12.6	11.2	10.8†	8.1	7.4

1 The number of unemployment-related benefit claimants as a percentage of the estimated total workforce (the sum of claimants, employee jobs self-employed, participants on work related government training programmes and HM Forces) at mid-year.

2 Seasonally adjusted and excluding claimants under 18, consistent with current coverage.

3 New geographical boundaries were introduced in May 1997 for UK regions and the data are only available from March 1986. Data for Standard Statistical Regions are available on request.

Source: ONS, Labour Market Statistics: 020 7533 6094

7.14 Claimant count[1,3]

Analysis by Government Office Regions[2]

Thousands, seasonally adjusted

	North East	North West	Yorkshire and the Humber	East Midlands	West Midlands	East	London	South East	South West	Wales	Scotland	Great Britain	Northern Ireland	United Kingdom
	DPDG	IBWA	DPAX	DPAY	DPBC	DPDJ	DPDK	DPDL	DPBB	DPBE	DPBF	DPAG	DPBG	BCJD
1984 Jan	..	..	260.0	174.3	318.6	..	341.9	..	172.7	155.2	304.6	2 741.2	108.9	2 850.1
Apr	..	..	263.5	177.0	319.7	..	345.5	..	175.7	158.1	304.7	2 764.9	109.4	2 874.3
Jul	..	..	267.4	180.0	322.4	..	352.6	..	178.5	160.2	308.9	2 803.1	109.6	2 912.7
Oct	..	..	274.0	184.4	326.3	..	361.9	..	184.3	164.3	312.1	2 856.8	108.8	2 956.6
1985 Jan	..	..	275.2	185.0	327.1	..	369.2	..	186.9	165.8	313.9	2 884.2	109.6	2 993.8
Apr	..	..	279.5	186.5	327.7	..	374.4	..	190.2	167.5	321.4	2 916.8	110.9	3 027.7
Jul	..	..	278.9	185.7	326.4	..	376.6	..	189.7	168.2	322.4	2 913.3	111.4	3 024.7
Oct	..	..	281.6	185.4	326.3	..	380.6	..	191.7	168.9	324.1	2 927.1	113.5	3 040.6
1986 Jan	..	..	289.4	188.3	327.0	..	385.3	..	194.3	170.8	323.7	2 958.1	116.5	3 074.6
Apr	198.9	..	292.2	188.8	328.9	193.1	393.9	249.0	196.5	170.6	326.9	2 985.8	120.6	3 106.4
Jul	198.0	..	294.3	190.2	329.5	192.4	396.4	248.0	197.8	169.3	333.4	3 000.0	122.6	3 122.6
Oct	194.1	..	290.4	188.2	325.2	188.5	391.0	243.4	194.6	164.4	335.9	2 956.4	124.1	3 080.5
1987 Jan	192.9	..	285.9	186.1	319.1	187.3	384.4	240.3	190.2	160.6	339.0	2 916.4	123.7	3 040.1
Apr	188.9	..	279.7	180.6	306.4	175.6	367.1	224.9	182.1	153.5	336.3	2 815.5	122.1	2 937.6
Jul	180.9	..	265.6	171.0	289.8	162.8	351.3	207.3	170.6	146.9	321.8	2 670.8	120.4	2 791.2
Oct	173.7	..	251.3	160.9	273.7	150.9	333.3	191.0	159.8	140.6	307.3	2 524.9	118.4	2 643.3
1988 Jan	166.0	..	239.8	152.2	258.0	139.4	315.8	173.1	151.9	134.0	298.0	2 393.0	114.8	2 507.8
Apr	160.1	..	231.3	144.4	243.3	125.0	301.8	155.9	141.4	129.8	286.9	2 268.6	112.6	2 381.2
Jul	153.0	..	218.1	135.4	226.2	111.8	279.5	141.2	130.9	122.0	273.6	2 125.1	110.8	2 235.9
Oct	147.4	..	208.9	128.5	211.7	102.3	267.2	131.8	123.5	118.1	268.0	2 029.6	109.0	2 138.6
1989 Jan	141.3	..	193.8	117.4	192.0	90.1	243.9	115.2	110.0	108.7	256.0	1 870.9	108.0	1 978.9
Apr	133.2	..	180.0	107.5	173.0	81.2	222.8	104.5	101.1	100.3	241.8	1 729.9	105.6	1 835.5
Jul	125.6	..	172.8	102.7	164.3	78.9	217.0	101.5	96.7	94.1	229.2	1 657.2	103.0	1 760.2
Oct	118.7	..	165.6	98.4	156.6	77.0	205.6	99.8	91.7	88.9	219.6	1 580.3	100.4	1 680.7
1990 Jan	112.8	..	159.7	94.5	151.1	75.7	200.2	96.9	87.5	84.4	208.2	1 514.8	97.2	1 612.0
Apr	109.8	..	155.8	94.0	147.0	77.7	198.5	98.6	88.7	83.2	204.1	1 497.6	96.3	1 593.9
Jul	110.2	..	157.6	97.0	148.7	83.9	204.8	107.0	94.9	84.7	198.6	1 529.5	95.0	1 624.5
Oct	113.0	..	165.5	104.4	156.7	97.4	226.2	126.0	105.6	88.8	198.9	1 631.5	93.7	1 725.2
1991 Jan	117.4	..	177.5	114.3	171.6	114.0	257.4	151.3	123.7	95.5	201.7	1 786.3	95.9	1 882.2
Apr	126.9	..	199.0	133.2	205.2	140.1	306.0	193.1	148.9	109.3	215.1	2 065.8	98.1	2 163.9
Jul	131.9	..	213.9	146.7	227.5	159.8	344.6	222.2	167.0	117.0	223.7	2 264.4	99.6	2 364.0
Oct	134.3	..	221.0	155.2	240.9	173.2	370.5	241.8	179.5	120.0	226.2	2 382.1	101.2	2 483.3
1992 Jan	135.1	..	224.6	162.3	252.7	185.4	393.4	261.2	190.6	122.3	229.8	2 484.1	102.6	2 586.7
Apr	136.8	..	230.1	169.8	262.8	198.0	413.9	281.0	201.7	124.4	234.9	2 590.8	103.7	2 694.5
Jul	138.0	..	234.5	173.7	268.4	206.7	429.3	290.8	208.3	125.6	238.4	2 652.7	105.4	2 758.1
Oct	142.3	..	241.8	179.9	279.1	220.2	450.3	310.3	217.0	130.0	244.8	2 761.3	105.9	2 867.2
1993 Jan	147.6	..	249.8	187.1	288.6	231.4	468.9	328.5	224.6	132.6	247.7	2 856.7	106.0	2 962.7
Apr	149.1	..	248.8	185.4	287.6	230.2	473.3	324.5	222.5	131.4	246.3	2 848.1	104.6	2 952.7
Jul	149.9	..	243.2	182.6	281.3	225.3	470.0	316.5	216.6	130.6	244.5	2 803.1	103.6	2 906.7
Oct	148.3	..	239.5	179.1	273.3	218.3	463.5	308.6	210.5	128.7	238.9	2 743.0	102.2	2 845.2
1994 Jan	147.1	328.1	235.7	176.3	264.2	211.8	455.1	299.7	204.8	127.6	237.8	2 685.5	100.0	2 785.5
Apr	143.3	319.0	228.7	171.6	253.0	202.9	441.7	284.3	195.6	123.8	233.9	2 590.7	99.3	2 690.0
Jul	141.5	308.7	224.9	168.0	244.5	193.6	431.5	271.4	190.2	120.4	230.7	2 522.2	97.8	2 620.0
Oct	137.9	296.2	217.6	161.6	232.6	183.4	419.1	254.3	181.0	113.8	220.4	2 414.4	94.3	2 508.7
1995 Jan	135.5	285.4	212.7	154.4	220.1	175.1	404.5	241.8	171.9	108.8	210.6	2 309.9	91.4	2 401.3
Apr	130.9	272.5	208.1	149.0	212.2	168.1	397.1	231.1	167.4	106.9	202.3	2 245.8	89.0	2 334.8
Jul	129.1	268.1	206.3	146.6	208.6	166.4	393.4	227.2	164.0	107.6	198.2	2 215.5	87.6	2 303.2
Oct	127.0	262.6	202.4	143.3	202.6	162.0	386.1	220.6	160.8	106.0	195.7	2 169.1	85.8	2 254.9
1996 Jan	124.4	258.3	199.6	141.4	198.6	158.4	379.7	214.7	156.6	104.9	195.3	2 132.0	86.4	2 218.4
Apr	122.8	256.6	197.0	138.6	195.5	154.4	369.8	209.1	153.4	105.5	197.1	2 099.7	86.6	2 186.3
Jul	117.8	250.2	190.6	133.0	189.2	148.1	360.4	200.8	148.5	102.6	194.8	2 036.1	86.4	2 122.6
Oct	110.9	239.9	182.5	125.8	178.7	139.8	344.1	186.7	139.2	98.7	188.7	1 934.8	81.6	2 016.3
1997 Jan	101.4	218.5	167.0	112.0	160.3	123.4	312.8	163.2	124.9	90.3	174.2	1 748.1	71.1	1 819.3
Apr	95.9	202.0	155.3	102.4	148.0	110.9	286.0	145.0	112.3	83.1	163.3	1 604.2	65.7	1 669.9
Jul	92.2	188.3	148.1	94.7	137.7	102.1	263.9	130.6	101.2	77.9	152.6	1 489.2	60.9	1 550.0
Oct	90.3	178.4	142.7	88.1	131.8	95.1	247.8	121.1	93.9	73.4	147.1	1 409.8	60.3	1 470.0
1998 Jan	87.7	169.6	136.8	82.4	125.8	87.9	233.9	111.4	86.6	70.6	141.0	1 333.6	60.1	1 393.8
Apr	84.5	165.6	134.1	79.7	123.1	85.2	229.6	108.3	85.0	69.7	139.4	1 304.0	58.6	1 362.6
Jul	81.5	162.7	132.6	79.1	120.6	82.8	223.6	103.9	84.0	68.1	137.5	1 276.5	56.7	1 333.2
Oct	81.6†	161.3	131.1†	79.9†	121.1†	82.4†	220.0†	103.0†	82.1†	68.1†	136.8†	1 267.6†	55.7†	1 323.3†
1999 Jan	82.6	159.2	129.5	78.6	122.3	79.7	215.3	100.4	79.5	67.8	135.8	1 250.6	56.1	1 306.7
Apr	82.6	157.7	127.0	78.4	123.4	79.2	208.4	99.2	78.6	67.4	134.7	1 236.6	55.1	1 291.7
Jul	80.2	152.3	121.3	75.4	119.5	75.8	203.0	93.8	74.8	63.3	127.6	1 184.1	49.7	1 233.8
Oct	76.7	150.5	118.4	73.8	116.3	74.1	196.9	91.8	71.9	61.2	125.8	1 157.4	46.6	1 204.0

1 The figures are based on the number of claimants receiving unemployment related benefits and are adjusted for seasonality and discontinuities to be consistent with current coverage.

2 New geographical boundaries were introduced in May 1997 for UK regions and data are only available from March 1986.

3 The latest national and regional seasonally adjusted claimant count figures are provisional and subject to revision in the following month.

Source: ONS, Labour Market Statistics: 020 7533 6094

7.15 Economic activity
United Kingdom
Age group by numbers and rates

Numbers in thousands, rates in percentage, Spring each year, nsa

	All aged 16 and over	16-59/64	16-17	18-24	16-24	25-34	35-49	50-64 (m) 50-59 (f)	65+ (m) 60+ (f)
	Numbers								
All									
	BEAM	BEIP	BEIS	BEIV	BEIY	BEJB	BEJE	BEJH	BEJK
1990	28 909	28 087	923	5 083	6 006	7 313	9 627	5 142	822
1991	28 813	27 995	882	4 828	5 709	7 438	9 763	5 085	818
1992	28 581	27 734	771	4 508	5 279	7 501	9 890	5 064	847
1993	28 447	27 640	663	4 334	4 997	7 610	9 966	5 068	806
1994	28 433	27 625	683	4 084	4 767	7 678	10 030	5 151	807
1995	28 426	27 614	706	3 916	4 622	7 693	10 114	5 185	813
1996	28 552	27 764	778	3 816	4 594	7 671	10 242	5 257	788
1997	28 716	27 892	819	3 695	4 514	7 680	10 233	5 465	824
1998	28 713	27 920	801	3 613	4 414	7 579	10 269	5 659	793
1999	28 992	28 159	792	3 606	4 397	7 455	10 461	5 846	833
Men									
	BEAN	BEIQ	BEIT	BEIW	BEIZ	BEJC	BEJF	BEJI	BEJL
1990	16 483	16 175	486	2 795	3 281	4 265	5 342	3 287	307
1991	16 401	16 099	459	2 641	3 100	4 341	5 412	3 246	302
1992	16 187	15 871	403	2 467	2 870	4 363	5 439	3 200	316
1993	16 021	15 754	338	2 382	2 721	4 390	5 473	3 170	267
1994	15 996	15 722	352	2 256	2 608	4 434	5 492	3 187	274
1995	15 981	15 686	364	2 162	2 525	4 428	5 549	3 184	296
1996	15 992	15 716	410	2 098	2 507	4 385	5 590	3 234	276
1997	16 023	15 743	411	2 038	2 449	4 366	5 580	3 348	280
1998	15 997	15 715	407	1 981	2 388	4 308	5 586	3 433	283
1999	16 120	15 824	410	1 971	2 381	4 201	5 700	3 542	296
Women									
	BEAO	BEIR	BEIU	BEIX	BEJA	BEJD	BEJG	BEJJ	BEJM
1990	12 427	11 912	437	2 288	2 725	3 047	4 285	1 855	515
1991	12 412	11 897	423	2 186	2 609	3 097	4 351	1 839	516
1992	12 395	11 863	368	2 041	2 409	3 138	4 452	1 864	532
1993	12 426	11 887	324	1 952	2 276	3 220	4 492	1 898	539
1994	12 436	11 904	331	1 828	2 158	3 244	4 538	1 963	533
1995	12 445	11 928	342	1 754	2 096	3 265	4 566	2 001	517
1996	12 561	12 048	368	1 719	2 087	3 286	4 653	2 023	512
1997	12 692	12 149	408	1 657	2 065	3 314	4 653	2 117	544
1998	12 716	12 206	394	1 631	2 025	3 272	4 682	2 226	510
1999	12 872	12 335	381	1 635	2 016	3 254	4 762	2 304	537
	Rates[1]								
All									
	BEJN	BEAY	BEJQ	BEJT	BEJW	BEJZ	BEKC	BEKF	BEKI
1990	64.1	80.2	62.0	82.4	78.4	83.3	85.9	69.7	8.1
1991	63.7	79.8	62.0	80.3	76.8	83.0	85.8	69.5	8.1
1992	63.1	78.8	55.9	76.7	72.7	82.6	86.0	69.1	8.4
1993	62.7	78.4	50.0	76.3	71.3	82.8	85.5	68.5	7.9
1994	62.5	78.2	52.4	74.5	70.3	83.0	85.2	68.6	7.9
1995	62.4	78.0	52.4	74.2	69.8	83.0	85.0	68.1	8.0
1996	62.4	78.1	54.5	75.2	70.7	82.9	84.9	68.2	7.7
1997	62.6	78.2	55.8	74.8	70.5	83.6	84.5	68.5	8.1
1998	62.3	78.0	54.9	73.8	69.5	83.8	84.3	68.8	7.7
1999	62.7	78.4	54.7	73.6	69.3	84.3	84.9	69.4	8.1
Men									
	BEJO	BEAZ	BEJR	BEJU	BEJX	BEKA	BEKD	BEKG	BEKJ
1990	75.6	88.3	63.5	88.7	83.8	96.1	95.4	75.5	8.8
1991	75.0	87.7	62.9	86.0	81.6	95.7	95.1	75.0	8.6
1992	73.8	86.3	56.9	82.3	77.4	94.9	94.6	73.9	8.9
1993	72.9	85.6	49.7	82.1	76.0	94.4	94.0	72.8	7.5
1994	72.5	85.2	52.6	80.5	75.1	94.5	93.3	72.3	7.6
1995	72.2	84.7	52.6	80.0	74.4	94.0	93.2	71.5	8.2
1996	71.9	84.6	56.0	80.8	75.3	93.2	92.5	71.8	7.6
1997	71.7	84.4	54.7	80.5	74.6	93.4	91.9	72.2	7.6
1998	71.3	83.9	54.4	79.1	73.4	93.5	91.4	72.0	7.6
1999	71.5	84.1	55.3	78.6	73.3	93.3	92.1	72.6	7.9
Women									
	BEJP	BEBJ	BEJS	BEJV	BEJY	BEKB	BEKE	BEKH	BEKK
1990	53.3	71.3	60.3	75.8	72.8	70.2	76.5	61.4	7.8
1991	53.1	71.0	61.1	74.3	71.8	69.9	76.4	61.4	7.8
1992	53.0	70.6	54.7	70.9	67.9	70.0	77.3	62.1	8.1
1993	53.1	70.6	50.2	70.2	66.4	71.0	77.1	62.5	8.2
1994	53.1	70.6	52.2	68.2	65.2	71.2	77.1	63.4	8.1
1995	53.1	70.6	52.1	68.1	64.9	71.6	76.8	63.4	7.9
1996	53.5	71.1	53.0	69.5	65.8	72.2	77.2	63.0	7.8
1997	53.9	71.4	57.0	68.8	66.1	73.4	77.0	63.4	8.3
1998	53.8	71.5	55.3	68.3	65.4	73.6	77.2	64.3	7.8
1999	54.4	72.1	54.1	68.4	65.1	75.0	77.6	64.9	8.2

1 Total economically active (see chapter text for definitions) as a percentage of all persons in the relevant age group.

Source: ONS, Labour Force Survey

7.16 Economically inactive
United Kingdom
Age group by numbers and rates

Numbers in thousands, rates in percentage, Spring each year, nsa

	All aged 16 and over	16-59/64	16-17	18-24	16-24	25-34	35-49	50-64 (m) 50-59 (f)	65+ (m) 60+ (f)
Numbers									
All									
	BEAV	BEKL	BEKO	BEKR	BEKU	BEKX	BELA	BELD	BELG
1990	16 198	6 931	566	1 089	1 655	1 465	1 578	2 233	9 267
1991	16 413	7 108	539	1 188	1 727	1 525	1 620	2 236	9 305
1992	16 728	7 439	609	1 368	1 978	1 581	1 616	2 264	9 289
1993	16 954	7 602	663	1 349	2 012	1 576	1 687	2 327	9 352
1994	17 033	7 683	620	1 399	2 018	1 572	1 740	2 352	9 350
1995	17 148	7 793	643	1 361	2 003	1 578	1 789	2 423	9 355
1996	17 172	7 770	649	1 256	1 905	1 584	1 825	2 456	9 402
1997	17 182	7 787	647	1 246	1 893	1 507	1 876	2 510	9 396
1998	17 343	7 886	659	1 280	1 938	1 470	1 907	2 571	9 456
1999	17 220	7 773	655	1 291	1 946	1 386	1 859	2 581	9 447
Men									
	BEAW	BEKM	BEKP	BEKS	BEKV	BEKY	BELB	BELE	BELH
1990	5 318	2 136	279	356	636	173	260	1 068	3 182
1991	5 470	2 251	271	430	700	193	277	1 080	3 219
1992	5 737	2 511	305	532	837	235	312	1 127	3 226
1993	5 964	2 661	342	518	860	262	352	1 187	3 304
1994	6 053	2 734	317	548	865	258	392	1 218	3 320
1995	6 151	2 825	328	539	868	281	407	1 270	3 325
1996	6 240	2 865	322	500	822	319	454	1 269	3 376
1997	6 317	2 917	340	493	833	307	489	1 288	3 400
1998	6 444	3 024	341	524	865	298	522	1 338	3 420
1999	6 422	2 994	332	535	867	302	488	1 337	3 428
Women									
	BEAX	BEKN	BEKQ	BEKT	BEKW	BEKZ	BELC	BELF	BELI
1990	10 880	4 794	287	732	1 019	1 291	1 318	1 165	6 086
1991	10 942	4 857	269	758	1 027	1 332	1 343	1 155	6 085
1992	10 991	4 928	305	837	1 141	1 347	1 304	1 136	6 063
1993	10 989	4 941	321	830	1 152	1 315	1 336	1 140	6 048
1994	10 979	4 949	303	851	1 154	1 314	1 347	1 134	6 030
1995	10 997	4 968	314	821	1 136	1 297	1 381	1 154	6 030
1996	10 932	4 906	327	756	1 083	1 265	1 371	1 187	6 026
1997	10 865	4 870	307	753	1 060	1 201	1 387	1 222	5 995
1998	10 898	4 862	318	756	1 074	1 171	1 384	1 233	6 036
1999	10 798	4 779	324	756	1 080	1 084	1 371	1 244	6 019
Rates[1]									
All									
	BELJ	BELM	BELP	BELS	BELV	BELY	BEMB	BEME	BEMH
1990	35.9	19.8	38.0	17.6	21.6	16.7	14.1	30.3	91.9
1991	36.3	20.2	38.0	19.7	23.2	17.0	14.2	30.5	91.9
1992	36.9	21.2	44.1	23.3	27.3	17.4	14.0	30.9	91.6
1993	37.3	21.6	50.0	23.7	28.7	17.2	14.5	31.5	92.1
1994	37.5	21.8	47.6	25.5	29.7	17.0	14.8	31.4	92.1
1995	37.6	22.0	47.6	25.8	30.2	17.0	15.0	31.9	92.0
1996	37.6	21.9	45.5	24.8	29.3	17.1	15.1	31.8	92.3
1997	37.4	21.8	44.2	25.2	29.5	16.4	15.5	31.5	91.9
1998	37.7	22.0	45.1	26.2	30.5	16.2	15.7	31.2	92.3
1999	37.3	21.6	45.3	26.4	30.7	15.7	15.1	30.6	91.9
Men									
	BELK	BELN	BELQ	BELT	BELW	BELZ	BEMC	BEMF	BEMI
1990	24.4	11.7	36.5	11.3	16.2	3.9	4.6	24.5	91.2
1991	25.0	12.3	37.1	14.0	18.4	4.3	4.9	25.0	91.4
1992	26.2	13.7	43.1	17.7	22.6	5.1	5.4	26.1	91.1
1993	27.1	14.4	50.3	17.9	24.0	5.6	6.0	27.2	92.5
1994	27.5	14.8	47.4	19.5	24.9	5.5	6.7	27.7	92.4
1995	27.8	15.3	47.4	20.0	25.6	6.0	6.8	28.5	91.8
1996	28.1	15.4	44.0	19.2	24.7	6.8	7.5	28.2	92.4
1997	28.3	15.6	45.3	19.5	25.4	6.6	8.1	27.8	92.4
1998	28.7	16.1	45.6	20.9	26.6	6.5	8.6	28.0	92.4
1999	28.5	15.9	44.7	21.4	26.7	6.7	7.9	27.4	92.1
Women									
	BELL	BELO	BELR	BELU	BELX	BEMA	BEMD	BEMG	BEMJ
1990	46.7	28.7	39.7	24.2	27.2	29.8	23.5	38.6	92.2
1991	46.9	29.0	38.9	25.7	28.2	30.1	23.6	38.6	92.2
1992	47.0	29.4	45.3	29.1	32.1	30.0	22.7	37.9	91.9
1993	46.9	29.4	49.8	29.8	33.6	29.0	22.9	37.5	91.8
1994	46.9	29.4	47.8	31.8	34.8	28.8	22.9	36.6	91.9
1995	46.9	29.4	47.9	31.9	35.1	28.4	23.2	36.6	92.1
1996	46.5	28.9	47.0	30.5	34.2	27.8	22.8	37.0	92.2
1997	46.1	28.6	43.0	31.2	33.9	26.6	23.0	36.6	91.7
1998	46.2	28.5	44.7	31.7	34.6	26.4	22.8	35.7	92.2
1999	45.6	27.9	45.9	31.6	34.9	25.0	22.4	35.1	91.8

1 Total economically active (see chapter text for definitions) as a percentage of all persons in the relevant age group.

Source: ONS, Labour Force Survey

7.17 Economically inactive
United Kingdom
All persons[1] by whether wants job, job search & availability

Thousands, Spring each year, not seasonally adjusted

	Total economically inactive	Does not want job	Wants job[2] but not seeking in last 4 weeks								Wants job[2] and seeking work but not available to start[3]		
				Availability to start work in next 2 weeks									
			Total	Available	Not available[3]	Discouraged workers[4]	Long term sick/ disabled	Looking after family/ home	Students	Other	All	Students	Other
All													
	BEKL	BEMK	BEMN	BEMQ	BEMT	BEMW	BEMZ	BENC	BENF	BENI	BENL	BENO	BENR
1993	7 602	5 440	1 839	831	1 008	143	409	738	225	323	323	172	151
1994	7 683	5 403	2 004	883	1 121	132	497	780	246	350	275	155	121
1995	7 793	5 496	2 015	889	1 126	105	518	763	256	373	282	172	110
1996	7 770	5 435	2 108	863	1 245	101	574	765	279	388	227	139	88
1997	7 787	5 374	2 165	749	1 416	88	685	733	287	371	248	144	104
1998	7 886	5 464	2 163	704	1 459	73	748	731	268	342	259	144	115
1999	7 773	5 428	2 093	659	1 434	70	748	668	261	347	252	140	111
Men													
	BEKM	BEML	BEMO	BEMR	BEMU	BEMX	BENA	BEND	BENG	BENJ	BENM	BENP	BENS
1993	2 661	1 875	639	281	358	85	254	42	118	140	146	85	61
1994	2 734	1 876	721	300	422	79	318	47	128	148	136	84	52
1995	2 825	1 967	725	298	428	62	320	49	137	157	133	84	49
1996	2 865	1 949	807	320	487	59	356	68	151	173	109	68	41
1997	2 917	1 959	839	253	586	51	413	68	150	158	118	79	40
1998	3 024	2 019	875	260	615	45	469	74	142	146	129	80	50
1999	2 994	2 030	852	256	596	41	458	70	136	146	112	68	44
Women													
	BEKN	BEMM	BEMP	BEMS	BEMV	BEMY	BENB	BENE	BENH	BENK	BENN	BENQ	BENT
1993	4 941	3 565	1 199	550	649	58	154	696	108	183	177	87	90
1994	4 949	3 527	1 283	584	699	53	179	733	117	202	139	70	69
1995	4 968	3 529	1 290	591	699	43	197	714	119	217	149	88	61
1996	4 906	3 486	1 301	543	758	42	218	697	129	215	118	71	47
1997	4 870	3 414	1 325	496	830	37	272	665	138	213	130	66	64
1998	4 862	3 445	1 288	444	843	28	279	658	127	196	130	65	66
1999	4 779	3 398	1 242	403	839	29	289	597	125	201	139	72	67

1 All persons of working age (men 16-64, women 16-59).
2 According to responses to LFS question.
3 Not available to start work in next two weeks including a few people who could not state whether or not they were available.
4 People whose reason for not seeking work was that they believed no jobs were available.

Source: ONS, Labour Force Survey

7.18 Labour disputes
United Kingdom

Thousands

SIC 1992		1994	1995	1996	1997	1998
Working days lost through all stoppages in progress: total	KBBZ	278	415	1 303	235	282
Analysis by industry						
Mining, quarrying, electricity, gas and water	DMME	1	1	2	2	–
Manufacturing	DMMF	58	65	97	86	34
Construction	DMMG	5	10	8	17	13
Transport, storage and communication	DMMH	110	120	884	36	139
Public administration and defence	DMMI	11	95	158	29	28
Education	DMMJ	70	67	128	28	6
Health and social work	DMMK	5	16	8	7	16
Other community, social and pesonal services	DMML	11	23	3	5	30
All other industries and services	DMMM	8	16	15	25	15
Analysis by number of working days lost in each stoppage						
Under 250 days	KBFC	11	11	14	12	8
250 and under 500 days	KBFJ	6	10	13	6	11
500 and under 1 000 days	KBFL	24	19	13	17	11
1 000 and under 5 000 days	KBFY	53	82	61	72	48
5 000 and under 25 000 days	KBFZ	68	195	123	101	118
25 000 and under 50 000 days	KBGS	–	29	54	26	–
50 000 days and over	KBGT	117	68	1 025	–	86
Working days lost per 1 000 employees all industries and services	KBHA	13	19	57†	10	12
Workers directly and indirectly involved: total	KBHB	107	174	364	130	93
Analysis by industry						
Mining, quarrying, electricity, gas and water	DMMN	–	2	1	–	1
Manufacturing	DMMO	23	33	34	28	14
Construction	DMMP	1	2	3	13	2
Transport, storage and communications	DMMQ	37	54	146	24	39
Public administration and defence	DMMR	8	28	32	20	4
Education	DMMS	29	30	122	15	4
Health and social work	DMMT	2	4	5	5	2
Other community, social and personal services	DMMU	2	10	2	1	22
All other industries and services	DMMV	5	11	21	23	4
Analysis by duration of stoppage						
Not more than 5 days	KBHM	75	142	208	108	57
Over 5 but not more than 10 days	KBHN	5	11	133	7	32
Over 10 but not more than 20 days	KBJQ	1	2	4	14	1
Over 20 but not more than 30 days	KBJR	6	2	3	–	–
Over 30 but not more than 50 days	KBJS	–	10	16	1	1
Over 50 days	KBJT	20	7	1	–	1
Numbers of stoppages in progress: total	KBLG	205	235	244	216	166
Analysis by industry						
Mining, quarrying, electricity, gas and water	DMMW	1	5	6	1	1
Manufacturing	DMMX	71	68	67	53	36
Construction	DMMY	4	9	11	11	13
Transport, storage and communications	DMMZ	54	56	72	68	57
Public administration and defence	DMNA	27	26	22	23	10
Education	DMNB	13	27	35	35	19
Health and social work	DMNC	7	17	9	7	6
Other community, social and personal services	DMND	15	19	12	8	17
All other industries and services	DMNE	14	14	11	12	7
Analysis of number of stoppages by duration						
Not more than 5 days	KBNH	176	199	196	184	130
Over 5 but not more than 10 days	KBNI	14	12	20	15	21
Over 10 but not more than 20 days	KBNJ	6	9	7	8	3
Over 20 but not more than 30 days	KBNK	5	6	6	2	4
Over 30 but not more than 50 days	KBNL	1	2	10	6	3
Over 50 days	KBNM	3	7	5	1	5

NOTES These figures exclude details of stoppages involving fewer than ten workers or lasting less than one day except any in which the aggregate number of working days lost is 100 or more.
There may be some under-recording of small or short stoppages; this would have much more effect on the total of stoppages than of working days lost.
Some stoppages which affected more than one industry group have been counted under each of the industries but only once in the totals.
Stoppages have been classified using *Standard Industrial Classification 1992.*

The figures for working days lost and workers involved have been rounded and consequently the sum of the constituent items may not agree with the totals.
Classifications by size are based on the full duration of stoppages where these continue into the following year.
Working days lost per thousand employees are based on the latest available mid-year (June) estimates of employee jobs.

Sources: Office for National Statistics,;
Labour Market Statistics: 01928 792825

7.18 Labour disputes United Kingdom

continued

Thousands

SIC 1980		1986	1987	1988	1989	1990	1991	1992	1993	1994
Working days lost through all stoppages in progress: total	KBBZ	1 920	3 546	3 702	4 128	1 903	761	528	649	278
Analysis by industry										
Coal extraction	KBCH	143	217	222	50	59	29	8	27	–
Other energy and water	KBCI	6	9	16	20	39	4	26	–	–
Metals, minerals and chemicals	KBCJ	192	60	70	42	42	27	14	6	8
Engineering and vehicles	KBCK	744	422	1 409	617	922	160	63	91	36
Other manufacturing industries	KBCL	135	115	151	91	106	35	16	13	15
Construction	KBCM	33	22	17	128	14	14	10	1	5
Transport and communication	KBCN	190	1 705	1 491	624	177	60	13	160	87
Public admin., sanitary services and education	KBCO	449	939	254	2 237	175	362	328	339	92
Medical and health services	KBDZ	11	6	36	151	345	1	1	2	1
All other industries and services	KBEZ	20	53	30	167	20	69	50	9	35
Analysis by number of working days lost in each stoppage										
Under 250 days	KBFC	48	48	33	30	28	15	7	10	11
250 and under 500 days	KBFJ	50	54	34	28	24	16	15	9	6
500 and under 1 000 days	KBFL	89	88	78	51	45	34	22	15	24
1 000 and under 5 000 days	KBFY	369	360	310	221	216	123	114	74	53
5 000 and under 25 000 days	KBFZ	381	388	325	365	286	205	156	78	68
25 000 and under 50 000 days	KBGS	258	118	127	234	216	190	115	149	–
50 000 days and over	KBGT	726	2 490	2 795	3 198	1 087	178	98	314	117
Working days lost per 1 000 employees all industries and services	KBHA	90	164	166	182	83	34	24	30	13
Workers directly and indirectly involved: total	KBHB	720	887	790	727	298	176	148	385	107
Analysis by industry										
Coal extraction	KBHC	87	98	92	25	15	6	3	14	–
Other energy and water	KBHD	2	2	2	10	18	2	6	–	–
Metals, minerals and chemicals	KBHE	17	9	10	7	5	2	2	2	2
Engineering and vehicles	KBHF	147	174	137	99	92	34	21	25	17
Other manufacturing industries	KBHG	30	19	29	12	11	15	3	3	4
Construction	KBHH	8	4	4	20	5	6	4	1	1
Transport and communication	KBHI	72	207	321	112	68	12	7	71	25
Public admin., sanitary services and education	KBHJ	348	361	161	414	70	87	92	261	39
Medical and health services	KBHK	4	4	31	9	10	–	2	–	1
All other industries and services	KBHL	6	10	4	19	3	11	9	7	18
Analysis by duration of stoppage										
Not more than 5 days	KBHM	369	308	381	194	185	133	111	364	75
Over 5 but not more than 10 days	KBHN	47	66	280	97	24	2	8	5	5
Over 10 but not more than 20 days	KBJQ	24	153	19	388	27	11	4	13	1
Over 20 but not more than 30 days	KBJR	58	25	22	8	21	3	1	–	6
Over 30 but not more than 50 days	KBJS	17	23	57	12	22	1	12	–	–
Over 50 days	KBJT	206	313	32	29	19	26	11	3	20
Numbers of stoppages in progress: total	KBLG	1 074	1 016	781	701	630	369	253	211	205
Analysis by industry										
Coal extraction	KBLR	351	296	154	146	87	32	10	5	–
Other energy and water	KBLS	10	6	6	7	7	3	6	–	–
Metals, minerals and chemicals	KBLT	62	36	50	40	36	17	16	15	14
Engineering and vehicles	KBLU	202	209	162	126	132	67	44	34	42
Other manufacturing industries	KBLV	85	98	75	63	45	30	19	21	16
Construction	KBLW	26	24	16	40	12	18	12	4	4
Transport and communication	KBLX	145	191	176	79	124	37	20	34	52
Public admin., sanitary services and education	KBLY	132	99	104	154	164	124	105	89	55
Medical and health services	KBLZ	35	24	21	18	13	6	6	3	2
All other industries and services	KBNG	34	41	31	31	16	35	16	6	21
Analysis of number of stoppages by duration										
Not more than 5 days	KBNH	836	784	587	553	476	270	188	176	176
Over 5 but not more than 10 days	KBNI	108	103	89	52	44	20	13	9	14
Over 10 but not more than 20 days	KBNJ	65	78	60	45	45	24	17	11	6
Over 20 but not more than 30 days	KBNK	23	23	17	18	20	15	6	2	5
Over 30 but not more than 50 days	KBNL	18	12	12	18	22	10	9	2	1
Over 50 days	KBNM	24	16	16	15	23	30	20	11	3

NOTES These figures exclude details of stoppages involving fewer than ten workers or lasting less than one day except any in which the aggregate number of working days lost is 100 or more.

There may be some under-recording of small or short stoppages; this would have much more effect on the total of stoppages than of working days lost.

Some stoppages which affected more than one industry group have been counted under each of the industries but only once in the totals.

Stoppages have been classified using *Standard Industrial Classification 1980.*

The figures for working days lost and workers involved have been rounded and consequently the sum of the constituent items may not agree with the totals.

Classifications by size are based on the full duration of stoppages where these continue into the following year.

Working days lost per thousand employees are based on the latest available mid-year (June) estimates of employee jobs.

Sources: Office for National Statistics,; Labour Market Statistics: 01928 792825

7.19 Average earnings and hours of manual employees by industry division
Full time employees on adult rates: pay unaffected by absence: Great Britain

At April

	Energy and water supply industries	Extraction minerals /ores other than fuels; manufacture of metals, mine products/ chemicals	Metal goods, engineering and vehicles industries	Other manufacturing industries	Construction	Distribution, hotels and catering; repairs	Transport and communication	Banking, finance, insurance, business services/ leasing	Other services	Manufacturing industries	Service industries	All industries and services
	1	2	3	4	5	6	7	8	9	2,3,4	6,7,8,9	0 - 9
Standard Industrial Classification: Revised 1980												
Men Weekly earnings	KINF	KING	KINH	KINI	KINJ	KINK	KINL	KINM	KINN	KINO	KINP	KINQ
1990	302.8	262.4	254.0	242.2	245.9	197.6	249.2	235.0	197.6	251.4	219.7	239.5
1991	334.1	273.1	263.2	254.2	257.1	208.7	266.5	250.9	216.7	261.8	236.4	253.1
1992	360.0	292.7	282.2	270.2	274.7	218.9	281.8	259.2	228.5	279.7	248.2	268.3
1993	369.0	302.9	289.6	278.9	274.3	224.7	290.7	256.8	237.8	287.9	254.2	274.3
1994	380.7	312.5	299.8	286.0	277.4	230.2	297.5	264.1	242.0	296.9	260.0	280.7
Hours worked	KIOD	KIOE	KIOF	KIOG	KIOH	KIOI	KIOJ	KIOK	KIOL	KIOM	KION	KIPO
1990	43.5	45.0	45.5	45.1	46.0	44.1	47.8	47.2	43.3	45.3	45.5	45.4
1991	43.8	44.0	43.3	44.1	45.4	43.5	46.6	47.0	43.1	43.7	44.9	44.4
1992	43.6	44.1	43.7	44.3	45.1	43.2	47.2	46.3	42.8	44.0	44.9	44.5
1993	43.0	44.2	43.3	44.2	44.7	43.4	46.8	46.6	42.6	43.8	44.7	44.3
1994	43.2	44.4	43.8	44.4	45.1	43.7	47.4	47.7	42.9	44.1	45.2	44.7
Hourly earnings	KIPB	KIPC	KIPD	KIPE	KIPF	KIPG	KIPH	KIPI	KIPJ	KIPK	KIPL	KIPM
1990	6.83	5.81	5.58	5.36	5.31	4.51	5.22	5.12	4.57	5.55	4.85	5.28
1991	7.50	6.19	6.08	5.74	5.63	4.83	5.71	5.42	5.03	5.98	5.28	5.70
1992	8.22	6.61	6.45	6.08	6.05	5.10	5.99	5.67	5.36	6.35	5.56	6.05
1993	8.54	6.81	6.68	6.29	6.12	5.21	6.23	5.60	5.62	6.56	5.72	6.21
1994	8.82	7.01	6.85	6.42	6.13	5.32	6.31	5.66	5.69	6.72	5.81	6.31
Women Weekly earnings	KIPZ	KIQA	KIQB	KIQC	KIQD	KIQE	KIQF	KIQG	KIQH	KIQI	KIQJ	KIQK
1990	..[1]	165.6	158.3	147.7	..[1]	128.7	201.2	178.1	138.7	152.8	143.6	148.4
1991	..[1]	176.3	167.2	157.2	..[1]	140.7	222.6	186.9	149.7	162.1	156.3	159.2
1992	..[1]	190.1	181.2	168.4	..[1]	143.7	244.4	189.4	161.6	174.4	166.0	170.1
1993	..[1]	198.0	187.4	177.3	..[1]	148.4	272.9	197.0	166.0	182.4	172.3	177.1
1994	..[1]	206.2	194.7	178.8	..[1]	155.5	271.8	195.2	170.7	186.4	177.6	181.9
Hours worked	KIQX	KIQY	KIQZ	KIRA	KIRB	KIRC	KIRD	KIRE	KIRF	KIRG	KIRH	KIRI
1990	..[1]	40.9	41.0	40.3	..[1]	39.3	41.5	39.6	38.8	40.5	39.3	40.0
1991	..[1]	40.3	39.9	40.0	..[1]	39.5	41.4	39.3	39.0	40.0	39.4	39.7
1992	..[1]	40.3	40.3	40.2	..[1]	39.3	42.3	40.5	38.9	40.2	39.4	39.8
1993	..[1]	40.4	40.4	40.3	..[1]	39.2	41.9	40.1	38.8	40.4	39.2	39.8
1994	..[1]	41.0	40.9	40.5	..[1]	39.6	42.2	40.1	38.9	40.6	39.5	40.1
Hourly earnings	KIRV	KIRW	KIRX	KIRY	KIRZ	KISA	KISB	KISC	KISD	KISE	KISF	KISG
1990	..[1]	4.06	3.86	3.67	..[1]	3.29	4.68	4.59	3.59	3.77	3.64	3.71
1991	..[1]	4.38	4.18	3.93	..[1]	3.60	5.21	4.95	3.87	4.06	3.97	4.01
1992	..[1]	4.71	4.50	4.19	..[1]	3.71	5.46	5.02	4.22	4.34	4.22	4.28
1993	..[1]	4.90	4.64	4.40	..[1]	3.83	5.74	5.04	4.32	4.53	4.32	4.42
1994	..[1]	5.04	4.78	4.42	..[1]	3.99	5.80	5.01	4.45	4.59	4.46	4.53
All Weekly earnings	KIST	KISU	KISV	KISW	KISX	KISY	KISZ	KITA	KITB	KITC	KITD	KITE
1990	300.6	250.3	240.7	214.2	245.2	181.3	245.6	229.6	176.4	231.9	203.6	223.3
1991	331.5	260.9	250.3	225.0	256.6	192.7	263.3	244.4	192.1	241.9	219.4	236.2
1992	357.6	280.1	268.3	240.4	274.0	201.9	279.1	252.5	204.2	258.9	230.9	250.7
1993	366.7	289.7	276.1	250.0	273.7	207.4	289.4	251.8	210.6	267.4	236.5	256.6
1994	378.7	298.6	285.7	255.8	276.8	213.8	295.5	258.4	215.3	275.4	242.5	262.7
Hours worked	KITR	KITS	KITT	KITU	KITV	KITW	KITX	KITY	KITZ	KIUO	KIUP	KIUQ
1990	43.4	44.5	44.9	43.6	46.0	42.9	47.4	46.5	41.7	44.3	44.2	44.4
1991	43.7	43.6	42.8	42.9	45.3	42.6	46.3	46.2	41.6	42.9	43.8	43.6
1992	43.5	43.7	43.3	43.0	45.0	42.4	46.9	45.8	41.4	43.2	43.8	43.7
1993	42.9	43.7	42.9	43.1	44.7	42.5	46.5	46.1	41.2	43.1	43.6	43.5
1994	43.1	43.9	43.4	43.3	45.0	42.9	47.1	47.1	41.5	43.4	44.1	43.9
Hourly earnings	KIVR	KIVS	KIVT	KIVU	KIVV	KIVW	KIVX	KIVY	KIVZ	KIWO	KIWP	KIWQ
1990	6.79	5.61	5.36	4.89	5.30	4.25	5.19	5.08	4.25	5.22	4.64	5.03
1991	7.46	5.97	5.84	5.22	5.63	4.57	5.68	5.38	4.64	5.62	5.04	5.43
1992	8.18	6.39	6.20	5.56	6.04	4.82	5.96	5.62	4.98	5.98	5.32	5.76
1993	8.51	6.58	6.43	5.77	6.11	4.93	6.21	5.56	5.18	6.19	5.46	5.92
1994	8.79	6.76	6.59	5.89	6.12	5.06	6.28	5.61	5.27	6.33	5.57	6.02

1 Denotes sample number too low or standard error too high for reliable estimate.

Source: ONS, New Earnings Survey: 01928 792077

7.19 Average earnings and hours of manual employees by industry division
Full time employees on adult rates: pay unaffected by absence: Great Britain

continued

At April

	Agriculture Hunting and forestry	Fishing	Mining and Quarrying	Manufacturing	Electricity Gas and Water Supply	Construction	Wholesale and Retail Trade; repair of motor vehicles, cycles, personal and household goods
Standard Industrial Classification: Revised 1992							
Full-time manual men							
Weekly earnings	KOTK	KOTL	KOTM	KOTN	KOTO	KOTP	KOTQ
1995	237.3	..	340.3	313.4	367.8	294.7	257.8
1996	241.2	..	367.8	323.6	399.7	308.2	264.1
1997	252.1	..	400.5	337.5	401.2	324.8	275.1
1998	260.3	..[1]	408.3	352.6	418.6	342.3	292.2
1999	272.4	..[1]	396.0	354.6	440.5	351.3	299.4
Hours worked	KOTZ	KPFV	KOUA	KOUB	KOUC	KOUD	KOUE
1995	48.2	..[1]	52.0	44.9	42.7	45.9	44.4
1996	47.5	46.5	50.8	44.2	42.6	45.8	44.4
1997	47.8	47.0	52.0	44.5	42.1	46.9	44.0
1998	47.0	..[1]	50.1	44.3	42.5	46.9	44.3
1999	47.5	..[1]	51.7	43.5	42.8	46.4	43.9
Hourly earnings	KOUN	KPFW	KOUO	KOUP	KOUQ	KOUR	KOUS
1995	4.93	..[1]	6.56	6.98	8.62	6.40	5.80
1996	5.06	..[1]	7.15	7.29	9.38	6.71	5.91
1997	5.27	..[1]	7.70	7.58	9.45	6.92	6.26
1998	5.54	..[1]	8.14	7.96	9.85	7.29	6.59
1999	5.75	..[1]	7.66	8.15	10.30	7.56	6.84
Full-time manual women							
Weekly earnings	KOYL	KPFX	KOYM	KOYN	KOYO	KOYP	KOYQ
1995	175.3	..[1]	..[1]	198.5	..[1]	..[1]	174.2
1996	177.9	..[1]	..[1]	205.0	..[1]	..[1]	185.4
1997	186.9	..[1]	..[1]	214.1	..[1]	..[1]	194.1
1998	185.7	..[1]	..[1]	224.2	..[1]	..[1]	203.6
1999	199.0	..[1]	..[1]	231.7	..[1]	..[1]	215.3
Hours worked	KOYZ	KPFY	KOZA	KOZB	KOZC	KOZD	KOZE
1995	41.6	..[1]	..[1]	40.9	..[1]	..[1]	39.6
1996	41.3	..[1]	..[1]	40.7	..[1]	..[1]	40.0
1997	40.9	..[1]	42.6	40.8	38.1	43.8	40.3
1998	42.0	..[1]	..[1]	40.7	..[1]	..[1]	40.3
1999	41.8	..[1]	..[1]	40.4	..[1]	..[1]	39.9
Hourly earnings	KOZN	KOZO	KOZP	KOZQ	KOZR	KOZS	KOZT
1995	4.21	..[1]	..[1]	4.86	..[1]	..[1]	4.39
1996	4.33	..[1]	..[1]	5.04	..[1]	..[1]	4.63
1997	4.50	..[1]	..[1]	5.26	..[1]	5.54	4.81
1998	4.43	..[1]	..[1]	5.52	..[1]	..[1]	5.06
1999	4.74	..[1]	..[1]	5.74	..[1]	..[1]	5.44
Full-time manual adults							
Weekly earnings	KPBO	KPFZ	KPBP	KPBQ	KPBR	KPBS	KPBT
1995	231.2	..[1]	338.9	290.7	366.6	293.4	246.0
1996	234.7	..[1]	366.5	300.8	398.5	307.0	253.5
1997	245.4	..[1]	398.9	314.6	399.6	323.4	264.3
1998	251.7	..[1]	403.8	329.9	416.2	340.9	280.3
1999	264.1	..[1]	392.5	333.7	439.5	350.9	287.7
Hours worked	KPCT	KPGA	KPCU	KPCV	KPCW	KPCX	KPCY
1995	47.6	45.6	50.8	44.1	..[1]	..[1]	43.7
1996	46.9	46.5	50.8	43.5	42.6	45.8	43.9
1997	47.1	47.0	51.9	43.8	42.0	46.8	43.5
1998	46.4	..[1]	49.9	43.7	42.4	46.9	43.8
1999	46.8	..[1]	51.6	43.0	42.7	46.4	43.4
Hourly earnings	KPDV	KPGB	KPDW	KPDX	KPDY	KPDZ	KPFA
1995	4.86	6.12	6.56	6.60	8.60	6.39	5.62
1996	5.00	..[1]	7.15	6.88	9.36	6.70	5.75
1997	5.20	5.42	7.69	7.18	9.42	6.90	6.08
1998	5.42	..[1]	8.07	7.56	9.81	7.27	6.40
1999	5.65	..[1]	7.61	7.77	10.29	7.56	6.66

1 Denotes sample number too low or standard error too high for reliable estimate.

Source: ONS, New Earnings Survey: 01928 792077

7.19 Average earnings and hours of manual employees by industry division
Full time employees on adult rates: pay unaffected by absence: Great Britain

continued

At April

	Hotels and restaurants	Transport, Storage and Communication	Financial Intermediation	Real Estate, Renting and Business	Public Administration and Defence; compulsory social security	Education	Health and Social work	Other community, social and personal service activities
Standard Industrial Classification: Revised 1992								
Full-time manual men								
Weekly earnings	KOTR	KOTS	KOTT	KOTU	KOTV	KOTW	KOTX	KOTY
1995	196.5	306.0	338.0	269.3	264.4	233.4	232.1	259.5
1996	203.3	314.9	373.8	275.6	275.4	243.5	241.3	264.4
1997	213.9	328.4	381.6	292.4	281.2	268.7	250.7	280.2
1998	227.0	344.0	394.8	302.7	289.0	277.9	264.4	289.0
1999	230.5	359.6	377.9	312.3	307.5	285.2	267.7	300.8
Hours worked	KOUF	KOUG	KOUH	KOUI	KOUJ	KOUK	KOUL	KOUM
1995	42.3	47.8	41.9	47.0	42.0	..[1]	41.9	44.5
1996	42.1	47.6	42.5	46.4	42.4	41.7	41.8	44.0
1997	41.9	48.5	42.1	46.7	42.3	41.5	42.4	44.3
1998	42.5	48.0	42.5	46.4	41.9	41.7	42.1	44.8
1999	42.1	47.5	39.9	45.5	42.2	40.8	41.8	44.8
Hourly earnings	KOUT	KOYE	KOYF	KOYG	KOYH	KOYI	KOYJ	KOYK
1995	4.65	6.38	8.07	5.74	6.31	5.55	5.55	5.86
1996	4.83	6.60	8.76	5.94	6.50	5.83	5.78	6.03
1997	5.10	6.77	9.07	6.27	6.66	6.48	5.92	6.33
1998	5.34	7.17	9.28	6.53	6.90	6.67	6.26	6.47
1999	5.48	7.56	9.46	6.86	7.30	7.00	6.41	6.72
Full-time manual women								
Weekly earnings	KOYR	KOYS	KOYT	KOYU	KOYV	KOYW	KOYX	KOYY
1995	147.2	279.9	..[1]	181.7	207.4	168.6	172.3	162.7
1996	156.7	286.7	..[1]	199.6	214.6	167.3	175.5	166.6
1997	160.5	278.3	..[1]	205.3	218.5	197.1	178.5	176.0
1998	170.8	292.7	..[1]	211.9	232.2	203.7	187.2	178.1
1999	180.5	311.5	..[1]	221.5	232.3	223.0	199.5	197.4
Hours worked	KOZF	KOZG	KOZH	KOZI	KOZJ	KOZK	KOZL	KOZM
1995	39.0	42.4	..[1]	40.4	39.7	38.7	39.4	39.9
1996	39.4	42.4	37.9	41.6	39.6	38.6	39.1	40.0
1997	39.2	42.2	37.5	41.8	39.4	38.8	39.1	39.7
1998	39.4	42.0	..[1]	41.4	39.6	38.3	39.3	39.6
1999	39.3	41.5	..[1]	40.4	38.6	38.0	39.1	39.8
Hourly earnings	KOZU	KOZV	KOZW	KOZX	KOZY	KOZZ	KPBN	KTDJ
1995	3.78	6.09	..[1]	4.54	5.23	4.34	4.38	4.07
1996	3.98	6.29	..[1]	4.81	5.43	4.35	4.49	4.17
1997	4.07	6.52	6.14	4.92	5.57	5.05	4.56	4.45
1998	4.30	6.98	..[1]	5.08	5.89	5.32	4.75	4.50
1999	4.55	7.52	..[1]	5.48	6.09	5.86	5.10	4.94
Full-time manual adults								
Weekly earnings	KPBU	KPBV	KPBW	KPBX	KPBY	KPBZ	KPCR	KPCS
1995	175.1	304.1	319.1	257.1	255.1	211.2	195.3	236.7
1996	182.4	312.5	349.7	264.6	264.7	220.0	199.7	241.8
1997	190.6	324.8	355.6	278.2	271.4	244.3	208.1	249.1
1998	203.6	340.0	367.3	289.1	280.7	254.8	218.9	258.3
1999	210.3	355.7	353.6	298.2	296.9	266.0	227.8	273.8
Hours worked	KPCZ	KPDO	KPDP	KPDQ	KPDR	KPDS	KPDT	KPDU
1995	40.9	47.5	..[1]	46.1	41.6	40.9	40.4	43.4
1996	40.9	47.3	41.7	45.7	41.9	40.8	40.1	43.0
1997	40.7	48.1	41.3	45.9	41.8	40.6	40.5	42.9
1998	41.2	47.6	41.6	45.6	41.6	40.7	40.5	43.3
1999	41.0	47.0	39.7	44.7	41.7	40.0	40.2	43.5
Hourly earnings	KPFB	KPFC	KPFD	KPFE	KPFF	KPFG	KPFT	KPFU
1995	4.29	6.37	7.76	5.60	6.14	5.16	4.85	5.47
1996	4.46	6.58	8.34	5.80	6.32	5.39	4.99	5.63
1997	4.67	6.76	8.61	6.07	6.50	6.02	5.14	5.82
1998	4.93	7.16	8.84	6.33	6.76	6.27	5.40	5.97
1999	5.12	7.56	8.92	6.67	7.15	6.66	5.66	6.29

1 Denotes sample number too low or standard error too high for reliable estimate.

2 It is now possible to recalculate data for previous years using SIC92. For further information contact the NES helpline on 01928 792077.

Source: ONS, New Earnings Survey: 01928 792077

7.20 Average weekly and hourly earnings and hours of full-time employees on adult rates: Great Britain

At April

	All Industries				Manufacturing industries			
			Average hourly earnings				Average hourly earnings	
	Average weekly earnings[1]	Average hours	including overtime	excluding overtime	Average weekly earnings	Average hours	including overtime	excluding overtime
	£		£	£	£		£	£
All adults								
	KBRZ	KBSA	KIUY	KIUZ	KBSD	KBSE	KIVO	KIVP
1994	325.7	40.1	8.03	8.03	321.6	41.7	7.62	7.58
1995	336.3	40.3	8.31	8.32	334.3	42.2	7.92	7.85
1996	351.7	40.2	8.71	8.72	349.2	41.9	8.29	8.22
1997	367.6	40.3	9.10	9.13	361.7	42.0	8.60	8.53
1998	384.5	40.2	9.53	9.54	384.5	41.8	9.17	9.10
1999	400.1	40.0	10.01	10.03	395.3	41.4	9.55	9.49
All men								
	KBSH	KBSI	KIVQ	KIWR	KBSL	KBSM	KIWS	KIWT
1994	362.1	41.6	8.61	8.65	350.9	42.5	8.16	8.12
1995	374.6	41.9	8.91	8.97	364.0	43.0	8.44	8.41
1996	391.6	41.7	9.34	9.39	380.0	42.7	8.86	8.81
1997	408.7	41.8	9.74	9.82	392.7	42.8	9.16	9.12
1998	427.1	41.7	10.20	10.26	416.8	42.6	9.75	9.72
1999	442.4	41.4	10.68	10.75	424.6	42.0	10.10	10.06
Manual men								
	KFHX	KFHY	KIWU	KIWV	KFJT	KFJU	KIWW	KIWX
1994	280.7	44.7	6.31	6.14	296.9	44.1	6.72	6.50
1995	291.3	45.2	6.44	6.25	313.4	44.9	6.98	6.74
1996	301.3	44.8	6.70	6.51	323.6	44.2	7.29	7.05
1997	314.3	45.1	6.97	6.79	337.5	44.5	7.58	7.34
1998	328.5	45.0	7.30	7.10	352.6	44.3	7.96	7.71
1999	335.0	44.4	7.54	7.36	354.6	43.5	8.15	7.92
Non-manual men								
	KFJX	KFJY	KIWY	KIWZ	KFMU	KFMV	KIXO	KIXP
1994	428.2	38.9	10.90	10.93	434.7	39.7	10.79	10.84
1995	443.3	39.0	11.33	11.36	449.2	39.9	11.24	11.29
1996	464.5	39.1	11.83	11.87	479.6	39.9	11.95	12.00
1997	483.5	39.1	12.33	12.39	489.2	39.8	12.28	12.32
1998	506.1	39.1	12.90	12.94	525.9	39.8	13.17	13.23
1999	525.5	39.0	13.49	13.52	541.6	39.6	13.68	13.73
All women								
	KBTF	KBTG	KIXQ	KIXR	KBTJ	KBTK	KIXS	KIXT
1994	261.5	37.6	6.89	6.88	226.8	39.1	5.76	5.72
1995	269.8	37.6	7.15	7.14	236.7	39.4	6.01	5.96
1996	283.0	37.6	7.51	7.50	246.7	39.3	6.27	6.23
1997	297.2	37.6	7.88	7.88	258.8	39.2	6.60	6.56
1998	309.6	37.6	8.23	8.22	274.5	39.2	7.01	6.97
1999	326.5	37.5	8.71	8.70	292.1	39.0	7.49	7.46
Manual women								
	KFMY	KFMZ	KIXU	KIXV	KFPS	KFPT	KIXW	KIXX
1994	181.9	40.1	4.53	4.45	186.4	40.6	4.59	4.49
1995	188.1	40.2	4.64	4.55	198.5	40.9	4.86	4.74
1996	195.2	40.2	4.81	4.72	205.0	40.7	5.04	4.92
1997	201.1	40.2	4.99	4.90	214.1	40.8	5.26	5.13
1998	210.8	40.2	5.23	5.14	224.2	40.7	5.52	5.39
1999	221.9	39.9	5.56	5.48	231.7	40.4	5.74	5.62
Non-manual women								
	KFRY	KFRZ	KIXY	KIYA	KFUY	KFUZ	KIYB	KIYC
1994	278.4	37.0	7.44	7.42	263.2	37.7	6.94	6.92
1995	288.1	37.0	7.76	7.75	275.0	37.8	7.26	7.23
1996	302.4	37.1	8.16	8.14	289.4	37.9	7.64	7.61
1997	317.8	37.1	8.56	8.55	300.0	37.8	7.94	7.92
1998	330.1	37.0	8.90	8.89	317.2	37.9	8.38	8.36
1999	346.9	37.0	9.37	9.36	341.5	37.9	9.02	9.01

1 Excluding those whose pay was affected by absence.

Source: ONS, New Earnings Survey: 01928 792077

7.21 Average earnings index: all employees: main industrial sectors

Great Britain

Analyses by industry based on Standard Industrial Classification 1992

1995 = 100

Unadjusted

	Annual averages	January	February	March	April	May	June	July	August	September[1]	October	November	December
Whole economy (Divisions 01 - 93)													
	LNMM												
1995	100.0	98.2	98.9	103.3	99.5	99.2	99.6	100.3	99.2	99.1	99.5	100.5	102.7
1996	103.6	101.0	102.4	106.8	103.0	102.4	103.0	104.0	102.7	103.3	103.0	104.1	107.1
1997	108.0	105.5	106.1	112.2	106.9	106.6	106.9	108.2	107.3	107.5	107.5	108.8	112.5
1998	113.5	110.7	111.7	118.1	113.1	113.2	112.6	114.0	112.4	112.8	112.6	113.7	117.2
1999	..	115.7	117.5	124.0	117.3	117.9	118.6	119.0	117.9†	117.8	..	..	..
Manufacturing industries (Divisions 15 - 37)													
	LNMN												
1995	100.0	97.4	98.7	102.5	99.6	99.5	99.7	100.9	98.5	99.1	100.2	100.8	102.9
1996	104.4	101.3	102.8	107.6	103.9	103.5	104.0	105.3	102.9	103.5	104.2	105.7	108.0
1997	108.8	105.4	107.3	111.1	108.0	107.9	108.3	109.2	107.5	107.8	108.9	110.8	112.9
1998	113.7	110.5	112.7	117.2	113.4	112.7	113.1	114.6	112.4	112.4	113.7	114.7	116.6
1999	..	115.0	116.6	121.3	117.4	116.6	117.0	118.7	117.0	117.4	..	..	..
Production industries (Divisions 10 - 41)													
	LNMO												
1995	100.0	97.5	98.8	102.3	99.6	99.5	99.7	101.0	98.6	99.2	100.2	100.7	102.9
1996	104.4	101.3	102.7	107.4	103.8	103.5	104.0	105.3	102.9	103.5	104.2	105.7	108.0
1997	108.5	105.3	107.1	110.9	107.7	107.9	108.1	109.0	107.2	107.5	108.6	110.6	112.7
1998	113.4	110.3	112.4	117.0	113.3	112.5	112.9	114.2	112.2	112.1	113.3	114.3	116.3
1999	..	114.6	116.2	120.9	117.2	116.3	116.6	118.1	116.4	116.8	..	..	..
Service industries (Divisions 50 - 93)													
	LNMP												
1995	100.0	98.5	99.1	103.9	99.4	99.1	99.4	100.0	99.3	98.8	99.2	100.4	102.8
1996	103.3	101.1	102.4	106.7	102.9	102.1	102.7	103.6	102.5	102.9	102.6	103.5	106.9
1997	107.9	105.8	106.0	112.8	106.8	106.3	106.6	107.9	107.2	107.2	107.1	108.2	112.5
1998	113.5	110.9	111.7	119.0	113.2	113.4	112.3	113.7	112.2	112.6	112.0	113.1	117.3
1999	..	116.0	117.9	125.2	117.2	118.3	119.2	119.1	118.1	117.8	..	..	..

Seasonally adjusted

	Annual averages	January	February	March	April	May	June	July	August	September[1]	October	November	December
Whole economy (Divisions 01 - 93)													
	LNMQ												
1995	100.0	98.8	99.2	99.5	99.5	99.5	99.8	99.9	100.1	100.5	101.0	101.2	101.4
1996	103.6	101.6	102.4	102.3	102.8	102.6	103.4	103.7	103.9	104.7	104.7	105.2	105.7
1997	108.0	106.1	106.2	107.0	106.6	106.8	107.5	107.9	108.6	109.0	109.6	110.2	110.9
1998	113.5	111.3	111.7	112.3	112.7	113.2	113.1	113.8	114.0	114.5	114.8	115.2	115.6
1999	..	116.4	117.2	117.8	117.2	117.8	119.0	118.9	119.6	119.8	..	..	..
Manufacturing industries (Divisions 15 - 37)													
	LNMR												
1995	100.0	98.3	98.9	99.0	99.5	99.5	99.8	100.3	100.2	100.8	101.3	101.1	101.7
1996	104.3	102.3	102.6	103.3	103.6	103.6	104.1	104.5	104.7	105.2	105.3	106.0	106.6
1997	108.8	106.5	107.1	107.1	107.5	108.1	108.4	108.6	109.4	109.7	110.0	111.1	111.5
1998	113.7	111.7	112.4	112.8	112.9	113.1	113.5	114.0	114.3	114.5	115.0	115.0	115.2
1999	..	116.2	116.2	116.7	116.9	117.0	117.4	118.0	118.9	119.4	..	..	..
Production industries (Divisions 10 - 41)													
	LNMS												
1995	100.0	98.3	98.9	99.0	99.4	99.5	99.8	100.3	100.3	100.8	101.3	101.1	101.5
1996	104.3	102.2	102.6	103.2	103.4	103.6	104.1	104.5	104.7	105.2	105.5	106.0	106.4
1997	108.5	106.4	106.8	106.9	107.3	108.0	108.2	108.5	109.1	109.4	109.9	110.9	111.1
1998	113.5	111.5	112.0	112.6	112.8	112.9	113.2	113.7	114.0	114.3	114.8	114.7	114.9
1999	..	115.7	115.8	116.3	116.6	116.7	117.0	117.5	118.3	118.8	..	..	..
Service industries (Divisions 50 - 93)													
	LNMT												
1995	100.0	98.9	99.2	100.0	99.4	99.6	99.7	99.8	100.0	100.3	100.9	101.3	101.4
1996	103.4	101.5	102.1	102.2	102.6	102.5	103.2	103.4	103.7	104.5	104.6	104.9	105.4
1997	107.9	106.0	105.8	107.3	106.4	106.7	107.2	107.7	108.4	108.7	109.4	109.9	110.8
1998	113.4	111.2	111.4	112.4	112.7	113.6	112.9	113.7	113.8	114.4	114.6	115.1	115.5
1999	..	116.4	117.4	117.8	117.1	118.3	119.6	119.2	119.9	120.0	..	..	..

1 September 1999 data are provisional.

Source: Office for National Statistics: 01928 792442

7.22 Average earnings index[1] excluding bonus payments: all employee jobs: by industry (unadjusted)

Great Britain

March 1996 = 100

	Agriculture[2], forestry and fishing	Mining and quarrying	Food products, beverages and tobacco	Textiles	Clothing, leather and footwear	Wood, wood products and other manufacturing n.e.c.	Pulp, paper products, printing and publishing	Chemicals and chemical products	Rubber and plastic products	Other non-metallic mineral products	Basic metals	Fabricated metal products (excluding machinery)	Machinery and equipment n.e.c.
SIC 1992 Class	(01,02,05)	(10-14)	(15,16)	(17)	(18-19)	(20,23,36,37)	(21,22)	(24)	(25)	(26)	(27)	(28)	(29)
	LOTJ	LOTK	LOTL	LOTM	LOTN	LOTO	LOTP	LOTQ	LOTR	LOTS	LOTT	LOTU	LOTV
1997	-	104.8	103.6	105.1	105.0	107.0	104.4	105.2	105.4	105.1	107.7	104.8	105.1
1998	-	108.8	108.1	107.3	109.2	111.6	108.5	111.5	110.5	109.4	113.0	108.3	109.4
1996 May	-	100.5	100.6	100.4	100.6	100.9	100.2	101.1	101.1	101.0	103.3	100.2	100.7
Jun	-	100.5	101.1	102.2	100.9	101.2	100.6	101.6	101.8	101.3	103.8	100.5	101.2
Jul	-	100.6	101.6	103.2	101.2	101.5	101.1	101.6	102.3	101.5	105.5	100.7	101.5
Aug	-	100.2	101.3	103.4	101.0	101.1	101.5	101.3	102.1	101.3	106.1	100.9	101.2
Sep	-	100.4	100.9	102.6	101.3	101.9	101.7	101.5	102.1	101.3	106.8	101.2	101.2
Oct	-	100.6	100.2	102.9	101.5	102.0	102.0	101.4	102.0	101.1	104.7	101.6	101.0
Nov	-	101.7	100.8	103.7	102.3	103.1	102.3	101.9	102.5	101.7	105.1	102.4	101.9
Dec	-	102.6	101.2	104.1	102.4	103.4	102.7	102.3	102.9	101.9	106.2	102.5	102.1
1997 Jan	-	103.3	101.6	103.7	103.0	103.7	102.4	102.7	103.1	101.9	106.3	102.3	102.3
Feb	-	103.8	101.4	103.4	103.5	103.8	102.2	103.0	103.2	102.0	106.5	102.4	102.6
Mar	-	103.9	101.6	103.1	104.3	104.3	101.8	103.0	103.6	102.5	106.0	102.9	102.9
Apr	-	104.5	102.4	103.6	104.8	105.1	102.4	103.7	104.0	103.5	106.3	103.8	104.1
May	-	104.7	103.1	104.0	105.2	106.2	102.9	104.2	104.4	104.4	107.0	104.0	104.5
Jun	-	104.7	103.3	105.1	105.5	106.7	103.8	105.0	104.8	105.2	107.1	104.4	105.6
Jul	-	105.0	103.8	105.6	105.6	107.1	104.2	105.1	105.5	105.4	108.1	104.8	105.6
Aug	-	104.8	103.8	105.8	105.2	106.5	105.0	105.4	105.8	105.5	107.4	104.7	105.7
Sep	-	104.6	103.9	105.4	104.9	106.8	105.6	105.5	106.1	105.5	108.6	105.1	105.2
Oct	-	104.0	103.9	105.6	104.8	107.4	106.4	105.9	106.0	106.0	108.0	105.3	105.2
Nov	-	104.8	104.6	106.6	105.2	109.1	106.5	106.4	106.7	106.6	109.0	106.3	105.7
Dec	-	106.1	105.6	107.0	105.3	110.0	106.4	107.5	107.1	107.2	109.2	106.7	106.8
1998 Jan	-	106.4	105.9	106.4	105.8	110.7	105.8	108.2	107.7	107.7	109.5	106.9	107.0
Feb	-	106.8	105.9	105.3	106.5	110.6	105.5	109.0	108.0	108.0	110.0	107.0	107.4
Mar	-	106.8	105.9	105.0	107.7	111.2	105.6	109.1	108.6	108.0	110.3	107.1	107.6
Apr	-	108.0	106.6	105.8	108.6	111.8	106.4	109.6	109.6	108.2	112.0	108.0	108.5
May	-	108.7	107.7	106.3	109.1	112.6	107.4	110.0	110.4	108.6	113.1	108.6	109.0
Jun	-	108.9	108.4	107.5	109.4	112.1	108.0	110.9	110.5	109.1	113.5	108.8	109.5
Jul	-	108.7	108.8	107.7	109.8	112.2	108.3	111.3	110.4	109.8	114.4	108.9	110.1
Aug	-	108.4	108.2	108.2	109.7	111.4	108.7	111.8	110.5	110.1	114.6	108.7	110.5
Sep	-	108.7	108.2	107.8	109.8	111.3	109.3	111.7	111.2	110.2	114.8	108.8	110.4
Oct	-	109.3	108.0	107.9	109.4	110.9	110.1	112.1	111.5	110.0	114.1	108.2	110.1
Nov	-	110.0	109.0	108.7	109.8	111.8	110.7	112.9	111.7	110.3	113.7	108.4	110.0
Dec	-	110.6	109.9	108.7	109.8	111.9	111.1	114.5	111.7	110.5	113.4	108.5	110.0
1999 Jan[5]	-	110.7	110.1	108.6	110.2	111.6	111.4	115.3	111.7	110.4	111.7	108.6	109.9
1999 Feb[5]	-	109.8	109.6	107.5	110.0	111.1	111.1	115.6	111.6	110.1	110.9	108.0	109.7
Mar	-	109.1	109.1	107.4	110.5	111.3	110.7	115.5	111.4	110.5	111.4	107.7	109.6
Apr	-	108.8	108.9	107.9	110.4	111.8	110.7	116.6	111.4	111.4	112.0	108.1	110.1
May	-	109.1	109.3	109.2	110.9	112.6	111.2	117.4	111.8	112.2	114.0	108.7	110.7
Jun	-	109.4	109.5	110.6	111.0	113.4	111.8	118.5	112.2	112.6	115.2	109.5	111.3
Jul	-	109.4	109.8	111.6	111.4	114.3	112.1	118.7	112.5	113.0	117.0	110.0	111.7
Aug	-	109.7	110.0	112.3	111.1	115.0	112.7	119.1	113.3	113.6	117.2	109.8	112.0
Sep[3]	-	109.7	110.3	112.5	111.7	116.0	113.5	119.7	114.2	114.1	117.5	110.0	112.0

1 Users should note that data contained in this table are not comparable with those previously published in Table 18.11 (prior to March 1999). Excluding bonuses and averaging the data over a three month period renders the data fundamentally different to the previous indices published for the same industries, but which included bonuses and related to single months only.

2 As a result of a discontinuity in the reporting of data for the agricultural sector, this series is not available.

3 Provisional.

4 The index for the sector Education, health and social work is based on a sample which excludes representatives of the private health and social work sector until June 1998. Monthly movements in the index for this sector therefore exclude health and social work up to May 1998.

5 As a result of a change in the survey questionnaire the series excluding bonuses are subject to a discontinuity between January and February 1999.

Source: Office for National Statistics: 01928 792442

7.22 Average earnings index[1] excluding bonus payments: all employee jobs: by industry (unadjusted)

continued

Great Britain

March 1996 = 100

	Electrical and optical equipment	Transport equipment	Electricity, gas and water supply	Constuction	Wholesale trade	Retail trade and repairs	Hotels and restaurants	Transport, storage and communication	Financial intermediation	Real estate, renting and business activities	Public administration	Education, health and social work[4]	Other services
SIC 1992 Class	(30-33)	(34,35)	(40,41)	(45)	(51)	(50,52)	(55)	(60-64)	(65-67)	(70-74)	(75)	(80-85)	(90-93)
	LOTW	LOTX	LOTY	LOTZ	LOUA	LOUB	LOUC	LOUD	LOUE	LOUF	LOUG	LOUH	LOUI
1997	105.7	101.6	101.1	103.4	104.9	97.9	106.3	104.2	106.7	104.1	101.0	104.6	106.1
1998	110.1	106.3	103.8	110.4	110.8	101.8	110.8	108.3	113.3	110.2	103.6	107.7	114.7
1996 May	100.8	98.0	100.9	99.0	100.4	96.2	100.3	99.5	100.8	100.4	99.4	101.1	99.1
Jun	101.1	97.1	101.4	98.8	100.7	94.6	100.5	99.3	101.1	100.3	99.0	101.9	98.5
Jul	101.2	97.7	101.5	99.2	100.7	95.1	101.3	99.2	100.8	100.1	99.4	102.7	99.2
Aug	101.2	97.6	101.8	99.2	101.2	95.5	101.7	99.2	100.9	99.7	99.6	103.4	99.8
Sep	101.4	97.3	101.9	99.3	101.4	95.6	101.8	99.9	101.2	99.7	100.3	104.1	100.9
Oct	101.5	96.8	102.5	99.3	101.6	95.1	101.7	100.5	101.5	99.8	100.1	103.8	100.9
Nov	101.9	97.8	102.4	100.0	101.5	94.7	101.5	101.3	101.8	100.5	100.3	103.0	101.9
Dec	102.7	99.0	103.2	100.5	102.0	94.7	103.6	101.5	102.3	101.2	99.9	102.5	102.4
1997 Jan	102.9	99.5	101.8	100.9	102.1	95.1	104.5	102.1	102.9	101.8	100.1	102.6	103.1
Feb	103.4	99.8	100.9	101.2	102.5	95.5	105.2	102.0	103.4	102.3	100.2	102.9	102.8
Mar	103.6	99.9	99.2	101.5	102.6	96.9	104.3	101.8	103.7	102.4	100.3	103.0	103.9
Apr	104.5	100.6	99.4	101.7	103.4	97.3	103.9	102.3	104.5	102.7	100.2	103.5	104.1
May	105.0	100.9	100.4	102.1	104.1	98.1	104.5	103.1	105.5	103.3	100.0	103.8	104.6
Jun	105.6	100.9	101.2	102.4	104.5	97.6	104.9	103.8	106.5	103.8	99.8	104.1	103.4
Jul	106.4	101.4	101.6	103.1	104.9	98.2	106.2	104.2	107.2	104.3	99.9	104.6	103.7
Aug	106.6	101.2	101.0	103.1	105.2	98.6	106.9	104.4	107.7	104.2	100.5	105.6	105.1
Sep	106.5	101.5	101.0	103.7	105.5	99.0	107.1	105.1	107.9	104.3	101.1	106.5	107.0
Oct	106.2	101.7	101.1	104.2	105.7	98.8	107.2	105.2	108.1	104.4	101.9	106.3	108.4
Nov	106.6	103.0	102.1	105.3	105.9	98.4	107.1	105.8	108.2	104.8	102.5	105.4	109.1
Dec	107.1	104.1	102.5	106.0	106.9	98.2	108.8	106.3	108.7	105.6	102.9	104.8	110.0
1998 Jan	107.1	104.6	102.5	106.4	107.4	98.5	109.9	107.0	109.3	106.8	102.6	104.8	110.6
Feb	107.5	104.8	102.2	106.8	107.9	99.0	110.4	107.3	110.0	108.0	102.5	104.9	110.8
Mar	107.7	105.0	102.3	107.1	107.8	99.2	109.5	107.3	110.7	108.7	102.8	104.5	111.5
Apr	108.6	105.7	103.2	107.7	108.8	99.8	109.3	108.6	111.3	109.2	102.9	104.9	112.8
May	109.0	106.0	103.9	108.3	109.9	101.1	109.9	109.8	112.3	109.9	102.9	105.6	113.7
Jun	109.4	106.6	104.1	109.0	111.0	102.1	110.4	108.8	113.2	110.3	102.9	106.7	113.6
Jul	110.0	107.0	104.3	110.1	111.6	102.9	110.9	107.4	113.8	110.5	103.4	107.7	114.0
Aug	110.4	107.0	104.3	111.0	111.9	103.0	110.8	106.5	113.8	110.0	103.8	108.9	115.5
Sep	110.7	106.6	104.6	111.9	112.0	103.3	110.8	108.0	114.0	110.2	103.7	109.8	116.5
Oct	111.0	106.2	104.5	112.4	112.1	102.8	110.7	108.3	114.4	110.2	104.1	109.9	116.9
Nov	111.6	106.4	104.5	113.3	112.2	102.5	111.0	109.3	114.9	111.1	104.3	109.5	116.9
Dec	112.5	107.1	104.5	113.5	112.5	102.6	112.7	108.9	115.4	111.8	104.7	109.5	117.2
1999 Jan[5]	112.9	107.3	103.7	113.6	112.5	103.4	113.7	109.3	115.8	112.9	104.8	109.8	117.1
1999 Feb[5]	113.2	107.5	102.5	113.0	112.4	103.1	113.8	109.5	115.7	113.5	104.8	110.2	117.0
Mar	113.5	107.7	101.4	113.0	112.4	102.1	112.9	109.8	115.9	114.0	105.0	109.9	117.2
Apr	114.0	108.8	102.2	113.0	113.1	101.6	113.4	109.8	116.5	114.6	105.0	110.3	117.6
May	114.6	109.6	103.8	113.3	113.6	102.1	115.2	110.2	117.6	115.4	105.1	111.1	118.3
Jun	115.1	110.2	104.9	113.4	113.8	103.2	117.1	110.9	118.2	116.0	105.6	112.7	119.5
Jul	116.0	110.5	103.7	113.9	113.8	103.2	118.0	111.9	118.6	116.0	105.9	113.5	121.2
Aug	116.9	111.1	102.2	114.4	114.0	103.7	119.0	111.8	118.6	115.2	106.1	114.5	122.5
Sep[3]	118.0	111.5	101.0	115.5	114.3	104.0	118.9	112.4	118.7	114.6	105.9	114.8	123.7

See footnotes on previous page.

Source: Office for National Statistics: 01928 792442

7.23 Gross weekly and hourly earnings of full-time adult employees

Great Britain

At April

	Gross weekly earnings					Gross hourly earnings				
	Lowest decile	Lower quartile	Median	Upper quartile	Highest decile	Lowest decile	Lower quartile	Median	Upper quartile	Highest decile
	£	£	£	£	£	£	£	£	£	£
All men										
	KBYD	KBYE	KBYF	KBYG	KBYH	KIYD	KIYE	KIYF	KIYG	KIYH
1991	160.7	206.9	277.5	376.5	507.8	3.87	4.88	6.53	9.12	12.99
1992	170.2	219.3	295.9	401.9	544.1	4.12	5.17	6.97	9.79	14.04
1993	174.7	226.0	304.6	417.3	567.2	4.22	5.35	7.25	10.28	14.55
1994	179.9	231.1	312.8	427.3	581.7	4.33	5.44	7.39	10.51	14.96
1995	182.0	237.1	323.2	442.7	600.8	4.33	5.55	7.60	10.90	15.60
1996	189.0	245.2	334.9	461.0	633.1	4.50	5.75	7.91	11.41	16.36
1997	198.6	256.4	349.7	480.0	656.9	4.71	6.00	8.24	11.89	17.01
1998	203.9	265.3	362.8	499.0	685.1	4.88	6.24	8.57	12.34	17.75
1999	211.1	274.5	374.3	517.3	711.6	5.09	6.50	8.92	12.96	18.63
Manual men										
	KFZY	KFZZ	KGAK	KGAL	KGAM	KIYN	KIYO	KIYP	KIYQ	KIYR
1991	149.3	186.0	235.4	298.2	375.5	3.60	4.36	5.40	6.70	8.10
1992	157.6	197.4	250.7	316.7	397.3	3.82	4.62	5.72	7.16	8.68
1993	160.9	201.2	256.4	324.0	407.1	3.87	4.72	5.89	7.35	8.98
1994	165.9	206.1	261.8	331.6	414.2	3.99	4.82	5.95	7.46	9.10
1995	166.9	211.1	271.8	347.1	437.5	3.97	4.86	6.09	7.66	9.43
1996	173.7	218.4	280.0	358.9	452.5	4.11	5.01	6.32	7.99	9.83
1997	182.4	228.8	292.5	373.9	472.8	4.32	5.25	6.60	8.28	10.17
1998	190.0	238.0	305.0	392.7	494.2	4.50	5.48	6.87	8.67	10.72
1999	194.9	244.8	312.8	399.0	500.6	4.68	5.69	7.11	8.95	11.04
Non-manual men										
	KGAS	KGAT	KGAU	KGAV	KGAW	KIYX	KIYY	KIYZ	KIZA	KIZB
1991	178.7	242.1	332.2	440.2	602.0	4.49	6.12	8.50	11.71	15.95
1992	188.0	257.6	353.4	473.3	641.2	4.77	6.51	9.05	12.57	17.11
1993	194.9	266.3	365.9	489.7	671.8	4.93	6.77	9.47	13.03	17.83
1994	200.4	274.4	376.0	503.9	688.5	5.03	6.90	9.67	13.30	18.20
1995	204.1	280.1	384.6	519.5	712.7	5.11	7.06	9.93	13.83	19.07
1996	211.1	287.9	399.5	545.9	755.1	5.28	7.28	10.29	14.42	19.90
1997	220.7	301.0	415.9	565.8	780.8	5.49	7.60	10.72	14.94	20.81
1998	225.5	309.1	430.7	586.1	823.4	5.62	7.79	11.09	15.54	21.74
1999	233.5	321.1	449.1	612.2	862.6	5.90	8.14	11.59	16.24	22.90
As percentages of the corresponding median										
All men										
	KBZH	KBZI	KBZJ	KBZK	KBZL	KBZM	KBZN	KBZO	KBZP	KBZQ
1991	57.9	74.6	100.0	135.7	183.0	59.3	74.8	100.0	139.6	198.8
1992	57.5	74.1	100.0	135.8	183.9	59.0	74.2	100.0	140.3	201.3
1993	57.4	74.2	100.0	137.0	186.2	58.2	73.8	100.0	141.8	200.8
1994	57.5	73.9	100.0	136.6	186.0	58.5	73.6	100.0	142.3	202.5
1995	56.3	73.4	100.0	137.0	185.9	57.0	73.0	100.0	143.4	205.3
1996	56.4	73.2	100.0	137.6	189.0	56.9	72.7	100.0	144.3	206.9
1997	56.8	73.3	100.0	137.3	187.8	57.2	72.8	100.0	144.3	206.4
1998	56.2	73.1	100.0	137.6	188.8	56.9	72.7	100.0	143.9	207.1
1999	56.4	73.3	100.0	138.2	190.1	57.1	72.9	100.0	145.3	208.9
Manual men										
	KGBS	KGBT	KGBU	KGBV	KGBW	KGBX	KGBY	KGBZ	KGCH	KGCI
1991	83.4	79.0	100.0	126.7	159.5	66.7	80.7	100.0	124.0	149.9
1992	82.8	78.7	100.0	120.3	158.5	66.9	80.8	100.0	125.2	151.8
1993	62.7	78.5	100.0	126.4	158.7	65.8	80.2	100.0	124.9	152.6
1994	63.4	78.7	100.0	126.7	158.2	67.1	80.9	100.0	125.3	152.9
1995	61.4	77.7	100.0	127.7	161.0	65.2	79.8	100.0	125.8	154.8
1996	62.1	78.0	100.0	128.2	161.6	65.0	79.3	100.0	126.4	155.6
1997	62.4	78.2	100.0	127.8	161.6	65.4	79.5	100.0	125.3	154.0
1998	62.3	78.0	100.0	128.8	162.0	65.5	79.8	100.0	126.2	156.1
1999	62.3	78.3	100.0	127.6	160.0	65.8	80.0	100.0	125.9	155.3
Non-manual men										
	KGCJ	KGCQ	KGCR	KGCS	KGCT	KGCU	KGCV	KGCW	KGCX	KGCY
1991	53.8	72.9	100.0	132.5	181.2	52.8	71.9	100.0	137.7	187.5
1992	53.2	72.9	100.0	133.9	181.4	52.7	71.9	100.0	138.9	189.0
1993	53.3	72.8	100.0	133.6	183.6	52.1	71.5	100.0	137.6	188.3
1994	53.3	73.0	100.0	134.0	183.1	52.0	71.4	100.0	137.5	188.2
1995	53.1	72.8	100.0	135.1	185.3	51.5	71.1	100.0	139.3	192.0
1996	52.8	72.1	100.0	136.6	189.0	51.3	70.7	100.0	140.2	193.5
1997	53.1	72.4	100.0	136.1	187.7	51.2	70.9	100.0	139.3	194.1
1998	52.4	71.8	100.0	136.1	191.2	50.7	70.2	100.0	140.2	196.0
1999	52.0	71.5	100.0	136.3	192.1	50.9	70.2	100.0	140.1	197.6

Source: ONS, New Earnings Survey: 01928 792077

7.23 Gross weekly and hourly earnings of full-time adult employees

Great Britain

continued

At April

	Gross weekly earnings					Gross hourly earnings				
	Lowest decile	Lower quartile	Median	Upper quartile	Highest decile	Lowest decile	Lower quartile	Median	Upper quartile	Highest decile
	£	£	£	£	£	£	£	£	£	£
All women										
	KCBK	KCBL	KCBM	KCBN	KCBO	KIZC	KIZD	KIZE	KIZF	KIZG
1991	120.8	150.6	195.7	271.6	353.3	3.20	3.96	5.16	7.20	9.97
1992	129.1	161.4	211.3	295.9	387.1	3.43	4.26	5.57	7.81	10.90
1993	134.0	168.2	221.6	309.1	402.3	3.55	4.44	5.86	8.19	11.30
1994	139.1	174.6	229.4	320.1	417.8	3.66	4.58	6.02	8.39	11.59
1995	141.1	179.5	237.2	332.5	430.7	3.70	4.69	6.24	8.87	12.30
1996	146.7	186.8	248.1	347.3	449.8	3.83	4.87	6.55	9.31	12.98
1997	154.5	196.1	260.5	364.7	472.1	4.03	5.11	6.84	9.79	13.54
1998	161.0	203.6	270.0	379.1	493.5	4.22	5.32	7.10	10.20	14.16
1999	169.5	213.3	284.0	398.2	521.1	4.47	5.59	7.47	10.75	14.98
Manual women										
	KGCZ	KGDA	KGDB	KGDC	KGDD	KIZM	KIZN	KIZO	KIZP	KIZQ
1991	100.5	119.9	147.4	185.5	233.8	2.68	3.12	3.76	4.59	5.67
1992	105.7	126.6	156.6	199.2	253.5	2.87	3.33	4.00	4.92	6.00
1993	109.6	130.5	162.2	205.8	263.7	2.95	3.41	4.12	5.12	6.25
1994	114.1	134.7	165.4	212.0	272.0	3.05	3.51	4.20	5.23	6.39
1995	114.8	137.4	171.0	222.1	282.6	3.02	3.56	4.32	5.37	6.61
1996	119.0	142.8	178.7	230.8	291.4	3.15	3.68	4.49	5.61	6.85
1997	122.4	148.7	185.4	237.3	300.0	3.25	3.83	4.66	5.81	7.16
1998	128.9	156.3	193.9	248.6	312.5	3.43	4.02	4.87	6.08	7.44
1999	140.4	164.7	201.4	260.9	327.8	3.73	4.27	5.08	6.37	7.93
Non-manual women										
	KGDJ	KGDK	KGDL	KGDM	KGDN	KIZW	KIZX	KIZY	KIZZ	KJAN
1991	131.0	162.8	211.1	289.8	365.4	3.52	4.37	5.62	7.82	10.55
1992	140.5	174.4	227.6	315.9	401.6	3.78	4.68	6.07	8.42	11.60
1993	146.4	182.5	238.4	329.6	414.1	3.93	4.90	6.35	8.83	11.98
1994	152.0	188.9	247.0	342.5	430.3	4.06	5.05	6.52	9.04	12.25
1995	155.6	194.7	256.2	355.5	445.3	4.13	5.22	6.83	9.61	13.09
1996	161.7	203.0	268.7	372.3	471.4	4.31	5.44	7.16	10.08	13.96
1997	171.4	212.8	283.3	390.7	495.1	4.53	5.68	7.53	10.61	14.38
1998	176.2	219.3	291.5	405.6	513.5	4.68	5.86	7.77	11.01	14.88
1999	184.0	230.3	305.0	422.3	540.9	4.92	6.14	8.16	11.52	15.78

As percentages of the corresponding median

	Lowest decile	Lower quartile	Median	Upper quartile	Highest decile	Lowest decile	Lower quartile	Median	Upper quartile	Highest decile
All women										
	KGEO	KGEP	KGEQ	KGER	KGES	KGET	KGEU	KGEV	KGEW	KGEX
1991	*61.7*	*76.9*	*100.0*	*138.8*	*180.6*	*62.0*	*76.7*	*100.0*	*139.4*	*193.2*
1992	*61.1*	*76.4*	*100.0*	*140.1*	*183.3*	*61.6*	*76.5*	*100.0*	*140.2*	*195.6*
1993	*60.5*	*75.9*	*100.0*	*139.5*	*181.5*	*60.6*	*75.8*	*100.0*	*139.8*	*192.9*
1994	*60.6*	*76.1*	*100.0*	*139.5*	*182.1*	*60.9*	*76.2*	*100.0*	*139.5*	*192.7*
1995	*59.5*	*75.6*	*100.0*	*140.2*	*181.5*	*59.3*	*75.2*	*100.0*	*142.1*	*197.1*
1996	*59.1*	*75.3*	*100.0*	*140.0*	*181.3*	*58.6*	*74.4*	*100.0*	*142.2*	*198.3*
1997	*59.3*	*75.3*	*100.0*	*140.0*	*181.2*	*58.8*	*74.7*	*100.0*	*143.0*	*197.8*
1998	*59.6*	*75.4*	*100.0*	*140.4*	*182.8*	*59.4*	*74.9*	*100.0*	*141.7*†	*191.5*†
1999	*59.7*	*75.1*	*100.0*	*140.2*	*183.5*	*59.8*	*74.8*	*100.0*	*143.9*	*200.5*
Manual women										
	KGDT	KGDU	KGDV	KGDW	KGDX	KGDY	KGDZ	KGEA	KGEB	KGEC
1991	*68.2*	*81.4*	*100.0*	*125.9*	*158.7*	*71.3*	*82.9*	*100.0*	*122.1*	*150.8*
1992	*67.5*	*80.8*	*100.0*	*127.2*	*161.8*	*71.7*	*83.1*	*100.0*	*123.0*	*149.8*
1993	*67.6*	*80.5*	*100.0*	*126.9*	*162.5*	*71.7*	*82.8*	*100.0*	*124.2*	*151.8*
1994	*69.0*	*81.4*	*100.0*	*128.2*	*164.4*	*72.6*	*83.5*	*100.0*	*124.4*	*151.9*
1995	*67.1*	*80.3*	*100.0*	*129.9*	*165.2*	*69.9*	*82.4*	*100.0*	*124.3*	*153.0*
1996	*66.5*	*79.9*	*100.0*	*129.1*	*163.0*	*70.1*	*82.0*	*100.0*	*125.0*	*152.6*
1997	*66.0*	*80.2*	*100.0*	*128.0*	*161.8*	*69.7*	*82.2*	*100.0*	*124.6*	*153.6*
1998	*66.5*	*80.6*	*100.0*	*128.2*	*161.2*	*70.4*	*82.6*	*100.0*	*124.9*	*152.7*
1999	*69.7*	*81.8*	*100.0*	*129.5*	*162.8*	*73.4*	*84.1*	*100.0*	*125.4*	*156.1*
Non-manual women										
	KGED	KGEE	KGEF	KGEG	KGEH	KGEI	KGEJ	KGEK	KGEL	KGEM
1991	*62.1*	*77.1*	*100.0*	*137.3*	*173.1*	*62.7*	*77.7*	*100.0*	*139.1*	*187.6*
1992	*61.7*	*76.6*	*100.0*	*138.8*	*176.5*	*62.3*	*77.1*	*100.0*	*138.9*	*191.3*
1993	*61.4*	*76.5*	*100.0*	*138.2*	*173.7*	*61.9*	*77.1*	*100.0*	*139.0*	*188.5*
1994	*61.5*	*76.5*	*100.0*	*138.6*	*174.2*	*62.3*	*77.5*	*100.0*	*138.7*	*188.0*
1995	*60.7*	*76.0*	*100.0*	*138.8*	*173.8*	*60.5*	*76.4*	*100.0*	*140.7*	*191.7*
1996	*60.2*	*75.6*	*100.0*	*138.6*	*175.4*	*60.2*	*76.0*	*100.0*	*140.7*	*192.1*
1997	*60.5*	*75.1*	*100.0*	*137.9*	*174.8*	*60.2*	*75.5*	*100.0*	*141.0*	*191.0*
1998	*60.4*	*75.2*	*100.0*	*139.1*	*176.2*	*60.2*	*75.5*	*100.0*	*141.7*	*191.5*
1999	*60.3*	*75.5*	*100.0*	*138.5*	*177.3*	*60.3*	*75.3*	*100.0*	*141.2*	*193.4*

Source: ONS, New Earnings Survey: 01928 792077

7.24 Gross weekly and hourly earnings of full-time adults
Northern Ireland
April of each year

	Gross weekly earnings[1]					Gross hourly earnings[1]				
	Lowest decile	Lower quartile	Median	Upper quartile	Highest decile	Lowest decile	Lower quartile	Median	Upper quartile	Highest decile
	£	£	£	£	£	p	p	p	p	p
All men	KCIF	KCIG	KCIH	KCII	KCIJ	KCIK	KCIL	KCIM	KCIN	KCIO
1991	128.7	170.9	237.9	337.5	460.3	318.8	409.2	544.0	808.7	1 042.2
1992	139.5	185.6	259.0	372.8	509.4	348.0	439.6	597.2	872.7	1 148.9
1993	147.1	195.0	274.3	393.6	534.2	368.3	465.6	636.0	964.2	1 256.6
1994	152.8	198.6	279.4	405.3	546.4	382.1	473.5	652.2	1 006.9	1 288.2
1995	161.0	205.3	285.6	419.4	548.5	394.6	491.2	674.6	1 043.5	1 371.6
1996	163.0	213.7	296.9	424.0	549.8	399.4	512.3	695.7	1 073.9	1 402.8
1997	171.1	222.2	307.1	449.6	575.8	414.8	522.1	724.4	1 106.2	1 452.0
1998	175.4	227.4	322.1	455.1	596.1	435.8	548.8	763.0	1 137.2	1 497.5
Manual men	KCIP	KCIQ	KCIR	KCIS	KCIT	KCIU	KCIV	KCIW	KCIX	KCIY
1991	122.4	153.9	195.9	254.5	323.0	308.7	377.4	463.7	571.5	734.7
1992	130.0	163.6	211.7	275.7	353.9	330.9	400.0	498.0	619.8	778.0
1993	135.9	172.2	218.2	281.9	356.9	339.8	412.2	513.3	627.0	761.8
1994	141.6	175.4	222.0	291.0	366.2	349.9	418.8	513.9	634.9	799.9
1995	149.7	181.5	230.9	299.4	383.3	359.7	435.5	527.0	664.6	826.5
1996	150.6	187.2	240.7	310.1	389.9	369.9	448.0	563.7	684.6	860.6
1997	158.4	192.9	249.4	321.4	402.7	385.0	460.0	570.5	709.5	882.0
1998	165.0	200.9	262.9	340.5	422.8	408.3	487.3	595.3	767.3	940.7
Non-manual men	KCIZ	KCJO	KCJP	KCJQ	KCJR	KCJS	KCJT	KCJU	KCJV	KCJW
1991	141.6	215.4	309.4	421.1	523.4	353.9	519.1	785.0	974.6	1 270.1
1992	161.2	237.2	337.6	467.4	573.4	404.0	577.0	851.7	1 078.3	1 356.5
1993	168.6	251.9	356.5	488.7	611.5	435.6	636.0	923.2	1 176.4	1 476.6
1994	174.8	253.8	366.4	501.5	615.2	445.9	641.7	956.2	1 222.5	1 461.7
1995	183.4	260.7	383.4	511.0	628.9	470.3	660.6	987.7	1 292.8	1 601.8
1996	192.5	268.4	387.1	506.4	617.2	482.4	682.5	1 013.0	1 322.7	1 612.2
1997	198.3	277.4	403.0	533.1	666.6	491.4	690.9	1 048.2	1 363.0	1 638.2
1998	200.5	284.2	418.4	543.8	685.9	500.0	715.7	1 074.7	1 404.5	1 739.5
As percentages of the corresponding median										
All men	KCJX	KCJY	KCJZ	KCKA	KCKB	KCKC	KCKD	KCKE	KCKF	KCKG
1991	*54.1*	*71.8*	*100.0*	*141.9*	*193.5*	*58.6*	*75.2*	*100.0*	*148.7*	*191.6*
1992	*53.9*	*71.7*	*100.0*	*143.9*	*196.7*	*58.3*	*73.6*	*100.0*	*146.1*	*192.4*
1993	*53.6*	*71.1*	*100.0*	*143.5*	*194.8*	*57.9*	*73.2*	*100.0*	*151.6*	*197.6*
1994	*54.7*	*71.1*	*100.0*	*145.1*	*195.6*	*58.6*	*72.6*	*100.0*	*154.4*	*197.5*
1995	*56.4*	*71.9*	*100.0*	*146.8*	*192.1*	*58.5*	*72.8*	*100.0*	*154.7*	*203.3*
1996	*54.9*	*72.0*	*100.0*	*142.8*	*185.2*	*57.4*	*73.6*	*100.0*	*154.4*	*201.6*
1997	*55.7*	*72.4*	*100.0*	*146.4*	*187.5*	*57.3*	*72.1*	*100.0*	*152.7*	*200.4*
1998	*54.5*	*70.6*	*100.0*	*141.3*	*185.1*	*57.1*	*71.9*	*100.0*	*149.0*	*196.3*
Manual men	KCKH	KCKI	KCKT	KCKK	KCKL	KCKM	KCKN	KCKO	KCKP	KCKR
1991	*62.5*	*78.6*	*100.0*	*129.9*	*164.9*	*66.6*	*81.4*	*100.0*	*123.2*	*158.4*
1992	*61.4*	*77.3*	*100.0*	*130.2*	*167.2*	*66.5*	*80.3*	*100.0*	*124.5*	*156.2*
1993	*62.3*	*78.9*	*100.0*	*129.2*	*163.6*	*66.2*	*80.3*	*100.0*	*122.2*	*148.4*
1994	*63.8*	*79.0*	*100.0*	*131.1*	*165.0*	*68.1*	*81.5*	*100.0*	*123.5*	*155.7*
1995	*64.8*	*78.6*	*100.0*	*129.7*	*166.0*	*68.3*	*82.6*	*100.0*	*126.1*	*156.8*
1996	*62.6*	*77.8*	*100.0*	*128.8*	*162.0*	*65.6*	*79.5*	*100.0*	*121.4*	*152.7*
1997	*63.5*	*77.3*	*100.0*	*128.9*	*161.5*	*67.5*	*80.6*	*100.0*	*124.4*	*154.6*
1998	*62.8*	*76.4*	*100.0*	*129.5*	*160.8*	*68.6*	*81.9*	*100.0*	*128.9*	*158.0*
Non-manual men	KCKQ	KCKS	KFZW	KCKU	KCKV	KCKW	KCKX	KCKY	KCKZ	KCLA
1991	*45.8*	*69.6*	*100.0*	*136.1*	*169.2*	*45.1*	*66.1*	*100.0*	*124.2*	*161.8*
1992	*47.8*	*70.3*	*100.0*	*138.5*	*169.9*	*47.4*	*67.8*	*100.0*	*126.6*	*159.3*
1993	*47.3*	*70.7*	*100.0*	*137.1*	*171.5*	*47.2*	*68.9*	*100.0*	*127.4*	*159.9*
1994	*47.7*	*69.3*	*100.0*	*136.9*	*167.9*	*46.6*	*67.1*	*100.0*	*127.8*	*152.9*
1995	*47.8*	*68.0*	*100.0*	*133.3*	*164.0*	*47.6*	*66.9*	*100.0*	*130.9*	*162.2*
1996	*49.7*	*69.3*	*100.0*	*130.8*	*159.4*	*47.6*	*67.4*	*100.0*	*130.6*	*159.2*
1997	*49.2*	*68.8*	*100.0*	*132.3*	*165.4*	*46.9*	*65.9*	*100.0*	*130.0*	*156.3*
1998	*47.9*	*67.9*	*100.0*	*130.0*	*163.9*	*46.5*	*66.6*	*100.0*	*130.7*	*161.9*

1 Those whose pay in the survey period was not affected by absence. Weekly earnings figures refer to April in each year and are gross before deductions excluding, generally, the value of incomes in kind, but including bonus, overtime and commission payments for the pay period.

Sources: Department of Economic Development (Northern Ireland); 028 9052 9429

7.24 Gross weekly and hourly earnings of full-time adults
Northern Ireland

continued

April of each year

	Gross weekly earnings[1]					Gross hourly earnings[1]				
	Lowest decile	Lower quartile	Median	Upper quartile	Highest decile	Lowest decile	Lower quartile	Median	Upper quartile	Highest decile
	£	£	£	£	£	p	p	p	p	p
All women										
	KCDS	KCDT	KCDU	KCDV	KCDW	KCDX	KCDY	KCDZ	KCFF	KCFG
1991	101.3	128.5	176.2	263.8	336.6	264.9	327.9	446.7	626.9	866.7
1992	113.3	141.2	193.1	291.2	385.7	293.4	350.0	482.0	670.2	937.4
1993	118.0	147.6	202.9	298.1	392.7	311.5	383.2	528.9	753.5	1 074.1
1994	120.4	149.0	201.5	303.7	399.0	321.3	398.0	545.5	809.4	1 138.4
1995	130.2	162.9	217.4	320.5	423.2	345.1	424.3	583.8	885.5	1 203.8
1996	133.3	165.5	226.2	331.9	434.8	354.5	431.0	595.8	893.6	1 256.9
1997	137.5	170.0	231.3	334.4	450.9	360.0	441.6	602.6	906.2	1 285.5
1998	148.1	181.4	243.9	348.0	465.2	390.2	471.5	652.5	950.9	1 304.5
Manual women										
	KCFH	KCFI	KCFJ	KCFK	KCFL	KCFM	KCFN	KCFO	KCFP	KCFQ
1991	91.8	107.2	129.0	158.9	201.4	251.9	278.7	335.0	401.8	507.8
1992	102.6	117.5	140.6	170.9	217.2	274.4	305.8	359.2	429.7	528.2
1993	103.9	121.6	142.9	173.7	218.7	276.2	314.6	371.0	431.0	533.1
1994	101.7	122.2	146.8	175.1	213.7	279.5	322.2	376.7	444.1	545.4
1995	109.1	131.5	161.0	193.7	247.7	287.6	343.0	401.7	483.0	616.2
1996	110.9	132.5	161.1	197.1	248.7	293.9	345.7	408.0	496.2	599.7
1997	113.4	138.3	160.0	201.8	250.8	308.3	354.4	413.7	500.0	593.0
1998	120.0	146.5	172.9	213.6	258.4	309.0	385.7	440.0	549.0	631.2
Non-manual women										
	KCFR	KCFS	KCFT	KCFU	KCFV	KCFW	KCFX	KCFY	KCFZ	KCHA
1991	111.0	145.1	200.6	292.2	349.1	291.7	373.9	493.6	684.7	909.7
1992	120.0	162.1	220.0	322.5	400.3	315.3	414.3	539.7	748.3	975.0
1993	129.6	169.2	229.4	332.3	402.2	334.8	442.0	595.3	837.0	1 127.7
1994	132.3	169.7	227.1	334.4	411.9	363.2	458.5	606.7	893.3	1 197.2
1995	139.1	176.9	241.4	346.3	435.8	375.2	473.9	650.1	943.0	1 269.6
1996	145.3	182.5	253.0	364.7	437.0	385.0	490.2	675.6	997.3	1 302.6
1997	149.7	188.6	254.0	364.3	451.3	389.0	501.6	679.4	996.7	1 342.9
1998	156.6	200.2	270.9	383.8	475.5	412.8	530.7	725.0	1 028.2	1 373.9

As percentages of the corresponding median

	Lowest decile	Lower quartile	Median	Upper quartile	Highest decile	Lowest decile	Lower quartile	Median	Upper quartile	Highest decile
All women										
	KCHB	KCHC	KCHD	KCHE	KCHF	KCHG	KCHH	KCHI	KCHJ	KCHK
1991	*57.5*	*72.9*	*100.0*	*149.7*	*191.0*	*59.3*	*73.4*	*100.0*	*140.3*	*194.0*
1992	*58.7*	*73.1*	*100.0*	*150.8*	*199.7*	*60.9*	*72.6*	*100.0*	*139.1*	*194.5*
1993	*58.2*	*72.8*	*100.0*	*146.9*	*193.5*	*58.9*	*72.5*	*100.0*	*142.5*	*203.1*
1994	*59.8*	*73.9*	*100.0*	*150.7*	*198.0*	*58.9*	*73.0*	*100.0*	*148.4*	*208.7*
1995	*59.9*	*74.9*	*100.0*	*147.4*	*194.7*	*59.1*	*72.7*	*100.0*	*151.7*	*206.2*
1996	*58.9*	*73.2*	*100.0*	*146.7*	*192.2*	*59.5*	*72.3*	*100.0*	*150.0*	*211.0*
1997	*59.4*	*73.5*	*100.0*	*144.6*	*194.9*	*59.7*	*73.3*	*100.0*	*150.4*	*213.3*
1998	*60.7*	*74.4*	*100.0*	*142.7*	*190.7*	*59.8*	*72.3*	*100.0*	*145.7*	*199.9*
Manual women										
	KCHL	KCHM	KCHN	KCHO	KCHP	KCHQ	KCHR	KCHS	KCHT	KCHU
1991	*71.2*	*83.1*	*100.0*	*123.2*	*156.1*	*75.2*	*83.2*	*100.0*	*119.9*	*151.6*
1992	*73.0*	*83.6*	*100.0*	*121.6*	*154.5*	*76.4*	*85.1*	*100.0*	*119.6*	*147.1*
1993	*72.7*	*85.1*	*100.0*	*121.6*	*153.0*	*74.4*	*84.8*	*100.0*	*116.2*	*143.7*
1994	*69.3*	*83.2*	*100.0*	*119.3*	*145.6*	*74.2*	*85.5*	*100.0*	*117.9*	*144.8*
1995	*67.8*	*81.7*	*100.0*	*120.3*	*153.9*	*71.6*	*85.4*	*100.0*	*120.2*	*153.4*
1996	*68.8*	*82.2*	*100.0*	*122.3*	*154.4*	*72.0*	*84.7*	*100.0*	*121.6*	*147.0*
1997	*70.9*	*86.4*	*100.0*	*126.1*	*156.8*	*74.5*	*85.7*	*100.0*	*120.9*	*143.3*
1998	*69.4*	*84.7*	*100.0*	*123.5*	*149.4*	*70.2*	*87.7*	*100.0*	*124.8*	*143.5*
Non-manual women										
	KCHV	KCHW	KCHX	KCHY	KCHZ	KCIA	KCIB	KCIC	KCID	KCIE
1991	*55.3*	*72.3*	*100.0*	*145.7*	*174.0*	*59.1*	*75.7*	*100.0*	*138.7*	*184.3*
1992	*54.6*	*73.7*	*100.0*	*146.6*	*182.0*	*58.4*	*76.8*	*100.0*	*138.7*	*180.7*
1993	*56.5*	*73.8*	*100.0*	*144.9*	*175.3*	*56.2*	*74.2*	*100.0*	*140.6*	*189.4*
1994	*58.3*	*74.7*	*100.0*	*147.2*	*181.4*	*59.9*	*75.6*	*100.0*	*147.2*	*197.3*
1995	*57.6*	*73.3*	*100.0*	*143.5*	*180.5*	*57.7*	*72.9*	*100.0*	*145.1*	*195.3*
1996	*57.4*	*72.1*	*100.0*	*144.2*	*172.7*	*57.0*	*72.6*	*100.0*	*147.6*	*192.8*
1997	*58.9*	*74.3*	*100.0*	*143.4*	*177.7*	*57.3*	*73.8*	*100.0*	*146.7*	*197.7*
1998	*57.8*	*73.9*	*100.0*	*141.7*	*175.5*	*56.9*	*73.2*	*100.0*	*141.8*	*189.5*

See footnotes on the first part of this table.

Sources: Department of Economic Development (Northern Ireland); 028 9052 9429

7.25 Average earnings by age group of full-time employees whose pay for the survey pay period was not affected by absence: Great Britain

At April 1999

	Gross weekly earnings								Average weekly hours		Average hourly earnings excluding overtime pay
	Average		As a percentage of the median			As a percentage of the median		Percentage earning under £250			
	Total	Overtime pay	Lowest decile	Lower quartile	Median	Upper quartile	Highest decile		Total	Overtime	
	£	£	per cent	per cent	£	per cent	per cent				£
All males											
Under 18	142.8	8.8	*58.1*	*77.2*	136.7	*125.2*	*149.6*	*95.4*	40.9	1.8	3.43
18 to 20	209.9	15.5	*70.0*	*82.3*	191.9	*126.2*	*159.7*	*78.1*	41.3	2.3	5.00
21 to 24	307.3	20.6	*63.7*	*78.4*	267.9	*128.6*	*166.3*	*42.5*	41.1	2.4	7.41
25 to 29	368.8	25.3	*60.6*	*77.4*	332.8	*129.6*	*170.0*	*22.9*	41.4	2.7	8.89
30 to 39	458.6	28.7	*57.8*	*74.4*	394.9	*135.1*	*184.7*	*14.5*	41.5	2.8	11.12
40 to 49	495.9	27.7	*55.8*	*73.0*	427.8	*134.9*	*185.8*	*12.2*	41.1	2.7	12.17
50 to 59	473.3	26.1	*57.0*	*73.7*	388.5	*140.1*	*197.6*	*16.1*	41.5	2.8	11.50
60 to 64	385.7	25.2	*60.9*	*75.9*	318.8	*135.4*	*189.2*	*27.3*	42.0	2.9	9.21
All ages	438.3	26.4	*55.5*	*73.0*	370.6	*138.7*	*191.3*	*19.8*	41.4	2.7	10.64
Manual males											
Under 18	138.6	8.6	*59.9*	*78.8*	130.0	*127.3*	*153.8*	*96.4*	41.4	1.9	3.30
18 to 20	208.9	18.2	*68.2*	*81.2*	191.9	*127.7*	*159.7*	*76.8*	42.3	2.8	4.84
21 to 24	273.4	29.4	*64.2*	*78.7*	254.1	*125.9*	*157.1*	*47.9*	43.3	3.6	6.17
25 to 29	318.9	40.7	*63.6*	*78.8*	298.5	*126.6*	*157.8*	*30.4*	44.4	4.6	7.01
30 to 39	349.7	47.5	*63.3*	*78.9*	325.8	*128.0*	*158.7*	*22.3*	44.7	5.1	7.63
40 to 49	358.3	49.7	*64.4*	*79.4*	331.5	*128.2*	*160.8*	*20.5*	44.7	5.2	7.82
50 to 59	338.0	43.6	*63.9*	*79.0*	319.0	*125.9*	*154.7*	*24.0*	44.5	4.9	7.44
60 to 64	302.6	36.3	*65.9*	*80.2*	280.7	*128.3*	*157.3*	*35.1*	43.9	4.4	6.74
All ages	331.1	43.2	*61.2*	*77.5*	309.6	*128.1*	*161.1*	*28.2*	44.4	4.8	7.28
Non-manual males											
Under 18	155.0	9.5	*54.7*	*71.9*	153.3	*118.9*	*153.3*	*92.4*	39.4	1.5	3.86
18 to 20	211.2	11.7	*71.7*	*83.1*	191.9	*124.4*	*159.5*	*79.8*	39.9	1.6	5.23
21 to 24	334.6	13.6	*63.9*	*78.9*	278.5	*129.9*	*175.6*	*38.2*	39.4	1.5	8.46
25 to 29	404.9	14.1	*59.9*	*76.5*	364.4	*131.0*	*170.5*	*17.5*	39.2	1.3	10.33
30 to 39	539.5	14.8	*55.7*	*74.0*	467.0	*133.7*	*186.1*	*8.7*	39.0	1.2	13.85
40 to 49	588.9	12.8	*56.1*	*75.9*	510.6	*132.5*	*184.8*	*6.6*	38.6	1.0	15.30
50 to 59	590.5	11.1	*52.0*	*71.3*	492.4	*136.4*	*194.9*	*9.2*	38.8	0.9	15.23
60 to 64	498.9	10.0	*52.7*	*72.7*	400.3	*142.9*	*201.4*	*16.5*	39.5	0.9	12.75
All ages	522.2	13.2	*51.7*	*71.1*	445.6	*136.7*	*192.9*	*13.2*	39.0	1.1	13.44
All females											
Under 18	145.6	5.5	*62.7*	*83.7*	143.2	*114.9*	*139.7*	*96.6*	38.8	1.0	3.71
18 to 20	189.2	6.9	*72.2*	*84.7*	178.6	*122.1*	*147.1*	*87.6*	38.6	1.0	4.86
21 to 24	258.1	8.1	*67.2*	*80.3*	235.7	*128.8*	*157.7*	*56.1*	38.2	1.0	6.73
25 to 29	319.4	8.4	*63.1*	*77.7*	290.4	*129.3*	*165.0*	*35.6*	37.7	0.8	8.44
30 to 39	364.1	8.1	*57.1*	*73.0*	323.3	*136.6*	*178.1*	*29.5*	37.6	0.8	9.69
40 to 49	345.5	7.0	*58.0*	*73.5*	299.2	*145.4*	*181.8*	*35.8*	37.1	0.8	9.33
50 to 59	318.0	6.5	*60.9*	*76.5*	274.6	*146.1*	*187.4*	*40.9*	37.2	0.8	8.56
60 to 64	282.0	6.7	*60.5*	*74.8*	249.6	*127.4*	*178.4*	*50.7*	37.5	0.8	7.31
All ages	324.4	7.5	*59.2*	*74.8*	282.2	*140.2*	*183.8*	*39.7*	37.5	0.8	8.64
Manual females											
Under 18	144.9	6.5	*67.2*	*83.7*	143.3	*110.6*	*139.4*	*96.9*	40.0	1.5	3.59
18 to 20	175.9	7.4	*68.7*	*81.9*	169.7	*120.6*	*144.9*	*90.6*	39.5	1.2	4.40
21 to 24	207.6	11.8	*70.3*	*82.3*	192.1	*126.1*	*154.9*	*77.6*	40.3	1.8	5.10
25 to 29	236.9	13.2	*66.7*	*79.5*	217.6	*129.1*	*158.6*	*65.2*	40.0	1.7	5.83
30 to 39	235.8	15.5	*67.1*	*80.7*	214.5	*131.1*	*166.9*	*65.5*	40.1	2.1	5.80
40 to 49	222.8	14.7	*70.3*	*82.2*	203.7	*127.7*	*160.5*	*70.8*	39.8	2.0	5.51
50 to 59	217.5	13.7	*71.2*	*82.5*	197.9	*129.9*	*160.0*	*72.8*	39.7	1.9	5.39
60 to 64	206.6	11.0	*73.5*	*82.9*	184.9	*124.0*	*167.4*	*79.3*	39.2	1.7	5.07
All ages	220.5	13.5	*69.4*	*81.6*	200.2	*129.8*	*163.3*	*71.8*	39.9	1.9	5.45
Non-manual females											
Under 18	145.9	4.9	*59.7*	*83.7*	143.2	*119.5*	*144.1*	*96.3*	38.1	0.8	3.78
18 to 20	193.2	6.7	*73.7*	*84.5*	182.3	*121.4*	*147.7*	*86.7*	38.4	0.9	5.00
21 to 24	267.3	7.4	*67.6*	*80.0*	245.7	*126.9*	*155.3*	*52.1*	37.8	0.8	7.04
25 to 29	331.7	7.7	*64.0*	*77.7*	303.4	*128.1*	*161.4*	*31.2*	37.4	0.7	8.85
30 to 39	384.5	7.0	*59.2*	*74.4*	342.5	*134.0*	*175.1*	*23.8*	37.2	0.6	10.34
40 to 49	370.1	5.5	*58.7*	*72.8*	327.1	*139.5*	*174.0*	*28.9*	36.5	0.5	10.15
50 to 59	343.2	4.7	*62.5*	*76.8*	300.0	*145.3*	*180.3*	*32.9*	36.5	0.5	9.40
60 to 64	311.3	5.1	*63.7*	*80.3*	268.9	*131.0*	*175.9*	*39.6*	36.8	0.5	8.22
All ages	344.7	6.3	*60.0*	*75.3*	302.7	*139.2*	*178.7*	*33.4*	37.1	0.6	9.30

Source: ONS, New Earnings Survey: 01928 792077

7.26 Trade unions[1,2]
United Kingdom
At end of year

per cent

		1987	1988	1989	1990	1991	1992	1993	1994	1995	1996	1997
Number of trade unions	KCLB	330	315	309	287	275	268	254	243	238	226	219
Analysis by number of members:												
Under 100 members	KCLC	*16.1*	*17.5*	*15.5*	*15.3*	*12.7*	*12.7*	*13.4*	*11.9*	*10.5*	*10.2†*	*12.8*
100 and under 500	KCLD	*25.8*	*24.1*	*23.6*	*22.6*	*24.0*	*21.6*	*22.8*	*20.2*	*21.0*	*23.0†*	*18.7*
500 and under 1 000	KCLE	*7.9*	*8.3*	*6.1*	*7.7*	*7.6*	*9.3*	*7.1*	*9.5*	*10.5*	*8.0†*	*11.4*
1 000 and under 2 500	KCLF	*15.8*	*14.9*	*15.9*	*15.0*	*15.6*	*17.2*	*16.9*	*16.5*	*17.2*	*19.9†*	*16.0*
2 500 and under 5 000	KCLG	*6.7*	*8.3*	*8.7*	*9.4*	*10.2*	*9.3*	*9.4*	*10.7*	*9.7*	*8.4†*	*9.6*
5 000 and under 10 000	KCLH	*4.8*	*4.7*	*5.8*	*5.2*	*5.5*	*6.3*	*7.1*	*7.8*	*7.1*	*6.2†*	*6.4*
10 000 and under 15 000	KCLI	*1.8*	*1.3*	*1.6*	*2.1*	*1.8*	*1.9*	*2.4*	*2.5*	*3.4*	*3.1*	*3.2*
15 000 and under 25 000	KCLJ	*3.3*	*3.5*	*3.9*	*3.1*	*3.3*	*3.7*	*3.1*	*2.9*	*2.5*	*2.2†*	*2.3*
25 000 and under 50 000	KCLK	*7.3*	*7.6*	*7.8*	*8.7*	*7.6*	*7.1*	*7.9*	*8.2*	*8.4*	*9.3†*	*9.1*
50 000 and under 100 000	KCLL	*2.1*	*2.5*	*2.3*	*2.4*	*3.3*	*3.4*	*3.1*	*2.9*	*2.9*	*2.2*	*2.7*
100 000 and under 250 000	KCLM	*3.9*	*4.1*	*4.2*	*4.9*	*4.4*	*4.1*	*3.5*	*4.1*	*3.4*	*3.5*	*3.7*
250 000 and over	KCLN	*3.3*	*3.2*	*3.2*	*3.1*	*3.6*	*3.4*	*3.1*	*2.9*	*3.4*	*4.0*	*4.1*
Membership unknown	KCLO	*1.2*	–	*1.3*	*0.3*	–	–	–	–	–	–	–
All sizes	KCLP	*100*	*100*	*100*	*100*	*100*	*100*	*100*	*100*	*100*	*100*	*100*
Membership[3,4]												
Analysis by size of union:												
Under 100 members	KCLQ	–	–	–	–	*0.02*	*0.02*	*0.02*	–	–	–	–
100 and under 500	KCLR	*0.2*	*0.2*	*0.2*	*0.2*	*0.2*	*0.2*	*0.2*	*0.2*	*0.2*	*0.2*	*0.1*
500 and under 1 000	KCLS	*0.2*	*0.2*	*0.1*	*0.2*	*0.2*	*0.2*	*0.2*	*0.2*	*0.2*	*0.1†*	*0.2*
1 000 and under 2 500	KCLT	*0.8*	*0.8*	*0.8*	*0.7*	*0.8*	*0.9*	*0.9*	*0.8*	*0.8*	*0.9*	*0.8*
2 500 and under 5 000	KCLU	*0.8*	*0.9*	*0.9*	*1.0*	*1.1*	*1.0*	*1.0*	*1.1*	*1.0*	*0.9†*	*1.0*
5 000 and under 10 000	KCLV	*1.0*	*1.0*	*1.2*	*1.0*	*1.2*	*1.3*	*1.5*	*1.6*	*1.5*	*1.2†*	*1.2*
10 000 and under 15 000	KCLW	*0.7*	*0.5*	*0.6*	*0.7*	*0.6*	*0.6*	*0.9*	*0.9*	*1.3*	*1.2*	*1.1*
15 000 and under 25 000	KCLX	*2.1*	*2.0*	*2.3*	*1.7*	*1.7*	*2.0*	*1.6*	*1.5*	*1.3*	*1.1†*	*1.1*
25 000 and under 50 000	KCLY	*8.4*	*8.5*	*8.6*	*9.0*	*7.8*	*7.7*	*8.4*	*8.9*	*9.2*	*9.2†*	*8.5*
50 000 and under 100 000	KCLZ	*5.3*	*5.5*	*4.9*	*4.9*	*6.2*	*6.9*	*6.7*	*6.0*	*6.0*	*3.9*	*4.5*
100 000 and under 250 000	KCMA	*20.3*	*20.2*	*20.0*	*22.4*	*19.8*	*18.9*	*16.4*	*19.3*	*16.4*	*15.2*	*15.3*
250 000 and over	KCMB	*60.3*	*60.1*	*60.4*	*58.2*	*60.5*	*60.2*	*62.4*	*59.5*	*62.1*	*66.0*	*66.2*
All sizes	KCMC	*100*	*100*	*100*	*100*	*100*	*100*	*100*	*100*	*100*	*100*	*100*
Total membership (Thousands)	KCMD	10 475	10 376	10 158	9 947	9 585	9 048	8 700	8 278	8 089	7 980†	7 835

1 The statistics relate to all organisations of employees known to Certification Officer with head offices in the United Kingdom.
2 Figures are confined to organisations which appear to satisfy the statutory definition of a trade union in Section 28 of the Trade Union and Labour Relations Act 1974 and more recently section 1 of the Trade Union and Labour Relations (Consolidation) Act 1992.
3 Included in the data are home and overseas membership figures of contributory and non-contributory members. Employment status of members is not provided and the figures may therefore include some people who are self-employed, unemployed or retired.
4 The 1997 figures are provisional and subject to revision as later information becomes available. Figures published in earlier years have been revised in line with the latest information.

Source: Department of Trade and Industry: 020 7215 5780

7.27 Vacancies at jobcentres in the United Kingdom[1,2]

Thousands, seasonally adjusted

	January	February	March	April	May	June	July	August	September	October	November	December
Numbers of vacancies remaining unfilled DPCB												
1989	230.8	227.5	222.6	220.2	217.9	223.9	222.1	218.3	220.8	218.1	213.7	198.2
1990	198.3	194.1	191.0	195.9	194.4	186.6	173.5	168.3	163.2	149.9	139.9	128.6
1991	140.0	138.9	133.9	119.8	110.2	105.6	106.0	109.2	112.5	109.6	111.8	117.4
1992	117.0	118.2	118.2	117.1	118.8	119.8	120.2	118.6	113.1	112.9	114.4	117.0
1993	119.0	120.3	124.3	123.6	125.9	124.8	129.2	129.8	129.5	133.2	136.3	138.2
1994	140.4	142.7	143.4	146.2	149.1	154.9	158.8	165.0	165.4	175.3	176.9	177.6
1995	175.2	174.3	177.5	186.0	185.4	182.9	181.8	181.7	184.5	181.7	185.2	186.7
1996	191.5	191.9	199.1	202.7	211.5	221.2	231.5	234.8	244.8	253.6	263.9	266.2
1997	267.8	275.2	277.5	277.8	277.9	284.1	285.2	290.1	296.0	305.1	284.6	281.9
1998	273.7	282.2	284.2	286.9	295.9	297.6	298.4	297.5	301.6	312.8†	314.1	309.0
1999	305.0	301.3	298.1	296.8	300.4	301.5	305.5	310.7	316.4	340.8	..	..
Inflow of vacancies DRYW												
1989	221.4	233.4	228.5	220.8	220.1	232.5	229.7	228.7	228.2	228.3	224.1	216.9
1990	201.4	224.5	217.1	215.1	214.2	200.9	197.6	198.3	196.7	186.9	185.2	176.4
1991	185.4	167.8	168.0	182.8	182.2	163.9	165.6	168.7	170.7	168.3	164.9	167.1
1992	166.4	166.9	171.7	166.0	166.2	176.4	173.3	164.0	167.1	171.3	164.7	173.8
1993	180.2	175.9	181.5	179.3	180.2	183.5	190.5	183.4	191.5	189.6	193.9	197.2
1994	199.2	199.5	199.1	203.7	205.2	212.2	208.3	223.6	216.6	219.1	223.1	226.8
1995	218.2	219.5	215.2	205.4	227.7	223.1	225.6	230.9	226.5	231.9	232.9	222.1
1996	224.0	221.3	220.8	230.9	220.0	220.1	225.1	222.5	222.0	203.9	230.9	230.5
1997	210.3	238.3	244.9	238.1	234.8	226.7	225.8	218.8	228.1	228.1	216.6	213.2
1998	198.5	222.4	224.3	221.5	209.4	222.9	217.8	217.6	223.0	236.8†	222.7	229.6
1999	229.9	226.4	225.9	231.6	216.4	224.0	227.2	230.1	232.8	240.6	..	..
Outflow of vacancies DRZL												
1989	231.2	237.9	233.2	225.0	224.5	224.7	229.0	232.3	227.6	230.5	230.1	223.9
1990	204.7	228.7	219.4	216.9	216.3	208.1	211.6	203.4	202.1	200.4	197.0	180.4
1991	176.4	168.7	172.4	197.5	197.1	168.3	165.0	164.9	168.7	170.2	159.9	160.8
1992	169.7	165.6	170.5	168.4	168.6	174.1	172.5	164.1	170.5	169.8	160.8	170.6
1993	178.8	173.6	176.6	180.1	181.0	184.0	185.6	182.0	189.8	186.4	191.0	195.3
1994	196.8	196.7	198.3	202.3	203.4	205.8	202.8	217.1	214.8	210.8	221.6	227.0
1995	219.0	220.6	214.1	195.1	229.0	225.0	224.6	230.4	225.5	237.1	229.7	219.2
1996	223.3	222.6	214.4	223.2	209.3	210.9	212.9	218.6	214.5	197.4	219.7	233.2
1997	215.0	234.0	248.3	234.2	233.2	219.8	223.1	214.1	217.1	222.1	232.6	222.3
1998	215.1	215.6	218.9	217.5	201.9	218.5	215.1	217.5	218.8	224.0†	220.7	228.8
1999	233.6	231.1	224.2	234.2	208.5	222.0	222.4	224.5	229.1	219.9	..	..
Number of placings DTQR												
1989	160.3	164.9	161.3	156.1	155.6	157.2	157.5	160.8	156.9	157.1	158.4	155.1
1990	144.9	159.9	153.7	151.4	151.1	144.7	148.5	145.9	144.9	144.7	142.6	131.4
1991	129.4	122.5	127.6	148.4	148.4	123.9	123.2	120.9	122.1	122.7	114.8	115.9
1992	124.6	119.7	123.6	122.6	123.0	126.5	126.5	119.6	125.9	127.6	120.6	130.2
1993	134.3	132.1	131.7	134.0	134.9	136.3	139.0	136.1	144.1	140.6	146.4	148.2
1994	149.5	150.6	151.5	156.6	157.7	161.9	157.7	169.6	166.4	162.3	170.3	173.2
1995	166.2	169.1	164.6	145.8	179.8	173.6	174.4	178.5	172.5	183.5	179.7	167.4
1996	166.0	163.8	153.5	155.4	150.0	147.0	148.3	152.5	148.7	134.3	150.4	161.6
1997	147.1	157.4	166.7	165.8	150.6	141.4	136.0	124.0	126.1	120.5	115.5	114.8
1998	121.9	116.8	120.6	117.5	109.1	112.9	110.4	112.8	117.4	119.1†	115.4	117.8
1999	126.3	121.5	118.2	129.3	110.7	117.8	118.4	120.1	122.9	120.3	..	..

Note: Vacancies notified to and placings made by jobcentres do not represent the total number of vacancies/engagements in the economy. Latest estimates suggest that about a third of all vacancies are notified to jobcentres: and about a quarter of all engagements are made through jobcentres. Inflow, outflow and placings figures are collected for four or five-week periods between count dates; the figures in this table are converted to a standard 4 $^1/_3$ week month.

1 Excluding vacancies on government programmes (except vacancies on Enterprise Ulster and Action for Community Employment (ACE) which are included in the seasonally adjusted figures for Northern Ireland). Note that Community Programme vacancies handled by jobcentres were excluded from the seasonally adjusted series when the coverage was revised in September 1985. The coverage of the seasonally adjusted series is therefore not affected by the cessation of C.P. vacancies with the introduction of Employment Training in September 1988. For further details, see the October 1985 *Employment Gazette* (now Labour Market Trends), 143.

2 The latest national seasonally adjusted vacancy figures are provisional and subject to revision in the following month.

Source: ONS, Labour Market Statistics: 020 7533 6094

7.28 Vacancies unfilled in Northern Ireland[1]

Unadjusted

Number

	January	February	March	April	May	June	July	August	September	October	November	December
Adults KAZZ												
1984	1 109	1 203	1 294	1 331	1 543	1 779	1 756	1 746	1 628	1 701	1 776	1 419
1985	1 216	1 349	1 609	1 669	1 876	1 912	1 766	1 651	1 662	1 627	1 516	1 502
1986	1 545	1 836	1 924	2 206	2 160	2 202	2 151	2 180	2 097	2 100	2 005	1 723
1987	1 766	1 952	1 985	2 162	2 147	2 223	2 130	2 111	2 148	2 222	2 326	2 728
1988	2 891	2 848	2 824	3 003	2 800	2 795	2 836	2 646	2 562	2 667	2 741	2 781
1989	2 748	3 654	3 511	3 820	3 671	3 722	3 837	3 739	4 302	4 326	3 997	3 585
1990	3 673	4 150	4 126	4 752	5 179	5 370	4 764	4 617	4 819	4 665	4 369	3 921
1991	3 809	4 228	3 949	4 108	4 317	4 539	4 177	4 076	4 496	3 953	3 696	3 477
1992	3 539	3 912	4 115	4 116	4 406	4 353	4 198	4 314	4 646	4 965	4 596	4 142
1993	4 416	4 558	4 444	4 422	4 993	5 371	5 510	5 393	5 663	6 246	5 782	5 381
1994	5 225	5 514	5 460	5 934	6 205	6 516	6 571	6 544	7 163	7 803	7 663	7 409
1995	7 370	7 135	7 087	7 218	7 141	7 337	7 296	7 371	8 315	8 254	7 651	7 253
1996	6 918	6 641	6 831	6 835	6 610	6 908	6 691	6 286	7 341	7 551	7 957	7 137
1997	6 371	6 253	6 209	6 024	6 337	6 773	6 530	6 623	7 530	7 910	7 829	7 570
1998	7 160	7 398	7 366	7 909	8 483	8 966	9 162	9 266	10 165	10 637	10 620	9 855
1999	8 436	8 108	10 221[3]	..[2]	..	..	..	..	..	..	..	..
Young Persons KBAZ												
1984	309	330	377	400	469	595	542	576	574	673	744	746
1985	709	766	797	837	887	967	758	628	691	651	605	489
1986	445	480	530	576	592	695	572	589	662	675	662	567
1987	546	598	703	635	737	893	814	806	846	978	921	819
1988	777	787	812	996	1 196	1 096	1 005	993	1 038	1 238	1 235	1 140
1989	1 083	1 150	1 271	1 373	1 300	1 284	1 240	1 250	1 491	1 491	1 487	1 302
1990	1 175	1 119	1 064	553	512	466	377	379	454	452	407	342
1991	318	256	277	258	290	287	280	238	302	306	295	319
1992	303	306	332	338	329	374	324	302	376	409	393	405
1993	396	428	519	549	639	686	659	643	672	724	699	587
1994	536	579	580	635	628	608	745	675	777	901	978	862
1995	829	861	833	785	786	715	635	644	752	733	711	659
1996	578	590	639	684	762	799	800	777	931	963	1 066	988
1997	881	886	860	843	854	895	906	859	1 002	1 085	1 158	1 074
1998	1 008	918	948	973	1 071	1 330	1 312	1 268	1 480	1 473	1 440	1 215
1999	1 063	999	1 247[3]	..[2]	..	..	..	..	..	..	..	..

1 The figures refer to vacancies notified to the local offices of the Training and Employment Agency, and remaining unfilled on the day of the count.
2 The publication of vacancy figures for Northern Ireland was temporarily suspended from May 1999 due to a problem caused by the introduction of a new computer system for processing vacancies notified to Training and Employment Agency Offices.
3 Provisional.

Sources: Training and Employment Agency (Northern Ireland); Research & Evaluation Branch: 028 9025 7621

Personal income, expenditure and wealth

8

Family expenditure survey
(Tables 8.1, 8.3 - 8.5)

The Family Expenditure Survey, introduced in 1957, covers all types of private households in the United Kingdom. It is a continuous survey with fieldwork carried out in every month of the year. In 1998-99 around 6,600 households in the UK provided information. The main purpose of the survey is to provide a source for the weighting pattern for the Index of Retail Prices, so it is primarily concerned with household expenditure on goods and services. However, it does have several other important uses.

Although the survey is primarily concerned with the expenditure of private households, much additional information is collected about income and the characteristics of co-operating households. Consequently, the survey provides a unique fund of important economic and social data.

Like all surveys based on a sample of the population, its results are subject to sampling error, and to some bias due to non-response.

The results of the survey are published in an annual report, the latest being *'Family Spending 1998-99'* (The Stationery Office). There are two significant quality improvements to the results of the survey for 1998-99. One is the inclusion of information from the expenditure diaries kept by children aged between 7 and 15. The other is the use of re-weighted data, which compensates for lower response rates among some types of household compared with others. The report includes a list of definitions used in the survey, items on which information is collected and a brief account of the fieldwork procedure.

Distribution of total incomes *(Table 8.2)*

The information shown in Table 8.2 comes from the Survey of Personal Incomes for the financial years 1994/95, 1995/96, 1996/97 and 1997/98. This is an annual survey that covers approximately 80 000 individuals across the whole of the UK. It is based on administrative data held by tax offices on individuals who could be liable to UK tax.

8.1 Average incomes of households before and after taxes and benefits[1,2], 1997/98

United Kingdom

	Retired households [3]		Non-retired households								
	1 adult	2 or more adults	1 adult	2 adults	3 or more adults	1 adult with children [4]	2 adults with 1 child [4]	2 adults with 2 children [4]	2 adults with 3 or more children [4]	3 or more adults with children [4]	All house-holds
Number of households in the population (Thousands)	3 477	2 787	3 692	4 909	2 283	1 453	1 887	2 249	882	937	24 556
Average per household (£ per year)											
Original income	3 219	9 693	13 793	27 241	35 439	4 943	28 337	30 047	24 437	31 721	19 680
Disposable income	7 132	14 100	12 183	22 699	30 628	9 408	23 646	24 934	23 007	28 604	18 402
Post-tax income	5 893	11 035	9 775	18 270	24 214	7 164	18 989	20 219	18 409	21 721	14 685

1 Original income is the total income in cash and kind of the household before the deduction of taxes or the addition of state benefits. The addition of cash benefits (retirement pensions, child benefit, etc) and the deduction of income tax, council tax, water charges, domestic rates and employees' national insurance contributions give disposable income. By further allowing for taxes paid on goods and services purchased, such as VAT, an estimate of 'post-tax' income is derived. These income figures are derived from estimates made by the Office for National Statistics, based largely on information from the Family Expenditure Survey (FES), and published each year in *Economic Trends*.

2 Redistribution of Income, included in the Effects of taxes and benefits upon household income, 1997/98, article published in the April 1999 edition of *Economic Trends*.

3 A retired household is defined as one where the combined income of retired members amounts to at least half the total gross income of the household, where a retired person is defined as anyone who describes themselves as 'retired' or anyone over the minimum NI pension age describing themselves as 'unoccupied' or 'sick or injured but not intending to seek work'.

4 Children are defined as persons aged under 16 or aged between 16 and 18, unmarried and receiving non-advanced further education.

Source: Office for National Statistics: 020 7533 5772

8.2 Distribution of total incomes before and after tax

United Kingdom

Years ended 5 April

1994/95 Annual Survey

Lower limit of range of income	Number of individuals (000s)	£ million: Total income before tax	£ million: Total tax[5]	£ million: Total income after tax
All incomes[1,2,3,4]	25 300	387 800	69 400	318 400
Income before tax £				
3 445	674	2 520	36	2 480
4 000	660	2 820	84	2 730
4 500	791	3 760	160	3 600
5 000	781	4 100	200	3 900
5 500	850	4 900	280	4 610
6 000	1 570	10 200	760	9 450
7 000	1 670	12 500	1 090	11 400
8 000	3 060	27 500	3 070	24 500
10 000	2 750	30 200	4 040	26 200
12 000	3 340	44 900	6 770	38 100
15 000	3 920	67 800	11 500	56 300
20 000	3 530	84 000	15 600	68 400
30 000	1 220	45 100	10 500	34 500
50 000	387	25 600	7 530	18 100
100 000 and over	118	21 850	7 680	14 130
Income after tax £				
3 445	812	3 080	52	3 030
4 000	807	3 570	130	3 450
4 500	938	4 680	230	4 440
5 000	952	5 320	310	5 010
5 500	995	6 130	410	5 710
6 000	2 010	14 400	1 260	13 100
7 000	2 040	17 100	1 740	15 300
8 000	3 740	38 400	4 850	33 500
10 000	3 080	39 600	5 820	33 700
12 000	3 470	55 700	9 130	46 600
15 000	3 490	73 100	13 100	60 000
20 000	2 140	63 600	13 100	50 500
30 000	627	31 500	8 480	23 000
50 000	189	18 300	5 880	12 400
100 000 and over	48	13 410	4 840	8 580

1995/96 Annual Survey

Lower limit of range of income	Number of individuals (000s)	£ million: Total income before tax	£ million: Total tax	£ million: Total income after tax
All incomes[1,2,3,4]	25 800	407 900	74 400	333 500
Income before tax £				
3 525	547	2 070	24	2 040
4 000	582	2 460	76	2 390
4 500	714	3 410	120	3 290
5 000	885	4 640	200	4 440
5 500	861	4 950	280	4 670
6 000	1 610	10 400	760	9 680
7 000	1 730	12 900	1 090	11 900
8 000	3 040	27 300	2 990	24 300
10 000	2 720	29 800	3 970	25 900
12 000	3 470	46 500	7 070	39 500
15 000	3 930	67 800	11 600	56 200
20 000	3 790	90 700	17 000	73 700
30 000	1 360	50 300	11 700	38 600
50 000	455	29 700	8 770	21 000
100 000 and over	126	24 700	8 720	15 950
Income after tax £				
3 525	689	2 640	39	2 600
4 000	676	2 980	110	2 870
4 500	861	4 300	190	4 110
5 000	1 110	6 130	320	5 810
5 500	1 090	6 700	450	6 250
6 000	1 920	13 700	1 170	12 500
7 000	2 140	17 700	1 750	16 000
8 000	3 740	38 400	4 800	33 600
10 000	3 130	40 300	5 980	34 300
12 000	3 550	56 900	9 410	47 500
15 000	3 580	75 500	13 700	61 800
20 000	2 360	70 500	14 500	56 000
30 000	730	36 800	9 980	26 900
50 000	200	19 300	6 180	13 100
100 000 and over	55	16 030	5 800	10 230

See footnotes on the second part of this table.

Source: Board of Inland Revenue: 020 7438 7412

8.2 Distribution of total incomes before and after tax

continued

United Kingdom

Years ended 5 April

1996/97 Annual Survey

Lower limit of range of income	Number of individuals (000s)	£ million: Total income before tax	£ million: Total tax	£ million: Total income after tax
All incomes[1,2,3,4]	25 700	427 000	75 800	351 100
Income before tax £				
3 765	259	1 010	6	1 000
4 000	567	2 400	49	2 360
4 500	589	2 800	100	2 700
5 000	896	4 710	180	4 530
5 500	810	4 660	220	4 430
6 000	1 640	10 700	680	9 990
7 000	1 570	11 800	910	10 900
8 000	3 080	27 800	2 690	25 100
10 000	2 830	31 100	3 770	27 300
12 000	3 470	46 700	6 630	40 100
15 000	3 980	69 000	10 900	58 000
20 000	3 770	90 800	16 000	74 800
30 000	1 550	57 500	12 600	44 800
50 000	491	32 400	9 270	23 100
100 000 and over	154	33 800	11 840	22 000
Income after tax £				
3 765	326	1 280	9	1 270
4 000	705	3 070	77	2 990
4 500	707	3 520	150	3 370
5 000	1 040	5 720	240	5 470
5 500	1 020	6 200	350	5 850
6 000	1 920	13 500	1 000	12 500
7 000	1 920	15 800	1 390	14 400
8 000	3 820	38 700	4 310	34 400
10 000	3 180	40 300	5 400	34 900
12 000	3 560	56 300	8 550	47 800
15 000	3 660	75 700	12 800	62 900
20 000	2 640	77 800	14 900	62 900
30 000	861	42 500	10 900	31 600
50 000	237	22 700	7 100	15 600
100 000 and over	72	23 900	8 570	15 300

1997/98 Annual Survey

Lower limit of range of income	Number of individuals (000s)	£ million: Total income before tax	£ million: Total tax	£ million: Total income after tax
All incomes[1,2,3,4]	26 200	461 000	79 500	381 000
Income before tax £				
4 045	509	2 170	21	2 150
4 500	560	2 660	74	2 580
5 000	678	3 580	127	3 450
5 500	841	4 840	190	4 650
6 000	1 640	10 700	600	10 100
7 000	1 540	11 600	820	10 800
8 000	3 050	27 400	2 440	25 000
10 000	2 850	31 200	3 520	27 700
12 000	3 530	47 500	6 270	41 200
15 000	4 260	73 800	11 100	62 700
20 000	4 260	102 700	17 200	85 500
30 000	1 730	63 700	13 100	50 600
50 000	582	38 400	10 600	27 700
100 000	135	18 000	5 850	12 100
200 000 and over	51	22 400	7 560	14 900
Income after tax £				
4 045	629	2 710	33	2 680
4 500	683	3 350	112	3 240
5 000	812	4 470	186	4 280
5 500	1 000	6 050	281	5 770
6 000	1 960	13 600	885	12 700
7 000	1 900	15 600	1 270	14 300
8 000	3 750	37 600	3 840	33 800
10 000	3 230	40 500	5 100	35 400
12 000	3 760	58 800	8 390	50 400
15 000	4 030	82 800	13 300	69 500
20 000	3 070	89 100	15 900	73 200
30 000	1 030	50 100	12 300	37 900
50 000	291	27 700	8 460	19 200
100 000	63	12 800	4 350	8 420
200 000 and over	23	15 500	5 150	10 400

1 These tables relate only to those individuals who have some tax liability. The distributions cover only incomes as computed for tax purposes and above a level which for each year corresponds approximately to the single person's allowance. Incomes below these levels are not shown because the information about them is incomplete.

2 All figures have been independently rounded.

3 Investment income from which tax has been deducted at source is not always known to local tax offices. Estimates of missing bank and building society interest and dividends from United Kingdom companies are included in these tables. The missing investment income is distributed in a manner consistent with information from the Family Expenditure Survey and the National Accounts, to individuals where there is no investment income already reported by the tax office.

4 Superannuation contributions are estimated and included in total income. They have been distributed among earners in the Survey of Personal Incomes sample by a method consistent with information about the number of employees who are contracted in or out of the State Earnings Related Pension Scheme and the proportion of their earnings contributed.

5 Before 1994-95 estimates of tax liability allow for mortgage interest relief at source (at basic rate).

Source: Board of Inland Revenue: 020 7438 7412

8.3 Sources of gross household income[1]
United Kingdom

		1988	1989	1990	1991	1992	1993	1994 /95	1995 /96	1996 /97	1997 /98	1998[3] /99
Number of households supplying data	KPDA	7 265	7 410	7 046	7 056	7 418	6 979	6 853	6 797	6 415	6 409	6 630
Average weekly household income by source (£)												
Wages and salaries	KPCB	176.80	189.40	212.50	222.00	222.70	228.30	237.90	245.00	256.30	280.20	309.20
Self-employment	KPCC	29.60	28.50	32.10	29.70	29.60	29.20	35.30	32.90	37.50	32.90	37.20
Investments	KPCD	13.30	14.00	19.00	23.70	20.80	18.00	16.20	18.10	17.80	18.70	18.80
Annuities and pensions (other than social security benefits)	KPCE	13.80	14.40	15.50	17.30	19.60	21.90	23.50	26.00	26.00	28.90	30.30
Social security benefits	KPCF	31.60	32.70	35.30	40.10	45.00	48.90	49.90	52.40	54.10	55.00	55.80
Imputed income from owner/rent-free occupancy[2]	KPCG	15.80	16.60	18.60	24.50	..	..	..	..	..	..	..
Other sources	KPCH	3.00	4.50	4.50	5.30	5.20	6.70	6.50	6.60	5.30	5.20	5.70
Total	KPCI	283.90	300.00	337.60	362.70	342.90	353.00	359.30	380.90	396.90	420.80	457.00
Sources of household income as a percentage of total household income												
Wages and salaries	KPCJ	*62*	*62*	*63*	*61*	*65*	*65*	*64*	*64*	*65*	*67*	*68*
Self-employment	KPCK	*10*	*9*	*10*	*8*	*9*	*8*	*10*	*9*	*9*	*8*	*8*
Investments	KPCL	*5*	*5*	*6*	*7*	*6*	*5*	*4*	*5*	*5*	*4*	*4*
Annuities and pensions (other than social security benefits)	KPCM	*5*	*5*	*5*	*5*	*6*	*6*	*6*	*7*	*7*	*7*	*7*
Social security benefits	KPCN	*11*	*11*	*11*	*11*	*13*	*14*	*14*	*14*	*14*	*13*	*12*
Imputed income from owner/rent-free occupancy[2]	KPCO	*6*	*6*	*6*	*7*	..	..	..	..	..	..	..
Other sources	KPCP	*1*	*2*	*1*	*2*	*2*	*2*	*2*	*2*	*1*	*1*	*1*
Total	KPCQ	*100*	*100*	*100*	*100*	*100*	*100*	*100*	*100*	*100*	*100*	*100*

1 Information derived from the Family Expenditure Survey (FES).
2 Imputed income is the weekly equivalent of the rateable value: this is adjusted to allow for general increases in rents since date of valuation, and is included in income of households living rent-free. 1985 assessments of rateable values in Scotland were used from 1 April 1985 in the calculation of imputed income and housing expenditure of owner-occupiers and those living rent-free (1987 FES Report definitions 14(h),(g)). From 1992 imputed income from owner-occupied and rent-free accommodation has not been estimated.
3 Income based on weighted data.

Source: Office for National Statistics: 020 7533 5756

8.4 Availability in households of certain durable goods[1]
United Kingdom

Percentage

		1988	1989	1990	1991	1992	1993	1994 /95	1995 /96	1996 /97	1997 /98	1998[2] /99
Number of households supplying data	KPDA	7 265	7 410	7 046	7 056	7 418	6 979	6 853	6 797	6 415	6 409	6 630
Car	KPDB	*66*	*66*	*67*	*68*	*68*	*69*	*69*	*70*	*69*	*70*	*72*
One	KPDC	*45*	*45*	*44*	*45*	*45*	*46*	*45*	*47*	*43*	*44*	*44*
Two	KPDD	*18*	*18*	*19*	*19*	*19*	*19*	*20*	*19*	*22*	*21*	*23*
Three or more	KPDE	*4*	*4*	*4*	*4*	*4*	*4*	*4*	*4*	*5*	*5*	*5*
Central heating, full or partial	KPDF	*77*	*78*	*79*	*81*	*82*	*83*	*84*	*85*	*87*	*89*	*89*
Washing machine	KPDG	*85*	*86*	*86*	*87*	*88*	*89*	*89*	*91*	*91*	*91*	*92*
Refrigerator or fridge/freezer	KPDH	*98*	*98*	*98*	*98*	*99*	*99*	*99*	*99*	..	..	..
Fridge/freezer or deep freezer	KPDI	*75*	*79*	*80*	*82*	*84*	*87*	*86*	*87*	*91*	*90*	*92*
Refrigerator	KPDJ	..	..	..	..	..	..	..	..	*49*	*51*	*52*
Television	KPDK	*98*	*98*	*98*	*98*	*98*	..	..	..	..	..	..
Telephone	KPDL	*85*	*86*	*87*	*88*	*88*	*90*	*91*	*92*	*93*	*94*	*95*
Home computer	KPDM	*17*	*17*	*17*	*18*	*19*	..	..	..	*27*	*29*	*33*
Video recorder	KPDN	*50*	*57*	*61*	*65*	*69*	*73*	*76*	*79*	*82*	*84*	*85*

1 Information derived from the Family Expenditure Survey (FES).
2 Percentages based on grossed number of households.

Source: Office for National Statistics: 020 7533 5756

8.5 Households and their expenditure[1]
United Kingdom

		1988	1989	1990	1991	1992	1993	1994 /95	1995 /96	1996 /97	1997[6] /98	1998[6,7] /99
Number of households supplying data	KPDA	7 265	7 410	7 046	7 056	7 418	6 979	6 853	6 797	6 415	6 409	6 630
Total number of persons	KPEA	18 280	18 590	17 437	17 089	18 174	17 291	16 617	16 586	15 732	15 430	16 218
Total number of adults[2]	KPEB	13 640	13 850	12 939	12 934	13 563	12 792	12 365	12 219	11 495	11 429	11 886
Household percentage distribution by tenure												
Rented unfurnished	KPED	*30*	*30*	*29*	*28*	*28*	*28*	*28*	*29*	*28*	*29*	*28*
Rented furnished	KPEE	*3*	*3*	*3*	*3*	*4*	*4*	*4*	*4*	*5*	*3*	*3*
Rent-free	KPEF	*2*	*1*	*1*	*2*	*2*	*2*	*1*	*1*	*2*	*2*	*2*
Owner-occupied	KPEG	*65*	*65*	*66*	*67*	*66*	*67*	*67*	*66*	*66*	*67*	*67*
Average number of persons per household												
All persons	KPEH	2.5	2.5	2.5	2.4	2.5	2.5	2.4	2.4	2.5	2.4	2.4
Males	KPEI	1.2	1.2	1.2	1.2	1.2	1.2	1.2	1.2	1.2	1.2	1.2
Females	KPEJ	1.3	1.3	1.3	1.3	1.3	1.3	1.3	1.3	1.3	1.3	1.2
Adults[2]	KPEK	1.9	1.9	1.8	1.8	1.8	1.8	1.8	1.8	1.8	1.8	1.8
Persons under 65	KPEL	1.5	1.5	1.5	1.5	1.5	1.5	1.4	1.4	1.4	1.4	1.5
Persons 65 and over	KPEM	0.4	0.4	0.4	0.4	0.4	0.4	0.4	0.4	0.4	0.3	0.4
Children[2]	KPEN	0.6	0.6	0.6	0.6	0.6	0.6	0.6	0.6	0.7	0.6	0.5
Children under 2	KPEO	0.1	0.1	0.1	0.1	0.1	0.1	0.1	0.1	0.1	0.1	0.1
Children 2 and under 5	KPEP	0.1	0.1	0.1	0.1	0.1	0.1	0.1	0.1	0.1	0.1	0.1
Children 5 and under 18	KPEQ	0.5	0.4	0.4	0.4	0.4	0.4	0.4	0.5	0.5	0.4	0.4
Persons economically active[3]	KPER	1.2	1.2	1.2	1.2	1.2	1.2	1.2	1.1	1.2	1.1	1.2
Persons not economically active[3]	KPES	1.3	1.3	1.3	1.3	1.3	1.3	1.3	1.3	1.3	1.3	1.2
Men 65 and over, women 60 and over	KPET	0.4	0.4	0.4	0.4	0.4	0.4	0.4	0.4	0.4	0.4	0.4
Others	KPEU	0.9	0.9	0.9	0.8	0.9	0.9	0.9	0.9	0.9	0.9	0.8
Average weekly household expenditure on commodities and services (£)												
Housing[4,5]	KPEV	35.80	38.40	44.40	50.20	47.40	44.90	46.40	48.30	49.10	51.50	57.20
Fuel and power	KPEW	10.50	10.60	11.10	12.30	13.00	13.20	13.00	12.90	13.40	12.70	11.70
Food	KPEX	38.30	41.70	44.80	46.10	47.70	50.00	50.40	52.90	55.20	55.90	58.90
Alcoholic drink	KPEY	9.20	9.50	10.00	10.80	11.10	12.00	12.30	11.40	12.40	13.30	14.00
Tobacco	KPEZ	4.50	4.80	4.80	5.20	5.40	5.60	5.60	5.80	6.10	6.10	5.80
Clothing and footwear	KCWC	14.50	15.30	16.00	15.80	16.40	17.40	17.10	17.20	18.30	20.00	21.70
Household goods[5]	KCWH	15.00	19.20	20.00	20.10	21.90	23.10	22.70	23.50	26.70	26.90	29.60
Household services[5]	KCWI	9.80	9.70	12.30	13.00	13.40	15.40	15.10	15.10	16.40	17.90	18.90
Personal goods and services	KCWJ	8.10	8.50	9.50	10.00	10.20	11.00	10.80	11.60	11.60	12.50	13.30
Motoring expenditure	KCWK	25.30	30.40	33.80	34.10	35.70	36.30	36.20	37.00	41.20	46.60	51.70
Fares and other travel costs	KCWL	4.90	5.40	6.20	5.60	7.20	7.00	6.60	6.20	7.50	8.10	8.30
Leisure goods	KCWM	9.70	11.00	11.30	12.10	13.30	13.30	13.90	13.20	15.20	16.40	17.80
Leisure services[5]	KCWN	18.10	19.00	21.50	22.20	27.60	25.60	31.20	32.10	34.00	38.80	41.90
Miscellaneous	KCWO	0.80	0.90	1.40	1.60	1.80	2.10	2.30	2.40	2.20	2.00	1.20
Total	KCWP	204.40	224.30	247.20	259.00	271.80	276.70	289.90	289.90	309.10	328.80	352.20

1 Information derived from the Family Expenditure Survey (FES).
2 Adults and children are:
Adults = all persons 18 and over and married persons under 18.
Children = all unmarried persons under 18.
3 Definitions of economic activity changed in 1990: see Family Spending 1990 for details.
4 Until 1992 excludes mortgage payments but includes imputed rent of owner-occupancy and of rent-free occupancy. Imputed expenditure is the weekly equivalent of the rateable value which is adjusted to allow for general increases in rents since date of valuation. 1985 assessments of rateable values in Scotland were used from 1 April 1985 in the calculation of imputed income and housing expenditure of owner-occupiers and those living rent-free (1987 FES Report definitions 14(h),(g)). In 1992 values of income and expenditure pertaining to households in owner-occupied and rent-free housing were no longer imputed. From 1992 onwards the interest element of mortgage payments is recorded as part of housing expenditure.
5 Expenditure on certain items was recorded on a retrospective basis from 1989 and 1990.
6 Rented unfurnished includes Local Authority and Housing Association (furnished and unfurnished) and privately rented (unfurnished). Rented furnished only includes privately rented (furnished). Owner-occupied includes shared owners (who part own and part rent).
7 Averages based on grossed number of households.

Source: Office for National Statistics: 020 7533 5756

8.5 Households and their expenditure[1]
United Kingdom

continued

		1988	1989	1990	1991	1992	1993	1994 /95	1995 /96	1996 /97	1997[6] /98	1998[6,7] /99
Expenditure on commodity or service as a percentage of total expenditure												
Housing[4,5]	KPFH	18	17	18	19	17	16	16	17	16	16	16
Fuel and power	KPFI	5	5	5	5	5	5	5	5	4	4	3
Food	KPFJ	19	19	18	18	18	18	18	18	18	17	17
Alcoholic drink	KPFK	5	4	4	4	4	4	4	4	4	4	4
Tobacco	KPFL	2	2	2	2	2	2	2	2	2	2	2
Clothing and footwear	KPFM	7	7	7	6	6	6	6	6	6	6	6
Household goods[5]	KCWQ	7	9	8	8	8	8	8	8	9	8	8
Household services[5]	KCWR	5	4	5	5	5	6	5	5	5	5	5
Personal goods and services	KCWS	4	4	4	4	4	4	4	4	4	4	4
Motoring expenditure	KCWT	12	14	14	13	13	13	13	13	13	14	15
Fares and other travel costs	KCWU	2	2	3	2	3	3	2	2	2	3	2
Leisure goods	KCWV	5	5	5	5	5	5	5	5	5	5	5
Leisure services[5]	KCWW	9	9	9	9	10	9	11	11	11	12	12
Miscellaneous	KPFR	–	–	1	1	1	1	1	1	1	1	–
Total	KPFS	100	100	100	100	100	100	100	100	100	100	100

See footnotes on the first part of this table.

Source: Office for National Statistics: 020 7533 5756

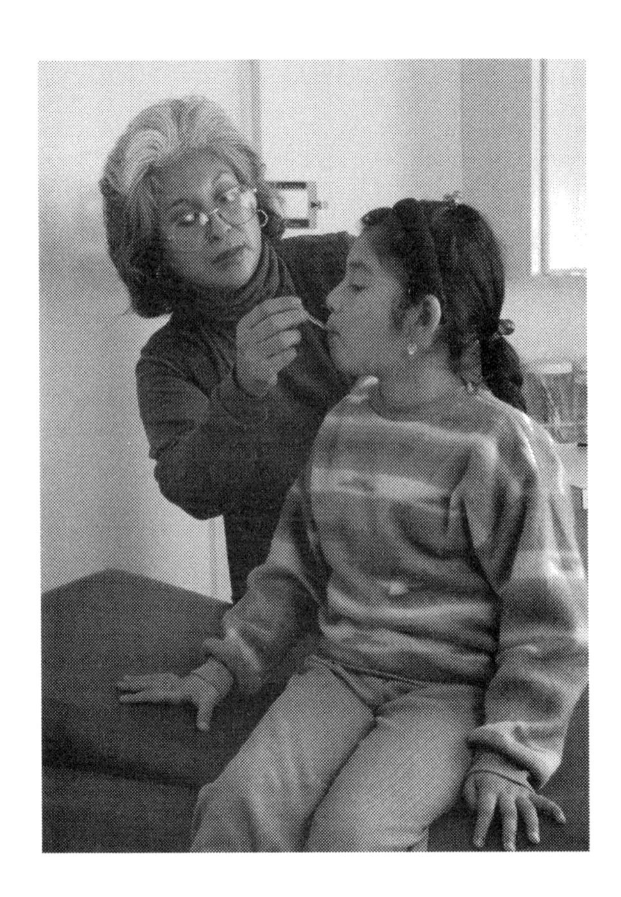

Health

Injuries at work *(Table 9.8)*

Employers are required to report injuries arising from workplace activities to HSC enforcing authorities under the Reporting of Injuries, Diseases and Dangerous Occurrences Regulations 1995 (RIDDOR 95). This includes fatal injuries, non-fatal major injuries, as defined by the Regulations, and other injuries causing incapacity for work for more than 3 days. Statistics from 1996/97 were compiled from reports made under RIDDOR 95 which updated previous legislation (RIDDOR 85) on 1 April 1996. The main differences in the updated Regulations are that the definition of a major injury to a worker is now wider and injuries caused by acts of physical violence at work are reportable. Therefore, statistics from 1996/97 cannot be compared with those for previous years.

Details of fatal, non-fatal major and over-3-day injuries involving employees and self-employed people are shown in the table. The introduction of RIDDOR 95 had no substantial effect on the number of workplace fatalities in 1996/97. However, it is estimated that approximately 70 per cent of the increase in the total number of non-fatal major injuries between 1995/96 and 1996/97 was a result of expanded definitions under RIDDOR 95. The expansion also resulted in a decrease in the total number of over-3-day injuries, since some of the non-fatal major injuries in 1996/97 would have been previously reportable as over-3-day injuries. However, since injuries caused by violence at work also became reportable, the decrease in the number of over-3-day injuries appears to be relatively small.

Not all employers make reports of non-fatal injuries as they should. In 1990 the HSC commissioned a health and safety supplement to the Labour Force Survey (LFS) in order to establish the true levels of workplace injury. This has been continued annually since 1993/94. The results of the LFS suggest levels of non-fatal injury reporting for employees to be about 47 per cent in 1997/98 compared with 34 per cent in 1989/90. For the self-employed, reporting levels were 5 per cent in 1989/90 and less than 10 per cent in 1997/98.

9.1 Hospital and family health services
England and Wales

			England 1994	England 1995	England 1996	England 1997	England 1998		Wales 1994	Wales 1995	Wales 1996	Wales 1997	Wales 1998
Hospital services[3]													
Average daily number of available beds	KNMY	Thousands	212	206	199	194	..	KNHY	16.8	16.0	15.6	15.2	14.2
Average daily occupation of beds:													
All departments	KNMX	"	..	..	162	157	..	KNHX	12.9	12.4	12.2	12.0	..
Psychiatric departments	KNMW	"	..	..	41	39	..	KNGZ	3.4	3.1	3.0	2.9	..
Persons waiting for admission at 31 March[2]	KNMV	"	1 065	1 044	1 048	1 158	1 298	KNGY	59.2	60.3†	65.0	69.9	61.8
Finished consultant episodes	KNLZ		..	..	..	..	..						
Day case admissions	KNLY	"	2 474	2 774†	2 907	3 071	..	KNBZ	240.0	261.6	286.4	309.7	328.0
Ordinary admissions	KIBS	"	8 065	8 264†	8 369	8 459	..	KNEO	513.4	516.0	516.4	521.1	522.7
Out-patients													
New cases	KNLX	"	10 363	10 988†	11 293	11 529	..	KNBY	638.3	658.7	667.1	693.3	691.3
Total attendances	KNLW	"	39 305†	40 116	40 872	41 635	..	KNBX	2 587.7	2 570.0	2 569.5	2 643.3	2 667.4
Accident and Emergency:													
New cases	KOTH	"	11 943	12 462†	12 484	12 794	..	KTCO	763.3	804.1	798.4	826.6	831.0
Total attendances	KOTI	"	13 812	14 234	14 126†	14 364	..	KTCP	960.6	995.8	975.3	997.9	982.0
Ward attendances	KOTJ	"	980	1 013	1 027	1 034	..	KTCQ	..	..	..	..	..
Family health services													
Medical services:													
Doctors on the list[4]	KNKX	Number	26 567	26 702	26 855	27 099	..	KNBR	1 710	1 719	1 736	1 753	1 749
Number of UPEs[13]	LQZZ		..	..	..	..	27 392						
Number of patients per doctor	KNKW	"	1 900	1 887	1 885	1 878	1 866	KNBQ	1 739	1 730	1 724	1 706	1 713
Paid to doctors[5,6]	KNKV	£ million	2 690.6	2 851.0	2 857.0	3 131.0	3 242.0	KNBP	317.4	371.0	387.5	466.6	470.6
Pharmaceutical services:[1,7]													
Number of prescription forms	KWUK	Million	275.4	284.4	282.9	287.4	291.1						
Number of prescription items	KWUL	"	456.0	473.3	484.9	500.2	513.2	KNBO	35.4	36.3	37.2	38.9	40.0
Total cost	KWUM	£ million	3 896.2	4 193.4	4 534.2	4 919.8	5 231.4	KNBN	281.0	303.8	325.6	351.9	366.4
Average total cost per prescription	KWUN	£	8.54	8.86	9.35	9.84	10.20	KNBK	7.95	8.37	8.75	9.00	9.16
Income from patients	KWUO	£ million	235.3	249.6	246.1	265.7	282.2	KNBM	18.3	18.8	18.0	20.4	15.7
Dental services:													
Principals on an FHSA/HA list at 30 September[8]	KIAZ	Number	15 084	15 064	15 280	15 509	15 820	KIBG	801	817	834	863	879
Number of adult courses of treatments	KIBA	Thousands	24 913	24 752	24 580	25 268	26 171	KIBH	1 369	1 350	1 357	1 446	1 514
Number of adult patients accepted into continuing care provision at 30 September	KIBB	"	21 050	19 994	19 524	19 383	16 721	KIBI	1 254	1 197	1 177	1 184	1 043
Number of children accepted into capitation at 30 September	KIBC	"	7 367	7 292	7 270	7 367	6 775	KIBJ	432	426	426	434	405
Gross expenditure[9]	KIBD	£ thousands	1 279 412	1 289 547	1 323 074	1 347 577	1 437 727	KIBK	70 416	70 743	73 554	78 083	83 500
Paid by patients[9]	KIBE	"	383 309	381 186	382 995	388 436	419 621	KIBL	19 625	19 616	19 789	21 384	22 681
Paid out of public funds[9]	KIBF	"	896 103	908 361	940 078	959 141	1 018 106	KIBM	50 791	51 127	53 765	56 699	60 819
General ophthalmic services:													
Sight tests[10,11]	KNJL	Thousands	6 383	6 512	6 808	6 991†	6 992	KNBD	421	439	464	477	477
Pairs of spectacles for which NHS vouchers redeemed[11]	KNJK	"	3 741	3 815	3 967	3 935	3 777	KNBC	268	275	287	300	288
Cost of services (gross)[9]	KNJJ	£ million	212.7	223.5	237.2	241.4	239.6	KNBA	15.0	16.0	17.1	17.9	17.9
Paid out of public funds:[9]													
For sight testing	KNJH	"	87.3	90.1	97.1	102.5	103.3	KMZZ	5.9	6.2	6.7	7.2	7.3
For cost of vouchers	KNHZ		125.3	132.6	139.8	138.6	136.0	KMZX	9.1	9.8	10.4	10.6	10.5

1 Data given are for calendar year for England and financial year for Wales. Data shown reflect data for the year commencing the year in the heading (for example the figure under 1994 reflects 1994 - 95 data). Financial year is from 1 April to 31 March.
2 People awaiting elective admission at NHS Trusts in England and Wales, as an inpatient or a day case.
3 Data collection changed in 1987 from calendar year to a financial year basis. Data shown reflect data for the year commencing the year in the heading (for example the figure under 1994 reflects 1994 - 95 data). Outpatient figures do not include accident and emergency figures or ward attenders which are given separately as follows: Information on general practitioner maternity clinics was not collected separately from 1992-93 in England but included for Wales.
4 Principals providing unrestricted services as at 1 October.
5 Provisional data subject to change.
6 Includes PFMA but excludes GPFH drugs and payments to providers.
7 The data cover all prescription items dispensed by community pharmacists and appliance contractors, dispensing doctors and prescriptions submitted by prescribing doctors for items personally administered. Total cost refers to the cost of the drug less discounts and includes on cost allowance, dispensing fees, container allowance, oxygen payments and VAT. Income from patients only covers community pharmacists and appliance contractors. The data does not include income from prescriptions dispensed by dispensing doctors or prepayment certificates.
8 Principals only. Assistants and vocational trainees are not included. Some dentists may have a contract with more than one Family Health Service Authority/Health Authority. These dentists have been counted once only.
9 Figures are for financial year and are based on the Appropriation account. 1997/98 figures are provisional.
10 Number of NHS sight tests paid for by FHSAs/HAs in the period.
11 Data given are for financial year and reflect the year commencing the year in the table heading (e.g. the figures for 1998 represents 1998/99 data).
12 To ensure consistency with PIPS data these figures are extracted from cash accounts, not accruals.
13 UPE's include Unrestricted Principals, PMS Contracted GP's and PMS salaried GP's. UPE data are as at 1 October.

Sources: NHS Executive: 020 7972 2231; National Assembly for Wales: 029 2082 5080

9.2 Hospital and primary care services
Scotland

		Unit	1988	1989	1990	1991	1992	1993	1994	1995	1996	1997	1998
Hospital and community services													
In-patients[1,2]:													
Average available staffed beds	KDEA	Thousands	54.5	53.4	52.1	50.6	48.7	46.7	44.2	42.4	40.6	38.4	36.8
Average occupied beds:													
All departments	KDEB	"	44.4	43.5	42.5	41.2	39.5	38.1	35.9	34.3	32.8	30.9	29.5
Psychiatric and learning disability	KDEC	"	18.2	17.6	17.0	16.3	15.5	14.6	13.2	12.6	11.7	10.8	10.0
Discharges or deaths[3]	KDED	"	871	882	896	912	920	942	952	960	973	965	977
Outpatients[2,4]:													
New cases	KDEE	"	2 275	2 325	2 381	2 403	2 426	2 457	2 503	2 577	2 666	2 675	2 715
Total attendances	KDEF	"	5 767	5 815	5 925	5 971	6 005	6 086	6 145	6 241	6 338	6 272	6 331
Medical and dental staff[5,6,7]:	JYXO	Number	7 769	7 777	7 936	7 890	8 069	8 078	8 317	8 524	8 774	9 098	9 171
Whole-time	KDEG	"	5 345	5 368	5 475	5 481	5 625	5 767	5 937	6 102	6 433	6 707	6 730
Part-time	KDEH	"	1 850	1 835	1 881	1 862	1 883	1 730	1 827	1 854	1 819	1 886	1 936
Honorary	JYXN	"	574	574	580	547	561	582	563	579	534	522	505
Professional and technical staff[5]:													
Whole-time	KDEI	"	9 200	9 320	9 291	9 387	9 759	9 953	10 062	10 452	10 584	10 740	10 884
Part-time	KDEJ	"	2 327	2 601	2 727	2 782	3 127	3 434	3 753	4 075	4 370	4 738	4 928
Nursing and midwifery staff[5,8,9,10]:													
Whole-time	KDEK	"	34 676	34 959	34 364	34 302	34 053	33 284	32 956	32 693	32 560	32 218	32 156
Part-time	KDEL	"	27 665	28 164	28 781	29 887	30 709	30 459	30 532	30 580	29 917	29 736	29 178
Administrative and clerical staff[5]:													
Whole-time	KDEM	"	12 250	12 648	13 223	13 718	14 447	15 125	15 723	15 815	15 155	14 708	14 564
Part-time	KDEN	"	4 652	4 769	4 985	5 276	5 815	6 113	6 624	7 005	6 986	7 174	7 265
Domestic, transport, etc. staff[5]:													
Whole-time	KDEO		17 141	15 214	12 468	11 622	10 805	10 205	9 574	9 037	8 596	8 187	8 090
Part-time	KDEP		18 521	16 059	16 685	16 167	15 753	15 403	14 464	14 105	13 554	13 082	12 716
Cost of services (gross)[11]	KDEQ	£ million	1 828.1	2 004.5	2 195.6	2 503.6	2 768.0	2 940.5	3 050.5	3 269.5	3 430.6	3 610.3	3 917.8
Payments by patients[11,23]	KDER	"	1.40	1.40	1.30	1.50	2.00	2.30	0.40	0.02	0.01	0.01	0.01
Payments out of public funds[11,23]	KDES	"	1 826.70	2 003.10	2 194.30	2 502.10	2 766.00	2 938.20	3 050.10	3 269.52	3 430.59	3 610.29	3 917.79
Primary care services													
Medical services													
Doctors on the list[12]:													
Principals[13]	KDET	Number	3 354	3 390	3 359	3 379	3 421	3 456	3 490	3 524	3 573	3 625	3 660
Assistants	KDEU	"	13	10	16	24	31	22	25	28	22	22	27
Average number of patients per principal doctor[14]	KDEV	"	1 606	1 590	1 593	1 580	1 556	1 542	1 524	1 506	1 488	1 468	1 449
Payments to doctor[15]	KDEW	£ million	165.7	183.3	222.2	247.2	263.8	275.3	291.7	311.9	333.2	356.4	362.0
Pharmaceutical services[16]													
Prescriptions dispensed	KDEX	Million	39.50	41.02	42.42	44.31	46.10	48.18	49.27	51.08	54.62	56.64	58.52
Payments to pharmacists (gross)	KDEY	£ million	244.02	269.99	298.16	329.82	370.20	406.07	434.85	474.22	543.40	588.38	627.18
Average gross cost per prescription	KDEZ	£	6.180	6.580	7.030	7.440	8.030	8.430	8.825	9.283	9.950	10.390	10.717
Dental services[17]													
Dentists on list[18]	KDFA	Number	1 523	1 585	1 645	1 676	1 702	1 772	1 763	1 764	1 772	1 798	1 854
Number of courses of treatment completed	KDFB	Thousands	3 054	3 143	3 057	2 485	2 569	2 647	2 723	2 711	2 825	3 406	3 349
Payments to dentists (gross)[19]	KDFC	£ million	86.9	97.7	99.9	132.8	133.6	128.4	136.1	137.3	139.2	154.9	157.5
Payments by patients	KDFD	"	33.9	46.4	48.0	41.6	40.5	40.0	42.2	41.7	41.4	45.9	47.4
Payments out of public funds[19]	KDFE	"	53.0	51.2	51.8	91.3	93.2	88.4	93.9	95.6	97.8	109.0	110.1
Average gross cost per course	KDFF	£	28.5	30.8	33.1	40.0	39.0	37.0	38.0	38.0	40.1	36.5	38.0
General ophthalmic services[24]													
Number of sight tests given[20,24]	KDFG	Thousands	1 141	716	417	477	537	568	614	618	635	656	..
Number of pairs of glasses supplied[21,24]	KDFH	"	322	313	320	374	404	463	473	461	474	488	..
Cost of services (gross)[22]	KDFI	£ million	17.9	14.5	13.7	16.4	19.7	22.2	24.4	25.8	27.7	29.1	29.8
Payments by patients	KDFJ	"	–	–	–	..	..	..	..	..	..	..	..
Payments out of public funds:													
For sight testing and dispensing	KDFK	"	17.9	14.5	13.7	16.4	19.7	22.2	24.4	25.8	27.7	29.1	29.8

1 Excludes joint user and contractual hospitals.
2 In year to 31 March.
3 Includes transfers out.
4 Including attendances at accident and emergency consultant clinics.
5 At 30 September.
6 Figures exclude officers holding honorary locum appointments. Part-time includes maximum part-time appointments.
7 There is an element of double counting of "heads" in this table as doctors can hold more than one contract. For example, they may hold contracts of different type, e.g. part time and honorary. Doctors holding two or more contracts of the same type, e.g. part time, are not double counted. Doctors, whose sum of contracts amounts to whole time, are classed as such.
8 Excludes nurse teachers, nurses in training and students on "1992". courses.
9 The apparent drop in whole-time nurses numbers between 1989 and 1990 is due largely to reclassification.
10 Between 1992 and 1993 the criterion for excluding nursing staff on low hours was changed from 0.25 hours to 2 hours. This partially accounts for any decrease in staff numbers between those two years.
11 Estimated from financial year figures.
12 At 1 October.
13 Unrestricted principals in post.
14 Unrestricted principals: establishment.
15 Provisional for 1997/98. Data relates to financial year, eg 1997 data is for year ending 31 March 1998. As 1994/95 data are unavailable for Dumfries & Galloway Health Board, 1993/94 data have been substituted for that board only.
16 For prescriptions dispensed in calendar year by all community pharmacists, (including stock orders) dispensing doctors and appliance suppliers.
17 From 1990, when the new dental contract came into effect, the statistics available centrally changed significantly. From 1991, data refers to financial year (eg 1991 data is for year ending 31 March 1992).
18 Assistants are excluded.
19 Before 1991 excludes capitation payments; from 1991 includes capitation, continuing care, weighted entry and item of service fees.
20 This figure represents sight tests paid for by health boards, hospital eye service referrals and GOS(s) ST (v) claimants.
21 Supply of spectacles restricted from 1 April 1985, includes hospital eye service.
22 Universal free eyesight testing ended in March 1989.
23 These figures are for Health Boards only and do not include the 2 NHS Trusts in 92/93, 17 in 93/94, 39 in 94/95, 47 in 95/96.
24 No data available for 1998.

Source: Scottish Health Service Common Services Agency: 0131 551 8899

9.3 Hospital and general health services
Northern Ireland

		Unit	1988	1989	1990	1991	1992	1993	1994	1995	1996	1997	1998
Hospital services[10]													
In-patients:													
Beds available[1]	KDGA	Number	14 684[10]	14 154[10]	13 487[10]	12 600[10]	11 712[10]	10 878[10]	10 356[10]	10 054[10]	9 475[10]	9 007[10]	8 819[10]
Average daily occupation of beds	KDGB	*Per cent*	*79.8*[10]	*79.2*[10]	*77.6*[10]	*77.1*[10]	*77.8*[10]	*77.9*[10]	*78.1*[10]	*77.7*[10]	*78.9*[10]	*80.8*[10]	*81.9*[10]
Discharges or deaths[15]	KDGC	Thousands	260[10]	264[10]	265[10]	269[10]	275[10]	290[10]	298[10]	304[10]	300[10]	306[10]	338[10]
Out-patients[2]:													
New cases	KDGD	"	761[10]	781[10]	782[10]	799[10]	819[10]	836[10]	873[10]	937[10]	933[10]	952[10]	962[10]
Total attendances	KDGE	"	1 974[10]	1 989[10]	1 954[10]	1 963[10]	1 969[10]	1 967[10]	2 051[10]	2 087[10]	2 072[10]	2 083[10]	2 091[10]
General health services													
Medical services													
Doctors (principals) on the list[3]	KDGF	Number	927	933	928	928	942	953	991	1 005	1 028	1 039	1 042
Number of patients per doctor	KDGG	"	1 808	1 794	1 811	1 922	1 808	1 779	1 743	1 731	1 698	1 690	1 693
Payments to doctors	KDGH	£ thousand	43 235	48 651	47 850	53 675	55 715	59 253	61 742	64 027	66 733	69 889	71 340
Pharmaceutical services													
Prescription forms dispensed	KDGI	Thousands	8 915	9 175	9 285	10 144	10 597	10 899	11 439	12 018	12 402	12 855	13 116
Number of prescriptions	KDGJ	"	14 606	15 726	15 404	16 754	17 496	18 043	18 561	19 894	20 581	21 430	22 171
Gross cost[13]	KDGK	£ thousand	88 688	97 721	108 144	133 134	146 925	160 420	173 064	197 580	213 258	229 852	241 973
Charges[13]	KDGL	"	3 900	4 185	4 893	4 745	5 582	6 017	6 529	6 255	6 028	6 607	6 853
Net Cost[13]	KDGM	"	84 787	93 536	103 751	119 694	141 343	154 403	172 412	191 325	207 230	223 245	235 120
Average gross cost per prescription	KDGN	£	6.10	6.50	7.00	7.70	8.40	8.90	9.44	9.93	10.36	10.73	10.91
Dental services													
Dentists on the list[4]	KDGO	Number	460	467	519	513	523	541	569	581	596	610	633
Number of courses of paid treatments	KDGP	Thousands	965	972	986	..	..	..	..	..	..	..	44 713
Gross cost[14]	KDGQ	£ thousand	27 442	31 697	35 026	47 210	47 637	43 904	47 076	48 781	51 512	53 735	56 835
Patients[14]	KDGR	"	5 976	8 009	8 732	9 922	11 036	10 378	11 417	11 530	11 870	12 433	13 686
Contributions (Net cost)[14]	KDGS	"	21 447	23 688	26 295	37 258	36 601	33 526	33 659	37 250	39 642	41 302	43 149
Average gross cost per paid treatment	KDGT	£	28.40	32.60	35.50	..	..	..	..	..	..	..	1.27
Ophthalmic services													
Number of sight tests given[5]	KDGU	Thousands	281	192	115	135	142	169	182	197	212	221	238
Number of optical appliances supplied[12]	KDGV	"	–[11]	–[11]	96[11]	110[11]	114[11]	133[11]	139[11]	146[11]	153[11]	156[11]	161[11]
Cost of service (gross)[6]	KDGW	£ thousand	5 135	4 583	4 158	5 607	5 555	7 363	7 127	8 568	9 555	10 114	10 469
Payments by patients	KDGX	"	–	–	–	–	..	..	..	..	..	..	..
Payments out of public funds	KDGY	"	5 135	4 583	4 158	5 607	5 555	..	..	..	9 555	10 114	10 469
Health and social services[7]													
Medical and dental staff:													
Whole-time	KDGZ	Number	2 091	2 218	2 238	2 262	2 226	2 209	2 358	2 053	2 107	2 156	2 196
Part-time	KDHA	"	719	674	564	560	511	554	694	1 154	1 094	1 041	1 009
Nursing and midwifery staff:													
Whole-time	KDHB	"	14 637	14 425	14 071	13 451	12 381	11 445	11 047	10 896	10 578	10 114	10 117
Part-time	KDHC	"	6 207	6 477	6 841	7 144	7 417	7 952	8 662	9 169	8 943	9 015	8 287
Administrative and clerical staff:													
Whole-time	KDHD	"	5 175	5 409	5 843	6 398	6 573	6 696	7 006	7 078	7 055	6 915	7 019
Part-time	KDHE	"	1 445	1 583	1 680	1 831	1 924	2 050	2 186	2 306	2 518	2 708	2 776
Professional and technical staff:													
Whole-time	KDHF	"	2 477	2 522	2 563	2 673	2 695	2 695	2 786	2 862	2 939	2 933	3 014
Part-time	KDHG	"	453	475	513	559	636	710	804	921	985	1 060	1 146
Social services staff(excluding casual home helps):													
Whole-time	KDHH	"	3 601	3 646	3 649	3 641	3 607	3 550	3 480	3 470	3 441	3 349	3 262
Part-time	KDHI	"	1 467	1 546	1 687	1 695	1 707	1 766	1 933	2 110	2 250	2 394	2 241
Ancillary and other staff:													
Whole-time	KDHJ	"	7 551	6 960	6 433	5 551	4 954	4 679	4 364	3 982	3 812	3 569	3 423
Part-time	KDHK	"	5 591	5 480	4 989	4 728	4 862	4 767	4 609	3 685	3 558	3 482	3 558
Cost of services (gross)	KDHL	£ thousand	695 515[8]	750 918[8]	822 795[8]	889 422[8]	947 354[8]	1 002 326[8]	1 043 745[8]	1 111 507[8]	1 120 563[8]	1 153 741[8]	..
Payments by recipients	KDHM	"	10 812	11 125	11 414	12 322	12 569	13 189	20 629	32 685	40 725	49 498	..
Payments out of public funds	KDHN	"	684 703	739 793	811 381	877 100	934 785	989 137	1 023 116	1 078 822	1 079 838	1 104 243	..

1 Average available beds in wards open overnight during the year.
2 Includes consultant outpatient clinics and Accident & Emergency departments.
3 At 30 September.
4 At 1st October.
5 Excluding sight tests given in hospitals and under the school health service.
6 Figures from 1977 include payments of superannuation.
7 Manpower figures refer to 31 December. From 1981, figures for medical and dental staff exclude some joint appointees, information for whom is not held on the computer payroll.
8 Figures relate to the cost of the hospital, community health, and personal social services, and have been estimated from financial year data.
9 The decrease in figures is due to legislation introduced on 1 July 1985 which allowed the issue of optical appliances by unregistered suppliers under the 'Voucher Scheme'.
10 Financial year.
11 Relates to the number of vouchers supplied.
12 Due to the change in the Dental Contract which came into force in October 1990, dentists are now paid under a combination of headings relating to Capitation and Continuing care patients. Prior to this, payment was simply on an item of service basis, which made statistics such as 'Number of course of treatment completed' and 'Average gross cost per course' relevant and meaningful. This is no longer the case.
13 Prior to 1992, headings read, 'Payments to Pharmacists (Gross)', 'Payments by Patients', 'Payments out of Public funds'.
14 Prior to 1992, headings read, 'Payments to Dentists (Gross)', 'Payments by Patients', 'Payments out of Public Funds'.
15 Includes transfers to other hospitals.

Sources: Central Service Agency Northern Ireland: 028 9032 4431; Dept of Health & Social Services Northern Ireland: 028 9052 2509; (For figures on Hospital Services: 028 9052 2800)

9.4 Health and personal social services: workforce summary

Great Britain

Number or whole-time equivalent

		1988	1989	1990	1991	1992	1993	1994	1995	1996	1997	1998
Health service staff and practitioners at 30 Sep:												
Medical staff: total[1]	KDBC	48 873	50 290	51 792	53 019	53 987	55 632	56 736	60 172	62 176	64 316	67 408
Hospital medical staff: total	KDBD	45 438	46 906	48 435	49 620	50 793	52 476	53 787	57 299	59 592	61 937	65 088
Consultant and senior hospital medical officer with allowance	KDBE	15 974	16 517	17 101	17 435	17 846	18 251	18 808	20 246	21 066	21 699	23 139
Staff grade[2]	KADJ	..	33	262	516	819	1 230	1 536	2 037	2 440	2 785	3 458
Associate specialist	KDBF	859	877	944	949	960	1 032	1 013	1 128	1 223	1 340	1 439
SpR/Sen Reg/ Reg[11]	KWUG	..	..	..	..	..	..	..	..	11 898	12 435	12 863
Senior registrar	KDBG	3 351	3 488	3 778	3 912	3 994	4 130	4 281	4 540	..	..	..
Registrar	KDBH	7 385	7 297	7 345	7 237	7 132	7 103	7 273	7 294	..	..	..
Senior house officer including post-registration house officer	KDBI	11 623	12 407	12 737	13 311	13 863	14 586	14 942	15 661	16 616	17 353	17 760
Pre-registration house officer	KADK	3 566	3 637	3 748	3 781	3 798	3 792	3 781	4 003	4 025	4 163	4 287
Hospital practitioner	KDBL	231	233	219	208	183	183	178	197	212	198	220
Part-time medical officer (clinical assistant)	KDBM	2 426	2 402	2 285	2 255	2 189	2 162	1 964	2 182	2 094	1 924	1 907
Other staff[3]	KDBK	23	17	15	17	10	8	11	11	18	38	16
Community health medical staff[4]	KDBN	3 435	3 384	3 357	3 399	3 194	3 156	2 948	2 873	2 584	2 379	2 320
Dental staff: total	KDBO	3 128	3 092	3 072	2 954	2 971	2 973	2 947	3 070	3 127	3 078	3 193
Hospital dental staff: total[1]	KDBP	1 465	1 475	1 496	1 478	1 546	1 573	1 597	1 687	1 737	1 696	1 807
Consultant and senior hospital dental officer with allowance	KDBQ	511	512	497	498	541	521	537	549	556	524	570
Staff grade[2]	KADL	..	–	2	5	13	24	39	54	67	86	99
Associate specialist	KDBR	84	81	82	71	70	69	62	69	65	62	68
Registrar Group[11]	LQMZ	..	..	..	..	..	..	..	..	..	..	309
Senior registrar	KDBS	118	129	135	123	122	135	124	123	138	169	..
Registrar	KDBT	216	204	197	189	188	184	178	186	183	125	..
Senior house officer	KDBU	223	246	260	266	300	353	394	446	490	491	531
Dental house officer	KDBV	138	129	145	145	143	126	101	86	68	58	59
Hospital practitioner	KDBX	12	14	17	15	15	17	20	17	23	22	21
Part-time medical officer (clinical assistant)	KDBY	160	158	160	165	154	142	141	158	145	152	144
Other staff[3]	KDBW	4	3	1	1	2	2	2	–	3	6	5
Community health dental staff[4]	KDBZ	1 663	1 617	1 576	1 476	1 425	1 400	1 350	1 383	1 390	1 382	1 386
Total non-medical staff	KWUH	920 608	912 481	910 455	912 028	906 236	899 234	876 630	859 213	851 185	894 309	855 013
Nursing and midwifery staff:[8,9,10] (excluding agency): total	KDCA	489 574	490 545	487 012	483 507	467 723	446 056	429 214	353 842	356 109	353 933	409 738
qualified	KSBR	294 828	299 527	297 320	298 299	300 698	295 245	291 070	292 248	264 744	262 732	299 010
unqualified	KSBS	117 454	116 088	114 273	114 474	115 431	116 360	116 138	116 053	89 998	90 595	107 262
learners	KSBT	77 284	74 929	71 645	59 615	44 879	27 560	13 945	7 580	..	..	9 904
unknown	KSBV	..	..	3 774	11 115	6 714	6 891	8 061	5 768	937	606	541
All Professional & Technical staff[8,9] (excluding works)	KSBM	95 521	97 149	100 226	103 572	107 367	109 463	111 597	100 350	105 496	107 158	123 833
Health care assistants	KWUI	..	..	..	..	..	..	..	..	18 025	19 260	21 887
Support staff	KWUJ	..	..	..	..	..	..	..	..	75 876	72 514	71 025
Ancillary, Works & Maintenance[8,9]	KSBN	167 185	156 277	145 573	132 321	123 100	118 867	110 577	102 669	..	..	36 265
Administrative & Clerical staff[5,6,8,9]	KSBO	137 899	144 041	157 519	165 467	179 300	180 820	186 446	180 132	178 461	177 957	159 351
Ambulance staff[8]	KSBP	22 302	22 443	21 698	21 369	21 744	21 424	21 379	15 681	16 330	16 424	18 526
Others[7,8,9]	KSBQ	..	..	..	5 696	7 165	639	1 250	8 759	6 834	3 337	3 014

1 Whole-time equivalent. Figures exclude locums and general medical practitioners participating in Hospital Staff Fund.
2 New grade introduced in 1989.
3 Figures include Senior Hospital Medical/Dental Officers (SHMO) without an allowance and other ungraded staff.
4 Whole-time equivalent. Figures exclude locums and occasional seasonal staff.
5 Figures exclude ambulance officers.
6 Includes Family Practitioner Service administrative and clerical staff and General Managers.
7 Due to changes in the collection procedure in 1991 a category of 'other staff' was introduced for these locally determined payscales. In previous years these staff were included in their respective main staff groups. Includes learners from 1996.
8 A new system for classifying NHS non-medical staff was used for the first time in September 1995 non-medical workforce census classifying staff according to what they do (known as new occupation codes). However in order to provide comparative information with earlier years, payscale based estimates were produced using data from organisations providing both payscale and occupation code information for at least 95 per cent of their non-medical staff. These estimates only provide a broad indication of 1995 levels of staff and therefore should be used with caution. For further information on the 1995 non-medical workforce see the Department of Health's statistical bulletin on non-medical staff in England (1985 - 95). These are England and Wales figures only from 1995.
9 Prior to 1987 figures for 'other statutory authorities' (eg Public Health Laboratory Service and Health Education Authority) were not collected in the annual non-medical workforce census.
10 Data for Scotland excludes bank nurses.
11 Includes Specialist Registrar (SpR), Senior Registrar and Registrar. The SpR grade was introduced formally on 1 April 1996.

Sources: Scottish Health Service Common Services Agency; NHS Executive: 020 7972 2231; National Assembly for Wales

9.4 Health and personal social services: workforce summary
Great Britain

continued

Number or whole-time equivalent

		1988	1989	1990	1991[7]	1992	1993	1994	1995	1996	1997	1998
Family Health services:[3]												
General medical practitioners:												
Analysis by type of practitioners at 1 October												
Total												
All Practitioners	LQZN	32 890	33 308	33 056	33 462	33 834	34 134	34 421	34 594	34 825	35 205	35 611
Unrestricted Principals & equivalents (UPE's)[1]	LQZO	30 275	30 629	30 616	30 712	31 065	31 447	31 767	31 945	32 164	32 477	32 801
of which GMS UP's	LQZP	30 275	30 629	30 616	30 712	31 065	31 447	31 767	31 945	32 164	32 477	32 414
PMS contracted	LQZQ	..	..	..	..	..	..	..	..	..	..	324
PMS salaried	LQZP	30 275	30 629	30 616	30 712	31 065	31 447	31 767	31 945	32 164	32 477	32 414
Restricted principals	LQZS	172	179	163	151	152	161	156	137	126	111	106
Assistants	LQZT	277	261	221	471	521	539	626	684	890	891	750
GP Registrars[2]	LQZU	2 166	2 239	2 040	2 102	2 067	1 956	1 841	1 790	1 605	1 678	1 830
of which GMS[3]	LQZV	2 166	2 239	2 040	2 102	2 067	1 956	1 841	1 790	1 605	1 678	1 796
Salaried Doctors (Para 52 SFA)	LQZW	..	..	..	..	..	..	..	..	..	..	60
PMS Other[4]	LQZX	..	..	..	..	..	..	..	..	..	..	12
Associates[5]	LQZY	..	..	16	26	29	31	31	38	40	48	52
General dental practitioners: total[15]	KDCQ	17 440	17 830	18 011	18 037	18 019	18 467	18 600	18 736	19 147	19 598	20 216
Principals	KDCR	17 144	17 436	17 539	17 440	17 384	17 699	17 642	17 670	17 897	18 174	18 557
Assistants and Vocational trainees	KDCS	296	394	472	597	635	768	958	1 066	1 250	1 424	1 659
Ophthalmic medical practitioners[7,13]	KDCT	930	882	882	862	837	789	735	752	766	830	855
Ophthalmic opticians[7,13]	KDCU	6 381	6 557	6 685	6 841	6 997	7 124	7 179	7 445	7 652	7 847	8 160
Personal Social Services staff[10]:												
Total	KDDE	235 969	238 754	240 342	237 280	235 240	232 911	237 752	233 861	233 655	229 439	223 500
Management, administration and ancillary staff[14]	KADS	24 993	25 801	27 006	28 049	29 433	..	..	..	..	..	..
Home help service	KSBU	60 435	60 411	59 772	58 591	58 004	57 593	59 391	56 961	55 430	53 573	50 417
Field Social Workers[16]	KSBX	22 924	23 702	24 457	25 537	26 509	28 850	29 820	31 926	32 140	32 990	33 400
Day care establishments staff	KADV	27 962.0	28 857.0	28 612.0	28 442.0	28 596.0	30 221.3	31 270.1	31 109.2	31 605.0	30 839.0	30 300.0
Residential care staff	KADW	88 496.0	88 285.0	87 858.0	83 001.0	78 918.0	74 038.3	72 155.3	68 650.5	67 975.0	65 422.0	62 100.0
All other staff[14]	KADX	3 502	3 932	4 284	5 355	5 049	..	..	..	..	..	..

1 UPE's includes Unrestricted Principals, PMS Contracted GP's and PMS salaried GP's.
2 GP Registrars were formerly referred to as Trainees.
3 GP Registrars in GMS Partnerships.
4 PMS other includes 2 Salaried Restricted Principals who are working in Scotland.
6 Figures relate to 1 October.
7 Figures for England and Wales relate to 31 December. Figures for Scotland relate to 31 March of the following year.
8 In Scotland this is the Dental Practice Division (DPD).
9 The Prescription Pricing Authority in England is synonymous with the Prescription Pricing Division in Scotland. Figures for the Prescription Pricing Division in Scotland relate to 30 November.
10 Figures are for England only.
11 Due to changes in the collection procedure in 1991 a category of 'other staff' was introduced for these locally determined payscales. In previous years these staff were included in their respective occupation groups.
12 1991 figure revised.
13 OMPs and OOs holding contracts with FHSAs/HAs and/or Scottish Health Boards to carry out NHS sight tests. OOs with contracts with more than one Health Board and/or a Health Board and a FHSA/HA will be counted more than once.
14 Breakdown for England not available in this form from 1993.
15 Some dentists may have a contract with more than one Family Health Service Authority/Health Authority. These dentists have been counted only once.
16 Includes care managers from 1983.

Sources: Scottish Health Service Common Services Agency; NHS Executive: 020 7972 2231; National Assembly for Wales

9.5 Notifications of infectious diseases

Number

		1988	1989	1990	1991	1992	1993	1994	1995	1996	1997	1998
United Kingdom												
Measles	KHQD	90 635	31 045	15 642	11 723	12 318	12 018	23 517	9 017	6 866	4 844	4 540
Mumps	KWNN	..	24 504	5 297	3 836	3 169	2 726	3 143	2 400	2 182	2 264	1 917
Rubella	KWNO	..	31 849	15 736	9 702	9 150	12 300	9 650	7 674	11 720	4 205	4 064
Whooping cough	KHQE	5 879	13 550	16 862	6 279	2 750	4 718	4 837	2 399	2 721	3 669	1 902
Scarlet fever	KHQC	7 087	10 386	9 505	6 876	5 978	7 341	8 031	6 863	6 101	4 639	4 708
Dysentery	KHQG	3 946	3 807	3 042	11 527	20 620	7 577	7 539	5 498	2 643	2 427	1 934
Food poisoning	KHQH	45 749	59 241[†]	59 721	59 497	72 139	76 711	91 129	92 604	94 923	105 579	105 060
Typhoid and Paratyphoid fevers	KHQB	369	273	294	288	298	277	390	386	291	249	252
Hepatitis	KWNP	6 438	8 304	9 864	9 856	9 616	6 142	4 284	3 823	2 876	3 601	3 781
Tuberculosis	KHQI	5 778	6 059	5 898	6 078	6 442	6 565	6 229	6 176	6 238	6 367	6 605
Malaria	KWNQ	..	..	1 565	1 652	1 253	1 281	1 219	1 363	1 743	1 549	1 163
England and Wales[1]												
Measles	KHRD	86 001	26 222	13 302	9 680	10 268	9 612	16 375	7 447	5 614	3 962	3 728
Mumps	KWNR	..	20 713	4 277	2 924	2 412	2 153	2 494	1 936	1 747	1 914	1 587
Rubella	KWNS	..	24 570	11 491	7 174	6 212	9 724	6 326	6 196	9 081	3 260	3 208
Whooping cough	KHRE	5 117	11 646	15 286	5 201	2 309	4 091	3 964	1 869	2 387	2 989	1 577
Scarlet fever	KHRC	5 949	8 295	7 187	5 217	4 645	5 855	6 193	5 296	4 873	3 569	3 339
Dysentery	KHRG	3 692	3 278	2 756	9 935	16 960	6 841	6 956	4 651	2 312	2 274	1 813
Food poisoning	KHRH	39 713	52 557	52 145	52 543	63 347	68 587	81 833	82 041	83 233	93 901	93 932
Typhoid and Paratyphoid fevers	KHRB	354	252	271	281	282	268	370	370	276	241	243
Viral hepatitis	KWNT	5 063	7 071	9 005	8 860	8 993	5 557	3 722	3 296	2 437	3 186	3 183
Tuberculosis[2]	KHRJ	5 164	5 432	5 204	5 436	5 799	5 921	5 591	5 608	5 654	5 859	6 087
Malaria	KWNU	1 271	1 478	1 493	1 553	1 189	1 198	1 139	1 300	1 659	1 476	1 110
Total meningitis	KHRO	2 987	2 722	2 572	2 760	2 571	2 082	1 800	2 285	2 686	2 345	2 072
Meningococcal meningitis	KHRP	1 304	1 133	1 138	1 117	1 067	1 053	938	1 146	1 164	1 220	1 152
Meningococcal septicaemia	KWNV	..	229	277	273	277	398	430	707	1 129	1 440	1 509
Ophthalmia neonatorum	KHRI	374	427	440	433	424	340	268	245	246	224	198
Scotland												
Measles	KHSE	2 258	3 359	2 006	1 701	1 747	1 911	6 192	1 307	1 055	762	700
Mumps	KWNW	..	3 095	833	723	601	458	546	371	368	282	251
Rubella	KWNX	..	6 628	3 702	2 171	2 645	2 048	2 916	1 258	2 449	818	745
Whooping cough	KHSF	356	659	1 291	838	236	493	639	399	186	545	225
Scarlet fever	KHSD	658	1 345	1 546	1 084	808	911	1 319	1 065	750	645	883
Dysentery	KHSH	196	154	235	1 526	3 486	607	446	575	176	124	103
Food poisoning[3]	KHSI	5 734	6 156[†]	6 757	6 318	7 877	7 170	8 291	9 297	10 234	10 144	9 186
Typhoid and Paratyphoid fevers	KHSB	14	18	20	7	15	7	18	16	14	6	6
Viral hepatitis	KWNY	848	831	546	556	319	290	296	405	360	359	490
Tuberculosis[4]	KHSL	528	533	563	546	559	554	546	478	509	433	457
Malaria	KWUC	62	60	68	91	50	75	74	58	70	57	30
Meningococcal infection	KWUD	149	190	216	178	207	207	201	190	201	271	313
Erysipelas	KHSC	112	152	125	155	128	130	118	125	84	95	66
Northern Ireland												
Measles	KHTD	2 376	1 464	334	342	303	495	950	263	197	120	112
Mumps	KHTR	532	696	187	189	156	115	103	93	67	68	79
Rubella[5]	KHTQ	124	651	543	357	293	528	408	220	190	127	111
Whooping cough	KHTE	406	1 245	285	240	205	134	234	131	148	135	100
Scarlet fever	KHTC	480	746	772	575	525	575	519	502	478	425	486
Dysentery	KHTG	58	375	51	66	174	129	137	272	155	29	18
Food poisoning	KHTH	302	501	819	636	915	954	1 005	1 266	1 456	1 534	1 942
Typhoid and Paratyphoid fevers	KHTB	1	3	3	–	1	2	2	–	1	2	3
Infective hepatitis	KHTO	527	402	313	440	304	295	266	122	79	56	108
Tuberculosis	KHTI	86	94	131	96	84	90	92	90	75	75	61
Malaria	KWUE	..	..	4	8	14	8	6	5	14	16	23
Acute encephalitis/meningitis	KHTM	156	131	158	172	118	122	144	116	105	91	64
Meningococcal septicaemia	KWUF	..	..	2	23	27	34	39	42	67	56	87
Gastro-enteritis (children under 2 years)	KHTP	869	1 243	1 157	1 091	1 070	1 379	887	1 072	745	896	1 371

1 The figures show the corrected number of notifications, incorporating revisions of diagnosis, either by the notifying medical practitioner or by the medical superintendent of the infectious diseases hospital. Cases notified in Port Health Authorities are included.

2 Formal notifications of new cases only. The figures exclude chemoprophylaxis.

3 Scotland's food poisoning includes 'otherwise ascertained' for the first time in 1995.

4 Figures include cases of tuberculosis not notified before death.

5 From July 1988 only.

Sources: Scottish Health Service Common Services Agency; Department of Health and Social Services (Northern Ireland); Communicable Diseases Surveillance Centre: 020 8200 6868

9.6 Deaths due to occupationally related lung disease[1]
Great Britain

		1987	1988	1989	1990	1991	1992	1993	1994	1995	1996	1997
Asbestosis (without mesothelioma)[2]	KADY	143	151	155	164	163	150	173	174	166	196	189
Mesothelioma[2]	KADZ	814	872	909	895	1 023	1 098	1 143	1 241	1 317	1 304	1 330
Pneumoconiosis (other than asbestosis)	KAEA	279	281	318	328	287	274	281	276	287	223	230
Byssinosis	KAEB	25	22	25	19	16	21	11	7	6	3	5
Farmer's lung and other occupational allergic alveolitis	KAEC	16	9	8	6	8	4	12	10	10	1	5
Total	KAED	1 277	1 335	1 415	1 412	1 497	1 547	1 620	1 708	1 786	1 727	1 759

1 The data in this table are derived from death certificates. For asbestosis and mesothelioma, the figure shown is the number of certificates mentioning the disease, those mentioning both diseases being counted as mesothelioma only. For the other diseases the figure is the number of deaths coded to the disease as underlying the case.

2 By definition every case of asbestosis is due to asbestos; the association with mesothelioma is also very strong, though there is thought to be a low natural background incidence.

Sources: Office for National Statistics;
Health and Safety Executive: 0151 951 4540

9.7 New cases of occupational diseases diagnosed under the Industrial Injuries and Pneumoconiosis, Byssinosis and Miscellaneous Diseases benefit schemes

Great Britain

		Disablement Benefit[1]										
		1988	1989	1990	1991	1992	1993	1994	1995	1996	1997	1998
A. Occupational lung diseases and deafness												
Pneumoconiosis (except asbestosis)[9]	KAEE	412	437	417[6]	447	428	456	657	433	362	249	554
Asbestosis	KAEF	202	268	306[6]	330	354	418	376	427	479	344	316
Byssinosis	KAEG	13	15	18[6]	7	4	5	2	6	4	1	2
Farmer's lung	KAEH	15	13	7	5	5	3	9	6	6	4	4
Occupational asthma	KAEI	222	220	216	293[3]	553	510	506	514	410	298	222
Mesothelioma	KAEJ	479	441	462	519	551	608	583	685	642	553	590
Lung cancer (asbestos)	KAEK	59	54	58	55	54	72	77	55	51	26	42
Lung cancer (other prescribed agents)[4]	KAEL	–	4	6	6	6	3	–	4	2	1	3
Pleural thickening[12]	KAEM	114	125	146	149	160	172	196	188	168	156	227
Other occupational lung disease	KAEN	5	–	4	6	5	2	–	–	1	–	–
Chronic bronchitis and/or emphysema[4,7]	KIUS	..	..	..	..	..	1 560	2 594	268	269	3 030	3 423
Occupational deafness[1]	KAEO	1 261	1 170	1 128	1 041	972	901	882	763	531	413	258
Total: Occupational lung disease and deafness	KAEP	2 782	2 747	2 768[6]	2 858	3 092	4 710	5 882	3 349	2 925	5 075	5 641

		Disablement Benefit[2,5]										
		1987/88	1988/89	1989/90	1990/91	1991/92	1992/93	1993/94	1994/95	1995/96	1996/97	1997/98
B. Other diseases												
Dermatitis	KEAI	368	285	301	434	411	419	392	368	328	336	271
Tenosynovitis	KEAJ	322	294	423	556	649	911	800	787	548	513	415
Vibration white finger	KEAK	1 673	1 056	2 601	5 403	2 369	1 447	1 425	1 747	3 016	3 288	3 033
Beat conditions	KEAL	171	112	95	187	317	256	257	205	199	153	104
Carpal Tunnel Syndrome[4]	KIUV	..	..	..	..	..	20	267	277	265	297	400
Viral hepatitis	KEAM	3	1	1	2	4	1	3	3	4	4	2
Tuberculosis	KEAN	3	5	–	3	3	6	10	9	10	6	11
Leptospirosis	KEAO	–	–	2	–	1	1	1	–	–	–	–
Other infections	KEAP	–	1	2	5	4	3	2	5	2	3	6
Poisonings[10]	KEAQ	2	16	6	4	8	9	15	9	10	7	10
Occupational cancers[11]	KEAR	29	11	18	17	23	30	43	26	30	44	37
Other conditions	KEAS	20	21	33	72	68	132	144	133	147	119	98
Allergic rhinitis[8]	KIUW	19	15	22	13	75	494	528	589	720	352	199
Total: Other diseases	KEAT	2 610	1 817	3 504	6 696	3 932	3 729	3 887	4 158	5 279	5 122	4 586

1 Awards for occupational deafness are made only for disablement at 20% or more and cases with lesser degrees of disabilities are excluded from the figures.
2 From October 1986 disablement benefit for most diseases was paid only for cases with disibility assessed at 14% or more; the figures for 1987/88 and subsequent years include cases not qualifying for benefit but assessed at 1-13%. Exceptions are claims made before October 1986, but assessed in subsequent years and all claims made for pneumoconiosis, asbestosis, mesothelioma and byssinosis, where the 1% threshold still applies, and deafness. (see note 1).
3 Series affected by changes in prescription benefit rules.
4 The following diseases were first prescribed during the period covered by the table (Year benefit first payable shown in brackets): Lung cancer (other prescribed agents except nickel)(1993) Chronic bronchitis and/or emphysema (1993) Carpal Tunnel Syndrone (1993)
5 Years ending in September.
6 Figures from 1990 do not include cases awarded by Medical Appeals Tribunals.
7 Prescribed from September 1993 for coal miners with a specified level of lung function impairment, and a minimum 20 years employment underground. The sharp fall in numbers in 1995 is typical of the pattern seen after a disease has been newly prescribed. The larger numbers in the two preceeding years probably represents the clearance of a backlog of pre-existing cases. The sharp rise in numbers in 1997 and 1998 results from relaxations in eligibility rules; future numbers are expected to fall.
8 Before March 1996 defined as inflammation or ulceration of upper respiritory tract or mouth. The recent increase in numbers for this disease represents increases in claims following a take-up campaign by trade unions rather than a true increase in incidence.
9 Figures for 1987 to 1994 include a small number of cases assessed under the Pneumoconiosis, Byssinosis and Miscellaneous Diseases Benefit Scheme, which are assumed to be all pneumoconiosis. Figures for 1995 onwards from this scheme are not presently available.
10 Excluding poisonings by nitrous fumes, beryllium or cadmium, which are included under 'other occupational lung disease'.
11 Excluding mesothelioma and lung cancer, see section A of table.
12 From 9 April 1997, the definition of this disease was changed to admit unilateral diffuse pleural thickening. Previously this was required to be bilateral.

Sources: Health and Safety Executive: 0151 951 4540; Department of Social Security

9.8 Injuries to workers by industry and severity of injury

Great Britain

As reported to all enforcing authorities

		Section	SIC (92)	Fatal 1995/96	Fatal 1996[7]/97	Fatal 1997/98		Major[1] 1995/96	Major[1] 1996[7]/97	Major[1] 1997/98		Over 3 Day[2] 1995/96	Over 3 Day[2] 1996[7]/97	Over 3 Day[2] 1997/98
Agriculture, hunting, forestry and fishing[3]	KSYS	A,B	01,02,05	40	55	40	KSZN	476	778	745	KTAZ	1 373	1 514	1 382
Energy and water supply industries[4]	KSYT	C,E	10-14,40/41	18	9	18	KSZO	525	694	684	KTBH	3 187	3 039	3 339
Mining and quarrying[4]	KSYU	C	10-14	16	7	14	KSZP	341	417	440	KTBI	1 404	1 476	1 920
Mining and quarrying of energy producing materials[4]	KSON	CA	10-12	10	6	10	KSZQ	226	228	297	KTBJ	779	861	1 359
Mining and quarrying except energy producing materials	KSOO	CB	13/14	6	1	4	KSZR	115	189	143	KTBK	625	615	561
Electricity, gas and water supply	KSOP	E	40/41	2	2	4	KSZS	184	277	244	KTBL	1 783	1 563	1 419
Manufacturing	KSOQ	D	15-37	43	59	61	KSZT	5 281	8 346	8 864	KTBM	42 301	40 159	41 742
of food products; beverages and tobacco	KSOR	DA	15/16	4	4	6	KSZU	983	1 501	1 556	KTBN	10 051	9 433	9 918
of textile and textile products	KSOS	DB	17/18	1	1	4	KSZV	236	338	353	KTBO	1 942	1 644	1 606
of leather and leather products	KSOT	DC	19	–	–	1	KSZW	16	33	43	KTBP	198	196	156
of wood and wood products	KSOU	DD	20	3	5	7	KSZX	231	390	373	KTBQ	1 235	1 144	1 168
of pulp, paper and paper products; publishing and printing	KSOV	DE	21/22	5	2	3	KSZY	348	552	598	KTBR	2 829	2 664	2 695
of coke, refined petroleum products and nuclear fuel	KSOW	DF	23	–	–	1	KSZZ	33	44	39	KTBS	182	140	138
of chemicals, chemical products and man-made fibres	KSOX	DG	24	3	4	2	KTAE	386	562	518	KTBT	2 309	2 187	2 054
of rubber and plastic products	KSOY	DH	25	1	3	2	KTAF	386	612	677	KTBU	3 140	3 270	3 618
of other non-metallic mineral products	KSOZ	DI	26	5	3	6	KTAG	268	433	483	KTBV	2 470	2 193	2 287
of basic metals and fabricated metal products	KSYV	DJ	27/28	10	13	14	KTAH	963	1 585	1 833	KTBW	6 268	5 911	6 407
of machinery and equipment n.e.c.	KSYW	DK	29	2	9	5	KTAI	499	780	765	KTBX	3 581	3 456	3 441
of electrical and optical equipment	KSYX	DL	30-33	3	1	2	KTAJ	244	442	513	KTBY	2 174	2 273	2 467
of transport equipment	KSYY	DM	34/35	6	10	5	KTAK	440	729	742	KTBZ	4 382	4 258	4 297
Manufacturing n.e.c.	KSYZ	DN	36/37	–	4	3	KTAL	248	345	371	KTCA	1 540	1 390	1 490
Construction	KSZA	F	45	79	90	80	KTAM	2 477	4 054	4 326	KTCB	9 695	9 666	10 265
Total service industries	KSZB	G-Q	50-99	78	74	75	KTAN	8 275	15 448	15 383	KTCC	72 963	75 190	79 045
Wholesale and retail trade, and repairs	KSZC	G	50-52	23	17	17	KTAO	1 918	3 493	3 527	KTCD	15 554	13 846	16 353
Hotel and restaurants	KSZD	H	55	1	4	1	KTAP	490	947	942	KTCE	2 482	2 696	3 234
Transport, storage and communication[5]	KSZE	I	60-64	22	18	27	KTAQ	1 295	2 906	2 771	KTCF	16 525	16 792	17 351
Financial intermediation	KSZF	J	65-67	–	–	–	KTAR	81	169	131	KTCG	591	658	623
Real estate, renting and business activities	KSZG	K	70-74	11	10	13	KTAS	375	1 015	1 013	KTCH	1 994	3 214	3 791
Public administration and defence	KSZH	L	75	10	8	3	KTAT	1 298	2 182	2 153	KTCI	14 495	14 744	13 895
Education	KSZI	M	80	2	3	–	KTAU	1 089	1 564	1 589	KTCJ	4 702	4 705	4 850
Health and social work	KSZJ	N	85	–	4	2	KTAV	1 130	1 871	1 966	KTCK	12 996	13 925	14 209
Other community, social and personal services activities	KSZK	O-Q	90-99	9	10	12	KTAW	599	1 301	1 291	KTCL	3 624	4 610	4 739
Unclassified	KSZL			–	–	–	KTAX	700	–	–	KTCM	3 457	–	–
All industries	KSZM			258	287	274	KTAY	17 734	29 320	30 002	KTCN	132 976	129 568	135 773

1 As defined in RIDDOR 85 prior to 1996/97 and RIDDOR 95 from 1996/97.
2 Injuries causing incapacity for normal work for more than 3 days.
3 Excludes sea fishing.
4 Includes the number of injuries in the oil and gas industry collected under offshore installations safety legislation.
5 Injuries arising from shore based services only. Excludes incidents reported under merchant shipping legislation.
6 n.e.c. not elsewhere classified.
7 1996/97 injury figures cannot be compared to previous years figures as they were reported under different legislation (RIDDOR 85 and 95).

Source: Health and Safety Executive (HSE): 0151 951 4842

Social protection

10

Social protection

Social security
(Tables 10.2 - 10.8 and 10.10 to 10.16)

Tables 10.2 to 10.6 and 10.10 to 10.16 give details of contributors and beneficiaries under the National Insurance and Industrial Injury Acts, supplementary benefits and war pensions.

There are three types of contributor:

Class 1 Employed people, that is, people working for employers. Their contributions are paid partly by themselves and partly by their employers. They are covered for all benefits.

Class 2 Self-employed people, that is, people working on their own account. They are covered for all benefits other than unemployment and industrial injuries.

Class 3 Non-employed people, that is, people who do not work for gain. These people pay contributions on a voluntary basis. They are covered for benefits other than unemployment, sickness, industrial injuries and maternity allowances.

Class 4 Payable, in addition to Class 2 by self-employed people, and the amount payable is proportionate to profits or gains between a lower and upper limit in any one year.

An employer must pay a contribution for every employee whose earnings exceed a base level. Most employed people pay the full employee's contribution, but retirement pensioners working for an employer do not and some married women and some widows who are working need not, unless they so wish, contribute except for industrial injuries benefit. Thus the total numbers in the analysis by benefit for which the contributions were payable are less than the total numbers in the analysis by class of contributor.

Sickness benefit, Invalidity benefit and Incapacity benefit (Tables 10.7 and 10.8)

Incapacity Benefit replaced Sickness Benefit and Invalidity Benefit from 13 April 1995. The first condition for entitlement to these contributory benefits is that the claimants are incapable of work because of illness or disablement. Secondly, that they satisfy the contribution conditions which depend on contributions paid as an employed (Class 1) or self-employed person (Class 2). Under Sickness and Invalidity Benefits the contribution conditions were automatically treated as satisfied if a person was incapable of work because of an industrial accident or prescribed disease. Under Incapacity Benefit those who do not satisfy the contribution conditions in this case do not have them treated as satisfied. Class 1A contributions paid by employers are in respect of the benefit of cars provided for the private use of employees, and the free fuel provided for private use. These contributions do not provide any type of benefit cover.

Since 6 April 1983, most people working for an employer and paying National Insurance contributions as employed persons, receive Statutory Sick Pay (SSP) from their employer when they are off work sick. SSP was payable for a maximum of 8 weeks until 5 April 1986, and 28 weeks thereafter. People who do not work for an employer, and employees who are excluded from the SSP scheme, or those who have run out of SSP before reaching the maximum of 28 weeks and are still sick can claim benefit. Any period of SSP is excluded from the tables.

From 14 September 1980, spells of incapacity of 3 days or less do not count as periods of interruption of employment, and are excluded from the tables. Exceptions are where people are receiving regular weekly treatment by dialysis, or treatment by radiotherapy, chemotherapy or plasmapheresis where 2 days in any 6 consecutive days make up a period of interruption of employment, and those whose incapacity for work ends within 3 days of the end of SSP entitlement.

At the beginning of a period of incapacity, benefit is subject to 3 waiting days, except where there was an earlier spell of incapacity of more than 3 days in the previous 8 weeks. Employees entitled to SSP for less than 28 weeks and who are still sick can get Sickness Benefit or Incapacity Benefit Short Term (Low) until they reach a total of 28 weeks provided they satisfy the conditions. After 28 weeks SSP and/or Sickness Benefit (SB), Invalidity Benefit (IVB) was payable up to pension age for as long as the incapacity lasts. From pension age Invalidity Benefit was paid at the person's Retirement Pension rate, until entitlement ceases when RP is paid or at deemed pension age (70 for a man, 65 for a woman). For people on Incapacity Benefit under State pension age there are two short-term rates: the lower rate is paid for the first 28 weeks of sickness and the higher rate for weeks 29 to 52. From week 53 the Long Term rate Incapacity Benefit is payable. The Short Term rate Incapacity Benefit is based on Retirement Pension entitlement for people over State pension age and is paid for up to a year if incapacity began before pension age.

The long-term rate of Incapacity Benefit applies to people under State pension age who have been sick for more than a year. People with a terminal illness or who are receiving the higher rate care component of Disability Living Allowance will get the Long Term rate. The Long Term rate is not paid for people over pension age.

Under Incapacity Benefit, for the first 28 weeks of incapacity, people previously in work will be assessed on the 'own occupation' test - the claimant's ability to do their own job. Otherwise, incapacity will be based on a new 'all work' test which will assess ability to carry out a range of work-related activities. The test will apply after 28 weeks of incapacity or from the start of the claim for people who did not previously have a job. Certain people will be exempted from this test.

The tables exclude all men aged over 65 and women aged over 60 who are in receipt of Retirement Pension, and all people over deemed pension age (70 for a man and 65 for a woman), members of the Armed Forces, mariners while at sea, and married women and certain widows who have chosen not to be insured for Sickness Benefit. In addition, employees of the Post Office are excluded prior to 29 March 1986, and the remainder of Civil Service Departments, British Telecom and fringe bodies prior to 25 August 1986. The tables include a number of individuals who were unemployed prior to incapacity.

The Short Term (Higher) and Long Term rates of incapacity Benefit are treated as taxable income.

There were transitional provisions for people who were on Sickness or Invalidity Benefit on 12 April 1995. They were automatically transferred to Incapacity Benefit, payable on the same basis as before. Former IVB recipients continue to get Additional Pension entitlement, but frozen at 1994 levels. Also their IVB is not subject to tax. If they were over State pension age on 12 April 1995 they may get Incapacity Benefit for up to 5 years beyond pension age.

Government expenditure on social services and housing *(Table 10.17 - 10.22)*

The tables of general government expenditure on the social services and housing in the United Kingdom comprise a summary table followed by separate tables for each of the social services and housing. The definition of government expenditure used in these tables follows that in Table 9.4 of *United Kingdom National Accounts 1994 Edition*, the CSO Blue Book, and covers both current and capital expenditure of the central government (including the National Insurance Fund) and local authorities. The housing table also includes the capital expenditure of public corporations concerned with housing. As in the Blue Book, government expenditure is measured after deducting fees and charges for services. Expenditure on administration includes the cost of common services (accommodation, stationery and printing, superannuation, etc) some of which is not directly borne by the departments administering each service. Transfers from one part of government to another have been eliminated to avoid double counting. The figures relate to years ended 31 March. Figures for the latest two years are the most recent estimates available and are subject to revision.

It should be noted that the figures no longer include imputed rents for the use of fixed assets owned and used by general government. In the Blue Book, imputed rents have been replaced by capital consumption. Capital consumption, however, cannot be allocated to individual services and is therefore not included in these tables.

The following notes give brief descriptions of each of the main services shown in the tables.

Education (Table 10.18)
This covers expenditure by the Education Departments, local education authorities and the University Grants Committee on education in schools, training colleges, technical institutions and universities. It includes expenditure on school meals and milk.

The education statistics in Table 10.18 are provided by the Department for Education. They are not normally able to provide a full set of estimates, from their sources, for the latest financial year. The ONS has in the past made estimates from a variety of sources but, following a review of their accuracy, it has been decided that they do not give a reliable guide to the final outcome and have therefore been withdrawn.

National Health Service (Table 10.19)
This covers expenditure by central government on hospital and community health, family practitioner and other health services. The expenditure by local authorities on the provision of health centres, health visiting, home nursing, ambulance services, vaccination and immunisation, etc. was transferred to central government on 1 April 1974. Only the net costs of providing these services are included in total government expenditure, receipts from patients being shown separately.

Personal social services (Table 10.20)
This covers local authority expenditure on the aged, handicapped and homeless, child care, care of mothers and young children, mental health, domestic help, etc. Also included are central government grants to voluntary approved schools.

Welfare foods (Table 10.20)
This covers the cost of providing welfare foods at reduced prices to children and expectant mothers. Only the net costs of providing these services are included in the total government expenditure, payments by the recipients of the services being shown separately.

Social security (Table 10.21)
This comprises both benefits under the Social Security schemes and non-contributory benefits and allowances, administered by the Department of Social Security. The analysis by type of Income Support is not exact; the estimates are derived from average numbers in receipt of benefit and average amounts paid. Unified housing benefit (rent rebates and allowances) is also included as social security expenditure and not as housing expenditure. This is now mainly administered by local authorities who receive grants from central government.

Housing (Table 10.22)
The table shows, in addition to government expenditure on housing, the capital expenditure of public corporations and the total expenditure of the public sector on housing. The government expenditure figures cover subsidies paid by the housing departments towards the provision of housing by local authorities, new town development corporations and housing associations; subsidies by local authorities to their housing revenue accounts; rent rebates for tenants of housing owned by local authorities and new towns; rent allowances for tenants of privately-owned housing; grants to persons for the reduction of mortgage

interest payments; capital expenditure on the provision of houses for letting; capital grants to housing associations; grants by local authorities towards the cost of conversion and improvement of privately-owned houses; net lending by the central government and local authorities for private house purchase and improvement and loans for first time purchasers. The public corporations' figures cover capital expenditure on the provision of houses for letting and lending by the Housing Corporation to housing associations.

10.1 National Insurance Fund

Years ended 31 March

£ thousands

		1988 /89	1989 /90	1990 /91	1991 /92	1992 /93	1993 /94	1994 /95	1995 /96	1996 /97	1997 /98
Receipts											
Opening balance	KJFB	7 481 237	10 634 672	10 579 741	12 161 227	8 637 594	3 577 671	4 672 345	7 042 118	8 044 941	7 869 121
Contributions		27 928 396	29 970 364	31 480 868	33 156 696	34 331 672	35 090 000	38 712 218	40 874 579	42 806 429	46 754 687
Grant from Consolidated Fund	KOTF	..	..	..	..	..	7 785 000	6 445 000	3 680 000	1 951 500	966 200
Compensation for SSP/SMP	KJQM	..	..	..	1 089 811	1 124 500	1 158 289	563 400	474 700	541 800	600 500
Consolidated Fund Supplement	KJFD	1 686 175	..	..	..	..	..	..	..	..	..
Reimbursement for industrial Injury Benefit payments	KJQN	..	..	436 514	..	..	..	..	..	..	..
Transfers from GB	KOTG	185 000	210 000	225 000	125 000	40 000	40 000	145 000	125 000	75 000	150 000
Income from investments	KJFE	794 860	1 060 702	1 022 506	1 160 026	970 862	488 484	365 096	459 217	488 728	474 296
Other receipts	KJFF	1 081	1 018	944 680	19 359	24 813	41 989	60 625	76 487	85 126	97 206
Redundancy receipts[1]	KIBQ	..	..	2 024	19 091	26 942	25 688	25 288	23 899	26 193	24 720
Total		38 076 748	41 876 756	44 691 332	47 731 208	45 156 384	48 990 048	50 988 972	52 756 000	54 019 717	56 936 730
Expenditure											
Total benefits		26 352 108	30 187 768	30 590 420	37 430 152	39 823 984	42 366 510	42 270 821	35 575 105	44 517 872	45 321 420
Unemployment	KJFH	1 147 832	764 593	898 963	1 641 928	1 800 589	1 689 604	1 330 922	1 131 373	605 095	148
Sickness	KJFI	202 078	214 228	226 968	286 455	378 136	378 407	350 700	14 711	..	..
Invalidity	KJFJ	3 500 571	3 993 974	4 613 004	5 701 239	6 452 717	7 341 161	8 009 286	599 829	..	..
Incapacity	JYXL	..	..	..	..	..	..	..	7 622 940	7 992 748	7 739 450
Maternity	KETY	27 873	31 140	34 590	32 071	32 660	34 230	28 230	29 897	34 283	36 715
Widows' pensions	KEWU	880 456	883 822	920 652	1 044 749	1 044 397	1 075 990	1 057 395	1 050 693	1 016 822	1 020 713
Guardians' allowances[2} Child's special allowance	KJFK	1 149	1 452	1 643	1 562	2 091	1 111	1 111	1 842	1 569	1 637
Retirement pensions[3]		19 951 518	23 634 712	23 305 444	28 604 996	29 995 404	31 720 928	31 366 872	32 619 571	34 735 720	36 396 421
Disablement[4]	KAAU	464 466	483 320	419 756	..	..	..	..	..	..	..
Death	KAAV	60 760	61 469	47 920	..	..	..	..	..	..	..
Pensioners' lump sum payments	KAAW	111 710	114 780	118 259	117 154	117 990	125 079	126 305	127 189	131 635	126 336
Other benefits	KAAX	3 695	4 278	3 222	..	..	..	..	..	..	..
Other payments	KAAZ	8 262	7 322	9 623	9 682	10 523	35 045	8 879	14 174	17 392	19 037
Administration	KABE	896 458	891 925	1 056 787	1 243 488	1 379 212	1 604 510	1 323 631	1 218 772	1 065 338	1 073 350
Transfers to Northern Ireland	KABF	185 000	210 000	225 000	125 000	40 000	40 000	145 000	125 000	75 000	150 000
Redundancy payments	KIBR	..	..	..	285 291	324 994	271 636	198 523	155 068	133 621	119 961
Total		27 441 828	31 297 016	31 881 832	39 093 616	41 578 712	44 317 704	43 946 854	37 088 119	45 809 223	46 683 768
Accumulated funds	KABH	10 634 672	10 579 741	12 530 598	8 637 594	3 577 671	4 672 345	7 042 118	8 044 941	7 869 121†	9 763 346

1 The assets of the Redundancy Fund became those of the National Insurance Fund on 31 January 1991.
2 Including figures of Child's special allowance for Northern Ireland.
3 Includes figures for Personal Pensions.
4 From 1/4/90, industrial injury benefits are no longer paid for out of the National Insurance Fund; this amount is reimbursement for expenditure incurred by the fund for the period 1/4/90 to 31/12/90.

Sources: Department of Social Security: 01253 856123 Ext 62436; Department of Health and Social Services (Northern Ireland)

10.2 Persons who paid National Insurance contributions in a tax year ending April[1]

Millions

		Total					Men					Women			
		1994	1995	1996	1997		1994	1995	1996	1997		1994	1995	1996	1997
Total[2]	KABI	23.99	24.27†	24.57	24.77	KEYF	13.69†	13.80	13.90	13.90	KEYP	10.29†	10.50	10.71	10.80
Class 1 Standard rate	KABJ	21.38†	21.66	21.99	22.28	KEYG	11.58	11.68†	11.81	11.94	KEYQ	9.80†	9.99	10.19	10.33
Not contracted out[3]	KABK	12.53†	12.76	12.96	13.28	KEYH	6.37	6.62	6.82†	7.04	KEYR	5.98†	6.15	6.14	6.23
Contracted out	KABL	7.51	7.28	7.26†	7.35	KEYI	4.68	4.44	4.32†	4.23	KEYS	2.83	2.84	2.84†	3.12
Mixed contracted in/out	KABM	1.04	1.21	1.43†	1.36	KEYJ	0.53	0.62	0.67†	0.67	KEYT	0.51	0.59	0.76	..
Class 1 Reduced rate (including standard rate)	KABO	0.48	0.41	0.34	0.29	KEYL	–	–	–	–	KEYV	0.48	0.41	0.35†	0.29
Class 2	KABP	2.08†	2.08	2.08	2.05	KEYM	1.73†	1.72	1.71	1.67	KEYW	0.34	0.36†	0.37	0.38
Mixed Class 1 and Class 2	KABQ	0.33	0.33†	0.33	0.35	KEYN	0.25†	0.25	0.24	0.26	KEYX	0.08	0.08	0.09†	0.09
Class 3[4]	KABR	0.20†	0.20	0.17	0.09	KEYO	0.13†	0.13	0.11	0.06	KEYY	0.07†	0.07	0.06	0.03

1 The tax year commences on 6 April and ends on 5 April of the following year. The years shown at the head of the column refer to the end of the tax year.
2 Not all figures agree because of rounding.
3 Includes those persons with an Appropriate Personal Pension (such persons pay contributions at the not contracted out rate but then receive a rebate).
4 Persons who paid a mixture of Class 3 contributions and others are not included in this category.

Source: Department of Social Security: 0191 225 7373

10.3 Weekly rates of principal social security benefits

£

		1989 Apr	1990 Apr	1991 Apr	1992 Apr	1993 Apr	1994 Apr	1995 Apr	1996 Apr	1997 Apr	1998 Apr	1999 Apr
Unemployment benefit:[1]												
Men and women	KJNA	34.70	37.35	41.40	43.10	44.65	45.45	46.45	48.25	–[6]	–[6]	–[6]
Jobseeker's allowance:[6]												
Personal allowances												
Single												
Depending on their	KXDH	..	..	..	..	..	..	..	..	29.60	30.30	30.95
circumstances	KXDI	..	..	..	..	..	..	..	..	38.90	39.85	40.70
Aged 18 - 24	KXDJ	..	..	..	..	..	..	..	..	38.90	39.85	40.70
Aged 25 or over	KXDK	..	..	..	..	..	..	..	..	49.15	50.35	51.40
Couple												
Both aged under 18	KXDL	..	..	..	..	..	..	..	..	58.70	60.10	61.90
One or both aged 18 or over	KXDM	..	..	..	..	..	..	..	..	77.15	79.00	80.65
Dependant children and young people												
Aged under 11	KXDN	..	..	..	..	..	..	..	..	16.90	17.30	20.20
Aged 11 - 16	KXDO	..	..	..	..	..	..	..	..	24.75	25.35	25.90
Aged 16 - 18	KXDP	..	..	..	..	..	..	..	..	29.60	30.30	30.95
Sickness benefit:[1,4]												
Men and women	KJNB	33.20	35.70	39.60	41.20	42.70	43.45	–[4]	–[4]	–[4]	–[4]	–[4]
Invalidity benefit:[4]												
Invalidity pension	KJNC	43.60	46.90	52.00	54.15	56.10	57.60	–[4]	–[4]	–[4]	–[4]	–[4]
Invalidity allowance:[4]												
High rate	KJND	9.20	10.00	11.10	11.55	11.95	12.15	12.40	12.90	13.15	13.60	14.05
Middle rate	KJNE	5.80	6.20	6.90	7.20	7.50	7.60	7.80	8.10	8.30	8.60	8.90
Low rate	KJNF	2.90	3.10	3.45	3.60	3.75	3.80	3.90	4.05	4.15	4.30	4.45
Increase for dependants:[4]												
Adult	KJNG	26.20	28.20	31.25	32.55	33.70	34.50	35.25	36.60	37.35	38.70	39.95
Each child	KJNH	8.95	9.65	10.70	10.85	10.95	11.00[3]	11.05[3]	11.15[3]	11.20[3]	11.30[3]	11.35[3]
Incapacity benefit:												
Short term (Lower) Under pension age	KOSB	..	..	..	..	..	..	44.40	46.15	47.10	48.80	50.35
Increase for adult dependant	KOSC	..	..	..	..	..	..	27.50	28.55	29.15	30.20	31.15
Short term (Lower) Over pension age	KOSD	..	..	..	..	..	..	56.45	58.65	59.90	62.05	66.75
Increase for adult dependant	KOSE	..	..	..	..	..	..	33.85	35.15	35.90	37.20	39.95
Short term (Higher)	KOSF	..	..	..	..	..	..	52.50	54.55	55.70	57.70	59.55
Increase for dependants:												
Adult	KOSG	..	..	..	..	..	..	27.50	28.55	29.15	30.20	31.15
Child	KOSH	..	..	..	..	..	..	11.05[3]	11.15[3]	11.20[3]	11.30[3]	11.35[3]
Long term	KOSI	..	..	..	..	..	..	58.85	61.15	62.45	64.70	66.75
Increase for dependants:												
Adult	KOSJ	..	..	..	..	..	..	35.25	36.60	37.35	38.70	39.95
Child	KOSK	..	..	..	..	..	..	11.05[3]	11.15[3]	11.20[3]	11.30[3]	11.35[3]
Incapacity age addition:[5]												
Higher rate	KOSL	..	..	..	..	..	..	12.40	12.90	13.15	13.60	14.05
Lower rate	KOSM	..	..	..	..	..	..	6.20	6.45	6.60	6.80	7.05
Attendance allowance:												
Higher rate	KJNI	34.90	37.55	41.65	43.35	44.90	45.70	46.70	48.50	49.50	51.30	52.95
Lower rate	KJNJ	23.30	25.05	27.10	28.95	30.00	30.55	31.20	32.40	33.10	34.30	35.40
Mobility allowance[2]	KJNK	24.40	26.25	29.80	–	–	–	–	–	–	–	–
Disability living allowance:												
care component												
Higher rate	KXDC	..	..	..	43.35	44.90	45.70	46.70	48.50	49.50	51.30	52.95
Middle rate	KXDD	..	..	..	28.95	30.00	30.55	31.20	32.40	33.10	34.30	35.40
Lower rate	KXDE	..	..	..	11.55	11.95	12.15	12.40	12.90	13.15	13.60	14.05
mobility component												
Higher rate	KXDF	..	..	..	30.30	31.40	31.95	32.65	33.90	34.60	35.85	37.00
Lower rate	KXDG	..	..	..	11.55	11.95	12.15	12.40	12.90	13.15	13.60	14.05

1 Persons under the age of 18 are entitled to the appropriate adult rate.
2 Mobility allowance was introduced from 1 January 1976 at the £5.00 rate. Disability living allowance has replaced Mobility allowance from April 1992.
3 For the first child only, the Child Dependency increase is reduced by £1.30 to £9.90 because of child benefit.
4 Incapacity benefit introduced from 13 April 1995, has replaced sickness benefit and invalidity benefit.
5 The rate of age addition depends on age at date of onset of incapacity: Higher rate = under age 35 Lower rate = age 35-44.
6 Jobseeker's allowance introduced 7 October 1996 has replaced Unemployment benefit and is for the unemployed.

Source: Department of Social Security: 0191 225 7373

10.3 Weekly rates of principal social security benefits

continued

£

		1989 Apr	1990 Apr	1991 Apr	1992 Apr	1993 Apr	1994 Apr	1995 Apr	1996 Apr	1997 Apr	1998 Apr	1999 Apr
Maternity benefit:												
Maternity allowances for insured women												
Higher rate[6]	KOSN	–	–	–	–	–	–	52.50	54.55	55.70	57.70	59.55
Lower rate[7]	KJNL	33.20	35.70	40.60	42.25	43.75	44.55	45.55	47.35	48.35	50.10	51.70
Death grant[2]	KJNM	–	–	–	–	–	–	–	–	–	–	–
Guardian's allowance	KJNN	8.95	9.65	10.70	10.85	10.95	11.00	11.05	11.15	11.20	11.30	..
Widow's benefit:												
Widow's pension	KJNO	43.60	46.90	52.00	54.15	56.10	57.60	58.85	61.15	62.45	64.70	66.75
Widowed mother's allowance	KJNP	43.60	46.90	52.00	54.15	56.10	57.60	58.85	61.15	62.45	64.70	66.75
Addition for each child	KJNQ	8.95	9.65	10.70	10.85	10.95	11.00	11.05	11.15	11.20	11.30	11.35
Retirement pension:[3]												
Single person	KJNR	43.60	46.90	52.00	54.15	56.10	57.60	58.85	61.15	62.45	64.70	66.75
Married couple	KJNS	69.80	75.10	83.25	86.70	89.80	92.10	94.10	97.75	99.80	103.40	106.70
Non-contributory retirement pension:												
Man or woman	KJNT	26.20	28.20	31.25	32.55	33.70	34.50	35.25	36.60	37.35	38.70	39.95
Married woman	KJNU	15.65	16.85	18.70	19.45	20.15	20.65	21.10	21.90	22.35	23.15	23.90
Industrial injuries benefit												
Injury benefit[5]	KJNV	–	–	–	–	–	–	–	–	–	–	–
Disablement pension at 100 per cent rate	KJNW	71.20	76.60	84.90	88.40	91.60	93.20	95.30	99.00	101.10	104.70	108.10
Widow's or widower's pension	KJNX	43.60	46.90	52.00	54.15	56.10	57.60	58.85	61.15	62.45	..	..
Increase for dependants:[4]												
Adult	KJNY	21.40	23.05	25.55	26.60	27.55	28.05	28.65	29.75	–	–	..
Each child	KJNZ	–	–	–	–	–	–	–	–	–	–	..
Child benefit:[8]												
First child	KJOA	7.25	7.25	8.25	9.65	10.00	10.20	10.40	10.80	11.05	11.45	14.40
Subsequent children	KETZ	–	–	–	7.80	8.10	8.25	8.45	8.80	9.00	9.30	9.60
Family Credit[9,10] (maximum awards payable):[1]												
Families with 1 child												
Birth to September following 11th birthday	KJOB	40.90	44.60	48.00	51.40	53.25	55.15	56.50	58.20	59.70	61.15	64.95
From September following 11th birthday to September following 16th birthday	KJOC	46.50	50.50	54.40	58.25	60.35	62.45	64.00	65.90	67.60	69.25	70.70
From September following 16th birthday to day before 19th birthday	KJOD	49.95	54.15	58.35	62.45	64.70	67.00	68.55	70.60	72.45	74.20	75.95
Increase for each additional child												
Birth to September following 11th birthday	KJOF	7.30	8.25	9.70	10.40	10.75	11.15	11.40	11.75	12.05	12.35	15.15
From September following 11th birthday to September following 16th birthday	KJOG	12.90	14.15	16.10	17.25	17.85	18.45	18.90	19.45	19.95	20.45	20.90
From September following 16th birthday to day before 19th birthday	KJOH	16.35	17.80	20.05	21.45	22.20	23.00	23.45	24.15	24.80	25.40	25.95
War pension:												
Ex-private (100 per cent assessment)	KJOJ	71.20	76.60	84.90	89.00	97.20	98.90	101.10	105.00	107.20	111.10	114.70
War widow	KJOK	56.65	60.95	67.60	70.35	72.90	74.70	76.35	79.35	81.00	83.90	86.60
Single householder:[11]												
Ordinary rate	KJON	–	–	–	–	–	–	–	–	–	–	..
Long-term rate	KJOO	–	–	–	–	–	–	–	–	–	–	..

1 Maximum award does not include the 30 hour credit.
2 Death grant is not payable in respect of the death of a person who on 4 July 1948 was aged 65 or over (men) and 60 or over (women). This benefit was abolished from April 1987 and replaced by payments from Social Fund.
3 Retirement pensioners over 80 receive 25p addition.
4 An allowance for one adult dependent is payable, where appropriate, with unemployment benefit, sickness benefit, retirement pension, injury benefit and maternity allowance. Allowances for dependent children are payable with any of these benefits. Changes in these increases take effect from the same date as the main benefit.
5 Injury benefit ceased on 6 April 1983.
6 Following an EU Directive employees' maternity benefit is aligned with the state benefit they would receive if off work sick.
7 Women who are either not employed or self employed will receive the lower rate.
8 From October 1991 first child receives £9.25 and each subsequent child £7.50.
9 Family credit superseded Family Income Supplement from April 1988.
10 Age bandings for children's personal allowances have been revised from 7 April 1997. Some children have protected rights. Rates are as follows: Age 11 before 7 October 1997 £20.45, age 16 before 7 October 1997 £25.40 and age 18 before 7 October 1997 £35.55.
11 Including any single person who is directly responsible for rent.

Source: Department of Social Security: 0191 225 7373

10.3 Weekly rates of principal social security benefits

continued

£

		1989 Apr	1990 Apr	1991 Apr	1992 Apr	1993 Apr	1994 Apr	1995 Apr	1996 Apr	1997 Apr	1998 Apr	1999 Apr
Non-householder aged:												
18 or over												
Ordinary rate	KJOP	–	–	–	–	–	–	–	–	–	–	–
Long-term rate	KJOQ	–	–	–	–	–	–	–	–	–	–	–
16-17												
Ordinary rate	KJOR	–	–	–	–	–	–	–	–	–	–	–
Long-term rate	KEUZ	–	–	–	–	–	–	–	–	–	–	–
Increase for children aged:												
11-15	KJOS	–	–	–	–	–	–	–	–	–	–	–
Under 10	KJOV	–	–	–	–	–	–	–	–	–	–	–
Income Support:												
Personal allowances[1]												
Single												
aged 16-17 either	KJOW	20.80	21.90	23.65[2]	25.55	26.45	27.50	28.00	28.85	29.60	30.30	30.95
or depending on their circumstances	KABS	–	28.80	31.15[3]	33.60	34.80	36.15	36.80	37.90	38.90	39.85	40.70
aged 18-24	KJOX	27.40	28.80	31.15	33.60	34.80	36.15	36.80	37.90	38.90	39.85	40.70
aged 25 or over	KJOY	34.90	36.70	39.65	42.45	44.00	45.70	46.50	47.90	49.15	50.35	51.40
Couple												
both aged under 18	KJOZ	41.60	43.80	47.30	50.60	52.40	54.55	55.55	57.20	58.70	60.10	61.90
one or both 18 or over	KJPA	54.80	57.60	62.25	66.60	69.00	71.70	73.00	75.20	77.15	79.00	80.65
Lone parent												
aged 16-17 either	KJPB	20.80	21.90	23.65	25.55	26.45	27.50	28.00	28.85	29.60	30.30	30.95
or depending on their circumstances	KABT	..	28.80	31.15[3]	33.60	34.80	36.15	36.80	37.90	38.90	39.85	40.70
aged 18 or over	KJPC	34.90	36.70	39.65	42.45	44.00	45.70	46.50	47.90	49.15	50.35	51.40
Dependant children and young people[5]												
aged under 11	KJPD	11.75	12.35	13.35[4]	14.55	15.05	15.65	15.95	16.45	–	–	..
aged 11-15	KJPE	17.35	18.25	19.75	21.40	22.15	23.00	23.40	24.10	–	–	..
aged 16-17	KJPF	20.80	21.90	23.65	25.65	26.45	27.50	28.00	28.85	–	–	..
aged 18	KABU	27.40	28.80	31.15	33.60	34.80	36.15	36.80	37.90	–	–	..
Birth to September following 11th birthday	KXDQ	..	..	..	..	..	..	..	..	16.90	17.30	20.20
From September following 11th birthday to September following 16th birthday	KXDR	..	..	..	..	..	..	..	..	24.75	25.35	25.90
From September following 16th birthday to day before 19th birthday	KXDS	..	..	..	..	..	..	..	..	29.60	30.30	30.95

1 In addition to personal allowances a claimant may also be entitled to premium(s). The types of premium are family, lone parent, pensioner, higher pensioner, disability, severe disability and disabled child.
2 From October 1991 the rate is £23.90.
3 From October 1991 the rate is £31.40.
4 From October 1991 each rate increases by 25p.
5 Age bandings for childrens personal allowances have been revised from 7 April 1997. Some children have protected rights. 1997 rates are as follows: age 11 before 7 April 1997 £24.75, age 16 before 7 April 1997 £29.60 and age 18 before 7 April 1997 £38.90. Rates for 1998: age 11 before 7 April 1997 £25.35, age 16 before 7 April 1997 £30.30 and age 18 before 7 April 1997 £39.85.

Source: Department of Social Security: 0191 225 7373

10.4 National Insurance contributions

	Class 1												Others	
	Not contracted out					Contracted out								
	Employee					Employee			Employer: Standard and reduced		Employer only: Standard only			
	Standard[1]		Reduced[2]	Employer: Standard and reduced	Employer only[3]: Standard only	Standard		Reduced						
	Up to LEL[4]	Over LEL				Up to LEL[4]	Over LEL		Up to LEL[4]	Over LEL	Up to LEL[4]	Over LEL	Class 2	Class 3
	%	%	%	%	%	%	%	%	%	%	%	%	£	£
6 April 1991														
Weekly earnings (£)														
52.00 - 84.99	*2.0*	*9.0*	*3.85*	*4.6*	*4.6*	*2.0*	*7.0*	*3.85*	*4.6*	*0.8*	*4.6*	*0.8*		
85.00 - 129.99	*2.0*	*9.0*	*3.85*	*6.6*	*6.6*	*2.0*	*7.0*	*3.85*	*6.6*	*2.8*	*6.6*	*2.8*		
130.00 - 184.99	*2.0*	*9.0*	*3.85*	*8.6*	*8.6*	*2.0*	*7.0*	*3.85*	*8.6*	*4.8*	*8.6*	*4.8*		
185.00 - 390.00	*2.0*	*9.0*	*3.85*	*10.4*	*10.4*	*2.0*	*7.0*	*3.85*	*10.4*	*6.6*	*10.4*	*6.6*		
over 390.00				*10.4*	*10.4*				*10.4*	*6.6*	*10.4*	*6.6*		
Classes 2 and 3													5.15	5.05
6 April 1992														
Weekly earnings (£)														
54.00 - 89.99	*2.0*	*9.0*	*3.85*	*4.6*	*4.6*	*2.0*	*7.0*	*3.85*	*4.6*	*0.8*	*4.6*	*0.8*		
90.00 - 134.99	*2.0*	*9.0*	*3.85*	*6.6*	*6.6*	*2.0*	*7.0*	*3.85*	*6.6*	*2.8*	*6.6*	*2.8*		
135.00 - 189.99	*2.0*	*9.0*	*3.85*	*8.6*	*8.6*	*2.0*	*7.0*	*3.85*	*8.6*	*4.8*	*8.6*	*4.8*		
190.00 - 405.00	*2.0*	*9.0*	*3.85*	*10.4*	*10.4*	*2.0*	*7.0*	*3.85*	*10.4*	*6.6*	*10.4*	*6.6*		
over 405.00				*10.4*	*10.4*				*10.4*	*6.6*	*10.4*	*6.6*		
Classes 2 and 3													5.35	5.25
6 April 1993														
Weekly earnings (£)														
56.00 - 94.99	*2.0*	*9.0*	*3.85*	*4.6*	*4.6*	*2.0*	*7.2*	*3.85*	*4.6*	*1.6*	*4.6*	*1.6*		
95.00 - 139.99	*2.0*	*9.0*	*3.85*	*6.6*	*6.6*	*2.0*	*7.2*	*3.85*	*6.6*	*3.6*	*6.6*	*3.6*		
140.00 - 194.99	*2.0*	*9.0*	*3.85*	*8.6*	*8.6*	*2.0*	*7.2*	*3.85*	*8.6*	*5.6*	*8.6*	*5.6*		
195.00 - 420.00	*2.0*	*9.0*	*3.85*	*10.4*	*10.4*	*2.0*	*7.2*	*3.85*	*10.4*	*7.4*	*10.4*	*7.4*		
over 420.00				*10.4*	*10.4*				*10.4*	*7.4*	*10.4*	*7.4*		
Classes 2 and 3													5.55	5.45
6 April 1994														
Weekly earnings (£)														
57.00 - 99.99	*2.0*	*10.0*	*3.85*	*3.6*	*3.6*	*2.0*	*8.2*	*3.85*	*3.6*	*0.6*	*3.6*	*0.6*		
100.00 - 144.99	*2.0*	*10.0*	*3.85*	*5.6*	*5.6*	*2.0*	*8.2*	*3.85*	*5.6*	*2.6*	*5.6*	*2.6*		
145.00 - 199.99	*2.0*	*10.0*	*3.85*	*7.6*	*7.6*	*2.0*	*8.2*	*3.85*	*7.6*	*4.6*	*7.6*	*4.6*		
200.00 - 430.00	*2.0*	*10.0*	*3.85*	*10.2*	*10.2*	*2.0*	*8.2*	*3.85*	*10.2*	*7.2*	*10.2*	*7.2*		
over 430.00				*10.2*	*10.2*				*10.2*	*7.2*	*10.2*	*7.2*		
Classes 2 and 3													5.65	5.55
6 April 1995														
Weekly earnings (£)														
58.00 - 104.99	*2.0*	*10.0*	*3.85*	*3.0*	*3.0*	*2.0*	*8.2*	*3.85*	*3.0*	*0.0*	*3.0*	*0.0*		
105.00 - 149.99	*2.0*	*10.0*	*3.85*	*5.0*	*5.0*	*2.0*	*8.2*	*3.85*	*5.0*	*2.0*	*5.0*	*2.0*		
150.00 - 204.99	*2.0*	*10.0*	*3.85*	*7.0*	*7.0*	*2.0*	*8.2*	*3.85*	*7.0*	*4.0*	*7.0*	*4.0*		
205.00 - 440.00	*2.0*	*10.0*	*3.85*	*10.2*	*10.2*	*2.0*	*8.2*	*3.85*	*10.2*	*7.2*	*10.2*	*7.2*		
over 440.00				*10.2*	*10.2*				*10.2*	*7.2*	*10.2*	*7.2*		
Classes 2 and 3													5.75	5.65
6 April 1996														
Weekly earnings (£)														
61.00 - 109.99	*2.0*	*10.0*	*3.85*	*3.0*	*3.0*	*2.0*	*8.2*	*3.85*	*3.0*	*0.0*	*3.0*	*0.0*		
110.00 - 154.99	*2.0*	*10.0*	*3.85*	*5.0*	*5.0*	*2.0*	*8.2*	*3.85*	*5.0*	*2.0*	*5.0*	*2.0*		
155.00 - 209.99	*2.0*	*10.0*	*3.85*	*7.0*	*7.0*	*2.0*	*8.2*	*3.85*	*7.0*	*4.0*	*7.0*	*4.0*		
210.00 - 455.00	*2.0*	*10.0*	*3.85*	*10.2*	*10.2*	*2.0*	*8.2*	*3.85*	*10.2*	*7.2*	*10.2*	*7.2*		
over 455.00				*10.2*	*10.2*	*2.0*	*8.2*	*3.85*	*10.2*	*7.2*	*10.2*	*7.2*		
Classes 2 and 3													*6.05*	*5.95*
6 April 1997														
Weekly earnings (£)														
62.00 - 109.99	*2.0*	*10.0*	*3.85*	*3.0*	*3.0*	*2.0*	*8.4*	*3.85*	*3.0*	*0.0*	*3.0*	*0.0*		
110.00 - 154.99	*2.0*	*10.0*	*3.85*	*5.0*	*5.0*	*2.0*	*8.4*	*3.85*	*5.0*	*2.0*	*5.0*	*2.0*		
155.00 - 209.99	*2.0*	*10.0*	*3.85*	*7.0*	*7.0*	*2.0*	*8.4*	*3.85*	*7.0*	*4.0*	*7.0*	*4.0*		
210.00 - 465.00	*2.0*	*10.0*	*3.85*	*10.0*	*10.0*	*2.0*	*8.4*	*3.85*	*10.0*	*7.0*	*10.0*	*7.0*		
over 455.00				*10.0*	*10.0*	*2.0*	*8.4*	*3.85*	*10.0*	*7.0*	*10.0*	*7.0*		
Classes 2 and 3													*6.15*	*6.05*

1 For those who are under pension age (65 men/60 women), but excluding those married women or widows who are liable for contributions at the reduced rate.

2 For employees who are married women or widows and liable for contributions at the reduced rate.

3 For all employees over pension age. Applicable also to employees who had made other arrangements to pay Class 1 contributions.

4 From 6 April 1991 LEL £52.00 a week,
from 6 April 1992 LEL £54.00 a week,
from 6 April 1993 LEL £56.00 a week,
from 6 April 1994 LEL £57.00 a week,
from 6 April 1995 LEL £58.00 a week,
from 6 April 1996 LEL £61.00 a week.
from 6 April 1997 LEL £62.00 a week.

Source: Department of Social Security: 0191 225 7373

10.5 Social Security Acts: number of persons receiving benefit[1]

United Kingdom

At any one time — Thousands

		1988	1989	1990	1991	1992	1993	1994	1995	1996	1997	1998
Persons receiving:												
Unemployment benefit[11]	KJHA	630.2	380.8	331.4	569.5	685.2	671.9	553.5	426.5	397.8	–	–
Jobseeker's allowance[24]	JYXM	..	..	..	..	..	..	..	..	..	1 520.7	1 344.8
Sickness and Invalidity benefit[2,3]	KJHB	1 278.0	1 394.7	1 515.6	1 479.5	1 646.3	1 805.0	1 893.9	1 987.7	–	–	–
Incapacity benefit[17,18]	KXDT	..	..	..	..	..	..	..	..	1 910.0	1 772.7	1 677.3
Attendance allowance[20]	KXDU	742.0	795.0	871.0	957.0	1 092.0	945.0	1 018.0	1 109.0	1 161.0	1 229.0	1 283.0
Mobility allowance[9,21]	KXDV	525	570	615	659	718	–	–	–	–	–	–
Disability living allowance[22]	KXDW	–	–	–	–	–	1 187.0	1 376.0	1 579.0	1 786.0	1 963.0	2 218.4
Guardians' allowances	KJHE	2.6	1.9	2.0	2.2	2.1	2.2	2.3	2.3	2.3	2.4	–
Widows' benefits[4]	KJHF	388.0	371.0	365.2	362.3	351.0	345.3	335.0	323.1	311.6	284.6	277.6
National Insurance retirement pensions[3]:												
Males	KJHH	3 479.1	3 481.7	3 553.9	3 512.8	3 613.7	3 623.6	3 657.5	3 728.7	3 836.6	3 926.7	4 010.6
Females	KJHL	6 523.5	6 520.4	6 625.7	6 515.3	6 683.0	6 680.0	6 727.1	6 777.2	6 948.5	6 983.2	7 031.1
Total	KJHG	10 001.6	10 002.2	10 179.6	10 028.1	10 296.8	10 303.6	10 384.7	10 505.9	10 785.1	10 909.9	11 041.9
Non-contributory retirement pensions[3]:												
Males	KJHI	6.0	6.7	6.7	6.1	6.1	6.1	6.1	6.5	6.1	5.9	5.7
Females	KJHJ	32.6	31.5	29.4	25.3	25.7	24.9	24.1	24.5	23.5	22.5	21.3
Total	KJHK	39.3	38.2	36.0	31.4	31.8	31.0	30.2	31.0	29.6	28.3	27.0
Industrial Injuries disablement pensions assessments[5]	KJHN	189.0[9]	193.0[9]	196.9	200.2[9]	204.3[9]	212.4[9]	232.8[16]	235.2[9]	248.9[9]	271.7[16]	285.6[16,9]
Reduced earnings allowance/ Retirement allowance assessments[5]	KEYC	150.5[9]	154.9[9]	160.0	163.7[9]	159.7[9]	156.0[9]	156.9[16]	152.0[9]	154.9[9]	158.5[16]	160.9[16,9]
Child benefit[3]												
Families receiving benefit	KJHO	6 923.0	6 695.0	6 949.5	7 024.4	7 078.3	7 136.4	7 179.9	7 222.2	7 251.6	7 195.7	7 202.3
Family income supplement	KJHP	–	–	–	–	–	–	–	–	–	–	–
Family Credit[6]	KHYH	313.1	311.9	331.7	360.4	404.6	505.7	564.5	630.9	716.3	801.5	766.9
Supplementary benefits[7]	KJHQ	–	–	–	–	–	–	–	–	–	–	–
Income Support[7]	KABV	4 536.0	4 350.0	4 376.0	4 683.0	5 292.9	5 858.4	5 897.6	5 896.5	5 778.3	4 329.4	4 018.4
Housing Benefit and Council Tax Benefit[8,9]												
Rent rebate	KABY	3 132.0[12]	2 971.5[12]	2 928.1[12]	2 944.0[12]	3 033.9[12]	3 052.5[12]	3 016.1[12]	3 075.2[12]	2 898.3[12]	2 792.2[12]	2 663.7[12]
Rent allowance[19]	KABZ	968.7[12]	958.1[12]	1 067.2[12]	1 082.4[12]	1 291.6[12]	1 480.5[12]	1 633.9[12]	1 793.1[12]	1 877.5[12]	1 846.9[12]	1 810.6[12,13]
Rate rebate	KACA	5 225.4[12]	4 299.8[12]	..[12,13]	..[12]	..[12]	..[12]	..[12]	..[12]	..[12]	..[12]	..[12]
Community charge benefit[10]	KACB	..	–[3]	6 518.4[13]	6 290.7[13]	6 563.9[13]	..[13]	..[13]	..[13]	..[13]	..[13]	..[13]
Community charge rebate[10]	KEZU	..	876.1[13]	..[13]	..[13]	..[13]	..[13]	..[13]	..[13]	..[13]	..[13]	..[13]
Council tax benefit[14,15]	KJPO	..	..	..	..	..	5 251.2	5 496.6	5 623.8	5 614.2	5 497.9	5 324.5
War pensions[3]	KJHR	258.3	252.2	248.0	249.6	260.1	292.9	309.2	315.4	327.5	320.7	302.0

1 Caseload counts at a specific date in the year which varies from benefit to benefit.
2 A relatively small number of claims do not result in the payment of benefit but are included here because they indicate notified incapacity for work.
3 Includes overseas cases.
4 Excluding widows' allowances paid during the first twenty-six weeks of widowhood. The number of such allowances does not exceed 35 000 in a six month period. Widows' allowance was replaced by widows' payment on 11 April 1988.
5 Industrial injuries disablement pension, reduced earnings allowance retirement allowance assessments starting 1 October up to 1986/87: first Monday in April thereafter.
6 Family Income Supplement was replaced by Family Credit from April 1988.
7 From April 1988 Supplementary Benefit was replaced by Income Support. From 9 October 1996 Income Support for the unemployed was replaced by Income-based Jobseeker's Allowance. Figures in this table up to and including 1996 include unemployed Income Support claimants. Figures from 1997 exclude unemployed who will be counted in the Jobseeker's Allowance claims.
8 Data prior to 1988 is not available.
9 Great Britain only.
10 Claimants with partners are treated as one recipient.
11 Figures are given at May each year.
12 The housing benefit scheme was reformed in April 1988.
13 Rate rebate in Scotland was replaced by community charge rebate (CCR) in April 1989. In April 1990 rate rebate in England and Wales, and CCR in Scotland was replaced by Community Charge Benefit.
14 Community Charge Benefit was replaced by Council Tax Benefit in April 1993.
15 Excluding Second Adult Rebate Claims.
16 Includes an allowance for late returns.
17 Incapacity Benefit replaced Sickness Benefit and Invalidity Benefit from 13 April 1995.
18 Incapacity Benefit figures are taken at the last day in February.
19 Rent Allowance figures include housing association tenants.
20 Attendance Allowance figures are based at 31 March until 1995 then at the end of February.
21 Mobility Allowance figures are based at 31 December until 1991 and then 31 March 1992.
22 Disability Living Allowance figures are based at the end of February each year.
23 Figures provided by the Child Benefit Centre Management Information Statistics as a new scan is being developed.
24 Jobseeker's Allowance introduced 7 October 1996, replacing unemployment benefit.

Sources: Department of Social Security: 0191 225 7373; Department of Health and Social Services (Northern Ireland): 028 9052 2062

10.6 Unemployed benefit/jobseekers allowance claimants analysed by benefit entitlement

United Kingdom at November

Thousands

		Jobseekers Allowance[1] 1996	Jobseekers Allowance May 1997	Jobseekers Allowance May 1998
All Persons				
All with benefit - total	KXDX	1 643.2	1 520.7	1 230.0
Contribution-based JSA only	KXDY	293.9	188.8	158.6
Contribution based JSA & income-based JSA	KXDZ	41.6	29.9	23.9
Income-based JSA only payment	KXEA	1 307.7	1 302.1	1 047.5
No benefit in payment	KXEB	158.6	165.6	120.7
Total	KXEC	1 801.8	1 686.3	1 301.3
Males				
All with benefit - total	KXED	1 264.7	1 190.8	922.1
Contribution-based JSA only	KXEE	189.7	124.5	104.6
Contribution based JSA & income-based JSA	KXEF	37.1	26.6	21.1
Income-based JSA only payment	KXEG	1 037.9	1 039.7	799.9
No benefit in payment	KXEH	105.7	111.9	79.5
Total	KXEI	1 370.3	1 302.7	997.9
Females				
All with benefit - total	KXEJ	378.6	330.0	272.1
Contribution-based JSA only	KXEK	104.2	64.3	54.2
Contribution based JSA & income-based JSA	KXEL	4.4	3.4	2.8
Income-based JSA only payment	KXEM	269.9	262.4	215.2
No benefit in payment	KXEN	52.9	53.6	41.3
Total	KXEO	431.5	383.6	313.3

1 Jobseeker's Allowance replaced Unemployment Benefit and Income Support for unemployed claimants on 7 October 1996. It is a unified benefit with two routes of entry: contribution-based which depends mainly upon national insurance contributions and income-based which depends mainly upon a means test. Some claimants can qualify by either route. In practice they receive income-based JSA but have underlying entitlement to the contribution-based element.

2 Figures quoted as at May of each year apart from 1996 which is November.

Sources: Department of Social Security: 0191 225 7373; Department of Health and Social Services (Northern Ireland): 028 9052 2062

10.6 Unemployed benefit/jobseekers allowance claimants analysed by benefit entitlement

continued

United Kingdom at November

Thousands

		Unemployment Benefit										
		1985[1]	1986	1987	1988	1989	1990	1991[2]	1992[2]	1993[2]	1994	1995[4]
All Persons												
Unemployment benefit in payment - total	KJIB	872.0	955.0	676.0	521.6	304.4	356.1	641.7	671.1	598.3	471.6	386.5
Unemployment benefit only	KJIC	679.0	768.0	549.0	402.2	220.3	294.0	520.2	548.3	495.6	333.6	266.3
Unemployment benefit and Income Support[3]	KJIF	193.0	187.0	127.0	119.3	82.1	62.0	121.5	122.8	102.7	138.0	120.2
Income Support only in payment[3]	KJIG	1 690.0	1 715.0	1 407.0	1 143.0	995.5	1 044.4	1 464.0	1 746.8	1 823.1	1 714.2	1 515.0
Neither Unemployment benefit nor Income Support in payment[3]	KJIH	461.0	500.0	425.0	347.3	252.0	2 529.0	306.6	374.2	316.9	250.1	215.7
Total	KJIA	3 023.0	3 170.0	2 508.0	2 012.8	1 551.9	1 653.3	2 412.3	2 792.1	2 738.3	2 435.9	2 117.1
Males												
Unemployment benefit in payment - total	KJIJ	541.0	596.0	442.0	328.2	200.5	254.1	472.1	488.4	425.8	326.0	265.4
Unemployment benefit only	KJIK	375.0	435.0	331.0	227.5	128.9	197.3	362.0	378.6	335.3	212.8	168.2
Unemployment benefit and Income Support[3]	KJIN	166.0	161.0	111.0	100.7	71.6	56.9	110.1	109.8	90.5	113.1	97.3
Income Support only in payment[3]	KJIO	1 281.0	1 308.0	1 140.0	886.0	778.2	825.4	1 171.7	1 410.3	1 465.1	1 366.3	1 198.8
Neither Unemployment benefit nor Income Support in payment[3]	KJIP	252.0	275.0	279.0	209.9	157.6	166.8	205.3	255.3	215.2	167.0	144.5
Total	KJII	2 074.0	2 179.0	1 861.0	1 424.1	1 136.3	1 246.3	1 849.4	2 154.0	2 106.1	1 859.2	1 608.7
Females												
Unemployment benefit in payment - total	KJIR	331.0	359.0	250.0	278.0	1 039.0	101.9	169.7	182.7	172.6	145.8	121.1
Unemployment benefit only	KJIS	304.0	333.0	231.0	174.8	92.5	96.8	158.1	169.8	160.4	120.8	98.2
Unemployment benefit and Income Support[3]	KJIV	27.0	26.0	19.0	18.6	11.5	5.1	11.5	12.9	12.2	25.0	22.9
Income Support only in payment[3]	KJIW	409.0	407.0	354.0	258.0	217.3	219.0	291.9	336.5	358.0	347.9	316.2
Neither Unemployment benefit nor Income Support in payment[3]	KJIX	209.0	225.0	184.0	138.3	95.4	86.1	101.4	119.0	101.8	83.1	71.2
Total	KJIQ	949.0	991.0	788.0	588.7	416.7	407.0	562.9	638.1	632.4	576.7	508.4

Note : Figures are based on a five per cent sample. For Northern Ireland, figures are based on a twenty per cent sample.

1 Figures are for Great Britain only. Due to industrial action November 1985 figures for Northern Ireland are not available.
2 Figures for Northern Ireland are at May.
3 Income Support replaced Supplementary Benefit from April 1988.
4 No Northern Ireland data are available for 1995.

Sources: Department of Social Security: 0191 225 7373; Department of Health and Social Services (Northern Ireland): 028 9052 2062

10.7 Sickness benefit, invalidity benefit and incapacity benefit
Claimants analysed by age and duration of spell

At end of statistical year[1]

Thousands

		1988	1989	1990	1991	1992	1993	1994	1995[2]	1996[3]	1997[3]	1998[4]
Age at 31 March												
Males												
All durations: All ages	KJJA	978.3	1 046.3	1 122.6	1 225.1	1 366.7	1 510.2	1 544.4	1 629.9	1 685.7	1 638.1	1 536.1
Under 20	KJJB	4.2	5.1	6.1	5.9	6.3	7.3	6.8	7.8	9.1	9.2	10.7
20-29	KJJC	41.0	48.6	51.1	63.3	77.6	88.2	97.2	106.0	123.6	123.3	124.1
30-39	KJJD	75.5	81.2	86.7	105.0	129.1	149.7	160.4	189.7	210.0	216.9	225.6
40-49	KJJE	138.4	148.4	160.7	175.6	201.0	234.8	245.2	260.4	285.2	283.6	283.4
50-59	KJJF	295.0	315.0	337.4	359.2	398.2	439.0	441.7	468.7	493.1	492.1	493.3
60-64	KJJG	290.8	292.4	298.8	309.1	329.3	354.7	356.6	362.5	364.3	362.2	357.2
65 and over	KJJH	134.2	156.0	181.8	207.1	225.2	236.6	236.5	234.8	200.7	150.7	97.0
Over six months: All ages	KJJI	821.7	895.6	972.4	1 057.3	1 183.4	1 311.8	1 361.5	1 442.1	1 492.7	1 436.0	1 401.4
Under 20	KJJJ	1.00	2.00	2.30	1.80	2.90	2.80	2.90	3.00	4.11	3.80	4.70
20-29	KJJK	23.1	27.8	31.0	39.0	51.5	60.9	65.7	74.3	89.4	85.8	87.7
30-39	KJJL	54.5	60.4	65.2	74.2	97.5	116.3	126.5	152.1	171.0	173.0	184.2
40-49	KJJM	107.2	118.6	126.9	143.3	165.0	190.6	208.6	219.7	245.0	241.3	244.1
50-59	KJJN	243.6	265.2	292.9	310.7	342.0	378.9	389.2	418.2	442.5	440.0	445.2
60-64	KJJO	260.0	266.5	273.1	281.7	300.0	326.4	332.6	340.2	341.3	342.2	338.8
65 and over	KJJP	133.2	155.2	181.1	206.6	224.5	235.8	236.0	234.6	199.4	150.1	96.7
Females												
All durations: All ages	KJJQ	347.3	400.3	452.5	511.6	592.0	670.2	704.7	776.5	815.5	832.4	814.8
Under 20	KJJR	6.2	8.2	8.2	9.2	9.4	9.6	11.0	12.0	13.0	12.9	13.6
20-29	KJJS	43.7	49.4	55.5	58.4	73.4	78.5	74.9	79.4	93.5	93.8	92.8
30-39	KJJT	61.8	65.6	71.0	77.0	90.8	99.8	108.5	122.4	142.4	145.7	153.7
40-49	KJJU	86.0	96.4	108.5	125.1	145.7	168.4	176.0	200.7	210.5	215.3	219.1
50-59	KJJV	128.7	154.4	175.7	200.1	220.8	253.1	266.6	290.7	297.1	316.3	337.1
60 and over	KJJW	22.0	26.3	33.6	41.8	51.9	60.7	67.7	71.3	58.9	48.5	33.9
Over six months: All ages	KJJX	267.0	322.3	370.2	432.9	501.9	574.2	611.4	673.1	704.6	718.2	734.7
Under 20	KJJY	1.0	2.4	3.1	4.0	3.4	3.6	4.7	4.2	6.5	6.8	5.7
20-29	KJJZ	24.4	30.5	34.7	40.5	49.0	55.9	55.9	60.3	69.9	69.4	69.2
30-39	KJKA	44.7	50.9	56.0	60.9	73.9	83.0	92.3	103.2	119.9	124.3	128.2
40-49	KJKB	68.2	79.0	89.4	106.8	124.5	143.6	151.8	170.8	184.0	187.2	191.6
50-59	KJKC	107.8	134.2	154.4	179.7	200.1	228.6	239.9	263.7	267.7	284.8	306.5
60 and over	KJKD	21.9	25.4	32.5	41.2	51.1	59.5	66.8	70.9	56.9	47.5	33.3

Note Figures for Great Britain are based on a 1 per cent sample up to 1995 and 5% sample thereafter.

1 For Great Britain the end of the statistical year is normally the Saturday before the first Monday in April.
2 For Great Britain the statistical year for 1994/95 was extended to 12/4/95, the day before the introduction of the new Incapacity Benefit which replaced Sickness and Invalidity Benefit from 13/4/95.
3 Age at 1 March.
4 Age at 28 February.

Sources: Department of Social Security: 0191 225 7373; Department of Health and Social Services (Northern Ireland): 028 9052 2062

10.8 Sickness, invalidity and incapacity benefit: days of certified incapacity
Analysis by age at end of period

Years starting on first Monday in April

Millions

		1987 /88	1988 /89	1989 /90	1990 /91	1991 /92	1992 /93	1993 /94	1994 /95	1995[1] /96	1996[1] /97	1997[1] /98
Age at 31 March												
Males: All ages	KJKH	294.8	315.5	338.5	367.7	415.2	458.9	468.8	507.9	612.8	593.9	588.0
Under 20	KJKI	1.1	1.3	1.4	1.6	1.7	1.5	1.6	1.8	2.7	2.4	3.7
20 - 29	KJKJ	11.7	13.4	14.3	17.3	21.9	24.9	27.0	30.4	43.1	41.3	45.4
30 - 39	KJKK	22.1	23.6	25.4	29.1	37.2	42.6	46.6	56.0	73.8	75.1	81.7
40 - 49	KJKL	40.3	43.6	46.5	51.0	59.0	66.7	72.7	78.9	100.4	99.1	103.5
50 - 59	KJKM	84.8	90.6	99.3	104.1	116.4	125.6	129.8	141.4	172.7	173.0	179.3
60 - 64	KJKN	86.6	87.2	88.2	92.2	99.2	105.6	107.3	112.6	133.6	134.0	131.9
65 and over	KJKO	48.3	55.8	63.3	72.2	79.8	83.0	83.9	86.8	86.5	68.8	43.7
Females: All ages	KJKP	101.9	117.7	133.2	153.0	177.2	197.6	211.4	237.5	289.8	296.1	309.8
Under 20	KJKQ	1.5	1.8	2.2	2.3	2.2	2.1	2.4	2.6	4.0	3.4	4.7
20 - 29	KJKR	13.7	14.5	16.0	17.4	20.4	22.9	22.1	23.9	32.3	32.0	34.2
30 - 39	KJKS	17.7	20.0	21.3	23.0	26.9	29.9	32.7	37.6	50.0	52.0	55.4
40 - 49	KJKT	24.2	27.9	30.9	36.2	42.6	48.1	51.5	58.9	73.3	74.5	79.8
50 - 59	KJKU	36.4	44.2	50.8	58.7	65.9	73.6	79.1	88.4	105.0	112.1	121.3
60 and over	KJKV	8.4	9.4	12.1	15.3	19.0	21.0	23.6	26.1	24.9	22.0	15.1

See *note* to Table 10.7.
1 Age at 1 March.

Sources: Department of Social Security: 0191 225 7373; Department of Health and Social Services (Northern Ireland): 028 9052 2062

10.9 Widows' benefit (excluding widows' allowance: widows' payment[1])

United Kingdom

Number in payment analysed by type of benefit and age of widow

Thousands

		September										
		1988	1989	1990	1991[3]	1992[3]	1993	1994	1995[3]	1996	1997	1998
All widows' benefit (excluding widows' allowance)												
All ages	KJGA	388.0	371.0	365.2	357.2	351.0	345.3	335.0	323.1	311.6	293.2	277.6
Under 30	KJGB	3.0	2.0	2.4	2.1	2.1	2.0	1.9	1.6	1.4	1.2	1.1
30 - 39	KJGC	17.0	16.0	16.2	15.5	15.9	15.0	14.5	13.9	13.5	13.0	12.3
40 - 49	KJGD	65.0	62.0	63.7	61.0	57.8	56.0	53.0	52.8	50.1	44.9	41.3
50 - 59	KJGE	258.0	241.0	239.6	230.8	224.2	219.9	216.7	206.9	202.5	197.3	187.7
60 and over	KJGF	46.0	49.0	43.3	47.7	51.1	52.5	48.8	47.8	44.0	36.8	35.2
Widowed mothers' allowance - with dependant children												
All ages	KJGG	59.0	53.0	54.4	52.4	53.2	53.2	53.3	52.8	51.7	49.5	47.1
Under 30	KJGH	2.0	2.0	2.3	2.0	2.0	1.9	1.7	1.6	1.4	1.2	1.1
30 - 39	KJGI	15.0	14.0	14.5	14.1	14.5	13.8	13.9	13.3	12.9	12.4	11.8
40 - 49	KJGJ	26.0	25.0	25.7	25.3	25.8	27.1	26.9	27.5	26.7	24.6	33.3
50 - 59	KJGK	15.0	12.0	11.5	10.8	10.6	10.2	10.7	10.1	10.2	10.7	10.5
60 and over	KJGL	–	–	0.3	0.2	0.2	0.2	0.2	0.4	0.3	0.3	0.2
Widowed mothers' allowance - without dependant children												
All ages	KJGM	20.0	21.0	17.8	14.2	10.7	8.3	6.7	6.2	5.6	4.6	3.4
Under 30	KJGN	–	–	0.1	0.2	0.1	–	0.1	0.1	–	–	–
30 - 39	KJGO	2.0	2.0	1.8	1.6	1.3	1.1	0.8	0.6	0.6	0.5	0.5
40 - 49	KJGP	8.0	9.0	8.0	6.4	5.1	3.9	3.4	3.3	2.8	2.3	1.9
50 - 59	KJGQ	9.0	10.0	7.8	5.8	4.0	2.9	2.2	1.9	2.1	1.7	1.0
60 and over	KJGR	–	–	0.3	0.3	0.2	0.2	0.2	0.1	0.1	0.1	–
Widows' pension												
All ages	KJGS	186.0	176.0	157.9	145.9	134.4	122.8	108.1	95.1	84.3	71.9	63.4
40 - 49	KJGT	–	–	–	–	–	–	–	–	–	–	–
50 - 59	KJGU	146.0	133.0	119.2	103.5	89.1	76.5	65.8	54.6	48.4	43.0	37.1
60 and over	KJGV	40.0	43.0	38.6	42.6	45.3	46.3	42.1	40.5	36.0	28.9	26.3
Age-related widows' pension[2]												
All ages	KJGW	124.0	121.0	135.0	144.6	152.7	161.0	166.9	168.9	170.0	167.1	163.7
40 - 49	KJGX	31.0	25.0	30.0	29.2	27.0	25.0	22.7	21.9	20.4	17.8	16.0
50 - 59	KJGY	88.0	87.0	100.9	110.8	120.4	130.3	137.9	140.2	141.9	141.6	139.0
60 and over	KJGZ	6.0	6.0	4.0	4.6	5.2	5.7	6.3	6.8	7.6	7.6	8.7

1 This is an especially high rate of benefit which is payable for the first 26 weeks of widowhood, provided that the widow is under pensionable age (age 60) or, if she is over that age, provided that her husband was not entitled to retirement pension. Widows' allowance was replaced by widows' payment on 11.4.88.

2 Figures for widows' basic pension are included in age-related widows' pension.

3 Northern Ireland data as at March.

Sources: Department of Social Security: 0191 225 7373; Department of Health and Social Services (Northern Ireland): 028 9052 2062

10.10 Child benefits

At 31 December

Thousands

		1988	1989	1990	1991	1992	1993	1994	1995	1996	1997[1]	1998[1,2]
Families receiving allowances:												
Total	KJMU	6 923	6 695	6 950	7 024	7 079	7 136	7 180	7 222	7 252	7 195	7 202
With 1 child	KJMV	2 956	2 872	2 955	2 977	2 986	3 001	3 024	3 054	3 068	..	..
2 children	KJMW	2 788	2 699	2 788	2 808	2 829	2 850	2 859	2 863	2 875	..	..
3 children	KJMX	878	844	894	918	934	948	960	968	969	..	..
4 children	KJMY	222	209	232	238	242	247	244	247	251	..	..
5 or more children	KJMZ	79	71	82	83	87	88	91	91	88	..	..

1 Figures provided by Child Benefit Centre Management Information Statistics as a new scan is being developed.

2 Figures for Northern Ireland January 1998

Sources: Department of Social Security: 0191 225 7373; Department of Health and Social Services (Northern Ireland): 028 9052 2062

10.11 Contributory and non-contributory retirement pensions
United Kingdom

Numbers in payment analysed by age-group[1]

Thousands (percentages in italics)

		At 30 September[2]										
		1988	1989[3]	1990[4]	1991	1992	1993	1994	1995	1996	1997	1998
Men:												
Age-groups:												
65-69	KJSB	1 178.5	1 194.0	1 163.8	1 124.6	1 095.5	1 063.9	1 053.6	1 079.1	1 142.5	1 200.5	1 249.9
Percentage	KJSC	*33.8*	*34.2*	*32.7*	*31.3*	*30.3*	*29.3*	*28.8*	*28.9*	*29.8*	*30.5*	*31.1*
70-74	KJSD	959.2	936.9	999.2	1 046.0	1 104.0	1 155.3	1 196.2	1 155.0	1 146.6	1 138.2	1 131.1
Percentage	KJSE	*27.5*	*26.9*	*28.1*	*29.1*	*30.5*	*31.8*	*32.7*	*31.0*	*29.9*	*28.9*	*28.2*
75-79	KJSF	748.2	750.5	761.4	761.7	738.5	706.3	681.8	748.8	787.3	831.3	876.6
Percentage	KJSG	*21.5*	*21.5*	*21.4*	*21.2*	*20.4*	*19.5*	*18.6*	*20.1*	*20.5*	*21.1*	*21.8*
80-84	KJSH	403.0	407.2	427.1	437.3	448.9	458.2	468.0	477.4	481.8	471.2	454.5
Percentage	KJSI	*11.6*	*11.7*	*12.0*	*12.2*	*12.4*	*12.6*	*12.8*	*12.8*	*12.6*	*12.0*	*11.3*
85-89	KJSJ	152.0	155.0	164.0	172.9	180.5	189.6	197.4	204.4	209.4	217.2	225.0
Percentage	KJSK	*4.4*	*4.4*	*4.6*	*4.8*	*5.0*	*5.2*	*5.4*	*5.5*	*5.5*	*5.5*	*5.6*
90 and over	KJSL	44.0	44.0	46.0	49.1	53.8	57.2	60.6	64.5	69.4	75.3	79.3
Percentage	KJSM	*1.3*	*1.3*	*1.3*	*1.4*	*1.5*	*1.6*	*1.7*	*1.7*	*1.8*	*1.9*	*2.0*
Total all ages	KJSA	3 485.8	3 488.0	3 560.5	3 591.5	3 621.3	3 630.4	3 657.5	3 729.2	3 837.0	3 933.8	4 016.3
Women:												
Age-groups:												
60-64	KJSO	1 117.6	1 099.0	1 138.7	1 124.2	1 121.1	1 099.1	1 090.6	1 094.3	1 200.6	1 245.0	1 296.2
Percentage	KJSP	*17.0*	*16.8*	*17.1*	*16.9*	*16.7*	*16.4*	*16.2*	*16.1*	*17.3*	*17.8*	*18.4*
65-69	KJSQ	1 565.3	1 597.8	1 533.3	1 484.6	1 472.7	1 447.0	1 436.6	1 437.6	1 461.2	1 458.9	1 442.8
Percentage	KJSR	*23.9*	*24.4*	*23.0*	*22.3*	*21.9*	*21.5*	*21.4*	*21.2*	*21.0*	*20.8*	*20.5*
70-74	KJSS	1 257.2	1 225.6	1 285.8	1 336.9	1 403.8	1 469.3	1 515.9	1 447.3	1 423.7	1 398.6	1 379.6
Percentage	KJST	*19.2*	*18.7*	*19.3*	*20.1*	*20.9*	*21.9*	*22.5*	*21.3*	*20.5*	*20.0*	*19.6*
75-79	KJSU	1 161.2	1 164.4	1 170.3	1 161.5	1 126.0	1 068.0	1 023.0	1 100.6	1 148.7	1 197.2	1 255.7
Percentage	KJSV	*17.7*	*17.8*	*17.6*	*17.5*	*16.8*	*15.9*	*15.2*	*16.2*	*16.5*	*17.1*	*17.8*
80-84	KJSW	827.7	831.8	855.6	864.2	868.0	873.9	882.0	892.9	888.4	861.0	814.6
Percentage	KJSX	*12.6*	*12.7*	*12.9*	*13.0*	*12.9*	*13.0*	*13.1*	*13.2*	*12.8*	*12.3*	*11.6*
85-89	KJSY	432.4	438.8	465.7	438.8	499.2	514.1	521.9	534.3	542.2	545.9	552.3
Percentage	KJSZ	*6.6*	*6.7*	*7.0*	*6.6*	*7.4*	*7.7*	*7.8*	*7.9*	*7.8*	*7.8*	*7.8*
90 and over	KJTA	194.6	195.8	205.7	219.1	231.0	244.9	256.8	271.9	285.3	300.4	311.4
Percentage	KJTB	*3.0*	*3.0*	*3.1*	*3.3*	*3.4*	*3.6*	*3.8*	*4.0*	*4.1*	*4.3*	*4.4*
Total all ages	KJSN	6 556.0	6 551.9	6 655.1	6 648.2	6 721.7	6 716.3	6 727.1	6 779.0	6 950.2	7 006.8	7 052.5

1 Including pensions payable to persons residing overseas.
2 Northern Ireland as at 30 November for years 1983 to 1987.
3 Great Britain as at 31 March.
4 Northern Ireland figures are at 31 March.

Sources: Department of Social Security: 0191 225 7373; Department of Health and Social Services (Northern Ireland): 028 9052 2062

10.12 Family credit
United Kingdom

At 31 December[2]

Thousands

		1990[2]	1991[2]	1992	1993	1994	1995	1996	1997	1998
Families receiving payments:										
Total	KJTO	331.7	360.3	15.8	536.5	600.6	669.7	742.7	808.4	808.8
Two-parent families: total	KJTP	201.1	221.5	11.0	303.1	339.2	371.4	405.6	421.1	399.5
With 1 child	KJTQ	44.5	52.4	1.9	75.6	82.9	92.6	101.6	103.2	98.6
2 children	KJTR	71.3	79.3	3.3	113.9	127.0	139.9	151.2	155.5	147.2
3 children	KJTS	49.1	52.4	2.8	72.0	80.2	87.2	95.2	100.0	93.3[4]
4 children	KJTT	22.7	23.5	1.8	28.8	34.4	36.1[3]	40.3[3]	42.1	39.0
5 children	KJTU	8.5	9.1	0.7	9.6	9.5	10.2	11.7	13.6	12.5
6 or more children	KJTV	5.0	4.8	0.4	5.2	5.3	5.6	5.5	6.7	5.8
One-parent families: total	KJTW	130.2	138.4	4.5	233.5	261.3	298.2	337.2	387.3	409.3
With 1 child	KJTX	72.4	74.7	2.7	121.3	138.4	157.2	176.3	203.0	237.5
2 children	KJTY	41.7	45.9	1.2	80.9	88.2	101.5	114.3	128.8	139.8
3 or more children	KJTZ	16.1	13.6	0.6	31.3	34.7	39.6	46.5	55.2	58.8

1 For N.I. data 1988 figures are at 1 November 1989, 1990 and 1997 figures are at 31 October, 1992 figures are at 30 April.
2 N.I. figures for two-parent families and one-parent families exclude 200 families in 1990, 385 families in 1991 and 340 families in 1992.
3 Includes 4 or more children for N.I.
4 Includes 3 or more children for N.I.

Sources: Department of Social Security: 0191 225 7373; Department of Health and Social Services (Northern Ireland): 028 9052 2062

10.13 Income support: number of beneficiaries receiving weekly payment[1]

Great Britain

On a day in May

Thousands

		1992	1993	1994	1995	1996	1997[3]	1998[3]
All income support	KACC	3 259	3 561	3 689	3 889	3 963	3 958	3 853
All aged 60 and over	KACD	1 626	1 760	1 749	1 770	1 753	1 720	1 659
Retirement pensioners	KACE	1 318	1 426	1 407	1 417	1 404	1 383	1 338
In receipt of other NI benefit	KACF	75	89	96	107	107	100	94
Others	KACG	234	245	246	245	242	237	228
All under 60	KACH	1 633	1 801	1 940	2 120	2 211	2 239	2 194
Disabled with contributory benefit	KACK	72	83	101	133	148	167	182
Disabled without contributory benefit	KFBJ	337	413	493	583	617	660	699
Lone parent premium not in other groups[2]	KACL	938	988	1 017	1 040	1 044	1 014	961
Others	KACM	287	317	328	363	402	398	353

1 Great Britain data are extracted from the Annual Statistical Enquiry undertaken in May 1988, 1989 , 1990, 1991, 1992 and 1993, and the Quarterly Statistical Enquiries May 1994 and May 1998.
2 Figures relate to one-parent families headed by a man or a woman.
3 Figures exclude unemployed claimants who transferred to Jobseeker's Allowance from 7 October 1996.

Source: Department of Social Security: 0191 225 7373

10.14 Income support

Great Britain[1,2]

On a day in May

Thousands

		1992	1993	1994	1995	1996	1997	1998
Number of regular weekly payments	KACN	3 259	3 561	3 689	3 889	3 963	3 958	3 853
Total number of persons provided for	KACO	5 618	6 146	6 465	6 858	7 004	6 977	6 770
Number of dependants	KACP	1 984	2 173	2 296	2 440	2 494	2 467	2 380
Partners	KACQ	374	442	480	529	546	551	537
Total children under 16 years	KACR	1 869	2 044	2 156	2 292	2 340	2 301	2 223
Under 11	KACS	1 444	1 557	1 623	1 708	1 728	1 683	1 605
11 - 15 years	KACT	425	487	533	584	612	618	618
16 - 17 years	KACU	97	104	112	119	129	134	129
Other dependants								
18 years and over	KACV	18	24	28	28	25	28	27

1 Great Britain data are extracted from the Annual Statistical Enquiry undertaken in May 1992-1993 and the Quarterly Statistical Enquiries undertaken in May 1994 and to May 1998.
2 Figures exclude unemployed claimants in receipt of Income Support prior to the introduction of Jobseeker's Allowance in October 1996.

Source: Department of Social Security: 0191 225 7373

10.15 Income support: average weekly amounts of benefit
Great Britain[1]
May

Thousands

		1992	1993	1994	1995	1996	1997[3]	1998[3]
All income support	KACW	51.77	54.75	56.16	56.29	57.26	58.03	58.72
All aged 60 and over	KJUB	40.79	43.59	42.45	41.18	41.69	42.24	42.29
Retirement pensioners	KACX	35.42	38.28	36.14	34.61	34.66	34.90	34.50
In receipt of other NI benefit	KJUD	39.13	38.40	41.75	40.60	43.93	46.74	49.10
Others	KACY	71.64	76.36	78.82	79.36	81.45	83.19	85.27
All under 60	KACZ	62.71	65.65	68.51	68.91	69.61	70.17	71.14
Disabled with contributory benefit	KADC	28.37	27.28	29.63	31.95	33.00	33.73	33.30
Disabled without contributory benefit	KADD	52.29	58.33	63.28	66.04	67.78	69.65	72.18
Lone parent premium not in other groups[2]	KADE	67.59	70.31	74.93	76.63	78.19	79.21	79.67
Others	KADF	67.60	70.73	68.44	64.92	63.59	63.24	65.38

1 Great Britain data are extracted from the Annual Statistical Enquiry undertaken in May 1988, 1989, 1990, 1991, 1992 and 1993, and from the Quarterly Statistical Enquiries May 1994 and 1998.

2 Figures relate to one-parent families headed by a man or a woman.

3 Figures exclude unemployed claimants who transferred to Jobseeker's Allowance from 7 October 1996.

Source: Department of Social Security: 0191 225 7373

10.16 War pensions
Estimated number of pensioners
At 31 March in each year

Thousands

		1988	1989	1990	1991	1992	1993	1994	1995	1996	1997	1998
Total	KADG	263.86	256.25	249.96	247.67	249.95	266.92	296.30	309.84	323.74	324.64	317.65
Disablement -												
1914 war; 1939 war and later service	KADH	201.70	196.45	192.47	192.04	196.19	214.53	245.44	260.30	265.37	264.59	259.16
Widows and dependants -												
1914 war; 1939 war and later service	KADI	62.16	59.80	57.48	55.83	53.76	52.39	50.86	49.54	58.37	60.05	58.49

Source: Department of Social Security: 0191 225 7373

10.17 Summary of government expenditure on social services and housing

Years ended 31 March

£ millions

		1988 /89	1989 /90	1990 /91	1991 /92	1992 /93	1993 /94	1994 /95	1995 /96	1996 /97	1997 /98	1998 /99
Education[1]	KJAA	22 137	24 664	26 728	29 550	31 576	33 544	35 367†	36 171	36 986	..	..
National health service	KJAB	23 080	25 029	27 911	31 842	35 413	37 259	39 879	40 691	42 383	43 878	45 579
Welfare services	CSWL	4 451	4 985	5 745	4 873	6 683	7 700	9 016	10 127	10 581	11 075	12 037
Social security benefits	KJAE	49 377	52 434	58 261	69 121	78 846	85 805	87 941	90 534†	92 217	92 107	93 808
Housing	KJAF	4 379	5 273	6 065	5 547	6 984	5 919	5 634	5 445	4 593†	3 415	3 767
Total government expenditure	KJAG	103 424	112 385	124 710	140 933	159 502	170 227	177 863	189 736†	186 760	..	..
Current expenditure	KJAH	98 463	105 356	117 243	133 848	151 010	163 289	171 303	183 297†	182 793	..	..
Capital expenditure	KJAI	4 961	7 029	7 467	7 085	8 492	6 938	6 560	6 439†	3 967	..	..
Total government expenditure	KJAG	103 424	112 385	124 710	140 933	159 502	170 227	177 863	189 736†	186 760	..	..
Central government	KJAK	75 090	80 577	90 224	103 304	116 357	126 407	130 828	136 993†	136 525	..	..
Local authorities	KJAL	28 334	31 808	34 486	37 629	43 145	43 820	47 035	52 743†	50 235	..	..
Total government expenditure	KJAG	103 424	112 385	124 710	140 933	159 502	170 227	177 863	189 736†	186 760	..	..
Total government expenditure on social services and housing as a percentage of GDP[2]		21.59	21.51	22.16	23.89	26.04	26.30	25.93	26.27†	24.37	..	..

1 Includes school meals and milk.
2 GDP adjusted to take account of change from rates to community charge.

Source: Office for National Statistics: 020 7533 5990

10.18 Government expenditure on education

Years ended 31 march

£ millions

			1986 /87	1987 /88	1988 /89	1989 /90	1990 /91	1991 /92	1992[1] /93	1993 /94	1994 /95	1995 /96	1996[15] /97
Current expenditure													
Nursery schools	KJBA	KEZN	81	4 743	5 259	5 889	6 458	7 247	8 262	8 712	9 094	9 349	9 676
Primary schools	KJBB		4 157										
Secondary schools	KJBC		5 583	5 991	6 437	6 832	7 147	7 787	8 347	8 615	8 875	8 987	9 253
Special schools	KJBD		727	812	888	1 008	1 121	1 245	1 354	1 420	1 451	1 492	1 567
Universities[3]	KJBG		1 654	1 824	1 958	2 104	2 265	2 437	3 361	..	..	..	..
Other Higher, Further and adult education[4]	KJBE		2 716	2 971	3 277	3 729	4 128	4 454	4 136	..	..	..	..
Higher Education Funding Council[4,5]	CSWM		..	..	..	..	..	..	..	4 908	5 192	5 455	5 738
Further Education Funding Council[6]	CSWO		..	..	..	..	..	..	..	3 072	3 200	3 374	3 712
Continuing Education	CSWP		..	..	..	..	..	..	..	380	294	266	273
Other education expenditure	KJBH		871	900	1 009	1 191	1 325	1 361	1 009	905	963	1 045	1 153
Related current expenditure:													
Training of teachers:													
residence[7]	KJBI		19	..	..	..	..	..	..	..	..	..	..
School welfare[8]	KJBJ		40	57	70	91	155	214	270	346	359	330	344
Meals and milk 9	KJBK		559	547	469	485	506	556	161	149	147	148	148
Youth service and physical training	KJBL		211	246	277	328	348	360	393	392	401	399	390
Maintenance grants and allowances to pupils and students[10,11]	KJBM		806	843	900	934	1 028	1 379	1 705	1 972	2 204	2 180	2 026
Transport of pupils	KJBN		290	302	311	335	393	442	417	444	486	506	569
Miscellaneous expenditure	KJBO		5	6	10	17	2	3	8	10	31	31	49
Total current expenditure[12]	KJBQ		17 719	19 240	20 865	22 943	24 876	27 485	29 424	31 325	32 697	33 562	34 898
Capital expenditure													
Nursery schools	KJBR	KEZP	7	192	246	314	353	376	384	414	517	529†	525
Primary schools	KJBS		180										
Secondary schools	KJBT		187	213	195	397	465	493	518	485	565	574	588
Special schools	KJBU		23	11	33	38	36	35	32	31	37	51	52
Universities[3]	KJBX		156	158	172	196	211	231	236	..	..	..	..
Other Higher, Further and adult education[4]	KJBV		152	177	180	248	230	293	285	..	..	..	..
Higher Education Funding Council[4,5]	CSWQ		..	..	..	..	..	..	..	406	412	453	74
Further Education Funding Council[6]	CSWR		..	..	..	..	..	..	..	194	201	213	19
Continuing Education	CSWS		..	..	..	..	..	..	..	6	8	10	4
Other education expenditure	KJBY		10	32	22	41	39	33	25	12	45	25	17
Related capital expenditure	KJBZ		24	21	25	32	26	20	17	23	25	17	18
Total capital expenditure[12]	KJCA		739	804	873	1 266	1 359	1 481	1 496	1 571	1 810	1 862†	1 287
VAT refunds to local authorities	KJBP		345	357	399	455	493	584	656	648	860†	747	801
Total expenditure													
Central government	KJCB		2 567	2 746	3 025	4 337	4 582	4 288	4 722	8 186	9 490	10 269†	10 426
Local authorities	KJCC		16 235	17 655	19 112	20 325	22 146	25 261	26 853	25 354	25 900	26 118	25 877
Total government expenditure on education[13]	KJAA		18 803	20 401	22 137	24 664	26 728	29 550	31 576	33 544	35 367†	36 171	36 986
Total government education expenditure as a percentage of GDP[14]			4.87	4.78	4.66	4.77	4.80	5.08	5.20	5.24	5.21†	5.07	4.92

1 Includes 1991/92 data for Wales.
2 Includes 1993/94 data for Wales.
3 Includes expenditure on University departments of Education for England and Wales.
4 Including tuition fees.
5 Includes expenditure on Higher Education Institutions in Northern Ireland.
6 Includes expenditure on Further Education Institutions in Northern Ireland.
7 With effect from 1987/8 included with maintenance grants and allowances.
8 Expenditure on the school health service is included in the National Health Service.
9 From 1992/93 expenditure on meals and milk in England has been recharged across other expenditure headings
10 From 1986/87, excludes the secondment of teachers on further training.
11 From 1990/91, includes student loans expenditure.
12 Due to rounding constituent figures may not sum to totals.
13 Excludes additional adjustment to allow for Capital consumption made for National Accounts purposes amounting to £1013m in 1995/96.
14 GDP includes adjustments to remove the distortion caused by the abolition of domestic rates which have led to revisions to the historical series.
15 Provisional Data. Includes 1995/96 data for Wales and Scotland.

Sources: Department for Education and Employment; Office for National Statistics: 020 7533 5990

10.19 Government expenditure on the National Health Service

Years ended 31 March

£ millions

		1988 /89	1989 /90	1990 /91	1991 /92	1992 /93	1993 /94	1994 /95	1995 /96	1996 /97	1997 /98	1998 /99
Current expenditure												
Central government:												
Hospitals and Community Health Services[1] and Family Health Services[2]	KJQA	21 110	22 197	25 276	29 061	32 195	35 567	37 698	38 514	39 425	40 993	43 600
Administration[3]	KJQB	682	855	979	1 119	1 258	–	–	–	–	–	–
less Payments by patients:												
Hospital services	KJQC	–347	–407	–510	–540	–505	–368	–111	–42	–42	–48	–84
Pharmaceutical services	KJQD	–202	–242	–247	–270	–297	–324	–342	–383	–376	–396	–390
Dental services	KJQE	–282	–340	–441	–477	–470	–440	–464	–494	–447	–475	–470
Ophthalmic services	KJQF	–	–	–	–	–	–	–	–	–	–	–
Total	KJQG	–831	–989	–1 198	–1 287	–1 272	–1 132	–917	–919	–865	–919	–944
Departmental administration	KJQH	206	202	268	293	319	270	256	242	265	245	227
Other central services	KJQI	604	693	738	865	1 301	1 651	2 304	2 538	3 124	3 242	3 357
Total current expenditure	KJQJ	21 771	22 958	26 063	30 051	33 801	36 356	39 341	40 375	41 949	43 561	45 346
Capital expenditure												
Central government	KJQK	1 309	2 071	1 848	1 791	1 612	903	538	316	434	317	233
Total government NHS Expenditure												
Central government	KJAB	23 080	25 029	27 911	31 842	35 413	37 259	39 879	40 691	42 383	43 878	45 579
Total NHS expenditure as a percentage of GDP[4]		4.82	4.79	4.96	5.40	5.78	5.76	5.81	5.63	5.53†	5.39	5.36

1 Including the school health service.
2 General Medical Services have been included in the expenditure of the Health Authorities. Therefore, Hospitals and Community Health Services and Family Practitioner Services (now Family Health Services) are not identifiable separately.
3 Administration costs are not separately identifiable from 1993/94.
4 GDP adjusted to take account of change from rates to community charge.

Source: Office for National Statistics: 020 7533 5990

10.20 Government expenditure on welfare services[1]

Years ended 31 March

£ millions

		1988 /89	1989 /90	1990 /91	1991 /92	1992 /93	1993 /94	1994 /95	1995 /96	1996 /97	1997 /98	1998 /99
Personal social services												
Central government current expenditure	KJCG	127	143	163	190	202	216	197	140	101	73	53
Local authorities current expenditure:												
Running expenses	CTKQ	3 971	4 395	5 022	5 725	6 122	7 113	8 400	9 716†	10 947	11 105	11 439
Capital expenditure	KJCI	163	213	227	200	189	190	235	229	221	213	191
Total	KJAC	4 346	4 877	5 626	4 732	6 513	7 519	8 832	10 085†	11 269	11 391	11 683
Welfare foods service												
Central government current expenditure on welfare foods (including administration)	KJCK	105	108	120	142	171	182	185	228	264	306	355
less Receipts from the public	KJCL	–	–	–1	–1	–1	–1	–1	–1	–1	–1	–1
Total	KJAD	105	108	119	141	170	181	184	227	263	305	354
Total government expenditure	CSWL	4 451	4 985	5 745	4 873	6 683	7 700	9 016	10 127	10 581	11 075	12 037
Total government expenditure as a percentage of GDP[2]		0.93	0.95	1.02	0.83	1.09	1.19	1.31	1.40	1.38	1.36	1.41

1 School meals and milk are included in Table 10.18.
2 GDP adjusted to take account of change from rates to community charge.

Source: Office for National Statistics: 020 7533 5990

10.21 Government expenditure on social security benefits

Years ended 31 March

£ millions

		1988 /89	1989 /90	1990 /91	1991 /92	1992 /93	1993 /94	1994 /95	1995 /96	1996 /97	1997 /98	1998 /99
Government current expenditure												
National insurance fund:												
Retirement pensions	EKXK	18 679	20 757	22 725	25 691	27 076	28 481	28 925	30 162†	32 174	33 640	35 714
Lump sums to pensioners	KJDB	109	112	114	114	115	122	123	124	129	118	120
Widows and guardians allowances	CSDH	881	896	893	884	1 014	1 041	1 034	1 018	974	992	973
Unemployment benefit	CSDI	1 318	752	892	1 627	1 761	1 623	1 277	1 099	588	–	–1
Jobseeker's allowance[3]	CJTJ	–	–	–	–	–	–	–	–	351†	590	475
Sickness benefit	CSDJ	217	209	222	278	365	294	426	12	–	–	–
Invalidity benefit	CSDK	3 820	3 935	4 544	5 461	6 198	7 146	8 042	271	–	–	–
Incapacity benefit[4]	CUNL	–	–	–	–	–	–	–	7 615	7 668†	7 471	7 295
Maternity benefit	CSDL	45	32	35	40	42	32	17	28	32	36	39
Death grant	CSDM	–	–	–	–	–	–	–	–	–	–	–
Injury benefit	CSDN	–	–	–	–	–	–	–	–	–	–	–
Disablement benefit	CSDO	504	475	526	–	–	–	–	–	–	–	–
Industrial death benefit	CSDP	70	61	62	–	–	–	–	–	–	–	–
Statutory sick pay	CSDQ	892	996	966	725	688	688	24	24	24†	28	28
Statutory maternity pay	GTKZ	263	292	344	396	416	440	498	476†	500	516	552
Payments in lieu of benefits foregone	GTKV	–	–	–	–	–	–	–	–	–	–	–
Total	ACHH	26 798	28 474	31 323	35 216	37 675	39 867	40 366	40 829	42 440†	43 391	45 195
Maternity Fund	GTKO	–	–	–	–	–	–	–	–	–	–	–
Redundancy Fund	GTKN	80	71	130	276	321	110	208	128	108	88	116
Social Fund	GTLQ	76	101	123	130	175	189	183	216	203†	152	216
Non-contributory benefits:												
War pensions	KJDP	572	641	688	844	976	913	1 083	1 247	1 352	1 284	1 262
Family benefits:												
Child benefit	KJDQ	4 720	4 751	4 840	5 433	5 950	6 347	6 294	6 332	6 645†	7 095	7 328
One parent benefit	KJDR	180	200	227	249	275	282	289	310	317	9	–
Family credit	KAAA	397	424	466	626	929	1 208	1 441	1 739	2 084	2 338	2 430
Family income supplement	KJDS	–	–	–	–	–	–	–	–	–	–	–
Maternity grants	KJDT	–	–	–	–	–	–	–	–	–	–	–
Income support/Supplementary benefits:												
Supplementary pensions	KJDU	–	–	–	–	–	–	–	–	–	–	–
Supplementary allowances	KJDV	–	–	–	–	–	–	–	–	–	–	–
Income support	KAAB	7 810	8 257	9 106	12 325	15 578	16 997	16 387	16 650	14 438	11 998†	11 793
Other non-contributory benefits:												
Old persons' pensions	KJDX	37	35	38	36	36	36	35	36	30	29	29
Lump sums to pensioners	KJDY	7	9	8	11	13	14	13	15	15	17	17
Attendance allowance	KJDZ	1 315	1 262	1 698	1 706	1 553	1 795	1 963	2 194	2 393	2 640	2 682
Invalid care allowance	KJEA	152	184	229	285	345	442	526	617	736	745	783
Mobility allowance	KJEB	675	770	895	1 063	68	–	–	–	–	–	–
Disability living allowance	EKXL	–	–	–	–	1 973	2 772	3 125	3 802	4 498	5 018	5 367
Disability working allowance	EKYE	–	–	–	–	3	7	11	19	34	44	49
Severe disablement allowance	KJEC	320	347	407	596	640	703	776	820	906	1 007	984
Industrial injury benefits	EKXM	–	–	142	655	668	687	706	731	744	754	–
RPI Adjustment	KAAC	–	–	–	–	–	–	–	–	–	–	–
Housing benefit[1]	KJED	3 711	4 095	4 735	6 053	7 670	9 163	10 345	10 773†	11 276	11 328	11 208
Administration	KJEE	2 527	2 813	3 206	3 617	3 998	4 273	4 190	4 076	3 998	4 170	4 349
Total government benefit expenditure	KJAE	49 377	52 434	58 261	69 121	78 846	85 805	87 941	90 534†	92 217	92 107	93 808
Total government benefit expenditure as a percentage of GDP[2]		10.31	10.04	10.35	11.72	12.87	13.26	12.82	12.53†	12.03	11.31	11.03

1 From 1981-82 comprises expenditure by the Department of Health and Social Security under the unified housing benefit scheme.

2 GDP adjusted to take account of change from rates to community charge.

3 Jobseeker's allowance was introduced in October 1996 to replace Unemployment benefit and Income Support for the unemployed.

4 Sickness benefit and Invalidity benefit were replaced by a single incapacity benefit in 1995

Source: Office for National Statistics: 020 7533 5990

10.22 Government and other public sector expenditure on housing

Years ended 31 March

£ millions

		1988 /89	1989 /90	1990 /91	1991 /92	1992 /93	1993 /94	1994 /95	1995 /96	1996 /97	1997 /98	1998 /99
Government expenditure												
Current expenditure												
Central government housing subsidies:												
to local authorities	CTMN	603	714	1 273	1 144	992	874	814	687	717	711†	737
to public corporations	ADVA	304	285	284	260	208	214	228	228	236	244†	239
to housing associations	KJRC	43	53	67	40	135	180	207	235	215	216†	–46
grants to housing associations	KJRD	8	7	85	117	59	13	3	1	1	–	–
Local authorities housing subsidies	KJVK	554	487	3	–	1	–	–	3	1	1	1
Grants under the option mortgage scheme	KJRF	–	–	–	–	–	–	–	–	–	–	–
Administration, etc	KGVL	250	249	320	373	394	364	405	402	447†	509	571
Total current expenditure	KJRH	1 763	1 794	2 032	1 934	1 789	1 645	1 657	1 556	1 617†	1 681	1 502
Capital expenditure												
Investment in housing by local authorities	KGVM	79	1 387	1 322	822	913	780	842	1 009	937†	–52	64
Capital grants to housing associations	KJRJ	956	552	1 495	2 045	2 583	2 136	1 774	1 329	1 393†	1 102	1 021
Improvement grants	ADCE	809	885†	863	1 152	1 240	1 287	1 105	846	980	1 157	950
Net lending for house purchase	KJRL	–148	–259	–596	–782	–211	–402	–141	–136	–719	–816†	–175
Capital grants to public corporations	KJRM	65	241	774	482	631	439	387	837	386	400	415
Net lending to public corporations	KJRN	855	672	176	–104	39	34	10	7	–1	–57	–10
Total capital expenditure	KJRO	2 616	3 479	4 033	3 613	5 195	4 274	3 977	3 889	2 976†	1 734	2 265
Total expenditure												
Central government	KJRP	3 023	2 547	3 835	3 697	4 595	3 839	3 409	3 313	2 362†	1 962	2 356
Local authorities	KJRQ	1 356	2 726	2 230	1 850	2 389	2 080	2 225	2 132	2 231†	1 453	1 411
Total government expenditure	KJAF	4 379	5 273	6 065	5 547	6 984	5 919	5 634	5 445	4 593†	3 415	3 767
Public corporations' capital expenditure												
Investment in housing	KGVN	285	451	519	673	549	568	560	580	574†	450	364
Net lending to private sector	AAFR	–8	–2	–3	–4	3	1	–1	–5	–1†	–1	–1
Total	KJRU	277	449	516	669	552	569	559	575	569†	449	363
Total public sector housing expenditure[1]	KJRV	3 736	4 809	5 631	5 838	6 866	6 015	5 796	5 176	4 777†	3 521	3 725
Total public sector housing expenditure as a percentage of GDP[2]		0.78	0.92	1.00	0.99	1.12	0.93	0.85	0.72	0.62†	0.43	0.44

1 Total government expenditure *less* grants and loans to public corporations *plus* public corporations' capital expenditure.
2 GDP adjusted to take account of change from rates to community charge.

Source: Office for National Statistics: 020 7533 5990

Crime and justice

11

Crime and justice

There are differences in the legal and judicial systems of England and Wales, Scotland and Northern Ireland which make it impossible to provide tables covering the United Kingdom as a whole in this section. These differences concern the classification of offences, the meaning of certain terms used in the statistics, the effects of the several Criminal Justice Acts and recording practices.

Recorded crime statistics *(Table 11.1)*

Notifiable Offences recorded by the police provide a measure of the amount of crime committed and include all indictable and triable-either-way offences together with a few summary offences which are closely linked to these offences. The statistics are based on counting rules, revised with effect from 1 April 1998, which are standard for all the police forces in England and Wales. The new rules have changed the emphasis of measurement to one crime per victim, and have also increased the coverage of offences. These changes have particularly impacted on the offence groups of violence against the person, fraud and forgery, drugs offences and other offences.

For a variety of reasons many offences are either not reported to the police or not recorded by them. The changes in the number of offences recorded do not necessarily provide an accurate reflection of changes in the amount of crime committed.

For further information please see the Home Office *Statistical Bulletin 18/99 on Recorded Crime Statistics.*

Court proceedings and police cautions *(Tables 11.3 - 11.8, 11.13 - 11.17, 11.20 - 11.22)*

The statistical basis of the tables of court proceedings is broadly similar in England and Wales, Scotland and Northern Ireland; the tables show the number of persons found guilty, recording a person under the heading of the principal offence of which he is found guilty, excluding additional findings of guilt at the same proceedings. A person found guilty at a number of separate court proceedings is included more than once.

The statistics on offenders cautioned, cover only those who, on admission of guilt, were given a formal caution by, or on the instructions of, a senior police officer as an alternative to prosecution (excluding Scotland). Written warnings by the police for motor offences and persons paying fixed penalties for certain motoring offences are excluded. There are no statistics on cautioning available for Northern Ireland.

In Scotland there are three criminal courts, the High Court of Justiciary, the Sheriff Court and the District Court. The High Court deals with serious solemn (ie Jury) cases and has unlimited sentencing power. The Sheriff Court is limited to imprisonment of 3 years for solemn cases, or 3 months (6 months when specified in legislation for second or subsequent offences and 12 months for certain statutory offences) for summary cases. The District Court deals only with summary cases and is limited to 60 days imprisonment and level 4 fines. Stipendiary Magistrates sit in Glasgow District Court and have the summary sentencing powers of a Sheriff.

Indictable offences are offences which are (a) triable on indictment only (these are the most serious crimes such as murder and rape) and are tried at the Crown Court in England and Wales and the High Court in Scotland; (b) triable either way offences which may be tried at the Crown Court or Magistrates Court in England and Wales or the High Court, Sheriff Court or District Court in Scotland; summary offences are those for which a defendant in England and Wales or accused in Scotland would normally be dealt with at the Magistrates Courts and the Sheriff or District Court in Scotland.

On 1 October 1988, Section 123 of the Criminal Justice Act 1988 replaced youth custody and detention centre orders by detention in a young offender institution. The Criminal Justice Act 1988 on 20 October 1988 also reclassified certain triable either way offences as summary offences (England and Wales only).

The Criminal Justice Act 1991 led to the following main changes in the sentences available to the courts: (a) introduction of combination orders, (b) introduction of the "unit fine scheme" at Magistrates' courts, (c) abolishing the sentence of detentions in a young offender institution for 14 year old boys and changing the minimum and maximum sentence lengths for 15 to 17 year olds to 10 and 12 months respectively, and (d) abolishing partly suspended sentences of imprisonment and restricting the use of a fully suspended sentence.

The Criminal Justice Act 1993 abolished the "Unit Fine Scheme" in Magistrates' courts which had been introduced under the Criminal Justice Act 1991.

A *charging standard for assault* was introduced on 31 August 1994 with the aim to promote consistency between the police and prosecution on the appropriate level of charge to be brought.

The Criminal Justice and Public Order Act 1994 created several new offences, mainly in the area of Public Order, but also including male rape. The Act also (a) extended the provisions of section 53 of the Children and Young Persons Act 1993 for 10 to 13 year olds, (b) increased the maximum sentence length for 15 to 17 year olds to 2 years, (c) increased the upper limit from £2 000 to £5 000 for offences of criminal damage proceeded against as if triable only summarily, (d) introduced provisions for the reduction of sentences for early guilty pleas, and (e) increased the maximum sentence length for certain firearm offences.

Provisions within the Crime (Sentences) Act 1997 included: a) an automatic life sentence for a second serious violent or sexual offence unless there are exceptional circumstances; b) a minimum sentence of seven years for an offender convicted for a third time of a class A drug trafficking offence unless the court considers this to be unjust in all the circumstances, and c) the new section 38A of the Magistrates' Courts' Act 1980 extending the circumstances in which a magistrates' court may commit a person convicted of an offence triable either way to the Crown Court for sentence - it was implemented in conjunction with section 49 of the Criminal Procedure and Investigations Act 1996, which involves the magistrates' courts in asking defendants to indicate plea before the mode of trial decision is taken and compels the court to sentence or commit for sentence any defendant who indicates a guilty plea.

The Crime and Disorder Act 1998 created provisions in relation to reprimands and final warnings, new offences and orders which have been implemented nationally or piloted in certain areas

Section 45 of the Criminal Justice (Scotland) Act 1980 was implemented on 15 November 1983; this introduced unified custodial sentencing for under 21 offenders and abolished borstal training.

The system of Magistrates' courts and Crown Courts in Northern Ireland operates in a similar way to that in England and Wales.

Previous convictions of prisoners *(Table 11.8)*

Information on previous convictions of prisoners published prior to 1995 was based upon Prison Service records. However, details of a prisoners previous conviction were often not recorded (eg this information was missing for 44 per cent of the 1990 male receptions under sentence). To overcome this problem the Home Office Offenders Index (a computerised database containing details of convictions for standard list offences) is now being used to provide information on prisoners previous convictions. Unfortunately, this means that the most up-to-date information on previous convictions is not directly comparable with that previously published. Standard list offences include indictable offences and some of the more serious summary offences so the coverage is not as complete. The published information also does not relate to 'prison receptions' but to those sentenced to immediate custody for standard list offences (which accounted for 82 per cent of those sentenced to custody in 1993).

The problems with non-availability of previous history information are much less acute using the Offenders Index data source. Previous convictions were found for 95 per cent of the 1993 prison population sample. Some of the cases where information is missing, would be accounted for by prisoners who are not sentenced for a standard list offence and have no previous record for such offences.

11.1 Recorded crime statistics
England and Wales

Thousands

		Old counting rules												New counting rules
		1988	1989	1990	1991	1992	1993	1994	1995	1996	1997	1998[2] /99		1998[2] /99
Violence against the person	BEAB	158.2	177.0	184.7	190.3	201.8	205.1	218.4	212.6	239.3	250.8	230.8	LQMP	502.8
Sexual offences	BEAC	26.5	29.7	29.0	29.4	29.5	31.3	32.0	30.3	31.4	33.2	34.9	LQMQ	36.2
Burglary	BEAD	817.8	825.9	1 006.8	1 219.5	1 355.3	1 369.6	1 256.7	1 239.5	1 164.6	1 015.1	951.9	LQMR	953.2
Robbery	BEAE	31.4	33.2	36.2	45.3	52.9	57.8	60.0	68.1	74.0	63.1	66.2	LQMS	66.8
Theft and handling stolen goods	BEAF	1 931.3	2 012.8	2 374.4	2 761.1	2 851.6	2 751.9	2 564.6	2 452.1	2 383.9	2 165.0	2 126.7	LQMT	2 191.4
Fraud and forgery	BEAG	133.9	134.5	147.9	174.7	168.6	162.8	145.3	133.0	136.2	134.4	173.7	LQMU	279.5
Criminal damage	BEAH	593.9	630.1	733.4	821.1	892.6	906.7	928.3	914.0	951.3	877.0	834.4	LQMV	879.6
Drug offences[1]	LQMO	..	..	..	..	..	..	..	..	..	..	21.3	LQYT	135.9
Other offences[1]	BEAI	22.7	27.6	31.1	34.6	39.4	41.0	47.7	50.7	55.8	59.8	42.0	LQYU	63.6
Total	BEAA	3 715.8	3 870.7	4 543.6	5 276.2	5 591.7	5 526.3	5 253.0	5 100.2	5 036.6	4 598.3	4 481.8	LQYV	5 109.1

1 Prior to 1/4/98 th offence of drug trafficking was included in the "Other offences" group. From 1/4/99, under the new counting rules, drug trafficking became part of a new "Drug offences" group which, with the expanded coverage, now includes possession and other drug offences. For 1998/99, under the old counting rules, drug trafficking - the only drugs offence counted - has been listed under drugs offences.

2 Provisional figures.

Source: Home Office: 020 7273 2583

11.2 Police forces: strength

End of year

Number

		1988	1989	1990	1991	1992	1993	1994	1995[1]	1996[1]	1997[1]	1998[1]
England and Wales												
Regular police												
Strength:												
Men	KERB	109 900	110 466	110 790	110 396	111 027	108 967	108 030	107 022	106 549	105 691	104 606
Women	KERC	13 007	13 695	14 352	14 898	15 841	16 571	17 263	17 688	18 501	19 124	19 659
Seconded:[2]												
Men	KERD	1 730	1 815	1 787	1 670	1 766	1 938	1 881	1 896	1 864	1 814	2 007
Women	KERE	122	134	161	163	167	182	184	202	209	233	232
Additional officers:[3]												
Men	KERF	69	75	85	73	68	63	97	105	111	200	267
Women	KERG	–	1	5	3	3	12	17	22	57	158	514
Special constables												
Enrolled strength:												
Men	KERH	10 578	10 390	10 483	11 592	12 251	13 240	12 772	12 751	12 594	12 483	11 331
Women	KERI	5 210	5 199	5 419	6 480	6 992	7 326	7 060	6 904	6 857	6 680	5 965
Scotland												
Regular police												
Strength[4]:												
Men	KERK	12 498	12 656	12 583	12 566	12 629	12 580	12 634	12 341†	12 347	12 495	12 479
Women	KERL	1 020	1 158	1 258	1 357	1 465	1 559	1 679	1 665†	1 856	2 019	2 202
Central service:[4,5]												
Men	KERM	65	69	67	64	70	73	79	96†	94	85	85
Women	KERN	3	4	6	6	5	5	4	7†	8	4	6
Seconded:[4,6]												
Men	KERO	88	92	98	103	120	109	110	108†	105	101	101
Women	KERP	7	8	8	6	10	16	16	16	16†	13	10
Additional regular police:												
Strength	KERR	64	64	64	93	90	58	91	90†	72	72	97
Special constables												
Strength:												
Men	KERS	1 514	1 521	1 475	1 436	1 454	1 466	1 518	1 411	1 336	1 286	1 289
Women	KERT	233	279	312	342	393	431	474	467	450	437	444
Northern Ireland												
Royal Ulster Constabulary												
Strength:												
Men	KERU	7 568	7 571	7 535	7 510	7 687	7 646	7 640	7 528	7 531	7 562	7 527
Women	KERV	659	688	708	707	791	818	853	887	897	923	933
Royal Ulster Constabulary Reserve												
Strength:												
Men	KERW	4 268	4 200	4 097	4 069	4 060	4 027	4 052	3 976	3 727	3 587	3 469
Women	KERX	386	425	449	491	533	545	638	709	675	719	705

1 Figures for England and Wales are as at 30 September. Figures for Scotland are as at 31 March.
2 Regional Crime Squads, other inter-force units and officers in central service.
3 Police Officers on loan to organisations outside the British Police Service such as the Royal Hong Kong police.
4 'Strength' includes central service and seconded police.
5 Instructors at Training Establishments, etc formerly shown as secondments.
6 Scottish Crime Squad, officers on courses, etc.

Sources: Home Office: 020 7273 2583; The Scottish Executive Justice Department: 0131 244 2225; Royal Ulster Constabulary: 028 9865 0222 ext 24135

11.3 Offenders found guilty: by offence group Magistrates' courts and the Crown Court

England and Wales

Thousands

		1988	1989	1990	1991	1992	1993	1994	1995	1996	1997	1998
All ages[1]												
Indictable offences												
Violence against the person:	KJEJ	53.5	55.6	52.5	47.2	43.6	38.9	37.6	29.1	30.0	34.6	35.7
Murder	KESB	0.2	0.2	0.2	0.2	0.2	0.2	0.2	0.2	0.3	0.3	0.3
Manslaughter	KESC	0.3	0.3	0.2	0.2	0.3	0.2	0.2	0.2	0.3	0.3	0.3
Wounding	KESD	50.9	53.7	50.9	45.5	42.0	37.4	36.1	27.4	28.3	32.7	35.2
Other offences of violence against the person	KESE	2.1	1.4	1.2	1.2	1.1	1.0	1.0	1.2	1.2	1.3	1.3
Sexual offences	KESF	7.2	7.3	6.6	5.5	5.0	4.3	4.5	4.7	4.4	4.5	4.6
Burglary	KESG	48.4	43.4	43.6	46.1	44.3	40.3	38.0	35.3	32.2	31.7	30.8
Robbery	KESH	4.3	4.6	4.8	4.8	5.1	5.1	4.9	5.2	5.9	5.6	5.5
Theft and handling stolen goods	KESI	163.4	134.5	134.3	133.5	127.9	121.6	121.6	116.1	114.5	118.4	125.7
Fraud and forgery	KESJ	22.7	22.3	21.9	21.2	20.0	17.5	18.4	17.2	16.3	17.0	19.8
Criminal damage	KESK	11.8	9.4	11.2	10.2	9.8	9.4	10.0	9.6	9.8	10.5	10.9
Drugs	KBWX	18.8	22.6	24.6	23.5	22.7	21.9	27.8	31.6	34.1	40.7	48.8
Other offences (excluding motoring)	KESL	24.9	28.5	32.3	34.4	36.0	37.8	39.4	42.2	43.5	47.6	49.6
Motoring offences	KESM	31.3	11.3	11.1	11.3	10.7	10.8	12.0	11.2	9.9	9.5	9.0
Total	KESA	386.2	339.5	342.8	337.6	324.9	307.6	314.1	302.2	300.6	320.1	341.7
Summary offences[2]												
Assaults[3]	KESO	11.5	15.1	17.3	16.5	18.0	19.0	21.9	29.3	30.0	32.0	35.3
Betting and gaming	KESP	–	0.1	–	–	–	–	–	–	–	–	–
Offences with pedal cycles	KBWY	3.1	2.8	2.8	1.9	1.4	1.2	1.0	1.1	1.3	1.5	2.1
Other Highways Acts offences	KBWZ	9.5	8.6	7.9	5.9	4.5	3.6	3.4	2.6	2.8	3.2	3.1
Breach of local or other regulations	KESQ	11.0	10.1	12.6	8.5	8.2	10.5	9.4	6.7	5.9	6.4	5.8
Intoxicating Liquor Laws:												
Drunkenness	KESR	45.3	42.9	37.8	29.4	23.8	18.8	20.2	19.8	24.2	28.8	30.8
Other offences	KESS	3.8	2.7	2.2	1.5	1.2	0.8	0.7	0.7	0.5	0.6	0.6
Education Acts	KEST	2.8	2.8	3.0	2.8	2.0	2.3	2.8	3.1	3.5	3.7	5.0
Game Laws	KESU	1.3	0.1	0.1	1.0	0.8	0.6	0.6	0.4	0.4	0.3	0.4
Labour Laws	KESV	0.1	1.2	1.1	0.1	0.1	0.1	0.1	0.1	–	0.1	0.1
Summary offences of criminal damage and malicious damage	KESW	35.2	36.5	33.9	28.5	24.6	21.6	22.7	22.6	23.4	24.7	26.5
Offences by prostitutes	KESX	9.4	11.1	11.5	10.9	9.8	8.2	7.7	6.8	6.6	6.6	6.0
Railway offences	KESY	9.2	9.9	8.3	5.4	4.7	4.0	5.6	6.2	9.1	11.4	12.6
Revenue Laws[2]	KESZ	113.4	108.5	104.9	115.6	121.6	123.0	126.2	123.8	139.1	143.5	174.7
Vagrancy Acts	KETB	1.0	1.8	1.9	2.1	1.8	1.7	1.9	1.6	2.0	2.0	2.2
Wireless Telegraphy Acts[2]	KETC	124.3	123.7	126.4	138.7	170.3	168.7	162.9	113.8	164.9	77.0	76.6
Other summary offences	KETD	74.5	94.4	95.4	85.6	78.9	69.0	67.8	71.5	74.7	74.7	80.9
Motoring offences (summary)[2]	KETA	713.9	722.7	704.6	713.1	723.1	664.7	638.7	642.4	649.0	649.3	665.2
Total	KESN	1 169.1	1 195.0	1 171.8	1 167.5	1 194.8	1 117.7	1 093.5	1 052.4	1 137.4	1 065.8	1 128.0
Persons aged under 18[2,4]												
Indictable offences												
Violence against the person:	KETF	8.0	7.5	6.5	5.6	5.2	5.1	5.8	4.7	5.3	5.9	5.9
Murder	KBXA	–	–	–	–	–	–	–	–	–	–	–
Manslaughter	KBXB	–	–	–	–	–	–	–	–	–	–	–
Wounding	KBXC	7.9	7.4	6.4	5.5	5.2	5.1	5.7	4.7	5.3	5.8	5.9
Other offences of violence against the person	KCAA	0.1	0.1	0.1	0.1	–	–	–	–	–	0.1	0.1
Sexual offences	KETG	0.5	0.5	0.5	0.4	0.4	0.4	0.4	0.4	0.4	0.5	0.5
Burglary	KETH	14.2	12.0	11.2	10.4	9.5	8.7	8.9	9.1	8.6	8.6	8.5
Robbery	KETI	1.3	1.4	1.4	1.4	1.4	1.5	1.7	2.0	2.4	2.3	2.2
Theft and handling stolen goods	KETJ	30.4	20.0	19.1	17.0	15.2	14.1	14.4	18.2	19.0	19.6	21.9
Fraud and forgery	KETK	1.1	1.0	0.9	0.8	0.6	0.4	0.5	0.6	0.7	0.8	1.0
Criminal damage	KETL	2.5	2.0	2.1	1.8	1.6	1.7	2.0	2.1	2.2	2.3	2.3
Drugs	KCAB	0.6	1.0	1.2	1.2	1.0	0.8	1.1	1.3	1.6	1.8	2.7
Other offences (excluding motoring)	KETM	2.2	2.3	2.5	2.8	2.7	2.4	2.8	3.3	3.8	4.2	4.2
Motoring	KETN	2.3	0.6	0.6	0.7	0.5	0.3	0.3	0.4	0.4	0.4	0.4
Total	KETE	63.2	48.2	46.1	42.0	38.1	35.4	37.9	42.2	44.4	46.4	49.7
Summary offences[2]												
Offences with pedal cycles	KETP	0.6	0.5	0.5	0.3	0.2	0.1	0.1	0.2	0.2	0.2	0.3
Breach of local or other regulations	KETR	0.6	0.5	0.4	0.3	0.2	0.1	0.1	0.2	0.3	0.2	0.2
Summary offences of criminal damage and malicious damage	KETS	5.2	4.7	3.9	2.9	2.5	2.2	2.9	3.4	3.9	4.4	5.2
Railway offences	KETT	0.5	0.5	0.4	0.3	0.2	0.2	0.3	0.4	0.4	0.5	0.5
Other summary offences	KETU	12.5	18.3	16.4	14.1	11.3	8.3	9.7	7.2	8.8	10.1	12.1
Motoring offences (summary)[2]	KCAC	18.3	17.9	16.4	13.7	10.8	8.9	8.6	9.3	10.8	10.8	11.3
Total	KETO	37.8	42.4	38.1	31.6	25.2	19.9	21.7	25.6	30.3	22.0	36.8

1 Includes 'Companies', etc.

2 It is estimated that in 1995 there was a shortfall of 75 100 offenders found guilty for certain summary offences.

3 A new charging standard was introduced for assault in 1994 (see Introduction - Law enforcement).

4 Figures for persons aged under 18 are included in the totals above.

Source: Home Office: 020 8760 8271

11.4 Offenders cautioned: by offence group
England and Wales

Thousands

		1988	1989	1990	1991	1992	1993	1994	1995	1996	1997	1998
All ages[1]												
Indictable offences												
Violence against the person	KELB	12.7	14.7	16.8	19.4	23.5	24.1	23.6	20.4	21.8	23.6	23.5
Murder	KCAD	–	–	–	–	–	–	–	–	–	–	–
Manslaughter	KCAE	–	–	–	–	–	–	–	–	–	–	–
Wounding	KCAF	12.2	14.5	16.5	19.1	23.2	23.8	23.2	20.1	21.4	23.3	22.9
Other offences of violence against the person	KCAG	0.4	0.2	0.2	0.2	0.3	0.3	0.3	0.3	0.4	0.4	0.6
Sexual offences	KELC	3.6	3.5	3.4	3.3	3.4	3.3	3.0	2.3	2.0	1.9	1.7
Burglary	KELD	12.2	12.0	14.3	13.3	14.4	12.8	11.5	10.5	10.2	9.4	8.4
Robbery	KELE	0.2	0.4	0.6	0.6	0.6	0.7	0.6	0.6	0.6	0.7	0.6
Theft and handling stolen goods	KELF	92.6	81.9	99.8	108.5	130.3	117.2	104.8	104.9	93.6	82.8	83.6
Fraud and forgery	KELG	4.0	4.0	4.7	5.6	7.5	8.1	7.6	7.9	7.5	7.2	7.4
Criminal damage	KELH	4.3	3.7	4.2	3.8	4.0	4.1	4.3	3.8	3.1	2.8	2.7
Drugs	KCAI	9.1	13.1	18.7	21.2	27.6	35.1	44.4	48.2	47.5	56.0	58.7
Other offences (excluding motoring)	KELI	1.9	2.9	3.9	4.1	4.8	4.2	4.0	4.0	4.4	5.0	5.0
Motoring[2]	KCAJ	..	..	..	..	..	..	..	..	..	..	..
Total	KELA	140.7	136.0	166.3	179.9	216.2	209.6	209.8	202.6	190.8	189.4	191.7
Summary offences												
Assaults[3]	KELK	0.9	1.1	1.4	1.6	2.4	3.1	4.2	8.1	9.1	9.1	–
Betting and gaming	KELL	0.1	–	–	–	–	–	–	–	–	–	–
Offences with pedal cycles	KCAK	3.3	2.9	2.6	1.7	1.4	0.9	0.8	0.8	0.9	0.9	0.8
Other Highways Acts offences	KCAL	1.8	1.6	1.5	1.2	1.2	1.0	0.9	0.9	0.8	0.8	0.8
Breach of local or other regulations	KELM	1.5	1.2	1.2	1.3	1.5	1.1	1.1	0.9	0.8	0.9	0.9
Intoxicating Liquor Laws:												
Drunkenness	KELN	48.6	49.9	48.6	46.0	45.0	41.1	37.7	22.9	25.9	25.7	22.8
Other offences	KELO	4.0	3.6	2.8	1.7	1.7	1.2	1.0	1.0	0.9	0.9	0.7
Education Acts	KELP	–	–	–	–	–	–	–	–	–	–	–
Game Laws	KELQ	0.2	0.1	0.2	0.2	0.1	0.2	0.1	0.1	0.1	0.1	0.1
Labour Laws	KELR	0.1	–	–	–	–	–	–	–	–	–	–
Summary offences of criminal damage and malicious damage	KELS	13.5	14.3	16.2	17.3	20.5	22.1	23.1	25.1	27.7	27.6	28.3
Offences by prostitutes	KELT	5.2	5.3	4.5	4.1	4.2	4.0	3.6	3.3	3.5	3.5	3.5
Railway offences	KELU	0.4	0.6	0.4	0.3	0.3	0.2	0.2	0.3	0.2	0.1	–
Revenue Laws	KELV	0.7	0.6	0.5	0.7	0.8	0.5	0.2	0.2	0.1	0.1	0.1
Vagrancy Acts	KELX	0.6	0.4	0.5	0.7	1.3	1.5	1.0	1.0	0.6	0.6	1.2
Wireless Telegraphy Acts	KELY	–	–	–	–	–	–	–	–	–	–	–
Other summary offences	KELZ	13.9	20.3	22.5	22.1	25.0	24.9	24.6	24.2	24.7	22.3	37.0
Motoring offences[2]	KCAM	..	..	..	..	..	..	..	..	..	..	..
Total	KELJ	94.7	102.1	102.8	98.9	105.1	101.8	98.7	88.7	95.4	92.7	96.2
Persons aged 18 and under												
Indictable offences												
Violence against the person	KEMB	6.9	7.8	8.5	9.2	10.8	10.8	11.1	9.4	9.4	9.6	9.5
Murder	KCAN	–	–	–	–	–	–	–	–	–	–	–
Manslaughter	KCAO	–	–	–	–	–	–	–	–	–	–	–
Wounding	KCAP	6.6	7.8	8.5	9.2	10.7	10.8	11.1	9.4	9.4	9.6	9.4
Other offences of violence against the person	KCCE	0.2	–	–	–	–	–	–	–	–	–	–
Sexual offences	KEMC	1.9	1.8	1.6	1.4	1.3	1.1	1.1	0.8	0.7	0.7	0.6
Burglary	KEMD	11.3	11.1	13.1	11.7	12.0	10.3	9.6	8.5	8.2	7.5	6.7
Robbery	KEME	0.2	0.3	0.5	0.6	0.6	0.6	0.6	0.5	0.6	0.6	0.5
Theft and handling stolen goods	KEMF	62.4	51.6	63.1	62.3	69.8	58.9	58.6	57.4	48.2	40.9	44.0
Fraud and forgery	KEMG	1.6	1.4	1.5	1.5	1.7	1.5	1.4	1.6	1.5	1.4	1.6
Criminal damage	KEMH	3.6	3.1	3.2	2.6	2.6	2.6	2.8	2.4	2.0	1.8	1.7
Drugs	KCCF	1.3	2.3	4.0	5.0	5.4	6.7	8.5	8.7	7.9	9.7	11.0
Other offences (excluding motoring)	KEMI	0.8	0.9	1.0	1.2	1.4	1.4	1.4	1.3	1.3	1.5	1.5
Motoring[2]	KCCG	..	..	..	..	..	..	..	..	..	..	..
Total	KEMA	89.9	80.4	96.6	95.5	105.6	94.1	95.1	90.6	79.9	73.7	77.2
Summary offences												
Offences with pedal cycles	KEMK	1.8	1.6	1.5	0.9	0.7	0.4	0.5	0.4	0.5	0.5	0.4
Breach of local or other regulations	KEMM	0.7	0.5	0.5	0.5	0.6	0.3	0.4	0.3	0.3	0.3	0.3
Summary offences of criminal damage and malicious damage	KEMN	9.3	9.5	10.4	10.1	10.7	11.4	12.5	12.8	13.8	13.5	14.2
Railway offences	KEMO	0.2	0.2	0.2	0.1	0.1	0.1	0.1	0.1	0.1	0.1	–
Other summary offences	KEMP	10.8	16.6	18.0	15.5	16.0	15.3	15.6	10.3	10.8	9.1	13.8
Motoring offences[2]	KCCH	..	..	..	..	..	..	..	..	..	..	..
Total	KEMJ	22.8	28.4	30.5	27.1	28.1	27.5	29.2	30.0	33.2	30.8	32.5

1 Includes 'Companies', etc.
2 Not applicable as motoring offences may attract written warning.
3 A new charging standard was introduced for assault in 1994 (see Introduction - Law enforcement).

Source: Home Office: 020 8760 8271

11.5 Offenders found guilty of offences: by age and sex
England and Wales

Magistrates' courts and the Crown Court

Thousands

		1988	1989	1990	1991	1992	1993	1994	1995	1996	1997	1998
Males												
Indictable offences												
All ages	KEFA	337.5	293.7	295.7	293.5	282.8	268.2	273.2	263.2	261.1	276.5	292.9
10 and under 14 years	KEFB	3.7	2.8	2.6	2.3	2.3	2.3	2.9	2.9	2.6	2.9	3.5
14 and under 18 years	KEFC	53.6	40.4	38.4	35.1	31.6	29.4	32.7	34.3	36.6	37.9	39.8
18 and under 21 years	KEFD	73.1	63.7	65.8	65.3	58.9	53.0	50.3	47.4	46.3	48.4	51.8
21 years and over	KEFE	207.1	186.8	188.8	190.8	190.1	183.4	187.4	178.6	175.6	187.3	197.9
Summary offences[1]												
All ages	KEFF	974.5	996.2	968.9	956.3	959.6	892.0	871.0	862.0	903.6	880.9	929.0
10 and under 14 years	KEFG	0.6	0.9	0.9	0.8	0.7	0.7	0.9	1.1	1.0	1.2	1.6
14 and under 18 years	KEFH	34.8	39.1	34.8	28.8	22.7	17.9	19.1	22.3	26.4	27.8	30.8
18 and under 21 years	KEFI	120.0	128.9	125.1	109.8	97.6	85.7	82.9	84.0	88.4	91.0	96.3
21 years and over	KEFJ	819.1	827.3	808.0	816.9	838.5	787.7	768.2	754.6	787.9	761.0	800.3
Females												
Indictable offences												
All ages	KEFK	46.1	43.0	44.0	41.9	40.0	37.8	39.5	37.5	38.0	42.2	47.3
10 and under 14 years	KEFL	0.3	0.2	0.2	0.2	0.2	0.2	0.3	0.3	0.3	0.3	0.5
14 and under 18 years	KEFM	5.6	4.8	4.9	4.4	4.0	3.6	4.4	4.7	4.9	5.3	5.9
18 and under 21 years	KEFN	8.8	8.2	8.3	8.1	7.3	6.3	6.2	5.7	5.7	6.3	7.1
21 years and over	KEFO	31.4	29.8	30.6	29.2	28.5	27.7	28.6	26.8	27.2	30.4	33.7
Summary offences[1]												
All ages	KEFP	176.4	179.2	185.3	194.1	219.0	213.3	211.5	180.5	222.9	174.9	188.3
10 and under 14 years	KEFQ	–	–	–	–	–	–	0.1	0.1	0.1	0.1	0.2
14 and under 18 years	KEFR	2.3	2.4	2.4	2.0	1.7	1.3	1.7	2.1	2.9	3.7	4.2
18 and under 21 years	KEFS	11.1	12.4	12.6	12.1	11.2	10.0	9.6	10.4	12.1	11.1	12.1
21 years and over	KEFT	163.0	164.3	170.4	179.8	206.0	201.9	200.2	167.9	207.9	160.0	171.7
Companies, etc												
Indictable offences	KEFU	2.7	2.9	3.1	2.2	2.1	1.7	1.4	1.5	1.5	1.3	1.5
Summary offences[1]	KEFV	18.2	19.6	17.6	17.2	16.2	12.5	10.9	9.9	10.9	10.0	10.7

1 It is estimated that in 1995 there was a shortfall of 75 100 offenders found guilty for certain summary offences.

Source: Home Office: 020 8760 8271

11.6 Persons cautioned by the police: by age and sex
England and Wales

Thousands

		1988	1989	1990	1991	1992	1993	1994	1995	1996	1997	1998
Males												
Indictable offences												
All ages	KEGA	107.0	102.8	124.2	131.4	155.0	153.6	153.6	149.3	142.6	143.3	142.9
10 and under 14 years	KEGB	22.9	20.3	22.5	21.0	22.5	19.9	19.9	18.1	15.4	14.1	15.2
14 and under 18 years	KEGC	48.6	43.3	52.1	50.5	53.7	48.6	47.9	46.4	42.7	40.8	40.5
18 and under 21 years	KEGD	9.5	10.9	15.1	18.4	23.8	24.5	25.0	24.8	24.3	25.2	25.7
21 years and over	KEGE	26.1	28.3	34.6	41.4	55.1	60.6	60.7	60.0	60.2	63.2	61.5
Summary offences												
All ages	KEGF	80.6	86.9	88.2	85.3	90.0	86.3	83.6	73.8	79.2	75.7	76.9
10 and under 14 years	KEGG	4.0	5.3	5.5	5.2	5.6	5.7	6.3	5.9	6.1	6.0	6.6
14 and under 18 years	KEGH	16.0	19.8	21.1	18.5	18.8	18.1	19.0	19.6	22.2	20.0	20.1
18 and under 21 years	KEGI	8.9	9.1	9.4	9.7	11.3	11.4	11.3	11.1	13.0	12.9	13.2
21 years and over	KEGJ	51.8	52.7	52.1	51.9	54.3	51.0	47.0	37.1	37.9	36.9	37.0
Females												
Indictable offences												
All ages	KEGK	33.6	33.2	42.1	48.5	61.1	55.9	56.2	53.3	48.2	46.0	48.8
10 and under 14 years	KEGL	5.3	4.5	5.9	6.3	8.3	7.5	9.1	8.2	6.6	5.4	6.7
14 and under 18 years	KEGM	13.2	12.2	16.2	17.8	21.2	18.0	18.6	17.9	15.1	13.4	14.8
18 and under 21 years	KEGN	2.6	3.0	4.2	5.5	7.3	6.7	6.1	6.0	5.6	5.7	5.9
21 years and over	KEGO	12.5	13.5	15.8	19.0	24.4	23.6	22.4	21.1	20.9	21.5	21.4
Summary offences												
All ages	KEGP	14.1	15.2	14.6	13.6	15.1	18.5	15.1	14.8	16.2	17.0	19.2
10 and under 14 years	KEGQ	0.4	0.4	0.5	0.5	0.6	0.6	0.7	0.8	0.8	0.8	1.1
14 and under 18 years	KEGR	2.5	2.9	3.4	2.9	3.0	6.0	3.2	3.6	4.1	4.0	4.8
18 and under 21 years	KEGS	2.2	2.4	2.1	1.9	2.0	2.1	1.9	1.9	2.1	2.3	2.6
21 years and over	KEGT	9.1	9.5	8.7	8.3	9.5	9.8	9.2	8.6	9.1	9.9	10.8
Companies, etc												
Indictable offences	KEGU	–	–	–	–	–	–	–	–	–	–	–
Summary offences	KEGV	–	–	–	–	–	–	–	–	–	–	–

Source: Home Office: 020 8760 8271

11.7 Sentence or order passed on offenders sentenced for indictable offences: by sex

England and Wales

Magistrates' courts and the Crown Court

Percentages

		1988	1989	1990	1991	1992	1993	1994	1995	1996	1997	1998
Males												
Sentence or order												
Absolute discharge	KEJB	0.6	0.7	0.7	0.7	0.8	0.9	0.8	0.8	0.8	0.7	0.7
Conditional discharge	KEJC	11.1	12.4	14.2	16.1	18.0	18.6	17.4	16.2	15.6	15.5	15.3
Probation order	KEJD	8.3	8.5	9.1	9.3	9.1	9.3	10.2	10.0	9.9	10.0	10.0
Supervision order	KEJE	1.8	1.6	1.5	1.4	1.4	2.0	2.4	2.7	2.9	2.7	2.7
Fine	KEJF	39.4	40.4	40.1	36.0	34.2	33.8	31.8	30.0	28.6	28.2	28.4
Community service order	KEJG	8.5	7.6	8.5	9.5	10.3	11.4	11.1	10.6	9.9	9.5	9.3
Attendance centre order	KEJH	2.2	1.8	1.9	1.9	1.8	1.9	2.0	2.0	1.9	1.8	1.7
Combination orders	KIJW	..	..	..	..	0.3	2.1	2.7	3.0	3.5	3.7	3.8
Curfew order	LUJP	..	..	..	..	..	..	..	..	0.1	0.1	0.2
Care order	KEJJ	0.1	0.1	0.1	–	..	..	..	..	..	..	..
Sec 53	LUJQ	0.1	–	–	–	–	0.1	0.1	0.1	0.2	0.3	0.2
Young offender institution[1]	KEJK	6.7	5.4	4.5	4.6	4.3	0.8	4.9	5.6	6.1	6.1	6.0
Secure training order	LUJR	..	..	..	..	..	..	..	..	..	..	–
Imprisonment												
Fully suspended	KEJL	7.3	6.4	6.3	6.4	5.4	0.9	0.7	0.7	0.8	0.8	0.7
Partly suspended[2]	KAFN	0.7	0.6	0.4	0.3	0.2	..	..	..	..	..	..
Unsuspended	KEJM	11.7	11.6	10.7†	11.4	11.6	12.0	13.6	16.0	17.2	17.9	18.2
Other sentence or order	KEJN	1.5	2.0	2.1†	2.2	2.5	2.5	2.3	2.2	2.4	3.0	2.6
Total number of offenders (thousands) = 100 per cent	KEJA	337.5	293.2	294.7	291.9	282.5	267.5	272.6	262.9	260.8	275.4	292.4
Females												
Sentence or order												
Absolute discharge	KEKB	0.8	0.9	1.0	0.9	1.0	0.9	0.9	0.8	0.9	0.8	0.7
Conditional discharge	KEKC	28.1	30.2	33.2	35.9	37.5	35.4	34.4	32.4	30.6	29.4	28.7
Probation order	KEKD	18.1	17.4	17.5	17.0	15.9	15.4	17.5	18.0	19.0	19.1	19.1
Supervision order	KEKE	1.2	1.1	1.1	1.1	1.3	1.8	2.4	2.7	2.9	2.9	3.1
Fine	KEKF	34.4	33.2	30.8	27.2	26.2	29.2	25.6	24.1	22.5	21.8	21.3
Community service order	KEKG	3.8	3.3	4.1	4.6	5.3	6.1	6.4	6.6	6.5	6.5	6.5
Attendance centre order	KEKH	0.4	0.3	0.3	0.4	0.4	0.5	0.8	1.0	1.0	1.0	0.9
Combination orders	KIJX	..	..	..	..	0.2	1.4	2.1	2.4	3.0	3.2	3.4
Curfew order	LUJT	..	..	..	..	..	..	..	..	–	0.1	0.1
Care order	KEKJ	0.1	0.1	0.1	–	..	..	..	..	..	..	..
Sec 53	LUJU	–	–	–	–	–	–	–	0.1	0.1	0.1	–
Young offender institution[1]	KEKK	1.3	1.1	0.8	0.9	0.8	1.1	1.1	1.5	1.8	1.9	2.2
Secure training order	LUJV	..	..	..	..	..	..	..	..	..	..	–
Imprisonment												
Fully suspended	KEKL	5.7	6.0	5.3	5.9	4.8	1.2	1.1	1.4	1.5	1.6	1.5
Partly suspended[2]	KAFP	0.6	0.5	0.3	0.3	0.2	..	..	..	..	..	..
Unsuspended	KEKM	4.2	4.2	3.5	4.1†	4.4	5.0	5.9	7.4	8.4	9.4	10.0
Other sentence or order	KEKN	1.4	1.8	1.9	1.8†	2.2	1.9	1.7	1.8	2.0	2.2	2.5
Total number of offenders (thousands) = 100 per cent	KEKA	46.1	43.0	43.9	41.9	40.0	37.7	39.5	37.5	38.0	42.1	47.2

1 Includes detention centre orders and youth custody (both abolished October 1988).

2 Abolished October 1992.

Source: Home Office: 020 8760 8271

11.8 Offenders sentenced to immediate custody for standard list offences[1]

England and Wales

Between 1993 and 1997: by gender and number of previous convictions

Percentage[2]

	Number of previous convictions[3]						
	Nil	1 - 2	3 - 6	7 - 10	11 and over	Total sentenced for standard list offences	Total sentenced to custody for all offences
Year and gender							
1993							
Males	19	16	24	18	23	45 789	55 879
Females	38	25	17	7	13	2 323	2 520
Total	20	16	23	17	23	48 112	58 399
1994							
Males	19	14	24	18	25	51 874	66 036
Females	35	21	17	10	18	2 842	3 149
Total	20	14	23	18	25	54 716	69 185
1995							
Males	18	14	22	18	27	58 592	75 369
Females	41	20	14	9	15	3 429	3 768
Total	20	14	22	18	27	62 021	79 137
1996[4]							
Males	18	16	21	17	27	78 392 (65 656)	80 240
Females	36	18	18	11	17	4 273 (4 057)	4 374
Total	20	16	21	17	27	82 665 (69 713)	84 614
1997[4]							
Males	16	16	23	16	29	85 580	87 620
Females	31	18	22	12	17	5 345	5 473
Total	17	16	23	16	28	90 934	93 093

1 These consist of all the indictable offences plus some of the more serious summary offences.
2 The percentages are based on samples of 2 994, 3 304, 2 975, 6 994 and 7 562 males and 149, 163, 160, 379 and 459 females in the years 1993, 1994, 1995, 1996 and 1997 respectively. Percentages are rounded and therefore may not add to 100.
3 Counting one conviction per court appearance.
4 From 1 January 1996 a number of summary motoring offences became standard list offences. The sentencing figures shown in brackets for 1996 exclude those sentenced for these offences. Excluding the new standard list offences from the analysis would slightly alter the percentages with previous convictions.

Source: Home Office: 020 7273 3177

11.9 Population in Prison Service establishments under sentence[1]

England and Wales

On 30 June 1993, 1994, 1995, 1996 and 1997 : by gender and number of previous convictions

Percentage[2]

	Number of previous convictions[3]						
	Previous convictions not found[3]	Nil	1 - 2	3 - 6	7 - 10	11 and over	Number of prisoners
Year and gender							
1993							
Males	5	17	16	24	18	20	31 375
Females	11	38	20	17	7	6	1 125
Total	5	18	16	24	18	20	32 500
1994							
Males	5	15	16	23	18	22	33 960
Females	10	36	18	18	8	9	1 266
Total	5	16	16	23	18	22	35 226
1995							
Males	5	16	16	21	16	25	37 479
Females	15	34	18	15	9	9	1 456
Total	6	17	16	21	16	25	38 935
1996[4]							
Males	9	22	16	19	14	20	41 187
Females	17	37	17	14	7	8	1 727
Total	9	22	16	19	14	19	42 914
1997[4]							
Males	9	15	16	20	15	24	46 611
Females	15	36	16	15	8	11	2 063
Total	9	16	16	20	15	24	48 674

1 Excludes fine defaulters.
2 The percentages are based on samples of 6 677, 6 675, 9 105, 11882 and 11800 males and 1 116, 1 243, 1 414 , 1 804 and 2032 females in the years 1993, 1994, 1995, 1996 and 1997 respectively. Percentages are rounded and therefore may not add to 100.
3 In some cases it was not possible to find details on previous convictions. This can happen when a prisoner is not sentenced for a standard list offence and has no previous record for such offences.
4 See footnote 1 of Table 11.8. Excluding the new standard list offences from the analysis would only alter the results very slightly.

Source: Home Office: 020 7273 3177

11.10 Receptions and average population in custody
England and Wales

Number

		1988	1989	1990	1991	1992	1993	1994	1995	1996	1997	1998
Receptions												
Type of inmate:												
Untried	KEDA	57 876	58 789	53 135	54 676	49 869	53 565	57 079	55 287	58 888	62 066	64 697
Convicted, unsentenced	KEDB	17 280	17 800	20 410	19 927	21 250	30 098	34 563	32 039	34 987	36 424	43 387
Sentenced	KEDE	81 836	76 430	67 510	72 313	69 832	72 966	83 657	89 173	82 861	87 168	91 282
Immediate custodial sentence	KEDF	65 019	59 445	50 851	53 340	50 006	50 563	61 188	69 016	74 306	80 832	85 908
Young offenders	KEDG	21 631	17 674	14 380	15 028	13 174	13 205	14 956	16 244	17 593	18 743	19 599
Up to 18 months	KEDH	18 268	14 764	11 758	12 447	10 862	11 114	12 739	13 783	14 156	14 893	15 965
Over 18 months up to 4 years	KEDJ	2 966	2 519	2 307	2 270	1 982	1 752	1 882	2 129	2 913	3 254	3 153
Over 4 years (including life)	KEDL	397	391	315	311	330	339	335	332	524	596	481
Adults	KFBO	43 388	41 771	36 471	38 312	36 832	37 358	46 232	52 772	56 713	62 089	66 309
Up to 18 months	KEDV	31 005	29 444	25 363	27 159	25 872	27 643	35 520	40 638	42 673	46 727	50 844
Over 18 months up to 4 years	KEDW	9 584	9 270	8 253	8 199	7 967	6 864	7 744	8 811	10 119	10 972	11 126
Over 4 years (including life)	KEDX	2 799	3 057	2 855	2 954	2 993	2 851	2 968	3 323	3 921	4 390	4 339
Committed in default of payment of a fine	KEDY	16 817	16 985	16 659	18 973	19 826	22 403	22 469	20 157	8 555	6 336	5 374
Young offenders	KEEA	3 968	3 671	3 522	4 209	4 282	3 353	3 268	2 846	885	555	568
Adults	KAFQ	12 849	13 314	13 137	14 764	15 544	19 050	19 201	17 311	7 670	5 781	4 806
Non-criminal prisoners	KEDM	3 032	3 021	2 314	2 791	3 109	5 073	4 507	3 789	3 128	3 204	3 290
Immigration Act 1971	KEDN	1 333	1 448	916	1 225	1 272	1 837	1 641	1 825	1 857	2 122	2 348
Others	KEDO	1 699	1 573	1 398	1 566	1 837	3 236	2 866	1 964	1 271	1 082	942
Average population[3]												
Total in custody	KEDP	49 949	48 610	45 636	45 897	45 817	44 565	48 794	51 047	55 281	61 114	65 298
Total in prison service establishments	KFBQ	48 872	48 500	44 975	44 809	44 718	44 551	48 621	50 962	55 281	61 114	65 298
Police cells[2]	KFBN	1 077	110	661	1 088	1 098	14	173	85	–	–	–
Untried	KEDQ	8 798	8 576	7 625	7 545	7 387	7 960	9 047	8 352	8 374	8 453	8 157
Convicted, unsentenced	KEDR	1 660	1 820	1 815	1 930	1 987	2 700	3 181	2 954	3 238	3 678	4 411
Remanded for medical examination[1]	KEDS	32	39	20	17	14	12	15	9	6	8	9
Others	KEDT	1 628	1 781	1 795	1 913	1 973	2 688	3 166	2 945	3 232	3 670	4 402
Sentenced	KEDU	38 187	37 885	35 336	35 034	35 037	33 317	35 753	39 040	43 043	48 413	52 176
Immediate custodial sentence	KFBR	37 714	37 427	34 972	34 665	34 707	32 825	35 308	38 636	42 863	48 272	52 045
Young offenders	KFBS	8 260	7 160	6 173	5 754	5 382	5 054	5 258	5 752	6 700	7 820	8 490
Up to 18 months	KFBU	4 523	3 638	3 143	3 095	2 808	2 671	2 736	2 911	2 930	3 267	3 507
Over 18 months up to 4 years	KFBV	2 947	2 749	2 328	2 033	1 999	1 800	1 902	2 159	2 606	3 019	3 155
Over 4 years (including life)	KFBW	790	773	699	623	575	583	620	682	1 164	1 534	1 828
Adults	KFCO	29 454	30 267	28 799	28 911	29 326	27 771	30 050	32 884	36 162	40 451	43 556
Up to 18 months	KFCP	8 229	7 646	7 001	7 194	7 170	7 054	8 051	8 845	8 199	9 724	10 308
Over 18 months up to 4 years	KFCQ	11 004	11 191	9 751	9 333	9 416	8 445	9 164	10 184	10 320	10 777	11 707
Over 4 years (including life)	KFCR	10 222	11 430	12 047	12 384	12 740	12 272	12 835	13 855	17 644	19 950	21 541
Committed in default of payment of a fine	KFCS	473	458	364	369	330	492	446	403	180	141	131
Young offenders	KFEW	78	71	80	85	65	77	62	54	22	13	15
Adults	KFEX	395	387	284	284	265	415	384	349	158	128	116
Non-criminal prisoners	KEEB	227	220	200	300	308	574	640	615	626	571	554
Immigration Act 1971	KEEC	146	157	144	222	227	431	487	483	516	485	476
Others	KEED	81	61	56	78	81	143	153	132	111	87	78

1 Under Section 30, Magistrates' Courts Act 1980.
2 Mostly untried prisoners.
3 Components may not add to totals because they have been rounded independently.

Source: Home Office: 020 7217 5567

11.11 Prison population serving sentences: analysis by age and offence[1]

England and Wales

Number

	Total	Age in years							
		14 - 16	17 - 20	21 - 24	25 - 29	30 - 39	40 - 49	50 - 59	60 and over
At 30 June 1992									
Offences									
Males									
Total	34 389	236	4 356	7 667	7 904	8 476	3 818	1 490	442
Violence against the person	6 779	23	539	1 304	1 671	1 896	870	362	114
Sexual offences	3 055	6	115	275	472	878	769	363	177
Burglary	5 447	72	1 142	1 733	1 330	867	234	55	14
Robbery	4 133	25	616	1 083	1 112	982	258	51	6
Theft, handling, fraud and forgery	3 819	36	476	791	861	980	471	179	25
Other offences	6 404	38	655	1 343	1 438	1 805	780	285	60
Offence not known	4 752	36	813	1 138	1 020	1 068	436	195	46
Females									
Total	1 175	8	111	178	274	378	166	48	12
Violence against the person	187	1	23	23	39	57	31	10	3
Sexual offences	10	-	1	-	2	4	2	1	-
Burglary	49	2	8	11	17	8	3	-	-
Robbery	53	2	13	8	12	15	2	1	-
Theft, handling, fraud and forgery	249	1	15	43	77	73	28	12	-
Other offences	422	-	23	48	86	167	76	17	5
Offence not known	205	2	28	45	41	54	24	7	4
At 30 June 1994									
Offences									
Males									
Total	34 474	815	3 942	6 919	7 704	8 954	4 019	1 599	522
Violence against the person	7 744	99	669	1 376	1 738	2 270	1 055	397	140
Sexual offences	3 270	15	96	228	454	925	823	494	235
Burglary	5 136	257	1 037	1 524	1 178	887	180	64	9
Robbery	5 094	126	617	1 183	1 451	1 338	313	60	6
Theft, handling, fraud and forgery	4 005	167	479	784	795	965	573	195	47
Other offences	7 292	104	614	1 414	1 672	2 154	944	323	67
Offence not known	1 933	47	430	410	416	415	131	66	18
Females									
Total	1 289	27	105	238	295	382	175	53	14
Violence against the person	279	9	31	48	56	71	44	15	5
Sexual offences	12	-	-	1	2	4	4	1	-
Burglary	41	-	10	8	5	12	6	-	-
Robbery	96	5	13	30	21	23	4	-	-
Theft, handling, fraud and forgery	298	6	14	48	83	91	38	16	2
Other offences	466	5	28	78	108	157	65	18	7
Offence not known	97	2	9	25	20	24	14	3	-

	Total	Age in years							
		15 - 17	18 - 20	21 - 24	25 - 29	30 - 39	40 - 49	50 - 59	60 and over
At 30 June 1995									
Offences									
Males									
Total	37 897	957	4 702	7 202	8 261	10 021	4 386	1 792	576
Violence against the person	8 515	120	871	1 400	1 915	2 529	1 122	414	144
Sexual offences	3 658	19	121	231	456	1 029	936	575	291
Burglary	5 938	257	1 220	1 689	1 418	1 086	197	63	8
Robbery	5 267	184	795	1 132	1 314	1 448	332	52	10
Theft, handling, fraud and forgery	4 613	183	561	827	932	1 187	617	263	43
Drugs offences	3 863	11	188	511	874	1 316	677	241	45
Other offences	4 421	112	578	1 072	1 022	1 083	389	142	23
Offence not known	1 622	71	368	340	330	343	116	42	12
Females									
Total	1 482	31	152	233	324	443	224	59	16
Violence against the person	290	7	32	46	63	69	50	18	5
Sexual offences	12	1	2	1	2	3	2	1	-
Burglary	57	2	12	14	16	9	4	-	-
Robbery	108	9	27	27	19	22	4	-	-
Theft, handling, fraud and forgery	379	5	19	54	96	122	57	22	4
Drugs offences	398	-	36	51	79	143	74	12	3
Other offences	145	3	12	20	27	59	18	4	2
Offence not known	93	4	12	20	22	16	15	2	2

1 Includes persons committed in default of payment of a fine.

Source: Home Office: 020 7217 5567

11.11 Prison population serving sentences: analysis by age and offence[1]
England and Wales

continued

Number

	Total	Age in years 15 - 17	18 - 20	21 - 24	25 - 29	30 - 39	40 - 49	50 - 59	60 and over
At 30 June 1996									
Offences									
Males									
Total	41 323	1 262	5 101	7 569	8 719	11 259	4 723	2 002	688
Violence against the person	9 236	177	937	1 547	1 986	2 747	1 174	501	167
Sexual offences	3 939	32	120	246	437	1 175	946	628	355
Burglary	6 351	373	1 287	1 753	1 484	1 199	202	49	4
Robbery	5 594	268	978	1 171	1 316	1 445	350	62	4
Theft, handling, fraud and forgery	4 709	180	540	798	893	1 311	678	266	43
Drugs offences	5 273	11	293	746	1 239	1 787	850	286	61
Other offences	4 729	119	600	1 050	1 102	1 248	419	160	31
Offence not known	1 492	102	346	258	262	347	104	50	23
Females									
Total	1 732	57	195	263	373	534	223	77	10
Violence against the person	355	16	51	55	60	89	57	22	5
Sexual offences	12	-	2	3	1	2	4	-	-
Burglary	80	5	23	25	12	11	4	-	-
Robbery	124	17	28	23	32	20	3	1	-
Theft, handling, fraud and forgery	433	7	29	48	115	141	62	27	4
Drugs offences	486	-	35	65	106	192	68	19	1
Other offences	166	5	23	31	32	52	16	7	-
Offence not known	76	7	4	13	15	27	9	1	-
At 30 June 1997									
Offences									
Males									
Total	46 739	1 620	6 078	8 472	9 939	12 503	5 046	2 299	782
Violence against the person	10 045	216	1 039	1 631	2 181	3 014	1 234	549	181
Sexual offences	4 069	39	121	232	465	1 169	934	726	383
Burglary	7 983	448	1 574	2 184	1 938	1 515	245	71	8
Robbery	6 278	437	1 227	1 352	1 370	1 471	337	73	11
Theft, handling, fraud and forgery	5 052	167	605	865	1 020	1 417	638	272	68
Drugs offences	6 486	29	356	892	1 494	2 242	1 027	371	75
Other offences	1 746	142	437	296	324	321	141	64	21
Offence not known	5 080	142	719	1 020	1 147	1 354	490	173	35
Females									
Total	2 066	53	198	372	449	615	271	91	17
Violence against the person	391	18	43	72	61	105	62	22	8
Sexual offences	8	-	1	1	1	3	2	-	-
Burglary	101	4	13	29	31	17	6	1	-
Robbery	161	18	50	29	29	28	5	2	-
Theft, handling, fraud and forgery	455	3	25	78	100	148	71	25	5
Drugs offences	691	4	43	111	173	237	88	31	4
Other offences	68	-	8	12	15	25	4	4	-
Offence not known	191	6	15	40	39	52	33	6	-
At 30 June 1998									
Offences									
Males									
Total	49 902	1 627	5 807	8 780	10 590	14 109	5 485	2 608	896
Violence against the person	10 530	235	1 028	1 670	2 215	3 239	1 345	594	204
Sexual offences	4 781	51	140	224	560	1 366	1 113	872	455
Burglary	8 541	432	1 502	2 256	2 179	1 799	298	65	10
Robbery	6 452	449	1 244	1 319	1 427	1 587	350	61	15
Theft, handling, fraud & forgery	5 193	190	556	906	1 071	1 535	587	300	48
Drugs offences	7 103	38	308	944	1 616	2 569	1 092	441	95
Other offences	5 277	137	690	1 087	1 132	1 461	509	215	46
Offence not known	2 025	95	339	374	390	553	191	60	23
Females									
Total	2 367	62	210	425	501	709	332	116	12
Violence against the person	420	11	44	68	80	116	68	27	6
Sexual offences	16	1	1	1	2	5	5	1	-
Burglary	118	6	17	27	32	28	6	2	-
Robbery	177	24	45	33	36	28	8	3	-
Theft, handling, fraud & forgery	514	6	36	97	107	169	72	26	1
Drugs offences	794	3	32	134	181	267	131	44	2
Other offences	218	8	22	42	41	65	30	8	2
Offences not known	110	3	13	23	22	31	12	5	1

1 Includes persons committed in default of payment of a fine.

Source: Home Office: 020 7217 5567

11.12 Expenditure on prisons
England and Wales

Operating cost and total capital employed, years ended 31 March

£ thousand

		1993 /94	1994 /95	1995 /96	1996 /97	1997 /98	1998 /99
Expenditure							
Staff costs	KWUV	833 900	866 600	895 700	948 700	935 900	995 200
Accommodation costs	KXCO	163 800	127 100	122 400	113 900	116 200†	163 400
Other operating costs	KXCP	295 500	323 400	369 000	373 900	472 600†	551 900
Depreciation	KXCQ	4 100	90 300	159 300	123 600	160 200	143 700
Cost of capital	KXCR	–	201 800	211 500	225 500	231 300	251 700
Total expenditure	KXCS	1 297 300	1 609 200	1 757 900	1 785 600	1 920 000	2 105 900
Income							
Contributions from industries	KXCT	–4 700	–4 600	–6 400	–6 600	–7 900	–8 300
Other operating income	KXCU	–4 300	–3 300	–4 900	–5 300	–8 500	–8 500
Total income	KXCV	–9 000	–7 900	–11 300	–11 900	–16 400	–16 800
Net operating costs	KXCW	1 288 300	1 601 300	1 746 600	1 773 700	1 903 600	2 089 100
Total capital employed	KXCX	..	3 452 000	3 580 700	3 920 900	4 116 900	4 340 500

1 For 1993/94 figures are not directly comparable with subsequent years. No figure is available for cost of capital or capital employed as no balance sheet was produced.

Source: Home Office: 020 7217 5567

11.13 Crimes and offences recorded by the police
Scotland

Thousands

		1988	1989	1990	1991	1992	1993	1994	1995	1996	1997	1998
Non-sexual crimes of violence against the person	KAFR	17.9	18.4	18.2	21.7	23.3	19.4	19.8	21.1	21.5	19.2	21.1
Serious assault, etc[1]	KAFS	7.7	7.3	6.3	7.0	7.7	6.5	6.7	6.9	7.0	6.1	6.6
Handling offensive weapons	KAFT	4.5	4.8	5.1	6.2	6.5	5.2	5.3	6.5	6.8	6.0	6.7
Robbery	KAFU	4.2	4.4	4.7	6.2	6.8	5.6	5.3	5.3	5.3	4.5	5.0
Other	KAFV	1.5	1.9	2.2	2.4	2.2	2.1	2.5	2.4	2.5	2.6	2.8
Crimes involving indecency	KAFW	5.1	5.7	6.0	5.8	6.2	6.0	6.0	5.5	5.7	7.1	7.4
Sexual assault	KAFX	1.3	1.5	1.5	1.4	1.6	1.6	1.6	1.6	1.7	2.0	2.2
Lewd and indecent behaviour	KAFY	2.4	2.6	2.6	2.6	2.6	2.7	2.7	2.4	2.5	3.0	3.0
Other	KAFZ	1.4	1.7†	2.0	1.8	2.0	1.7	1.7	1.5	1.5	2.2	2.3
Crimes involving dishonesty	KAGA	344.5	355.5	385.2	430.2	415.0	374.9	350.3	321.2	295.4	267.2	275.4
Housebreaking	KAGB	91.4	93.7	101.7	116.1	113.2	97.8	88.4	74.2	64.5	55.5	56.6
Theft by opening a lockfast places	KAGC	77.5	84.7	92.4	102.8	92.2	84.8	74.9	66.5	60.5	51.1	51.8
Theft of a motor vehicle	KAGD	26.4	29.1	36.1	44.3	47.4	42.8	42.0	37.5	34.2	28.6	28.4
Shoplifting	KAGE	25.5	26.3	26.8	30.1	29.7	26.7	26.6	28.0	26.9	26.3	29.6
Other theft	KAGF	91.2	92.7	97.8	104.6	98.8	93.3	88.9	87.7	82.6	79.6	80.1
Fraud	KAGG	22.4	19.3	19.6	22.0	22.6	19.1	17.7	17.1	16.1	15.7	18.6
Other	KAGH	10.1	9.7	10.8	10.3	11.1	10.3	12.0	10.2	10.8	10.3	10.2
Fire-raising, vandalism, etc	KAGI	73.5	79.1	86.5	89.7	92.2	84.2	88.5	86.5	89.0	81.0	79.2
Fire-raising	KAGJ	3.7	4.4	4.3	4.8	4.7	4.1	3.6	3.3	3.3	2.8	2.5
Vandalism, etc	KAGK	69.9	74.7	82.1	84.9	87.6	80.1	85.0	83.2	85.7	78.2	76.6
Other crimes[2]	KAGL	17.0	19.5	22.6	25.5	28.2	32.7	35.4	41.3	40.3	46.1	48.5
Crimes against public justice[2]	KAGM	11.6	12.3	12.9	13.3	14.4	14.5	16.0	16.4	16.1	16.6	16.9
Drugs	KAGN	5.2	7.0	9.6	12.0	13.6	18.0	19.3	24.8	24.0	29.4	31.5
Other	KAGO	0.2	0.2	0.2	0.2	0.2	0.2	0.2	0.2	0.1	0.1	0.1
Total crimes[2]	KAGQ	457.9	478.2	518.5	572.9	564.9	517.2	500.1	475.7	452.0	420.6	431.6
Miscellaneous offences	KAGR	124.9	124.9	127.0	122.3	127.5	126.6	133.2	134.4	146.1	155.9	153.7
Petty assault[1]	KAGS	36.6	37.7	39.6	41.0	42.5	41.3	45.1	46.6	47.6	50.1	51.0
Breach of the peace	KAGT	55.9	56.3	57.7	55.3	60.0	61.4	65.5	66.1	70.8	73.1	71.7
Drunkenness	KAGU	12.2	12.2	11.7	10.4	10.4	10.1	10.3	9.7	9.6	9.7	8.5
Other	KAGV	20.2	18.8	18.0	15.6	14.6	13.7	12.3	11.9	18.0	23.1	22.6
Motor vehicle offences	KAGW	260.7	283.8	296.2	305.6	306.4	315.1	330.7	317.5	305.9	331.0	362.1
Dangerous and careless driving	KAGX	23.5	25.8	25.3	23.1	22.5	20.0	21.1	18.7	17.3	16.3	15.8
Drunk driving	KAGY	11.4	10.9	11.4	11.0	11.3	10.9	10.8	10.7	11.8	11.2	10.6
Speeding	KAGZ	64.8	81.5	90.0	100.1	93.6	85.4	85.8	85.1	82.4	91.9	115.5
Unlawful use of a motor vehicle	KAHA	68.7	67.6	70.6	75.7	79.9	85.8	88.7	83.4	79.1	79.1	75.5
Vehicle defect offences	KAHB	47.7	48.2	47.5	46.8	47.8	51.4	56.9	56.3	53.5	60.1	63.6
Other	KAHC	44.7	49.8	51.5	48.9	51.5	61.7	67.4	63.2	61.8	72.3	81.2
Total offences	KAHD	385.6	408.6	423.2	427.9	433.9	441.7	463.9	451.9	452.0	486.9	515.8
Total crimes and offences[2]	KAHE	843.5	886.9	941.7	1 000.8	998.8	959.0	964.0	927.6	903.9	907.5	947.3

1 The definition of serious assault was changed in January 1990 to improve consistency between forces. It is estimated that the number of serious assaults that would have been recorded in 1989, using the revised definition, is some 1 150 fewer than actually recorded, with a corresponding rise in petty assaults.

2 Data from 1983 onwards has been revised as a result of a legislative change which came into force on 1 April 1996. From this date "offending while on bail" is no longer regarded as an offence in its own right.

Source: The Scottish Executive Justice Department: 0131 244 2225

11.14 Persons proceeded against
Scotland

Number of persons

		1987[1]	1988[1]	1989[1]	1990[1]	1991[1]	1992[1]	1993[1]	1994[1]	1995[1]	1996	1997
Non-sexual crimes of violence	KEHC	4 394	4 290	4 198	4 392	4 356	5 141	5 327	4 910	4 969	5 741	5 633
Homicide	KEHD	123	116	119	100	106	148	150	122	159†	159	119
Serious assault, etc	KEHE	1 569	1 412	1 432	1 512	1 425	1 546	1 592	1 508	1 298	1 421	1 485
Handling offensive weapons	KEHF	1 340	1 444	1 430	1 427	1 504	2 028	2 028†	2 001	2 357	2 946	2 882
Robbery	KEHG	981	906	786	846	885	949	1 006	990	855	929	848
Other violence	KEHH	381	412	431	507	436	470	371	289	302	286	299
Crimes of indecency	KEHI	1 464	1 366	1 371	1 825	1 660	1 563	1 683	1 587	1 452	1 154	1 379
Sexual assault	KEHJ	227	220	225	233	200	203	229	195	182	194	212
Lewd and libidinous practices	KEHK	484	451	438	440	429	363	370	399	350	376	399
Other indecency	KEHL	753	695	708	1 152	1 031	997	1 084	993	920	584	768
Crimes of dishonesty	KEHM	45 633	43 054	40 808	40 329	40 765†	40 950	38 601	37 026	35 301	33 695	31 931
Housebreaking	KEHN	9 076	8 266	7 673	7 408	7 258	7 105	6 370	6 126	5 452	4 639	4 054
Theft by opening lockfast place	KEHO	3 867	3 832	3 847	3 586	3 906†	3 698	3 640	3 565	3 111	2 906	2 665
Theft of motor vehicle	KEHP	3 381	3 363	3 082	3 226	3 476†	3 688	3 480	3 494	3 449	3 497	3 259
Shoplifting	KEHQ	7 169	6 526	6 646	7 098	7 921	8 469	8 262	7 260	7 185	7 840	7 959
Other theft	KEHR	13 644	12 975	11 574	10 765	9 944†	9 485	8 565	7 532	7 485	7 170	7 113
Fraud	KEHS	3 881	3 514	3 531	3 753†	3 438	3 326	3 519	4 191	4 073	3 566	3 202
Other dishonesty	KEHT	4 615	4 578	4 455	4 493†	4 822	5 179	4 765	4 858	4 546	4 077	3 679
Fire-raising, vandalism, etc	KEHU	7 421	7 393	6 937	6 757	6 593	6 651	6 029	5 644	5 808	6 198	5 917
Fire-raising	KEHV	279	235	229	214	192	213	188	203	177	173	149
Vandalism, etc	KEHW	7 142	7 158	6 708	6 543	6 401	6 438	5 841	5 441	5 631	6 025	5 768
Other crime	KEHX	10 041	9 169	9 639	10 749	11 990†	12 126	13 751	15 108	16 018	17 243	15 515
Crime against public justice	KFBK	7 120	6 485	6 854	7 272	7 537†	7 377	8 407	8 892	9 424	9 748	7 248
Drugs offences	KFBL	2 842	2 638	2 756	3 436	4 413	4 713	5 313	6 185	6 556	7 454	8 219
Other	KFBM	79	46	29	41	40	36	31	31	38	41	48
Total crimes	KEHB	68 953	65 272	62 953	64 052	65 364†	66 431	65 391	64 275	63 548	64 031	60 375
Miscellaneous offences	KEHZ	64 440	67 674	63 559	60 781	58 962†	60 589	54 497	48 383	50 780	52 714	51 837
Petty assault	KEIA	14 353	16 146	16 858	15 990	15 116†	14 715	14 452	13 962	14 809	15 577	15 875
Breach of the peace	KEIB	30 826	27 397	24 830	23 483†	21 955	21 148	20 353	19 102	20 613	22 377	22 336
Drunkenness	KEIC	3 776	3 465	3 028	2 930	2 376	2 253	1 803	1 483	1 364	1 160	983
Other miscellaneous offences	KEID	15 485	20 666	18 843	18 378†	19 515	22 473	17 889	13 836	13 994	13 600	12 643
Motor vehicle offences	KEIE	66 561	65 627	67 392	73 784	76 566	72 055	64 580	66 009	62 840	58 712	60 344
Dangerous and careless driving	KEIF	6 701	7 363	8 603	8 862	7 840	7 177	5 883	5 162	5 273	5 193	5 150
Drunk driving	KEIG	11 438	10 680	8 875	8 997	8 679	8 314	7 812	7 601	7 798	8 313	8 900
Speeding	KEIH	9 803	11 216	13 959†	18 202	21 807	19 519	15 405	16 787	15 159	12 645	11 805
Unlawful use of vehicle	KEII	26 639	22 820	20 002	20 881	21 742	22 033	21 951	22 947	21 528	21 124	22 213
Vehicle defect offences	KEIJ	4 740	5 197	5 071	5 035	4 489	4 111	3 308	3 560	3 683	3 548	3 775
Other motor vehicle offences	KEIK	7 240	8 351	10 882†	11 807	12 009	10 901	10 221	9 952	9 399	7 889	8 501
Total offences	KEHY	131 001	133 301	130 951	134 565	135 528†	132 644	119 077	114 392	113 620	111 426	112 181
Total crimes and offences	KEHA	199 954	198 573	193 904	198 617	200 892	199 075	184 468	178 667	177 168	175 457	172 556

1 Figures for 1986-1995 have been revised.

Source: Scottish Executive Justice Department: 0131 244 2229

11.15 Persons called to court
Scotland

Number of persons

		1987[1]	1988[1]	1989[1]	1990[1]	1991[1]	1992[1]	1993[1]	1994[1]	1995[1]	1996	1997
Court procedure												
High Court[2]	KEIQ	1 252	1 062	1 056	1 099	1 227	1 419	1 601	1 228	1 390	1 420	1 356
Sheriff Court	KEIU	104 060	130 866	103 150	101 678	100 013	100 109	96 391	97 579	98 306	100 928	98 867
District Court	KEIV	78 591	81 063	80 153	83 591	87 186	84 704	74 920	68 183	66 037	62 485	61 648
Stipendiary Magistrate Court	KEIW	15 431	11 785	8 745	11 300	11 522	11 732	10 472	10 893	10 649	9 973	9 943
Total called to court[3]	KEIZ	199 334	197 805	193 162	197 722	200 002	198 038	183 674	178 067	176 423	174 844	171 932

1 Figures for 1986-1995 have been revised.
2 Including cases remitted to the High Court from the Sheriff Court.
3 Includes court type not known.

Source: Scottish Executive Justice Department: 0131 244 2229

11.16 Persons with charge proved: by main penalty
Scotland

Number of persons

		1987[1]	1988[1]	1989[1]	1990[1]	1991[1]	1992[1]	1993[1]	1994[1]	1995[1]	1996	1997
Main penalty												
Absolute discharge	KEXA	457	669	699	869	887	967	989	839	939	1 008	1 137
Admonition or caution	KEXB	15 678	15 983	15 552	16 558	17 137	17 441	16 976	16 243	15 857	15 859	15 039
Probation	KEXC	3 131	3 343	3 840	4 268	4 877	5 385	5 722	6 145	6 145	6 435	6 814
Remit to children's hearing	KEXD	68	59	67	52	81[†]	72	83	124	172	193	219
Community service order	KEXE	3 287	3 423	4 056	4 739	5 190	5 473	5 079	5 320	5 339	5 711	5 707
Fine	KEXF	140 752	139 283	133 817	135 273	135 479	131 842	116 918	112 748	110 337	105 384	103 861
Compensation order	KEXG	1 616	1 703	1 771	1 678	1 591	1 575	1 578	1 535	1 527	1 415	1 304
Insanity, hospital, guardianship order	KYAN	133	160	151	152	148	133	138	133	136	159	162
Prison	KEXI	9 293	9 669	9 091	8 796	9 222	10 085	10 832	11 583	11 561	12 134	11 621
Young offenders' institution	KEXJ	2 937	3 333	4 531	4 150	4 318	4 488	4 461	4 472	4 646	4 744	4 557
Detention centre	KEXL	1 881	1 466	–	–	–	–	–	–	–	–	–
Detention of child	KEXM	43	28	19	23	35	22	30	36	48	45	29
Total persons with charge proved	KEXO	179 276	179 119	173 594	176 558	178 965[†]	177 483	162 806	159 178	156 707	153 087	150 450

1 Figures for 1986-1995 have been revised.

Source: Scottish Executive Justice Department: 0131 244 2229

11.17 Persons with charge proved: by age and sex
Scotland

Number of persons

		1987[1]	1988[1]	1989[1]	1990[1]	1991[1]	1992[1]	1993[1]	1994[1]	1995[1]	1996	1997
Males	KEWA	158 376	154 360	148 157	150 105	150 482	147 692	136 127	136 533	133 330	130 961	129 461
Under 16	KEWB	307	276	215	173	189	138	138	171	180	149	137
16 to 20	KEWC	48 160	44 358	41 206	39 720	38 284	36 494	32 589	30 708	30 113	31 703	31 310
21 to 30	KEWD	58 568	58 375	56 509	59 114	59 532	59 561	56 052	56 517	54 184	52 217	50 684
Over 30	KEWE	47 343	48 317	46 602	48 390	49 802	49 093	45 701	47 788	47 648	45 799	46 281
Not known	KEWF	3 998	3 034	3 625	2 708	2 675	2 406	1 647	1 349	1 205	1 093	1 049
Females	KEWG	19 116	22 791	23 186	24 400	26 340	28 051	25 405	21 650	22 412	21 308	20 246
Under 16	KEWH	23	20	9	11	18	9	6	8	17	12	11
16 to 20	KEWI	3 690	3 982	3 700	3 903	3 927	3 885	3 589	2 939	3 098	3 302	3 424
21 to 30	KEWJ	6 830	9 184	9 110	10 077	10 958	11 998	10 557	9 190	9 284	8 461	8 095
Over 30	KEWK	6 964	8 760	9 261	9 507	10 519	11 409	10 611	9 002	9 439	8 894	8 266
Not known	KEWL	1 609	845	1 106	902	918	750	642	511	574	639	450
Males and Females	KEWM	177 492	177 151	171 343	174 505	176 822	175 743	161 532	158 183	155 742	152 269	149 707
Under 16	KEWN	330	296	224	184	207	147	144	179	197	161	148
16 to 20	KEWO	51 850	48 304	44 906	43 623	42 211	40 379	36 178	33 647	33 211	35 005	34 734
21 to 30	KEWP	65 398	67 559	65 619	69 194	70 490	71 559	66 609	65 707	63 468	60 678	58 779
Over 30	KEWQ	54 307	57 077	55 863	57 897	60 321	60 502	56 312	56 790	57 087	54 693	54 547
Not known	KEWR	5 607	3 879	4 731	3 610	3 593	3 156	2 289	1 860	1 779	1 732	1 499
Companies	KEWS	1 784	1 869	2 212	2 024	2 126	1 717	1 263	991	961	812	737
Total persons with charge proved[2]	KEWT	179 276	179 119	173 594	176 558	178 965	177 483	162 806	159 178	156 707	153 087	150 450

1 Figures for 1986-1995 have been revised.
2 Includes sex not known.

Source: Scottish Executive Justice Department: 0131 244 2229

11.18 Penal establishments: average daily population and receptions
Scotland

Number

		1988	1989	1990	1991	1992	1993	1994	1995	1996[5]	1997	1998
Average daily population												
Male	KEPB	5 057	4 838	4 587	4 696	5 099	5 466	5 408	5 451	5 673	5 900	5 825
Female	KEPC	172	147	137	143	158	171	177	175	189	184	193
Total	KEPA	5 229	4 986	4 724	4 839	5 257	5 637	5 585	5 626	5 862	6 084	6 018
Analysis by type of custody												
Remand	KEPD	844	770	751	770	876	948	1 015	998	1 000	947	938
Persons under sentence: total	KEPE	4 381	4 209	3 961	4 056	4 375	4 686	4 569	4 624	4 861	5 134	5 077
Adult prisoners	KEPF	3 434	3 341	3 201	3 322	3 552	3 795	3 785	3 823	4 026	4 282	4 281
Young offenders	KEPI	817	813	708	684	769	819	720	719	770	787	712
Detention centre inmates	KEPM	85	–	–	–	–	–	–	–	–	–	–
Persons recalled from supervision/licence	KEPN	28	33	39	39	32	40	37	44	46	46	67
Others	KEPO	17	21	13	12	21	32	28	38	18	19	17
Persons sentenced by court martial	KEPP	3	6	11	12	6	2	1	3	–	1	2
Civil prisoners	KEPQ	1	1	1	1	1	1	1	–	1	1	1
Receptions to penal establishments												
Total[1]												
Remand	KEPR	15 000	14 276	15 168	13 127	13 546	13 412	14 922	14 253	14 977	14 826	14 995
Male	KEPS	14 225	13 486	14 323	12 360	12 722	12 478	13 985	13 377	13 976	13 850	13 887
Female	KEPT	775	790	845	767	824	934	937	876	1 001	976	1 108
Persons under sentence: total	KEPU	20 540	19 484	17 134	18 226	19 966	22 157	21 111	19 030	22 155	23 202	20 755
Male	KEPV	19 427	18 466	16 235	17 033	18 856	20 741	19 697	17 737	20 869	21 936	19 517
Female	KEPW	1 113	1 018	899	1 193	1 110	1 416	1 414	1 293	1 286	1 266	1 238
Imprisoned:												
directly	KEPX	7 984	7 619	7 551	7 951	8 543	9 444	9 349	8 730	10 040	9 698	9 607
in default of fine	KEPY	7 234	6 801	5 182	6 336	6 603	7 956	7 377	6 299	7 432	8 873	6 862
in default of compensation order	KEPZ	46	49	23	6	40	41	26	13	..[4]	..[4]	..[4]
Sentenced to young offenders' institution:												
directly	KEQA	2 160	2 651	2 719	2 356	3 041	3 052	2 855	2 772	3 111	2 784	2 748
in default of fine	KEQB	2 466	2 353	1 653	1 573	1 736	1 660	1 498	1 210	1 567	1 847	1 538
in default of compensation order	KEQC	6	9	3	2	2	4	6	4	..[4]	..[4]	..[4]
Sentenced to detention centre[2]	KEQE	642	–	–	–	–	–	–	–	–	–	–
Persons recalled from supervision/licence	JYYD	2	2	3	2	1	–	–	2	5	–	–
Persons sentenced by court martial	KEQH	10	8	10	1	2	7	5	4	4	4	5
Civil prisoners[3]	KEQI	8	15	21	25	34	37	27	25	32	23	11

1 Total receptions cannot be calculated by adding together receptions in each category because there is double counting. This arises because a person received on remand and then under sentence in relation to the same set of charges, is counted in both categories.

2 From 1 November 1988 detention centre sentences were no longer available.

3 For 1995 and 1996 data is estimated.

4 From 1996 compensation orders are included in the figures for default of fine.

5 Please note that, with effect from 1 April 1996, there was a change in the method of collection of statistical information from Scottish Prison Service establishments. The historical method of manual submission of aggregate population and reception data was replaced by electronic transmission of detailed data from the Scottish Prison Service Prisoner Records System (an operational system used to record all prisoner details). This change in the data collection procedures has resulted in some changes to the population and reception time series. In summary, there are changes in the population categorisation of prisoners, the coverage of receptions and the main crime/offence classification for reception purposes.

Source: The Scottish Executive Justice Department: 0131 244 2225

11.19 Expenditure on penal establishments
Scotland

Years ended 31 March

£ thousands

		1988 /89	1989 /90	1990 /91	1991 /92	1992 /93	1993 /94	1994 /95	1995 /96	1996 /97	1997 /98	1998 /99
Expenditure												
Manpower and Associated Services	KPHC	96 694	103 394	110 454	124 898	129 597	135 301	140 009	135 941	143 107	137 890	144 660
Prisoner and Associated Costs	KPHD	6 834	8 307	8 446	8 504	10 274	10 795	11 679	12 373	13 377	16 313	18 891
Capital Expenditure	KPHE	11 420	15 551	12 866	13 032	13 681	11 845	15 636	15 377	22 577	22 136	23 697
Gross Expenditure	KPHF	114 948	127 252	131 766	146 434	153 552	157 941	167 324	163 691	179 061	176 339	187 248
Less Receipts:-	KPHG	4 793	4 302	3 819	3 711	3 542	3 598	3 042	2 800	2 600	2 810	8 160
Net Operating Costs	KPHH	110 155	122 950	127 947	142 723	150 010	154 343	164 282	160 891	176 461	173 529	179 088

Source: The Scottish Executive Justice Department: 0131 244 2225

11.20 Disposals given to those convicted by court
Northern Ireland

Number

		1987	1988	1989	1990	1991	1992	1993	1994	1995	1996	1997
Magistrates court - all offences												
Prison	KYAO	1 463	1 222	1 006	1 009	960	830	1 027	945	1 046	1 003	989
Young offenders centre	KYAP	452	401	368	370	502	588	575	499	483	443	430
Training school	KYAQ	220	170	168	153	177	120	125	193	169	147	148
Total immediate custody	KYAR	2 135	1 793	1 542	1 532	1 639	1 538	1 727	1 637	1 698	1 593	1 567
Prison suspended	KYAS	1 858	1 647	1 544	1 513	1 379	1 420	1 529	1 558	1 674	1 722	1 506
YOC suspended	KYAT	386	310	319	310	432	507	447	447	385	444	461
Attendance centre	KYAU	107	139	108	118	90	66	94	89	101	91	66
Probations/supervision	KYAV	700	763	778	854	742	849	881	1 017	1 137	1 134	1 155
Community supervision order	KYAW	519	523	435	575	547	464	536	551	547	591	561
Fine	KYAX	27 774	28 019	26 306	26 644	19 569	23 418	25 166	24 390	22 726	20 612	21 313
Recognizance	KYAY	600	442	412	399	514	713	858	961	1 001	1 203	1 267
Conditional discharge	KYAZ	2 598	2 707	2 022	1 982	2 102	1 965	2 021	1 830	1 928	1 679	1 597
Absolute discharge	KYBA	1 657	1 656	1 305	1 274	845	732	690	661	608	509	424
Disqualification	KYBB	4 734	4 810	4 106	4 335	4 211	640	6	6	2	5	2
Other	KYBC	39	36	20	28	24	12	7	11	8	10	6
Total	KYBD	43 107	42 845	38 897	39 564	32 094	32 324	33 962	33 158	31 815	29 593	29 925
Crown court - all offences												
Prison	KYBE	681	536	472	493	493	447	555	471	533	469	475
Young offenders centre	KYBF	199	112	111	106	125	119	130	87	76	106	111
Training school	KYBG	16	5	10	4	13	5	2	5	6	–	4
Total immediate custody	KYBH	896	653	593	603	631	571	687	563	615	575	590
Prison suspended	KYBI	389	313	318	295	238	249	211	277	265	253	220
YOC suspended	KYBJ	172	93	73	78	46	63	37	43	63	71	60
Attendance centre	KYBK	–	1	2	–	–	–	–	1	–	–	–
Probation/supervision	KYBL	86	80	72	85	103	95	73	58	60	49	47
Community supervision order	KYBM	141	106	71	105	89	79	48	59	60	54	37
Fine	KYBN	50	35	39	33	23	17	33	23	27	39	40
Recognizance	KYBO	14	7	3	8	7	9	5	16	–	7	10
Conditional discharge	KYBR	55	65	51	37	53	36	19	15	64	30	31
Absolute discharge	KYBS	3	5	1	2	5	8	3	2	1	–	1
Disqualification	KYBT	7	4	8	10	6	2	–	–	–	–	–
Other	KYBU	2	5	3	–	8	6	6	1	2	3	3
Total	KYBV	1 815	1 367	1 234	1 256	1 209	1 135	1 122	1 058	1 157	1 081	1 039

Source: Northern Ireland Office: 028 9052 7538

11.21 Persons found guilty at all courts by offence group
Northern Ireland

Number

		1987	1988	1989	1990	1991	1992	1993	1994	1995	1996	1997
Violence against the person	KYCT	1 534	1 491	1 424	1 750	1 634	1 558	1 674	1 498	1 685	1 597	1 594
Sexual offences	KEVG	210	228	226	275	193	184	126	148	182	184	130
Burglary	KYBW	1 869	1 553	1 363	1 362	1 208	1 149	1 114	979	951	801	715
Robbery	KYBX	314	296	273	220	162	202	159	168	195	161	166
Theft	KYBY	4 108	3 614	3 298	3 399	3 429	3 158	3 254	3 044	3 128	2 765	2 596
Fraud and forgery	KYBZ	728	688	783	699	648	683	633	568	533	467	491
Criminal damage	KYCA	1 230	1 048	939	1 054	1 019	967	1 145	1 134	1 008	1 076	1 163
Offences against the state	KYCB	321	226	216	233	195	187	184	137	166	147	165
Other indictable	KYCC	319	429	380	287	368	448	606	669	863	899	739
Total indictable[1]	KYCD	10 633	9 573	8 902	9 279	8 856	8 536	8 895	8 345	8 711	8 097	7 759
Summary	KYCE	6 072	6 071	5 113	4 293	4 278	4 115	4 307	4 369	4 137	4 402	4 435
Motoring[2]	KYCF	28 218	28 568	26 116	27 248	20 169	20 808	21 882	21 502	20 124	18 177	18 770
All offences	KYCG	44 923	44 212	40 131	40 820	33 303	33 459	35 084	34 216	32 972	30 676	30 964

1 Excludes indictable motoring offences
2 Includes indictable motoring offences

Source: Northern Ireland Office: 028 9052 7538

11.22 Juveniles found guilty at all courts by offence group
Northern Ireland

Number

		1987	1988	1989	1990	1991	1992	1993	1994	1995	1996	1997
Violence against the person	KYCH	55	58	33	44	38	46	43	49	51	75	49
Sexual offences	KAHF	12	15	14	17	8	11	7	8	7	4	8
Burglary	KYCI	274	251	228	232	194	165	155	180	170	137	124
Robbery	KYCJ	11	19	11	10	7	8	4	9	22	13	18
Theft	KYCK	421	419	308	329	328	247	280	283	345	338	334
Fraud and forgery	KYCL	21	6	10	10	10	16	14	14	21	14	11
Criminal damage	KYCM	165	114	119	90	92	82	94	117	116	121	136
Offences against the state	KYCN	16	19	14	6	5	6	1	8	9	6	10
Other indictable	KYCO	8	11	15	10	11	8	2	6	14	24	10
Total indictable[1]	KYCP	983	912	752	748	693	589	600	674	755	732	700
Summary	KYCQ	262	190	186	138	111	113	125	131	180	182	198
Motoring[2]	KYCR	129	100	81	92	71	40	44	74	74	58	57
All offences[3]	KYCS	1 374	1 202	1 019	978	875	742	769	879	1 009	972	955

1 Excludes indictable motoring offences.
2 Includes indictable motoring offences.
3 Juveniles are aged 10 - 16 years inclusive.

Source: Northern Ireland Office: 028 9052 7538

11.23 Prisons and Young Offenders Centres
Northern Ireland
Receptions and average population

Number

		1988	1989	1990	1991	1992	1993	1994	1995	1996	1997	1998
Receptions:												
Reception of untried prisoners	KEOA	1 853	1 795	1 773	1 851	1 987	2 045	2 043	2 003	2 292	2 188	2 284
Reception of sentenced prisoners:												
Imprisonment under sentence of immediate custody[1]	KEOB	1 136	1 029	953	974	941	1 135	1 029	1 070	1 070	1 062	949
Imprisonment in default of payment of a fine	KEOC	1 493	1 291	1 264	1 282	1 352	1 221	1 190	1 248	1 374	1 513	1 530
Total	KEOD	2 629	2 320	2 217	2 256	2 293	2 356	2 219	2 318	2 444	2 575	2 479
Reception into Young Offender Centres:												
Detention under sentence of immediate custody	KEOE	437	422	371	348	371	416	346	371	362	331	347
Detention in default of payment of a fine	KEOF	457	378	309	356	364	353	276	351	373	366	385
Total	KEOG	894	800	680	704	735	769	622	722	735	697	732
Other receptions:												
Civil committals	KEOI	50	46	21	17	10	21	13	45	27	42	70
Total	KEOL	50	46	21	17	10	21	13	45	27	42	70
Daily average population:												
Total	KEOM	1 901	1 815	1 785	1 796	1 810	1 934	1 899	1 762	1 639	1 632	1 507
Unconvicted[2]	KEON	270	313	361	350	414	427	440	322	337	376	383
Convicted[3]	KEOP	1 631	1 502	1 424	1 446	1 396	1 507	1 459	1 440	1 302	1 256	1 124

1 Includes those detained under Section 73 of the Children and Young Persons (NI) Act 1968.
2 Prisoners on remand or awaiting trial and prisoners committed by civil process.
3 Includes those sentenced to immediate custody and fine defaulters.

Source: Northern Ireland Office: 028 9052 7538

Lifestyles

12

Lifestyles

Cinema statistics *(Tables 12.3)*

These estimates represent the motion picture projection activity for all individual legal units (that is companies, sole proprietorships, partnerships etc.) whose main activity is motion picture projection (i.e. heading 92.13 of the UK Standard Industrial Classification 1992). However, where such activity can be identified, the estimates also include the film exhibition activity of a number of legal units where motion picture projection is not their main activity. The Office for National Statistics does not maintain registers of all sites where film exhibition takes place so that the estimates will not be completely comprehensive. It is believed, however, that the estimates cover the great majority of units engaged in cinema activity in Great Britain (i.e. England, Scotland and Wales but not Northern Ireland, the Isle of Man, the Channel Islands or the Republic of Ireland).

These figures are produced by combining information obtained by grossing data from a voluntary quarterly panel with estimates derived from the larger sample approached in the annual inquiry to the service trades.

All financial figures are exclusive of Value Added Tax.

12.1 Expenditure by the Department for Culture, Media and Sport

£ millions

	Museums and galleries (England)	Libraries (UK)	Museums library archives (UK)	The arts (England)	Sport (UK)	Historic buildings, monuments, and sites (England)	The Royal Parks (UK)	Tourism (UK)	Broadcasting and media (UK)	Central administration (England)	National Lottery Commission	European regional development fund	Commemorative services and royal funerals
	KWFN	KWFO	LQYX	KWFP	KWFQ	KWFR	LQYY	KWFS	KWFT	KWFU	LQYZ	JYXQ	JYXP
1991/92	204	129	..	212	47	..	..	44	22	8	..	..	..
1992/93	217	133	..	235	50	..	..	46	79	17	..	..	..
1993/94	212	114	9	235	54	164	23	46	85	22	2	..	..
1994/95	223	146	9	195	53	164	24	44	93	21	3	..	3
1995/96	228	171	9	200	54	164	25	45	98	20	2	6	3
1996/97	214	114	10	195	52	162	23	46	97	21	2	16	..
1997/98	211	104	12	196	50	156	22	45	43	21	2	26	4
1998/99[1]	204	89	12	199	49	144	21	45	99	24	2	6	..
1999/00[2]	220	90	16	228	52	148	26	48	104	26	2	6	..
2000/01[2]	223	91	20	238	52	144	22	48	104	26	2	6	..
2001/02[2]	247	95	19	253	52	142	22	47	105	26	2	6	..

1 Data are Estimated Outturn.
2 Data are plans.

Source: Department for Culture, Media and Sport

12.2 Employment in tourism related industries
Great Britain

At June in each year

Thousands, not seasonally adjusted

	Hotels and other tourist accommodation	Restaurants, cafes etc.	Bars, pubs, clubs	Travel agents, tour operators	Libraries, museums, culture	Sport and other recreation	All	Total employees in tourist related industries	Estimated self employment
	KWFV	KWFW	KWFX	KWFY	KWFZ	KWGA	LQZA	KWGB	KWGC
1989	299.2	283.4	428.2	64.9	82.8	294.7	1 644.2	1 453.2	191.0
1990	314.4	303.0	445.8	70.0	80.0	311.5	1 714.7	1 524.7	190.0
1991	307.9	297.7	435.0	69.7	75.6	316.5	1 685.4	1 502.4	183.0
1992	311.0	303.0	414.2	69.2	74.8	320.8	1 671.0	1 493.0	178.0
1993	317.6	298.0	370.6	69.3	75.6	316.5	1 643.6	1 447.6	196.0
1994	375.5†	372.3†	399.4†	83.6†	77.4†	356.0†	1 872.8	1 664.2†	208.6†
1995	385.9	386.6	445.4	90.7	80.3	363.1	1 752.0	1 536.6	215.4
1996	370.8	394.5	437.9	96.8	78.4	355.8	1 949.5	1 734.1	215.4
1997	344.1	410.4	479.5	108.9	80.0	363.1	2 003.8	1 786.0	217.8
1998	332.6	413.7	467.3	110.0	86.4	357.8	1 951.7	1 767.8	183.9
1999	306.0	357.7	428.8	114.2	84.5	337.7	1 787.5	1 628.9	158.6

Source: Office for National Statistics, short term employment survey

12.3 GB Cinema exhibitor statistics

	Sites	Screens	Total no. of admissions	Gross box office takings	Amount paid out for films	Revenue per admission	Revenue per screen
	number	number	millions	£ million	£ million	£	£ thousand
	CKNU	CKNV	CKQE	CKQF	CKQG	CKQH	CKQI
1988	495	1 117	75.2	142.2	55.1	1.89	126.9
1989	481	1 177	82.9	169.5	65.4	2.04	143.7
1990	496	1 331	78.6	187.7	69.5	2.39	143.8
1991	537	1 544	88.9	229.7	78.5	2.58	152.4
1992	480	1 547	89.4	243.7	81.9	2.73	153.6
1993	495	1 591	99.3	271.3	95.9	2.73	171.0
1994	505	1 619	105.9	293.5	104.6	2.77	183.5
1995	475	1 620	96.9	286.5	96.3	2.96	176.7
1996	483†	1 738†	118.7†	373.5†	144.9†	3.15†	215.5†
1997[1]	504	1 886	128.2	443.2	165.9	3.46	255.0
1998[1]	481	1 975	123.4	449.5	162.6	3.64	227.6

1 Provisional.

Source: Office for National Statistics: 01633 812264

12.4 Films

	Production of UK films		Expenditure on feature films in UK				
	Films produced in the UK	Production costs	UK box office	Video rental	Video retail	Subscriptions to movie channels	Box office, video, subscription channels
	number	£ million 1998 prices	£ million current prices	£ million current prices	£ million current prices	£ million current prices	£ million current prices
	KWGD	KWGE	KWHU	KWHV	KWHW	KWHX	KWHY
1988	48	264.0†	193	371†	184†	..	748†
1989	30	145.2	227	416	345	..	988
1990	60	271.8	273	418	374	47	1 112
1991	59	285.3	295	407	440	121	1 263
1992	47	208.6	291	389	506	283	1 469
1993	67	252.9	319	350	643	350	1 662
1994	84	502.7	364	339	698	540	1 941
1995	78	441.0	385	351	789	721	2 246
1996	128	785.0	426	382	803	1 319	2 930
1997	116	581.9	506	369	858	..	1 733
1998	90	508.2	515	437	940	..	1 892

Sources: Department for Culture, Media and Sport; (Screen Finance/ONS/BVA/EDI/BSKYB)

12.5 Tourism

International tourism

	Visits to the UK by overseas residents	Spending in the UK by overseas residents	Spending in the UK by overseas residents	Visits overseas by UK residents	Spending overseas by UK residents	Spending overseas by UK residents
	Thousands	£ million current prices	£ million constant 1995 prices	Thousand	£ million current prices	£ million constant 1995 prices
	KWGF	KWGG	KWGH	KWGI	KWGJ	KWGK
1988	15 799	6 184	9 142	28 828	8 216	12 515
1989	17 338	6 945	9 567	31 030	9 357	12 861
1990	18 013	7 748	9 853	31 150	9 886	12 021
1991	17 125	7 386	8 627	30 808	9 951	11 775
1992	18 535	7 891	8 784	33 836	11 243	12 678
1993	19 863	9 487	10 188	36 720	12 972	13 184
1994	20 794	9 786	10 050	39 630	14 365	14 852
1995	23 537	11 763	11 763	41 345	15 386	15 386
1996	25 163	12 290	11 954	42 050	16 223	15 897
1997	25 515	12 244	11 542	45 957	16 931	18 652
1998	25 745	12 671	11 573	50 872	19 489	21 847

UK Domestic tourism: one or more nights

	Number of trips	Number of nights spent	Expenditure at current prices	Average nights spent	Average expenditure per trip
	Millions	Millions	£ million	Numbers	£
	KWGL	KWGM	KWGN	KWGO	KWGP
1990	95.3	399.1	10 460	4.2	109.8
1991	94.4	395.6	10 470	4.2	110.9
1992	95.6	399.7	10 665	4.2	111.6
1993	90.9	375.9	12 430	4.1	136.7
1994	109.8	416.5	13 215	3.8	120.4
1995	121.0	449.8	12 775	3.7	105.6
1996	127.0	454.6	13 895	3.6	109.4
1997	133.6	473.6	15 075	3.6	112.9
1998	122.3	437.6	14 030	3.6	114.7

Holidays taken by GB residents

	Taking a holiday in GB	Taking a holiday abroad	Taking only one holiday (4+ nights)	Taking two holidays (4+ nights)	Taking 3 or more holidays (4+ nights)
	Percentage	Percentage	Percentage	Percentage	Percentage
	KWGQ	KWGR	KWGS	KWGT	KWGU
1988	39	30	37	16	8
1989	37	30	37	15	7
1990	38	30	36	15	8
1991	41	30	36	16	8
1992	36	32	35	15	9
1993	37	35	36	16	9
1994	35	36	34	16	10
1995	36	37	35	17	10
1996	35	34	35	15	9
1997	33	35	31	16	10
1998	31	38	34	16	9

UK Trends in number of visits to tourist attractions (index 1988 = 100)

	Historic properties	Gardens	Museums	Wildlife sites	Other attractions	Any types of attraction
	LQZB	LQZC	LQZD	LQZE	LQZF	LQZG
1988	100	100	100	100	100	100
1989	104	112	101	103	106	104
1990	106	114	107	104	109	107
1991	102	119	108	98	108	106
1992	102	118	111	90	107	106
1993	104	125	110	92	110	108
1994	103	130	111	90	112	109
1995	107	136	109	89	116	111
1996	110	140	112	93	118	115
1997	112	142	112	95	119	116
1998	111	132	113	95	116	114

Sources: Department for Culture, Media and Sport; Office for National Statistics; British Tourist Authority

12.6 Participation in leisure activities[1]

Percentages

		1987	1990	1993	1996
Sports, games and physical activities					
Walking	KWHB	37.9	40.7	40.8	44.5
Swimming	KWHC	13.1	14.8	15.4	14.8
Cue sports	KWHD	15.1	13.6	12.2	11.3
Keep fit/yoga	KWHE	8.6	11.6	12.1	12.3
Cycling	KWHF	8.4	9.3	10.2	11.0
Darts	KWHG	8.8	7.1	5.6	..
Weight lifting)					1.3
Weight lifting/training)	KWHH	4.5	4.8	5.5	
Weight training)					5.6
Golf	KWHI	3.9	5.0	5.3	4.7
Jogging	KWHJ	5.2	5.0	4.6	4.5
Football	KWHK	4.8	4.6	4.5	4.8
Any activity other than walking	KWHL	44.7	47.8	47.3	45.6
Selected leisure activities					
Watching TV	KWHM	99	99	99	99
Visiting/entertaining friends or relatives	KWHN	95	96	96	96
Listening to the radio	KWHO	88	89	89	88
Listening to tapes or records	KWHP	73	76	77	78
Reading books	KWHQ	60	62	65	65
Gardening	KWHR	46	48	48	48
DIY	KWHS	43	43	42	42
Dressmaking, needlework, knitting	KWHT	27	23	22	22

1 Percentage aged 16 and over participating in each activity in the four weeks before interview.

Source: Office for National Statistics

12.7 Gambling

		1986 /87	1987 /88	1988 /89	1989 /90	1990 /91	1991 /92	1992 /93	1993 /94	1994 /95
Money spent on gambling (£ million at 1994/95 prices)										
Lotteries (not including National Lottery)	KWHZ	32	31	28	29	31	60	49	44	39
Casinos	KWIA	2 382	2 437	2 296	2 329	2 178	2 063	2 154	2 291	2 461
Bingo clubs	KWIB	816	886	789	765	746	771	822	811	–

Source: Gaming Board for Great Britain

12.8 Private households with usual residents[1]: Census 1991[2]

Thousands

	Number of households				
		Selected tenures of households in permanent buildings			
			Rented		
	All households	Owner occupied	From council or new town	Unfurnished	With no car
Household and family type					
Great Britain					
All households	21 441	14 288	4 589	765	7 095
Households with no family	6 344	3 297	1 778	363	3 936
One person	5 643	2 922	1 652	327	3 657
Two or more persons	701	375	126	36	280
Households with one family	14 898	10 856	2 760	398	3 113
Married couple family	11 847	9 279	1 719	273	1 889
With no children	5 208	3 994	796	156	1 145
With dependent child(ren)	4 802	3 813	651	85	526
With non-dependent child(ren) only	1 837	1 471	272	32	218
Cohabiting couple family	1 139	720	219	57	224
With no children	718	502	68	41	117
With dependent child(ren)	374	187	139	15	100
With non-dependent child(ren) only	47	31	12	1	7
Lone parent family	1 912	858	823	68	1 000
With dependent child(ren)	1 122	391	567	40	689
With non-dependent child(ren) only	790	467	256	28	311
Households with two or more families	199	134	50	4	46
England and Wales					
All households	19 458	13 252	3 838	717	6 253
Households with no family	5 725	3 053	1 498	341	3 500
One person	5 083	2 703	1 393	306	3 254
Two or more persons	642	351	105	34	247
Households with one family	13 551	10 071	2 298	373	2 713
Married couple family	10 779	8 589	1 411	254	1 645
With no children	4 762	3 714	661	146	1 015
With dependent child(ren)	4 358	3 517	534	78	448
With non-dependent child(ren) only	1 659	1 358	215	30	182
Cohabiting couple family	1 064	682	192	55	199
With no children	672	474	58	40	104
With dependent child(ren)	348	178	124	14	89
With non-dependent child(ren) only	43	29	10	1	6
Lone parent family	1 708	800	694	64	868
With dependent child(ren)	1 004	368	484	38	601
With non-dependent child(ren) only	703	432	210	26	267
Households with two or more families	183	127	42	3	40
Scotland					
All households	1 983	1 036	751	48	843
Households with no family	619	244	281	22	436
One person	560	219	260	20	403
Two or more persons	60	25	21	2	33
Households with one family	1 348	785	463	25	401
Married couple family	1 068	690	307	19	244
With no children	446	280	135	10	130
With dependent child(ren)	444	296	116	7	78
With non-dependent child(ren) only	178	113	56	2	36
Cohabiting couple family	75	38	27	2	25
With no children	45	28	10	1	13
With dependent child(ren)	26	9	15	1	11
With non-dependent child(ren) only	3	2	2	0	1
Lone parent family	204	58	129	4	132
With dependent child(ren)	118	23	83	2	88
With non-dependent child(ren) only	86	35	46	2	44
Households with two or more families	16	7	8	0	6

1 Number of households and family type by selected tenures of households in permanent buildings and with no car.
2 Later figures not available.

Sources: Office for National Statistics: 0151 951 4540; General Register Office (Scotland)

Environment, water and housing

13

Environment, water and housing

Environment *(Tables 13.1 to 13.17)*

Greenhouse gases *(Table 13.1)*

Estimates of the relative contribution to global warming of the main greenhouse gases, or classes of gases is presented weighted by their global warming potential. This is the format required by the Intergovernmental Panel on Climate Change (IPCC).

Air emissions *(Tables 13.2 to 13.7)*

Emissions of air pollutants arise from a wide variety of sources. Except for large combustion plants, information is not available on actual emissions from specific individual sources. Consequently, the National Atmospheric Emissions Inventory (NAEI), prepared annually for DETR (Department of the Environment, Transport and the Regions) by the National Environmental Technology Centre (NETCEN), is compiled as estimates derived from statistical information and from research on emission factors for stationary and mobile sources. Although for any given year considerable uncertainties surround the emission estimates for each pollutant, trends over time are likely to be more reliable.

UK national emission estimates are updated annually and any developments in methodology are applied retrospectively to earlier years. Adjustments in the methodology are made to accommodate new technical information and to improve international comparability.

UNECE: United Nations Economic Commission for Europe
CORINAIR: part of the CORINE (Co-ordination of Information on the Environment) work plan relating to providing an air emission inventory for Europe

Water quality *(Tables 13.8 and 13.9)*

The chemical quality of river and canal waters in the UK is monitored in a series of separate national surveys. The General Quality Assessment (GQA) Scheme used in the surveys provides a rigorous and objective method for assessing the basic chemical quality of rivers and canals based on three determinants – dissolved oxygen, biochemical oxygen demand (BOD), and ammoniacal nitrogen. The GQA grades river stretches into six categories (A-F) of chemical quality and these in turn have been grouped into four broader groups – good (classes A and B), fair (C and D), poor (E) and bad (F). In Scotland river and canal water quality is assessed using the following classes: unpolluted, fairly good, poor and grossly polluted. These classes are not directly comparable with those used in England and Wales and Northern Ireland.

To provide a more comprehensive picture of the health of rivers and canals, biological testing was carried out in parallel with chemical testing in the latest two quinquennial river quality surveys in England and Wales, Scotland and Northern Ireland. The biological grading is based on the monitoring of tiny animals (invertebrates) which live in or on the bed of the river. Research has shown that there is a relationship between species composition and water quality. Using a procedure known as the River Invertebrate Prediction and Classification System (RIVPACS), species groups recorded at a site were compared with those which would be expected to be present in the absence of pollution, allowing for the different environmental characteristics in different parts of the country. In England and Wales and in Northern Ireland two different summary statistics (known as ecological quality indices (EQI)) were calculated and then the biological quality was assigned to one of six bands based on a combination of these two statistics. In Scotland, a third EQI was also calculated by SEPA and the grading system based on a combination of all three statistics or EQIs. The results for Scotland are not directly comparable for this reason.

Bathing waters *(Table 13.12)*

Under the EC Bathing Water Directive, eleven physical, chemical and microbiological parameters are measured including total and faecal coliforms which are generally considered to be the most important indicators of the extent to which water is contaminated by sewage. The mandatory value for total coliforms is 10 000 per 100 ml, and for faecal coliforms 2 000 per 100 ml. For a bathing water to comply with the coliform standards, the Directive requires that at least 95 per cent of samples taken for each of these parameters over the bathing season are less than equal to the mandatory values. In the UK a minimum of 20 samples are normally taken at each site. In practice this means that where 20 samples are taken, a maximum of only one sample may exceed the mandatory value for the bathing water to comply, and where less than 20 samples are taken none may exceed the mandatory value for the bathing water to comply.

Radioactive wastes *(Table 13.13)*

Solid radioactive wastes are not discharged to the environment but stored and conditioned by processes such as supercompaction, cementation or turning into glass. Such wastes cover a wide range of materials and can be classified, according to the nature and quantity of radioactivity associated with them, as high level wastes (HLW), intermediate level wastes (ILW) or low level wastes (LLW). For explanation of what these three types of waste comprise see the footnotes to the table. HLW result from the reprocessing of nuclear fuel and are highly radioactive. Although small in volume, they contain over 95 per cent of all the radioactivity in wastes from nuclear establishments. Final disposal of HLW and ILW will be to deep underground repositories when these become available.

The table shows recent trends in the volume of radioactive waste stocks for particular groups of nuclear sites. Data for HLW and ILW are presented in two physical states – as-stored and conditioned - and conditioned for disposal for LLW (see footnotes to Table). There is also no simple relationship between as-stored and conditioned volumes of waste, since the effects of conditioning can vary with different wastes. Estimates of conditioned wastes are indicative only and should be interpreted with care.

Noise complaints *(Table 13.14)*
The table shows trends in the number of complaints received by local authority Environmental Health Officers. The figures are from those authorities making returns and are calculated per million people based on the population of the authorities making returns.

Recycling (*Table 13.16)*
The ratios shown reflect the amount of secondary material (scrap) used in the UK in a year as a proportion of consumption in that year. This gives some indication of the extent to which secondary material is displacing raw materials from the feedstock. This definition takes no account of the lifetime of the article being recovered ie that material which is being recycled now and related to current consumption might originally have been delivered to an end market some years ago.

Protected areas *(Table 13.17)*
National Parks, Areas of Outstanding Natural Beauty (AONBS) in England, Wales and Northern Ireland and National Scenic Areas in Scotland are the major areas which have been designated to protect their landscape importance.

Further details can be found in DETR's annual *Digest of Environmental Statistics.* If you would like to discuss the tables, Adrian Redfern can be contacted at the Department of the Environment, Transport and the Regions on 020 7890 6497.

13.1 Estimated total emissions[1] of UK greenhouse gases on IPCC basis
United Kingdom

Million tonnes (CO_2 equivalent)

		1990	1991	1992	1993	1994	1995	1996	1997
Emissions weighted by global warming potential									
Carbon dioxide (CO_2(asCO_2))	JZCK	616†	620	605	590	584	575	594	568
Methane (CH_4)	JZCL	76.37†	75.33	73.36	65.91	60.96	60.38	59.08	57.26
Nitrous oxide (N_2O)	JZCM	66.10†	64.15	57.12	53.44	58.07	55.80	57.80	59.53
Hydrofluorocarbons (HFC)	JZCN	11.37†	11.86	12.35	12.90	13.81	15.21	16.29	18.45
Perfluorocarbons (PFC)	JZCO	2.28†	1.79	0.96	0.81	0.98	1.09	0.91	0.66
Sulphur hexafluoride (SF_6)	JZCP	0.72†	0.76	0.80	0.84	1.00	1.05	1.18	1.17
Total	JZCQ	773.0†	774.0	750.0	723.7	718.7	708.3	729.3	704.8

1 Emissions inventories based on the methodology developed by the Intergovernmental Panel on Climate Change (IPCC) are used to report UK emissions to the Climate Change Convention.

Source: NETCEN

13.2 Estimated emissions of sulphur dioxide (SO_2) by source
United Kingdom

Thousand tonnes

By source category (UNECE/CORINAIR94)		*Percentage of total in 1997*	1970	1975	1980	1985	1990	1991	1992	1993	1994	1995	1996	1997
Power stations	JZCR	*62*	2 913	2 941	3 007	2 627	2 723	2 535	2 434	2 084†	1 762	1 590	1 319	1 025
Refineries	JZCS	*10*	227†	237	250†	136†	151†	160	178	203	180	175	178	174
Combustion in fuel extraction & transformation	JZCT	*1*	306	120	73	55	44	40	17†	14	14	12	11	14
Domestic	JZCU	*4*	522	301	226	202	108	115	103	113†	91	65	68	63
Commercial, public & agricultural combustion	JZCV	*3*	320	214	218	133	90	85	89	95	81	61	59	48
Iron and steel	JZCW	*4*	433	272	128	80	88	86	66†	75	68	65	61	62
Other industrial combustion	JZCX	*11*	1 383†	1 012	774†	334†	352†	369	417	412	343	256	216	178
Production processes	JZCY	*1*	91†	87	74†	55†	44†	38	33	31	29	23	24	21
Road transport	JZCZ	*2*	44	48	42	45†	63	58	62†	59	63	51	37	28
Other transport	JZDA	*25*	92	83	61	49†	47†	50	49	48	46	44	44	42
Other sources	JZDB	-	9	9	10	14	21	10	12	9	10	8	8†	7
Total	JZDC	*100*	6 341†	5 325	4 863†	3 729†	3 731†	3 548	3 459	3 143	2 687	2 351	2 025	1 660

Source: NETCEN

13.3 Estimated emissions of black smoke by source
United Kingdom

Thousand tonnes

By source category (UNECE/CORINAIR94)		*Percentage of total in 1997*	1970	1975	1980	1985	1990	1991	1992	1993	1994	1995	1996	1997
Power stations	JZDD	*6*	32	32	29	28	35	34	36	32	38	22	21	18
Domestic	JZDE	*20*	770	426	316	285	137	144	129	127	93	63	67	64
Commercial, public & agricultural combustion	JZDF	*1*	12	8	8	6	4	4	4	4	4	3	3	3
Industrial combustion[1]	JZDG	*3*	44	29	21	14	14	15†	15	14	13	11	10	9
Road transport	JZDH	*59*	100†	108	117	141	207	208†	216	211	214	203	197	184
Other transport	JZDI	*10*	48†	46	41†	39	35†	36	36	35	34	32	34	33
Waste treatment and disposal	JZDJ	-	44	44	44	44	37	36	33	32	24	6	5	1
Other sources	JZDL	*1*	19†	17	22†	19†	13†	12	9	3	3	3	3	3
Total	JZDM	*100*	1 069†	710	598†	576†	482†	488	477	458	424	343	339	314

1 Comprises Iron and Steel and Other Industrial Combustion.

Source: NETCEN

13.4 Estimated emissions of PM_{10}[1] by source
United Kingdom

Thousand tonnes

By source category (UNECE/CORINAIR94)		Percentage of total in 1997	1970	1975	1980	1985	1990	1991	1992	1993	1994	1995	1996	1997
Power stations	JZDN	12	67†	65	76†	64†	71†	70	67	56	51	37	34	23
Domestic	JZDO	15	216	127	98	89	48	51	46	47	38	28	30	28
Commercial, public & agricultural combustion	JZDP	3	19	11	11	10	8	8	7	7	7	6	6	6
Industrial combustion[2]	JZDQ	11	74	43	32†	26†	26	27†	29	28	26	24	22	20
Production processes[3]	JZDR	25	52†	50	43†	45†	53†	49	47	48	50	48	46	46
Road transport[4]	JZDS	26	46	50	56†	58	68†	68	65	63	62	58	54	48
Other transport	JZDT	2	6†	6	5†	5†	5†	5	5	5	4	4	5	5
Waste treatment and disposal	JZDU	1	5†	5	7†	6†	5†	5	5	5	5	2	2	1
Other sources	JZDV	4	21	9	8†	7†	7†	7	7	7	7	7	8	8
Total	JZGA	100	507†	366	335†	309†	290†	290	277	264	249	215	207	184

1 Particles which pass through a size selective inlet with a 50 per cent efficiency cut-off at 10µm aerodynamics diameter.
2 Comprises emissions from Iron and Steel and Other Industrial Combustion.
3 Emissions from construction, mining, quarrying and industrial processes.
4 Includes emissions from tyre and brake wear.

Source: NETCEN

13.5 Estimated emissions[1] of nitrogen oxides (NOx)[2] by source
United Kingdom

Thousand tonnes

By source category (UNECE/CORINAIR94)		Percentage of total in 1997	1970	1975	1980	1985	1990	1991	1992	1993	1994	1995	1996	1997
Power stations	JZGB	20	812†	797	861†	753†	781	683	672	568†	526	493	448	370
Refineries	JZGC	3	40	41	42†	38	41†	43	44	46	46	48	49	48
Combustion in fuel extraction & transformation	JZGD	3	68†	49	49†	71†	67†	67	68	69	80	53	51	50
Domestic	JZGE	4	62†	58	64	68	64	71	69	72	68	66	75	69
Commercial, public & agricultural combustion	JZGF	2	57	41	46	44	37	40	39	38	38	37†	39	36
Iron and steel	JZGG	1	75	47	25	20	22	22	22	23	24	24†	24	25
Other industrial combustion	JZGH	8	291†	259	220†	172†	171†	167	163	160	162	155	154	145
Production processes	JZGI	-	14†	15	14†	16†	11†	10	9	8	7	5	5	5
Road transport	JZGJ	48	742†	835	926†	988†	1 233†	1 227	1 176	1 124	1 076	1 005	958	883
Other transport	JZGK	11	250†	251	229†	225†	226†	225	222	217	205	198	207	200
Waste treatment and disposal	JZGL	-	5	5	11	9	8	8	7	7	7	8	6†	4
Agriculture and managed forestry	JZGM	0	10†	10	15†	14	9	8	6	–	–	–	–	–
Other sources	JZGN	-	–	–	–	1	1	1	1	–	1	1	1	1
Total	JZGO	100	2 427†	2 409	2 502†	2 419†	2 673†	2 571	2 497	2 333	2 242	2 093	2 018	1 835

1 Most of the figures in this table are based on a single NOx emission factor for each fuel which are held constant over time.
2 Expressed as nitrogen dioxide equivalent.

Source: NETCEN

13.6 Estimated emissions[1] of carbon monoxide (CO) by source
United Kingdom

Thousand tonnes

By UNECE source category		Percentage of total in 1997	1970	1975	1980	1985	1990	1991	1992	1993	1994	1995	1996	1997
Power stations	JZGP	1	117†	111	121†	105†	114†	113	110	100	106	104	102	70
Combustion in fuel extraction & transformation	JZGQ	-	29†	18	14†	17†	15†	15	15	15	16	19	18	18
Domestic	JZGR	5	1 251†	780	622†	579†	358†	382	347	369	320	258	265	250
Commercial, public & agricultural combustion	JZGS	-	45†	28	26†	26†	22†	22	21	21	20	19	19	19
Industrial combustion[2]	JZGT	3	320†	220	138†	153†	168†	158	160	158	163	165	167	173
Production processes	JZGU	8	288†	324	235†	329†	377†	356	343	349	362	368	381	389
Extraction and distribution of fossil fuels	LQMK	-	2	2	2	4	7	3	3	2	3	3	3	3
Road transport	JZGV	75	6 005†	5 990	5 913†	5 584†	5 834†	5 695	5 386	5 031	4 731	4 414	4 132	3 797
Other transport	JZGW	7	504†	465	440†	402†	369†	383	390	377	367	348	358	344
Waste treatment and disposal	JZGX	-	2	3	43	28	29	27	26	26	33	24	24†	19
Agriculture and managed forestry	JZGY	0	288†	294	449†	416	266	228	165	4	–	–	–	–
Other sources	JZGZ	-	6	6	6†	9	12	8	9	8	8	9	9	9
Total	JZHA	100	8 854†	8 240	8 008†	7 649†	7 566†	7 386	6 972	6 457	6 126	5 727	5 475	5 090

1 Most of the figures in this table are based on a single carbon monoxide emissions factor for each fuel which are held constant over time.
2 Comprises emissions from Iron and Steel and Other Industrial Combustion.

Source: NETCEN

13.7 Estimated emissions[1] of volatile organic compounds (VOCs)[2] by source
United Kingdom

Thousand tonnes

By source category (UNECE/CORINAIR94)		Percentage of total in 1997	1970	1975	1980	1985	1990	1991	1992	1993	1994	1995	1996	1997
Power stations	JZHB	-	7†	7	8†	7†	7†	7	7	7	8	8	9	8
Domestic	JZHC	2	293†	168	129†	117†	64†	67	61	61	48	36	39	37
Production processes	JZHD	14	371†	361	362†	348†	360†	352	353	343	328	323	307	295
Extraction and distribution of fossil fuels	JZHE	14	89	114	214†	285†	282†	283	271	267	287	279	294	296
Solvent use	JZHF	27	646†	642	621†	621†	680	650†	620	615	615	599	582	582
Road transport[3]	JZHG	30	687†	731	819†	872†	1 019†	1 006	960	898	840	774	713	641
Other transport	JZHH	2	62†	59	56†	54†	50†	51	52	51	50	48	49	48
Waste treatment and disposal	JZHI	1	10†	11	54†	41†	42†	39	38	37	44	37	35	31
Agriculture and managed forestry	JZHJ	-	37†	37	58†	54	35	30	22	–	–	–	–	–
Forests[4]	JZHK	8	178†	178	178†	178†	178†	178	178	178	178	178	178	178
Other sources	JZHL	1	17†	17	15†	15	16†	15	15	15	16	14	14	14
Total	JZHM	100	2 396†	2 326	2 514†	2 591†	2 733†	2 678	2 576	2 471	2 413	2 295	2 221	2 130

1 Most of the figures in this table are based on a single VOC emission factor for each source held constant over time.
2 Excluding methane.
3 Includes evaporative emissions from the petrol tank and carburettor of petrol-engined vehicles.
4 An order of magnitude estimate of natural emissions from managed and unmanaged forests.

Source: NETCEN

13.8 Chemical quality of rivers and canals[1]

	Years	Length[2] in km: Good A	B	Fair C	D	Poor E	Bad F	Total	Percentage of total: Good or fair	Poor or bad
England and Wales, Northern Ireland										
EA Region										
North West	1988-90	720	620	560	430	650	200	3 180	*73*	*27*
	1995-97	1 540	1 750	990	620	680	170	5 750	*85*	*15*
North East	1988-90	840	1 650	570	470	630	190	4 350	*81*	*19*
	1995-97	1 620	1 900	1 110	630	730	130	6 120	*86*	*14*
Midlands	1988-90	450	1 510	1 620	970	960	130	5 640	*81*	*19*
	1995-97	780	2 280	2 180	860	680	60	6 850	*89*	*11*
Anglian	1988-90	40	760	1 690	1 190	790	100	4 560	*81*	*19*
	1995-97	200	1 170	1 560	960	880	40	4 810	*81*	*19*
Thames	1988-90	290	990	1 100	520	620	40	3 560	*81*	*19*
	1995-97	360	1 160	1 060	620	570	10	3 800	*84*	*15*
Southern	1988-90	240	710	670	290	240	30	2 180	*88*	*12*
	1995-97	300	830	570	240	240	30	2 220	*87*	*12*
South West	1988-90	1 640	2 700	1 300	700	360	70	6 770	*94*	*6*
	1995-97	2 130	2 500	970	280	180	10	6 070	*97*	*3*
Welsh	1988-90	1 780	1 420	430	230	130	40	4 030	*96*	*4*
	1995-97	3 690	880	280	90	80	-	5 040	*98*	*2*
England and Wales	1988-90	6 000	10 360	7 930	4 800	4 380	810	34 280	*85*	*15*
	1995-97	10 620	12 470	8 740	4 320	4 050	453	40 650	*89*	*11*
Northern Ireland	1989-91	100	640	680	170	60	20	1 680	*95*	*5*
	1994-96	190	850	750	340	220	20	2 370	*90*	*10*

1 Based on the GQA chemical classification system.
2 Lengths are rounded to the nearest 10km and may not sum to totals.

Sources: Environment Agency; Environment and Heritage Service

	Year	Length[1] in km: Unpolluted	Fairly good	Poor	Grossly polluted	Total	Percentage of total: Unpolluted or fairly good	Poor or grossly polluted
Scotland								
Former River Purification Board								
Highland	1990	14 020	50	10	-	14 070	*100*	-
	1995	13 780	20	10	10	13 830	*100*	-
North East	1990	8 620	50	10	-	8 680	*100*	-
	1995	8 620	40	-	-	8 660	*100*	-
Tay	1990	6 230	60	-	-	6 290	*100*	-
	1995	6 110	30	-	-	6 150	*100*	-
Forth	1990	2 930	340	150	40	3 470	*94*	*6*
	1995	2 830	350	130	30	3 350	*95*	*5*
Tweed	1990	2 780	10	-	-	2 790	*100*	-
	1995	2 750	30	-	-	2 770	*100*	-
Solway	1990	4 890	70	10	10	4 980	*100*	-
	1995	4 940	20	-	-	4 960	*100*	-
Clyde	1990	9 990	630	60	20	10 690	*99*	*1*
	1995	9 860	590	70	10	10 530	*99*	*1*
Scotland	1990	49 450	1 200	240	70	50 960	*99*	*1*
	1995	48 890	1 090	220	60	50 260	*99*	*1*

1 Lengths are rounded to the nearest 10km and may not sum to totals. Figures for 1990 and 1995 include lengths of rivers on islands which were not included in previous surveys.

Source: Scottish Executive

13.9 Biological water quality of rivers and canals[1]

	Year	Length[2] in km: Good A	Good B	Fair C	Fair D	Poor E	Bad F	Total	Percentage of total: Good or fair	Poor or bad
England and Wales, Northern Ireland										
EA Region										
North West	1990	430	1 070	690	330	620	880	4 020	*63*	*37*
	1995	960	1 530	900	500	810	260	4 970	*78*	*22*
North East	1990	1 340	1 170	520	410	380	310	4 130	*83*	*17*
	1995	2 260	1 200	730	520	480	270	5 460	*86*	*14*
Midlands	1990	330	810	1 210	770	470	220	3 810	*82*	*18*
	1995	970	1 680	1 790	920	360	120	5 840	*92*	*8*
Anglian	1990	470	1 420	1 520	480	210	70	4 170	*93*	*7*
	1995	1 050	2 020	1 170	350	120	20	4 730	*97*	*3*
Thames	1990	720	1 000	690	350	240	110	3 090	*89*	*11*
	1995	1 130	1 120	790	350	160	20	3 570	*95*	*5*
Southern	1990	400	450	400	130	30	10	1 420	*97*	*3*
	1995	980	720	300	160	30	-	2 190	*98*	*2*
South West	1990	2 140	2 110	830	220	150	90	5 550	*96*	*4*
	1995	3 330	1 900	530	100	50	20	5 940	*99*	*1*
Welsh	1990	1 400	1 440	610	240	100	20	3 810	*97*	*3*
	1995	2 320	1 680	680	150	30	-	4 860	*99*	*1*
England and Wales	1990	7 210	9 480	6 470	2 940	2 200	1 710	30 000	*87*	*13*
	1995	13 000	11 860	6 890	3 040	2 040	720	37 560	*93*	*7*
Northern Ireland	1991	710	950	410	100	10	-	2 190	*100*	-
	1995	850	910	440	120	10	-	2 330	*100*	-
	1996	720	960	450	190	10	-	2 330	*100*	-

1 Based on the new GQA biological classification system. Figures for 1990 and 1995 are not directly comparable.
2 Lengths are rounded to the nearest 10km and may not sum to totals.

Sources: Environment Agency; Environment and Heritage Service

	Year	Length[1] in km of rivers and canals biologically classified: Good A	Moderate B	Poor C	Very poor D	Total	Percentage of total: Good or moderate	Poor or very poor
Scotland[2]								
Former River Purification Board								
Highland	1990	2 010	470	-	-	2 480	*100*	-
	1995	2 730	70	-	-	2 800	*100*	-
North East	1990	1 540	420	70	30	2 050	*95*	*5*
	1995	2 190	30	40	20	2 270	*97*	*3*
Tay	1990	1 240	160	20	-	1 420	*98*	*2*
	1995	2 870	130	40	10	3 060	*98*	*2*
Forth	1990	680	240	60	40	1 030	*90*	*10*
	1995	1 320	390	120	50	1 880	*91*	*9*
Tweed	1990	730	100	20	-	840	*98*	*2*
	1995	1 350	60	-	-	1 410	*100*	-
Solway	1990	1 070	130	10	-	1 210	*99*	*1*
	1995	1 540	40	10	-	1 590	*99*	*1*
Clyde	1990	1 260	460	60	70	1 850	*93*	*7*
	1995	2 970	640	60	30	3 700	*97*	*3*
Scotland	1990	8 530	1 960	230	140	10 870	*97*	*3*
	1995	14 960	1 360	280	110	16 710	*98*	*2*

1 Lengths are rounded to the nearest 10km and may not sum to totals.
2 The Scottish biological survey covered about 21 per cent of the freshwaters covered by the chemical classification system.

Source: Scottish Executive

13.10 Water pollution incidents and prosecutions
United Kingdom

Water region		Incidents reported									Incidents substantiated					
		1990	1991	1992	1993	1994	1995	1996	1997		1992	1993	1994	1995	1996	1997
North West	JZHN	4 171	4 090	4 206	4 842	4 776	5 014	4 000	3 433	JZIA	3 270	3 656	3 532	3 717	2 818	2 160
North East	JZHO	3 826	3 652	4 048	4 715	4 761	4 364	4 060	3 955	JZIB	3 059	3 642	3 243	2 576	2 143	2 404
Midlands	JZHP	6 081	5 298	6 301	6 689	6 633	6 310	5 953	5 848	JZKR	4 420	4 876	4 895	4 259	4 305	4 411
Anglian	JZHQ	1 647	3 004	3 369	3 504	3 693	3 416	3 318	3 246	JZKS	2 462	2 625	2 819	2 156	2 417	2 411
Thames	JZHR	3 441	3 416	3 595	3 538	3 763	3 977	3 782	3 525	JZKT	1 955	2 071	2 006	1 972	1 959	1 917
Southern	JZHS	1 789	1 735	1 578	1 853	1 719	2 389	2 507	2 357	JZKU	1 089	1 355	1 316	1 235	1 189	1 174
South West	JZHT	4 442	4 847	4 890	5 142	5 508	5 975	5 273	5 101	JZKV	4 278	4 129	4 340	4 558	3 042	2 847
Welsh	JZHU	2 746	3 330	3 686	4 013	4 438	4 445	3 516	3 234	JZKW	2 798	2 945	3 264	2 990	2 285	2 247
England and Wales	JZHV	28 143	29 372	31 673	34 296	35 291	35 890	32 409	30 699	JZKX	23 331	25 299	25 415	23 463	20 158	19 571
Scotland[1,2]	JZHW	..	..	..	..	..	..	..	..	JZKY	3 020	3 081	3 170	2 752	2 878	..
Northern Ireland	JZHX	1 623	1 959	1 845	1 877	2 216	2 380	2 881	..	JZKZ	..	..	..	..	2 055	..

1 1991 figure includes broad estimates for unavailable data.
2 Data for 1992, 1993 and 1994 refer to financial years 1992/93, 1993/94 and 1994/95.

Sources: Environment Agency; Scottish Environment Protection Agency; Environment and Heritage Service

13.11 Estimated abstractions from all surface and groundwater sources by purpose[1]
England and Wales

Ml/day

		1987	1988	1989	1990	1991	1992	1993	1994	1995	1996	1997
Public water supply[2]	JZLA	17 240	17 597	18 205	18 337	17 562	17 953	16 651	16 735	17 346	17 453	16 820
Spray irrigation[3]	JZLB	101	133	298	377	366	269	164	283	352	369	292
Agriculture (excl spray irrigation)	JZLC	121	120	115	128	131	129	139	115	103	136	108
Electricity supply industry[4]	JZLD	25 824	28 868	33 302	30 768	30 359	37 693	26 581	27 732	25 805	31 294	33 307
Other industry	JZLE	7 736	7 875	8 854	9 275	5 469	5 326	6 017	4 292	7 513	4 960	4 352
Mineral washing[5]	JZLF	..	..	..	..	173	213	198	222	262	250	297
Fish farming, cress growing, amenity ponds[6]	JYXG	1 100	1 262	1 111	1 323	3 882	4 479	3 818	3 985	4 268	4 338	4 211
Private water supply[7]	JZLG	..	..	..	..	..	53	81	82	98	171	162
Other[2,8]	JZLH	..	..	..	..	1 254	1 794	93	194	223	531	408
Total	JZLI	52 122	55 855	61 885	60 208	59 196	67 909	53 742	53 640	55 970	59 503	59 957

1 From 1991 data were collected on a different basis. Figures are therefore not strictly comparable with those in previous years. Some regions report licensed and actual abstracts for financial rather than calendar years. As figures represent an average for the whole year expressed as daily amounts, differences between amounts reported for financial and calendar years are small.
2 The 1991 figures include some private water supply.
3 Includes small amounts of non-agricultural spray irrigation.
4 In South West region in 1991 and 1992, Hydroelectric licences were classified as "Other", in 1993 they were classified as "Other industry" and in 1994 and 1995 they were classified correctly as "Electricity supply".
5 "Mineral washing" included for the first time as a separate category in 1991 survey.
6 Category amended to include amenity ponds, but exclude miscellaneous in 1991.
7 Private abstractions for domestic use and individual households. New category in 1992. In the 1991 survey private water supply was included in either the "public supply" or "other" categories.
8 "Other" included for the first time as a separate category in 1991 survey. In 1991 "Other" included some private domestic water supply wells and boreholes, public water supply transfer licences and frost protection use. From 1992 "private water supply" shown as separate category.

Source: Environment Agency

13.12 Bathing water surveys
United Kingdom

Compliance with EC Bathing Water Directive[1] coliform standards[2] during the bathing season

	Identified bathing waters					Numbers complying					*Percentage complying*				
	1994	1995	1996	1997	1998	1994	1995	1996	1997	1998	1994	1995	1996	1997	1998
Northumbrian[3]	34	34	34	34	34	29	33	29	32	26	*85*	*97*	*85*	*94*	*76*
Yorkshire[3]	22	22	22	22	22	20	20	20	19	21	*91*	*91*	*91*	*86*	*95*
Anglian	33	34	35	35	36	27	30	34	35	36	*82*	*88*	*97*	*100*	*100*
Thames	3	3	3	3	3	2	3	2	3	3	*67*	*100*	*67*	*100*	*100*
Southern	67	67	69	75	77	53	62	62	67	75	*79*	*93*	*90*	*89*	*97*
Wessex[3]	42	42	42	43	43	40	40	41	39	40	*95*	*95*	*98*	*91*	*93*
South West[3]	133	134	138	137	140	111	127	126	125	127	*83*	*95*	*91*	*91*	*91*
Welsh	51	56	57	64	68	39	49	52	60	64	*76*	*88*	*91*	*94*	*94*
North West	33	33	33	34	34	24	15	20	17	21	*73*	*45*	*61*	*50*	*62*
England and Wales	418	425	433	447	457	345	379	386	397	413	*83*	*89*	*89*	*89*	*90*
Scotland	23	23	23	23	23	16	19	21	18	12	*70*	*83*	*91*	*78*	*52*
Northern Ireland	16	16	16	16	16	15	15	16	14	15	*94*	*94*	*100*	*88*	*94*
United Kingdom	457	464	472	486	496	376	413	423	429	440	*82*	*89*	*90*	*88*	*89*

1 76/160/EEC
2 At least 95 per cent of samples must have counts not exceeding the mandatory limit values for total and faecal coliforms.
3 In 1993, Northumbria and Yorkshire were amalgamated into one region, now called North East; and Wessex and South West were amalgamated into one region, now called South West. Results for the old regions are given for comparison with earlier years.

Sources: Environment Agency; Scottish Environment Protection Agency; Environment and Heritage Service

13.13 Radioactive waste stocks and arisings[1]
Great Britain

m^3

	Stocks						
	1/1/86	1/1/87	1/1/88	1/1/89	1/1/91	1/4/94	1/4/98
High level waste[2]							
As stored[3]	1 351	1 430	1 463	1 575	1 686	1 639	1 804
Conditioned[4]	436	517	594	674	681	653	717
Intermediate level waste[5]							
As stored	41 887	43 602	47 783	45 313	51 558	61 494	70 948
Conditioned[6]	56 211	59 479	66 020	81 762	78 512	66 102	74 131
Low level waste[7,8]							
As stored	2 429	2 343	1 002	13 752	6 252	7 882	7 983

1 Excludes waste from defence establishments before 1991. Such intermediate and low level wastes will add no more than 20 per cent to the volumes of radioactive waste from civil sources before 1991.
2 "High level wastes" (HLW) come from the reprocessing of irradiated nuclear fuel and are intensely radioactive. They contain over 95 per cent of all the radioactivity in wastes from the nuclear cycle. HLWs are of relatively small volume, but have a high heat output as a result of the energy from radioactive decay.
3 "As stored" is the form in which the waste is currently stored. For the majority of low level waste, storage is short term prior to disposal.
4 Conditioned HLW is the form suitable for long-term storage (i.e. vitrified).
5 "Intermediate level waste" (ILW) have a lower radioactivity and heat output than HLW but radioactivity content that exceeds the upper limits for low level wastes. Examples of ILW include the irradiated metal cladding for nuclear reactor fuel, reactor components, and chemical process residues and filters.
6 "Conditioned" ILW is the form suitable for ultimate disposal to a deep repository.
7 "Low level wastes" (LLW) include concrete, rubble and soil from building demolition, discarded protective clothing and worn out or damaged plant and equipment. Unlike HLW and ILW, LLW does not normally require shielding against radiation emissions during handling and transport.
8 LLW volumes are reported in the form in which the waste will be disposed of in existing facilities or a future repository. For 1991 and 1994, the volumes are for the wastes after supercompaction.

Source: Electrowatt Engineering Services (UK) Ltd

13.14 Noise complaints received by Environmental Health Officers[1]

Number per million people

		1984/85	1985/86	1986/87	1987/88	1988/89	1989/90	1990/91	1991/92	1992/93	1993/94	1994/95	1995/96	1996/97
England and Wales														
Not controlled by the Act:[2]														
Road traffic[3]	JZLJ	41	40	41	41	42	47	46	65	59	59	60	66	62
Aircraft[4]	JZLK	15	25	19	20	47	48	34	72	73	64	111	48	58
Other	JZLL	32	27	54	48	35	59	60	99	105	84	91	90	74
Total	JZLM	88	92	114	109	124	154	140	236	237	207	262	204	194
Controlled by the Act:[2]														
Industrial/commercial premises	JZLN	636	601	654	670	713	811	913	1 037	1 108	1 120	1 320	1 466	1 455
Road works, construction and demolition	JZLO	98	103	153	196	233	180	252	148	191	168	300	229	242
Domestic premises	JZLP	1 244	1 276	1 269	1 579	1 620	1 855	2 264	2 627	3 137	3 468	3 949	4 895	5 051
Vehicles, machinery & equipment in streets[5]	JZLQ	..	..	..	..	..	..	..	..	..	..	249	225	206
Total	JZLR	1 978	1 980	2 076	2 445	2 566	2 846	3 429	3 812	4 436	4 756	5 818	6 815	6 954
Controlled by Section 62 of the Act:[6]														
Noise in streets[7]	JZLS	48	45	55	52	69	60	75	100	76	92	89	109	111
Total complaints received	JZLT	2 114	2 117	2 245	2 606	2 759	3 060	3 644	4 148	4 749	5 055	6 169	7 128	7 259

		1984	1985	1986	1987	1988	1989	1990	1991	1992[8]/93	1993/94	1994/95	1995/96	1996/97
Scotland														
Not controlled by the Act:[2]														
Road traffic[3]	JZLU	26	28	28	32	36	39	54	51	35	..	..	..	24
Aircraft[4]	JZLV	9	7	4	4	5	7	17	8	14	..	..	..	14
Other[9]	JZLW	22	9	6	4	6	7	87	128	172	..	..	..	304
Total	JZLX	57	44	38	40	47	53	158	187	221	..	..	..	342
Controlled by the Act:[2]														
Industrial/commercial premises	JZLY	392	427	437	411	436	501	537	519	560	..	..	..	845
Road works, construction and demolition	JZLZ	59	55	90	85	100	129	162	177	89	..	..	..	206
Domestic premises	JYWN	99	176	169	202	148	161	210	302	431	..	..	..	754
Vehicles, machinery & equipment in streets[5]	JYXH	..	..	..	..	..	..	..	..	..	..	..	..	12
Total	JYWO	550	658	696	698	684	791	909	998	1 080	..	..	..	1 817
Controlled by Section 62 of the Act:[6]														
Noise in streets[7]	JYXI	47	35	32	27	23	24	25	61	33	..	..	..	27
Total complaints received	JYWP	654	737	766	765	754	868	1 092	1 246	1 334	..	..	..	2 186

1 Figures relate to those authorities making returns.
2 Control of Pollution Act 1974 until 1990/91; Environmental Protection Act 1990 for 1991/92 onwards.
3 Most complaints about traffic noise are usually addressed to highways authorities or Department of Transport Regional Directors, and will not necessarily be included in these figures.
4 Complaints about noise from civil aircraft are generally received by aircraft operators, the airport companies, the Department of Transport or Civil Aviation Authority. Complaints about military flying are dealt with either by Station Commanding Officers or by Ministry of Defence headquarters. The figures in the table will not necessarily include these complaints.
5 Noise & Statutory Nuisance Act 1993.
6 Control of Pollution Act 1974.
7 Primarily includes the chimes of ice-cream vendors and the use of loudspeakers other than for strictly defined purposes.
8 1992/93 figures cover a period of 15 months from January 1992 to March 1993 due to the change from calendar year to financial year in order to come into line with England and Wales.
9 Complaints about barking dogs were included for the first time in 1990 figures for Scotland.

Sources: The Chartered Institute of Environmental Health; Royal Environmental Health Institute of Scotland

13.15 Estimated total annual waste arisings by sector
United Kingdom

Sector	Annual arisings (million tonnes)	Date of estimate	Status[1]	Source	*Percentage of total arisings*
Agriculture[2]	80	1991	NC	MAFF	*19*
Mining & quarrying					
colliery[3,4]	18	1996	NC	DETR	*4*
slate[4,5]	6	1996	NC	DETR	*1*
china clay[4,6]	20	1996	NC	DETR	*5*
quarrying[4,7]	30	1996	NC	DETR	*7*
Sewage sludge[8,9]	35	1996	C	MAFF/WSA	*8*
Dredged spoils[10]	51	1996	C	MAFF	*12*
Municipal waste[11]	28	1996/97	C	DETR	*7*
of which household	26	1996/97	C	DETR	*6*
Commercial[12]	15	..	C	DETR	*4*
Demolition & construction[13]	70	1990	C	DETR	*17*
Industrial					
blast furnace & steel slag	6	1990	C	DETR	*1*
power station ash	13	1990	C	DETR	*3*
other[14]	50	..	C	DETR	*12*
Total	423				100

1 NC = Not classed as a controlled waste under the terms of the Environmental Protection Act (Controlled Waste Regulations) 1992; C = controlled wastes under the terms of the Environmental Protection Act (Controlled Waste Regulation) 1992.
2 Refers to wastes from housed livestock only. Wet weight. Figure is not comparable with previously published estimates which included excreta from all livestock (both housed and grazing animals) and an allowance for other wastes including straw, plastics and packaging, animal carcasses and slurries.
3 Excludes wastes from opencast coal mining. Assumes ratio of spoil to saleable coal is 1:2 (1:3 in Scotland).
4 Provisional.
5 Assumes that the ratio of waste to final product is 20:1.
6 Assumes that the ratio of waste to china clay extracted is 9:1.
7 Includes wastes from primary aggregate extraction comprising the overburden and any contaminating materials found in the quarry, most of which is used as infill when the quarry is abandoned. Waste arisings are estimated at 10% of annual aggregate production.
8 Wet weight arisings, estimated on the basis of 4% solid content on average.
9 WSA = Water Services Association.
10 The data are for all UK waters, i.e. external and internal waters; provisional figure.
11 UK estimates based on returns made by Waste Disposal/Unitary Authorities in England and Wales to a DETR/Welsh Office survey.
12 Rounded to nearest 5 million tonnes.
13 Figure revised to include excavated soil and miscellaneous material as well as hard materials, e.g. brick, concrete and road planings.
14 Rounded to nearest 10 million tonnes.

Sources: Ministry of Agriculture, Fisheries and Food; Department of the Environment, Transport and the Regions; Water Services Association

13.16 Recycling[1] of selected materials
United Kingdom

Scrap reused as a percentage of consumption

		1984	1985	1986	1987	1988	1989	1990	1991	1992	1993	1994	1995	1996	1997
Ferrous	JYWQ	46	39	36	38	34	36	38	36	45	42	42	40	44	45
Aluminium	JYWR	41	40	34	34	41	44	39	41	39	29	39	53	44	40
Copper	JYWS	43	43	44	47	49	49	50	49	35	35	32	34	36	37
Lead	JYWT	67	69	66	77	73	71	67	62	64	67	74	71	73	69
Zinc	JYWU	24	22	24	24	24	23	24	24	21	21	20	20	19	19
Paper & board	JYWV	28	28	27	27	27	28	32	34	34	32	34	37	38	38
Glass cullet	JYWW	8	12	..	14	14	17	21	21	26	28	27	26	26	25

1 The ratios shown reflect the amount of secondary material used (scrap collected less exported scrap plus imported scrap) in the UK in a year as a proportion of consumption in that year.

Sources: Iron and Steel Statistics Bureau; Aluminium Federation; Customs and Excise; World Bureau of Metal Statistics; British Paper & Board Industry Federation; British Glass Manufacturers Confederation

13.17 Designated areas[1] by region at December 1997

	National Parks[3]		Areas of Outstanding Natural Beauty[3]		Defined Heritage Coasts Length (km)
	Area ('000 ha)	*Percentage of total area in region*	Area ('000 ha)	*Percentage of total area in region*	
Region					
Northern	362	*23*	226	*15*	128
Yorkshire and Humberside	315	*20*	92	*6*	82
East Midlands	92	*6*	52	*3*	-
East Anglia	-	-	91	*7*	121
South East	-	-	662	*24*	72
South West	165	*7*	712	*30*	638
West Midlands	20	*2*	127	*10*	-
North West	10	*1*	78	*11*	-
England	963	*7*	2 039	*16*	1 041
Wales	413	*20*	83	*4*	496
England and Wales	1 376	*9*	2 122	*15*	1 539
Scotland	-	-	1 002	*13*	-
Northern Ireland	-	-	285	*20*	-

1 Some areas may be in more than one category.
2 Estimated.
3 National Scenic Areas in Scotland.

Sources: Countryside Commission; Environment & Heritage Service, Northern Ireland; Countryside Council for Wales; Scottish Natural Heritage

13.18 Stock of dwellings
Great Britain

		1988	1989	1990	1991	1992	1993	1994	1995	1996	1997	1998
Estimated annual gains and losses (Thousands)												
Gains: New construction	KGAA	232.4	211.2†	195.5	184.7	172.5	179.0	186.6	191.4	180.9	180.6	168.9
Other	KGAB	16.7	16.6	14.5	14.5	10.7	10.8	10.4	11.2	10.7	..	..
Losses: Slum clearance	KGAC	8.2	7.9	9.4	4.7	4.9	5.9	4.6	6.3	6.6	..	..
Other	KGAD	11.9	9.2	9.5	12.5	7.7	7.3	9.4	8.1	7.0	..	..
Net gain	KGAE	234.5	216.2†	196.6	207.4	170.6	176.6	183.0	188.2	178.0	..	..
Stock at end of year[1]	KGAF	22 527	22 743	22 939†	23 141	23 314	23 489	23 672	23 860	24 037	24 216	..
Estimated tenure distribution at end of year (percentage)												
Owner occupied	KGAG	*64.0*	*65.2*	*65.8*	*66.3*	*66.3*	*66.6*	*66.8*	*66.8†*	*67.0*	*67.2*	..
Rented: From local authorities and new towns	KGAH	*24.0*	*22.8*	*21.9*	*21.2*	*20.5*	*19.9*	*19.1*	*18.4*	*17.9*	*17.0*	..
From housing associations	KGAI	*2.7*	*2.9*	*3.1*	*3.1*	*3.3*	*3.6*	*4.0*	*4.3*	*4.6*	*4.9*	..
From private owners including other tenures	KGAJ	*9.2*	*9.1*	*9.2*	*9.6*	*9.8*	*9.9*	*10.1*	*10.4*	*10.6†*	*10.6*	..

Note: For statistical purposes the stock estimates are expressed to the nearest thousand, but should not be regarded as accurate to the last digit.

1 Estimates are based on data from the 1971 and 1981 Censuses.

Sources: Department of the Environment, Transport and the Regions: 0117 987 8055; Scottish Executive; National Assembly for Wales

13.19 Renovations
Great Britain

Number of dwellings

		1988	1989	1990[5]	1991[6]	1992[7]	1993	1994	1995	1996	1997	1998
England												
Local authorities and new towns[1]	KGBA	204 070†	238 957	231 558	179 584	171 202	244 582	297 613	311 640	321 802	..	..
Housing associations[2]	KGBB	11 237	13 024	10 657	6 425†	7 532	6 012	6 881	..	..	..	..
Private owners: grants paid[3]	KGBC	105 309†	89 686	90 864	84 696	82 866	80 734	91 233	93 090	95 230	46 193	6 650
Wales												
Local authorities[1]	KGBE	8 333	8 444	10 843	10 513	9 491	23 264†	20 397	16 665	4 705	..	..
Housing associations[2]	KGBF	867	812	399	305	322	300	287	163	147	131	275
Private owners: grants paid[3]	KGBG	20 187	20 074	26 522	21 228†	16 642	16 061	14 619	17 314	17 345	4 911	377
Scotland												
Local authorities and new towns[4]	KGBI	68 921	52 549	84 261	70 361	80 753	90 519	117 497	75 845†	88 279	114 877	..
Housing associations[2]	KGBJ	1 225	1 122	816	1 680	1 785	1 524	1 229	1 328	767	337	..
Private owners: grants paid[3]	KGBK	31 512	26 915	23 586	23 478	37 934	21 194	19 697	19 027†	15 800	13 410	..
Great Britain												
Local authorities and new towns[1,4]	KGBM	281 324†	299 950	326 662	260 458	261 446	358 365	435 507	404 150	380 633	..	..
Housing associations[2]	KGBN	13 329	14 958	11 872	8 410†	9 639	7 836	8 397	..	..	..	..
Private owners: grants paid[3]	KGBO	157 008	136 675	140 972†	129 402	137 442	117 989	125 549	129 431	128 375	64 514	..

1 Work completed. Includes improvement for sale. Figures for Wales not available before 1984.

2 Figures for England are of work completed funded by the Housing Corporation and Local Authorities. Figures for Wales are for work completed by Housing Corporation funded housing associations only. Figures for Scotland are of work approved under specific housing association legislation. Figures from Scottish Homes not available for 1991.

3 Including grants paid to housing associations under private owner grant legislation. Figures include a small number of grants to tenants in both public and private sectors.

4 Work approved in Scotland.

5 Of the 123 190 renovation grants paid in 1990 to private owners and tenants in England and Wales, 114 192 were paid under the Housing Act 1985 and 8 998 were paid under the Local Government and Housing Act 1989.

6 Of the 106 016 renovation grants paid in 1991 to private owners and tenants in England and Wales, 36 304 were paid under the Housing Act 1985 and 69 712 were paid under the Local Government and Housing Act 1989.

7 Of the 101 627 renovation grants paid in 1992 to private owners and tenants in England and Wales, 5 070 were paid under the Housing Act 1985 and 96 557 under the Local Government and Housing Act 1989.

Sources: Department of the Environment, Transport and the Regions: 0117 987 8055;
Scottish Executive;
National Assembly for Wales

13.20 Slum clearance: dwellings demolished or closed

Number

		1987 /88	1988 /89	1989 /90	1990 /91	1991 /92	1992 /93	1993 /94	1994 /95	1995 /96	1996 /97	1997 /98
England and Wales: total	KGCA	6 683	5 509	6 157	6 210	2 383	2 072	4 061†	3 997	2 047	1 629	1 289
Demolished: In clearance areas	KGCB	3 617	2 290	3 433	4 421	1 221	1 221	1 461	2 044†	1 180	861	847
Elsewhere	KGCC	1 605	1 456	2 031	1 465	459	347	261	1 953	493†	391	332
Closed[1]	KGCD	1 461	1 763	713†	324	703	504	2 339	574	374	377	110
Scotland[2]: total	KGCE	1 801	1 175	1 007	2 224	1 816	3 685	4 311	4 371	5 542†	3 437	5 014
Unfit[3]	KGCF	794	373	447	..	..	..	..	..	..	..	..
Other[1,3]	KGCG	1 007	802	560	..	..	..	..	..	..	..	..

1 Less dwellings included in numbers demolished which had previously been reported as closed.

2 Action under the Housing Acts, Town and Country Planning Acts and other specific statutory powers and other action. Unfit houses comprise houses failing to meet the tolerable standard introduced by the Housing (Scotland) Act 1969 and houses dealt with under earlier statutory provisions as being unfit for human habitation.

3 Figures after 31 March 1990 no longer include information about unfit or other dwellings demolished, as data collection on above and below tolerable standard demolitions ceased.

Sources: Department of the Environment, Transport and the Regions: 0117 987 8055;
Scottish Executive;
National Assembly for Wales

13.21 Permanent dwellings completed

Number

	United Kingdom				England and Wales			
	Total	For local housing authorities[1]	For private owners	Housing Associations and other[2]	Total	For local housing authorities[1]	For private owners	Housing Associations and other[2]
	KAAD	KAAE	KAAF	KAAG	KAAH	KAAI	KAAJ	KAAK
1975	321 936	150 526	154 528	16 882	278 694	122 857	140 381	15 456
1976	324 769	151 824	155 229	17 716	278 660	124 152	138 477	16 031
1977	314 093	143 250	143 905	26 938	276 011	121 246	128 688	26 077
1978	288 603	112 340	152 166	24 097	254 001	96 752	134 578	22 671
1979	251 816	88 495	144 055	19 276	220 722	77 192	125 306	18 224
1980	241 986	87 974	131 974	22 038	224 919	78 012	116 164	20 743
1981	206 636	68 050	118 590	19 996	179 794	58 126	104 012	17 656
1982	182 863	39 960	129 022	13 881	159 407	33 430	113 893	12 084
1983	209 033	38 921	153 038	17 074	181 399	31 391	134 901	15 107
1984	220 465	37 381	165 574	17 510	191 163	31 160	145 282	14 721
1985	207 465	30 303	163 395	13 767	178 284	24 259	142 020	12 005
1986	223 803	25 070	178 017	13 505	187 758	20 303	156 065	11 390
1987	226 234	21 095	191 250	13 889	198 732	16 836	169 895	12 001
1988	242 359	21 129	207 423	13 807	214 156	16 687	185 733	11 736
1989	221 521	18 685	187 542	15 294	191 048	14 694	163 344	13 010
1990	203 380†	17 651†	167 473	18 256	175 099	14 483	144 849	15 767
1991	191 604	11 154	159 269†	21 181†	165 197	8 469	138 572†	18 156†
1992	180 246	5 473	147 322	27 451	154 088	3 414	127 020	23 654
1993	186 182	3 351	146 673	36 158	157 592	1 583	123 255	32 754
1994	193 591	2 882	153 341	37 368	165 210	1 321	130 066	33 823
1995	200 271	3 447	157 649	39 175	166 953	969	132 549	33 435
1996	188 076	1 616	154 350	33 181	159 507†	505†	129 309	29 693
1997	188 410	1 480	161 021	28 313	158 161	297	134 805	23 059
1998	..	..	..	..	149 165	287	127 458	21 420

	Scotland				Northern Ireland			
	Total	For local housing authorities[1]	For private owners	Housing Associations and other[2]	Total	For local housing authorities[1]	For private owners	Housing Associations and other[2]
	KAAL	KAAM	KAAN	KAAO	KAAP	KAAQ	KAAR	KAAS
1975	34 323	22 784	10 371	1 168	8 919	4 885	3 776	258
1976	36 527	21 154	13 704	1 669	9 582	6 518	3 048	16
1977	27 320	14 328	12 132	860	10 762	7 676	3 085	1
1978	25 759	9 907	14 443	1 409	8 843	5 681	3 145	17
1979	23 782	7 857	15 175	750	7 312	3 436	3 574	302
1980	20 611	7 455	12 242	914	6 456	2 507	3 568	381
1981	20 015	7 065	11 021	1 929	6 827	2 859	3 557	411
1982	16 423	3 716	11 523	1 184	7 033	2 814	3 606	613
1983	17 929	3 486	13 166	1 277	9 705	4 044	4 971	690
1984	18 838	2 633	14 115	2 090	10 464	3 588	6 177	699
1985	18 411	2 811	14 435	1 165	10 770	3 233	6 940	597
1986	18 637	2 187	14 870	1 580	10 197	2 580	7 082	535
1987	17 707	2 495	13 904	1 308	9 795	1 764	7 451	580
1988	18 272	2 730	14 179	1 363	9 931	1 712	7 511	708
1989	20 190	2 283	16 287	1 620	10 283	1 708	7 911	664
1990	20 362†	1 869†	16 461	2 032	7 919	1 299	6 163	457
1991	19 529	1 732	15 533	2 264	6 878	953	5 164	761
1992	18 443	1 010	14 389	3 044	7 715	1 049	5 913	753
1993	21 392	958	17 711	2 723	7 198	810	5 707	681
1994	21 404	661	17 753	2 990	6 977	900	5 522	555
1995	24 486	1 173	18 310	5 003	8 832	1 305	6 790	737
1996	21 402	258	18 462†	2 683	8 239†	853†	6 579†	805†
1997	22 485	108	17 870	4 507	10 168	1 075	8 346	747
1998	..	..	..	..	10 075	683	8 581	813

1 Including the Commission for the New Towns and New Towns Development Corporations, the Scottish Special Housing Association, the Northern Ireland Housing Trust and the Northern Ireland Housing Executive.

2 Dwellings provided by housing associations other than the Scottish Special Housing Association and the Northern Ireland Housing Trust and provided or authorised by government departments for the families of police, prison staffs, the Armed Forces and certain other services.

Sources: Department of the Environment, Transport and the Regions: 0117 987 8055; Scottish Executive; National Assembly for Wales; Department of the Environment, Northern Ireland

13.22 Water industry expenditure[1,2,3,4]
England and Wales

£ millions

		1989 /90	1990 /91	1991 /92	1992 /93	1993 /94	1994 /95	1995 /96	1996 /97	1997 /98	1998 /99
Operating expenditure											
Water supply	KQQX	..	2 377.4	1 953.8	2 024.4	2 108.0	2 209.0	2 347.0	2 304.2	2 331.0	2 386.1
Sewerage services	KQQY	..	1 963.1	1 580.9	1 635.4	1 757.9	1 755.0	1 763.0	1 780.7	1 856.1	1 971.4
Capital expenditure											
Water supply	KQSX	632.9	986.4	1 307.0	1 358.2	1 386.7	1 081.0	1 074.5	1 314.3	1 467.2	1 294.1
Sewerage	KQSY	269.3	475.8	559.0	455.3	410.7	381.0	375.6	479.9	455.2	507.5
Sewage treatment and disposal	KQSZ	470.9	759.2	968.0	906.0	735.3	763.0	773.0	959.4	1 296.3	1 374.5

1 Data is taken from the annual and regulatory accounts of the water and sewerage companies and water companies of England and Wales.
2 Figures are given based on current cost rather than historical cost accounting principles.
3 The elements which make up operating expenditure are as follows:
Manpower costs
Other costs of employment
Power
Local Authority rates
Water charges
Local Authority sewerage agencies
Materials and consumables
Hired and contracted services
Charge for bad and doubtful debts
Depreciation
Infrastructure renewals expenditure
Infrastructure renewals accrual
Exceptional items
Other operating costs.
4 Capital expenditure figures are the addition to tangible fixed assets including management and general expenditure but excluding infrastructure renewals expenditure.

Source: Ofwat: 0121 625 1388

Transport and communications

14

14.1 Goods transport in Great Britain

			1988	1989	1990	1991	1992	1993	1994	1995	1996	1997	1998[1]
		Tonne kms (thousand millions)											
Road[2]	KCTA		130.2	137.8	136.3	130.0	126.5	134.5	143.7	149.6	153.9	157.1	159.5
Rail[6]	KCTB	"	18.2†	17.3	15.8	15.3	15.5	13.8	13.0	13.3	15.1	16.9	17.4
Water:[3] Coastwise oil	KCTC	"	34.2	34.1	32.1	31.2	29.4	28.9	28.9	31.4	38.7	33.8	36.4
Water:[3] Other	KCTD	"	25.1	23.8	23.6	26.5	25.5	22.3	23.3	21.7†	16.6	14.3	20.8
Pipelines[4]	KCTE	"	11.1	9.8	11.1[5]†	11.1	11.0	11.6	12.0	11.1	11.6	11.2	11.2
Total	KCTF	"	218.8	222.8	218.9†	214.1	207.9	211.1	220.9	227.1	235.9	233.3	245.3
		Million tonnes											
Road[2]	KCTG		1 758	1 812	1 749	1 600	1 555	1 615	1 689	1 701	1 730	1 740	1 727
Rail[6]	KCTH	"	150	146†	140	136	122	103	97	101	102	105	102
Water:[3] Coastwise oil	KCTI	"	47	46	44	44	43	42	43	47	54	52	55
Water:[3] Other	KCTJ	"	109	109	108	100	97	92	97	96†	88	90	94
Pipelines[4]	KCTK	"	99	93	121[5]	105	106	125	161	168	157	148	148
Total	KCTL	"	2 163	2 206	2 162†	1 985	1 923	1 977	2 087	2 113	2 131	2 135	2 126

1 Water and pipeline figures are provisional for 1998.
2 All road freight by goods vehicles (over 3.5 tonnes gross vehicle weight) and small commercial vehicles.
3 Oil comprises crude oil and all petroleum products. 'Coastwise' includes all sea traffic within the UK, Isle of Man and Channel Islands. 'Other' means coastwise plus inland waterway traffic and one-port traffic (largely) crude oil direct from rigs.
4 Excluding movements of gases by pipelines.
5 The increase as compared with the corresponding figure for 1989 is believed to be largely due to changes in coverage.
6 Rail figures are in financial years e.g. 1997 = 1997/98.

Source: Department of the Environment, Transport and the Regions: 020 7890 4847

14.2 Passenger transport in Great Britain: estimated passenger kilometres

Thousand million passenger kilometres

		1988	1989	1990	1991	1992	1993	1994	1995	1996	1997	1998
Air[1]	KCTM	5	5	5	5	5	5	5†	6	6	7	7
Rail[2]	KCTN	41	39	39	38	38	36	35	36	38	41	42
Road:												
Public service vehicles[3]	KCTP	46	47	46	44	43	44	44	44	44	44†	43[5]
Cars, vans and taxis[4]	KCTQ	536	581	588	582	583	584	591	596	606†	614	616[5]
Motor cycles[4]	KCTR	6	6	6	6	6†	5	4	4	4	4	4
Pedal cycles	KCTS	5	5	5	5	5	4†	5	4	4	4	4
Total	KCTT	640†	683	689	681	679	678	684	690	703	714	716

1 Domestic scheduled and non-scheduled services, including Northern Ireland, Isle of Man and Channel Islands. Excludes air taxi services, private flying and passengers paying less than 25% of the full fare on scheduled and non-scheduled services.
2 Financial year. Consists of national rail, London Underground and other Urban Rail Systems.
3 Changes to figures for buses and coaches are estimated by deflating passenger receipts by the most appropriate price indices available.
4 Based on statistics of vehicle mileage derived from the traffic counts and estimates of average numbers of persons per vehicle, derived from the National Travel surveys. In 1998 occupancy rates were 1.57 for cars and taxis and 1.06 for motorcycles.
5 Provisional figures.

Source: Department of the Environment, Transport and the Regions: 020 7890 4847

14.3 Length of public roads in Great Britain

At 1 April in each year

Kilometres

		1988	1989	1990	1991	1992	1993	1994	1995	1996	1997	1998
Motorway[1]	KCTU	2 993	2 995	3 070	3 102	3 133	3 139	3 168	3 189	3 226	3 294	3 303
Trunk	KCTV	12 581	12 715	12 674	12 322	12 295	12 203	12 111	12 108	12 359	12 269	12 230
Principal	KCTW	34 938	35 039	35 149	35 580	35 641	35 715	35 791	35 957	35 856	35 834	36 002
Other[2]	KCTX	303 803	305 854	307 142	308 962	311 241	313 155	313 897	315 744	317 380	318 470	320 068
Total	KCTY	354 315	356 602	358 034	359 966	362 310	364 212	364 966	366 999	368 821	369 867	371 603

1 Including local authority motorways, the percentage of which is small, less than 3%.
2 Excluding unsurfaced roads and green lanes.

Source: Department of the Environment, Transport and the Regions: 020 7890 3095

14.4 Estimated traffic on all roads in Great Britain

Billion vehicle kilometres

		1988	1989	1990	1991	1992	1993	1994	1995	1996	1997	1998
All motor vehicles	KCVZ	375.7	406.9	410.8	411.6	412.1	412.2	422.6	430.9	442.5	452.5†	459.4
Cars and taxis[1]	KCWA	305.4	331.3	335.9	335.2	338.0	338.5	345.7	353.2	362.4	370.9†	375.9
Two-wheeled motor vehicles	KCWB	6.0	5.9	5.6	5.4	4.5	4.1	4.1	4.1	4.2	4.1†	4.0
Buses and coaches	KDZS	4.3	4.5	4.6	4.8	4.6	4.6	4.7	4.7	4.8	4.9	5.0
Light vans[2]	KDZT	32.0	35.4	35.7	37.2	36.7	36.5	38.3	39.1	40.4	40.8†	42.5
Other goods vehicles	KDZU	27.9	29.8	29.1	29.0	28.3	28.5	29.6	29.8	30.7	31.9†	32.1
Total goods vehicles	KDZV	60.0	65.1	64.8	66.2	65.0	64.9	68.0	68.9	71.1	72.7†	74.6
Pedal cycles	KDZW	5.2	5.2	5.3	5.2	4.7	4.5	4.5	4.5	4.3	4.1†	4.0

1 This category includes three-wheeled cars; excludes all vans whether licensed for private or for commercial use.
2 Not exceeding 3 500 Kgs gross vehicle weight.

Source: Department of the Environment, Transport and the Regions: 020 7890 3095

14.5 Motor vehicles currently licensed[1,2]
Great Britain

Thousands

		1988	1989	1990	1991	1992[9]	1993	1994	1995[10]	1996	1997	1998
Total	BMBI	23 302	24 196	24 673	24 511	24 577	24 826	25 231	25 369	26 302	26 974	27 538
Private and light goods[3]												
Private cars[3]	BMBJ	18 432	19 248	19 742	19 737	19 870	20 102	20 479	20 505	21 172	21 681	22 115
Other vehicles[3]	BMBK	2 095	2 199	2 247	2 215	2 198	2 187	2 192	2 217	2 267	2 317	2 362
Total[3]	KCTZ	20 527	21 447	21 989	21 952	22 068	22 289	22 672	22 722	23 439	23 998	24 477
Motor cycles, etc:												
Up to 50c.c.	KCUA	312	280	248	207	173	147	129	112	105	96	102
Other	KCUB	600	595	585	543	512	503	502	482	504	530	582
Total	BMBB	912	875	833	750	685	650	630	594	609	626	684
Public road passenger vehicles												
Buses, coaches, taxis, etc												
Not over 8 passengers	KCUC	59	49	42	38	35	33	33	8[11]	..	..	..
Over 8 passengers	KCUD	73	73	73	71	72	73	75	74	77	79	80
Total	BMBE	132	122	115	109	107	106	107	82	77	79	80
Goods[4,5]	BMBD	503	505	482	449	432	428	434	421	413	414	412
Agricultural tractors, etc[6]	BMBC	383	384	376	346	324	318	309	..[12]	..	..	..
Special concession group	KSBY	..	..	..	..	..	..	..	274	254	249	243
Other licensed vehicles[7]	BMBF	83	77	71	65	59	55	50	..	..	..	..
Special vehicles group	KSBZ	..	..	..	..	..	..	..	28[13]	48	48	47
Other licensed vehicles	KSCA	..	..	..	..	..	..	..	44	40	38	37
Exempt from licence duty:[8]												
Crown vehicles	KCUE	38	38	37	36	36	34	34	20[14]	15	22	15
All other exempt vehicles	KCUF	722	747	770	804	867	945	996	1 150[14]	1 409	1 500	1 543
Total[8]	BMBL	761	785	807	840	903	979	1 030	1 169	1 424	1 522	1 558

1 Since 1978, censuses have been taken annually on 31 December and are obtained from a full count of licensing records held at DVLA Swansea. Prior to this, figures were compiled from a combination of DVLA data and records held at Local Taxation offices.
2 Excludes vehicles officially registered by the Armed Forces.
3 Includes all vehicles used privately. Mostly consists of private cars and vans. However, from October 1990, goods vehicles less than 3 500 kgs gross vehicle weight are now included in this category.
4 Mostly Goods vehicles over 3 500 kgs gross vehicle weight but includes farmers' and showmen's vehicles that are less than 3 500 kgs.
5 Includes agricultural vans and lorries and showmen's goods vehicles licensed to draw trailers.
6 Includes combine harvesters, mowing machines, digging machines, mobile cranes and works trucks.
7 Includes three wheelers, pedestrian controlled vehicles and showmen's haulage and recovery vehicles. Recovery vehicle tax class introduced January 1988.
8 From 1980 - 1995, includes electric vehicles which were exempt from vehicle excise duty.
9 Source 1982 - 91: DVLA: Annual Vehicle Census analyses. Source 1992 - 93: DOT Directorate of Statistics: Vehicle Information Database.
10 Various revisions to the vehicles taxation system were introduced on 1 July 1995 and on 29 November 1995. Separate taxation classes for farmers' goods and showman's goods vehicles were abolished on 1 July 1995, with vehicles in these classes moving mainly to general HGV classes or to the special vehicles group.
11 From 1 July 1995 separate taxation of public transport vehicles with 8 or fewer seats was abolished. Remaining vehicles shown in this category are taxed on licences obtained before 1 July and will move to the private and light goods class on relicensing.
12 The old agricultural and special machines was abolished at the end June 1995. The special concession group includes agricultural and mowing machines, snow ploughs and gritting vehicles. Electric vehicles are also included in this group and are no longer exempt from VED. Steam propelled vehicles were added to this group from November 1995.
13 The special vehicles group was created on 1 July 1995 and consists of various vehicle types over 3.5 tonnes gross weight but not required to pay VED as heavy goods vehicles. The group includes mobile cranes, works trucks, digging machines, road rollers and vehicles previously taxed as showman's goods and haulage.
14 From 29 November 1995 vehicles over 25 years old and previously taxed as PLG or motorcycles or tricycles became exempt from payment of VED. Vehicles must still display a valid tax disc, but a nil rate of duty is payable.

Source: Department of the Environment, Transport and the Regions: 020 7944 3077

14.6 New vehicle registrations by taxation class
Great Britain

Thousands

		1988	1989	1990	1991	1992	1993	1994	1995[7]	1996	1997	1998
Total[1]	BMAX	2 723.5	2 828.9	2 438.7	1 921.5	1 901.8	2 074.0	2 249.0	2 306.5	2 410.1	2 597.2	2 740.3
Private and light goods[2]												
Private cars	BMAA	2 154.7	2 241.2	1 942.3	1 536.6	1 528.0	1 694.6	1 809.1	1 828.3	1 888.4	2 015.9	2 123.5
Other vehicles	BMAE	282.3	294.0	237.6	171.9	166.4	158.8	182.6	195.7	205.0	228.4	244.5
Total	KCUG	2 403.6	2 494.0	2 179.9	1 708.5	1 694.4	1 853.4	1 991.7	2 024.0	2 093.4	2 244.3	2 368.0
Motor cycles, etc:												
Up to 50c.c.	KCUH	24.7	22.9	18.8	13.1	9.1	6.5	6.9	6.3	8.9	14.2	22.6
Other	KCUI	65.4	74.4	75.9	63.6	56.5	51.9	57.7	62.6	80.7	107.1	120.7
Total	BMAD	90.1	97.3	94.4	76.5	65.6	58.4	64.6	68.9	89.6	121.3	143.3
Public road passenger vehicles												
Buses, coaches, taxis, etc												
Not over 8 seats	KCUJ	4.2	3.0	2.9	2.2	2.0	1.8	2.5	1.3[8]	..	..	..
Over 8 seats	KCUK	5.0	5.2	4.8	3.0	3.1	3.6	4.2	5.2	6.5	6.6	7.4
Total	BMAG	9.2	8.2	7.7	5.2	5.1	5.4	6.7	6.5	6.5	6.6	7.4
Heavy general goods[2] and farmers[3]												
Goods vehicles: by weight	KCUL	63.4	64.5	44.4	28.6	28.7	32.8	41.1	48.0	45.5	41.8	49.1
Privately owned agricultural tractors and engines[4]	BMAH	45.2	42.5	34.2	26.1	24.1	30.0	35.3	..	..	..	..
Special concession group	KSCB	..	..	..	..	..	..	..	33.0[9]	25.7	21.7	15.2
Other licensed vehicles[5]	KCUM	3.4	2.6	2.1	0.7	1.6	1.4	1.3	1.0	1.0	1.5	1.5
Special vehicles group	DMNR	..	..	..	..	..	..	..	3.3[10]	8.1	8.6	7.6
Exempt from license duty												
Crown vehicles	KCUN	4.1	4.4	4.0	3.2	3.7	3.1	4.1	3.3	1.2	0.7	1.1
All other exempt vehicles[1,6]	KCUO	66.0	74.2	72.2	72.6	78.7	89.4	104.3	118.1	139.1	150.7	146.6
Total	KCUP	70.1	78.6	76.2	75.8	82.3	92.4	108.4	121.4	140.3	151.4	147.7

1 Including personal and direct export vehicles.
2 For years up to 1990 retrospective counts within these new taxation classes have been estimated.
3 Owned by a farmer and available for hauling produce and requisites for his own farm.
4 Agricultural tractors are excluded unless driven on public roads.
5 Includes three wheelers, pedestrian controlled vehicles, general haulage and showmen's tractors and recovery vehicles. Recovery vehicle tax class introduced January 1988.
6 Between 1980 and 30 June 1995 electric vehicles were exempt from duty. From 1 July 1995 electric vehicles pay VED as part of the special conession group.
7 Various revisions to the vehicle taxation system were introduced on 1 July 1995 and on 29 November 1995. Separate taxation classes for farmers' goods vehicles were abolished on 1 July 1995; after this date new vehicles of this type were registered as general HGVs. The total includes 5 900 vehicles registered between 1 January and 30 June in the (now abolished) agricultural and special machines group in classes which were not eligible to register in the special concession group.
8 From 1 July 1995 separate taxation of public transport vehicles with 8 or fewer seats was abolished. After this date new vehicles of this type were registered as PLG.
9 The old agricultural and special machines taxation group was abolished at end June 1995 and replaced by the special concession group from 1 July 1995. The group includes agricultural and mowing machines, snow ploughs and gritting vehicles. Electric vehicles are also included in this group and are no longer exempt from VED. Steam propelled vehicles were added to this group from November 1995.
10 The special vehicles group was created on 1 July 1995 and consists of various vehicle types over 3.5 tonnes gross weight but not required to pay VED as heavy goods vehicles. The group includes mobile cranes, works trucks, digging machines, road rollers and vehicles previously taxed as showman's goods and haulage. Figure shown for 1995 covers period from 1 July to 31 December only.

Source: Department of the Environment, Transport and the Regions: 020 7944 3077

14.7 Driving tests
Great Britain

Applications and results

Thousands

		1988	1989	1990	1991	1992	1993	1994	1995	1996	1997	1998
Applications received	KCUQ	2 170.2	2 117.9	2 048.8	1 805.5	1 704.1	1 611.9	1 607.9	1 631.4	1 741.3	1 205.7	1 286.1
Tests conducted	KCUR	2 001.9	1 939.3	1 992.8	1 803.9	1 650.4	1 511.0	1 482.9	1 489.0	1 685.4	1 121.9	1 165.8
Tests passed	KCUS	1 028.1	1 002.2	1 033.0	917.8	831.6	730.0	697.4	684.2	748.4	526.0	535.1
Percentage of passes	KCUT	*51*	*52*	*52*	*51*	*50*	*48*	*47*	*46*	*44*	*47*	*46*

Source: Department of the Environment, Transport and the Regions: 020 7890 3077

14.8 Vehicles with licences current[1]
Northern Ireland

Number

		1988[5]	1989	1990	1991	1992	1993	1994	1995	1996[6]	1997	1998
Private cars, etc	KNKA	443 081	456 611	481 090	498 471	516 194	515 185	514 760	521 605	540 083	575 923	584 706
Cycles and tricycles	KNKB	8 957	9 460	10 167	9 684	9 023	8 634	8 775	9 142	10 026	10 932	11 663
Public road passenger vehicles:												
Taxis up to 4 seats	KNKD	1 128	741	603	656	494	462	623	739	..	..	..
Buses, coaches, over 4 seats	KNKE	2 205	2 221	2 183	2 231	2 280	2 217	2 455	1 353	2 090	2 144	2 175
Total	KNKC	3 333	2 962	2 786	2 887	2 744	2 679	3 078	2 092	2 090	2 144	2 175
General (HGV) goods vehicles:												
Unladen weight												
Not over 1 525kgs	KNKG	..	..	..	..	..	..	..	..	..	..	..
Over 1 525 and not over 12 000kgs	KNKH	17 524	18 424	..	..	..	..	..	..	..	..	..
Over 12 000kgs	KNKI	..	..	..	..	..	..	..	..	..	..	..
Total	KNKF	22 547	23 514	16 191	13 907	14 286	14 576	14 810	15 844	17 401	18 172	18 312
Farmers' goods vehicles[2]	KNKJ	4 858	4 948	4 962	4 994	5 315	5 498	5 904	2 854	..	..	..
Tractors for general haulage	KNKK	144	120	..	..	..	..	..	..	..	..	..
Tower wagons	KNKL	21	22	..	..	..	..	..	..	..	..	..
Agricultural tractors and engines, etc[2]	KNKM	7 640	7 965	8 021	7 199	6 892	7 201	7 317	7 318	5 911	6 378	5 906
Other	KNKN	..	..	513	403	343	329	354	658	1 019	1 188	1 193
Vehicles exempt from duty:												
Government owned	KNKP	5 021	5 142	5 211	5 120	5 004	4 828	4 818	3 872	3 753	3 705	3 785
Other:												
Ambulances	KNKQ	49	56	74	98	97	101	104	250	371	389	425
Fire engines	KNKR	210	234	162	250	251	205	194	301	292	291	285
Other exempt[3]	KNKS	10 147	11 712	13 937	15 305	18 163	27 089	35 837	47 626	58 340	64 447†	66 981
Total	KNKO	15 427	17 144	19 384	20 773	23 515	27 395	36 135	48 177	59 003	68 832†	71 476
Total	KNKT	500 985	517 656	543 114	558 318	578 312	586 325	595 951	611 562	639 286	683 569	695 431

1 Licences current at any time during the quarter ended December.
2 Owned by a farmer and available for hauling produce and requisites for his farm. From 1 July 1995 farmers goods taxation classes have been abolished.
3 Changes in the Mobility Allowance (DSS) have contributed to the increase in Other exempt.
4 Licences current at 31 March 1988.
5 Licences current at 31 December 1988.
6 Due to a revision of taxation classes, 1996 data are not directly comparable with previous years.

Source: Department of the Environment for Northern Ireland: 028 9054 0808

14.9 New vehicles registrations
Northern Ireland

Number

			1988	1989	1990	1991	1992	1993	1994	1995	1996	1997	1998
Private cars, etc	KNLA		66 153	67 112	68 918	63 739	62 777	65 360	70 765	73 718	77 817	83 968	91 141
Cycles and tricycles	KNLB		1 870	2 163	2 343	2 218	1 993	1 885	1 943	2 362	2 803	3 376	4 307
Public road passenger vehicles:													
Taxis[1]	KNLD		61	74	..	..	..	..	..	..	..	..	..
Buses, coaches etc	KNLE		536	528	..	..	..	..	..	..	..	..	..
Total	KNLC		597	602	606	620	551	466	1 143	622	724	714	486
Goods vehicles:													
General haulage vehicles:													
Unladen weight:													
Not over 1.5 tons	KNLH		3 847	3 813	4 301[5]	6 273	6 236	6 468	6 908	7 357	7 232	8 468	10 107
Over 1.5 tons and not over 3 tons[2]	KNLI		..	..	..	..	..	..	..	..	..	..	..
Over 3 tons	KNLJ		3 911	4 117	4 630[6]	2 619	2 471	2 593	2 668	2 935	3 492	3 521	3 572
Total	KNLG		7 758	7 930	..	..	..	..	..	..	..	..	..
Agricultural vans and lorries[3]	KNLK		476	466	..	..	..	..	..	..	..	..	..
Tractors for general haulage	KNLL		12	11	..	..	..	..	..	..	..	..	..
Goods vehicles: Total	KNLF		8 246	8 407	..	..	..	..	..	..	..	..	..
Agricultural tractors and engines, etc[4]	KNLM		1 731	1 805	1 610[7]	1 177	1 184	1 658	1 558	1 619	1 292	1 364	971
Vehicles exempt from duty:													
Ambulances	KNLO	KGVE	–	9	332[8]	330	..	..	..	..	..	..	..
Fire engines	KNLP		2	–									
Road construction vehicles	KNLQ		4	2	..	..	..	..	..	..	..	..	..
Other exempt	KNLR		1 934	2 225	2 181	2 006	2 463	4 550	6 423	8 333	10 520	10 885	10 718
Total	KNLN		1 940	2 236	..	..	..	..	..	..	..	..	..
Total	KNLS		80 537	82 325	84 921	78 982	77 675	82 980	91 408	96 946	103 880	112 296	121 302

1 From 1990 figure is for Hackneys.
2 Goods vehicles 1.5 - 3 tons from 1988.
3 Owned by a farmer and available for hauling produce and requisites for his farm.
4 Agricultural tractors are excluded unless driven on public roads.
5 From 1990 figure is for light goods.
6 From 1990 figure is for heavy goods.
7 From 1990 figure is for tractors only.
8 From 1990 figure is for crown vehicles.

Source: Department of the Environment for Northern Ireland: 028 9054 0808

14.10 Buses and Coaches[1]
Great Britain

			1987 /88	1988 /89	1989 /90	1990 /91	1991 /92	1992 /93	1993 /94	1994 /95	1995 /96	1996 /97	1997 /98
Number of vehicles[2]	KNAA	Thousands	71.7	72.0	72.5	71.9	71.4	72.7	74.8	75.3	75.7	75.7	76.2
Single deckers:		"											
Up to 16 seats	KILH	"	8.0	7.7	8.4	8.1	7.9	8.7	9.4	9.3	8.8	10.0	10.5
17 - 35 seats	KILI	"	7.8	9.8	10.3	11.5	12.4	13.6	14.5	16.0	16.5	16.6	13.6
36 plus seats	KILJ	"	31.8	31.0	31.0	30.2	29.8	29.5	30.8	30.4	30.8	30.5	35.0
All single deckers	KNAB	"	47.6	48.5	49.8	49.8	50.1	51.7	54.7	55.6	56.1	57.1	59.1
All double deckers	KNAC	"	24.1	23.5	22.7	22.2	21.3	20.9	20.1	19.7	19.6	18.6	17.1
Vehicle kilometres	KNAJ	Millions	3 664	3 738	3 835	3 838	3 879	3 864	4 026	4 077	4 105	4 141	4 191
Local bus services													
London	KILK	"	276	285	292	304	316	330	343	356	353	342	362
English metropolitan areas	KILL	"	616	634	654	650	662	679	693	720	695	692	700
English shire counties	KILM	"	1 015	1 027	1 041	1 035	1 035	1 040	1 058	1 080	1 102	1 116	1 085
Scotland	KILN	"	329	325	336	336	355	347	361	368	350	368	368
Wales	KILO	"	105	118	119	123	120	119	130	125	123	120	117
All local bus services	KILP	"	2 342	2 390	2 442	2 448	2 488	2 515	2 585	2 649	2 623	2 638	2 632
Other (non-local) services	KILQ	"	1 322	1 349	1 394	1 390	1 391	1 349	1 440	1 428	1 482	1 503	1 559
Local bus passenger journeys	KILR	"	5 292	5 215	5 074	4 850	4 665	4 480	4 385	4 420	4 383	4 350	4 337
London	KILS	"	1 207	1 211	1 188	1 178	1 149	1 129	1 117	1 167	1 205	1 242	1 294
English metropolitan areas	KILT	"	1 732	1 695	1 648	1 547	1 478	1 383	1 337	1 331	1 292	1 246	1 237
English shire counties	KILU	"	1 550	1 501	1 474	1 396	1 333	1 307	1 274	1 277	1 265	1 265	1 248
Scotland	KILV	"	647	647	613	585	571	532	525	513	494	467	438
Wales	KILW	"	156	161	151	145	133	129	133	132	127	130	120
Passenger receipts[3]	KNAX	£ million	2 389	2 514	2 723	2 866	2 969	3 036	3 222	3 334	3 431	3 558	3 778
Local bus services													
London	KILX	"	311	333	368	400	415	432	463	492	538	580	629
English metropolitan areas	KILY	"	499	523	556	585	616	626	643	656	657	672	720
English shire counties	KILZ	"	628	661	701	736	749	763	800	830	838	866	906
Scotland	KIMA	"	230	231	236	248	261	267	279	295	293	290	296
Wales	KIMB	"	62	68	69	69	69	70	76	78	81	83	81
All local bus services	KIMC	"	1 730	1 816	1 929	2 038	2 109	2 158	2 261	2 351	2 407	2 491	2 632
Other (non-local) services	KIMD	"	659	698	794	827	860	878	961	983	1 024	1 067	1 146

1 Including traditional trams.
2 For 1986/87, figures shown are annual averages; for all other years, they relate to the year end.
3 Passenger receipts on local bus services include concessionary fare reimbursement from local authorities. Receipts at current prices.

Source: Department of the Environment, Transport and the Regions: 020 7890 3076

14.11 Indices of local bus service fares
Great Britain

1992 = 100

		1988 /89	1989 /90	1990 /91	1991 /92	1992 /93	1993 /94	1994 /95	1995 /96	1996 /97	1997 /98	1998 /99
London	KNEP	71.2	78.5	86.6	94.9	102.2	110.3	116.7	122.7	127.9	132.7†	138.0
English metropolitan areas	KILD	70.6	76.4	84.9	94.8	101.5	106.3	110.4	116.2	122.4	129.7	135.8
English shire counties	KILE	75.1	81.3	89.7	96.7	100.9	103.9	108.3	112.9	118.3	124.4	130.3
Scotland	KILF	77.9	81.8	88.1	94.8	101.4	105.0	111.4	115.8	124.2	133.9	140.0
Wales[1]	KILG	..	..	..	..	100.8	104.6	108.6	112.3	116.5	122.8	129.7
All Great Britain	KNEU	74.1	80.0	88.0	95.8	101.3	105.9	110.9	116.0	121.9	128.4†	134.3
Retail prices index[2]	KNEV	78.6	84.8	93.0	97.4	100.4	102.2	105.0	108.4	111.0	114.7	118.5

1 Figures for Wales prior to 1992/93 are omitted because insufficient data are available.
2 The published Retail Prices Index has been rescaled to 1992 = 100 for comparability.

Source: Department of the Environment, Transport and the Regions: 020 7890 3076

14.12 Road accidents, vehicles involved and casualties
Great Britain

Number

		1988	1989	1990	1991	1992	1993	1994	1995	1996	1997	1998
Road accidents[1]	KKKA	246 994	260 759	258 441	235 889†	233 104	228 975	234 254	230 544	236 193	240 287	238 923
Vehicles involved:												
Pedal cycles	KKKB	26 561	29 327	27 108	25 439†	25 299	24 612	25 415	25 497	25 102	25 200	23 423
Motor vehicles	KKKC	404 571	429 237	427 625	391 890†	390 736	386 458	396 750	388 836	402 001	413 197	413 172
Two-wheeled motor vehicles	KKKD	44 279	43 995	40 404	31 722†	27 660	25 836	25 127	24 219	23 798	25 211	25 514
Cars and taxis	KKKE	303 693	325 213	330 181	308 076†	313 382	312 790	322 946	318 083	331 091	338 924	337 794
Light goods vehicles[2]	KKKF	24 671	25 793	24 052	21 802†	20 490	19 069	19 495	18 674	19 186	20 070	20 083
Heavy goods vehicles[3]	KKKG	16 376	17 894	16 524	15 241†	14 500	14 417	14 572	13 771	13 582	14 385	14 526
Buses and coaches	KKKH	12 086	12 711	12 200	11 417†	11 264	10 947	11 413	10 994	11 196	11 241	11 762
Other motor vehicles	KKKI	3 466	3 631	3 664	3 632†	3 440	3 399	3 197	3 095	3 148	3 366	3 493
Vehicles involved per hundred million kilometres travelled:												
Pedal cycles	KKKL	508	563	517	489†	538	547	565	567	584	615	591
All two- wheeled motor vehicles	KKKM	734	741	725	587†	615	630	613	591	567	615	646
Cars and taxis	KKKN	99	98	98	92	93†	93	94	90	91	92	90
Light goods vehicles[2]	KKKO	77	73	69	59	56	52	51	48	47	50	47
Heavy goods vehicles[3]	KKKP	59	60	57	53	51	51	49	46	44	45	45
Total casualties	KKKQ	322 305	341 592	341 141	311 368†	310 753	306 135	315 359	310 687	320 578	327 803	325 212
Killed[4]:												
Total	KKKR	5 052	5 373	5 217	4 568	4 229	3 813†	3 650	3 621	3 598	3 599	3 421
Pedestrians	KKKS	1 753	1 706	1 694	1 496	1 347	1 241	1 124	1 038	997	973	906
Pedal cycles	KKKT	227	294	256	242	204	186	172	213	203	183	158
All two-wheeled motor vehicles	KKKU	670	683	659	548	469	427	444	445	440	509	498
Cars and taxis	KKKV	2 142	2 426	2 371	2 053	1 978	1 760	1 764	1 749	1 806	1 795	1 696
Others	KKKW	260	264	237	229	231	199†	146	176	152	139	163
Killed and seriously injured[5]:												
By age group[6]												
0-4	KKKY	1 339	1 342	1 363	1 211†	1 141	1 010	993	968	831	826	823
5-9	KKKZ	3 122	3 094	3 217	2 663†	2 588	2 166	2 457	2 186	2 208	2 024	1 904
10-14	KKLA	3 871	3 943	3 832	3 376†	3 328	3 099	3 330	3 323	3 161	2 964	2 721
15-19	KKLB	12 866	12 616	11 595	8 755†	7 509	6 179	6 456	6 339	6 641	6 022	5 871
20-24	KKLC	11 629	11 512	10 583	8 840†	8 016	7 017	6 922	6 678	6 182	5 620	5 163
25-29	KKLD	7 026	7 404	7 240	6 455†	6 137	5 678	5 632	5 641	5 407	5 474	4 921
30-39	KKLE	7 999	8 047	7 905	6 983†	7 384	7 082	7 451	7 605	7 884	8 050	7 721
40-49	KKLF	5 892	5 873	5 737	5 137†	5 196	4 852	5 084	5 064	4 923	4 805	4 623
50-59	KKLG	4 620	4 532	4 425	3 860†	3 748	3 507	3 666	3 482	3 492	3 578	3 640
60 and over	KKLH	9 517	9 596	9 113	8 239†	7 749	7 531	7 406	7 163	6 649	6 552	6 220
By type of road user:												
Child pedestrians[7]	KIJS	5 897	5 836	5 914	5 097†	4 901	4 231	4 610	4 400	4 132	3 954	3 737
Adult pedestrians	KIJT	11 781	11 460	11 228	9 733†	9 125	8 260	8 114	7 716	7 300	6 925	6 592
Child pedal cyclists[7]	KIJU	1 576	1 623	1 490	1 345	1 195	1 146	1 234	1 249	1 231	1 016	915
Adult pedal cyclists	KIJV	3 272	3 477	3 075	2 800	2 752†	2 598	2 710	2 673	2 517	2 542	2 345
Moped riders	KKLN	1 903	1 729	1 417	989	807	682†	562	560	442	419	403
Motor scooter riders	KKLO	380	382	289	226	152	135	102	110	98	114	145
Motor scooter passengers	KCUU	46	44	24	20	11	11	10	13	10	8	4
Motor cycle riders	KCUV	9 364	9 414	8 576	6 601†	5 787	5 537	5 508	5 418	5 177	5 426	5 457
Motor cycle passengers	KCUW	924	884	783	651	569	502	473†	504	463	467	421
Car and taxi drivers	KCUX	17 576	17 834	17 403	15 630†	15 406	14 280	14 877	14 557	15 015	14 881	13 841
Car and taxi passengers	KCUY	11 770	11 850	11 717	9 765†	9 718	8 553	9 015	8 904	9 033	8 310	7 835
Users of buses and coaches	KCUZ	912	835	807	725	655	725	815	836†	695	601	631
Users of goods vehicles	KCVA	2 485	2 673	2 399	2 122	1 967	1 717	1 672	1 741	1 544	1 501	1 509
Users of other vehicles	KCVB	255	247	251	229	214	216†	213	237	201	226	204
All severities:												
Total	KCVC	322 305	341 592	341 141	311 368†	310 753	306 135	315 359	310 687	320 578	327 803	325 212
Pedestrians	KCVD	58 843	60 080	60 230	54 030†	51 612	48 128	48 695	47 083	46 450	45 601	44 886
Vehicle users	KCVE	263 462	281 512	280 911	257 336†	259 138	258 005	266 663	263 603	274 128	282 202	280 326
Breath tests on car drivers involved in accidents:												
All drivers	KCVG	303 693	325 213	330 181	308 076†	313 382	312 790	322 946	318 083	331 091	338 924	337 794
Tested	KCVH	60 798	81 771	91 661	90 138†	90 295	88 282	91 927	99 631	133 347	157 373	173 610
Failed test[8]	KCVI	8 549	8 508	8 073	7 356†	6 893	6 171	6 366	6 639	7 303	7 087	6 690

1 Accidents on public roads, involving injury, which are reported to the police within 30 days of occurrence.
2 1.5 tons unladen weight and under. The definition of LGV changed in 1994 to vehicles up to 3.5 tonnes maximum gross weight.
3 Over 1.5 tons unladen weight. The definition of HGV changed in 1994 to vehicles licensed as over 3.5 tonnes maximum permissible gross weight.
4 Died within 30 days of accident.
5 Hospital in-patients *plus* casualties with any fracture, internal injury, concussion, crushing, severe general shock, etc, *plus* deaths after 30 days.
6 These figures may not add up to total killed and seriously injured figures, due to the exclusion of road users whose age and casualty type was not reported.
7 Age 0 - 15.
8 Positive result, or refused to provide a specimen.

Source: Department of the Environment, Transport and the Regions: 020 7890 3078

14.13 Road accident casualty rates: by type of road user and severity
Great Britain

Rate per 100 million vehicle kilometres

		1988	1989	1990	1991	1992	1993	1994	1995	1996	1997	1998
Pedal cyclists:												
Killed	KNMA	4.3	5.7	4.9	4.7	4.3	4.1	3.8	4.7†	4.7	4.5	4.0
Killed or seriously injured	KNMB	93	99	87	81	84†	84	89	88	88	88	84
All severities	KNMC	494	552	501	477†	522	535	552	554	572	601	578
Two-wheel motor vehicle riders:												
Killed	KNMD	10.0	11.0	11.0	9.3	9.7	9.6†	9.7	10.0	9.9	12.0	12.0
Killed or seriously injured	KNME	193	195	184	145	150	155†	151	148	136	145	152
All severities	KNMF	657	668	646	525†	551	566	550	528	509	554	582
Car drivers:												
Killed	KNMG	0.4	0.5	0.4	0.4	0.4	0.3	0.3	0.3	0.3	0.3	0.3
Killed or seriously injured	KNMH	6.0	5.0	5.2	4.7	4.6	4.2	4.3	4.1	4.1	4.0	3.7
All severities	KNMI	33	33	34	32	34	34	35	34	36	36	36
Bus and coach drivers:												
Killed	KNMJ	0.1	–	0.1	–	0.1	0.1	–	–	–	–	–
Killed or seriously injured	KNMK	1.0	1.0	1.5	1.3	1.3†	1.3	1.3	1.8	1.3	1.5	1.4
All severities	KNML	13	15	14	14	15	14	16	16	17	17	18
Light goods vehicle drivers:												
Killed	KNMM	0.3	0.3	0.2	0.2	0.2	0.2	0.1	0.1	0.1	0.1	0.1
Killed or seriously injured	KNMN	4.0	3.0	3.0	2.6	2.3	2.1	2.2†	2.1	1.8	1.7	1.6
All severities	KNMO	20	19	18	16	15	14	15†	13	13	14	13
Heavy goods vehicle drivers:												
Killed	KNMP	0.2	0.2	0.2	0.2	0.2	0.2	0.1	0.2	0.2	0.1	0.2
Killed or seriously injured	KNMQ	2.0	2.0	2.1	2.0	2.0	1.9	1.7	1.9†	1.5	1.5	1.5
All severities	KNMR	11.0	11.0	11.0	10.0	10.0	10.0	10.0	10.0†	8.9	8.9	9.2
All road users[1]:												
Killed	KNMS	1.4	1.3	1.3	1.1	1.0	0.9	0.9	0.8	0.8	0.8	0.7
Killed or seriously injured	KNMT	19.0	17.0	16.0	13.5†	12.9	11.8	11.8	11.3	10.8	10.2	9.6
All severities	KNMU	89.0	84.0	82.0	74.7†	74.9	73.8	74.0	71.4	71.8	71.8	70.2

1 Includes other road users and road user not reported.

Source: Department of the Environment, Transport and the Regions: 020 7890 3078

14.14 Road goods transport
Great Britain

Analysis by mode of working and by gross weight of vehicle[1,2]

			1988	1989	1990	1991	1992	1993	1994	1995	1996	1997	1998
Estimated tonne kilometres		Thousand million											
Own account	KNNC	"	37.2	36.8	36.0	38.8	34.9	35.4	37.0	37.2	37.7	37.4	37.6
Public haulage	KNND	"	87.6	95.3	94.7	85.8	86.4	93.2	100.8	106.5	109.1	112.2	114.3
Total	KNNB	"	124.8	132.1	130.6	124.6	121.3	128.6	137.8	143.7	146.8	149.6	151.9
By gross weight of vehicle		Billion tonne kms											
Not over 25 tonnes	KNNF	"	30.1	30.6	29.2	28.0	26.3	25.9	26.6	24.7	25.3	24.3	22.5
Over 25 tonnes	KNNG	"	94.7	101.5	101.4	96.6	95.0	102.7	111.2	119.0	121.5	125.2	129.4
Estimated tonnes carried		Millions											
Own account	KNNJ	"	696	684	667	643	620	612	618	622	618	599	589
Public haulage	KCVJ	"	957	1 020	978	862	843	911	980	987	1 011	1 044	1 041
Total	KNNI	"	1 653	1 704	1 645	1 505	1 463	1 523	1 597	1 609	1 628	1 643	1 630
By gross weight of vehicle		Million tonnes											
Not over 25 tonnes	KCVL	"	682	672	632	578	544	541	527	467	447	419	382
Over 25 tonnes	KCVM	"	971	1 033	1 013	927	919	982	1 070	1 142	1 181	1 224	1 248

1 Includes the small amount of work performed in Northern Ireland by British registered goods vehicles.

2 Excludes work done by small goods vehicles not exceeding 3.5 tonnes gross vehicle weight.

Source: Department of the Environment, Transport and the Regions: 020 7890 4442

14.15 National Rail and Railtrack: assets and privately-owned freight vehicles operated at year end

Great Britain

		Unit	1988 /89	1989 /90	1990 /91	1991 /92	1992 /93	1993 /94	1994 /95	1995 /96	1996 /97	1997 /98	1998 /99
Permanent way and stations													
Route open for traffic: total	KNCZ	Kilometres	16 599	16 587	16 584	16 558	16 528	16 536	16 542	16 666	16 666	16 656	16 659
Electrified[6]	KNDA	"	4 376	4 546	4 912	4 886	4 910	4 968	4 970	5 163	5 176	5 166	5 166
Non-electrified[7]	KNDB	"	12 223	12 041	11 672	11 672	11 618	11 568	11 572	11 503	11 490	11 490	11 493
Open for passenger traffic	KNDC	"	14 309	14 318	14 317	14 291	14 317	14 357†	14 359	15 002	15 034	15 024	15 038
Track open for traffic: total	KNDD	"	37 868	37 849	37 810	37 757	37 731	37 740	..	..	..	..	..
Running lines	KNDE	"	32 628	32 623	32 604	32 551	32 525	32 534	..	..	..	..	..
Sidings	KNDF	"	5 240	5 226	5 206	5 206	5 206	5 206	..	..	..	..	..
Stations: total	KNDG	Number	2 596	2 598	2 615	2 551	2 543	2 553	2 565	2 497†	2 498	2 495	2 499
Passenger[8]	KNDH	"	2 470	2 471	2 488	2 468	2 482	2 493	2 506	2 497†	2 498	2 495	2 499
Parcel[9]	KNDI	"	1	2	2	–	–	–	–	–	–	–	–

1 Excludes multiple unit power cars, mainly passenger-carrying.
2 Includes shunting locomotives.
3 Includes passenger-carrying multiple unit power cars.
4 Excludes 1 narrow gauge vehicle sold to Brecon Mountain Railway in 1989. Includes power cars.
5 Includes brake vans and other operating vehicles up to and including 1991/2.
6 26km of route transferred to Greater Manchester Metro Ltd in 1991/2.
7 Excludes 19km of narrow gauge line sold to Brecon Mountain Railway Ltd in 1989.
8 Excludes 8 stations on narrow gauge line sold to Brecon Mountain Railway Ltd in 1989. From 1995/96 the number of stations shown include only those owned by Railtrack.
9 Parcel stations are those handling parcels traffic only. Stations accepting both parcels and passengers are included in Passenger stations. (236 at 31/3/94)
10 Includes depots owned by Freightliners Ltd from 1988/9.

Source: Department of the Environment, Transport and the Regions: 020 7890 3089

14.16 National Rail: passenger and freight receipts and traffic
Great Britain

		Units	1988 /89	1989 /90	1990 /91	1991 /92	1992 /93	1993 /94	1994[8] /95	1995 /96	1996 /97	1997 /98	1998 /99
Passenger receipts: total[2,3]	KNDL	£ million	1 802.9†	1 907.0	2 056.7	2 116.9	2 153.5	2 192.6	2 171.0	2 379.4	2 572.5	2 820.8	3 090.5
Ordinary fares	KGVF	"	1 290.9†	1 356.8	1 482.8	1 513.9	1 550.6	1 577.0	1 559.5	1 719.5	1 870.3	2 047.6	2 243.6
Season tickets	KNDO	"	512.0†	550.2	573.9	603.0	602.9	615.6	611.5	659.9	702.2	773.2	846.9
Traffic													
Passenger kilometres: total (estimated)	KNDZ	Millions	34 323.0†	33 323.0	33 191.0	32 466.0	31 718.0	30 357.0	28 655.0	30 039.0	32 135.0	34 190.0[9]	35 135.0[9]
Ordinary fares	KNEN	"	23 226.0†	22 424.0	22 803.0	22 435.0	22 280.0	21 332.0	20 659.0	22 152.0	23 389.0	24 877.0	25 310.0
Season tickets	KNEC	"	11 097.0†	10 899.0	10 389.0	10 031.0	9 438.0	9 025.0	7 996.0	7 886.0	8 746.0	9 312.0	9 825.0
Freight traffic originating: total[6,7]	BMHA	Million tonnes	149.5	143.1	138.2	135.8	122.4	103.2†	97.3	100.7	101.7	105.4	102.1
Coal and coke	BMHB	"	79.2	75.8	74.7	75.1	67.9	48.9†	42.5	45.2	52.2	50.2	45.2
Iron and steel	BMHC	"	20.6	18.9	18.0	17.8	15.9	15.8	16.9	..	..	..	..
Other[10]	BMHD	"	49.7†	48.4	45.4	43.0	38.5	38.5	37.9	55.5	49.5	55.2	56.9
Tonne kilometres: total (estimated)[6,7]	KNEJ	Millions	18 100.0	16 700.0	16 000.0	15 300.0	15 486.0	13 765.0	12 979.0	13 272.0	15 144.0	16 949.0†	17 369.0
Coal and coke	KNEK	"	4 826.0	4 638.0	5 047.0	4 988.0	5 429.0	3 945.0	3 272.0	3 636.0	3 880.0	4 414.0†	4 462.0
All other freight train	KNEL	"	13 274.0	12 062.0	10 953.0	10 312.0	10 057.0	9 820.0	9 706.0	9 636.0	11 264.0	12 535.0†	12 907.0

1 National Rail traffic was affected by the miners' strike and associated action by National Rail staff in 1984 and 1985.
2 Grants received by National Rail in respect of certain subsidised passenger services are not included in passenger receipts. Figures include various miscellaneous receipts (eg car park charges).
3 Receipts and passenger kilometres include through booked journeys which began on London Underground or other administrations' trains.
4 From 1989/90 includes non-rail haulage.
5 Season ticket journeys and passenger kilometres are calculated on a rate of 480 journeys per year per annual season ticket and 540 journeys for shorter periods. Return tickets are counted as two journeys.
6 Excluding free-hauled traffic on freight trains.
7 Excluding freight carried by coaching trains.
8 Rail traffic was affected by the signal workers' strike.
9 There is some underestimation of passenger kilometres in 1997/98 and 1998/99. The figures are therefore subject to revision and must be interpreted with caution.
10 Includes iron and steel from 1995/96.

Source: Department of the Environment, Transport and the Regions: 020 7890 3089

14.17 London Underground: receipts, operations and assets

		Unit	1988 /89	1989 /90	1990 /91	1991 /92	1992 /93	1993 /94	1994 /95	1995 /96	1996 /97	1997 /98	1998 /99
Receipts													
Passenger: total	KNOA	£ million	432	462	531	559	589	637	718	765	797	899	977
Ordinary[1]	KNOB	"	228	262	295	307	322	350	396	430	449	510	547
Season tickets	KNOD	"	204	200	236	252	267	287	322	335	348	389	430
Traffic													
Passenger journeys: total	KNOE	Million	815	765	775	751	728	735	764	784	772	832	866
Ordinary[1]	KNOF	"	363	380	399	368	365	376	398	416	418	448	463
Season tickets	KNOH	"	452	385	376	383	363	359	366	368	354	384	403
Passenger kilometres	KNOI	"	6 292	6 016	6 164	5 895	5 758	5 814	6 051	6 337	6 153	6 479	6 716
Operations													
Loaded train kilometres	KNOJ	Million	50.5	50.1	52.4	52.5	52.5	52.7	54.8	57.2	58.6	62.1	61.2
Place kilometres	KNOK	"	43.6	43.0	44.9	45.2	45.3	45.6	49.4	51.6	52.2	55.5	54.8
Rolling stock													
Railway cars	KNOL	Number	3 950	3 908	3 880	3 880	3 895	3 955	3 923	3 923	3 867	3 886	3 923
Seating capacity	KNOM	Thousand	169.8	171.6	171.6	166.9	166.9	168.5	167.7	165.4	164.5	164.0†	165.0
Permanent way and stations													
Route kilometres open for traffic	KNON	Kilometres	394	394	394	394	394	394	392	392	392	392	392
Stations	KNOO	Number	246	245	246	246	246	245	245	245	245	245	246

1 Includes one day travelcards and concessionary fares.

Source: Department of the Environment, Transport and the Regions: 020 7890 3076

14.18 Accidents on railways
Great Britain

Number

		1988	1989	1990	1991[1]	1991/92	1992/93	1993/94	1994/95	1995/96	1996[2]/97	1997/98	1998/99
Train accidents													
Number of accidents: total	KNIA	1 330	1 434	1 283	242	960	1 152	977	907	989	1 753	1 863†	1 728
Collisions	KNIB	296	329	290	37	187	154	135	125	123	120	127†	120
Derailments	KNIC	231	192	183	46	144	205	113	149	104	119	93	119
Running into level crossing gates and other obstructions	KNID	486	510	473	81	340	532	445	397	488	741	680†	589
Fires	KNIE	229	283	257	63	225	202	247	217	256	302	344	336
Missiles through cab windscreen[3]	KYAL	..	..	..	..	..	..	..	..	..	468	619	564
Miscellaneous	KNIF	88	120	80	15	64	59	37	19	18	3	–	–
Persons killed	KNIG	40	18	4	3	11	5	6	12	7	1	10	3
Passengers	KNIH	34	6	–	2	2	–	–	3	1	1	7	–
Railway staff	KNII	2	6	1	–	2	1	–	5	1	–	–	–
Others	KNIJ	4	6	3	1	7	4	6	4	5	–	3	3
Persons injured	KNIK	705	404	243	569	391	153	246	296	166	255	244	84
Passengers	KNIL	615	311	157	549	307	66	134	190	62	180	190	40
Railway staff	KNIM	68	71	73	18	65	74	95	83	75	61	39	31
Others	KNIN	22	22	13	2	19	13	17	23	29	14	15	13
Other accidents through movement of railway vehicles													
Persons killed	KNIO	53	41	69	13	49	27	26	27	17	20	32†	27
Passengers	KNIP	34	25	35	6	28	16	14	12	7†	13	15	16
Railway staff	KNIQ	11	8	19	5	9	5	3	3	2	2	3	1
Others	KNIR	8	8	15	2	12	6	9	12	8†	5	14	10
Persons injured	KNIS	2 810	2 769	2 777	412	2 360	2 558	2 373	2 417	3 078	828	883	982
Passengers	KNIT	2 721	2 698	2 658	394	2 254	2 414	2 209	2 215	2 860	559	617	688
Railway staff	KNIU	85	63	116	17	104	140	157	201	206	253	249	281
Others	KNIV	4	8	3	1	2	4	7	1	12	16	17	13
Other accidents on railway premises													
Persons killed	KNIW	4	10	5	–	8	7	8	3	4	4	6†	7
Passengers	KNIX	1	2	2	–	1	2	2	2	2	3	4†	3
Railway staff	KNIY	3	4	2	–	6	5	5	1	2	–	–	3
Others	KNIZ	–	4	1	–	1	–	1	–	–	1	2	1
Persons injured	KNJA	7 351	7 628	7 028	1 489	6 900	7 524	8 244	7 933	9 046	3 668†	4 186	4 172
Passengers	KNJB	4 035	4 429	3 647	626	3 419	3 883	4 468	4 385	4 762	1 710	1 940	1 963
Railway staff	KNJC	3 142	3 001	3 168	839	3 335	3 418	3 600	3 379	4 088	1 838	2 151	2 134
Others	KNJD	174	198	213	24	146	223	176	169	196	120	95	75
Trespassers and suicides													
Persons killed	KNJE	334	298	311	63	300	264	262	254	246	251	264	249
Persons injured	KNJF	129	92	123	23	119	89	97	85	82	106	136	153

1 Covering the period January - March 1991.
2 New accident reporting regulations (RIDDOR 95) came into force on 1 April 1996.
3 Category now reportable under RIDDOR 95.

Source: Department of the Environment, Transport and the Regions: 020 7890 3089

14.19 Railways: permanent way and rolling stock
Northern Ireland

At end of year

Number

		1988	1989	1990	1991	1992	1993	1994	1995	1996	1997	1998
Length of road open for traffic[1] (km)	KNRA	332	332	332	332	332	330	331	333	335	335	335
Length of track open for traffic (km)												
Total	KNRB	553	553	553	553	553	504	503	506	506	505	526
Running lines	KNRC	500	500	500	500	500	463	462	464	464	464	484
Sidings (as single track)	KNRD	53	53	53	53	53	41	41	42	42	42	42
Locomotives												
Diesel-electrics	KNRE	12	12	9	9	9	9	11	11	8	6	5
Passenger carrying vehicles												
Total	KNRF	115	115	115	115	114	112	112	112	112	112	120
Rail motor vehicles:												
Diesel-electric, etc	KNRG	31	30	30	30	30	30	30	30	30	30	28
Trailer carriages:												
Total locomotive hauled	KNRH	28	28	28	28	28	28	28	28	28	28	38
Ordinary coaches	KNRI	26	26	26	26	26	26	26	26	26	26	36
Restaurant cars	KNRJ	2	2	2	2	2	2	2	2	2	2	2
Rail car trailers	KNRK	56	56	56	56	56	54	54	54	54	54	54
Non-passenger carrying vehicles												
Post Office and luggage vans, etc	KNRL	-	-	-	-	-	-	-	-	-	-	-
Trucks and wagons owned												
Total	KNRM	17	17	17	17	18	-	-	-	-	-	-
Merchandise wagons:												
Open	KNRN	-	-	-	-	-	-	-	-	-	-	-
Covered	KNRO	-	-	-	-	-	-	-	-	-	-	-
Rail and timber trucks	KNRP	-	-	-	-	-	-	-	-	-	-	-
Brake vans	KNRQ	-	-	-	-	-	-	-	-	-	-	-
Special wagons	KNRR	17	17	17	17	18	-	-	-	-	-	-
Containers	KNRS	-	-	-	-	-	-	-	-	-	-	-
Rolling stock for maintenance and repair	KNRT	46	55	55	55	36	42	41	41	41	41	26

1 The total length of railroad open for traffic irrespective of the number of tracks comprising the road.

Source: Department of the Environment for Northern Ireland: 028 9054 0540

14.20 Operating statistics of railways
Northern Ireland

		Unit	1988	1989	1990	1991	1992	1993	1994	1995	1996	1997	1998
Maintenance of way and works													
Material used:													
Ballast	KNSA	Thousand m^2	28.3	33.0	25.0	22.5	30.2	16.0	33.2	22.5	27.0	51.3	38.5
Rails	KNSB	Thousand tonnes	1.72	1.60	0.91	1.68	2.26	2.00	1.80	1.76	2.12	0.37	2.50
Sleepers	KNSC	Thousands	23.20	18.00	19.00	23.20	31.23	14.60	22.40	22.90	27.50	5.10	32.00
Track renewed	KNSD	Km	17.20	9.70	8.80	10.40	14.00	16.00	12.00	16.00	20.00	2.40	22.50
New Track laid	KPGD	Km	..	..	..	..	..	..	3.2	2.5	-	-	-
Engine kilometres													
Total[1]	KNSE	Thousand Km	3 572	3 575	3 200	3 200	3 540	3 640	3 640	4 000	4 100	4 100	4 100
Train kilometres:													
Total	KNSF	"	3 394	3 391	3 844	3 410	3 210	3 210	3 210	3 570	3 670	3 670	3 670
Coaching	KNSG	"	3 391	3 393	3 840	3 406	3 206	3 206	3 206	3 566	3 666	3 666	3 666
Freight	KNSH	"	4	4	4	4	4	4	4	4	4	4	4

1 Including shunting, assisting, light, departmental, maintenance and repair.

Source: Department of the Environment for Northern Ireland: 028 9054 0540

14.21 Main output[1] of United Kingdom airlines

		1988	1989	1990	1991	1992	1993	1994	1995	1996	1997	1998
All services: total	KNTA	17 225	18 920	20 375	20 166	23 145	25 144	27 714	29 904	32 210	35 538	40 021
Percentage growth on previous year	KNTB	*8.7*	*9.9*	*7.7*	*-1.0*	*14.8*	*8.5*	*10.2*	*7.4*	*7.7*	*10.3*	*12.5*
Scheduled services: total	KNTC	12 405	13 427	15 274	15 188	17 065	18 605	20 360	22 016	23 793	26 504	29 756
Percentage growth on previous year	KNTD	*8.5*	*8.2*	*13.8*	*-0.6*	*12.4*	*9.0*	*9.4*	*8.1*	*8.1*	*11.4*	*12.3*
Non-scheduled services: total	KNTE	4 820	5 496	5 103	4 978	6 079	6 510	7 265	7 695	8 044	9 034	10 265
Percentage growth on previous year	KNTF	*9.0*	*14.0*	*-7.2*	*-2.5*	*22.0*	*7.1*	*13.0*	*5.9*	*4.5*	*7.3*	*13.3*

1 Available tonne kilometres (millions).

Source: Civil Aviation Authority: 020 7379 7311

14.22 Air traffic between the United Kingdom and abroad[1]

Aircraft flights and passengers carried

Thousands

		1988	1989	1990	1991	1992	1993	1994[3]	1995	1996	1997	1998
Flights												
United Kingdom airlines												
Scheduled services	KNUA	216.9	245.0	271.0	250.5	283.0	290.8	325.8	342.1	373.0	410.3	443.7
Non-scheduled services	KNUB	225.5	221.3	207.9	202.6	236.2	215.2	195.0	204.8	198.0	208.2	218.7
Overseas airlines[2]												
Scheduled services	KNUC	249.8	269.9	294.8	300.4	328.0	336.2	351.3	363.3	390.0	399.6	426.4
Non-scheduled services	KNUD	43.0	46.1	45.5	34.6	33.9	38.0	35.1	31.5	31.3	32.5	34.8
Total	KNUE	735.2	782.3	819.2	788.1	881.1	880.2	907.2	941.7	992.3	1 050.6	1 223.6
Passengers carried												
United Kingdom airlines												
Scheduled services	KNUF	19 237.6	21 863.2	25 316.8	23 271.6	27 138.8	29 798.2	32 578.0	34 934.7	37 902.2	41 854.7	46 747.7
Non-scheduled services	KNUG	23 062.1	22 558.8	19 679.1	19 715.2	23 347.6	24 777.1	19 501.0	20 434.5	26 304.4	28 699.5	31 616.6
Overseas airlines[2]												
Scheduled services	KNUH	25 029.3	26 517.8	28 224.4	26 719.4	29 240.9	31 163.4	35 134.5	34 568.5	36 992.1	39 900.7	42 554.5
Non-scheduled services	KNUI	4 086.5	4 325.6	4 187.9	3 078.8	3 222.6	3 638.0	3 509.5	4 244.2	4 416.3	4 413.0	4 569.7
Total	KNUJ	71 415.5	75 265.3	77 408.2	72 785.0	82 949.9	89 376.7	90 723.0	94 231.9	105 615.0	114 867.9	125 488.5

1 Excludes travel to and from the Channel Islands.
2 Includes airlines of overseas UK Territories.
3 Due to the introduction of European licencing, off shore helicopter movements are no longer included in this figure.

Source: Civil Aviation Authority: 020 7379 7311

14.23 United Kingdom airlines[1]

Operations and traffic on scheduled services: revenue traffic

		Unit	1988	1989	1990	1991	1992	1993	1994	1995	1996	1997	1998
All services													
Aircraft stage flights:													
Number	KNFA	Number	546 386	606 505	617 477	568 122	601 500	601 620	621 272	658 958	702 492	749 806	797 682
Average length	KNFB	Kilometres	811	801	848	876	916	971	1 023	1 032	1 047	1 079	1 111
Aircraft-kilometres flown	KNFC	Millions	443.2	485.7	523.8	497.7	551.2	584.3	663.2	679.9	735.3	809.2	886.3
Passengers uplifted	KNFD	"	31.4	35.2	38.4	34.6	38.2	40.1	43.9	47.5	51.1	56.3	61.7
Seat-kilometres used	KNFE	"	63 868.1	70 196.0	79 579.6	74 615.4	86 731.4	94 670.1	104 294.5	115 347.1	124 846.5	136 388.2	151 969.1
Cargo uplifted:[2]													
Total	KNFF	Tonnes	435 071	453 430	485 535	466 622	507 356	541 986	618 067	643 181	690 806	782 855	831 436
Tonne-kilometres used:		Millions											
Passenger	KNFH	"	5 869.8	6 604.9	7 465.6	7 007.9	8 135.2	8 905.3	9 789.2	11 171.5	12 189.6	13 287.2	14 754.9
Freight	KNFI	"	2 057.3	2 206.4	2 388.7	2 379.9	2 644.1	2 919.6	3 378.1	3 567.3	3 831.9	4 454.0	4 663.3
Mail	KNFJ	"	179.2	162.3	168.6	182.6	161.1	141.5	147.3	151.1	176.0	172.2	177.7
Total	KNFG	"	8 106.3	8 973.5	10 023.0	9 570.5	10 940.5	11 966.4	13 314.6	14 889.9	16 197.5	17 913.4	19 595.9
Domestic services													
Aircraft stage flights:													
Number	KNFK	Number	277 682	303 147	300 683	285 346	299 893	300 416	301 652	318 884	331 109	336 218	352 936
Average length	KNFL	Kilometres	279.0	281.0	288.0	301.0	305.7	311.3	314.8	317.0	320.0	330.0	333.1
Aircraft-kilometres flown	KNFM	Millions	77.4	85.2	86.5	86.0	91.6	93.5	94.9	101.1	105.8	111.0	117.6
Passengers uplifted	KNFN	"	11.2	12.2	12.7	11.6	11.6	12.1	13.0	14.0	15.0	15.9	16.6
Seat-kilometres used	KNFO	"	4 381.1	4 767.6	5 020.8	4 663.7	4 728.2	4 933.8	5 334.0	5 753.6	6 204.3	6 645.7	6 947.5
Cargo uplifted:[2]													
Total	KNFP	Tonnes	48 794	46 660	45 818	37 739	35 420	30 660	32 670	33 659	35 432	30 679	31 879
Tonne-kilometres used:		Millions											
Passenger	KNFR	"	355.2	392.0	412.0	382.3	387.4	405.2	417.3	485.0	527.8	568.9	592.6
Freight	KNFS	"	10.1	8.9	8.7	6.7	6.6	5.6	6.3	6.9	7.4	6.1	6.0
Mail	KNFT	"	6.3	7.1	7.6	7.4	7.0	6.5	6.7	6.6	6.4	6.0	6.0
Total	KNFQ	"	371.5	408.0	428.3	396.5	401.1	417.3	430.3	498.5	541.6	581.0	604.7
International services													
Aircraft stage flights:													
Number	KNFU	Number	268 704	303 358	316 794	282 776	301 607	301 204	319 620	339 714	371 400	413 588	444 746
Average length	KNFV	Kilometres	1 361	1 320	1 381	1 456	1 523	1 629	1 693	1 703	1 695	1 688	1 729
Aircraft-kilometres flown	KNFW	Millions	365.8	400.5	437.4	411.7	459.5	490.8	541.4	578.8	629.5	698.2	768.8
Passengers uplifted	KNFX	"	20.2	22.9	25.7	22.9	26.5	28.0	30.9	33.5	36.1	40.4	45.1
Seat-kilometres used	KNFY	"	59 487.1	65 428.3	74 558.8	69 951.7	82 003.1	89 736.3	99.0	109.6	118.6	129.7	145.0
Cargo uplifted:[2]													
Total	KNFZ	Tonnes	386 277.0	406 770.0	439 717.0	428 883.0	471 936.0	511 326.0	585.4	609.5	655.4	752.2	799.6
Tonne-kilometres used:		Millions											
Passenger	KNJX	"	5 514.6	6 212.9	7 053.6	6 625.6	7 747.8	8 500.1	9 352.3	10 636.4	11 661.9	12 718.2	14 162.3
Freight	KNJY	"	2 047.2	2 197.4	2 380.1	2 373.2	2 637.4	2 914.0	3 371.8	3 560.4	3 824.5	4 448.0	4 657.2
Mail	KNJZ	"	172.9	155.2	161.1	175.2	154.0	135.0	140.5	144.4	169.5	166.3	171.7
Total	KNJW	"	7 734.8	8 565.6	9 594.7	9 174.0	10 539.3	11 549.1	12 864.6	14 391.2	15 655.9	17 332.5	18 991.2

1 Includes services of British Airways and other UK private companies (including operations performed by Cathay Pacific Airways on their scheduled service London - Hong Kong from May 1981 until December 1984).
2 Cargo has re-defined as freight and mail.

Source: Civil Aviation Authority: 020 7379 7311

14.24 United Kingdom airlines

Accidents on scheduled fixed wing passenger-carrying services[1]

	Passenger casualties			Crew casualties						Fatal accidents		
	Number of fatal accidents	Killed	Seriously injured	Killed	Seriously injured	Thousand aircraft stage flights per fatal accident	Million aircraft-kms. flown per fatal accident	Thousand passengers carried per passenger killed	Million passenger kms. flown per passenger killed	per 100 000 aircraft stage flights	per hundred million aircraft-kms.	Passengers killed per hundred million passenger-kms.
1950-54	7	194	9	28	4	107.4	61.8	46.1	50.0	0.93	1.62	1.99
1955-59	7	123	28	29	8	158.3	92.1	155.2	158.5	0.63	1.09	0.63
1960-64	5	104	35	21	6	303.7	182.3	373.4	390.6	0.33	0.55	0.25
1965-69	6	273	2	32	2	282.7	194.9	222.2	255.2	0.35	0.52	0.39
1970-74	2	167	5	14	2	897.4	737.6	466.3	657.7	0.11	0.14	0.15
1975-79	1	54	6	9	-	1 797.2	1 481.6	1 697.0	3 240.0	0.06	0.07	0.03
1980-84	-	-	4	-	-	-	-	-	-	-	-	-
1985-89	2	47	77	1	8	1 220.0	1 014.5	3031.0	6 262.9	0.08	0.10	0.02
1990-94	-	-	-	-	-	-	-	-	-	-	-	-
	KCVN	KCVO	KCVP	KCVQ	KCVR							
1994	-	-	-	-	-)							
1995	1	9	-	3	-)							
1996	-	-	-	-	-)	3 331.0	3 596.4	28 722	68 947	0.03	0.03	0.001
1997	-	-	-	-	-)							
1998	-	-	-	-	-)							

1 Excluding accidents involving the deaths of third parties only.

Source: Civil Aviation Authority: 020 7832 5504

14.25 Activity at civil aerodromes

United Kingdom

		Unit	1989	1990	1991	1992	1993	1994	1995	1996	1997	1998
Movement of civil aircraft		Thousands										
Commercial												
Transport	KNQC	"	1 359	1 420	1 365	1 448	1 487	1 552	1 615	1 686	1 764	1 871
Other[1]	KNQD	"	128	127	115	112	109	112	124	128	143	162
Total	KNQB	"	1 487	1 547	1 481	1 561	1 596	1 664	1 739	1 814	1 907	2 033
Non-commercial[2]	KNQE	"	1 650	1 708	1 466	1 321	1 523	1 684	1 809	1 281	1 330	1 343
Total	KNQA	"	3 137	3 255	2 947	2 881	3 119	3 348	3 548	3 095	3 237	3 376
Passengers handled												
Terminal	KNQG	"	98 898	102 417	95 768	106 123	112 278	122 364	129 586	135 998	146 823	158 997
Transit	KNQH	"	1 655	1 725	1 525	1 680	1 620	1 578	1 514	1 508	1 427	1 251
Total	KNQF	"	100 553	104 142	97 293	107 804	113 898	123 942	131 100	137 506	148 250	160 248
Commercial freight handled[3]		Tonnes										
Set down	KNQJ	"	592 139	625 131	595 886	658 183	696 055	794 860	854 859	891 150	987 065	1 077 731
Picked up	KNQK	"	556 318	567 919	524 087	581 315	682 186	811 353	862 584	892 206	968 506	1 014 791
Total	KNQI	"	1 148 457	1 193 050	1 119 973	1 239 498	1 378 241	1 606 213	1 717 443	1 783 356	1 955 571	2 092 522
Mail handled												
Set down	KNQM	"	71 918	76 262	77 082	80 236	85 803	91 212	93 562	101 664	110 355	108 110
Picked up	KNQN	"	90 586	95 902	93 390	97 241	101 092	110 817	116 057	127 544	138 002	134 275
Total	KNQL	"	162 504	172 164	170 472	177 477	186 895	202 029	209 619	229 208	248 357	242 385

1 Local pleasure flights and non-transport charter flights for reward (for example: aerial survey work, crop dusting and delivery of empty aircraft).

2 Test and training flights, scheduled service positioning flights, private, aero-club and official flights, etc.

3 Figures include weight of vehicles carried on vehicle ferry services.

Source: Civil Aviation Authority: 020 7379 7311

14.26 United Kingdom and Crown Dependency registered trading vessels of 500 gross tons and over

Summary of tonnage by type

End of year

	1988	1989	1990	1991	1992	1993	1994	1995	1996	1997	1998
Number											
Tankers[1]	156	136	123	124	108	106	113	113	115	133	145
Bulk carriers[2]	49	42	39	32	30	22	14	18	26	27	26
Specialised carriers[3]	18	18	19	18	13	13	13	12	11	11	11
Container (FC)[4]	43	43	39	32	28	26	34	37	38	39	45
Ro-Ro[5]	99	93	93	96	91	86	84	83	88	89	92
Other general cargo[6]	108	110	105	99	84	82	93	90	87	81	86
Passenger[7]	9	8	9	8	9	9	9	12	12	12	11
All vessels	482	450	427	409	363	344	360	365	377	392	416
Thousand gross tons											
Tankers[1]	2 661	2 252	2 210	2 166	2 188	2 161	2 481	2 346	2 383	3 407	2 977
Bulk carriers[2]	1 301	1 253	828	489	446	293	294	485	819	831	854
Specialised carriers[3]	128	122	118	99	100	124	110	52	49	49	49
Container (FC)[4]	1 335	1 368	1 275	1 091	1 015	1 017	1 236	1 326	1 110	1 113	1 379
Ro-Ro[5]	586	510	555	604	632	657	874	910	1 068	1 093	1 123
Other general cargo[6]	332	277	257	242	174	145	212	282	269	254	307
Passenger[7]	259	242	269	271	276	272	281	360	360	361	358
All vessels	6 603	6 025	5 512	4 963	4 831	4 670	5 488	5 761	6 057	7 108	7 048
Thousand deadweight tonnes											
Tankers[1]	4 787	4 063	3 995	3 875	3 944	3 915	4 576	4 289	4 347	6 119	5 163
Bulk carriers[2]	2 318	2 238	1 425	825	749	479	528	884	1 501	1 519	1 563
Specialised carriers[3]	128	118	109	81	84	87	76	31	29	29	29
Container (FC)[4]	1 265	1 308	1 212	1 019	976	995	1 253	1 358	1 212	1 224	1 543
Ro-Ro[5]	310	221	283	279	273	272	281	273	332	337	364
Other general cargo[6]	489	418	372	349	260	221	299	375	360	335	414
Passenger[7]	47	44	49	49	49	47	47	55	55	55	56
All vessels	9 344	8 409	7 446	6 477	6 335	6 017	7 061	7 266	7 835	9 618	9 132

1 Tankers cover oil, gas, chemical and other specialised tankers.
2 Bulk carriers (large and small) including combination - ore/oil and ore/bulk/oil-carriers.
3 Includes vessels such as livestock carriers, car carriers, chemical carriers.
4 Fully cellular container ships only.
5 Ro-Ro passenger and cargo vessels.
6 Reefer vessels, general cargo/passenger vessels, and single and multi-deck general cargo vessels.
7 Cruise liner and other passenger.

Source: Department of the Environment, Transport and the Regions: 020 7890 4443

14.27 United Kingdom owned trading vessels of 500 gross tons and over

End of year

	1988	1989	1990	1991	1992	1993	1994	1995	1996	1997	1998
Number											
Tankers[1]	197	178	174	168	150	145	145	139	129	123	127
Bulk carriers[2]	79	65	61	61	61	51	41	41	42	35	29
Specialised carriers[3]	17	15	23	23	22	22	22	22	19	11	10
Container (FC)[4]	56	60	52	52	45	44	48	52	54	60	62
Ro-Ro[5]	105	97	96	99	96	96	85	85	87	85	91
Other general cargo[6]	154	154	161	170	167	158	177	187	168	156	148
Passenger[7]	13	12	14	13	14	14	14	14	15	16	19
All vessels	621	581	581	586	555	530	532	540	514	486	486
Thousand gross tons											
Tankers[1]	5 352	4 965	5 052	4 713	4 469	4 395	4 129	3 666	2 958	2 704	2 408
Bulk carriers[2]	2 125	1 760	1 700	1 615	1 619	1 338	1 393	1 648	1 775	1 408	1 230
Specialised carriers[3]	151	139	144	120	120	126	145	97	87	43	42
Container (FC)[4]	1 486	1 640	1 506	1 479	1 349	1 368	1 467	1 531	1 491	1 626	1 841
Ro-Ro[5]	603	520	514	570	614	627	724	780	834	827	991
Other general cargo[6]	548	524	563	589	574	517	647	758	681	654	526
Passenger[7]	338	320	356	355	360	355	401	455	484	548	541
All vessels	10 604	9 869	9 836	9 441	9 105	8 726	8 906	8 935	8 309	7 809	7 579
Thousand deadweight tonnes											
Tankers[1]	9 966	9 291	9 513	8 807	8 306	8 211	7 735	6 856	5 538	5 048	4 411
Bulk carriers[2]	3 726	3 117	3 033	2 872	2 878	2 367	2 502	3 011	3 255	2 575	2 254
Specialised carriers[3]	160	141	144	120	112	113	88	44	40	30	29
Container (FC)[4]	1 427	1 598	1 457	1 439	1 332	1 362	1 470	1 555	1 519	1 672	1 948
Ro-Ro[5]	342	238	225	249	258	269	236	243	251	243	285
Other general cargo[6]	800	765	818	858	825	749	893	1 010	928	887	713
Passenger[7]	67	65	70	67	67	66	72	75	81	81	86
All vessels	16 488	15 214	15 259	14 413	13 777	13 137	12 996	12 793	11 611	10 546	9 727

1 Tankers cover oil, gas, chemical and other specialised tankers.
2 Bulk carriers (large and small) including combination - ore/oil and ore/bulk/oil-carriers.
3 Includes vessels such as livestock carriers, car carriers, chemical carriers.
4 Fully cellular container ships only.
5 Ro-Ro passenger and cargo vessels.
6 Reefer vessels, general cargo/passenger vessels, and single and multi-deck general cargo vessels.
7 Cruise liner and other passenger.

Source: Department of the Environment, Transport and the Regions: 020 7890 4443

14.28 United Kingdom international passenger movement by sea and air

Arrivals plus departures by country of embarkation or landing

Thousands

		1988	1989	1990	1991	1992	1993	1994	1995	1996	1997	1998[9]
All passenger movements												
By sea	KMUO	24 994	28 967	30 263	31 625	33 076	34 943	37 038	34 595	34 828	36 288†	33 275
By air	KMUP	70 540	74 424	75 772	71 879	82 141	87 289	95 790	100 902	104 861	114 099	124 744
Irish Republic:												
By sea	KMUQ	2 434	2 737	2 773	3 037	3 123	3 407	3 478	3 632	3 887	4 066†	4 610
By air	KMUR	3 521	4 092	4 429	3 947	4 323	4 363	5 126	6 017	6 938	7 786	8 528
Total	KMUS	5 955	6 828	7 202	6 984	7 446	7 770	8 604	9 649	10 825	11 852†	13 138
European continent and Mediterranean Sea area[5]												
By sea												
Belgium	BMLB	3 232	3 444	3 588	3 510	3 239	2 958	2 878	2 480	2 053	2 075†	1 747
France[6]	BMLC	15 975	19 246	20 104	21 248	22 874	24 643	27 224	25 164	25 470	26 975†	23 908
Netherlands	BMLD	2 218	2 365	2 524	2 610	2 693	2 550	1 987	1 837	1 956	1 961†	1 767
Other European Union[7]	KMUT	815	855	922	886	838	1 010	1 052	1 056	1 014	1 009†	1 024
Other countries	BMLF	159	161	177	130	146	145	188	188	191	172	192
Total	BMLA	22 399	26 071	27 315	28 384	29 790	31 306	33 329	30 725	30 684	32 192†	28 639
By air												
Belgium and Luxembourg	BMLI	1 365	1 442	1 550	1 493	1 698	1 834	2 040	2 026	2 086	2 529	2 889
Denmark	BMLJ	696	780	862	910	1 047	1 150	1 283	1 390	1 598	1 673	1 696
Germany	KMQN	4 492	4 800	5 572	5 112	5 537	5 898	6 190	6 538	6 939	7 125	7 456
France	BMLL	4 883	5 702	6 202	5 915	6 566	6 754	7 260	6 567	6 399	6 443	7 073
Italy	BMLQ	3 095	3 311	3 392	3 076	3 586	3 687	4 188	4 693	4 941	5 234	5 895
Netherlands	BMLS	2 896	3 070	3 251	3 159	3 536	3 683	4 013	4 285	4 941	5 773	6 482
Norway	BMLT	821	792	858	767	916	959	1 074	1 166	1 283	1 498	1 624
Portugal[1]	BMLU	2 157	2 138	2 247	2 418	2 559	2 553	2 692	2 761	2 674	2 888	3 178
Sweden	BMLW	762	842	947	833	956	1 007	1 122	1 213	1 306	1 589	1 880
Switzerland	BMLX	2 489	2 655	2 730	2 536	2 595	2 629	2 647	2 717	2 870	3 101	3 230
Greece	BMLO	3 796	3 526	3 556	3 459	4 470	4 548	4 963	4 578	3 589	3 774	4 435
Spain[2]	BMLV	14 865	14 234	11 782	11 653	13 076	14 592	17 645	18 267	17 789	19 558	22 090
Yugoslavia	BMLZ	1 047	1 053	1 066	258	68	74	93	132	173	250	290
Eastern Europe	KMRR	751	852	975	1 081	1 212	1 492	1 591	1 762	1 964	2 204	2 400
Middle East countries[4]	KMRS	1 621	1 717	1 647	1 622	2 036	2 254	2 479	2 765	2 908	3 157	3 382
Austria	BMLH	707	851	904	845	966	1 091	1 143	1 173	1 161	1 152	1 191
Other Western Europe[3]	KMRU	3 493	3 796	3 763	3 518	4 844	5 181	5 413	5 851	5 774	6 151	6 558
Total	BMLG	49 935	51 560	51 304	48 656	55 668	59 392	65 836	67 889	68 395	74 099	81 750
Rest of World												
By sea												
United States of America	BMMG	30.7	29.3	18.0	29.6	22.5	33.7	31.1	29.9	20.4	26.6	24.7
Canada	BMMH	–	–	–	–	0.5	0.1	–	–	–	0.3	–
Australia	BMMI	1.1	0.4	1.6	1.7	1.6	1.0	1.0	1.2	1.0	0.6	0.2
New Zealand	BMMJ	0.3	0.2	0.3	0.4	0.1	0.2	0.3	0.2	0.4	0.1	0.1
South Africa	BMMK	0.6	0.1	0.8	0.8	0.8	0.6	0.6	0.6	0.4	0.7	0.1
West Africa	BMML	0.3	0.2	0.3	0.5	0.1	0.3	0.2	0.1	0.1	–	0.1
British West Indies and Bermuda	BMMM	–	–	–	0.1	–	–	–	–	0.3	0.7	–
Other countries	BMMN	0.6	1.0	0.5	0.9	0.7	1.1	1.1	0.6	1.3	0.4	0.8
Total	BMMF	33.6	31.2	21.6	33.9	26.5	37.0	34.2	32.6	23.9	29.4	26.1
By air												
United States of America	BMNC	8 574	9 449	10 143	9 699	11 430	12 015	12 173	13 249	14 402	15 652	17 152
Canada	BMMQ	1 877	1 984	2 057	1 846	1 979	2 038	2 096	2 292	2 543	2 868	3 140
Australasia	BMMP	583	488	644	752	784	832	867	969	872	948	1 085
North Africa	KMRX	932	930	845	604	988	1 075	1 079	1 149	1 148	1 296	1 140
Rest of Africa	KMRY	1 250	1 322	1 456	1 542	1 567	1 631	1 734	1 841	2 012	2 077	2 225
South America	BMNB	151	192	250	286	321	315	391	409	466	472	572
Latin America and Caribbean	KMRZ	642	847	709	726	746	979	1 170	1 308	1 587	2 067	2 217
Indian sub-continent and Indian Ocean Islands	KMRO	897	1 038	1 115	1 013	1 069	1 164	1 425	1 588	1 759	1 852	1 932
Japan	BMMY	552	703	838	821	1 035	1 031	1 186	1 241	1 348	1 462	1 440
Rest of Asia	KMRP	1 624	1 820	1 981	1 988	2 229	2 455	2 706	2 955	3 391	3 521	3 563
Total	BMMO	17 082	18 773	20 039	19 277	22 149	23 534	24 828	27 001	29 528	32 215	34 465
Pleasure cruises beginning and/or ending at United Kingdom seaports[8]	KMRQ	127	129	153	172	138	193	236	207	233	..	..

1 Includes Azores and Madeira.
2 Includes Canary Islands.
3 Includes Cyprus, Faroes, Finland, Gibraltar, Iceland, Malta, Turkey.
4 Includes Israel, Iran, Iraq, Jordan, Kuwait, Lebanon, Gulf States, Saudi Arabia, United Arab Emirates, Yemeni Arab Republic, Yemeni People's Republic.
5 Includes North Africa and Middle East Mediterranean countries.
6 Includes hovercraft passengers.
7 Portugal, Spain, Finland and Sweden have been included in this grouping throughout even though they did not join the EU until 1986 (Spain and Portugal) or 1995 (Finland and Sweden).
8 Cruise passengers, like other passengers, are included at both departure and arrival if their journeys begin and end at a UK sea port.
9 Sea figures provisional.

Sources: Department of the Environment, Transport and the Regions: 020 7271 3743; Civil Aviation Authority

14.29 Postal services and television licences
United Kingdom

Years ended 31 March[1]

		Unit	1989[5]	1990	1991	1992	1993	1994	1995	1996	1997	1998	1999
Letters and parcel post		Millions											
Letters, etc posted[2]	KMRA	"	13 741	15 293	15 902	16 038	16 364	16 651	17 468	18 322	18 101	18 389†	18 963
of which:													
Registered and insured	KMRB	"	*21.0*	*22.2*	*21.3*	*20.7*	*21.5*	*21.8*	*21.5*	*23.5*	*25.6*	*28.7*	*31.6*
Airmail (Commonwealth and foreign)[3]	KMRC	"	403.0	444.0	467.7	544.5	534.4	545.0	567.1	655.1	684.5	658.4	693.2
Business reply and freepost items[6]	KMRD	"	377.3	438.7	416.7	428.9	481.3	482.9	477.6	493.1	505.8	524.7	503.6
Postal orders													
Total issued[4]	KMRH	Thousands	43 882	42 281	39 644	39 867	38 401	39 089	37 901	35 542	33 404	31 907	30 289
Television licences													
in force on 31 March	KMQL	Thousands	19 396	19 645	19 546	19 631	20 067	20 413	20 732	21 105	21 305	21 723	22 240
of which:													
Colour	KMQM	"	*17 469*	*17 964*	*18 111*	*18 426*	*19 031*	*19 524*	*19 957*	*20 505*	*20 849*	*21 344*	*21 944*

1 Years ended 31 March for letter and parcel post, postal orders and telegrams sent. For all other items figures relate to 31 March in each year.
2 Including printed papers, newspapers, postcards and sample packets.
3 Including letters without special charge for air transport.
4 Excluding those issued on HM ships, in many British possessions and in other places abroad. For 1988 to 1998 includes Postal Orders issued Overseas and by Ministry of Defence.
5 Industrial Action during year.
6 Business reply and Freepost is now known as Response Services.

Sources: Royal Mail: 01246 547012;
Subscription Services Limited: 0126 217673;
Post Office Counters Limited: 020 7921 9384

National accounts

15

National accounts *(Tables 15.1 to 15.22)*

The tables which follow are based on those in the *Blue Book 1999 Edition.* Some of the figures are provisional and may have to be revised later; this applies particularly to the figures for 1997 and 1998.

The accounts are based on the European System of Accounts 1995 (ESA95). The *Blue Book* contains an introduction to the system of the UK accounts outlining some of the main concepts and principles of measurement used. It explains how key economic indicators are derived from the sequence of accounts and how the figures describing the whole economy are broken down by sector and by industry. A detailed description of the structure for the accounts is provided in a separate ONS publication *United Kingdom National Accounts: Concepts, Sources and Methods* (TSO 1998). Further information on the financial accounts is given in the *Financial Statistics Explanatory Handbook.*

In the tables in this chapter on national income, analyses by industry are based, as far as possible, on the Standard Industrial Classification Revised 1992. The first aggregate measured in these tables is the **Gross domestic product** (GDP). This is a concept of the value of the total economic activity taking place in UK territory. It can be viewed as incomes earned, as expenditures incurred, or as production. Adding all primary incomes received from the rest of the world and deducting all primary incomes payable to non-residents produces **Gross national income** (previously known as gross national product). This is a concept of the value of all incomes earned by UK residents.

ESA95, the new internationally compatible accounting framework, provides a systematic and detailed description of the UK economy. It includes the sector accounts which provide, by institutional sector, a description of the different stages of the economic process from production through income generation, distribution and use of income to capital accumulation and financing; and the input output framework, which describes the production process in more detail. It contains all the elements required to compile such aggregate measures as GDP, gross national income (GNI) and saving.

Gross domestic product and national income *(Tables 15.1, 15.2, 15.3).* Table 15.1 shows the main national accounts aggregates, at current prices and at constant 1995 prices.

Table 15.2 shows the various money flows which generate the gross domestic product and gross national income. The output approach to GDP shows the total output of goods and services, the use of goods and services in the production process (intermediate consumption) and taxes and subsidies on products. The expenditure approach to GDP shows consumption expenditure by households and government, gross capital formation and expenditure on UK exports by overseas purchasers. The sum of these items overstates the amount of income generated in the United Kingdom by the value of imports of goods and services; this item is therefore subtracted to produce gross domestic product at market prices. The income approach to GDP shows gross operating surplus, mixed income and compensation of employees (previously known as income from employment). Taxes are added and subsidies are deducted to produce the total of the income-based components at market prices.

Table 15.2 also shows the primary incomes received from the rest of the world, which are added to GDP and primary incomes payable to non-residents, which are deducted from GDP, to arrive at **Gross national income.** Property income comprises compensation of employees, taxes *less* subsidies on production and property and entrepreneurial income.

Table 15.3 shows the expenditure approach to GDP at constant 1995 prices. When looking at the change in the economy over time the main concern is usually whether more goods and services are actually being produced now than at some time in the past. Over time changes in current price GDP show changes in the monetary value of the components of GDP and, as these changes in value can reflect changes in both price and volume, it is difficult to establish how much of an increase in the series is due either to increased activity in the economy or to an increase in the price level. As a result, when looking at the real growth in the economy over time it is useful to look at volume (or constant price) estimates of GDP. In constant price series, for all years the transactions are re-valued to a constant price level using the average prices of a selected year, known as the base year, presently 1995.

Industrial analysis *(Tables 15.4, 15.5)* The analysis of gross value added by industry at current prices shown in Table 15.4 reflects the estimates based on the Standard Industrial Classification, Revised 1992 (SIC92). The table is based on current price data reconciled through the input-output process for 1992 to 1997. Figures for the latest year, 1998, are based on data reconciled through the national accounts balancing process. The estimates are valued at basic prices, that is, the only taxes included in the price will be taxes paid as part of the production process, such as business rates, and not any taxes specifically levied on the production of a unit of output, for example VAT.

Table 15.5 shows constant price estimates of gross value added at basic prices by industry. Constant price gross value added (output approach) provides the lead indicator of economic change in the short term.

Sector analysis - Distribution of income accounts and capital account *(Tables 15.6-15.13)*
The new national accounts accounting framework includes the sector accounts which provide, by institutional sector, a description of the different stages of the economic process from production through income generation, distribution and use of income to capital accumulation and financing.

Tables 15.6-15.12 show the allocation of primary income account and the secondary distribution of income account for the non-financial corporations, financial corporations, government and households sectors. Additionally, Table 15.12 shows the use of income account for the households sector and Table 15.13 provides a summary of the capital account. The full sequence of accounts is shown in the *Blue Book.*

The allocation of primary income account shows the resident units and institutional sectors as recipients rather than producers of primary income. It demonstrates the extent to which operating surpluses are distributed to the owners of the enterprises. The resources side of the allocation of primary income accounts includes the components of the income approach to measurement of GDP. The balance of this account is the gross balance of primary income (B.5g) for each sector, and if the gross balance is aggregated across all sectors of the economy the result is **gross national income.**

The secondary distribution of income account describes how the balance of income for each sector is allocated by redistribution; through transfers such as taxes on income, social contributions and benefits and other current transfers. The balancing item of this account is gross disposable income (B.6g). For the households sector, gross disposable income at constant prices is shown as real household disposable income.

Table 15.12 shows, for the households sector, the use of disposable income where the balancing item is saving (B.8g). For the non-financial corporations sector the balancing item of the secondary distribution of income account, gross disposable income (B.6g) is equal to saving (B.8g).

The summary capital account *(Table 15.13)* brings together the saving and investment of the several sectors of the economy. It shows saving, capital transfers, gross capital formation and net acquisition of non-financial assets for each of the four sectors.

Household and non-profit institutions serving households (NPISH) consumption expenditure at current and constant prices *(Tables 15.14-15.17).*
Household and NPISH consumption expenditure is a major component of the expenditure measure of gross domestic product at current prices *(Table 15.2)* and at constant prices *(Table 15.3).*

Household final consumption expenditure includes the value of income-in-kind and imputed rent of owner-occupied dwellings but excludes business expenditure allowed as deductions in computing income for tax purposes. It includes expenditure on durable goods, for instance motor cars, which from the point of view of the individual might more appropriately be treated as capital expenditure. The only exceptions are the purchase of land and dwellings and costs incurred in connection with the transfer of their ownership and expenditure on major improvements by occupiers, which are treated as personal capital expenditure.

The estimates of household consumption expenditure include purchases of second-hand as well as new goods, *less* the proceeds of sales of used goods.

The most detailed figures are published quarterly in *Consumer Trends.*

Change in inventories (previously known as value of physical increase in stocks and work in progress) *(Table 15.18).* This table gives a broad analysis by industry, and, for manufacturing industry, by asset, of the value of entries *less* withdrawals and losses of inventories (stocks).

Gross fixed capital formation *(Tables 15.19-15.22)* Gross fixed capital formation comprises expenditure on the replacement of, and additions to, fixed capital assets located in the United Kingdom, including all ships and aircraft of UK ownership.

15.1 UK national and domestic product
Main aggregates: index numbers and values

At current and constant 1995 prices

		1990	1991	1992	1993	1994	1995	1996	1997	1998
INDICES (1995=100)										
VALUES AT CURRENT PRICES										
Gross domestic product at current market prices ("money GDP")	YBEU	77.8	81.8	85.1	89.5	94.9	100.0	105.9	112.8†	118.4
Gross value added at current basic prices	YBEX	78.8	82.3	85.8	90.2	95.3	100.0	106.1	112.5†	117.9
VALUES AT 1995 PRICES										
Gross domestic product at 1995 market prices	YBEZ	92.4	91.0	91.1	93.2	97.3	100.0	102.6	106.2†	108.5
Gross national disposable income at 1995 market prices	YBFP	92.0	91.4	91.9	93.9	98.6	100.0	103.4	108.6†	111.3
Gross value added at 1995 basic prices	CGCE	92.0	90.9	91.1	93.2	97.4	100.0	102.5	106.0	108.6
PRICES										
Implied deflator of GDP at market prices (expenditure based, "total home costs per unit of output")	YBGB	84.2	89.9	93.5	96.1	97.5	100.0	103.3	106.3†	109.1
VALUES AT CURRENT PRICES (£ million)										
Gross measures, before deduction of fixed capital consumption										
at current market prices										
Gross domestic product at current market prices ("money GDP")	YBHA	554 486	582 946	606 582	637 817	676 036	712 548	754 601	803 889†	843 725
Employment, property and entrepreneurial income from the rest of the world (receipts *less* payments)	YBGG	–558	–1 953	2 115	685	7 770	5 976	8 111	11 170†	15 174
Subsidies (receipts) *less* taxes (payments) on production from/to the rest of the world	-QZOZ	–5 138	–3 512	–4 253	–4 510	–3 063	–4 927	–3 888	–2 598†	–3 437
Gross national income at current market prices	ABMX	548 790	577 481	604 444	633 992	680 743	713 597	758 824	812 461†	855 462
Current transfers from the rest of the world (receipts *less* payments)[1]	-YBGF	889	2 849	–562	–98	–1 605	–1 987	–636	–2 455†	–3 090
Gross national disposable income at current market prices[1]	NQCO	549 679	580 330	603 882	633 894	679 138	711 610	758 188	810 006†	852 372
adjustment to current basic prices										
Gross domestic product at current market prices	YBHA	554 486	582 946	606 582	637 817	676 036	712 548	754 601	803 889†	843 725
Adjustment to current basic prices (*less* taxes *plus* subsidies on products)	-NQBU	–54 744	–61 399	–62 678	–65 979	–71 874	–78 482	–82 029	–90 275†	–96 181
Gross value added at current basic prices	ABML	499 742	521 547	543 904	571 838	604 162	634 066	672 572	713 614†	747 544
Net measures, after deduction of fixed capital consumption	-NQAE	–67 762	–70 299	–68 255	–71 336	–74 296	–79 294	–83 969	–85 834†	–88 771
Net domestic product at current market prices	NHRK	486 724	512 647	538 327	566 481	601 740	633 254	670 632	718 055†	754 954
Net national income at current market prices	NSRX	481 028	507 182	536 189	562 656	606 447	634 303	674 855	726 627†	766 691
Net national disposable income at current market prices[1]	NQCP	481 917	510 031	535 627	562 558	604 841	632 315	674 218	724 172†	763 601
VALUES AT 1995 PRICES (£ million)										
Gross measures, before deduction of fixed capital consumption										
at 1995 market prices										
Gross domestic product at 1995 market prices	ABMI	658 480	648 639	648 975	664 018	693 177	712 548	730 767	756 430†	773 380
Terms of trade effect ("Trading gain or loss")	YBGJ	2 245	4 284	7 303	8 058	4 836	–	2 182	8 891†	14 436
Real gross domestic income	YBGL	660 726	652 922	656 278	672 076	698 013	712 548	732 949	765 321†	787 816
Real employment, property and entrepreneurial income from the rest of the world (receipts *less* payments)	YBGI	–684	–2 207	2 288	721	8 024	5 976	7 884	10 638†	14 195
Subsidies (receipts) *less* taxes (payments) on production from/to the rest of the world	-QZPB	–6 720	–3 580	–4 097	–4 795	–2 896	–4 927	–4 649	–996†	–1 310
Gross national income at 1995 market prices	YBGM	653 322	647 135	654 469	668 002	703 141	713 582	736 183†	774 963	800 701
Real current transfers from the rest of the world (receipts *less* payments)	-YBGP	1 040	3 206	–620	–102	–1 662	–1 987	–622	–2 338†	–2 891
Gross national disposable income at 1995 market prices	YBGO	654 408	650 425	653 865	667 926	701 479	711 610	735 565†	772 625	797 810
adjustment to 1995 basic prices										
Gross domestic product at 1995 market prices	ABMI	658 480	648 639	648 975	664 018	693 177	712 548	730 767	756 430†	773 380
Adjustment to 1995 basic prices (*less* taxes *plus* subsidies on products)	-NTAQ	–75 493	–72 335	–71 524	–73 383	–75 775	–78 482	–80 538	–84 002	..
Gross value added at 1995 basic prices	ABMM	583 525	576 523	577 540	590 757	617 402	634 066	650 229	671 887†	688 602
Net measures, after deduction of fixed capital consumption at 1995 prices	-YBFX	–72 709	–74 189	–74 866	–76 674	–78 336	–79 294	–81 223	–82 780†	–84 550
Net national income at 1995 market prices	YBET	580 623	572 685	579 359	591 047	624 805	634 303	654 961	692 183†	716 151
Net national disposable income at 1995 market prices[1]	YBEY	581 745	576 042	578 768	590 991	623 143	632 315	654 339	692 034	713 260

1 No values are calculated for this series for years before 1987.

Source: Office for National Statistics: 020 7533 6031

15.2 UK gross domestic product and national income

Current prices

£ millions

		1990	1991	1992	1993	1994	1995	1996	1997	1998
GROSS DOMESTIC PRODUCT										
Gross domestic product: output approach										
Gross value added, at basic prices										
Output of goods and services, at basic prices	NQAF	1 016 525	1 039 824	1 101 705	1 168 193	1 271 305	1 359 593	1 450 016	1 526 371†	1 599 339
less intermediate consumption, at purchasers' prices	-NQAJ	−516 783	−518 277	−557 801	−596 355	−667 143	−725 527	−777 444	−812 757†	−851 795
Total	ABML	499 742	521 547	543 904	571 838	604 162	634 066	672 572	713 614†	747 544
Value added taxes (VAT) on products	QYRC	33 846	38 415	40 710	42 757	46 113	48 467	51 623	55 658†	57 066
Other taxes on products	NSUI	27 124	29 206	28 888	30 651	33 311	37 724	39 457	42 661†	46 568
less subsidies on products	-NZHC	−6 226	−6 222	−6 920	−7 429	−7 550	−7 709	−9 051	−8 044†	−7 453
Gross domestic product at market prices	YBHA	554 486	582 946	606 582	637 817	676 036	712 548	754 601	803 889†	843 725
Gross domestic product: expenditure approach										
Final consumption expenditure										
Actual individual consumption										
Household final consumption expenditure	ABPB	336 492	357 785	377 147	399 108	419 262	438 453	467 841	498 307†	525 463
Final consumption expenditure of NPISH	ABNV	10 755	10 447	10 165	13 290	14 567	15 718	17 577	18 725†	19 661
Individual government final consumption expenditure	NNAQ	59 085	66 065	72 718	74 558	77 067	81 951	85 294	87 721†	91 266
Total actual individual consumption	ABRE	406 332	434 297	460 030	486 956	510 896	536 122	570 712	604 753†	636 390
Collective government final consumption expenditure	NNAR	50 450	54 734	55 551	56 007	58 469	58 455	60 817	60 052†	62 298
Total final consumption expenditure	ABKW	456 782	489 031	515 581	542 963	569 365	594 577	631 529	664 805†	698 688
Households and NPISH	NSSG	347 247	368 232	387 312	412 398	433 829	454 171	485 418	517 032†	545 124
Central government	NMBJ	65 565	72 292	77 374	80 606	83 728	86 142	89 692	90 784†	94 338
Local government	NMMT	43 970	48 507	50 895	49 959	51 808	54 264	56 419	56 989†	59 226
Gross capital formation										
Gross fixed capital formation	NPQX	114 314	104 680	100 278	101 230	107 390	116 360	125 675	134 153†	147 629
Changes in inventories	ABMP	−1 800	−4 927	−1 937	329	3 708	4 512	1 771	4 388†	3 621
Acquisitions less disposals of valuables	NPJO	−113	−86	36	−9	136	−92	−185	39†	573
Total gross capital formation	NPDN	112 401	99 667	98 377	101 550	111 234	120 780	127 261	138 580†	151 823
Exports of goods and services	KTMW	133 501	135 365	143 291	162 078	178 767	202 412	220 303	229 326†	224 202
less imports of goods and services	-KTMX	−148 198	−141 117	−150 667	−168 774	−183 330	−205 221	−224 492	−228 822†	−232 714
External balance of goods and services	KTMY	−14 697	−5 752	−7 376	−6 696	−4 563	−2 809	−4 189	504†	−8 512
Statistical discrepancy between expenditure components and GDP	GIXM	–	–	–	–	–	–	–	–†	1 726
Gross domestic product at market prices	YBHA	554 486	582 946	606 582	637 817	676 036	712 548	754 601	803 889†	843 725
Gross domestic product: income approach										
Operating surplus, gross										
Non-financial corporations										
Public non-financial corporations	NRJT	4 110	2 207	2 237	3 206	3 627	5 367	5 114	4 205†	4 307
Private non-financial corporations	NRJK	107 132	108 502	108 460	118 453	133 237	143 086	158 805	169 778†	171 367
Financial corporations	NQNV	11 711	7 836	15 657	18 393	23 616	20 077	18 970	18 151†	18 788
Adjustment for financial services	-NSRV	−19 347	−15 828	−20 997	−20 025	−26 410	−25 499	−25 557	−25 678†	−29 370
General government	NMXV	9 454	9 721	9 677	9 872	10 329	10 901	11 656	11 840†	12 518
Households and non-profit institutions serving households	QWLS	24 755	28 513	32 749	34 462	36 790	39 321	41 105	44 217†	45 602
Total operating surplus, gross	ABNF	137 815	140 951	147 783	164 361	181 189	193 253	210 093	222 513†	223 212
Mixed income	HAXH	33 030	31 635	33 872	36 096	38 336	40 239	41 570	41 665†	43 379
Compensation of employees	HAEA	315 208	333 850	347 036	356 323	369 960	385 397	404 521	432 388†	463 398
Taxes on production and imports	NZGX	74 659	82 732	84 811	88 466	94 101	101 368	107 468	115 367†	121 253
less subsidies	-AAXJ	−6 226	−6 222	−6 920	−7 429	−7 550	−7 709	−9 051	−8 035	..
Statistical discrepancy between income components and GDP	GIXQ	–	–	–	–	–	–	–	–†	−64
Gross domestic product at market prices	YBHA	554 486	582 946	606 582	637 817	676 036	712 548	754 601	803 889†	843 725
GROSS NATIONAL INCOME at market prices										
Gross domestic product at market prices	YBHA	554 486	582 946	606 582	637 817	676 036	712 548	754 601	803 889†	843 725
Compensation of employees										
receipts from the rest of the world	KTMN	543	551	551	595	681	887	911	1 007	777
less payments to the rest of the world	-KTMO	−653	−614	−600	−560	−851	−1 183	−818	−924	−701
Total	KTMP	−110	−63	−49	35	−170	−296	93	83	76
less Taxes on production paid to the rest of the world *plus* Subsidies received from the rest of the world	-QZOZ	−5 138	−3 512	−4 253	−4 510	−3 063	−4 927	−3 888	−2 598†	−3 437
Property and entrepreneurial income										
receipts from the rest of the world	HMBN	88 186†	84 419	74 549	78 825	82 639	97 086	101 669	106 898	110 588
less payments to the rest of the world	-HMBO	−88 634†	−86 309	−72 385	−78 175	−74 699	−90 814	−93 651	−95 811	−95 490
Total	HMBM	−448†	−1 890	2 164	650	7 940	6 272	8 018	11 087	15 098
Gross national income at market prices	ABMX	548 790	577 481	604 444	633 992	680 743	713 597	758 824	812 461†	855 462

Source: Office for National Statistics: 020 7533 6031

15.3 UK gross domestic product
Constant (1995) prices

£ millions

		1990	1991	1992	1993	1994	1995	1996	1997	1998
GROSS DOMESTIC PRODUCT										
Gross domestic product: expenditure approach										
Final consumption expenditure										
Actual individual consumption										
Household final consumption expenditure	ABPF	415 788	408 309	410 026	420 081	431 462	438 453	454 686	472 701†	488 505
Final consumption expenditure of non-profit institutions serving households	ABNU	11 621	11 656	11 769	13 963	15 119	15 718	15 936	16 235†	16 862
Individual government final consumption expenditure	NSZK	72 295	74 787	77 053	78 005	79 006	81 950	83 999	85 295†	85 967
Total actual individual consumption	YBIO	499 280	494 388	498 504	511 954	525 587	536 121	554 621	574 231†	591 334
Collective government final consumption expenditure	NSZL	60 691	62 101	60 501	58 383	59 250	58 456	58 821	55 529†	56 243
Total final consumption expenditure	ABKX	559 996	556 555	559 059	570 348	584 837	594 577	613 442	629 760†	647 577
Gross capital formation										
Gross fixed capital formation	NPQR	119 368	109 000	108 246	109 127	113 042	116 360	122 042	131 246†	144 184
Changes in inventories	ABMQ	−2 079	−5 349	−1 962	360	4 836	4 512	1 830	3 762†	3 643
Acquisitions less disposals of valuables	NPJP	−130	−90	39	−9	140	−92	−186	14†	533
Total gross capital formation	NPQU	117 847	104 505	106 940	109 888	118 018	120 780	123 686	132 366	..
Gross domestic final expenditure	YBIK	678 506	659 991	665 194	679 523	702 855	715 357	737 128	764 782†	795 937
Exports of goods and services	KTMZ	156 640	156 362	162 819	169 225	184 873	202 412	217 600	236 283†	241 123
Gross final expenditure	ABME	834 871	816 136	827 860	848 628	887 728	917 769	954 728	1 001 065†	1 037 060
less imports of goods and services	-KTNB	−176 377	−167 472	−178 879	−184 607	−194 551	−205 221	−223 961	−244 635†	−265 261
Statistical discrepancy between expenditure components and GDP	GIXS	–	–	–	–	–	–	–	–†	1 581
Gross domestic product at 1995 market prices	ABMI	658 480	648 639	648 975	664 018	693 177	712 548	730 767	756 430†	773 380
of which External balance of goods and services	KTNC	−19 737	−11 110	−16 060	−15 382	−9 678	−2 809	−6 361	−8 352†	−24 138

Source: Office for National Statistics: 020 7533 6031

15.4 Gross value added at current basic prices: by industry[1,2]

£ millions

		1990	1991	1992	1993	1994	1995	1996	1997	1998
Agriculture, hunting, forestry and fishing	EWSH	9 700	9 376	10 349	10 765	10 776	11 713	11 963	10 594†	9 656
Mining & quarrying	EWSL	13 404	13 410	13 167	13 305	14 550	16 116	19 447	17 643†	12 748
Manufacturing	EWSP	114 809	111 337	113 704	118 718	128 202	136 747	143 485	148 619†	147 306
Electricity, gas & water supply	EWST	11 618	14 553	14 721	16 049	15 932	15 562	16 120	16 230†	16 737
Construction	EWSX	34 555	31 995	29 965	29 144	31 347	32 948	34 563	36 927†	39 262
Wholesale and retail trade	EWTB	70 948	75 349	79 652	83 823	88 224	92 557	99 170	108 450†	113 070
Transport & communication	EWTF	41 934	43 909	45 714	47 294	50 708	52 297	53 994	57 916†	63 340
Financial intermediation	EWTJ	116 804	121 850	133 501	141 698	154 498	159 069	168 448	184 163†	206 347
Adjustment for financial services	-NSRV	–19 347	–15 828	–20 997	–20 025	–26 410	–25 499	–25 557	–25 678†	–29 370
Public administration and defence	EWTN	33 021	35 561	38 537	39 352	38 894	39 196	39 181	39 020†	40 495
Education, health & social work	EWTR	53 428	60 169	64 566	68 621	72 193	76 130	81 291	85 162†	89 041
Other services	EWTV	18 867	19 866	21 026	23 096	25 248	27 229	30 467	34 567†	38 912
All industries including adjustment for financial services	ABML	499 742	521 547	543 904	571 838	604 162	634 066	672 572	713 614†	747 544

1 The contribution of each industry to the gross domestic product before providing for consumption of fixed capital. The industrial composition in this table is consistent with the Input-Output analyses. Although the industrial composition for 1989-91 is based on Input-Output analyses also, there are improvements to the underlying data from 1992. Between 1989 and 1991, the data were compiled on a different basis, which lead to step changes in 1991 and 1992.

2 Components may not sum to totals due to rounding.

Source: Office for National Statistics: 020 7533 6031

15.5 Gross value added at 1995 basic prices: by industry

Index numbers

Indices 1995=100

	Weight per 1000[1] 1995		1990	1991	1992	1993	1994	1995	1996	1997	1998
Agriculture, hunting & forestry; fishing	18.5	GDQA	100.7	105.7	110.1	101.2	100.0	100.0	102.0	103.5†	102.6
Production											
Mining and quarrying											
Mining and quarrying of energy producing materials											
Mining of coal	1.9	CKZP	199.0	202.0	181.8	142.3	94.6	100.0	95.5	90.3†	76.5
Extraction of mineral oil and natural gas	21.3	CKZO	59.8	63.8	68.4	77.5	96.1	100.0	105.6	104.7†	107.5
Other mining and quarrying	2.3	CKZQ	112.3†	107.9	103.4	105.6	105.4	100.0	88.9	87.5	98.4
Total mining and quarrying	25.4	CKYX	73.3†	76.7	78.9	84.2	96.8	100.0	103.3	102.1	104.4
Manufacturing											
Food; beverages & tobacco	28.7	CKZA	97.2†	97.0	98.7	99.0	101.5	100.0	101.0	104.6	102.7
Textiles & textile products	10.1	CKZB	111.8†	101.0	101.9	101.3	103.6	100.0	99.6	96.4	88.9
Leather & leather products	1.4	CKZC	112.1†	97.1	94.5	97.8	97.3	100.0	99.9	103.8	89.2
Wood & wood products	2.9	CKZD	111.6†	99.4	98.2	100.4	108.2	100.0	97.6	95.8	96.6
Pulp, paper & paper products; publishing & printing	26.3	CKZE	96.4†	92.0	93.1	96.1	98.5	100.0	98.0	98.2	98.2
Coke, petroleum products & nuclear fuel	4.7	CKZF	77.4†	83.5	88.5	88.9	89.7	100.0	91.8	93.5	88.2
Chemicals, chemical products & man-made fibres	24.1	CKZG	83.5†	85.8	88.5	90.4	95.1	100.0	100.7	102.4	103.4
Rubber & plastic products	10.5	CKZH	88.2†	83.1	85.1	88.9	98.0	100.0	98.8	98.6	101.1
Other non-metallic mineral products	7.9	CKZI	109.4†	99.1	94.7	99.1	102.8	100.0	96.5	99.0	96.8
Basic metals & fabricated metal products	24.6	CKZJ	111.2†	101.0	96.0	95.0	97.3	100.0	99.7	101.2	98.6
Machinery & equipment not elsewhere classified	19.2	CKZK	110.6	98.8†	94.8	94.7	99.9	100.0	98.0	95.8	95.4
Electrical & optical equipment	26.9	CKZL	80.8†	77.6	78.9	83.2	93.3	100.0	104.0	105.9	113.3
Transport equipment	20.5	CKZM	108.8†	101.8	99.8	98.1	100.7	100.0	105.7	109.5	115.3
Manufacturing not elsewhere classified	7.6	CKZN	112.5†	98.5	98.0	99.4	102.4	100.0	100.2	101.7	100.0
Total manufacturing	215.7	CKYY	97.7†	92.8	92.8	94.1	98.5	100.0	100.4	101.7	102.1
Electricity, gas and water supply	24.5	CKYZ	86.5†	91.4	92.9	96.8	97.8	100.0	105.3	105.7	107.5
Total production	265.6	CKYW	94.1†	91.0	91.3	93.3	98.3	100.0	101.1	102.1	102.8
Construction	52.0	GDQB	111.3	102.3	98.3	97.1	100.8	100.0	101.5	104.7†	106.0
Service industries											
Wholesale & retail trade (including motor trade); repair of motor vehicles, personal & household goods	116.9	GDQC	88.4	86.3	87.5	92.5	97.5	100.0	103.4	107.3†	109.6
Hotels and restaurants	29.0	GDQD	107.5	100.4	96.4	98.5	101.2	100.0	102.4	103.6†	104.5
Transport, storage & communication											
Transport and storage	53.6	GDQF	87.9	84.4	87.5	89.3	96.3	100.0	100.7	106.5†	112.6
Communication	28.9	GDQG	80.1	79.6	79.5	83.6	90.7	100.0	110.8	126.2†	136.1
Total	82.5	GDQH	85.1	82.7	84.6	87.3	94.3	100.0	104.2	113.4†	120.8
Financial intermediation	67.4	GDQI	98.1	98.7	94.2	96.3	97.3	100.0	104.3	109.9	112.7
Adjustment for financial services	–40.2	GDQJ	95.5	94.0	90.9	91.6	95.8	100.0	105.5	112.8†	118.9
Real estate, renting & business activities											
Letting of dwellings, including imputed rent of owner occupiers	72.2	GDQL	93.3	94.8	95.3	96.5	97.8	100.0	101.4	103.5†	106.3
Other real estate, renting & business activities	111.3	GDQK	86.4	84.1	82.6	84.7	93.5	100.0	106.8	117.6†	127.3
Total	183.5	GDQM	89.0	87.9	86.8	88.9	95.2	100.0	104.7	112.0†	119.1
Public administration & defence	61.2	GDQO	103.7	105.6	105.0	102.9	101.3	100.0	98.5	96.7†	96.0
Education	56.0	GDQP	89.2	92.2	95.0	95.0	98.7	100.0	101.6	103.0†	103.8
Health and social work	64.6	GDQQ	80.6	86.1	88.6	92.6	95.0	100.0	104.4	108.3†	112.0
Other social & personal services, private households with employees and extra-territorial organisations	42.9	GDQR	81.5	81.4	83.2	90.1	95.7	100.0	103.7	106.7†	112.2
Total service industries	663.9	GDQS	89.5	89.3	89.7	92.5	96.8	100.0	103.3	107.8†	111.8
All industries	1 000.0	YBFR	92.3	90.9	91.1	93.1	97.5	100.0	102.6	106.1	108.9

1 The weights are in proportion to total gross value added in 1995. The GVA for sections L, M, and N in this table follows the SIC(92) and differs from that shown in table 2.3, which is based on Input-Output groups. Central government expenditure on teachers' pay is included in Education in table 2.4 but in PAD in table 2.3. The administation costs of the NHS are included in PAD in table 2.4 but are included in Health and social work in table 2.3.

2 The output analysis of gross value added is estimated in terms of change and expressed in index number form. It is therefore inappropriate to show as a statistical adjustment any divergence of an output measure of GDP derived from it from other measures of GDP. Such an adjustment does, however, exist implicitly.

Source: Office for National Statistics: 020 7533 6031

15.6 Non-financial corporations
Allocation of primary income account

£ millions

		1990	1991	1992	1993	1994	1995	1996	1997	1998
Resources										
Operating surplus, gross	NQBE	111 242	110 709	110 697	121 659	136 864	148 453	163 919	173 983†	175 674
of which UK continental shelf companies gross trading profits	CAGD	8 752	8 322	8 345	9 312	10 654	12 018	15 538	13 786	11 208
of which other private non-financial corporations' gross trading profits	CAED	98 304	94 969	93 641	102 082	117 351	126 313	134 776	146 576†	149 775
Property income, received										
Interest	EABC	16 109	13 907	11 263	7 883	7 902	9 772	9 989	10 468†	13 021
Distributed income of corporations	EABD	14 106	14 797	16 430	15 943	18 437	24 517	25 458	29 223†	27 653
Reinvested earnings on direct foreign investment	HDVR	8 120	5 446	4 530	7 444	11 673	10 772	12 713	13 810	..
Attributed property income of insurance policy-holders	FAOF	662	730	786	765	780	862	779	897	..
Rent	FAOG	111	109	108	106	108	109	115	120	..
Total	FAKY	39 126	35 004	33 126	32 165	38 931	46 121	49 123	52 496	..
Total resources	FBXJ	150 368	145 713	143 823	153 824	175 795	194 574	213 042	226 038†	229 003
Uses										
Property income, paid										
Interest	EABG	35 283	32 784	29 804	23 718	23 594	26 943	26 218	27 983†	33 853
Distributed income of corporations	NVCS	51 121	49 380	53 038	52 843	58 717	72 090	77 403	82 215†	79 035
of which private non-financial corporations' dividends	RVFT	29 772	30 171	33 403	33 727	38 549	48 671	54 793	59 253†	55 612
Reinvested earnings on direct foreign investment	HDVB	1 796	248	118	1 669	3 523	3 898	5 261	5 723	..
Rent	FBXO	758	697	682	734	693	719	753	752	..
Total	FBXK	88 956	83 099	83 633	78 967	86 537	103 597	109 551	118 446	..
Balance of primary incomes, gross	NQBG	61 412	62 614	60 190	74 857	89 258	90 977	103 491	110 719†	109 053
Total uses	FBXJ	150 368	145 713	143 823	153 824	175 795	194 574	213 042	226 038†	229 003

Source: Office for National Statistics: 020 7533 6031

15.7 Non-financial corporations
Secondary distribution of income account

£ millions

		1990	1991	1992	1993	1994	1995	1996	1997	1998
Resources										
Balance of primary incomes, gross	NQBG	61 412	62 614	60 190	74 857	89 258	90 977	103 491	110 719†	109 053
Social contributions										
Imputed social contributions	NSTJ	1 041	2 478	3 031	3 226	3 805	3 371	2 983	2 872†	3 298
Current transfers other than taxes, social contributions and benefits										
Non-life insurance claims	FCBP	8 564	10 464	11 147	9 834	9 286	10 287	10 121	9 613	..
Miscellaneous transfers	NRJY	–	37	112	256	420	494	537	557†	595
Total	NRJB	8 564	10 501	11 259	10 090	9 706	10 781	10 658	10 095†	10 545
Total resources	FCBR	71 017	75 593	74 480	88 173	102 769	105 129	117 132	117 513	..
Uses										
Current taxes on income, wealth etc.										
Taxes on income	FCBS	19 719	16 563	12 951	12 827	14 497	19 318	21 219	27 205	..
Social benefits other than social transfers in kind	NSTJ	1 041	2 478	3 031	3 226	3 805	3 371	2 983	2 872†	3 298
Current transfers other than taxes, social contributions and benefits										
Net non-life insurance premiums	FCBY	8 564	10 464	11 147	9 834	9 286	10 287	10 121	9 613	..
Miscellaneous current transfers	FDBI	241	238	240	272	311	343	1 455	402	363
Total, other current transfers	FCBX	8 805	10 702	11 387	10 106	9 597	10 630	11 576	10 015	..
Disposable income, gross[1]	NRJD	41 452	45 850	47 111	62 014	74 870	71 810	81 354	82 843†	83 288
Total uses	FCBR	71 017	75 593	74 480	88 173	102 769	105 129	117 132	117 513	..

1 Gross disposable income equals gross saving.

Source: Office for National Statistics: 020 7533 6031

15.8 General government
Allocation of primary income account

£ millions

		1990	1991	1992	1993	1994	1995	1996	1997	1998
Resources										
Operating surplus, gross	NMXV	9 454	9 721	9 677	9 872	10 329	10 901	11 656	11 840†	12 518
Taxes on production and imports, received										
Taxes on products										
Value added tax (VAT)	NZGF	29 182	35 157	36 651	38 286	42 992	43 622	47 152	52 261†	52 838
Taxes and duties on imports excluding VAT										
Import duties	NMBS	–	–	–	–	–	–	–	–	–
Taxes on imports excluding VAT and import duties	NMBT	–	–	–	–	–	–	–	–	–
Taxes on products excluding VAT and import duties	NMBV	25 230	27 242	26 891	28 422	31 079	35 211	37 113	40 290†	44 450
Total taxes on products	NVCC	54 412	62 399	63 542	66 708	74 071	78 833	84 265	92 551†	97 288
Other taxes on production	NMYD	13 689	15 111	15 213	15 058	14 677	15 177	16 388	17 048†	17 619
Total taxes on production and imports, received	NMYE	68 101	77 510	78 755	81 766	88 748	94 010	100 653	109 595†	114 907
less Subsidies, paid										
Subsidies on products	-NMYF	–4 806	–4 512	–5 117	–5 239	–5 260	–5 278	–6 124	–4 870†	–4 544
Other subsidies on production	-NMCC	–	–	–	–	–	–	–	–	–
Total	-NMRL	–4 806	–4 512	–5 117	–5 239	–5 260	–5 278	–6 124	–4 870†	–4 544
Property income, received										
Interest										
from general government	NMYI	4 733	4 774	4 606	4 055	3 866	3 767	3 901	4 004†	4 085
from other sectors	NMYJ	5 267	4 407	4 042	3 965	4 562	4 846	4 997	4 313†	4 198
Total	NMYL	10 000	9 181	8 648	8 020	8 428	8 613	8 898	8 317†	8 283
Distributed income of corporations	NMYM	2 754	2 609	2 184	1 518	1 313	1 642	1 565	1 620†	2 039
Property income attributed to insurance policy holders	NMYO	24	28	29	28	28	32	28	33	27
Rent										
from sectors other than general government	NMYQ	723	662	647	699	658	684	718	717	533
Total property income, received										
from general government	NMYS	4 733	4 774	4 606	4 055	3 866	3 767	3 901	4 004†	4 085
from other sectors	NMYT	8 768	7 706	6 902	6 210	6 561	7 204	7 308	6 683†	6 797
Total	NMYU	13 501	12 480	11 508	10 265	10 427	10 971	11 209	10 687†	10 882
Total resources	NMYV	86 250	95 199	94 823	96 664	104 244	110 604	117 394	127 252†	133 763
Uses										
Property income, paid										
Interest										
to general government	NMYW	4 733	4 774	4 606	4 055	3 866	3 767	3 901	4 004†	4 085
to other sectors	NMYX	20 900	18 659	18 737	19 981	22 963	26 305	27 975	29 935†	30 507
Total	NMYY	25 633	23 433	23 343	24 036	26 829	30 072	31 876	33 939†	34 592
Balance of primary incomes, gross	NMZH	60 617	71 766	71 480	72 628	77 415	80 532	85 518	93 313†	99 171
Total uses	NMYV	86 250	95 199	94 823	96 664	104 244	110 604	117 394	127 252†	133 763

Source: Office for National Statistics: 020 7533 6031

15.9 General government
Secondary distribution of income account

£ millions

		1990	1991	1992	1993	1994	1995	1996	1997	1998
Resources										
Balance of primary incomes, gross	NMZH	60 617	71 766	71 480	72 628	77 415	80 532	85 518	93 313†	99 171
Current taxes on income, wealth etc.										
Taxes on income	NMZJ	79 903	81 428	80 222	78 275	85 344	95 045	99 292	107 389†	123 758
Other current taxes	NVCM	13 231	10 496	10 299	10 551	11 140	11 937	12 795	13 820†	14 892
Total	NMZL	93 134	91 924	90 521	88 826	96 484	106 982	112 087	121 209†	138 650
Social contributions										
Actual social contributions										
Employers' actual social contributions	NMZM	21 251	22 573	23 185	24 615	25 000	26 141	27 580	29 327†	31 336
Employees' social contributions	NMZN	15 276	16 029	16 525	17 235	19 649	21 091	21 700	24 121†	25 804
Social contributions by self- and non-employed persons	NMZO	1 177	1 206	1 281	1 472	1 469	1 541	1 771	1 848†	1 842
Total	NMZP	37 704	39 808	40 991	43 322	46 118	48 773	51 051	55 296†	58 982
Imputed social contributions	NMZQ	3 966	4 943	5 426	5 397	5 419	5 279	5 300	5 356†	5 667
Total	NMZR	41 670	44 751	46 417	48 719	51 537	54 052	56 351	60 652†	64 649
Other current transfers										
Non-life insurance claims	NMZS	314	382	408	361	340	377	371	349†	363
Current transfers within general government	NMZT	39 290	48 859	54 527	55 891	57 736	58 587	59 458	59 506†	60 252
Current international cooperation	NMZU	1 752	4 616	1 907	2 558	1 752	1 233	2 424	1 739	1 384
from institutions of the EC	NMEX	1 714	2 506	1 898	2 558	1 752	1 233	2 424	1 739	1 384
Miscellaneous current transfers										
from sectors other than general government	NMZX	222	353	178	296	420	469	429	476†	390
Total, other current transfers										
from general government	NMZY	39 290	48 859	54 527	55 891	57 736	58 587	59 458	59 506†	60 252
from other sectors	NMZZ	2 288	5 351	2 493	3 215	2 512	2 079	3 224	2 564†	2 137
Total	NNAA	41 578	54 210	57 020	59 106	60 248	60 666	62 682	62 070†	62 389
Total resources	NNAB	236 999	262 651	265 438	269 279	285 684	302 232	316 638	337 244†	364 859
Uses										
Social benefits other than social transfers in kind	NNAD	65 999	81 753	94 766	102 585	106 187	109 877	112 568	116 302†	116 605
Other current transfers										
Net non-life insurance premiums	NNAE	314	382	408	361	340	377	371	349†	363
Current transfers within general government	NNAF	39 290	48 859	54 527	55 891	57 736	58 587	59 458	59 506†	60 252
Current international cooperation	NNAG	1 374	1 677	1 931	1 774	1 825	2 018	1 598	1 527†	1 503
to institutions of the EC	NMFA	–	–	–	2	7	8	8	31	–1
Miscellaneous current transfers										
to sectors other than general government	NNAI	4 561	4 965	5 096	9 055	10 760	11 369	13 386	14 531†	16 831
GNP based fourth own resource	NMFH	1	813	914	1 558	2 071	1 826	2 454	2 458	3 920
Total other current transfers										
to general government	NNAL	39 290	48 859	54 527	55 891	57 736	58 587	59 458	59 506†	60 252
to other sectors	NNAM	6 249	7 024	7 435	11 190	12 925	13 764	15 355	16 407†	18 697
Total	NNAN	45 539	55 883	61 962	67 081	70 661	72 351	74 813	75 913†	78 949
Disposable income, gross	NNAO	125 461	125 015	108 710	99 613	108 836	120 004	129 257	145 029†	169 305
Total uses	NNAB	236 999	262 651	265 438	269 279	285 684	302 232	316 638	337 244†	364 859

Source: Office for National Statistics: 020 7533 6031

15.10 Households and non-profit institutions serving households

Allocation of primary income account

£ millions

		1990	1991	1992	1993	1994	1995	1996	1997	1998
Resources										
Operating surplus, gross	QWLS	24 755	28 513	32 749	34 462	36 790	39 321	41 105	44 217†	45 602
Mixed income, gross	QWLT	33 030	31 635	33 872	36 096	38 336	40 239	41 570	41 665†	43 379
Compensation of employees										
Wages and salaries	QWLW	276 337	291 905	303 018	310 252	321 545	335 589	351 547	375 643†	401 634
Employers' social contributions	QWLX	38 761	41 882	43 969	46 106	48 245	49 512	53 067	56 828†	61 840
Total	QWLY	315 098	333 787	346 987	356 358	369 790	385 101	404 614	432 471†	463 474
Property income										
Interest	QWLZ	38 448	36 922	33 122	23 802	22 339	26 043	23 681	26 043†	29 903
Distributed income of corporations	QWMA	29 179	26 705	27 641	27 572	29 222	34 607	35 882	38 467†	38 828
Attributed property income of insurance policy holders	QWMC	32 444	33 782	34 765	35 267	37 001	42 078	47 564	51 800†	52 542
Rent	QWMD	99	98	99	95	96	99	103	104	103
Total	QWME	100 170	97 507	95 627	86 736	88 658	102 827	107 230	116 414†	121 376
Total resources	QWMF	473 053	491 442	509 235	513 652	533 574	567 488	594 519	634 767†	673 831
Uses										
Property income										
Interest	QWMG	53 146	49 861	45 796	36 253	36 985	39 887	38 376	42 045†	51 366
Rent	QWMH	203	200	200	194	197	201	211	217	216
Total	QWMI	53 349	50 061	45 996	36 447	37 182	40 088	38 587	42 262†	51 582
Balance of primary incomes, gross	QWMJ	419 704	441 381	463 239	477 205	496 392	527 400	555 932	592 505†	622 249
Total uses	QWMF	473 053	491 442	509 235	513 652	533 574	567 488	594 519	634 767†	673 831

Source: Office for National Statistics: 020 7533 6031

15.11 Households and non-profit institutions serving households

Secondary distribution of income account

£ millions

		1990	1991	1992	1993	1994	1995	1996	1997	1998
Resources										
Balance of primary incomes, gross	QWMJ	419 704	441 381	463 239	477 205	496 392	527 400	555 932	592 505†	622 249
Imputed social contributions	RVFH	114	288	360	412	491	448	383	427	342
Social benefits other than social transfers in kind	QWML	88 988	109 677	126 932	136 370	143 303	149 234	156 429	165 245†	170 191
Other current transfers										
Non-life insurance claims	QWMM	8 978	10 976	11 690	10 313	9 739	10 785	16 203	10 259†	10 482
Miscellaneous current transfers	QWMN	10 842	11 254	11 248	15 481	17 013	17 722	20 060	20 478†	21 391
Total	QWMO	19 820	22 230	22 938	25 794	26 752	28 507	36 263	30 737†	31 873
Total resources	QWMP	528 626	573 576	613 469	639 781	666 938	705 589	749 007	788 914†	824 655
Uses										
Current taxes on income, wealth etc										
Taxes on income	QWMQ	60 628	66 109	68 063	65 316	69 507	74 288	74 938	74 958†	88 551
Other current taxes	NVCO	13 231	10 496	10 299	10 551	11 140	11 937	12 795	13 820†	14 892
Total	QWMS	73 859	76 605	78 362	75 867	80 647	86 225	87 733	88 778†	103 443
Social contributions										
Actual social contributions										
Employers' actual social contributions	QWMT	33 534	33 909	34 867	36 795	38 233	40 158	44 161	47 711†	52 862
Employees' social contributions	QWMU	43 135	45 005	46 477	45 415	49 615	53 747	59 900	66 802†	70 425
Social contributions by self and non-employed	QWMV	1 177	1 206	1 281	1 472	1 469	1 541	1 771	1 848†	1 842
Total	QWMW	77 846	80 120	82 625	83 682	89 317	95 446	105 832	116 361†	125 129
Imputed social contributions	QWMX	5 227	7 973	9 102	9 311	10 012	9 354	8 906	8 911†	9 551
Total	QWMY	83 073	88 093	91 727	92 993	99 329	104 800	114 738	125 272†	134 680
Social benefits other than social transfers in kind	QWMZ	318	560	691	871	946	918	853	897	814
Other current transfers										
Net non-life insurance premiums	QWNA	8 978	10 976	11 690	10 313	9 739	10 785	16 203	10 259†	10 482
Miscellaneous current transfers	QWNB	5 787	6 675	6 802	7 618	8 114	8 287	8 199	9 067†	9 301
Total	QWNC	14 765	17 651	18 492	17 931	17 853	19 072	24 402	19 326†	19 783
Disposable income, gross	QWND	356 611	390 667	424 197	452 119	468 163	494 574	521 281	554 641†	565 935
Total uses	QWMP	528 626	573 576	613 469	639 781	666 938	705 589	749 007	788 914†	824 655
Read households' disposable income, at 1995 prices	RVGK	438 935	445 552	461 964	475 850	481 924	494 574	505 392	524 501†	524 660

Source: Office for National Statistics: 020 7533 6031

15.12 Households and non-profit institutions serving households

Use of disposable income account

£ millions

		1990	1991	1992	1993	1994	1995	1996	1997	1998
Resources										
Disposable income, gross	QWND	356 611	390 667	424 197	452 119	468 163	494 574	521 281	554 641†	565 935
Adjustment for the change in net equity of households in pension funds	NSSE	18 238	15 596	13 265	10 742	10 577	11 690	14 824	15 692†	16 567
Total resources	NSSF	374 849	406 263	437 462	462 861	478 740	506 264	536 105	570 333†	582 502
Uses										
Final consumption expenditure										
Individual consumption expenditure	NSSG	347 247	368 232	387 312	412 398	433 829	454 171	485 418	517 032†	545 124
Saving, gross	NSSH	27 602	38 031	50 150	50 463	44 911	52 093	50 687	53 301†	37 378
Total uses	NSSF	374 849	406 263	437 462	462 861	478 740	506 264	536 105	570 333†	582 502
Saving ratio (per cent)	RVGL	7.4	9.4	11.5	10.9	9.4	10.3	9.5	9.3†	6.4

Source: Office for National Statistics: 020 7533 6031

15.13 Summary capital accounts and net lending/net borrowing

£ millions

		1990	1991	1992	1993	1994	1995	1996	1997	1998
Non-financial corporations										
Gross saving[1]	RPJV	41 452	45 850	47 111	62 014	74 870	71 810	81 354	77 594	..
Capital transfers (net receipts)	GZQW	8 349	3 519	3 317	3 187	3 275	4 882	3 599	2 459†	1 709
Gross capital formation[2]	RQBZ	61 601	53 430	52 474	55 434	60 357	68 514	74 481	85 293†	93 769
Net acquisition of non-financial assets	RQAX	391	–39	41	254	214	301	117	195	200
Financial corporations										
Gross saving[1]	RPPS	7 908	3 196	10 593	9 407	16 695	13 534	11 474	11 803†	17 342
Capital transfers (net receipts)	GZQE	–	–	–	–88	–518	–	–	–	–
Gross capital formation[2]	RPYP	6 783	6 249	4 969	3 748	5 566	4 984	5 212	5 087†	7 536
Net acquisition of non-financial assets	RPYO	66	–8	–49	–203	86	–77	–1	–39	–46
General Government										
Gross saving[1]	RPQC	15 926	4 216	–19 559	–30 952	–26 700	–20 402	–16 854	–1 551	..
Capital transfers (net receipts)	GZQU	–10 260	–6 702	–6 170	–6 985	–5 929	–7 097	–5 040	–3 910†	–3 596
Gross capital formation[2]	RPZF	14 719	14 213	14 227	13 698	13 807	13 851	11 482	10 629	..
Net acquisition of non-financial assets	RPZE	–605	–396	–312	–497	–485	–143	–467	–372	–382
Households & NPISH										
Gross saving[1]	RPQL	27 602	38 031	50 150	50 463	44 911	52 093	50 687	53 301†	37 378
Capital transfers (net receipts)	GZQI	2 408	3 473	3 274	4 195	3 205	2 749	2 163	2 289†	2 326
Gross capital formation[2]	RPZV	24 053	21 304	21 024	23 377	26 190	27 813	31 029	32 920	..
Net acquisition of non-financial assets	RPZU	148	443	320	446	185	–81	337	250	246
Net lending(+)/net borrowing(-)[3]										
Non-financial corporations	RQAW	–17 436	–8 493	–7 770	4 220	12 260	2 259	5 298	–7 065	..
Financial corporations	RPYN	1 059	–3 045	5 673	5 774	10 525	8 627	6 263	6 755†	9 852
General government	RPZD	–8 448	–16 303	–39 644	–51 138	–45 951	–41 207	–32 909	–16 109	..
Households sector	RPZT	5 809	19 757	32 080	30 835	21 741	27 110	21 484	21 804†	3 939
Rest of the world[4]	RQCH	19 016	8 084	9 661	10 309	1 425	3 211	–136	–7 427†	–557
Statistical Discrepancy[5]	RVFE	–	–	–	–	–	–	–	–†	–1 789

1 Before providing for depreciation, inventory holding gains.

2 Comprises gross fixed capital formation and changes in inventories and acquisitions less disposals of valuables.

3 This balance is equal to gross saving *plus* capital transfers *less*gross fixed capital formation,*less* Net acquisition of non-financial assets, *less*changes in inventories.

4 Equals, the current balance of payments accounts, *plus* capital transfers.

5 Series is only available annually.

Source: Office for National Statistics: 020 7533 6031

15.14 Household final consumption expenditure: classified by commodity
At current market prices

£ millions

		1990	1991	1992	1993	1994	1995	1996	1997	1998
Durable goods:										
Cars, motorcycles and other vehicles	CCDT	19 034	16 977	16 470	18 063	19 778	20 749	23 467	26 731†	28 377
Other durable goods	ABZB	15 483	15 890	17 044	17 311	18 102	18 747	20 212	23 194†	24 250
Total durable goods	AEIT	34 517	32 867	33 514	35 374	37 880	39 496	43 679	49 925†	52 627
Non-Durable Goods:										
Food (household expenditure)	CCDW	41 817	44 044	45 193	46 334	47 122	49 274	52 513	53 188†	54 113
Alchohol and tobacco	CDFH	30 009	32 680	33 553	34 865	36 624	37 456	39 618	40 858†	41 577
Clothing and footwear	CDDE	21 212	22 209	23 404	24 777	26 893	28 347	29 564	31 115†	31 571
Energy products	CCEC	22 422	24 955	25 399	26 136	26 857	27 118	28 822	28 593†	28 660
Other goods	ABZN	39 659	42 171	44 461	46 738	48 935	51 947	57 094	61 921†	66 543
Total non-durable goods	ABZR	155 119	166 059	172 010	178 850	186 431	194 142	207 611	215 675†	222 464
Services:										
Rental and water charges	ABRG	36 543	42 411	48 637	52 405	56 297	59 798	62 197	66 421†	67 858
Catering	CDEY	30 076	30 671	32 287	35 154	36 459	37 727	40 786	42 652†	46 300
Transport and communication	ABOZ	30 969	32 843	34 610	37 840	39 603	41 360	43 211	46 247†	50 433
Financial services	CEGK	13 856	14 549	14 801	16 316	16 226	16 784	18 958	21 461†	23 993
Other services	AEJC	34 984	37 571	39 759	42 026	44 391	48 602	50 958	54 981†	59 697
Total services	AELL	146 428	158 045	170 094	183 741	192 976	204 271	216 110	231 762†	248 281
Total household final expenditure in the UK by resident and non-resident households (domestic concept)	ABQI	336 064	356 971	375 618	397 965	417 287	437 909	467 400	497 362†	523 372
Final consumption expenditure outside the UK by UK resident households	ABTA	9 284	9 385	10 605	11 890	13 096	13 768	14 433	14 945	16 799
Less Final consumption expenditure in the UK by households resident in the rest of the world	CDFD	-8 856	-8 571	-9 076	-10 747	-11 121	-13 224	-13 992	-14 000	-14 708
Final consumption expenditure by UK resident households in the UK and abroad (national concept)	ABPB	336 492	357 785	377 147	399 108	419 262	438 453	467 841	498 307†	525 463

Data for all series in this table are available in *Consumer Trends* or on the ONS Databank. Some of these quarterly data are published regularly in *UK Economic Accounts* in table A7.

Source: Office for National Statistics: 020 7533 6031

15.15 Household final consumption expenditure: classified by commodity
At 1995 market prices

£ millions

		1990	1991	1992	1993	1994	1995	1996	1997	1998
Durable goods:										
Cars, motorcycles and other vehicles	CCBJ	22 425	18 598	17 397	19 016	20 334	20 749	22 673	24 680†	25 953
Other durable goods	ABZD	15 968	15 894	16 980	17 196	18 249	18 747	20 028	23 273†	25 121
Total durable goods	AEIV	38 101	34 436	34 474	36 221	38 583	39 496	42 701	47 953†	51 074
Non-Durable Goods:										
Food (household expenditure)	CCBM	47 055	47 114	47 664	48 282	48 931	49 274	50 931	51 786†	51 972
Alchohol and tobacco	FCCA	41 654	40 258	38 415	37 861	38 441	37 456	38 007	37 533†	36 378
Clothing and footwear	FCCB	22 105	22 502	23 683	24 875	26 928	28 347	29 773	31 076†	31 734
Energy products	CCBS	27 389	28 281	27 961	28 123	27 754	27 118	28 210	27 427†	27 561
Other goods	ABZP	47 995	47 567	47 817	49 035	50 119	51 947	55 419	59 165†	62 434
Total non-durable goods	ABZT	184 983	184 760	185 041	187 926	192 173	194 142	202 340	206 987†	210 079
Services:										
Rental and water charges	ABRI	56 127	56 743	57 191	58 015	58 862	59 798	60 344	60 934†	61 595
Catering	CCHS	39 539	36 806	36 279	37 539	37 319	37 727	39 704	40 399†	42 927
Transport and communication	ABPD	36 320	35 121	35 704	38 473	40 011	41 360	42 188	43 387†	46 119
Financial services	CEGM	16 603	16 477	16 202	17 168	16 397	16 784	18 034	19 516†	19 968
Other services	AEJZ	45 577	44 808	44 736	45 183	46 008	48 602	48 863	50 371†	51 504
Total services	AELN	194 093	189 592	189 729	196 363	198 597	204 271	209 133	214 607†	222 113
Total household final expenditure in the UK by resident and non-resident households (domestic concept)	ABQJ	415 812	407 322	408 208	419 536	429 353	437 909	454 174	469 547†	483 266
Final consumption expenditure outside the UK by UK resident households	ABTC	11 262	11 081	11 953	12 093	13 525	13 768	14 125	16 357†	18 677
Less Final consumption expenditure in the UK by households resident in the rest of the world	CCHX	−11 249	−10 031	−10 086	−11 524	−11 416	−13 224	−13 613	−13 203	−13 438
Final consumption expenditure by UK resident households in the UK and abroad (national concept)	ABPF	415 788	408 309	410 026	420 081	431 462	438 453	454 686	472 701†	488 505

Data for all series in this table are available in *Consumer Trends* or on the ONS Databank. Some of these quarterly data are published regularly in *UK Economic Accounts* in table A7.

Source: Office for National Statistics: 020 7533 6031

15.16 Individual consumption expenditure at current market prices by households, NPISHs[1] and general government

Classified by function (COICOP/COPNI/COFOG)[1]

£ millions

		1990	1991	1992	1993	1994	1995	1996	1997	1998
FINAL CONSUMPTION EXPENDITURE OF HOUSEHOLDS										
Food and non-alcoholic beverages	ABZV	41 817	44 044	45 193	46 334	47 122	49 274	52 513	53 188†	54 113
Food	ABZW	37 380	39 363	40 465	41 630	42 392	44 049	47 034	47 631†	48 216
Non-alcoholic beverages	ADFK	4 437	4 681	4 728	4 704	4 730	5 225	5 479	5 557	5 897
Alcoholic beverages and tobacco	ADFL	30 009	32 680	33 553	34 865	36 624	37 456	39 618	40 858†	41 577
Alcoholic beverages	ADFM	21 360	23 032	23 481	24 399	25 616	26 039	27 774	29 109†	29 722
Tobacco	ADFN	8 649	9 648	10 072	10 466	11 008	11 417	11 844	11 749†	11 855
Clothing and footwear	ADFP	21 934	22 977	24 165	25 511	27 625	29 140	30 370	31 978†	32 479
Clothing	ADFQ	18 196	19 199	20 137	21 379	23 228	24 687	25 895	27 286†	27 988
Footwear	ADFR	3 738	3 778	4 028	4 132	4 397	4 453	4 475	4 692†	4 491
Housing, water, electricity, gas and other fuels	ADFS	56 729	65 270	71 951	75 651	79 772	83 473	87 440	91 855†	94 341
Actual rentals for housing	ADFT	10 748	12 829	15 155	17 122	18 238	18 925	19 703	20 706†	20 731
Imputed rentals for housing	ADFU	23 257	26 608	30 182	31 646	34 045	36 629	37 969	40 994†	42 263
Maintenance and repair of the dwelling	ADFV	7 455	8 114	8 329	7 939	7 834	7 744	8 297	9 628†	11 068
Water supply and miscellaneous dwelling services	ADFW	3 019	3 522	3 905	4 326	4 770	5 058	5 397	5 648†	5 818
Electricity, gas and other fuels	ADFX	12 250	14 197	14 380	14 618	14 885	15 117	16 074	14 879†	14 461
Furnishings, household equipment and routine maintenance of the house	ADFY	19 882	20 945	22 338	23 393	24 758	25 767	28 032	30 881†	31 999
Furniture, furnishings, carpets and other floor coverings	ADFZ	6 564	6 667	7 281	7 771	8 589	9 109	10 286	11 603†	11 652
Household textiles	ADGG	2 112	2 306	2 321	2 629	2 656	2 825	3 239	3 419†	3 483
Household appliances	ADGL	5 149	5 396	5 642	5 387	5 487	5 603	5 905	6 656†	7 190
Glassware, tableware and household utensils	ADGM	485	511	540	598	643	644	673	731†	792
Tools and equipment for house and garden	ADGN	1 100	1 154	1 213	1 247	1 302	1 317	1 380	1 525†	1 652
Goods and services for routine household maintenance	ADGO	4 472	4 911	5 341	5 761	6 081	6 269	6 549	6 947†	7 230
Health	ADGP	3 559	4 232	4 669	4 774	5 298	5 463	5 881	5 906†	6 186
Medical products, appliances and equipment	ADGQ	2 159	2 560	3 035	3 213	3 651	3 698	4 070	4 190†	4 384
Out-patient services	ADGR	1 195	1 436	1 398	1 336	1 411	1 514	1 556	1 466†	1 539
Hospital services	ADGS	205	236	236	225	236	251	255	250†	263
Transport	ADGT	51 767	52 083	53 148	56 811	59 788	62 224	66 851	74 047†	78 806
Purchase of vehicles	ADGU	21 247	19 754	19 509	21 024	22 551	23 192	25 621	28 722†	29 874
Operation of personal transport equipment	ADGV	19 054	20 560	21 109	22 325	23 270	23 972	25 085	27 639†	29 552
Transport services	ADGW	11 466	11 769	12 530	13 462	13 967	15 060	16 145	17 686†	19 380
Communications										
Communications	ADGX	6 485	7 041	7 332	7 831	8 594	9 092	9 432	9 918†	10 835
Recreation and culture	ADGY	35 733	36 739	38 145	40 409	41 750	45 236	48 247	53 334†	58 485
Audio-visual, photographic and information processing equipment	ADGZ	6 463	6 606	6 970	7 179	6 966	7 479	7 662	9 032†	10 174
Other major durables for recreation and culture	ADHL	2 652	2 381	2 093	2 309	2 532	2 850	3 279	3 697†	4 446
Other recreational items and equipment; flowers, garden and pets	ADHZ	6 867	7 195	7 839	8 164	8 705	9 232	10 094	11 582†	12 456
Recreational and cultural services	ADIA	12 242	12 780	13 176	13 756	14 614	16 683	17 380	18 683†	20 430
Newspapers, books and stationery	ADIC	6 922	7 178	7 435	8 313	8 220	8 250	9 028	9 499†	10 066
Package holidays	ADID	587	599	632	688	713	742	804	841†	913
Education										
Educational services	ADIE	3 221	4 028	4 788	5 186	5 677	6 392	6 984	7 734†	8 492
Restaurants and hotels	ADIF	24 762	24 671	25 475	28 113	29 450	31 196	34 344	36 254†	39 910
Catering services	ADIG	19 569	19 377	19 127	21 239	22 303	23 432	26 315	28 296†	29 359
Accommodation services	ADIH	5 193	5 294	6 348	6 874	7 147	7 764	8 029	7 958†	10 551
Miscellaneous goods and services	ADII	40 166	42 261	44 861	49 087	50 829	53 196	57 688	61 409†	66 149
Personal care	ADIJ	6 947	7 497	7 798	8 258	8 970	9 812	11 269	11 707†	12 360
Personal effects n.e.c.	ADIK	4 361	4 453	4 491	4 769	5 053	5 992	6 284	6 511†	6 935
Social protection	ADIL	5 839	6 462	7 303	7 550	7 631	7 343	7 389	7 498†	7 677
Insurance	ADIM	13 632	14 649	15 442	17 410	17 273	16 834	17 613	18 839†	19 835
Financial services n.e.c.	ADIN	4 594	4 540	4 948	5 864	6 168	6 677	7 895	8 972†	10 560
Other services n.e.c.	ADIO	4 793	4 660	4 879	5 236	5 734	6 538	7 238	7 882†	8 782
Final consumption expenditure in the UK by resident and non-resident households (domestic concept)	ABQI	336 064	356 971	375 618	397 965	417 287	437 909	467 400	497 362†	523 372
Final consumption expenditure outside the UK by UK resident households	ABTA	9 284	9 385	10 605	11 890	13 096	13 768	14 433	14 945	16 799
less Final consumption expenditure in the UK by households resident in the rest of the world	CDFD	−8 856	−8 571	−9 076	−10 747	−11 121	−13 224	−13 992	−14 000	−14 708
Final consumption expenditure by UK resident Households in the UK and abroad(national concept)	ABPB	336 492	357 785	377 147	399 108	419 262	438 453	467 841	498 307†	525 463

1 Non-profit institutions serving households.

Source: Office for National Statistics: 020 7533 6031

15.16 Individual consumption expenditure at current market prices by households, NPISHs[1] and general government

continued

Classified by function (COICOP/COPNI/COFOG)[1]

£ millions

		1990	1991	1992	1993	1994	1995	1996	1997	1998
CONSUMPTION EXPENDITURE OF UK RESIDENT HOUSEHOLDS										
Final consumption expenditure of UK resident households in the UK and abroad	ABPB	336 492	357 785	377 147	399 108	419 262	438 453	467 841	498 307†	525 463
FINAL INDIVIDUAL CONSUMPTION EXPENDITURE OF NPISH										
Final individual consumption expenditure of NPISH	ABNV	10 755	10 447	10 165	13 290	14 567	15 718	17 577	18 725†	19 661
FINAL INDIVIDUAL CONSUMPTION EXPENDITURE OF OF GENERAL GOVERNMENT										
Health	QYOT	26 340	29 583	33 263	35 518	37 356	39 018	41 498	42 683†	45 147
Recreation and culture	QYSU	2 817	3 109	3 311	3 480	3 183	3 119	3 119	3 015†	2 913
Education	QYSE	21 380	23 750	25 537	23 502	24 027	25 284	25 361	26 377†	27 243
Social protection	QYSP	8 548	9 623	10 607	12 058	12 501	14 530	15 316	15 646†	15 963
Housing	QYXO	–	–	–	–	–	–	–	–	–
Final individual consumption expenditure of of general government	NNAQ	59 085	66 065	72 718	74 558	77 067	81 951	85 294	87 721†	91 266
Total, individual consumption expenditure/	NQEO	406 332	434 297	460 030	486 956	510 896	536 122	570 712	604 753†	636 390
actual individual consumption	ABRE	406 332	434 297	460 030	486 956	510 896	536 122	570 712	604 753†	636 390

1 "Purpose" or "function" classifications are designed to indicate the "socio-economic objectives" that institutional units aim to achieve through various kinds of outlays. COICOP is the Classification of Industrial Consumption by Purpose and applies to households. COPNI is the Classification of the Purposes of Non-profit Institutions Serving Households and COFOG the Classification of the Functions of Government. The introduction of ESA95 coincides with the redefinition of these classifications and data will be available on a consistent basis for all European Union member states.

Source: Office for National Statistics: 020 7533 6031

15.17 Individual consumption expenditure at 1995 market prices by households, NPISHs[1] and general government

Classified by function (COICOP/COPNI/COFOG)[1]

£ millions at 1995 prices

		1990	1991	1992	1993	1994	1995	1996	1997	1998
FINAL CONSUMPTION EXPENDITURE OF HOUSEHOLDS										
Food and non-alcoholic beverages	ADIP	47 055	47 114	47 664	48 282	48 931	49 274	50 931	51 786†	51 972
Food	ADIQ	42 390	42 401	43 024	43 542	44 010	44 049	45 687	46 565†	46 605
Non-alcoholic beverages	ADIR	4 669	4 714	4 650	4 749	4 921	5 225	5 244	5 221†	5 367
Alcoholic beverages and tobacco	ADIS	41 654	40 258	38 415	37 861	38 441	37 456	38 007	37 533†	36 378
Alcoholic beverages	ADIT	28 133	27 088	25 956	25 967	26 689	26 039	26 906	27 278†	26 857
Tobacco	ADIU	13 678	13 342	12 603	11 959	11 752	11 417	11 101	10 255†	9 521
Clothing and footwear	ADIW	22 983	23 359	24 508	25 645	27 686	29 141	30 555	31 885†	32 542
Clothing	ADIX	18 934	19 451	20 348	21 461	23 267	24 688	25 997	26 979†	27 760
Footwear	ADIY	4 065	3 918	4 171	4 189	4 419	4 453	4 558	4 906†	4 782
Housing, water, electricity, gas and other fuels	ADIZ	80 247	82 003	81 951	82 700	82 955	83 470	85 190	86 178†	87 976
Actual rentals for housing	ADJA	16 984	17 366	17 987	18 672	18 876	18 925	19 069	18 871†	18 736
Imputed rentals for housing	ADJB	35 013	35 234	35 093	35 211	35 724	36 629	37 027	37 772†	38 665
Maintenance and repair of the dwelling	ADJC	9 073	8 936	8 767	8 203	8 006	7 743	8 038	9 065†	9 971
Water supply and miscellaneous dwelling services	ADJD	4 642	4 707	4 735	4 856	5 038	5 056	5 090	5 155†	5 040
Electricity, gas and other fuels	ADJE	14 301	15 381	15 129	15 598	15 311	15 117	15 966	15 315†	15 564
Furnishings, household equipment and routine maintenance of the house	ADJF	22 606	22 430	23 246	24 049	25 374	25 766	27 298	29 808†	30 575
Furniture, furnishings, carpets and other floor coverings	ADJG	7 633	7 331	7 759	8 203	9 005	9 109	9 797	10 794†	10 664
Household textiles	ADJH	2 135	2 305	2 349	2 692	2 695	2 825	3 170	3 333†	3 368
Household appliances	ADJI	5 450	5 440	5 599	5 306	5 494	5 603	5 929	6 791†	7 404
Glassware, tableware and household utensils	ADJJ	561	548	564	617	656	645	660	708†	759
Tools and equipment for house and garden	ADJK	1 225	1 209	1 240	1 260	1 318	1 316	1 367	1 514†	1 638
Goods and services for routine household maintenance	ADJL	5 587	5 555	5 694	5 944	6 206	6 268	6 375	6 668†	6 742
Health	ADJM	4 498	4 918	5 089	5 069	5 387	5 462	5 671	5 497†	5 480
Medical products, appliances and equipment	ADJN	2 693	2 928	3 248	3 366	3 684	3 698	3 907	3 901†	3 896
Out-patient services	ADJO	1 548	1 717	1 579	1 457	1 458	1 515	1 515	1 366†	1 358
Hospital services	ADJP	272	293	271	250	245	249	249	230†	226
Transport	ADJQ	63 127	58 411	57 501	60 050	61 626	62 211	64 480	67 295†	69 724
Purchase of vehicles	ADJR	24 749	21 537	20 456	22 076	23 106	23 179	24 780	26 638†	27 636
Operation of personal transport equipment	ADJS	24 945	24 519	23 968	23 896	24 066	23 972	24 025	24 760†	25 366
Transport services	ADJT	13 454	12 534	13 241	14 128	14 454	15 060	15 675	15 897†	16 722
Communications										
Communications	ADJU	6 661	6 663	6 757	7 141	8 155	9 092	9 599	10 369†	11 510
Recreation and culture	ADJV	41 660	40 229	40 347	41 709	42 602	45 250	46 972	51 335†	55 819
Audio-visual, photographic and information processing equipment	ADJW	6 194	6 208	6 622	6 825	6 803	7 480	7 769	9 454†	11 209
Other major durables for recreation and culture	ADJX	3 032	2 557	2 210	2 380	2 595	2 863	3 202	3 532†	4 174
Other recreational items and equipment; flowers, gardens and pets	ADJY	7 768	7 704	8 146	8 373	8 844	9 232	9 907	11 218†	12 077
Recreational and cultural services	ADJZ	15 495	14 877	14 507	14 649	15 029	16 683	16 784	17 553†	18 490
Newspapers, books and stationery	ADKM	8 594	8 297	8 152	8 743	8 601	8 250	8 529	8 783†	9 023
Package holidays	ADMI	773	719	708	733	730	742	781	795†	846
Education										
Educational services	ADMJ	4 653	5 309	5 731	5 855	6 205	6 392	6 398	6 490†	6 562
Restaurants and Hotels	ADMK	32 724	29 521	28 697	30 043	30 013	31 196	33 356	34 119†	36 650
Catering services	ADML	26 170	23 575	21 886	23 077	23 219	23 432	25 290	26 206†	26 162
Accommodation services	ADMM	6 597	5 984	6 795	6 958	6 794	7 764	8 066	7 913†	10 488
Miscellaneous goods and services	ADMN	49 531	48 761	49 389	51 818	51 978	53 199	55 717	57 252†	58 078
Personal care	ADMO	8 938	8 843	8 718	8 912	9 453	9 811	10 781	10 744†	10 738
Personal effects n.e.c.	ADMP	4 936	4 891	4 807	4 902	5 081	5 992	6 232	6 448†	6 849
Social protection	ADMQ	7 530	7 858	8 193	8 059	7 956	7 342	7 245	7 318†	7 361
Insurance	ADMR	15 748	15 901	16 053	17 754	17 259	16 835	17 298	17 272†	16 967
Financial services n.e.c.	ADMS	6 195	5 650	5 945	6 483	6 297	6 677	7 201	8 002†	8 369
Other services n.e.c.	ADMT	6 268	5 625	5 715	5 674	5 932	6 542	6 960	7 468†	7 794
Final consumption expenditure in the UK by resident and non-resident households (domestic concept)	ABQJ	415 812	407 322	408 208	419 536	429 353	437 909	454 174	469 547†	483 266
Final consumption expenditure outside the UK by UK resident households	ABTC	11 262	11 081	11 953	12 093	13 525	13 768	14 125	16 357†	18 677
less Final consumption expenditure in the UK by households resident in the rest of the world	CCHX	−11 249	−10 031	−10 086	−11 524	−11 416	−13 224	−13 613	−13 203	−13 438
Final consumption expenditure by UK resident households in the UK and abroad (national concept)	ABPF	415 788	408 309	410 026	420 081	431 462	438 453	454 686	472 701†	488 505

1 Non-profit institutions serving households.

Source: Office for National Statistics: 020 7533 6031

15.17 Individual consumption expenditure at 1995 market prices by households, NPISHs[1] and general government

continued

Classified by function (COICOP/COPNI/COFOG)[1]

£ millions at 1995 prices

		1990	1991	1992	1993	1994	1995	1996	1997	1998
CONSUMPTION EXPENDITURE OF UK RESIDENT HOUSEHOLDS										
Final consumption expenditure of UK resident households in the UK and abroad	ABPF	415 788	408 309	410 026	420 081	431 462	438 453	454 686	472 701†	488 505
FINAL INDIVIDUAL CONSUMPTION EXPENDITURE OF NPISH										
Final individual consumption expenditure of NPISH	ABNU	11 621	11 656	11 769	13 963	15 119	15 718	15 936	16 235†	16 862
FINAL INDIVIDUAL CONSUMPTION EXPENDITURE OF GENERAL GOVERNMENT										
Health	QYXJ	32 601	33 704	34 968	36 532	37 561	39 018	40 084	40 987	42 229
Recreation and culture	QYXK	3 362	3 488	3 582	3 644	3 233	3 119	3 046	2 898†	2 615
Education	QYXL	25 697	26 120	26 623	25 194	24 979	25 284	25 558	25 804†	26 046
Social protection	QYXM	10 635	11 475	11 880	12 635	13 233	14 529	15 311	15 606†	15 077
Housing	QYXO	–	–	–	–	–	–	–	–	–
Final individual consumption expenditure of general government	NSZK	72 295	74 787	77 053	78 005	79 006	81 950	83 999	85 295†	85 967
Total, individual consumption expenditure/ actual individual consumption	YBIO	499 280	494 388	498 504	511 954	525 587	536 121	554 621	574 231†	591 334

1 "Purpose" or "function" classifications are designed to indicate the "socio-economic objectives" that institutional units aim to achieve through various kinds of outlays. COICOP is the Classification of Individual Consumption by Purpose and applies to households. COPNI is the Classification of the Purposes of Non-profit Institutions Serving Households and COFOG the Classification of the Functions of Government. The introduction of ESA95 coincides with the redefinition of these classifications and data will be available on a consistent basis for all European Union member states.

Source: Office for National Statistics: 020 7533 6031

15.18 Change in inventories at constant 1995 prices[1,2]

£ millions

		1990	1991	1992	1993	1994	1995	1996	1997	1998
Mining and quarrying	FADO	–103	170	66	–45	–267	–123	–47	55†	349
Manufacturing industries										
Materials and fuel	FBID	–266	–861	–211	20	433	514	–105	403†	200
Work in progress	FBIE	–1 511	–1 358	–1 124	–981	639	1 144	–217	–1 332†	–400
Finished goods	FBIF	–136	–1 346	–47	–320	548	998	6	340†	855
Total	DHBH	–1 913†	–3 565	–1 382	–1 281	1 620	2 656	–316	–589	655
Electricity, gas and water supply	FADP	–129	200	–100	–270	–661	–205	15	103†	–119
Distributive trades										
Wholesale[3]	FAJM	–552	–612	117	802	1 332	597	681	1 499†	–3
Retail[3]	FBYH	181	–404	216	368	884	811	638	799†	784
Other industries[4]	DLWV	716	–420	–616	738	1 928	776	859	1 895†	1 977
Change in inventories	ABMQ	–2 079	–5 349	–1 962	360	4 836	4 512	1 830	3 762†	3 643

1 Estimates are given to the nearest £ million but cannot be regarded as accurate to this degree.
2 Components may not sum to totals due to rounding.
3 Wholesaling and retailing estimates exclude the motor trades.
4 Quarterly alignment adjustment included in this series. For description see notes.

Source: Office for National Statistics: 020 7533 6031

15.19 Gross fixed capital formation at current purchasers' prices Analysis by broad sector and type of asset

Total economy

£ millions

		1990	1991	1992	1993	1994	1995	1996	1997	1998
Private sector										
New dwellings, excluding land	EQBT	16 867	15 577	16 246	17 124	18 285	18 784	20 467	22 018†	24 589
Other buildings and structures	EQBU	24 323	21 347	18 378	15 743	16 213	17 841	20 408	24 352†	28 358
Transport equipment	EQBV	9 324	7 574	6 972	8 357	10 049	10 021	11 279	12 283†	13 846
Other machinery and equipment and cultivated assets	EQBW	33 123	32 387	31 553	32 616	34 968	41 532	46 606	49 980†	54 357
Intangible fixed assets	EQBX	3 206	3 591	3 290	3 122	2 982	3 179	3 278	3 395†	3 651
Costs associated with the transfer of ownership of non-produced assets	EQBY	7 409	5 729	3 926	5 089	5 208	5 222	7 065	7 599†	7 916
Total	EQBZ	94 252	86 205	80 365	82 051	87 705	96 579	109 103	119 627†	132 717
Public non-financial corporations										
New dwellings, excluding land	DEER	201	153	172	150	139	162	151	123†	114
Other buildings and structures	DEES	2 659	3 015	3 332	3 416	3 492	3 781	3 397	2 675†	2 614
Transport equipment	DEEP	496	383	678	516	512	354	225	191†	114
Other machinery and equipment and cultivated assets	DEEQ	2 084	442	873	720	761	857	765	756†	652
Intangible fixed assets	DLXJ	159	252	256	279	374	496	585	595†	605
Costs associated with the transfer of ownership of non-produced assets	DLXQ	–100	168	358	376	349	126	133	314†	182
Total	FCCJ	5 499	4 413	5 669	5 457	5 627	5 776	5 256	4 654†	4 281
General government										
New dwellings, excluding land	DFHW	3 980	2 609	2 407	2 618	2 809	2 642	2 148	1 786†	1 716
Other buildings and structures	EQCH	10 048	10 010	9 928	9 940	10 010	9 997	9 193	7 920†	8 463
Transport equipment	EQCI	773	711	747	721	768	680	659	689†	719
Other machinery and equipment and cultivated assets	EQCJ	2 610	2 246	2 264	2 230	1 964	2 075	1 845	1 493†	1 380
Intangible fixed assets	EQCK	206	220	236	247	257	264	273	259†	411
Costs associated with the transfer of ownership of non-produced assets	EQCL	–3 054	–1 734	–1 338	–2 034	–1 750	–1 653	–2 802	–2 275†	–2 058
Total	NNBF	14 563	14 062	14 244	13 722	14 058	14 005	11 316	9 872†	10 631
Total gross fixed capital formation	NPQX	114 314	104 680	100 278	101 230	107 390	116 360	125 675	134 153†	147 629

Source: Office for National Statistics: 020 7533 6031

15.20 Gross fixed capital formation at current purchasers' prices Analysis by type of asset

Total economy

£ millions

		1990	1991	1992	1993	1994	1995	1996	1997	1998
Tangible fixed assets										
New dwellings, excluding land	DFDK	21 048	18 339	18 825	19 892	21 233	21 588	22 766	23 927†	26 419
Other buildings and structures	DLWS	37 030	34 372	31 638	29 099	29 715	31 619	32 998	34 947†	39 435
Transport equipment	DLWZ	10 593	8 668	8 397	9 594	11 329	11 055	12 163	13 186†	14 679
Other machinery and equipment and cultivated assets	DLXI	37 817	35 075	34 690	35 566	37 693	44 464	49 216	52 206†	56 389
Total	EQCQ	106 488	96 454	93 550	94 151	99 970	108 726	117 143	124 266†	136 922
Intangible fixed assets	DLXP	3 571	4 063	3 782	3 648	3 613	3 939	4 136	4 249†	4 667
Costs associated with the transfer of ownership of non-produced assets	DFBH	4 255	4 163	2 946	3 431	3 807	3 695	4 396	5 638†	6 040
Total gross fixed capital formation	NPQX	114 314	104 680	100 278	101 230	107 390	116 360	125 675	134 153†	147 629

Source: Office for National Statistics: 020 7533 6031

15.21 Gross fixed capital formation at 1995 purchasers' prices[1]
Analysis by broad sector and type of asset
Total economy

£ millions at 1995 prices

		1990	1991	1992	1993	1994	1995	1996	1997	1998
Private sector										
New dwellings, excluding land	DFDP	18 746	16 603	17 311	18 353	19 076	18 784	19 903	20 824†	22 086
Other buildings and structures	EQCU	21 821	20 273	19 280	17 449	17 914	17 841	19 099	22 860†	25 645
Transport equipment	EQCV	11 101	8 460	7 678	9 242	10 597	10 021	10 916	12 148†	13 495
Other machinery and equipment and cultivated assets	EQCW	37 919	36 209	34 662	33 926	35 464	41 532	46 486	51 914†	59 435
Intangible fixed assets	EQCX	3 799	3 960	3 414	3 152	2 997	3 179	3 298	3 288†	3 584
Costs associated with the transfer of ownership of non-produced assets	EQCY	7 540	5 716	5 045	5 821	5 705	5 222	6 435	6 160†	5 416
Total	EQCZ	101 564	91 844	87 965	88 081	91 753	96 579	106 137	117 194†	129 661
Public non-financial corporations										
New dwellings, excluding land[2]	DEEW	..	..	..	..	150	162	147	119†	107
Other buildings and structures[2]	DEEX	..	..	..	..	3 847	3 781	3 154	2 417†	2 359
Transport equipment[2]	DEEU	..	..	..	..	529	354	223	181†	104
Other machinery and equipment and cultivated assets[2]	DEEV	..	..	..	..	766	857	768	764†	683
Intangible fixed assets	EQDE	189	281	274	295	387	496	571	553	554
Costs associated with the transfer of ownership of non-produced assets	EQDF	–103	182	351	368	355	126	151	327†	233
Total	EQDG	5 204	4 282	5 863	5 930	6 034	5 776	5 014	4 361†	4 040
General government										
New dwellings, excluding land	DFID	3 958	2 633	2 570	2 932	3 041	2 642	2 104	1 726†	1 605
Other buildings and structures	EQDI	8 759	9 211	10 296	11 174	11 162	9 997	8 511	7 313†	7 612
Transport equipment	EQDJ	806	742	804	785	801	680	638	653†	667
Other machinery and equipment and cultivated assets	EQDK	2 817	2 381	2 380	2 259	1 941	2 075	1 870	1 563†	1 498
Intangible fixed assets	EQDL	242	232	227	231	247	264	293	262†	437
Costs associated with the transfer of ownership of non-produced assets	EQDM	–3 103	–1 724	–1 755	–2 343	–1 937	–1 653	–2 525	–1 826†	–1 336
Total	EQDN	13 277	13 339	14 436	15 043	15 255	14 005	10 891	9 691†	10 483
Total gross fixed capital formation	NPQR	119 368	109 000	108 246	109 127	113 042	116 360	122 042	131 246†	144 184

1 For the years before 1994, totals differ from the sum of their components.
2 Data not available for years before 1994.

Source: Office for National Statistics: 020 7533 6031

15.22 Gross fixed capital formation at 1995 purchasers' prices[1]
Analysis by type of asset
Total economy

£ millions at 1995 prices

		1990	1991	1992	1993	1994	1995	1996	1997	1998
Tangible fixed assets										
New dwellings, excluding land	DFDV	23 434	19 587	20 041	21 491	22 267	21 588	22 154	22 669†	23 798
Other buildings and structures	EQDP	32 741	32 143	32 974	32 414	32 923	31 619	30 764	32 590†	35 616
Transport equipment	DLWJ	12 499	9 656	9 253	10 589	11 927	11 055	11 777	12 982†	14 266
Other machinery and equipment and cultivated assets	DLWM	43 181	39 086	38 031	36 958	38 171	44 464	49 124	54 241†	61 616
Total	EQDS	110 850	100 464	100 751	101 630	105 288	108 726	113 819	122 482†	135 296
Intangible fixed assets	EQDT	4 229	4 474	3 914	3 677	3 631	3 939	4 162	4 103†	4 575
Costs associated with the transfer of ownership of non-produced assets	DFDW	4 337	4 176	3 640	3 857	4 123	3 695	4 061	4 661†	4 313
Total gross fixed capital formation	NPQR	119 368	109 000	108 246	109 127	113 042	116 360	122 042	131 246†	144 184

1 For the years before 1994, totals differ from the sum of their components.

Source: Office for National Statistics: 020 7533 6031

Prices

16

Prices

Producer price index numbers *(Tables 16.1 and 16.2)*

The producer price indices were published for the first time in August 1983, replacing the former wholesale price indices. Full details of the differences between the two indices were given in an article published in *British Business*, 15 April 1983. The producer price indices are calculated using the same general methodology as that used by the wholesale price indices.

The high level index numbers in Tables 16.1 and 16.2 are constructed on a net sector basis. That is to say, they are intended to measure only transactions between the sector concerned and other sectors. Within sector transactions are excluded. Index numbers for the whole of manufacturing are thus not weighted averages of sector index numbers.

The index numbers for selected industries in Tables 16.1 and 16.2 are constructed on a gross sector basis i.e. all transactions are included in deriving the weighting patterns, including sales within the same industry.

All the index numbers are compiled exclusive of value-added tax. Excise duties on cigarettes, manufactured tobacco and alcoholic liquor are included as is the duty on hydrocarbon oils.

The indices relate to the average prices for a year. The movement in these prices are weighted to reflect the relative importance of the composite products in a chosen year (known as the base year) currently 1995.

Since July 1995, PPIs have been published fully reclassified to the 1992 version of the Standard Industrial Classification (SIC). Until 1998, high level PPIs on the 1980 SIC will also be shown in Tables 16.1 and 16.2.

Purchasing power of the pound *(Table 16.3)*

Changes in the internal purchasing power of a currency may be defined as the 'inverse' of changes in the levels of prices; when prices go up, the amount which can be purchased with a given sum of money goes down. Movements in the internal purchasing power of the pound are based on the consumers' expenditure deflator (CED) prior to 1962 and on the General index of retail prices (RPI) from January 1962 onwards. The CED shows the movement in prices implied by the national accounts estimates of consumers' expenditure valued at current and at constant prices, whilst the RPI is constructed directly by weighting together monthly movements in prices according to a given pattern of household expenditure derived from the Family Expenditure Survey. If the purchasing power of the pound is taken to be 100p in a particular month (quarter, year), the comparable purchasing power in a subsequent month (quarter, year) is:

$$100 \times \frac{\text{earlier period price index}}{\text{later period price index}}$$

, where the price index used is the CED for years 1946-1961 and the RPI for periods after 1961.

Index of retail prices *(Table 16.4)*

The retail prices index measures the change from month to month in the average level of prices of goods and services purchased by most households in the United Kingdom. The expenditure pattern on which the index is based is revised each year using information from the Family Expenditure Survey. The expenditure of certain higher income households and households of retired people dependent mainly on social security benefits is excluded.

The index covers a large and representative selection of more than 600 separate goods and services, for which price movements are regularly measured in 146 locations throughout the country. Around 120 000 separate price quotations are used in compiling the index.

The index of retail prices replaced the interim index from January 1956 (indices of the interim index of retail prices for the period 1952 to January 1956 were last published in *Annual Abstract of Statistics* No 103, 1965). A new set of weights was introduced, based on expenditure in 1953-54, valued at January 1956 prices. Between January 1962 and 1974 the weights have been revised each January on the basis of expenditure in the three years ended in the previous June, valued at prices obtained at the date of revision. From 1975 the weights have been revised on expenditure for the latest available year.

Following the recommendations of the Retail Prices Index Advisory Committee, the index has been re-referenced to make January 13, 1987 = 100. Calculations of price changes which involve periods spanning the new reference date are made as follows:

$$\%\text{ change} = \frac{\underset{\text{(Jan 1987 = 100)}}{\text{Index for later month}} \times \underset{\text{(Jan 1974 = 100)}}{\text{Index for Jan 1987}}}{\text{Index for earlier month (Jan 1974 = 100)}} - 100$$

Tax and price index (TPI) *(Table 16.5)*

The purpose and methodology of the TPI were described in an article in the August 1979 issue (No 310) of *Economic Trends* (The Stationery Office). The TPI measures the change in *gross* taxable income needed for taxpayers to maintain their purchasing power, allowing for changes in retail prices. The TPI thus takes account of the changes to direct taxes (and employees' National Insurance contributions) facing a representative cross-section of taxpayers as well as changes in the retail prices index (RPI).

When direct taxation or employees' National Insurance contributions change, the TPI will rise by less than or more than the RPI according to the type of changes made. Between Budgets, the monthly increase in the TPI is normally slightly larger than that in the RPI, since all the extra income needed to offset any rise in retail prices is fully taxed.

Index numbers of agricultural prices
(Tables 16.6 and 16.7)

The indices of agricultural prices for the United Kingdom are based on the calendar year 1990 and designed to provide an indication of movements in the purchase prices of the means of agricultural production and of the prices received by producers for their agricultural products. The methodology is comparable with that used for the other member states of the European Community and enables the compilation of indices for the Fifteen, which appear in the Communities Eurostat series of publications.

16.1 Producer price index numbers of materials and fuels purchased

All manufacturing and selected industries SIC(92)

United Kingdom

1995 = 100

			Annual averages							
			1991	1992	1993	1994	1995	1996	1997	1998
Net sector										
Materials and fuel purchased by manufacturing industry	PLKW	6292000000	86.6	86.3	90.2	91.9	100.0	98.8	90.6	82.5
Materials	PLKX	6292000010	86.1	83.8	88.9	90.3	100.0	99.6†	90.9	81.6
Fuels	PLKY	6292000020	100.0	104.6	106.1	104.6	100.0	91.7	88.3	89.2
Materials and fuels purchased by manufacturing industry-seasonally adjusted	PLKZ	6292008900	86.6	86.2	90.2	91.9	100.0	98.8	90.6	82.4
Materials and fuels purchased by manufacturing industry other than food, beverages, petroleum and tobacco	PLLA	6292990000	87.6	85.9	89.0	91.3	100.0	95.7†	89.7	85.9
Materials purchased by manufacturing industry other than food, drink and tobacco	RWCJ	6292990010	84.0	81.0	84.9	88.5	100.0	96.6	90.0	85.2
Materials purchased by manufacturing industry other than food and beverages	RWCK	6292990020	99.8	104.5	106.0	104.6	100.0	91.7	88.2	89.1
Materials and fuels purchased by manufacturing industries other than food, beverages, petroleum and tobacco-seasonally adjusted	PLLB	6292998900	87.6	85.9	89.0	91.3	100.0	95.7	89.7	85.8
Gross sector										
All manufacturing	RBBO	6192000000	89.6	89.5	92.0	94.0	100.0	100.3	97.0	93.2
Other mining & quarrying products	RABE	6112140000	89.2	90.2	93.1	95.7	100.0	101.9	102.9	104.0
Manufacture of food products	RBBQ	6192151600	85.4	87.2	93.1	94.2	100.0	99.8	94.2	90.3
Food products and beverages	RABF	6112150000	85.7	87.3	92.9	94.2	100.0	100.9	96.6	93.3
Tobacco products	RABG	6112160000	84.9	84.1	87.8	90.7	100.0	100.9	95.8	91.0
Manufacture of textiles	RBBR	6192171800	87.4	87.6	88.9	92.1	100.0	99.2	96.3	93.2
Textiles	RABH	6112170000	87.3	87.2	88.6	91.6	100.0	99.1	95.8	92.4
Wearing apparel	RABI	6112180000	87.6	88.3	89.6	92.9	100.0	99.5	97.0	94.6
Manufacture of leather	RBBS	6192190000	88.3	89.2	91.9	94.0	100.0	101.0	98.3	94.9
Manufacture of wood and wood products	RBBT	6192200000	84.7	84.4	88.6	93.7	100.0	99.0	98.1	95.3
Manufacture of pulp, paper, publishing & printing	RBBU	6192212200	87.0	85.1	85.4	88.4	100.0	99.2	93.1	90.4
Pulp and paper products	RABL	6112210000	86.5	85.0	85.5	88.3	100.0	99.1	93.1	90.5
Printed matter and recording material	RABM	6112220000	87.4	85.3	85.3	88.4	100.0	99.3	93.1	90.3
Manufacture of coke	RBBV	6192230000	101.9	97.4	101.5	94.7	100.0	118.1	107.4	78.9
Manufacture of chemical products	RBBW	6192240000	87.7	88.0	91.3	93.5	100.0	98.7	95.9	94.2
Manufacture of rubber products	RBBX	6192250000	84.9	81.8	84.3	87.4	100.0	94.7	90.2	84.3
Manufacture of other non-metallic mineral products	RBBY	6192260000	91.7	92.5	95.4	97.0	100.0	100.6	100.4	100.7
Manufacture of basic metals	RBBZ	6192272800	86.8	86.3	89.0	92.8	100.0	99.3	97.3	94.9
Basic metals	RABV	6112270000	86.1	85.2	88.1	92.2	100.0	98.8	96.3	93.3
Fabricated metal products	RABW	6112280000	87.7	87.7	90.2	93.5	100.0	99.9	98.5	96.9
Manufacture of machinery and equipment n.e.c.	RBCA	6192290000	89.6	90.6	92.7	95.2	100.0	100.8	100.5	99.6
Manufacture of electrical and optical equipment	RBCB	6192303300	102.2	99.2	96.8	98.0	100.0	97.8	93.6	88.6
Office machinery and computers	RABY	6112300000	120.9	110.6	103.0	102.6	100.0	95.3	87.7	78.8
Electrical machinery and apparatus n.e.c.	RACB	6112310000	90.3	90.3	91.3	94.0	100.0	99.4	97.6	95.0
Radio, television and communication equipment	RACC	6112320000	96.3	96.7	95.9	97.3	100.0	98.4	95.1	91.2
Medical, precision, optical instruments and clocks	RACD	6112330000	95.9	95.7	95.2	96.7	100.0	98.9	96.0	92.6
Manufacture of transport equipment	RBCC	6192343500	88.5	90.5	93.5	95.7	100.0	101.9	102.0	102.3
Motor vehicles, trailers and semi-trailers	RACE	6112340000	88.5	90.2	93.1	95.4	100.0	101.9	101.9	101.8

Source: Office for National Statistics: 01633 812106

16.1 Producer price index numbers of materials and fuels purchased

continued

All manufacturing and selected industries SIC(92)

United Kingdom

1995 = 100

			Annual averages							
			1991	1992	1993	1994	1995	1996	1997	1998
Other transport equipment	RACF	6112350000	88.8	91.3	94.6	96.5	100.0	101.9	102.4	103.8
Manufacturing n.e.c.	RBCD	6192363700	87.0	87.1	89.6	93.1	100.0	100.0	98.8	96.8
Electricity	PQMX	4010990000	93.7	99.9	103.8	103.8	100.0	96.6	90.7	90.6
Gas	PQMZ	4020990000	112.9	113.5	109.2	107.1	100.0	76.4	79.1	83.3
Collected and purified water	PQNB	4100000000	73.8†	83.3	91.3	96.8	100.0	104.9	109.6	115.3
Construction materials	RWDZ	6101451000	87.7	87.7	90.1	94.4	100.0	100.5	102.1	102.9
House building materials	RWEA	6101452000	87.4	87.7	90.2	94.6	100.0	100.7	102.2	103.0
SIC (80) Net sector										
Materials and fuels purchased by manufacturing industry	DZBR	6200000000	85.5	84.9	89.7	91.7	100.0	96.4	89.0	84.1

Source: Office for National Statistics: 01633 812106

16.2 Producer price index numbers of output
All manufacturing and selected industries SIC(92)
United Kingdom

1995 = 100

			Annual averages							
			1991	1992	1993	1994	1995	1996	1997	1998
Net sector										
Output of manufactured products	PLLU	7209200000	87.4	90.2	93.8	96.1	100.0	102.6	103.6	104.2
All manufacturing excl. duty	PVNP	7209200010	88.3	90.7	94.2	96.2	100.0	102.3	102.5	102.1
All manufacturing excl. duty - seasonally adjusted	PVNQ	7209200890	88.3	90.7	94.2	96.2	100.0	102.3	102.5	102.1
Products of manufacturing industries other than the food, beverages, petroleum and tobacco manufacturing industries -unadjusted	PLLV	7209299000	89.5	91.5	93.9	96.0	100.0	102.0	102.2	102.1
All manufacturing excl. food, beverages, tobacco & petroleum - seasonally adjusted	PLLW	7209299890	89.5	91.5	93.9	96.0	100.0	102.0	102.2	102.1
Gross sector										
Manufactured product	POKE	7109200000	93.3	93.2	95.6	96.6	100.0	102.1	101.5	99.2
Manufactured products excl. food, drink, tobacco & petroleum	POKF	7109299000	92.7	92.8	94.0	95.8	100.0	101.4	100.9	100.1
Other mining & quarrying products	POKW	7112140000	89.4	90.0	91.1	94.3	100.0	105.0	108.1	111.2
Food products, beverages & tobacco	POKH	7111151600	85.9	89.6	94.7	96.3	100.0	103.1	103.0	102.7
Food products, beverages & tobacco including duty	RBGA	7111151680	85.0	89.4	94.5	96.6	100.0	103.3	103.9	104.7
Food products & beverages	POKX	7112150000	86.5	90.0	95.2	96.8	100.0	103.1	102.9	102.3
Food products excluding beverages	RBGD	7112159900	87.3	90.8	95.9	97.3	100.0	103.1	102.1	101.0
Tobacco products	POKY	7112160000	69.6	74.0	79.4	83.5	100.0	101.2	105.6	114.5
Textiles & textile products	POKI	7111171800	91.0 B	93.5 B	94.6 B	95.9 B	100.0 B	102.6 B	104.7	105.7
Textiles	POKZ	7112170000	90.5 B	93.0 B	94.0 B	95.5 B	100.0 B	102.3	104.1	104.9
Wearing apparel: Furs	POLA	7112180000	92.3 B	94.5 B	96.0 B	97.0 B	100.0 B	103.3	106.2	107.7
Leather & leather products	POKJ	7111190000	89.3	91.0	93.8	97.4	100.0	102.7	103.3	102.8
Wood & wood products	POKK	7111200000	86.1 B	86.3 B	90.2 B	94.0 B	100.0 B	98.8 B	100.3	98.9
Pulp, paper & paper products, recorded media & printing services	POKL	7111212200	87.2	88.8	91.0	92.9	100.0	103.4	102.0	102.5
Pulp, paper & paper products	POLD	7112210000	80.8	81.3	82.8	85.6	100.0	101.7	97.0	95.8
Printed matter & recorded media	POLE	7112220000	90.5	92.6	95.2	96.6	100.0	104.3	104.6	106.0
Chemicals, chemical, products & manmade fibres	POKN	7111240000	88.0	88.8	91.1	93.4	100.0	98.9	97.3	94.0
Rubber & plastic products	POKO	7111250000	88.4	89.9	91.2	92.3	100.0	101.4	100.8	98.4
Other non metallic mineral products	POKP	7111260000	89.4	89.9	91.8	95.9	100.0	101.1	103.5	106.5
Base metals & fabricated metal products	POKQ	7111272800	88.9	89.2	90.4	93.6	100.0	101.4	100.7	100.1
Base metals	POLJ	7112270000	84.5	82.8	84.5	90.0	100.0	98.8	96.0	93.0
Fabricated metal products, except machinery and equipment	POLK	7112280000	92.0 B	93.6 B	94.5 B	96.0 B	100.0 B	103.1	104.0	105.0
Machinery and equipment n.e.c.	POKR	7111290000	90.2 B	93.0 B	95.5 B	97.5 B	100.0 B	102.9 B	105.7	106.7
Electrical and optical equipment	POKS	7111343500	119.2	110.3	104.6	103.8	100.0	98.1	93.3	87.1
Office machinery and computers	POLM	7112300000	171.7	138.8	120.8	115.5	100.0	93.7	81.7	67.6
Electrical machinery and apparatus n.e.c.	POLN	7112310000	88.1 B	92.6 B	94.0 B	96.1 B	100.0 B	101.6	102.2	101.4
Radio, television and communication equipment and apparatus	POLO	7112320000	104.1	103.6	99.6	100.2	100.0	95.8	90.1	83.2
Medical precision and optical instruments, watches and clocks	POLP	7112330000	90.1 B	93.6 B	97.7 B	99.2 B	100.0 B	104.1	104.9	106.1

B = These index values are considered less reliable due to lack of market coverage.

Source: Office for National Statistics: 01633 812106

16.2 Producer price index numbers of output
All manufacturing and selected industries SIC(92)
continued

United Kingdom

			Annual averages							
			1991	1992	1993	1994	1995	1996	1997	1998
Gross sector continued										
Transport equipment	POKT	7111343500	88.3	90.8	94.5	96.5	100.0	103.0	104.4	106.4
Motor vehicles, trailers and semi-trailers	POLQ	7112340000	88.7	90.6	93.9	96.2	100.0	103.0	104.3	105.9
Other transport	POLR	7112350000	86.9 B	91.7 B	96.3 B	97.4 B	100.0 B	102.9	104.9	108.2
Furniture: other manufactured goods n.e.c.	POLS	7112360000	90.6 B	93.2 B	95.5 B	96.9 B	100.0 B	103.7	105.1	107.3
SIC (80) Net sector										
Output of manufactured products	DZCV	7200000000	89.0	91.4	94.3	96.3	100.0	102.3	102.8	103.2

B = These index values are considered less reliable due to lack of market coverage.

Source: Office for National Statistics: 01633 812106

16.3 Internal purchasing power of the pound (based on RPI)[1,2]
United Kingdom

Pence

	Year in which purchasing power was 100p																		
	1980	1981	1982	1983	1984	1985	1986	1987	1988	1989	1990	1991	1992	1993	1994	1995	1996	1997	1998
	BAMM	BAMN	BAMO	BAMP	BAMQ	BAMR	BAMS	BAMT	BAMU	BAMV	BAMW	BASX	CZVM	CBXX	DOFX	DOHR	DOLM	DTUL	CDQG
1980	100	112	122	127	133	142	146	152	160	172	189	200	207	210	216	223	228	236	244
1981	89	100	109	114	119	127	131	136	143	154	169	179	185	188	193	199	204	211	218
1982	82	92	100	105	110	116	120	125	132	142	155	164	171	173	177	184	188	194	201
1983	79	88	96	100	105	111	115	120	126	136	148	157	163	166	170	176	180	185	192
1984	75	84	91	95	100	106	110	114	120	129	141	150	155	158	162	167	171	177	183
1985	71	79	86	90	94	100	103	108	113	122	133	141	146	149	152	158	161	166	172
1986	68	76	83	87	91	97	100	104	109	118	129	136	142	144	147	152	156	161	167
1987	66	73	80	83	88	93	96	100	105	113	124	131	136	138	141	146	150	155	160
1988	63	70	76	79	83	88	92	95	100	108	118	125	130	132	135	139	143	147	152
1989	58	65	71	74	77	82	85	88	93	100	109	116	120	122	125	129	133	137	141
1990	53	59	64	67	71	75	78	81	85	91	100	106	110	112	114	118	121	125	129
1991	50	56	61	64	67	71	73	76	80	86	94	100	104	105	108	112	114	118	122
1992	48	54	59	61	64	68	71	74	77	83	91	96	100	102	104	108	110	114	118
1993	48	53	58	60	63	67	70	72	76	82	90	95	98	100	102	106	109	112	116
1994	46	52	56	59	62	66	68	71	74	80	88	93	96	98	100	103	106	109	113
1995	45	50	54	57	60	63	66	68	72	77	85	90	93	94	97	100	102	106	109
1996	44	49	53	56	58	62	64	67	70	75	83	87	91	92	94	98	100	103	107
1997	42	47	52	54	57	60	62	65	68	73	80	85	88	89	92	95	97	100	103
1998	41	46	50	52	55	58	60	63	66	71	77	82	85	86	88	92	94	97	100

1 To find the purchasing power of the pound in 1993, given that it was 100 pence in 1980, select the column headed 1980 and look at the 1993 row. The result is 48 pence.

2 These figures are calculated by taking the inverse ratio of the respective annual averages of the Retail Prices Index.

Source: Office for National Statistics: 020 7533 5874

16.4 Retail Prices Index
United Kingdom

13 January 1987=100

	ALL ITEMS (RPI)	All items excluding: mortgage interest payments (RPIX)	All items excluding: mortgage interest payments and depreciation[1]	All items excluding: housing	All items excluding: food	All items excluding: seasonal food[2]	Food and catering	Alcohol and tobacco	Housing and household expenditure	Personal expenditure	Travel and leisure	Consumer durables	All items excluding mortgage interest payments & indirect taxes (RPIY)[3]
Weights													
	CZGU	CZGY	DOGZ	CZGX	CZGV	CZGW	CBVV	CBVW	CBVX	CBVY	CBVZ	CBWA	
1991	*1 000*	*924*	*924*	*808*	*849*	*976*	*198*	*109*	*353*	*101*	*239*	*128*	
1992	*1 000*	*936*	*936*	*828*	*848*	*978*	*199*	*116*	*344*	*99*	*242*	*127*	
1993	*1 000*	*952*	*952*	*836*	*856*	*979*	*189*	*113*	*336*	*97*	*265*	*127*	
1994	*1 000*	*956*	*956*	*842*	*858*	*980*	*187*	*111*	*326*	*95*	*281*	*127*	
1995	*1 000*	*958*	*928*	*813*	*861*	*978*	*184*	*111*	*356*	*93*	*256*	*123*	
1996	*1 000*	*958*	*929*	*810*	*857*	*978*	*191*	*113*	*353*	*92*	*251*	*116*	
1997	*1 000*	*961*	*932*	*814*	*864*	*981*	*185*	*114*	*351*	*96*	*254*	*122*	
1998	*1 000*	*955*	*923*	*803*	*870*	*982*	*178*	*105*	*359*	*95*	*263*	*121*	
1999	*1 000*	*958*	*928*	*807*	*872*	*980*	*179*	*100*	*358*	*95*	*268*	*127*	
Annual averages													
	CHAW	CHMK	CHON	CHAZ	CHAY	CHAX	CHBS	CHBT	CHBU	CHBV	CHBW	CHBY	CBZW
1990	126.1	122.1	122.1	119.2	127.4	126.4	120.9	120.6	139.0	117.6	119.8	111.3	121.4
1991	133.5	130.3	130.3	128.3	135.1	133.8	128.6	136.2	142.2	123.6	128.9	114.8	129.5
1992	138.5	136.4	136.4	134.3	140.5	139.1	132.6	146.8	144.2	126.9	136.8	115.5	135.1
1993	140.7	140.5	140.5	138.4	142.6	141.4	136.1	155.1	141.2	129.5	141.8	115.9	139.0
1994	144.1	143.8	143.8	141.6	146.5	144.8	138.5	161.4	144.4	131.9	145.7	115.5	141.3
1995	149.1	147.9	148.0	145.4	151.4	149.6	143.9	169.0	150.8	133.6	148.4	116.2	144.5
1996	152.7	152.3	152.3	149.3	154.9	153.4	148.9	175.9	153.0	135.1	152.8	117.1	148.2
1997	157.5	156.5	156.4	152.9	160.5	158.5	150.4	183.2	158.4	137.7	159.0	117.3	151.5
1998	162.9	160.6	160.3	156.2	166.5	163.8	153.4	192.3	166.2	139.9	162.8	115.9	154.5
Monthly figures													
1996 Sep	153.8	153.6	153.6	150.5	156.2	154.7	149.3	177.2	153.6	137.2	154.5	118.5	149.6
Oct	153.8	153.6	153.6	150.5	156.4	154.8	148.4	177.5	153.5	137.5	155.1	118.1	149.6
Nov	153.9	153.7	153.7	150.6	156.6	154.9	148.1	177.2	153.9	138.6	155.0	119.3	149.7
Dec	154.4	154.2	154.2	151.1	157.2	155.4	148.3	177.9	154.3	138.7	155.9	120.0	149.5
1997 Jan	154.4	153.9	153.9	150.7	157.0	155.3	149.3	179.7	154.1	133.7	156.7	114.2	149.3
Feb	155.0	154.5	154.5	151.3	157.7	156.0	149.2	180.7	154.8	135.0	157.3	115.5	149.9
Mar	155.4	154.9	154.9	151.7	158.4	156.5	148.7	180.8	155.6	137.0	157.4	117.9	150.3
Apr	156.3	155.8	155.9	152.2	159.3	157.4	149.3	181.9	156.9	138.3	157.6	117.8	150.8
May	156.9	156.3	156.4	152.7	159.8	157.9	150.2	182.9	157.3	138.7	157.9	118.3	151.3
Jun	157.5	156.7	156.7	153.0	160.3	158.4	151.3	183.2	158.2	138.3	158.5	117.9	151.8
Jul	157.5	156.4	156.3	152.6	160.4	158.4	151.0	183.9	158.6	134.5	159.5	114.4	151.0
Aug	158.5	157.1	157.1	153.5	161.5	159.4	151.1	184.8	159.8	136.4	160.4	116.1	151.8
Sep	159.3	157.8	157.6	154.1	162.5	160.3	151.1	185.0	160.7	139.8	160.8	118.4	152.6
Oct	159.5	157.9	157.8	154.2	162.8	160.5	151.4	185.4	161.0	139.7	161.0	117.9	152.9
Nov	159.6	158.0	157.8	154.2	163.0	160.6	151.0	184.9	161.5	140.8	160.7	119.0	152.9
Dec	160.0	158.3	158.1	154.5	163.5	161.0	151.1	185.6	162.4	140.7	160.7	119.7	152.8
1998 Jan	159.5	157.7	157.4	153.7	162.8	160.4	151.4	188.6	161.2	134.9	161.4	113.2	152.1
Feb	160.3	158.5	158.3	154.6	163.8	161.4	151.6	189.8	162.1	137.8	161.8	115.2	153.0
Mar	160.8	158.9	158.7	155.2	164.4	161.8	151.4	190.3	162.9	139.6	161.6	117.3	153.4
Apr	162.6	160.4	160.2	155.9	166.4	163.7	151.9	191.4	165.9	140.2	163.5	116.5	154.1
May	163.5	161.3	161.0	156.8	167.2	164.4	153.7	192.5	166.6	141.4	163.8	117.7	155.1
Jun	163.4	161.1	160.8	156.6	167.1	164.3	153.4	192.5	166.5	141.4	163.6	117.0	154.9
Jul	163.0	160.5	160.1	155.8	166.7	164.1	153.2	193.1	167.0	136.6	163.5	113.1	154.2
Aug	163.7	161.1	160.7	156.4	167.3	164.6	154.6	193.4	167.5	138.5	163.5	114.2	155.0
Sep	164.4	161.8	161.4	157.1	168.2	165.4	154.3	193.6	168.4	142.3	163.6	116.8	155.7
Oct	164.5	161.9	161.3	157.1	168.3	165.5	154.7	193.9	168.8	141.7	163.2	115.6	155.7
Nov	164.4	162.0	161.4	157.1	168.2	165.4	154.6	193.5	168.9	142.4	162.7	116.7	155.6
Dec	164.4	162.4	161.9	157.6	168.0	165.2	155.5	195.4	168.6	142.0	162.0	118.0	155.8
1999 Jan	163.4	161.8	161.2	156.8	166.7	164.2	156.1	198.2	165.9	136.3	162.8	110.6	155.1
Feb	163.7	162.3	161.7	157.4	167.0	164.5	156.4	198.6	165.9	138.5	162.6	112.3	155.8
Mar	164.1	163.2	162.6	158.4	167.7	165.0	155.8	200.8	165.5	139.8	164.0	114.2	156.0
Apr	165.2	164.3	163.7	159.0	169.1	166.3	155.4	201.9	166.8	140.3	166.1	113.1	156.9
May	165.6	164.7	164.1	159.4	169.5	166.5	156.1	202.6	167.3	140.8	166.2	114.0	157.4
Jun	165.6	164.7	164.0	159.2	169.6	166.6	155.7	203.4	167.3	140.9	165.9	113.1	157.3
Jul	165.1	164.1	163.4	158.6	169.1	166.3	155.1	204.2	166.9	136.9	166.3	109.6	156.7
Aug	165.5	164.5	163.8	158.9	169.7	166.8	154.7	204.5	167.6	138.3	166.6	110.5	157.2
Sep	166.2	165.2	164.4	159.6	170.6	167.4	154.6	204.5	168.5	141.6	166.5	112.7	157.8

1 This series has been constructed using the index for all items excluding mortgage interest payments prior to February 1995.

2 Seasonal food is defined as items of food the prices of which show significant seasonal variations. These are fresh fruit and vegetables, fresh fish, eggs and home-killed lamb.

3 There are no weights available for RPIY.

Source: Office for National Statistics: 020 7533 5874

16.5 Tax and Price Index
United Kingdom

	Tax and Price Index: January 1978 = 100					January 1987 = 100												
	BSAA					DQAB												
	1983	1984	1985	1986	1987	1987	1988	1989	1990	1991	1992	1993	1994	1995	1996	1997	1998	1999
January	170.7	177.9	184.7	192.9	198.0	100.0	101.4	107.1	113.9	123.6	128.1	128.7	132.1	137.2	141.6	143.6	147.1	150.5
February	171.6	178.8	186.4	193.7	..	100.5	101.8	108.0	114.7	124.3	128.8	129.6	132.9	138.2	142.3	144.2	147.9	150.8
March	171.9	179.4	188.4	194.0	..	100.7	102.3	108.5	115.9	124.9	129.3	130.2	133.4	138.8	143.0	144.6	148.4	151.2
April	171.8	178.8	190.2	192.5	..	99.7	101.4	109.8	118.2	125.4	129.6	131.3	135.3	140.3	141.7	143.8	149.7	151.2
May	172.6	179.6	191.2	192.9	..	99.8	101.9	110.5	119.4	125.8	130.2	131.8	135.8	141.0	142.0	144.4	150.6	151.7
June	173.1	180.1	191.7	192.8	..	99.8	102.3	110.9	119.9	126.5	130.2	131.7	135.8	141.2	142.1	145.0	150.5	151.7
July	174.2	179.9	191.3	192.1	..	99.7	102.4	111.1	120.0	126.2	129.6	131.4	135.1	140.4	141.5	145.0	150.1	151.1
August	175.1	181.8	191.8	192.9	..	100.0	103.7	111.4	121.4	126.5	129.7	132.1	135.8	141.3	142.2	146.0	150.8	151.5
September	176.0	182.2	191.7	194.0	..	100.4	104.3	112.2	122.7	127.0	130.3	132.7	136.1	142.0	143.0	146.9	151.5	152.3
October	176.7	183.5	191.4	194.3	..	100.9	105.4	111.7	123.8	127.5	130.8	132.6	136.4	141.2	143.0	147.1	151.6	..
November	177.5	184.1	192.1	196.3	..	101.5	106.0	112.8	123.4	128.1	130.6	132.4	136.5	141.2	143.1	147.2	151.5	..
December	178.0	183.9	192.4	197.1	..	101.4	106.3	113.1	123.3	128.2	130.1	132.7	137.2	142.1	143.6	147.6	151.5	..

	Retail Prices Index: January 1974 = 100					January 1987 = 100												
	CBAB					CHAW												
	1983	1984	1985	1986	1987	1987	1988	1989	1990	1991	1992	1993	1994	1995	1996	1997	1998	1999
January	325.9	342.6	359.8	379.7	394.5	100.0	103.3	111.0	119.5	130.2	135.6	137.9	141.3	146.0	150.2	154.4	159.5	163.4
February	327.3	344.0	362.7	381.1	..	100.4	103.7	111.8	120.2	130.9	136.3	138.8	142.1	146.9	150.9	155.0	160.3	163.7
March	327.9	345.1	366.1	381.6	..	100.6	104.1	112.3	121.4	131.4	136.7	139.3	142.5	147.5	151.5	155.4	160.8	164.1
April	332.5	349.7	373.9	385.3	..	101.8	105.8	114.3	125.1	133.1	138.8	140.6	144.2	149.0	152.6	156.3	162.6	165.2
May	333.9	351.0	375.6	386.0	..	101.9	106.2	115.0	126.2	133.5	139.3	141.1	144.7	149.6	152.9	156.9	163.5	165.6
June	334.7	351.9	376.4	385.8	..	101.9	106.6	115.4	126.7	134.1	139.3	141.0	144.7	149.8	153.0	157.5	163.4	165.6
July	336.5	351.5	375.7	384.7	..	101.8	106.7	115.5	126.8	133.8	138.8	140.7	144.0	149.1	152.4	157.5	163.0	165.1
August	338.0	354.8	376.7	385.9	..	102.1	107.9	115.8	128.1	134.1	138.9	141.3	144.7	149.9	153.1	158.5	163.7	165.5
September	339.5	355.5	376.5	387.8	..	102.4	108.4	116.6	129.3	134.6	139.4	141.9	145.0	150.6	153.8	159.3	164.4	166.2
October	340.7	357.7	377.1	388.4	..	102.9	109.5	117.5	130.3	135.1	139.9	141.8	145.2	149.8	153.8	159.5	164.5	..
November	341.9	358.8	378.4	391.7	..	103.4	110.0	118.5	130.0	135.6	139.7	141.6	145.3	149.8	153.9	159.6	164.4	..
December	342.8	358.5	378.9	393.0	..	103.3	110.3	118.8	129.9	135.7	139.2	141.9	146.0	150.7	154.4	160.0	164.4	..

	Percentage changes on one year earlier																
	1983	1984	1985	1986	1987	1988	1989	1990	1991	1992	1993	1994	1995	1996	1997	1998	1999
Tax and Price Index																	
January	*5.2*	*4.2*	*3.8*	*4.4*	*2.6*	*1.4*	*5.6*	*6.3*	*8.5*	*3.6*	*0.5*	*2.6*	*3.9*	*3.2*	*1.4*	*2.4*	*2.3*
February	*5.7*	*4.2*	*4.3*	*3.9*	*2.7*	*1.3*	*6.1*	*6.2*	*8.4*	*3.6*	*0.6*	*2.5*	*4.0*	*3.0*	*1.3*	*2.6*	*2.0*
March	*4.8*	*4.4*	*5.0*	*3.0*	*2.8*	*1.6*	*6.1*	*6.8*	*7.8*	*3.5*	*0.7*	*2.5*	*4.0*	*3.0*	*1.1*	*2.6*	*1.9*
April	*3.5*	*4.1*	*6.4*	*1.2*	*2.5*	*1.7*	*8.3*	*7.7*	*6.1*	*3.3*	*1.3*	*3.0*	*3.7*	*1.0*	*1.5*	*4.1*	*1.0*
May	*3.1*	*4.1*	*6.5*	*0.9*	*2.4*	*2.1*	*8.4*	*8.1*	*5.4*	*3.5*	*1.2*	*3.0*	*3.8*	*0.7*	*1.7*	*4.3*	*0.7*
June	*3.0*	*4.0*	*6.4*	*0.6*	*2.5*	*2.5*	*8.4*	*8.1*	*5.5*	*2.9*	*1.2*	*3.1*	*4.0*	*0.6*	*2.0*	*3.8*	*0.8*
July	*3.1*	*3.3*	*6.3*	*0.4*	*2.8*	*2.7*	*8.5*	*8.0*	*5.2*	*2.7*	*1.4*	*2.8*	*3.9*	*0.8*	*2.5*	*3.5*	*0.7*
August	*3.6*	*3.8*	*5.5*	*0.6*	*2.6*	*3.7*	*7.4*	*9.0*	*4.2*	*2.5*	*1.9*	*2.8*	*4.1*	*0.6*	*2.7*	*3.3*	*0.5*
September	*4.2*	*3.5*	*5.2*	*1.2*	*2.4*	*3.9*	*7.6*	*9.4*	*3.5*	*2.6*	*1.8*	*2.6*	*4.3*	*0.7*	*2.7*	*3.1*	*0.5*
October	*4.0*	*3.8*	*4.3*	*1.5*	*2.9*	*4.5*	*6.0*	*10.8*	*3.0*	*2.6*	*1.4*	*2.9*	*3.5*	*1.3*	*2.9*	*3.1*	..
November	*3.9*	*3.7*	*4.3*	*2.2*	*2.4*	*4.4*	*6.4*	*9.4*	*3.8*	*2.0*	*1.4*	*3.1*	*3.4*	*1.3*	*2.9*	*2.9*	..
December	*4.4*	*3.3*	*4.6*	*2.4*	*1.9*	*4.8*	*6.4*	*9.0*	*4.0*	*1.5*	*2.0*	*3.4*	*3.6*	*1.1*	*2.8*	*2.6*	..
Retail Prices Index																	
January	*4.9*	*5.1*	*5.0*	*5.5*	*3.9*	*3.3*	*7.5*	*7.7*	*9.0*	*4.1*	*1.7*	*2.5*	*3.3*	*2.9*	*2.8*	*3.3*	*2.4*
February	*5.3*	*5.1*	*5.4*	*5.1*	*3.9*	*3.3*	*7.8*	*7.5*	*8.9*	*4.1*	*1.8*	*2.4*	*3.4*	*2.7*	*2.7*	*3.4*	*2.1*
March	*4.6*	*5.2*	*6.1*	*4.2*	*4.0*	*3.5*	*7.9*	*8.1*	*8.2*	*4.0*	*1.9*	*2.3*	*3.5*	*2.7*	*2.6*	*3.5*	*2.1*
April	*4.0*	*5.2*	*6.9*	*3.0*	*4.2*	*3.9*	*8.0*	*9.4*	*6.4*	*4.3*	*1.3*	*2.6*	*3.3*	*2.4*	*2.4*	*4.0*	*1.6*
May	*3.7*	*5.1*	*7.0*	*2.8*	*4.1*	*4.2*	*8.3*	*9.7*	*5.8*	*4.3*	*1.3*	*2.6*	*3.4*	*2.2*	*2.6*	*4.2*	*1.3*
June	*3.7*	*5.1*	*7.0*	*2.5*	*4.2*	*4.6*	*8.3*	*9.8*	*5.8*	*3.9*	*1.2*	*2.6*	*3.5*	*2.1*	*2.9*	*3.7*	*1.3*
July	*4.2*	*4.5*	*6.9*	*2.4*	*4.4*	*4.8*	*8.2*	*9.8*	*5.5*	*3.7*	*1.4*	*2.3*	*3.5*	*2.2*	*3.3*	*3.5*	*1.3*
August	*4.6*	*5.0*	*6.2*	*2.4*	*4.4*	*5.7*	*7.3*	*10.6*	*4.7*	*3.6*	*1.7*	*2.4*	*3.6*	*2.1*	*3.5*	*3.3*	*1.1*
September	*5.1*	*4.7*	*5.9*	*3.0*	*4.2*	*5.9*	*7.6*	*10.9*	*4.1*	*3.6*	*1.8*	*2.2*	*3.9*	*2.1*	*3.6*	*3.2*	*1.1*
October	*5.0*	*5.0*	*5.4*	*3.0*	*4.5*	*6.4*	*7.3*	*10.9*	*3.7*	*3.6*	*1.4*	*2.4*	*3.2*	*2.7*	*3.7*	*3.1*	..
November	*4.8*	*4.9*	*5.5*	*3.5*	*4.1*	*6.4*	*7.7*	*9.7*	*4.3*	*3.0*	*1.4*	*2.6*	*3.1*	*2.7*	*3.7*	*3.0*	..
December	*5.3*	*4.6*	*5.7*	*3.7*	*3.7*	*6.8*	*7.7*	*9.3*	*4.5*	*2.6*	*1.9*	*2.9*	*3.2*	*2.5*	*3.6*	*2.8*	..

Note: The purpose and methodology of the Tax and Price Index were described in an article in the August 1979 issue of *Economic Trends*. The purpose is to produce a single index which measures changes in both direct taxes (including national insurance contributions) and in retail prices for a representative cross-section of taxpayers. Thus, while the Retail Prices Index may be used to measure changes in the purchasing power of after-tax income (and of the income of non-taxpayers) the Tax and Price Index takes account of the fact that taxpayers will have more or less to spend according to changes in direct taxation. The index measures the change in gross taxable income which would maintain their after tax-income in real terms.

Source: Office for National Statistics: 020 7533 5874

16.6 Index of purchase prices of the means of agricultural production
United Kingdom

Annual averages

1990 = 100

		Weights	1988	1989	1990	1991	1992	1993	1994	1995	1996	1997	1998
Goods and services currently consumed	BYEA	*100*	96.5	96.5	..	103.5	106.6	111.2	111.5	115.2	122.2	119.0	115.0
Seeds	BYEB	*4.3*	97.6	93.0	100.0	102.5	111.4	119.0	114.1	133.5	141.6	124.5	127.5
Animals for rearing and production	BYEC	*0.5*	96.0	99.7	100.0	93.9	100.5	88.1	111.6	139.1	139.2	133.4	127.8
Energy, lubricants	BYED	*6.8*	83.0†	91.4	100.0	103.1	105.8	106.6	104.8	108.0	116.4	115.4	110.5
Fuels for heating	KVBA	*2.0*	71.5	85.0	100.0	97.8	89.5	96.6	88.9	92.5	110.4	102.1†	82.2
Motor fuel	KVBB	*2.1*	90.6†	98.1	100.0	102.4	107.9	109.4	111.5	122.0	133.1	141.5	149.5
Electricity	KVBC	*2.4*	85.6	91.3	100.0	106.8	116.5	112.3	111.2	107.7	107.2†	103.6	99.0
Lubricants	KVBD	*0.4*	85.2	88.3	100.0	112.6	109.6	110.2	112.4	113.1	112.3	114.4†	115.9
Fertilisers and soil improvers	BYEE	*10.9*	91.0	98.7	100.0	96.6	91.1	85.6	91.0	102.0	112.8	97.9†	83.0
Straight fertilisers	KVBE	*4.4*	92.6	99.0	100.0	94.1	87.6	80.1	87.4	98.4	108.8	93.8†	78.9
Compound fertilisers	KVBF	*6.0*	90.2	99.2	100.0	98.1	92.6	87.9	92.0	103.7	115.0	98.6†	82.2
Other fertiliser	KVBG	*0.5*	86.6	90.7	100.0	100.6	102.8	106.3	111.4	113.2	122.2	124.3†	127.4
Plant protection products	BYEF	*6.6*	87.3	91.5	100.0	108.4†	110.1	111.5	117.5	115.9	123.5	124.4	116.1
Animal feedingstuffs	BYEG	*40.5*	94.3	99.8	100.0	101.7	104.7	112.3	109.0	110.0	118.4	114.1	108.6
Straight feedingstuffs	KVBH	*11.3*	97.4	101.2	100.0	103.8	104.8	113.0	110.2	110.0	119.6	115.9	106.8
Wheat	KVBI	*3.2*	93.3	96.2	100.0	110.6	110.9	118.0	112.4	115.5	121.0	116.4	111.6
Whole barley	KVBJ	*2.3*	92.9	96.2	100.0	107.8	106.1	113.3	109.5	111.2	112.4	113.7	108.2
Flaked maize	KVBK	*0.2*	94.8	94.7	100.0	115.8	112.1	116.7	113.1	117.2	116.7	122.6	128.5
Whole oats	KVBL	*0.1*	99.7	92.8	100.0	102.4	108.4	131.3	125.1	124.5	129.0	122.4	117.5
Oilcake	KVBM	*3.4*	100.4	110.5	100.0	97.7	100.2	115.8	114.1	109.4	129.0	121.7	104.3
White fish meal	KVBN	*0.7*	107.6	105.1	100.0	96.8	100.9	101.8	99.5	107.7	131.2	130.1	138.3
Sugar beet pulp	KVBO	*1.5*	101.3	97.2	100.0	99.5	101.1	100.0	102.1	97.6	101.1	97.9	83.1
Compound feedingstuffs	KVBP	*29.2*	93.1	99.2	100.0	100.9	104.7	112.1	108.5	110.0	118.0	113.4	109.4
for:													
Calves	KVBQ	*1.1*	88.1	96.3	100.0	100.5	106.7	114.6	114.1	115.4	122.6	116.3	112.4
Cattle	KVBR	*8.7*	92.4	98.5	100.0	98.5	103.5	113.5	107.7	107.5	116.2	109.2	102.7
Pigs	KVBS	*6.4*	93.2	98.8	100.0	103.4	107.6	115.8	110.8	113.0	117.6	114.2	111.5
Poultry	KVBT	*11.1*	94.4	100.8	100.0	102.2	104.5	109.0	106.6	108.7	116.9	113.9	110.7
Sheep	KVBU	*1.9*	91.5	96.2	100.0	96.1	100.2	109.2	112.9	115.8	130.8	125.6	123.1
Material and small tools	BYEH	*4.4*	88.0	93.2	100.0	111.2†	116.0	119.0	121.8	122.3	122.7	122.3	119.7
Maintenance and repair of plant	BYEI	*10.0*	87.2	91.9	100.0	110.6	119.0	126.2	129.0	132.2	139.3	145.2	149.8
Maintenance and repair of buildings	BYEJ	*4.9*	89.0†	95.2	100.0	102.1	102.1	105.0	109.9	116.4	117.0	118.8	119.8
Veterinary services	BYEK	*2.4*	92.7	95.3	100.0	103.0†	108.3	110.4	111.5	113.4	116.2	119.4	122.9
General expenses	BYEL	*8.7*	90.8	96.4	100.0	107.1	114.4	121.6	123.5	126.6	128.5	131.7†	133.2
Goods and services contributing to investment in agriculture	BYEM	*100*	88.7	94.5	100.0	105.9†	109.8	113.2	115.8	118.8	121.8	125.2	128.0
Machinery and other equipment	BYEN	*50.9*	90.4	95.0	100.0	106.6†	111.7	115.7	117.0	117.7	121.8	123.9	124.8
Machinery and plant for cultivation	KVBV	*6.2*	86.4	94.0	100.0	109.0†	117.2	118.9	119.8	121.2	125.0	128.1	130.7
Machinery and plant for harvesting	KVBW	*11.2*	89.9	95.7	100.0	106.3	110.8	113.8	113.8	113.8	113.8	113.8	113.8
Farm machinery and installations	KVBX	*4.7*	87.0	93.5	100.0	105.9	109.8†	111.3	112.3	113.9	115.3	116.9	118.6
Tractors	KVBY	*22.3*	92.6	95.4	100.0	106.4†	111.5	117.7	119.0	118.8	125.5	128.7	129.5
Other vehicles	KVBZ	*6.5*	92.0	95.2	100.0	106.4†	111.2	116.4	118.4	119.1	125.1	128.1	128.7
Buildings	BYEO	*49.1*	86.9	94.0	100.0	105.2†	107.8	110.6	114.5	120.0	121.7	126.5	131.3
Farm buildings	KVCA	*33.6*	88.2	94.3	100.0	104.5†	106.5	109.3	113.7	119.4	121.9	126.7	130.7
Engineering and soil improvement operations	KVCB	*15.5*	84.2†	93.1	100.0	106.8	110.8	113.5	116.2	121.4	121.2	126.0	132.7

The price indices shown are all on the 1990 = 100 series.

Source: Stats (C & S) H, Ministry of Agriculture, Fisheries and Food: 01904 455253

16.7 Index of producer prices of agricultural products
United Kingdom

Annual averages

1990 = 100

		Weights	1988	1989	1990	1991	1992	1993	1994	1995	1996	1997	1998
All products	BYEP	*100*	92.9	98.8	100.0	99.2	100.9	106.1	106.6	116.0	114.3	98.8†	90.0
All crop products	BYEQ	*40.4*	92.7	95.9	100.0	101.1	96.7	96.2	98.4	111.7	103.2	85.9†	87.6
Cereals	BYER	*15.9*	97.7	98.3	100.0	105.8	108.1	111.5	97.2	103.1	104.7	83.9	72.7
Wheat for:													
breadmaking	KVDA	*3.9*	102.8	98.6	100.0	111.4	116.6	122.5	102.3	106.6	108.6	92.5	83.5
other milling	KVDB	*0.5*	98.0†	98.0	100.0	109.2	110.6	111.4	96.4	101.9	103.6	83.7	72.7
feeding	KVDC	*6.6*	95.2	97.2	100.0	104.6	105.7	107.9	93.5	100.4	102.5	81.0	68.8
Barley for:													
feeding	KVDD	*2.9*	95.0	98.0	100.0	103.8	105.4	109.7	98.1	101.9	101.3	79.2	68.5
malting	KVDE	*1.7*	99.2	102.8	100.0	100.6	102.1	102.0	97.9	109.1	110.8	84.0	71.9
Oats for:													
milling	KVDF	*0.2*	102.6	94.3	100.0	102.2	108.2	120.6	100.6	92.8	100.3	78.5	63.4
feeding	KVDG	*0.1*	100.5	93.9	100.0	101.5	109.6	124.8	101.0	93.3	100.6	78.7	63.4
Root crops	BYES	*5.6*	82.9	93.2	100.0	99.5	86.0	80.6	121.7	177.3	105.2	70.4	107.6
Potatoes:													
early	KVDH	*0.3*	127.7	174.1	100.0	224.7	100.7	110.1	257.7	197.8	139.1	99.5	194.8
main crop	KVDI	*3.3*	73.6	84.5	100.0	87.7	72.4	58.3	119.2	214.4	94.4	51.4	112.7
Sugar beet	KVDJ	*2.1*	91.5	96.1	100.0	101.6	105.5	111.9	107.6	116.0	117.8	96.5	88.0
Fresh vegetables	BYET	*7.8*	91.0	90.7	100.0	99.1	90.5	97.4	101.9	109.3	108.4	96.6†	104.6
Cauliflowers	KVDK	*0.5*	84.8	93.8	100.0	111.5	95.7	104.4†	106.6	97.1	107.7	88.2	100.3
Lettuce	KVDL	*1.0*	93.3	97.1	100.0	98.3	121.2	113.4	124.3	122.8	106.1	116.7†	107.7
Tomatoes	KVDM	*0.7*	102.4	99.9	100.0	88.8	80.4	75.8	84.8	78.5	98.5	69.2	77.9
Carrots	KVDN	*0.8*	84.4	73.1	100.0	104.0	67.1	86.0	100.1	122.8	121.0	73.3	108.9
Cabbage	KVDO	*0.7*	85.2†	87.8	100.0	98.5	85.6	106.4	94.5	120.5	112.1	102.4	119.4
Beans	KVDP	*0.2*	68.1	63.2	100.0	68.0	57.2	82.8	93.1	90.8	87.6	92.9	69.8
Onions	KVDQ	*0.5*	71.8	75.8	100.0	89.9	78.2	88.1	108.6	122.4	75.5	84.2	108.5
Mushrooms	KVDR	*1.6*	101.1	95.3	100.0	104.2	96.4	106.0	104.3	107.9	111.6	98.2	100.1
Fresh fruit	BYEU	*2.1*	90.7	91.9	100.0	104.5	100.5	87.5	87.0	95.9	107.5	98.3	98.5
Dessert apples	KVDS	*0.7*	93.2	94.7	100.0	117.3	124.7	79.6	74.8	96.4	112.6	97.1	93.5
Dessert pears	KVDT	*0.1*	60.8	80.3	100.0	90.9	90.3	77.8	78.9	86.6	82.9	78.6	75.3
Cooking apples	KVDU	*0.3*	89.3	70.9	100.0	99.3	78.0	76.3	86.9	88.5	105.6	112.8	125.5
Strawberries	KVDV	*0.5*	107.0	102.0	100.0	109.9	105.3	101.2	103.0	91.3	94.1	91.0	85.3
Raspberries	KVDW	*0.2*	88.8	101.6	100.0	85.9	69.0	95.5	93.1	124.7	156.8	106.1	111.6
Seeds	BYEV	*1.1*	93.8	96.7	100.0	99.1	101.1	92.2	107.3	143.3	122.7	92.1	89.6
Flowers and plants	BYEW	*3.8*	101.1	102.3	100.0	97.1	99.3	97.9	104.4	107.2	114.6	114.8	109.0
Other crop products	BYEX	*4.1*	82.8	96.1	100.0	91.3	73.7	60.3	62.6	63.2	66.8	60.2	59.7
Animals and animal products	BYEY	*59.6*	93.1	100.8	100.0	97.9	103.8	112.8	112.1	119.0	121.8	107.5	91.6
Animals for slaughter	BYEZ	*33.8*	94.9	103.0	100.0	94.9	102.1	111.8	109.1	112.7	115.5	102.2	84.7
Calves	KVDX	*0.3*	135.0	141.1	100.0	92.5	114.8	150.7	145.4	130.9	93.6	96.2	82.0
Clean cattle	KVDY	*11.7*	101.9	107.2	100.0	100.5	102.7	120.2	114.6	115.1	98.2	89.7	79.6
Cows and bulls	KVDZ	*2.1*	110.5	114.7	100.0	97.4	131.8	110.0†	118.1	117.0	106.6	87.7	81.0
Clean pigs	KVEA	*7.7*	80.8	100.3	100.0	91.4	102.3	92.2	88.6	105.9	122.1	98.4	71.9
Sows and boars	KVEB	*0.3*	81.6	112.0	100.0	103.2	121.4	92.0	95.4	121.3	131.0	102.5	53.4
Sheep	KVEC	*4.3*	101.5	105.4	100.0	84.0	105.1	126.7	135.4	136.6	163.2	148.5	111.4
Ewe and ram	KVED	*0.3*	136.3	119.6	100.0	87.5	109.9	133.8	125.9	123.5	172.0	178.0	105.0
Poultry	KVEE	*7.0*	86.5	90.8	100.0	95.3	95.0	101.9	101.5	98.4	108.4	100.0	92.3
Chickens	KVEF	*5.0*	89.3	94.6	100.0	92.9	94.0	101.5	99.3	94.2	104.8	97.3	85.5
Turkeys	KVEG	*1.6*	76.6	77.4	100.0	100.8	95.9	102.3	107.2	108.6	117.4	103.2	105.4
Cows milk	BYFA	*21.7*	93.4	99.6	100.0	104.6	109.6	117.8	119.3	134.0	134.6	118.8	104.1
Eggs	BYFB	*3.6*	72.9	87.7	100.0	87.5	86.2	98.8	99.4	90.0	107.6	92.3	85.1
Other animal products:													
Wool (clip)	BYFC	*0.6*	104.4†	103.5	100.0	92.0	87.1	72.7	92.2	102.5	95.4	84.0	60.0

The price indices shown are all on the 1990 = 100 series.

Source: Stats (C & S) H, Ministry of Agriculture, Fisheries and Food: 01904 455253

16.8 Commodity price trends[1]
United Kingdom
Calendar years

			1988	1989	1990	1991	1992	1993	1994	1995	1996	1997	1998
Wheat *£ per tonne*	KVAA	Average ex-farm price[2]	105.2	104.9	109.8	116.5	120.5	124.2	106.4	115.6	112.5	88.7	..
Barley *£ per tonne*	KVAB	Average ex-farm price[2]	103.7	106.7	108.8	107.8	113.9	113.6	105.4	107.6	103.7	81.2	..
Oats *£ per tonne*	KVAC	Average ex-farm price[2]	104.8	98.5	106.4	105.9	115.1	128.8	108.4	101.4	107.4	83.5	..
Rye *£ per tonne*	KVAD	Average ex-farm price[2]	108.20	106.80	114.60	118.50	115.20	113.40	112.81	107.88	113.88	..	..
Hops *£ per tonne*	KVAE	Average farm-gate price	2 468	2 662	2 922	3 211	3 208	3 357	4 005	3 595†	3 360	3 550	3 664
Potatoes *£ per tonne*	KVAF	Average farm-gate price[3]	70.77†	89.83	99.13	92.99	74.39	64.30	126.79	185.84	101.06	65.49	121.36
Sugar beet *£ per tonne*	KVAG	Producer price[4]	25.68	27.87	31.47	31.42	38.02	36.47	34.74	38.34	37.72	32.86	31.85
Oilseed rape	KVAH	Average market price[5]	232.00	286.00	280.00	248.00	133.00	153.03	185.43	177.73	186.85	160.23	164.27
Apples *£ per tonne*	KPUE	Dessert average farm-gate price[6]	407.0	385.6	524.1	596.2	310.3	291.8	337.0†	407.3	430.1	409.2	349.0
	KVAI	Dessert average market price	424.0	383.0	500.0	589.0	414.0	370.0	447.4	465.7	474.7	538.1	..
	KVAJ	Culinary average market price	385.0	286.0	408.0	420.0	312.0	327.0	446.4	405.9	462.8	558.4	..
	KPUJ	Culinary average farm-gate price	278.8	199.0	268.3	278.8	189.6	229.4	236.1†	236.8	267.3	292.1	305.3
Pears *£ per tonne*	KPUG	Average farm-gate price	413.1	476.9	473.0	545.8	426.9	365.4	460.7	420.4	440.6	426.8	..
	KVAK	Average market price	321.0	402.0	485.0	468.0	461.0	380.0	408.3†	447.5	441.7	433.1	394.2
Tomatoes *£ per tonne*	LQMH	Average farm-gate price[6]	561.7	551.1	669.8	584.9	481.2	535.5	617.8	510.8	704.0	486.0	326.3
	KVAL	Average market price[6]	571.0	559.0	694.0	604.0	494.0	552.8	675.2	624.0	700.1	655.6	..
Cauliflowers *£ per tonne*	KPUI	Average farm-gate price[6]	186.4	196.5	236.2	243.1	198.1	213.4	225.7†	244.6	233.0	200.4	231.3
	KVAM	Average market price[6]	247.0	257.0	296.0	292.0	253.0	296.9	323.5	394.9	442.6	372.0	..
Cattle (rearing)	KVAN	1st quality Hereford/cross bull calves[7,13]	186.00	188.43	128.00	119.92	145.44	185.95	182.06	166.32	131.81	146.95	107.86
£ per head	KVAO	1st quality beef/cross yearling steers[13]	414.00	430.73	397.00	397.30	419.49	459.27	434.95	434.68	321.30	367.54	308.16
Cattle (fat) *p per kg liveweight*	KVAP	Clean cattle[8]	109.26	114.03	106.30	106.92	109.58	128.04	121.71	123.15	105.52	96.89	85.95

See footnotes on the second part of this table.

Source: Ministry of Agriculture, Fisheries and Food: 01904 455332

16.8 Commodity price trends[1]

United Kingdom

continued

Calendar years

			1988	1989	1990	1991	1992	1993	1994	1995	1996	1997	1998
Sheep (store) *£ per head*	KVAQ	1st quality lambs, hoggets and tegs[7]	37.50	36.10	33.80	35.50	37.90	35.56	40.02	44.46	46.83	53.42	31.28
Sheep (fat)	KVAR	Great Britain[11]	177.00	184.47	174.47	148.16	182.12	218.96	236.88	236.40	283.13	239.20†	192.47
p per kg estimated dressed carcase weight	KVAS	Northern Ireland[12]	208.00	202.50	170.94	179.41	172.37	199.50	221.95	214.41	260.46†	228.23	179.06
Pigs *£ per kg deadweight*	KVAT	Average price clean pigs	91.13	113.85	112.42	103.05†	115.02	103.03	99.65	118.95	137.77	111.06	81.01
Broilers *p per kg carcass weight*	KVAU	Average producer price	78.0	82.1	86.8	81.0	82.7	86.8	86.3	83.8†	90.7	85.9	76.0
Milk *p per litre*	KVAV	Average net return to producers[9]	17.65†	19.10	19.19	19.81	20.93	22.68	22.89	25.03	25.07	22.35	19.52
Eggs *p per dozen*	KVAW	Average producer price[10]	31.00	37.30	42.60	36.30	36.60	41.90	42.40	38.20	45.70	39.80	36.22
Wool *p per kg*	KHWQ	Average producer price for clip paid to producers by the British Wool Marketing Board	96.90	95.30	91.04	81.98	79.82	61.00	97.97	94.08	86.90†	74.90	45.90

1 This table gives indications of the movement in commodity prices at the first point of sale. The series do not always show total receipts by farmers; for some commodities additional premiums or deficiency payments are made to achieve support price levels.
2 Weighted average ex-farm prices of United Kingdom cereals.
3 Weighted average price paid to growers for early and main crop potatoes in the United Kingdom. (includes all potatoes and a value for sacks)
4 Returns to growers figures since 1986 prices per 'adjusted' tonne at 16% sugar content.
5 Typical contract price adjusted to delivered basis and 40 per cent oil content.
6 Weighted average wholesale prices for England and Wales. From 1982, for England only. Average farm-gate price for England and Wales, crop year (June-May).
7 Average prices at representative markets in England and Wales.
8 Based on Meat and Livestock Commission all clean cattle prices.
9 Derived by dividing total value of output by the total quantity of output available for human consumption.
10 Average price of all Class A eggs weighted according to quantity in each grade.
11 Average of Great Britain weekly market prices as used to determine the level of ewe premium.
12 Average of Northern Ireland weekly market prices used to determine the level of ewe premium.
13 Category change 1988: formerly 1st quality yearling steers beef/dairy cross, now consists of Hereford/cross, Charolais/cross, Limousin/cross, Simmental/cross, Belgian/cross, other continental cross, other beef/dairy cross, other beef/beef cross.

Source: Ministry of Agriculture, Fisheries and Food: 01904 455332

Government finance

17

Government finance

Public sector *(Tables 17.1 to 17.3 and 17.5)*

In Table 17.1 the term public sector describes the consolidation of central government, local government and public corporations. General government is the consolidated total of central government and local government. The table shows details of the key public sector finances' indicators, consistent with the European System of Accounts 1995 (ESA95).

The concepts in Table 17.1 are consistent with the new format for public finances in the *Economic and Fiscal Strategy Report (EFSR),* (published by HM Treasury 11 June 1998). The public sector surplus on current budget in the *EFSR* is equivalent to net saving in national accounts plus capital tax receipts. Net investment is gross capital formation, plus payments less receipts of investment grants, less depreciation. Net borrowing is net investment less surplus on current budget. Net borrowing differs from the net cash requirement (see below) in that it is measured on an accruals basis whereas the net cash requirement is mainly a cash measure which includes some financial transactions.

Details of the public sector net cash requirement and the contributions to the net cash requirement are given in Table 17.2. The net cash requirement indicates the extent to which the public sector borrows from other sectors of the economy and overseas to finance the balance of expenditure and receipts arising from its various activities. Part of the central government net cash requirement is needed to finance central government on lending to local government and public corporations. Their additional 'contributions' to the public sector are therefore equal to their net cash requirements *less* their direct borrowing from central government.

General government net cash requirement is the sum of the borrowing requirements of central government and local government *less* direct borrowing by local government from central government. The public sector net cash requirement is general government *plus* public corporations' net cash requirement *less* their direct borrowing from central government.

The net cash requirements for local government and public corporations are measured from the financing items rather than as the difference between receipts and payments because this information comes to hand more quickly.

Table 17.3 shows public sector net debt. Public sector net debt consists of the public sector's financial liabilities at face value minus its liquid assets- mainly foreign currency exchange reserves and bank deposits. General government gross debt (consolidated) in table 17.3 is consistent with the definition of general government gross debt reported to the European Commission under the requirements of the Maastricht Treaty. More information on the concepts in table 17.1, 17.2 and 17.3 can be found in a guide to monthly public sector finance statistics, *GSS Methodology Series No 12,* the ONS First Releases *Public Sector Finances* and *Public Sector Accounts* and *Financial Statistics Explanatory Handbook*.

Table 17.5 (Part 1) shows the central government net cash requirement on own account, which is the net measure of all receipts and expenditure on central government funds and accounts, after excluding transactions which do not cross the central government boundary with the rest of the economy. Part 2 shows central government net cash requirement. The table shows the main financial instruments through which central government borrows to finance the net cash requirement. The details of the changes in the financial liabilities and assets of central government in aggregate match the net cash requirement.

Rateable values *(Table 17.12)*

Major changes to local authority finance in England and Wales took effect from 1 April 1990. These included the abolition of domestic rating - replaced by the community charge (replaced in 1993 by the council tax), the revaluation of all non-domestic properties, and the introduction of the Uniform Business Rate. As a result the analysis presented here for 1990 differs substantially from that shown for earlier years. Rateable values in 1990 are about eight times those for 1989, on average, but there are wide variations about this average both for individual properties and for classes of property. Also in 1990, a new classification scheme was introduced. While the new classifications have been matched as far as possible to those used in earlier years there are differences in coverage. Further differences are caused by legislative changes which have changed the treatment of certain types of property. Comparisons between 1990 and earlier years should therefore be made with caution. With effect from 1 April 1995 all non-domestic properties were revalued. Although overall there was little change in rateable values, those for offices fell and there was a rise for all other main property types.

Central government

The central government embraces all bodies for whose activities a Minister of the Crown, or other responsible person, is accountable to Parliament. It includes, in addition to the ordinary government departments, a number of bodies administering public policy but without the substantial degree of financial independence which characterises the public corporations; it also includes certain extra-budgetary funds and accounts controlled by departments.

The government's financial transactions are handled through a number of statutory funds, or accounts. The most important of these is the Consolidated Fund which is the government's main account with the Bank of

England. Up to 31 March 1968 the Consolidated Fund was virtually synonymous with the term 'Exchequer' which was then the government's central cash account. From 1 April 1968 the National Loans Fund, with a separate account at the Bank of England, was set up by the National Loans Act, 1968. The general effect of this Act was to remove from the Consolidated Fund most of the government's domestic lending and the whole of the government's borrowing transactions and to provide for them to be brought to account in the National Loans Fund.

Revenue from taxation and miscellaneous receipts, including interest and dividends on loans made from Votes, continue to be paid into the Consolidated Fund. After meeting the ordinary expenditure on Supply Services and the Consolidated Fund Standing Services, the surplus or deficit of the Consolidated Fund (Table 17.4), is payable into or met by the National Loans Fund.

Table 17.4 also provides a summary of the transactions of the National Loans Fund. The service of the National Debt, previously borne by the Consolidated Fund, is now met from the National Loans Fund which receives (a) interest payable on loans to the nationalised industries, local authorities and other bodies, whether the loans were made before or after 1 April 1968 and (b) the profits of the Issue Department of the Bank of England, mainly derived from interest on government securities, which were formerly paid into the Exchange Equalisation Account. The net cost of servicing the National Debt after applying these interest receipts and similar items is a charge on the Consolidated Fund as part of the standing services. Details of National Loans Fund loans outstanding are shown in Table 17.7.

Details of borrowing and repayments of debt, other than loans from the National Loans Fund, are shown in Table 17.6.

17.1 Public sector finances[1]

United Kingdom

£ millions, not seasonally adjusted

		1988 /89	1989 /90	1990 /91	1991 /92	1992 /93	1993 /94	1994 /95	1995 /96	1996 /97	1997 /98	1998 /99
Surplus on current budget[2]												
General Government	ANLW	8 958	10 960	7 951	–7 221	–31 752	–37 764	–30 537	–24 946	–21 603†	–2 013	10 204
Public sector	ANMU	9 026	8 763	4 394	–9 919	–34 632	–39 879	–32 917	–25 454	–22 572†	–2 696	9 503
Net investment[3]												
General Government	-ANNV	4 090	11 889	14 982	13 596	14 793	12 925	14 468	11 537	7 566†	5 715	5 959
Public sector	-ANNW	2 048	6 840	8 416	11 389	12 898	10 942	10 360	9 651	5 159†	4 749	5 002
Net borrowing[4]												
General Government	-NNBK	–4 868	929	7 031	20 817	46 545	50 689	45 005	36 483	29 169†	7 728	–4 245
Public sector	-ANNX	–6 978	–1 923	4 022	21 308	47 530	50 821	43 277	35 105	27 731†	7 445	–4 501
Net cash requirement[5]												
General Government	RUUS	–12 544	–2 982	239	13 420	36 577†	46 070	38 096	34 489	24 152	1 830	–6 644
Public sector	RURQ	–14 471†	–6 905	–786	13 856	36 180	46 162	36 688	31 485	22 728	1 111	–7 037

1 National accounts entities as defined under the European System of Accounts 1995 (ESA95) consistent with the latest national accounts.
2 Net saving, plus capital taxes.
3 Gross capital formation, plus payments less receipts, of investment grants less depreciation.
4 Net borrowing = net investment minus surplus on current budget.
5 Previously called Public Sector Borrowing Requirement (PSBR).

Source: Office for National Statistics: 020 7533 5984

17.2 Contributions to the public sector net cash requirement

United Kingdom

£ millions

		1988 /89	1989 /90	1990 /91	1991 /92	1992 /93	1993 /94	1994 /95	1995 /96	1996 /97	1997 /98	1998 /99
Central government net cash requirement	RUUW	–6 959	–4 575	–2 635	13 020	36 230	49 699	39 123	35 315	25 156	3 542†	–4 537
of which: own account	RUUX	–12 884	–4 314	–3 092	11 830	42 312	48 850	39 057	35 628	24 995	2 650†	–6 172
Local government												
Direct borrowing from central government[1]	ABEC	4 941	2 262	1 472	639	–7 267	–654	–392	473	1 517	955	1 869
Net borrowing from other sources	ABEF	–4 553	–967	1 817	973	1 506	–2 025	–610	–1 611	–2 357	–1 761	–2 240
less Transactions in other public sector dept:												
Central government	ABEE	11	32	–32	19	–12	43	25	4	2	15	103
Public corporations	AAEJ	37	–69	–10	3	–14	58	–66	–3	1	–1	4
Net cash requirement	ABEG	340	1 332	3 331	1 590	–5 735	–2 780	–961	–1 139	–843	–820	–478
General government net cash requirement	RUUI	–11 560	–5 505	–776	13 971	37 762	47 573	38 554	33 703	22 796	1 767†	–6 884
Public corporations:												
Direct borrowing from central government	ABEI	984	–2 523	–1 015	551	1 185	1 503	458	–786	–1 356	–63	–234
Net borrowing from other sources	ABES	–2 491†	–1 778	–216	–357	–599	–845	–633	–419	1 187	–295	–843
less Transactions in other public sector debt:												
Central government	ABEK	428	–236†	–220	–225	865	531	1 267	2 495	1 340	356	–659
Local authorities	ABEL	–8	–142	14	–17	118	35	–34	–696	–85	5	–31
Net cash requirement	ABEM	–1 927†	–3 923	–1 025	436	–397	92	–1 408	–3 004	–1 424	–719	–387
Public sector net cash requirement	RURQ	–14 471†	–6 905	–786	13 856	36 180	46 162	36 688	31 485	22 728	1 111	–7 037
Public sector net cash requirement excluding privatisation proceeds	RURS	–7 402†	–2 686	4 559	21 779	44 364	51 592	43 121	33 920	27 162	2 881	–6 967

1 Excluding market transactions of central government in public sector debt; these transactions are included in 'Net borrowing from other sources.'

Source: Office for National Statistics: 020 7533 5984

17.3 Public sector net debt
United Kingdom

£ millions

		1991 /92	1992 /93	1993 /94	1994 /95	1995 /96	1996 /97	1997 /98	1998 /99
Central government gross debt:									
F20 Currency and sight deposits	RUTQ	3 059	3 337	3 779	4 070	4 073	4 346	4 645	5 187
F30 Other deposits	EYJZ	47 198	52 578	56 982	60 431	66 952	70 566	71 677	70 181
F40 Bills and short term bonds	RUTT	12 205	8 185	6 206	10 913	13 664	7 484	4 354	7 062
F50 Long term bonds	RUTX	125 693	165 450	212 110	235 357	264 208	292 431	297 790	292 391
F70 Short-term loans	ANVE	7 093	7 595	7 371	16 776	16 258	14 843	13 331	16 060
F80 Long-term loans	EYKD	2 749	2 805	2 771	2 516	2 431	2 246	2 083	1 513
F90 Insurance technical reserves	EYKE	–	–	–	–	–	–	–	–
Central government gross debt total	ANVB	197 997	239 950	289 219	330 063	367 586	391 916	393 880	392 394
Local government gross debt:									
F20 Currency and sight deposits	EYKH	–	–	–	–	–	–	–	–
F30 Other deposits	EYKI	–	–	–	–	–	–	–	–
F40 Bills and short term bonds	EYKJ	–	–	–	–	–	–	–	–
F50 Long term bonds	EYKK	148	160	279	452	641	721	685	710
F70 Short-term loans	EYKL	2 442	2 210	1 636	1 655	1 855	1 481	1 278	1 103
F80 Medium and long-term loans	EYKM	52 691	47 527	48 324	48 122	48 056	49 396	49 969	50 980
F90 Insurance technical reserves	EYKN	–	–	–	–	–	–	–	–
Local government gross debt total	EYKP	55 281	49 897	50 239	50 229	50 552	51 598	51 932	52 793
General government gross debt (unconsolidated)									
F20 Currency and sight deposits	ANUX	3 059	3 337	3 779	4 070	4 073	4 346	4 645	5 187
F30 Other deposits	EYKR	47 198	52 578	56 982	60 431	66 952	70 566	71 677	70 181
F40 Bills and short term bonds	ANUY	12 205	8 185	6 206	10 913	13 664	7 484	4 354	7 062
F50 Long term bonds	BKLK	125 841	165 610	212 389	235 809	264 849	293 152	298 475	293 101
F70 Short-term loans	BKLL	9 535	9 805	9 007	18 431	18 113	16 324	14 609	17 163
F80 Medium and long-term loans	EYKV	55 440	50 332	51 095	50 638	50 487	51 642	52 052	52 493
F90 Insurance technical reserves	EYKW	–	–	–	–	–	–	–	–
General government gross debt	BKLM	253 278	289 847	339 458	380 292	418 138	443 514	445 812	445 187
less:									
Central government holdings of local government debt	EYKZ	49 173	41 527	40 977	40 707	41 266	42 555	43 397	45 277
Local government holdings of central government debt	EYLA	93	81	124	149	153	155	170	274
General government gross debt (consolidated)[1]	BKLQ	204 012	248 239	298 357	339 436	376 719	400 804	402 245	399 636
as percentage of GDP	EZPW	33.9	39.5	44.8	48.2	50.7	50.7	48.1	46.5
Public corporations gross debt	EYYD	11 510	16 194	23 867	26 916	26 595	26 158	26 044	26 775
less:									
Central government holdings of public corporations debt	EYXY	10 313	15 227	22 950	26 279	25 981	25 664	25 668	26 440
Local government holdings of public corporations debt	EYXZ	29	11	69	3	–	1	–	4
Public corporations holdings of central government debt	BKPZ	1 348	2 617	2 854	3 503	5 724	7 107	7 467	6 336
Public corporations holdings of local government debt	EYXV	1 483	1 585	1 620	1 586	890	805	810	780
Public sector gross debt (consolidated)	BKQA	202 349	244 993	294 731	334 981	370 719	393 385	394 344	392 851
Public sector liquid assets									
Official reserves	AIPD	25 952	27 153	28 908	28 330	30 463	25 547	21 293	22 147
Bank and building society deposits	BKQO	1 656	1 631	1 775	1 816	1 808	2 067	2 286	1 635
Instalments due on BGS	BKQC	739	3 386	1 250	–	–	–	–	–
Central government liquid assets	BKQD	28 347	32 170	31 933	30 146	32 271	27 614	23 579	23 782
Bank deposits	BKQE	3 459	3 471	4 907	5 408	6 398	7 134	7 994	8 040
Building society deposits	BKQF	2 436	2 870	3 855	3 927	3 831	4 142	3 796	4 235
Other short term assets	BKQG	227	1 932	2 424	2 621	2 826	3 256	3 693	4 295
Local government liquid assets	BKQH	6 122	8 273	11 186	11 956	13 055	14 532	15 483	16 570
Bank deposits	BKNC	1 284	1 166	1 353	2 622	3 088	1 778	1 469	1 860
Other short term assets	BKQP	499	1 077	800	943	1 054	964	937	1 227
Public corporations liquid assets	BKQQ	1 783	2 243	2 153	3 565	4 142	2 742	2 406	3 087
Public sector liquid assets	BKQJ	36 252	42 686	45 272	45 667	49 468	44 888	41 468	43 439
Public sector net debt	BKQK	166 097	202 307	249 459	289 314	321 251	348 497	352 876	349 412
as percentage of GDP	RUTO	27.6	32.2	37.5	41.1	43.2	44.1	42.2	40.6

1 General government gross debt as reported to the European Commission under the terms of the Maastricht Treaty.

Source: Office for National Statistics: 020 7533 5984

17.4 Consolidated Fund and National Loans Fund: revenue and expenditure; receipts and payments

United Kingdom Years ended 31 March

£ millions

		1988 /89	1989 /90	1990 /91	1991 /92	1992 /93	1993 /94	1994 /95	1995 /96	1996 /97	1997 /98	1998 /99
Consolidated Fund												
Revenue												
Inland Revenue	KCWZ	68 812.7	76 674.3	82 320.7	79 509.6	76 345.9	77 270.6	87 230.4	97 100.9	103 892.4	117 632.8	128 249.8
Customs and Excise	KCXA	49 565.3	52 190.2	55 336.9	61 826.9	63 398.0	66 885.5	72 485.6	76 668.6	82 351.6	89 839.6	94 018.3
Motor vehicle duties	KCXB	2 810.8	2 920.0	2 971.4	2 945.3	3 196.0	3 752.3	3 805.9	4 043.6	4 217.5	4 543.0	4 666.3
National Non-Domestic Rates	KPOI	..	..	..	..	..	..	16 280.3	13 373.4	14 269.3	14 036.9	15 878.3
Miscellaneous receipts	KCXE	12 402.5	13 136.5	21 736.6	27 193.0	30 101.2	22 150.2	16 280.3	10 289.3	12 634.6	14 176.6	10 737.1
Total revenue	KCXF	133 592.8	144 921.6	162 366.1	171 474.8	173 041.2	170 058.6	191 272.0	201 475.8	217 365.4	240 228.9	253 549.8
Expenditure												
Supply services	KCXG	110 713.9	122 548.8	145 763.1	168 702.6	189 961.6	204 376.1	207 465.8	211 403.3	214 226.0	209 440.8	213 439.2
Debt interest[1]	KCXH	10 827.7	11 343.3	10 229.2	8 943.0	10 675.1	13 459.6	16 039.8	18 423.1	20 702.3	21 605.7	21 320.8
Payments to Northern Ireland	KCXI	2 383.3	2 683.1	2 927.7	2 944.9	2 632.1	3 090.3	4 052.6	3 903.5	3 685.2	4 581.5	4 709.3
Payments to the European Communities, etc	KCXJ	3 899.5	4 537.7	4 806.1	3 323.3	4 867.1	6 079.5	5 258.2	7 650.7	6 875.3	7 039.9	8 060.7
Other expenditure[2]	KCXL	177.8	380.7	503.7	276.3	117.7	–453.3	271.6	–14.0	876.2	–53.8	–62.1
Total expenditure	KCXM	128 002.2	141 493.6	164 229.8	184 189.7	208 253.6	226 552.2	233 088.0	241 366.5	246 365.0	242 614.1	247 467.9
Deficit met from the National Loans Fund	KCXN	–5 590.6	–3 428.0	1 863.7	12 715.5	35 212.4	56 493.6	41 816.0	39 890.7	28 999.6	2 385.2	–6 081.9
National Loans Fund												
Receipts												
Profits of the Issue Department of the Bank of England - income[5]	KZAW	1 309.2	1 237.1	2 247.3	1 816.5	1 350.8	996.6	955.2	1 275.5	1 200.9	1 604.8	1 600.9
Other miscellaneous receipts	KZAX	2.9	3.9	–	6.2	66.7	106.6	4.8	4.4	3.6	5.0	5.0
Interest on loans	KCXO	6 099.0	6 282.4	5 429.0	5 925.1	5 100.7	4 695.2	4 564.5	4 598.7	4 393.5	5 103.6	4 409.7
Service of the National Debt - balance met from the Consolidated Fund	KCXP	10 827.7	11 335.5	10 229.3	8 943.1	10 675.1	13 459.6	16 039.8	18 423.2	20 702.3	21 605.7	21 320.8
Gilt Edged Official Operations Account Net Income	KJDO	..	..	..	..	..	2.0	–	–	–	–	–
Total	KCXQ	18 238.8	18 858.9	17 905.6	16 690.9	17 193.3	19 260.0	21 564.3	24 301.8	26 300.3	28 319.1	27 336.4
Exchange Equalisation Account-sterling capital	KCXR	–	6 175.0	–1 050.0	300.0	3 574.5	400.0	1 250.0	2 750.0	2 150.0	3 650.0	1 880.0
Net borrowing[4]	KCXS	–	–	3 684.7	12 112.0	25 907.6	58 939.5	39 848.1	37 688.8	27 547.9	–	–
International Monetary Fund-maintenance of sterling holdings	KCXT	234.7	–	–	59.8	47.6	–	–	–	82.1	707.0	181.2
Profits of the Issue Department of the Bank of England: capital appreciation	KZAY	..	..	..	..	..	..	..	..	18.8	21.2	19.3
Reduction of National Debt Commissioners' Liability in respect of the National Savings Bank Investment Account	KCXU	–	–	–	–	–	–	–	–	–	–	–
Change in balances and other items	KCXV	–	–	–	–	–	–	–	–	–	–	–
NILO Gilt-Edged operations A/C	KJDL	–	–	–	–	–	2 500.0	2 500.0	2 500.0	2 000.0	2 500.0	2 500.0
Total	KCXW	18 473.5	25 033.9	20 540.3	29 162.7	46 723.0	81 099.5	65 162.4	42 938.8	58 099.1	35 197.3	31 916.9
Payments												
Service of the National Debt:												
Interest	KJDM	18 065.8	18 679.9	17 730.6	16 516.3	17 009.5	19 107.0	21 334.7	24 004.0	25 942.5	27 877.8	26 792.1
Management and expenses	KCXY	173.0	179.0	175.0	174.6	183.8	153.0	229.6	297.8	357.8	441.3	544.3
Total	KCXZ	18 238.8	18 858.9	17 905.6	16 690.9	17 193.3	19 260.0	21 564.3	24 301.8	26 300.3	28 319.1	27 336.4
Consolidated Fund deficit met from the National Loans Fund	KCYA	–5 590.6	–3 435.7	1 863.7	12 714.9	35 212.4	56 493.6	41 816.0	39 890.7	28 999.6	2 385.2	–6 081.9
Net repayment[4]	KCYB	1 846.8	7 923.3	–	–	–	–	–	–	–	2 482.1	3 558.2
Net lending[3]	KCYC	3 978.5	1 538.3	623.9	–243.1	–6 787.0	–286.6	–974.8	334.1	280.6	–4.2	1 766.3
International Monetary Fund-maintenance of value of sterling holding	KCYD	–	149.1	147.1	–	1 104.3	632.5	226.2	202.1	–	–	–
International Monetary Fund-additional subscription	KCYE	–	–	–	–	–	–	–	–	–	–	2 826.5
NILO Gilt-Edged Operations A/C	KJDN	–	–	–	–	–	–	2 500.0	2 500.0	2 500.0	2 000.0	2 500.0
Discharge of Treasury Liability to the Bank of England Issue Department	KPUK	..	..	–	–	–	–	30.7	11.9	18.6	15.1	11.4
Total	KCYF	18 473.5	25 033.9	20 540.3	29 162.7	46 723.0	81 099.5	65 162.4	42 938.8	58 099.1	35 197.3	31 916.9

1 Payment to National Loans Fund representing its payments for the service of the National Debt *less* its receipts of interest on loans outstanding, etc.

2 Includes net issues to Contingencies Fund.

3 Minus sign indicates a net issue repayment.

4 See Table 17.6.

5 Prior to 1996-97, receipts from the Bank of England for appreciation of the assets of the Issue Department were included in the total amount for the profits of the Issue Department.

Source: HM Treasury: 020 7270 1586

17.5 Central government net cash requirement
United Kingdom

Years ended 31 March

£ millions

		1988 /89	1989 /90	1990 /91	1991 /92	1992 /93	1993 /94	1994 /95	1995 /96	1996 /97	1997 /98	1998 /99
Inland revenue	ACAB	68 812	76 673	82 322	79 353	76 292	77 289	87 255	97 120	103 957	117 632	128 249
Customs and excise	ACAC	49 566	52 190	55 336	61 827	63 398	66 885	72 486	76 670	82 351	89 840	94 027
Social security contributions[1]	ABIA	30 702	31 020	33 013	34 391	35 420	36 985	40 659	42 882	45 121	49 255	53 227
Interest and dividends	RUUL	10 126	11 362	10 997	10 924	9 828	9 595	8 380	9 607	9 082	9 557†	9 455
Other receipts[2]	RUUM	3 894	3 107	14 798	16 052	16 529	17 870	17 082	19 289	20 906	20 848†	19 629
Total cash receipts	RUUN	163 100	174 352	196 466	202 547	201 467	208 624	225 862	245 568	261 417	287 132†	304 587
Interest payments	RUUO	17 618	18 090	17 465	16 538	17 528	19 160	21 998	24 452	27 554	27 807†	27 031
Privatisation proceeds	ABIF	–7 069	–4 219	–5 345	–7 923	–8 184	–5 430	–6 433	–2 435	–4 434	–1 770	–70
Net departmental outlays[3]	RUUP	139 667	156 167	181 254	205 762	234 435	243 744	249 354	259 179	263 292	263 745†	271 454
Total cash outlays	RUUQ	150 216	170 038	193 374	214 377	243 779	257 474	264 919	281 196	286 412	289 782†	298 415
Own account net cash requirement	RUUX	–12 884	–4 314	–3 092	11 830	42 312	48 850	39 057	35 628	24 995	2 650†	–6 172
On lending to local government	ABEC	4 941	2 262	1 472	639	–7 267	–654	–392	473	1 517	955	1 869
On lending to public corporations	ABEI	984	–2 523	–1 015	551	1 185	1 503	458	–786	–1 356	–63	–234
Central government net cash requirement	RUUW	–6 959	–4 575	–2 635	13 020	36 230	49 699	39 123	35 315	25 156	3 542†	–4 537

1 Excluding Northern Ireland contributions
2 Including some elements of expenditure not separately identified.
3 Net of certain receipts.

		1991 /92	1992 /93	1993 /94	1994 /95	1995 /96	1996 /97	1997 /98	1998 /99
Central government funds and accounts									
National Loans Fund:									
Total payments	ACAU	16 449	10 407	18 973	20 648	24 631	26 581	28 309	29 100
Total receipts	ACAQ	16 692	17 193	19 260	21 593	24 286	26 282	28 318	27 335
Less surplus from consolidated fund	ACAP	–12 715	–35 213	–56 494	–41 817	–39 889	–28 983	–2 386	6 083
Borrowing required	ACAX	12 471	28 426	56 205	40 871	40 235	29 279	2 376	–4 318
Surplus of National Insurance Funds[2]	ACAY	–3 121	–6 033	903	2 304	674	326	1 798	3 053
Departmental balances and miscellaneous	ACAZ	2 522	–1 615	5 585	–539	4 192†	3 856	–3 046	–2 945
Northern Ireland central government debt	ACBA	–50	156	–20	15	–52	59	–82	–111
Financing of the central government net cash requirement analysed by type of instrument									
Net cash requirement	RUUW	13 020	36 230	49 699	39 123	35 315	25 156	3 542†	–4 537
Liabilities									
Coin	-EYMW	47	33	97	121	74	136	145	220
Sterling Treasury bills	NAVG	–1 291	–4 322	–1 824	4 566	2 730	–5 786	–2 890	2 615
British government securities	ANTA	11 141	30 468	46 411	21 206	26 607	26 492	6 285	–8 909
National savings	-AACE	3 136	4 353	4 111	3 570	5 109	4 821	1 554	367
Tax instruments	-AACF	22	–329	–252	–521	–390	–369	–147	–134
Loans from MFIs	ANTB	269	–4 113	3 400	9 028	–292	–1 342	–1 468†	2 572
Northern Ireland cg debt	-AACH	–29	7	–27	–3	–7	–4	–21	–19
ECS liabilities	-AACI	–18	–50	–43	–24	–35	–9	4†	–4
Deposits with cg from other sectors	ANTC	38	693	1 242	1 406	1 393	990	156†	–1 642
Government foreign currency debt	-AACL	694	6 966	–1 639	–25	–401	–248	–1 803	–44
Other government overseas financing	-AACM	–74	–100	–94	–93	–94	–88	–91	–99
Assets									
NILO lending (except PWLB)	ANTD	290	388	–124	–122	–97	253	140†	–2
Net change in official reserves	AIPA	–847	2 184	–1 442	62	705	635	1 902†	5
Deposits with MFIs	ANSZ	–358	52	–117	–48	13	–325	–224†	537

1 This is equal to National Loans Fund borrowing and special transactions (net) *less* receipts from other central government funds, etc.
2 A negative item represents a surplus, a positive item a deficit.

Sources: HM Treasury; Office for National Statistics: 020 7533 5984

17.6 Borrowing and repayment of debt
United Kingdom

Years ended 31 March

£ millions

		1988/89	1989/90	1990/91	1991/92	1992/93	1993/94	1994/95	1995/96	1996/97	1997/98	1998/99
Borrowing												
Government securities: new issues	KQGA	4 045.6	–	3 269.3	15 920.7	33 601.1	55 959.2	32 137.5	34 150.3	40 800.8	28 484.4	12 048.0
National savings securities:												
National savings certificates	KQGB	2 139.4	1 256.2	2 658.1	3 549.1	3 453.4	3 381.5	2 187.9	3 425.9	3 695.5	4 435.2	3 028.7
Capital bonds	KQGC	190.1	288.0	188.7	282.1	524.8	556.2	521.5	504.8	450.8	619.0	469.6
Income bonds	KQGD	1 741.9	1 482.9	1 629.1	1 329.1	1 251.2	1 940.8	1 255.7	780.7	1 272.7	1 043.4	1 371.7
Deposit bonds	KQGE	143.7	2.4	–	–	–	–	–	–	–	–	–
British savings bonds	KQGF	–	–	–	–	–	–	–	–	–	–	–
Premium savings bonds	KQGG	320.0	256.5	185.9	246.5	378.6	931.5	1 837.2	2 040.6	2 552.5	3 158.8	3 652.8
Save As You Earn	KQGH	66.9	76.4	88.2	94.3	85.3	76.4	66.5	50.8	34.1	20.7	11.4
Yearly plan	KQGI	114.2	88.8	71.0	87.0	109.9	112.4	116.5	94.1	–	–	5.2
National savings stamps and gift tokens	KQGJ	2.8	0.5	11.4	–	–	–	–	–	–	–	–
National Savings Bank Investments	KQGK	1 596.7	1 434.7	1 491.2	1 215.1	1 227.9	1 370.5	1 368.1	1 312.1	1 478.7	1 282.3	1 085.0
Children's Bonus Bonds	KGVO	–	–	–	127.0	95.5	91.2	118.9	144.9	352.4	255.3	205.0
First Option Bonds	KIAR	–	–	–	–	859.4	1 026.0	812.0	826.4	1 139.8	1 152.9	1 001.8
Pensioners Guaranteed Income Bond	KJDW	..	..	..	..	..	786.1	1 190.3	2 104.2	2 863.8	1 126.9	201.0
Treasurer's account	KWNF	..	..	..	..	..	..	..	..	21.1	39.9	17.1
Certificate of tax deposit	KQGL	891.1	1 065.9	1 343.1	1 137.7	607.2	470.6	91.2	76.2	109.4	84.1	66.4
Nationalised industries', etc temporary deposits	KQGM	12 708.0	29 728.0	25 855.0	20 709.0	19 350.0	28 700.9	29 294.0	36 870.6	53 198.2	46 375.9	39 962.4
British Gas corporation deposits	KQGN	–	–	–	–	–	–	–	–	–	–	–
Sterling Treasury bills (net receipt)	KQGO	588.4	7 282.5	–	–	–	–	5 196.1	2 606.1	–	–	3 546.2
ECU Treasury bills (net receipt)	KQGP	1 562.1	625.3	429.2	–	–	–	–	–	–	–	–
ECU Treasury notes (net receipt)	KDZZ	–	–	–	1 066.4	1 396.0	2 073.2	429.8	–	–	–	–
Ways and means (net receipt)	KQGQ	2 329.6	1 753.9	4 107.5	2 530.5	–	9 007.0	8 891.7	1 162.3	511.1	–	183.6
Other debt : payable in sterling :												
Interest free notes	KQGR	142.2	323.6	319.1	161.2	1 045.6	796.6	382.6	247.4	99.2	32.4	2 130.9
Other debt : payable in external currencies	KHCY	–	588.1	1 939.8	40.0	33 147.7	9 243.9	–	–	2 565.2	–	–
Total receipts	KHCZ	28 582.7	46 253.7	43 586.6	48 495.7	97 133.6	116 524.0	85 897.5	86 397.4	111 145.3	88 111.2	68 986.8
Repayment of debt												
Government securities : redemptions	KQGS	8 537.0	17 068.4	6 806.0	7 841.5	8 351.8	7 390.5	9 333.3	4 652.7	14 488.4	20 678.9	18 575.5
Statutory sinking funds	KQGT	2.8	2.8	2.7	2.5	2.5	2.4	2.3	2.2	2.2	2.1	2.0
Terminable annuities:												
National Debt Commissioners	KQGU	–	–	–	–	–	–	–	–	–	–	–
National savings securities:												
National savings certificates	KQGV	2 514.3	2 879.1	2 187.4	1 653.6	1 092.5	2 208.0	1 612.6	2 258.4	3 263.7	4 058.5	3 449.0
Capital bonds	KQGW	–	5.9	23.5	14.6	21.1	398.0	647.7	509.0	698.3	1 160.5	888.3
Income bonds	KQGX	940.3	1 094.9	831.3	729.0	955.2	1 000.7	1 181.8	1 256.9	1 394.0	1 148.9	880.8
Deposit bonds	KQGY	135.3	202.5	138.4	86.8	80.9	71.8	72.3	72.5	64.8	72.6	84.2
Yearly Plan	KQGZ	17.0	81.0	127.5	107.1	99.8	88.9	77.0	101.9	96.3	113.2	120.0
British savings bonds	KQHA	–	–	–	–	–	–	–	–	–	–	–
Premium savings bonds	KQHB	134.0	185.7	194.0	154.3	153.4	181.0	333.0	590.1	869.3	1 203.1	1 398.4
Save As You Earn	KQHC	94.1	87.9	89.6	103.6	116.9	96.3	92.3	98.7	70.1	68.2	37.1
National savings stamps and gift tokens	KQHD	2.3	0.1	11.1	–	–	0.7	–	–	–	–	–
National Savings Bank Investments (repayments)	KQHE	1 858.4	2 056.6	1 775.3	1 796.1	2 039.0	1 646.3	1 745.3	1 755.8	1 837.0	2 175.7	2 027.0
Children's Bonus Bonds	KGVQ	–	–	–	0.1	1.0	1.2	4.3	0.1	257.8	187.9	183.2
First Option Bonds	KIAS	–	–	–	–	4.6	1 085.1	969.9	732.7	833.9	1 283.0	1 055.5
Pensioners Guaranteed Income Bond	KPOB	–	–	–	–	–	–	57.0	104.5	185.0	318.8	897.8
Treasurer's account	KWNG	..	..	..	..	..	..	..	..	1.2	11.8	13.7
Certificates of tax deposit	KQHF	1 526.3	797.1	1 192.3	1 116.5	936.9	722.6	612.1	466.0	478.9	229.0	199.9
Tax reserve certificates	KQHG	–	–	0.1	0.1	–	–	–	–	–	–	–
Nationalised industries', etc temporary deposits	KQHH	12 770.0	29 314.0	25 833.0	21 186.0	18 477.0	28 098.5	28 557.8	35 263.1	51 979.3	46 835.7	41 776.9
Debt to the Bank of England	KPOC	–	–	–	–	–	–	11.0	–	–	–	–
British Gas Corporation (Repayment of Deposits)	KQHI	–	–	–	–	–	–	–	–	–	–	–
Sterling Treasury bills (net repayment)	KQHJ	–	–	140.3	1 191.0	4 882.1	1 547.5	–	–	4 009.6	1 928.5	–
ECU Treasury bills (net repayment)	KJEG	..	..	..	..	..	73.3	–	–	–	–	–
ECU Treasury notes (net repayment)	KSPA	..	..	..	..	..	..	..	439.1	318.3	3.3	13.2
Ways and means (net repayment)	KQHK	–	–	–	–	6 786.3	–	–	–	–	5 815.4	–
Other debt: payable in sterling :												
Interest free notes	KQHL	330.0	170.8	169.6	268.3	155.8	101.3	225.2	301.3	87.6	1 215.5	850.5
Other	KQHM	0.1	0.1	0.1	0.1	–	–	–	–	–	–	–
Other debt : payable in external currencies	KQHN	1 567.6	229.3	379.7	132.5	27 068.7	12 870.4	514.5	103.6	2 661.7	2 082.7	92.0
Total payments	KQHO	30 429.5	54 177.0	39 901.9	36 383.7	71 226.0	57 584.5	46 049.4	48 708.6	83 597.4	90 593.3	72 545.0
Net borrowing	KQHP	–	–	3 684.7	12 112.0	25 907.6	58 939.5	39 848.1	37 688.8	27 547.9	–	–
Net repayment	KHDD	1 846.8	7 923.3	–	–	–	–	–	–	–	2 482.1	3 558.2

Source: HM Treasury: 020 7270 1586

17.7 Consolidated Fund and National Loans Fund: assets and liabilities

United Kingdom

At 31 March in each year

£ millions

		1989	1990	1991	1992	1993	1994	1995	1996	1997	1998
Consolidated Fund											
Total estimated assets	KQIA	8 654.9	6 820.0	14 951.0	19 852.9	18 872.0	26 914.5	33 992.4	33 809.0	36 177.4	35 490.6
Subscriptions and contributions to international financial organisations	KQIB	2 619.8	2 866.1	4 329.6	4 315.2	5 009.2	5 438.2	5 898.1	6 470.8	6 528.9	6 660.8
International Bank for Reconstruction and Development	KQIC	202.8	215.7	224.8	236.3	259.8	274.5	265.2	271.6	265.0	262.5
International Finance Corporation	KQID	37.0	41.6	39.3	39.4	53.1	60.9	62.0	73.1	74.9	72.9
International Development Association	KQIE	1 811.9	1 979.5	2 127.3	2 343.3	2 576.0	2 785.4	3 005.5	3 205.8	3 372.7	3 562.2
African Development Bank	KQIF	59.0	65.9	80.8	91.2	108.1	127.9	141.2	148.3	162.9	180.0
Asian Development Bank	KQIG	123.2	142.3	155.8	173.0	180.9	188.9	201.5	214.6	240.4	272.5
Caribbean Development Bank	KQIH	19.5	20.9	24.6	27.5	29.4	31.2	32.6	34.2	34.1	34.6
European Investment Bank	KQII	249.4	277.7	1 550.3	1 252.8	1 588.9	1 713.7	1 876.4	2 166.6	2 036.5	1 840.7
European Bank for Reconstruction and Development	KPOD	..	..	..	22.3	61.3	99.4	146.3	189.4	175.6	164.2
Inter-American Development Bank	KQIJ	85.2	87.5	89.9	89.4	109.8	112.5	107.2	119.6	117.5	219.3
International Fund for Agricultural Development	KQIK	28.7	31.8	34.7	36.9	38.5	40.2	42.7	44.2	46.1	48.8
Multilateral Investment Guarantee Agency	KQIL	3.1	3.2	3.0	3.0	3.5	3.5	3.2	3.4	3.2	3.1
Amounts due from overseas governments	KQIM	3.9	1.3	–	–	–	–	–	–	–	–
War of 1939-45	KQIN	–	–	–	–	–	–	–	–	–	–
Other	KQIO	3.9	1.3	–	–	–	–	–	–	–	–
Loans from Votes	KQIP	3 850.4	2 257.1	2 420.3	4 047.3	6 709.2	11 057.4	13 599.7	12 967.0	13 684.7	13 647.6
Issues of public dividend capital:	KQIQ	1 641.7	1 629.0	1 617.9	1 642.7	4 920.0	7 979.2	11 467.0	12 161.1	12 424.8	13 102.3
British Airways Board	KQIR	–	–	–	–	–	–	–	–	–	–
British Steel Corporation	KQIS	–	–	–	–	–	–	–	–	–	–
Royal Ordnance Factories	KQIT	–	–	–	–	–	–	–	–	–	–
National Enterprise Board	KQIU	–	–	–	–	–	–	–	–	–	–
Royal Mint	KQIV	7.0	7.0	7.0	7.0	7.0	7.0	7.0	7.0	7.0	7.0
Post Office	KQIW	22.0	22.0	–	–	–	–	–	–	–	–
Scottish Development Agency	KQIX	21.5	7.9	–	–	–	–	–	–	–	–
Welsh Development Agency	KQIY	18.6	15.5	15.5	13.5	12.6	11.6	10.9	9.8	8.9	8.8
British Aerospace	KQIZ	–	–	–	–	–	–	–	–	–	–
British Shipbuilders	KQJA	1 572.6	1 576.6	1 595.4	1 597.4	1 598.3	1 598.3	1 598.3	1 598.3	1 598.5	1 598.5
Patent Office	KIAT	–	–	–	6.3	6.3	6.3	6.3	6.3	6.3	6.3
NHS Trusts	KIAU	–	–	–	1 325.8	3 279.6	6 336.2	9 603.1	10 173.8	10 349.7	11 023.0
Companies House	KIAV	–	–	–	16.9	15.9	15.9	15.9	15.9	15.9	15.9
Central Office of Information	KIAW	–	–	–	0.3	0.3	0.3	0.3	0.3	0.3	0.3
Chessington Computer Centre	KPOE	–	–	–	–	–	3.5	3.5	3.5	–	–
Buying Agency	KWNH	..	..	..	..	..	..	0.1	0.4	0.4	0.4
Defence Evaluation and Research Agency	KWNI	..	..	..	..	..	..	128.8	253.0	274.5	274.5
Fire Service College	KWNJ	..	..	..	..	..	..	16.7	16.7	16.7	16.7
Land Registry	KWNK	..	..	..	..	..	..	55.4	55.4	61.5	61.5
Medicines Control Agency	KWNL	..	..	..	..	..	..	1.6	1.6	1.6	1.6
Meteorological Office	KZAZ	..	..	..	..	..	..	..	..	58.9	58.9
Registers of Scotland	KZBA	..	..	..	..	..	..	..	..	4.3	4.3
Vehicle Inspectorate	KWNM	..	..	..	..	..	..	19.1	19.1	20.3	20.3
Driving Standards Agency	LQMI	..	..	..	..	..	..	..	..	..	3.5
Queen Elizabeth II Conference Centre	LQMJ	..	..	..	..	..	..	..	..	..	0.8
Contingencies Fund - capital	KQJB	307.0	0.6	942.0	1 097.0	947.0	347.0	447.0	297.0	977.0	533.8
Balance on revenue accounts	KQJC	232.1	66.3	237.2	725.2	477.6	358.7	1 433.6	1 096.1	954.9	1 546.0
Privatisation receipts	KIAX	–	–	5 403.9	6 701.0	809.0	1 734.0	1 147.0	817.0	1 607.1	–
Total liabilities	KQJD	118 005.6	123 397.0	128 386.2	–	–	–	–	333 927.9	364 803.0	364 950.8
Liability to balance National Loans Fund	KQJE	115 545.2	121 170.9	125 807.1	142 306.7	182 102.2	240 351.0	286 055.9	331 164.9	362 506.5	362 582.5
Payment from Votes:	KQJF	69.0	68.3	67.6	66.9	66.1	65.2	64.3	63.4	62.4	61.3
Married quarters for Armed Forces	KQJG	69.0	68.3	67.6	66.9	66.1	65.2	64.3	63.4	62.4	61.3
Liability to Post Office Superannuation Fund	KQJH	–	–	–	–	–	–	–	–	–	–
Post-war credits outstanding and interest due - estimated	KQJI	46.3	46.2	46.1	46.1	46.0	46.0	46.0	45.9	45.9	45.9
Revenue paid over in advance of collection	KQJJ	10.3	0.5	–	8.3	–	83.7	37.0	–	28.2	13.8
Inland Revenue	KQJK	–	–	–	–	–	–	–	–	–	–
Customs and Excise	KQJL	–	0.5	–	–	–	–	–	–	28.2	–
Broadcast receiving licences	KQJM	–	–	–	–	–	–	–	–	–	–
Vehicle Excise Duty	KQJN	10.3	–	–	8.3	–	83.7	37.0	–	–	13.8
National insurance surcharge - Great Britain	KQJO	–	–	–	–	–	–	–	–	–	–
National insurance surcharge - Northern Ireland	KQJP	–	–	–	–	–	–	–	–	–	–

Source: HM Treasury: 020 7270 1586

17.7 Consolidated Fund and National Loans Fund: assets and liabilities
United Kingdom

continued

At 31 March in each year

£ millions

		1989	1990	1991	1992	1993	1994	1995	1996	1997	1998
Consolidated Fund (*continued*)											
Promissory notes issued by Minister of Overseas Development	KQJQ	2 334.8	2 111.1	2 465.1	770.3	979.5	1 002.6	996.7	1 005.8	1 021.9	822.1
International Development Association	KQJR	642.7	650.0	705.3	489.4	673.1	670.5	656.5	663.2	673.6	484.0
African Development Fund	KQJS	57.4	71.2	77.1	92.1	100.7	106.6	93.2	86.2	95.3	95.2
Asian Development Bank	KQJT	2.8	2.1	1.4	0.1	–	–	–	–	–	–
Asian Development Fund	KQJU	109.8	119.9	108.7	92.1	108.7	131.8	140.5	127.6	136.0	120.5
Caribbean Development Bank	KQJV	–	–	–	0.1	0.1	1.2	1.3	1.4	1.3	1.3
Special Development Fund	KQKC	6.2	9.5	8.1	8.3	6.9	8.0	9.5	10.6	13.2	15.8
European Community/International Development Association Special Action Account	KQJW	–	–	–	–	–	–	–	–	–	–
European Investment Bank	KQJX	96.0	86.6	61.2	–	–	–	–	–	–	–
Inter-American Development Bank	KQJY	2.8	1.6	1.7	1.7	2.6	2.2	1.3	2.0	1.8	1.9
Fund for special operations	KQJZ	24.2	23.4	22.5	22.5	15.4	16.4	13.1	11.3	8.8	6.3
International Fund for Agricultural Development	KQKA	18.5	15.3	16.2	17.8	16.1	18.2	15.8	14.2	12.3	14.2
International Bank for Reconstruction and Development	KQKB	58.9	48.2	34.2	31.9	30.2	24.2	39.2	60.7	71.7	81.9
European Bank for Reconstruction and Development	KIAY	–	–	–	14.3	20.4	19.9	20.9	21.0	6.1	–
United Nations Environment Programme	KJEH	–	–	–	–	5.3	3.6	5.4	7.6	1.8	1.0
Other contributions and instalments due in respect of international subscriptions, etc	KQYX	1 315.5	1 083.3	1 429.0	2 125.1	1 520.7	2 165.8	1 499.6	1 647.9	1 138.1	1 425.2
National Loans Fund											
Total assets	KQKD	197 319.8	192 981.4	198 703.4	214 527.6	248 643.5	306 872.6	349 159.5	390 681.8	419 548.9	418 431.0
Total National Loans Fund loans outstanding	KQKE	60 915.3	56 154.1	56 777.9	55 492.8	48 705.9	48 470.8	47 496.1	46 600.9	46 746.8	46 742.6
Loans to nationalised industries:											
Post Office	KQKF	–	–	–	–	–	–	–	–	–	–
British Coal	KQKG	1 908.6	1 145.9	387.4	257.7	180.0	–	–	–	–	–
Electricity Council	KQKH	1 335.4	–	–	–	–	–	–	–	–	–
North of Scotland Hydro-Electric Board	KQKI	416.7	–	–	–	–	–	–	–	–	–
Scottish Hydro-Electric	KQKJ	–	527.8	534.9	–	–	–	–	–	–	–
South of Scotland Electricity Board	KQKK	1 189.1	–	–	–	–	–	–	–	–	–
Scottish Power	KQKL	–	548.8	508.6	–	–	–	–	–	–	–
Scottish Nuclear Ltd	KQKM	–	205.0	203.2	201.3	199.2	196.8	194.1	96.1	–	–
British Gas Corporation	KQKN	–	–	–	–	–	–	–	–	–	–
British Steel Corporation	KQKO	–	–	–	–	–	–	–	–	–	–
British Airways Board	KQKP	–	–	–	–	–	–	–	–	–	–
Railtrack	KTCR	..	..	..	..	..	..	1 287.7	–	–	–
European Passenger Services	KTCS	..	..	..	..	..	..	99.0	761.2	–	–
Civil Aviation Authority	KQKQ	93.2	146.9	216.3	271.2	311.9	403.7	453.3	476.1	447.5	420.9
British Airports Authority	KQKR	–	–	–	–	–	–	–	–	–	–
British Railways Board	KQKS	233.0	202.8	938.5	1 040.8	1 923.9	2 445.6	749.2	718.3	601.2	573.7
British Transport Docks Board	KQKT	–	–	–	–	–	–	–	–	–	–
British Waterways Board	KQKU	24.2	22.8	21.6	20.0	19.2	18.6	18.3	18.2	18.2	18.2
National Freight Corporation	KQKV	–	–	–	–	–	–	–	–	–	–
National Bus Company	KQKW	–	–	–	–	–	–	–	–	–	–
Scottish Transport Group	KQKX	–	–	–	–	–	–	–	–	–	–
British National Oil Corporation	KQKY	–	–	–	–	–	–	–	–	–	–
British Aerospace	KQKZ	–	–	–	–	–	–	–	–	–	–
British Shipbuilders	KQLA	–	–	–	–	–	–	–	–	–	–
British Telecommunications	KQLB	–	–	–	–	–	–	–	–	–	–
Regional Water Authorities	KQLC	4 526.1	–	–	–	–	–	–	–	–	–
Loans to other public corporations:											
New Towns - Development Corporations and Commission	KQLD	2 731.6	2 481.0	1 714.6	1 441.5	1 303.9	1 130.1	1 008.8	314.3	122.2	36.2
Scottish Special Housing Association	KQLE	596.2	–	–	–	–	–	–	–	–	–
Scottish Homes	KQLF	–	796.1	786.4	654.5	401.2	399.4	397.3	395.0	392.5	259.8
Housing Corporation (Scotland)	KQLG	1 967.8	–	–	–	–	–	–	–	–	–
Housing Corporation (England)	KQLH	–	1 744.1	1 611.3	1 021.4	986.0	899.0	869.9	926.3	848.7	4.0
Housing for Wales	KQLI	–	139.9	125.8	96.6	86.7	86.4	69.2	59.0	–	–
Covent Garden Market Authority	KQLJ	1.0	0.4	–	–	–	–	–	–	–	–
National Enterprise Board	KQLK	–	–	–	–	–	–	–	–	–	–
Land Authority for Wales	KQLL	6.1	7.6	6.8	8.8	10.3	6.2	3.2	3.2	1.3	1.3
Scottish Enterprise	KQLM	6.3	5.7	5.4	4.6	2.7	1.9	1.3	0.6	0.5	0.1
Welsh Development Agency	KQLN	2.5	2.8	2.4	1.8	1.5	1.2	0.9	1.1	1.2	1.2
Land Registry Trading Fund	KPOF	..	–	–	–	–	54.9	–	–	–	–
Development Board for Rural Wales	KQLO	8.8	8.8	8.8	8.8	8.8	8.8	8.8	7.9	4.1	4.0
Royal Mint	KQLP	0.3	0.1	–	–	–	–	–	–	–	–
Royal Ordnance Factories	KQLQ	–	–	–	–	–	–	–	–	–	–
The Crown Suppliers	KQLR	23.7	23.7	–	–	–	–	–	–	–	–
Crown Agents	KQLS	2.9	2.8	2.5	2.5	2.3	2.1	2.0	1.9	–	–
Her Majesty's Stationery Office	KQLT	27.7	22.6	16.8	10.6	6.7	3.3	–	7.0	–	–
Urban Development Corporations	KQLU	0.9	0.8	0.4	0.8	0.7	1.0	–	–	–	–
Harbour Authorities	KQLV	52.4	42.3	36.3	33.4	7.1	1.5	0.8	0.7	0.6	0.5
Commonwealth Development Corporation	KQLW	–	–	–	–	–	–	–	–	–	–
UK Atomic Energy Authority	KQLX	68.0	64.0	106.7	111.0	149.8	125.0	147.3	141.0	–	–

Source: HM Treasury: 020 7270 1586

17.7 Consolidated Fund and National Loans Fund: assets and liabilities
United Kingdom

continued

At 31 March in each year

£ millions

		1989	1990	1991	1992	1993	1994	1995	1996	1997	1998
National Loans Fund (*continued*)											
Loans to other public corporations: (*continued*)											
Central Office of Information	KJEI	–	–	–	1.3	1.0	0.8	0.5	0.3	–	–
Registers of Scotland	KZBB	..	..	..	..	..	..	..	..	6.4	5.6
East of Scotland Water Authority	KZBC	..	..	..	..	..	..	..	..	163.0	229.0
North of Scotland Water Authority	KZBD	..	..	..	..	..	..	..	..	155.0	189.2
West of Scotland Water Authority	KZBE	..	..	..	..	..	..	..	..	185.0	304.9
Loans to local authorities	KQLY	44 333.0	46 567.0	48 087.1	48 742.3	41 464.6	40 979.2	40 440.7	40 969.2	42 134.0	42 951.1
Loans to private sector:											
Shipbuilding Industry Board	KQLZ	–	–	–	–	–	–	–	–	–	–
Shipowners (Ship credit scheme)	KGVR	–	–	–	–	–	–	–	–	–	–
Housing associations	KGVS	16.5	15.8	14.9	14.6	14.1	13.8	13.2	12.7	0.5	0.5
Building societies	KGVT	–	–	–	–	–	–	–	–	–	–
British Nuclear Fuels Ltd	KGVU	2.4	1.3	–	–	–	–	–	–	–	–
British Aerospace plc	KGVV	3.3	1.1	–	–	–	–	–	–	–	–
Loans within central government:											
Northern Ireland Exchequer	KGVW	1 268.6	1 357.6	1 373.1	1 480.3	1 558.3	1 626.2	1 666.3	1 627.4	1 602.5	1 681.1
Married quarters for Armed Forces	KGVX	69.0	68.3	67.6	66.9	66.1	65.2	64.3	63.4	62.4	61.3
Redundancy Fund	KGVY	–	–	–	–	–	–	–	–	–	–
Other assets:											
Exchange Equalisation Account - capital	KGVZ	16 200.0	10 025.0	11 075.0	10 774.5	7 200.0	6 800.0	5 550.0	2 800.0	650.0	–
Subscriptions and contributions to international financial organisations:											
International Monetary Fund	KGXE	4 396.8	4 904.6	4 785.5	4 885.3	6 893.0	7 066.4	7 172.2	7 102.6	6 241.2	5 895.6
Gilt-Edged Official Operations Account											
-advances outstanding	KPUF	..	–	–	–	–	2 500.0	2 500.0	2 500.0	3 000.0	2 500.0
-surplus not paid to the National Loans Fund	KPUH	..	–	–	–	–	2.0	–	–	–	141.6
Borrowing included in national debt but not brought to account by 31 March	KGXF	262.6	726.7	257.8	1 068.2	3 742.5	1 682.4	385.2	513.4	404.7	568.6
National Debt Commissioners' liability in respect of the National Savings Bank's Investment Fund	KCYG	–	–	–	–	–	–	–	–	–	–
Other	KCYH	–	–	–	–	–	–	–	–	–	–
Consolidated Fund liability	KCYI	115 545.1	121 170.9	125 807.1	142 306.7	182 102.2	240 351.0	286 055.9	331 164.9	362 506.2	362 582.7
Total liabilities											
National Loans Fund - national debt outstanding	KCYJ	197 319.8	192 981.4	198 703.4	214 527.6	248 643.5	306 872.6	349 159.5	390 681.8	419 548.9	418 431.0

Source: HM Treasury: 020 7270 1586

17.8 British government and government guaranteed marketable securities[1]
Nominal values of official and other holdings by maturity[2,3,4]

At 31 March in each year

£ millions

		1989	1990	1991	1992	1993	1994	1995	1996	1997	1998	1999
Total holdings	KQMO	139 464	125 018	122 437	133 121	162 642	209 507	232 486	262 262	290 259	297 366	291 788
Up to 5 years	KQMP	41 242	36 523	34 933	40 363	46 019	58 437	69 011	81 122	90 357	86 094	95 112
Over 5 and up to 15 years	KQMQ	61 577	53 395	56 309	61 239	72 612	94 308	101 960	111 510	125 401	131 758	124 603
Over 15 years (including undated)	KQMR	36 645	35 100	31 194	31 520	44 010	56 762	61 515	69 630	74 501	79 515	72 074
Official holdings:[4]												
Total	THAA	15 478†	15 085	13 633	8 799	4 138	6 239	6 614	7 186	6 858	6 345	6 394
Up to 5 years	THAB	6 844†	6 958	5 811	3 478	1 434	1 685	2 007	2 345	2 850	2 499	2 600
Over 5 and up to 15 years	THAC	7 154†	6 806	6 527	4 323	1 826	3 194	3 700	3 774	3 041	2 726	2 989
Over 15 years (including undated)	THAD	1 480†	1 321	1 295	998	879	1 359	907	1 068	967	1 120	805
Market holdings:												
Total	THAE	123 986†	109 933	108 804	124 323	158 503	203 268	225 872	255 075	283 402	291 021	285 394
Up to 5 years	THAF	34 398†	29 565	29 122	36 885	44 585	56 751	67 004	78 777	87 508	83 595	92 512
Over 5 and up tp 15 years	THAG	54 423†	46 589	49 782	56 916	70 786	91 115	98 260	107 736	122 360	129 032	121 614
Over 15 years (including undated)	THAH	35 165†	33 779	29 899	30 522	43 131	55 403	60 608	68 562	73 536	78 395	71 269

1 The government guaranteed securities of nationalised industries only. A relatively small amount of other government guaranteed securities is excluded.

2 Securities with optional redemption dates are classified according to the final redemption date.

3 The nominal value of index-linked BGS has been raised by the amount of accrued capital uplift.

4 Official holdings have changed following the introduction of the central bank sector into the UK national accounts. These holdings now principally include those of the National Investment and Loans Office and other government departments. The Issue and Banking Departments of the Bank of England are classified within the central bank sector, and so as part of the market.

Source: Bank of England: 020 7601 4068

17.9 National savings

Year ended 31 March

£ millions

		1988 /89	1989 /90	1990 /91	1991 /92	1992 /93	1993 /94	1994 /95	1995 /96	1996 /97	1997 /98	1998 /99
Department for National Savings												
Receipts												
Total[1]	KQNA	6 998.5	5 562.9	6 408.9	7 704.7	8 485.5	10 336.2	9 246.9	11 596.9	13 353.5	12 565.6	10 212.9
Capital bonds	KQNB	199.5	284.0	185.6	338.3	476.6	566.4	511.2	500.5	482.7	583.6	238.1
National savings certificates:												
Fixed interest	KQNC	1 383.1	632.6	696.7	2 119.6	1 894.7	1 929.1	1 083.6	1 519.8	1 805.9	1 753.4	1 174.4
Index-linked	KQND	785.2	634.6	1 503.2	1 500.8	1 600.0	1 434.9	1 083.2	1 949.8	1 871.8	2 640.4	1 734.1
Yearly plan	KQNE	119.3	90.1	74.2	89.3	115.7	118.2	117.5	97.0	0.7	–	–
Save as you earn:												
Fixed interest (1st, 2nd and 4th issues)	KQNF	63.3	76.2	83.2	94.1	85.5	76.9	66.7	49.9	33.4	20.3	11.2
Index-linked (3rd issue)	KQNG	2.9	0.2	–	–	–	–	–	–	–	–	–
Income bonds	KQNH	1 729.7	1 512.1	1 605.6	1 355.9	1 241.2	1 948.9	1 228.9	802.4	1 263.6	1 044.1	1 365.4
First Option bonds	KHYZ	–	–	–	–	885.0	247.0	95.6	194.3	398.3	149.2	110.1
Investment account[2]	KQNI	1 629.6	1 491.5	1 515.7	1 247.7	1 107.4	1 392.9	1 393.1	1 299.2	1 478.3	1 270.1	1 045.4
Premium savings bonds	KQNJ	313.5	249.4	185.7	257.5	384.5	998.2	1 826.2	2 075.7	2 578.1	3 199.6	3 622.9
Ordinary account	KQNK	637.6	592.2	559.0	566.0	604.4	636.1	635.6	627.3	643.1	653.0	622.3
Children's bonus bonds	KHCU	–	–	–	135.5	90.5	92.6	116.8	144.9	124.3	109.7	73.4
Pensioners' guaranteed income bonds	KJRB	..	..	..	..	..	895.0	1 088.5	2 336.1	2 651.4	1 104.2	199.1
Deposit bonds	KQNL	134.8	–	–	–	–	–	–	–	–	–	–
Indexed-income bonds	KQNM	–	–	–	–	–	–	–	–	–	–	–
British savings bonds	KQNN	–	–	–	–	–	–	–	–	–	–	–
Treasurer's account	KXCY	–	–	–	–	–	–	–	–	21.9	38.0	16.5
Repayments												
Total[1]	KQNO	8 488.0	9 432.0	7 339.8	6 495.7	5 853.8	8 051.0	7 759.8	8 756.9	11 048.5	13 201.7	11 665.7
Capital bonds	KQNP	–	5.0	23.6	14.9	21.8	345.3	678.3	453.1	776.6	1 143.5	649.7
National savings certificates:												
Fixed interest - Principal	KQNQ	2 117.3	2 498.2	1 455.3	1 193.4	647.9	1 292.6	843.0	840.1	1 838.8	2 256.2	1 934.3
- Accrued interest	KQNR	1 578.5	1 642.6	991.0	738.7	417.4	670.9	437.2	474.8	937.0	1 021.4	645.6
Index-linked - Principal	KQNS	409.4	394.4	412.9	418.8	448.2	913.5	761.7	1 419.7	1 432.7	1 803.5	1 520.1
- Accrued interest/ index-linking, bonuses and supplements	KQNT	329.0	370.6	403.3	396.6	378.5	598.1	425.8	672.0	613.6	703.1	527.5
Yearly plan:												
Principal	KQNU	16.6	82.4	124.2	106.5	98.4	89.5	77.0	102.0	96.3	119.5	111.8
Interest	KQNV	1.3	23.6	43.2	40.9	35.7	30.3	26.2	42.1	39.2	43.6	30.6
Save as you earn:												
Fixed interest - Principal	KQNW	38.3	40.9	59.3	74.4	88.7	86.1	85.0	96.7	69.1	67.4	35.9
- Accrued interest	KQNX	10.8	9.7	13.3	16.0	18.2	18.1	19.2	21.9	17.9	18.5	9.1
Index-linked - Principal	KQNY	51.3	45.0	24.2	29.4	26.5	9.3	7.4	1.8	0.9	0.8	0.6
- Accrued interest/ index-linking	KQNZ	28.5	33.7	27.1	38.5	38.9	14.3	12.2	3.3	1.6	1.5	1.2
Income bonds	KQOA	943.0	1 091.2	824.2	719.9	956.0	1 000.3	1 180.1	1 256.2	1 392.0	1 141.9	880.7
First Option bonds	KHYY	–	–	–	–	4.8	375.8	288.4	126.5	139.1	335.2	205.5
Investment account[2]	KQOB	1 957.3	2 074.3	1 904.0	1 803.9	1 780.0	1 677.1	1 777.2	1 797.2	1 856.0	2 177.1	1 981.6
Premium savings bonds	KQOC	133.5	186.4	193.4	153.7	153.4	181.0	333.0	590.4	869.0	1 203.1	1 398.4
Ordinary account	KQOD	727.8	733.2	700.9	663.7	658.1	674.7	676.8	676.5	692.3	703.4	684.6
Children's bonus bonds	KHCV	–	–	–	0.1	0.9	1.2	1.8	2.6	30.8	39.8	52.4
Pensioners' guaranteed income bonds	KJRE	..	..	..	..	..	..	56.7	104.8	172.5	331.4	903.1
Deposit bonds	KQOE	142.4	197.6	137.7	85.3	79.8	72.4	72.2	73.5	65.5	76.6	79.0
Indexed-income bonds	KQOF	3.0	3.2	2.2	1.0	0.6	0.5	0.6	1.7	6.5	2.6	–
British savings bonds:												
Before maturity	KQOG	–	–	–	–	–	–	–	–	–	–	–
On maturity	KQOH	–	–	–	–	–	–	–	–	–	–	–
Treasurer's account	KXCZ	–	–	–	–	–	–	–	–	1.1	11.6	14.0

1 Excludes Ulster Savings Certificates.
2 Investment Account contributes to the funding of PSBR/CGBR from January 1981.
3 Including National Savings stamps and gift tokens and securities on the National Savings register, which (except for British savings bonds) do not appear in the first part of the table.
4 Nominal value held in National Savings section.
5 Including accrued interest to date.
6 From September 1979 excludes £7.7 million deemed to be outstanding National Savings stamps written off on 15 August 1979.
7 From 1 April 1986 onwards the data series have been revised. The revisions, notably those for the estimates of accrued interest, coincided with the adoption of an improved transaction accounting system for National Savings.
8 Stock on the National Savings Register transferred to the Bank of England in July 1998.

Source: Department for National Savings: 020 7605 9314

17.9 National savings

Year ended 31 March

continued

£ millions

		1988 /89	1989 /90	1990 /91	1991 /92	1992 /93	1993 /94	1994 /95	1995 /96	1996 /97	1997 /98	1998 /99
Department for National Savings												
Interest accruing												
Capital bonds	KQOO	–	12.2	42.0	76.8	130.8	223.1	218.8	214.6	249.9	172.0	146.9
National savings certificates:												
Fixed interest	KQOP	810.0	675.0	525.7	443.3	521.2	604.2	622.2	703.7	787.6	557.8	485.3
Index-linking (including bonuses and supplements)	KQOQ	365.8	422.0	477.0	351.7	267.8	303.9	458.5	655.2	671.9	690.3	502.7
Yearly plan	KQOR	29.6	42.8	40.9	36.7	32.5	30.5	34.1	37.5	38.0	28.3	17.2
Save as you earn:												
Fixed interest (1st, 2nd and 4th issues)	KQOS	11.5	13.6	14.4	18.3	18.8	18.3	18.0	18.3	14.6	11.7	5.7
Index-linked (3rd issue)	KQOT	33.7	26.3	21.1	8.8	1.1	0.3	0.9	0.4	0.5	0.5	0.5
First Option bonds	KIAJ	–	–	–	–	–	70.7	40.4	31.8	40.4	54.8	48.4
Investment account[2]	KQOU	700.8	831.2	1 007.8	864.4	657.7	541.8	529.1	518.5	509.0	497.8	464.5
Ordinary account	KQOV	65.2	64.0	60.9	58.2	57.3	44.6	42.8	35.5	36.2	29.4	30.6
Childrens bonus bonds	KHCW	–	–	–	–	9.3	14.4	19.2	27.0	98.9	73.7	66.6
Deposit bonds	KQOW	87.0	97.4	95.4	77.9	59.8	46.4	41.4	39.4	38.3	37.2	31.5
Treasurer's account	KXDA	–	–	–	–	–	–	–	–	–	–	–
Total[1]	KQOX	2 103.6	2 184.5	2 285.2	1 936.1	1 756.3	1 989.2	2 025.4	2 281.9	2 485.3	2 153.5	1 799.9
Amounts remaining invested												
Capital bonds[5]	KQOY	199.5	490.7	694.7	1 094.9	1 680.5	2 124.7	2 176.4	2 438.4	2 394.4	2 006.5	1 741.8
National savings certificates												
Fixed interest - Principal	KQOZ	7 744.8	5 879.2	5 120.6	6 046.8	7 293.6	7 930.1	8 170.7	8 850.4	8 817.5	8 314.7	7 554.8
- Accrued interest	KQPA	3 626.3	2 658.7	2 193.4	1 898.0	2 001.8	1 935.1	2 120.1	2 349.0	2 199.6	1 736.0	1 575.7
Index-linked - Principal	KQPB	2 683.0	2 923.2	4 013.5	5 095.5	6 247.3	6 768.7	7 090.2	7 620.3	8 059.4	8 896.3	9 110.3
- Bonuses and supplements/ index-linking	KQPC	1 375.0	1 426.4	1 500.1	1 455.2	1 344.5	1 050.3	1 083.0	1 066.2	1 124.5	1 111.7	1 086.9
Yearly plan:												
Principal	KQPD	483.4	491.1	441.1	423.9	441.2	469.9	510.4	505.4	409.8	290.3	178.5
Interest	KQPE	56.2	75.4	73.1	68.9	65.7	65.9	73.8	69.2	68.0	52.7	39.3
Save As You Earn:												
Fixed interest - Principal	KQPF	149.6	184.9	208.8	228.5	225.3	216.1	197.8	151.0	115.3	68.2	43.5
- Accrued interest	KQPG	17.5	21.4	22.5	24.8	25.4	25.6	24.4	20.8	17.5	10.7	7.3
Index-linked - Principal	KQPH	161.0	116.2	92.0	62.6	36.1	26.8	19.4	17.6	16.7	15.9	15.3
- Accrued interest/ index-linking	KQPI	121.4	114.0	108.0	78.3	40.5	26.5	15.2	12.3	11.2	10.2	9.5
Stock on the National savings register (DNS only)[4]	KQPJ	846.4	868.4	963.1	1 026.2	1 180.3	1 198.6	1 392.7	1 530.7	1 541.1	1 468.5	–
Income bonds	KQPK	7 760.9	8 181.8	8 963.2	9 599.2	9 884.4	10 833.0	10 881.8	10 428.0	10 299.6	10 205.2	10 689.9
First Option bonds	KIAK	–	–	–	–	880.2	822.1	669.7	769.3	1 068.9	937.7	890.7
Investment account[5]	KQPL	7 758.6	8 007.0	8 626.5	8 934.7	8 919.8	9 177.4	9 322.4	9 342.9	9 474.2	9 065.0	8 593.3
Premium savings bonds	KQPM	2 271.0	2 334.0	2 326.3	2 430.1	2 661.2	3 478.4	4 971.6	6 456.9	8 166.0	10 162.5	12 387.0
Ordinary account[5]	KQPN	1 632.3	1 555.3	1 474.3	1 434.8	1 438.4	1 444.4	1 446.0	1 432.3	1 419.3	1 398.3	1 366.6
Children's bonus bonds[5]	KHCX	–	–	–	135.4	234.3	340.1	474.3	643.6	836.0	979.6	1 067.2
Pensioners' guaranteed income bonds	KJRG	–	–	–	–	–	895.0	1 926.8	4 158.1	6 637.0	7 409.8	6 705.8
Deposit bonds[5]	KQPO	879.9	779.7	737.4	730.0	710.0	684.0	653.2	619.1	591.9	552.5	505.0
Indexed-income bonds	KQPP	22.3	19.1	16.9	15.9	15.3	14.8	14.2	12.5	6.0	–	–
British saving bonds	KQPQ	–	–	–	–	–	–	–	–	–	–	–
National savings stamps and gift tokens[6]	KQPR	2.4	2.1	1.9	1.8	1.8	1.7	1.7	1.6	1.6	1.6	1.6
Treasurer's account	KXDB	–	–	–	–	–	–	–	–	20.8	47.2	49.7
Total administered by DNS[3]	KQPS	37 791.5	36 128.6	37 577.4	40 785.5	45 327.6	49 529.2	53 235.8	58 495.6	63 296.3	64 741.1	63 619.7

See footnotes on the first part of this table.

Source: Department for National Savings: 020 7605 9314

17.10 Income tax: allowances and reliefs
United Kingdom

£

		1989/90	1990/91	1991/92	1992/93	1993/94	1994/95	1995/96	1996/97	1997/98	1998/99	1999/00
Personal allowances												
Single person's allowance	KSOD	2 785	..	..	..	..	..	..	..	..	..	..
Married man's allowance[1]	KDZQ	4 375	..	..	..	..	..	..	..	..	..	..
Wife's earned income allowance[2]	KEBW	2 785	..	..	..	..	..	..	..	..	..	..
Age allowance:[3]												
Married (either partner between 65-74 but neither partner over 75)	KEBX	5 385	..	..	..	..	..	..	..	..	..	..
Married (either partner between 75-79 but neither partner 80 or over)	KSOE	5 565	..	..	..	..	..	..	..	..	..	..
Married (either partner 80 or over)	KEIX	5 565	..	..	..	..	..	..	..	..	..	..
Single (between 65 and 74)	KEJI	3 400	..	..	..	..	..	..	..	..	..	..
Single (between 75 and 79)	KEKI	3 540	..	..	..	..	..	..	..	..	..	..
Single (80 or over)	KSOF	3 540	..	..	..	..	..	..	..	..	..	..
Income limit	KSOG	11 400	..	..	..	..	..	..	..	..	..	..
Marginal fraction		1/2										
Personal allowance	KDZP	..	3 005	3 295	3 445	3 445	3 445	3 525	3 765	4 045	4 195	4 335
Married couples allowance[4]	KDZR	..	1 720	1 720	1 720	1 720	1 720	1 720	1 790	1 830	1 900	1 970
Age allowance:[3]												
Personal (aged 65-74)	KSOH	..	3 670	4 020	4 200	4 200	4 200	4 630	4 910	5 220	5 410	5 720
Personal (aged 75 or over)	KSOI	..	3 820	4 180	4 370	4 370	4 370	4 800	5 090	5 400	5 600	5 980
Married couple's (either partner between 65-74 but neither partner 75 or over)	KEDI	..	2 145	2 355	2 465	2 465	2 665	2 995	3 115	3 185	3 305	5 125
Married couples (either partner 75 or over)	KEIY	..	2 185	2 395	2 505	2 505	2 705	3 035	3 155	3 225	3 345	5 195
Income limit	KEOO	..	12 300	13 500	14 200	14 200	14 200	14 600	15 200	15 600	16 200	16 800
Marginal fraction		-	1/2	1/2	1/2	1/2	1/2	1/2	1/2	1/2	1/2	1/2
Additional personal allowance[5]	KEPG	1 590	1 720	1 720	1 720	1 720	1 720	1 720	1 790	1 830	1 900	1 970
Widow's bereavement allowance[6]	KEPH	1 590	1 720	1 720	1 720	1 720	1 720	1 720	1 790	1 830	1 900	1 970
Dependent relative allowance:[7]												
Maintained by single woman	KEPJ	..	..	..	..	..	..	..	..	..	..	..
Other cases	KEPK	..	..	..	..	..	..	..	..	..	..	..
Limit of relative's income	KETX	..	..	..	..	..	..	..	..	..	..	..
Daughter's or son's services allowance[8]	KRUA	15	18	6	9	10	11	11	12	..	..	..
Housekeeper allowance[9]	KEUK	..	..	..	..	..	..	..	..	..	..	..
Blind person's allowance[10]												
Single or married (one spouse blind)	KSOJ	540	1 080	1 080	1 080	1 080	1 200	1 200	1 250	1 280	1 330	1 380
Married (both spouses blind)	KSOK	1 080	2 160	2 160	2 160	2 160	2 400	2 400	2 500	2 560	2 660	2 760
Life Assurance Relief[11]												
(Percentage of gross premium)	KFDR	*12.5 or Nil*	*12.5 or Nil*	*12.5 or Nil*	*12.5 or Nil*	*12.5 or Nil*	*12.5 or Nil*	*12.5 or Nil*	*12.5 or Nil*	*12.5 or Nil*	*12.5 or Nil*	*12.5 or Nil*

1 The married man's allowance was that for a full year, payable instead of the personal allowance. In the year of marriage the allowance is reduced by one twelfth of the difference between the married and single personal allowances for each complete month (beginning on the sixth day of each calendar month) prior to the date of marriage.
2 The wife's earned income allowance has as its maximum value the amount shown. Where the earned income was less, the allowance is reduced to the actual amount of earned income. From 1990-91, under Independent Taxation, the allowance is abolished and wives obtain a personal allowance.
3 The age allowance replaces the single or married allowances, provided the taxpayer's income is below the limit shown. For incomes in excess of the limit, the allowance is reduced by £2 for each additional £3 of income until the ordinary single or married allowance is reached. From 1989-90 the allowance was reduced by £1 for each additional £2 of income. The relief was due where the taxpayer (or, prior to 1990-91, wife living with him) was aged 65 or over in the year of assessment. For 1987-88 *et seq* there is additional relief at a higher age. For 1987-88 and 1988-89 the increased relief was available to those aged 80 or over in the year of assessment. From 1989-90 the increased level of relief was made available to those aged 75 and over. The married age allowances were abolished in 1990-91, under Independent Taxation and replaced by age-related married couple's allowances.
4 Following the introduction of Independent Taxation from 1990-91 the married couple's allowance was introduced. It is payable in addition to the personal allowance and between 1990-91 and 1992-93 went to the husband unless the transfer condition was met. This was that a husband could only transfer part or all of the married couple's allowance to his wife if he was unable to make full use of the allowance himself. In 1993-94 all or half of the allowance may be transferred to the wife if the couple have agreed beforehand. The wife has the right to claim half the allowance. The married couple's allowance and allowances linked to it, are restricted to 20 per cent in 1994-95 and to 15 per cent from 1995-96.
5 The additional personal allowance may be claimed by a single parent (or by a married man if his wife is totally incapacitated) who maintains a resident child at his or her own expense.
6 Widow's bereavement allowance is due to a widow in the year of her husband's death and in the following year provided the widow has not remarried before the beginning of that year.
7 The dependent relative allowance was due to a taxpayer who maintained, wholly or partially, either (a) an aged or infirm relative, or (b) his or his wife's separated, divorced or widowed mother. The relative's income had to be below the limits shown in order for the full allowances to be given (the limit was equal to the basic National Insurance Retirement Pension). The allowance was reduced by the excess of the relative's income over the income limit.
8 The daughter's or son's services allowance could be claimed by an aged or infirm taxpayer who maintained a daughter or son on whose services the taxpayer or his wife were dependent.
9 The housekeeper allowance could be claimed by a widow or widower who had a resident housekeeper.
10 The blind person's allowance may be claimed by blind persons (in England and Wales, registered as blind by a local authority) and surplus blind person's allowance may be transferred to a husband or wife.
11 Relief on life assurance premiums is given by deduction from the premium payable. From 1984-85, it is confined to policies made before 14 March 1984.

Source: Board of Inland Revenue: 020 7438 7412

17.11 Rates of Income tax
United Kingdom

	1990/91		1991/92		1992/93		1993/94		1994/95	
	Slice of taxable income (£)[1]	Rate per cent	Slice of taxable income (£)[1]	Rate per cent	Slice of taxable income (£)[1]	Rate per cent	Slice of taxable income (£)[1]	Rate per cent	Slice of taxable income (£)[1]	Rate per cent
Lower rate	-	-	-	-	1 - 2000	*20*	1 - 2 500	*20*	1 - 3 00	*20*
Basic rate	1 - 20 700	*25*	1 - 23 700	*25*	2 001 - 23 700	*25*	2 501 - 23 700	*25*[2]	3 001 - 23 700	*25*[2]
Higher rate	over 20 700	*40*	over 23 700	*40*	over 23 700	*40*	over 23 700	*40*	over 23 700	*40*

	1995/96		1996/97		1997/98		1998/99		1999/00	
	Slice of taxable income (£)[1]	Rate per cent	Slice of taxable income (£)[1]	Rate per cent	Slice of taxable income (£)[1]	Rate per cent	Slice of taxable income (£)[1]	Rate per cent	Slice of taxable income (£)[1]	Rate per cent
Lower rate	1 - 3 200	*20*	1 - 3 900	*20*	1 - 4 100	*20*	1 - 4 300	*20*	1 - 1 500	*10*
Basic rate	3201 - 24 300	*25*[2]	3 901 - 25 500	*24*[3]	4 101 - 26 100	*23*[3]	4 301 - 27 100	*23*[3]	1 501 - 28 000	*23*[3]
Higher rate	over 24 300	*40*	over 25 500	*40*	over 26 100	*40*	over 27 100	*40*	over 28 000	*40*[4]

1 Taxable income is defined as gross income for income tax purposes less any allowances and reliefs available at the taxpayer's marginal rate.
2 The basic rate of tax on dividend income is 20%.
3 The basic rate of tax on dividends and savings income is 20%.
4 The higher rate of tax on dividends is 32.5%.

Source: Board of Inland Revenue: 020 7438 7412

17.12 Rateable values
England and Wales
At 1 April in each year

		Unit	1989	1990[2]	1991[2]	1992[2]	1993[2]	1994[2]	1995[2]	1996[2]	1997[2]	1998[2]	1999[2]
Number of properties		Thousands											
Total-all classes	KMIH	"	23 916	1 685	1 709	1 734	1 737	1 734	1 723	1 726	1 725	1 719	1 720
Domestic: total	KMII	"	20 058	..	..	..	..	..	..	..	..	..	..
Houses and flats with rateable values:													
Not over £75)	KMIJ	"											
Over £75 but not over £100 }	KHDG KMIK	"	19 748[1]	..	..	..	..	..	..	..	..	..	..
Over £100)	KMIL	"											
Agricultural dwelling-houses, etc	KMIM	"	310	..	..	..	..	..	..	..	..	..	..
Commercial: total	KMIN	"	3 210	1 190	1 211	1 227	1 230	1 227	1 223	1 228	1 225	1 223	1 279
Shops and cafes	KMIO	"	569	576	572	570	567	565	562	497†	491	488	484
Offices	KMIP	"	214	223	234	243	248	251	252	255	255	257	258
Other	KMIQ	"	2 427	392	405	413	415	411	409	476†	479	478	537
On-licensed premises: total	KMIR	"	55	55	55	55	55	55	60	59	59	59	60
Entertainment and recreational: total	KMIS	"	79	95	95	94	93	92	87	87	87	81	80
Cinemas	KMIT	"	1	1	1	1	1	1	1	1	1	1	1
Theatres and music-halls	KMIU	"	1	1	1	1	1	1	1	1	1	1	1
Other	KMIV	"	78	94	94	93	92	91	86	85	86	80	79
Public utility: total	KMIW	"	38	15	15	15	15	15	9	8	8	9	9
Educational and cultural: total	KMIX	"	40	42	42	42	42	41	41	41	41	41	41
Miscellaneous: total	KMIY	"	323	53	53	59	58	58	56	55	55	56	61
Industrial: total	KMIZ	"	114	234	237	241	244	247	248	249	249	250	250
Value of assessments		£ million											
Total-all classes	KMHA	"	8 231	32 916	34 214	35 608	35 263	34 129	33 912	34 245	34 299	33 909	33 649
Domestic: total	KMHB	"	4 032	..	..	..	..	..	..	..	..	..	..
Houses and flats with rateable values:													
Not over £75)	KMHC	"											
Over £75 but not over £100 }	KHDH KMHD	"	3 963[1]	..	..	..	..	..	..	..	..	..	..
Over £100)	KMHE	"											
Agricultural dwelling-houses, etc	KMHF	"	69	..	..	..	..	..	..	..	..	..	..
Commercial: total	KMHG	"	2 166	19 566	20 659	21 440	21 336	20 662	19 626	19 822	19 859	19 733	20 649
Shops and cafes	KMHH	"	656	7 229	7 382	7 472	7 335	7 068	7 780	6 094†	5 959	5 860	5 840
Offices	KMHI	"	815	6 787	7 545	8 093	8 239	8 027	5 587	5 630	5 641	5 624	5 575
Other	KMHJ	"	696	5 550	5 732	5 876	5 762	5 568	6 260	8 098†	8 259	8 249	9 234
On-licensed premises: total	KMHK	"	66	679	691	698	679	642	968	969	970	980	997
Entertainment and recreational: total	KMHL	"	91	951	994	1 025	1 022	988	1 009	1 018	1 033	1 040	1 045
Cinemas	KMHM	"	3	18	21	24	25	24	32	32	36	39	45
Theatres and music-halls	KMHN	"	2	21	21	22	21	19	21	21	21	21	20
Other	KMHO	"	86	912	951	979	975	946	956	965	975	979	980
Public utility: total	KMHP	"	435	3 485	3 493	3 518	3 505	3 424	3 455	3 469	3 488	3 380	3 361
Educational and cultural: total	KMHQ	"	235	1 981	1 994	1 970	1 883	1 813	1 873	1 883	1 894	1 773	1 672
Miscellaneous: total	KMHR	"	480	806	823	1 375	1 367	1 319	1 429	1 500	1 494	1 464	1 439
Industrial: total	KMHS	"	726	5 448	5 559	5 583	5 470	5 280	5 550	5 584	5 561	5 540	5 463

1 The detailed breakdown of the various rateable value bands for houses and flats is not available after 1985.
2 See chapter text.

Source: Board of Inland Revenue: 020 7438 6314

17.13 Local authorities: gross loan debt outstanding[1]

At 31 March in each year

£ millions

		1989	1990	1991	1992	1993	1994	1995	1996	1997	1998	1999
United Kingdom												
Total debt[2]	KQBR	58 834	59 982	57 985	56 331	55 943	55 869	..	..	..	..	..
Public Works Loan Board	KQBS	44 324	46 499	48 001	48 717	41 447	40 803	40 388	40 445	41 683	42 901	44 701
Northern Ireland Consolidated Fund[2]	KQBT	197	208	239	93	109	126	138	157	175	220	..
Other debt	KQBU	14 313	13 275	9 745	..	..	..	..	..	..	..	..
England and Wales												
Total debt	KQBV	50 091	51 689	49 914	48 187	47 357	46 695	..	..	..	..	..
of which Public Works Loan Board	KQBW	*37 563*	*39 969*	*41 025*	*41 501*	*34 082*	*33 132*	*32 448*	*32 617*	*33 660*	*34 606*	*36 049*
Scotland												
Total debt[3]	KQBX	8 650	8 192	7 948	8 005	8 429	8 999	9 984	10 850	..	..	..
of which Public Works Loan Board	KQBY	*6 761*	*6 530*	*6 976*	*7 216*	*7 365*	*7 671*	*7 940*	*7 828*	*8 023*	*8 295*	*8 652*
Northern Ireland												
Total debt[2]	KQBZ	93	101	123	139	157	175	186	183	211	225	..

1 The sums shown exclude inter-authority loans and debt transfers, and temporary loans and overdrafts obtained for the purpose of providing for current expenses. No deduction has been made in respect of sums held in sinking funds for the repayment of debt.
2 The Northern Ireland Loans Fund was abolished on 1 April 1983 and its functions transferred to the Northern Ireland Consolidated Fund.
3 The data for 1991 is taken from information supplied by the Chartered Institute of Public Finance and Accountancy. Data for 1996 is before transfer of responsibility for Water & Sewerage and the Scottish Childrens Reporters Administration.

Sources: Department of the Environment (1981 to 1990); Scottish Executive, Economic Advice & Statistics: 0131 244 7033; Public Works Loan Board: 020 7270 3874; Department of the Environment for Northern Ireland: 028 9054 0707; Welsh Office (1981 to 1990); Chartered Institute of Public Finance and Accountancy (England and Wales,1991)

17.14 Revenue expenditure of local authorities

	1996/97 outturn £m	1997/98 outturn[8] £m	1998/99 provisional outturn £m	1998/99 adjusted provisional outturn[6] £m	1999/2000 budget £m	Difference 1999/2000 over 1998/99 adjusted provisional outturn[9] £m	%
England							
Education	18 691	18 908	20 157	21 397	22 591	1 194	6
Personal social services	7 943	8 454	9 040	9 040	9 693	653	7
Police	6 261	6 498	6 754	6 754	7 083	330	5
Fire	1 281	1 338	1 414	1 414	1 496	82	6
Highway maintenance	1 679	1 628	1 690	1 690	1 736	46	3
EPC services	14 147	14 297	14 910	14 910	15 469	559	4
Comprising:							
Civil defence and other Home office	544	556	551	551	578	27	5
Magistrates courts	303	308	308	308	317	9	3
Parking and public transport	626	659	683	683	680	-2	-
Housing benefit and council tax administration[1]	5 625	5 531	5 624	5 624	5 817	193	3
Non HRA housing	363	367	411	411	430	19	5
Libraries and art galleries	739	745	907	907	937	30	3
Sports and recreation centres	539	537	536	536	533	-3	-
Local environmental services	3 372	3 474	4 602	4 602	4 848	246	5
Other	2 036	2 121[7]	1 289[7]	1 289[7]	1 330[7]	41	3
Net current expenditure	50 002	51 123	53 964	55 204	58 067	2 863	5
Capital charges	1 936	1 970	2 131	2 131	2 117	-14	-1
CERA[2]	757	712	810	810	744	-66	-8
Interest receipts	-898	-1 053	-979	-979	-758	221	-23
Other non-current[3]	3 777	3 897	4 003	2 224	2 291	67	3
Gross revenue expenditure[4]	55 574	56 650	59 929	59 390	62 461	3 071	5
Specific grants outside AEF	-8 991	-9 395	-8 938	-8 438	-8 694	-256	3
Other income[5]	-51						
Revenue expenditure	46 532	47 256	50 991	50 952	53 767	2 815	6
Specific & special grants inside AEF	-1 618	-1 825	-2 223	-2 253	-2 368	-115	5
Net revenue expenditure	44 914	45 431	48 769	48 700	51 399	2 699	6
Other adjustments	-18	-1	-59	-59	-50	9	-16
Use of reserves	-116	54	-679	-679	-810	-131	19
Budget requirement	44 781	45 483	48 030	47 961	50 539	2 578	5
SSA reduction grant	-219	-123	-102	-102	-68	34	-33
Police grant	-3 164	-3 243	-3 375	-3 375	-3 505	-130	4
Revenue support grant		-18 649	-19 480	-19 411	-19 875	-464	2
Central Support Protection Grant					-52		
Council Tax Benefit Subsid Limitation Scheme					32		
Non-domestic rates	-12 736	-12 027	-12 524	-12 524	-13 612	-1 088	9
Other items	-199	-201	-218	-218	-181	37	-17
Precepts/demand on collection fund	10 461	11 241	12 332	12 332	13 278	946	8
Scotland							
Net revenue expenditure on general fund	6 578	6 520	6 730	..	6 984	254	4

1 Administration and payment of housing benefit and administration of Council Tax.
2 Capital Expenditure charged to the Revenue Account.
3 Includes:
(i) Gross expenditure on council tax benefit.
(ii) Expenditure on council tax reduction scheme.
(iii) Discretionary (non-domestic) rate relief.
(iv) Flood defence payments to the National Rivers authority (now Environment Agency)
(v) Payments in respect of grant maintained schools.
(vi) Bad debt provision.
4 This measure of expenditure is referred to as "Gross Revenue Expenditure" on local authority returns, but is actually net of expenditure met by sales, fees, charges and interest receipts, and excludes most expenditure on the housing revenue account and on trading services.
5 From licence fees etc.
6 The adjustments, made to allow for introduction of the nursery voucher scheme, under 5's Specific Grant and additional community care responsibilities, result in figures on a comparable basis to those for 1998/99.
7 Includes Other income from licence fees etc.
8 The 1997/98 data for Scotland are provisional.
9 For Scotland only, the difference is between the 1999/2000 budget over 1998/99 provisional outturn.

Sources: Department of the Environment, Transport and the Regions: 020 7890 4163;
Scottish Executive, Economic Advice & Statistics: 0131 244 7033

17.14 Revenue expenditure of local authorities

continued

	1993/94 outturn	1994/95 outturn	1995/96 outturn	1996/97 outturn	1997/98 outturn	1998/99 budget	1999/2000 budget
Wales							
Education	1 265	1 253	1 276	1 283	1 292	1 355	1 427
Personal social services	352	426	482	526	558	578	616
Police	273	282	298	321	344	355	372
Fire	70	69	74	78	81	89	95
Highways maintenance	146	150	157	148	144	147	145
Other services	697	745	806	822	829	835	866
Comprising:							
Civil defence and other Home office	28	31	32	37	36	33	35
Magistrates courts	17	19	19	21	21	19	19
Parking and public transport	15	14	15	16	14	14	18
Housing benefit and council tax administration	189	218	237	251	250	265	260
Non HRA housing	15	15	17	17	18	17	17
Libraries and art galleries	35	35	37	37	38	47	50
Sports and recreation centres	50	49	53	50	51	53	55
Local environmental services	229	240	255	231	235	268	291
Other	119	125	141	161	166	120	120
Net current expenditure	2 803	2 925	3 093	3 178	3 248	3 360	3 521
Capital charges	188	187	202	210	230	255	259
CERA	37	43	63	14	19	20	19
Interest receipts	-39	-43	-55	-30	-39	-21	-20
Other non-current	81	98	106	120	128	136	111
Gross revenue expenditure	3 070	3 209	3 410	3 492	3 586	3 750	3 890
Specific grants outside AEF	-444	-451	-475	-499	-501	-488	-459
Other income	-2	-2	-2	-2	-	-	-
Revenue expenditure	2 624	2 756	2 933	2 991	3 085	3 262	3 431
Specific & special grants inside AEF	-209	-215	-70	-73	-78	-73	-68
Net revenue expenditure	2 415	2 541	2 863	2 918	3 007	3 189	3 362
Other adjustments	-1	-	-1	-3	-4	-	-
Use of reserves	47	16	-74	-7	-22	-30	-17
Budget requirement	2 461	2 557	2 787	2 908	2 982	3 158	3 345
SSA reduction grant	-	-	-	-45	-17	-31	-22
Police grant	-	-	-156	-164	-168	-174	-181
Revenue support grant	-1 666	-1 739	-1 717	-1 790	-1 731	-1 798	-1 890
Non-domestic rates	-470	-464	-520	-459	-584	-612	-656
Other items	-16	-18	-15	-	-	-	-
Precepts/demand on collection fund	309	337	379	449	482	543	596
Northern Ireland							
Total expenditure[1]							
Capital	56.4	65.3	47.4	73.7	65.2		
Other	214.1	227.8	251.7	294.6	304.3		
Total income	235.0	251.3	243.5	301.1	370.4		

1 "Capital" and "Other" figures cannot be added to give "Total" figure because of double counting of loan charges, which are recorded as both "Capital" and "Other" expenditure.

Sources: National Assembly for Wales: 029 2082 5355; Department of the Environment for Northern Ireland: 028 9054 0707

17.15 Funding of revenue expenditure

Years ended 31 March

£ millions

		Domestic rating system		Community charge system			Council tax system						
		1988 /89	1989 /90	1990 /91	1991 /92	1992 /93	1993 /94	1994 /95	1995 /96	1996 /97	1997 /98	1998[13,6] /99	1999[13,7] /00
England													
Revenue expenditure[1]													
1999/00 prices £m[5]	KRTM	47 194†	46 921	46 871†	48 593	50 072	48 182†	49 900	48 862	50 143	49 545	52 138	53 767
Cash £m	KRTN	31 240	33 282	35 851	39 472	42 020	41 506	43 602	44 827	46 532	47 256†	50 991	53 767
Government grants[2]													
Cash £m	KRTO	13 201	13 481	12 927	18 620	20 968	21 685	23 679	23 335	23 003	23 840†	25 180	25 867
Per cent	KRTP	42	41	36	47	50	52	54	52	49	50†	49	48
Non- domestic rates[3]													
Cash £m	KRTQ	8 819	9 595	10 429	12 408	12 306	11 584	10 692	11 361	12 743	12 034	12 531	13 619
Per cent	KRTR	28	29	29	32	29	28	25	25	27	25	25	25
Domestic rates, community charges & council taxes[4]													
Cash £m	KRTS	8 957	9 713	12 251	8 533	9 521	8 912	9 239	9 777	10 461	11 241	12 332	13 278
Per cent	KRTT	29	29	34	22	23	21	21	22	22	23	24	25
Wales[8]													
Revenue expenditure[9]													
Cash £m	JYXV	1 956	2 122	2 228	2 484	2 656	2 627	2 756	2 933	2 991†	3 083	3 262	3 431
Government grants[10]													
Cash £m	JYXW	1 249†	1 327	1 329	1 785	1 862	1 877	1 954	1 943	2 073	1 993†	2 077	2 162
Per cent	JYXZ	64†	63	60	72	70	71	71	66	69	65	64	63
Non- domestic rates[11]													
Cash £m	JYXX	369	407†	443	525	536	470	464	520	459	584	612	656
Per cent	JYYA	19	19	20	21	20	18	17	18	15	19	19	19
Domestic rates, community charges & council taxes[12]													
Cash £m	JYXY	337†	387	468	180	270	325	354	394	449	482	543	596
Per cent	JYYB	17†	18	21	7	10	12	13	13	15	16	17	17

1 1981/82 to 1989/90: net expenditure financed from relevant grants rates and balance. 1990/91 onwards: "Revenue expenditure to compare with TSS" ie expenditure financed from revenue support grant, specific grants within Aggregate External Finance, special grants, non domestic rates, community charges/council taxes and balances. Also includes spending met by community charge grant (1991/92), and additional grant for teacher pay (1992/93), SSA reduction grant (1994/95 to 1999/00), police grant (1995/96 to 1999/00) and Control Support Reduction Grant (1999/00). This line is not the total of the others. The difference is due to funding by balances and other adjustments.

2 1981/82 to 1989/90: Aggregate Exchequer Grant 1990/91 onwards: revenue support grants, specific and special grants within AEF, community charge grant (1991/92), teachers' pay award additional grant (1992/93), SSA reduction grant (1994/95 to 1999/00), police grant (1995/96 to 1999/00).

3 1981/82 to 1989/90 net of rate relief. 1990/91 onwards: distributable amount from non-domestic rate pool. 1993/94 onwards: includes City Offset.

4 1981/82 to 1989/90: gross of domestic rate rebates. 1990/91 to 1992/93: gross of community charge benefit and community charge transitional relief/community charge reduction scheme grant. 1993/94 onwards: gross of council tax benefit and council tax transitional reduction scheme.

5 Revenue expenditure at 1999/00 prices have been calculated using the GDP deflator. Major function changes include:
(i) Responsibilities for municipal bus service transferred to public transport companies in 1986;
(ii) Funding of Polytechnic and higher education colleges transferred to Polytechnics and colleges Funding Council (PCFC) on 1 April 1989;
(iii) Increasing responsibilities as a result of care in the community since 1 April 1993;
(iv) Funding of colleges of further education and sixth form colleges transferred to the Further Education Funding Council (FEFC) on 1 April 1993.

6 Provisional Outturn for England.

7 Budget Estimate for England.

8 This table provides a general indication of the trend in the financing of revenue expenditure. Comparisons between the rates (prior to 1990/91), community charge (1990/91 to 1992/93) and council tax systems must be made with caution since expenditure definitions and categories differ. For example, specific and supplementary grants within aggregate exchequer grant (included under government grants prior to 1990/91) are not comparable in coverage to those within aggregate external finance (included 1990/91 onwards). Also the split between domestic and non-domestic rates, and rebates relating to each, prior to 1990/91 has been estimated. The components do not add to the totals for technical reasons; these include rounding and authorities' use of balances.

9 Local authority revenue expenditure is defined here (i) prior to 1990/91 as relevant expenditure plus expenditure on certain non-relevant items such as net expenditure on mandatory student awards (ii) for 1990/91 onwards as revenue expenditure on a total standard spending basis.

10 Prior to 1990/91 comprises rates support grant, specific and supplementary grants within aggregate exchequer grant, domestic rate relief grant and rate rebate grants. For 1990/91 onwards comprises revenue support grant, specific and supplementary grants within aggregate external finance, community charge grant (1991/92), transitional relief (1990/91), community charge reduction scheme grant (1991/92 to 1992/93), council tax reduction scheme grant (1996/97 and 1997/98) and police damping grant (1997/98).

11 Prior to 1990/91 estimated non-domestic rate income, net of rebates. For 1990/91 onwards, the distributable amount.

12 Prior to 1990/91, estimated domestic rate income, including domestic rate relief grant and rate rebates. For 1990/91 to 1992/93, community charge, net of transitional relief/community charge reduction scheme grants and community charge grant but including community charge benefit grant and community council precepts. For 1993/94 onwards, council tax, net of council tax reduction scheme grant but including council tax benefit grant and community council precepts.

13 Budget for Wales.

Sources: Department of the Environment, Transport and the Regions: 020 7890 4163;
National Assembly for Wales: 029 2082 5355

17.16 Local authority capital expenditure and receipts

Years ended 31 March

£ millions

		1994	1995	1996	1997	1998
England		1994/95 final outturn	1995/96 final outturn	1996/97 final outturn	1997/98 final outturn	1998/99 provisional outturn
Expenditure						
Housing[1]	KRUB	2 039†	1 999	1 746	1 665	1 964
Transport	KRUC	1 436†	1 374	1 176	1 086	908
Education	KRUD	807†	788	804	837	959
Personal Social Services	KRUE	195	198	191†	148	141
Fire services	KRUF	72	63	49	50†	47
Agriculture, fisheries and food	KRUG	33	53	55	51†	30
Sport and recreation	KRUH	175	192	201†	194	235
Protective services[2]	KRUI	314	341	290†	272	297
Urban and regeneration programmes[3]	KRUJ	653†	564	725	649	552
Single regeneration budget - environment[4]	KRUK	609†	530	683	604	507
Single regeneration budget - other departments[4]	KRUL	44†	34	42	45	45
Other environment	KRUM	9	1	..	..	..
Other services						
Environment	KRUN	880†	898	730	847	1 001
Other departments	KRUO	69†	61	65	127	127
Housing Association grant	KRUP	331	364	338†	330	310
Total	KRUR	7 004†	6 895	6 370	6 256	6 571
Receipts						
Housing	KRUS	1 391†	1 203	1 049	1 289	1 446
Transport	KRUU	47†	14	49	10	10
Education	KRUT	128	75	120†	133	75
Personal Social Services	KRUV	45	40	57†	43	47
Fire services	KRUW	3	3	5	9†	3
Agriculture, fisheries and food	KTDH	16	9	12†	64	44
Sport and recreation	KRUX	13	15	9†	13	16
Protective services	KRUY	56	58	82†	71	75
Other services						
Environment	KRUZ	463	359	545†	544	675
Other departments	KRVA	69	44†	52	2	2
Total	KRVB	2 230†	1 821	1 980	2 178	2 393

1 Excludes expenditure on Estate Action and Housing Association Grant.
2 Includes police, probabtion and aftercare, civil defence and magistrates' courts.
3 Includes expenditure on Estate Action schemes.
4 Includes environmental services, consumer protection, careers and sheltered employment and museums, galleries and libraries.

Source: Department of the Environment, Transport and the Regions: 020 7890 4076

17.17 Capital expenditure and income

£ millions

	Expenditure			Income				
	Expenditure on land works, etc	Capital assigned to repayment of debt	All expenditure	Loans	Government grants	Miscellaneous	All income	Gross debt at end of year
Financial year								
	KRVC	KRVD	KRVE	KRVF	KRVG	KRVH	KRVI	KRVJ
1970/71	1 792	150	1 942	1 516	106	349	1 970	13 384
1971/72	1 938	170	2 109	1 605	116	421	2 143	14 450
1972/73	2 418	213	2 631	2 030	122	531	2 682	16 105
1973/74	3 286	225	3 511	2 781	143	619	3 544	18 300
1974/75[1]	3 712	127	3 839	3 209	128	498	3 835	18 884
1975/76	3 917	198	4 115	3 285	177	647	4 109	21 930
1976/77	3 783	312	4 095	3 097	249	803	4 149	24 534
1977/78	3 487	352	3 839	2 677	255	981	3 913	26 282
1978/79	3 621	390	4 011	2 627	351	1 139	4 117	27 103
1979/80	4 249	331	4 580	2 992	385	1 367	4 745	30 187
1980/81	4 476	413	4 889	2 900	492	1 864	5 256	32 076
1981/82	4 061	563	4 623	2 527	470	2 177	5 174	34 069
1982/83	5 090	634	5 724	3 358	416	3 100	6 874	36 231
1983/84	5 890	562	6 452	3 538	379	3 294	7 211	38 698
1984/85	6 352	515	6 867	3 381	327	3 283	6 991	40 554
1985/86	5 748	348	6 096	3 008	360	3 239	6 607	40 138
1986/87	5 899	328	6 227	2 814	388	3 878	7 081	43 033
1987/88	6 091	486	6 577	2 953	297	4 286	7 536	44 904
1988/89	7 166	658	7 824	2 985	270	6 122	9 376	47 295
1989/90	9 590	474	10 064	2 919	440	6 110	9 469	48 695

		Income							
	Gross capital expenditure	Credit approvals used	Government grants	Capital receipts	Other income	Total income	Capital receipts set aside[2,4]	Credit ceiling[3]	Provision for credit liabilities[3]
Financial year									
At 1 April 1990	-	-	-	-	-	-	4 241	42 167	4 241
	KRVK	KRVL	KRVM	KRVN	KRVO	KRVP	KRVQ	KRVR	KRVS
1990/91	6 869	2 786	907	3 165	542	7 400	2 022	41 125	5 677
1991/92	6 572	3 140	1 041	2 251	674	7 106	1 353	41 234	6 502
1992/93	6 567	3 229	1 210	2 110	619	7 168	908	37 051	6 282
1993/94	7 124	2 948	1 279	3 310	651	8 188	356	37 941	6 041
1994/95	6 950	2 722	1 176	2 458	724	7 080	1 409	37 673	6 921
1995/96	6 910	2 264	1 484	1 966	1 278	6 992	1 160	37 542	7 677
1996/97	6 420	2 120	1 388	2 183	1 132	6 823	1 039	37 157	8 520
1997/98	6 298	2 099	1 262	2 349	1 129	6 839	1 186	36 797	8 594

1 Reorganisation of local government in April 1974 transferred responsibility for various services to regional health and water authorities.

2 Excluding Social Housing Grant and European Regional Development Fund (ERDF) grants.

3 At end of year.

4 Gross receipts received in year.

Source: Department of the Environment, Transport and the Regions: 020 7890 4076

17.18 Expenditure of local authorities
Scotland

Year ended 31 March

£ thousands

		Out of revenue [1]										
		1987/88	1988/89	1989[8]/90	1990/91	1991/92	1992/93	1993/94	1994/95	1995/96	1996/97	1997[11]/98
Total	KQTA	5 721 849	6 198 024	6 794 154	7 429 771	7 985 137	8 560 071	8 713 471	9 111 751	9 690 424	9 196 125	9 591 951
General Fund Services:	KQTB	4 380 607	4 771 959	5 212 273	5 650 719	6 093 642	6 542 916	6 589 420	6 904 228	7 324 381	7 151 759	6 692 339
Education[2]	KQTC	1 817 823	1 977 117	2 174 652	2 390 134	2 564 363	2 757 058	2 537 582	2 563 049	2 654 158	2 629 961	2 513 340
Libraries, museums and galleries	KQTD	69 410	74 991	83 712	103 489	109 181	117 243	125 791	132 055	149 427	138 483	121 387
Social work[2]	KQTE	433 590	483 888	559 013	668 792	733 067	806 613	914 436	1 045 638	1 222 693	1 289 928	1 315 387
Law, order and protective services[2]	KQTF	480 487	514 793	580 977	653 197	704 943	763 917	804 616	828 480	866 567	816 315	931 795
Roads[3]	KQTG	439 473	467 912	506 562	561 522	589 564	633 874	650 555	684 452	676 116	716 570	441 689
Environmental services[4]	KQTH	317 334	336 953	221 857	264 507	283 968	305 645	322 534	330 039	351 689	329 674	343 565
Planning	KQTI	111 731	127 816	125 298	162 566	174 380	188 753	202 027	211 346	226 073	210 827	163 380
Leisure and recreation	KQTJ	217 527	237 221	276 331	338 532	362 560	383 081	396 874	405 083	451 952	426 422	364 853
Central administration[9]	KQTK	205 176	226 432	305 743	–	–	–	–	–	–	–	–
Other services[2,4]	KQTL	170 891	219 355	258 196	383 221	445 968	456 198	516 842	580 561	591 766	562 462	468 876
Other general fund expenditure[5]	KQTM	117 165	105 481	119 932	124 759	125 648	130 534	118 163	123 525	133 940	31 117	28 067
Housing	KQTN	1 257 988	1 325 226	1 330 796	1 473 950	1 573 590	1 672 566	1 738 427	1 806 022	1 924 930	2 000 684	1 658 935
Trading services:	KQTO	200 419	206 320	371 017	429 861	443 553	475 123	503 787	525 026	575 053	74 799	76 429
Water supply	KQTP	142 223	151 002	168 347	188 478	203 847	218 864	239 291	255 081	274 773	–	–
Sewerage[6]	KQTQ	–	–	154 436	176 449	185 216	194 576	200 036	209 323	233 525	–	–
Passenger transport	KQTR	149	188	263	560	603	859	681	685	794	2 849	1 510
Ferries	KQTS	3 577	4 158	5 358	6 596	4 508	8 336	7 203	7 574	7 744	6 831	7 512
Harbours, docks and piers	KQTT	38 879	36 967	25 757	23 428	28 008	27 811	16 215	16 804	15 301	13 482	13 261
Airports	KQTU	951	1 142	–	–	–	–	–	–	–	–	–
Road bridges	KQTV	7 378	8 459	7 942	9 277	9 629	9 508	10 948	7 579	11 755	12 759	16 064
Slaughterhouses	KQTW	3 169	3 121	3 366	2 737	1 427	1 282	985	976	1 000	794	940
Markets	KQTX	2 193	2 184	..	..	6 086	8 145	8 462	8 615	10 336	14 278	13 479
Other trading services	KQTY	1 900	–901	5 548	22 336	4 229	5 742	19 966	18 389	19 825	23 806	23 663
Loan charges:[11] Total	KQTZ	1 132 636	1 215 398	1 199 706	1 267 466	1 327 639	1 343 768	1 346 433	1 383 167	1 451 179	1 121 448	1 192 315
Allocated to :												
General Fund services	KMHV	502 383	538 518	538 693	567 143	607 867	631 479	649 212	677 377	710 801	639 380	715 293
Housing	KMHW	558 172	602 914	504 201	530 566	542 248	528 488	505 407	499 904	504 162	475 507	471 274
Trading services	KMHX	72 081	73 966	156 812	169 757	177 524	183 801	191 814	205 886	236 216	6 561	5 748

		On capital works [7]										
		1987/88	1988/89	1989[8]/90	1990/91	1991/92	1992/93	1993/94	1994/95	1995/96	1996/97	1997/98
Total	KQUA	1 079 366	1 142 074	1 262 076	1 249 744	1 276 006	1 344 250	1 428 545	1 518 362	1 528 167	889 572	813 900
General Fund Services:	KQUB	424 624	450 701	617 013	589 921	620 287	650 779	699 250	760 564	767 795	540 127	540 096
Education[2]	KQUC	60 201	72 144	81 643	75 166	81 007	92 584	85 619	113 121	114 128	101 898	112 753
Libraries, museums & galleries	KQUD	3 555	6 031	5 389	7 223	9 225	15 319	11 077	12 822	16 757	11 602	9 974
Social work[2]	KQUE	18 846	23 448	24 801	25 797	26 017	32 940	29 040	29 067	30 298	20 658	19 660
Law, order and protective services[2]	KQUF	27 121	27 257	29 789	23 378	19 903	28 230	32 951	33 635	35 847	41 326	37 701
Roads	KQUG	127 768	125 870	167 797	159 467	164 333	179 275	189 525	198 178	187 988	116 881	108 227
Environmental services[4]	KQUH	67 476	70 541	15 708	12 738	12 798	13 239	18 905	16 595	14 580	10 226	21 367
Planning	KQUI	34 600	30 076	30 897	55 899	84 900	69 065	76 477	102 152	103 221	51 182	69 648
Leisure and recreation	KQUJ	29 032	33 714	55 885	43 872	51 490	44 774	74 563	65 411	57 243	36 232	29 692
Administrative buildings & equipment[10]	KQUK	39 239	42 919	22 857	32 251	29 996	24 980	17 888	23 994	14 693	40 014	45 374
Other services[2,4]	KQUL	16 786	18 701	182 247	154 130	140 618	150 373	163 205	165 589	193 040	110 108	85 700
Housing	KQUM	603 511	629 816	501 082	504 378	485 131	460 140	467 811	497 997	517 593	345 713	270 005
Trading Services:	KQUN	51 231	61 557	143 981	155 445	170 588	233 331	261 484	259 801	242 779	3 732	3 799
Water supply	KQUO	42 989	51 470	61 991	63 872	77 000	109 655	128 081	123 564	107 064	–	–
Sewerage[6]	KQUP	–	–	65 417	77 096	88 035	113 998	127 047	130 634	131 684	–	–
Passenger transport	KQUQ	–	24	–	–	–	–	–	–	–	–	–
Ferries	KQUR	564	1 573	3 422	846	773	18	–	376	355	521	770
Harbours, docks and piers	KQUS	4 155	5 703	3 222	8 444	336	5 020	2 300	1 982	1 218	934	1 175
Airports	KQUT	86	262	59	14	2	2	184	173	763	1 149	439
Road bridges	KQUU	998	1 440	2 276	237	380	694	799	830	805	277	973
Slaughterhouses	KQUV	85	67	170	73	56	87	1 068	139	63	112	69
Markets	KQUW	171	313	–	–	–	–	–	–	–	–	–
Other trading services	KMHY	2 183	705	7 424	4 863	4 006	3 857	2 005	2 103	827	739	373

1 Gross expenditure less inter-authority and inter-account transfers.
2 From 1989/90 Education includes careers service (previously Other Services); Social work includes sheltered employment (previously Other Services); Other services includes district courts (previously Law, order and protective services), public analyst (previously Environmental services) and Housing loans and grants (previously Housing).
3 Including general fund support for transport (LA and NON-LA).
4 From 1989/90 onwards burial grounds and crematoria are included within Environmental services. From 1981/82 to 1988/89 (inclusive) they were included within Other services.
5 General fund contributions to Housing and Trading services (excluding transport), are also included in the expenditure figures for these services. From 1996/97 water and sewerage are excluded from other general fund expenditure.
6 From 1989/90 onwards Sewerage is shown as a trading service (previously included in Environmental services).
7 Expenditure out of loans, government grants and other capital receipts.
8 From 1980/81 to 1988/89 inclusive, revenue contributions to capital are recorded as expenditure in **both** the revenue and capital summaries. Table 17.19 details the amount of these contributions. From 1989/90 revenue contributions to capital are excluded from the capital summary and shown **only** in the revenue summary.
9 From 1990/91 Central Administration expenditure is included in the relevant service expenditure.
10 Prior to 1990-91 was called Central Administration.
11 From 1997/98 loan charges are not included within individual service totals.

Source: Scottish Executive, Economic Advice & Statistics: 0131 244 7033

17.19 Income of local authorities: classified according to source

Scotland

Year ended 31 March

£ thousands

		1987 /88	1988 /89	1989 /90	1990 /91	1991 /92	1992 /93	1993 /94	1994 /95	1995 /96	1996 /97	1997 /98
Revenue account												
Rates[1,2,3]	KQXA	1 942 864	2 050 666	1 265 922	1 351 627	1 414 346	1 336 395	1 258 863	1 198 575	1 310 721	1 313 531	1 326 129
Community charges	KQXB	–	–	867 519	890 645	802 005	932 994	–	–	–	–	–
Council tax	KPUC	–	–	–	–	–	–	822 830	918 502	976 465	968 153	1 070 405
Government grants												
RSG[4]:	KQXC	–	–	2 346 281	2 495 840	2 706 629	3 546 958	3 582 127	3 741 567	3 716 567	3 649 694	3 520 461
Needs element	KQXD	1 509 400	1 690 000	–	–	–	–	–	–	–	–	–
Resources element	KQXE	196 200	215 800	–	–	–	–	–	–	–	–	–
Domestic element	KQXF	90 800	91 900	–	–	–	–	–	–	–	–	–
Rate rebate grant[2]	KQXG	248 045	189 632	24 164	30 941	37 304	33 741	41 772	39 860	4 456	496	–
Community charge grant[5]	KIMJ	..	..	..	..	437 105	15 512	–	–	–	–	–
Community charge rebate grants	KQXH	–	–	187 182	192 378	112 904	153 231	–	–	–	–	–
Council tax rebate grants	KPUD	–	–	–	–	–	–	166 015	186 219	193 937	226 132	260 424
Other grants and subsidies[6]	KQXI	750 741	796 789	808 535	882 296	951 871	1 049 899	1 118 978	1 179 327	1 236 160	1 347 706	1 480 328
Sales	KQXJ	60 769	55 881	49 439	50 001	45 134	54 399	61 970	59 182	64 284	59 059	46 874
Fees and charges[7]	KQXK	936 928	1 070 254	1 185 303	1 288 728	1 410 355	1 435 951	1 421 565	1 471 320	1 528 270	1 539 611	1 625 549
Other income[6,7]	KQXL	110 219	126 990	123 212	132 009	161 791	184 581	194 151	209 819	207 005	162 825	193 399
Capital account												
Sale of fixed assets	KQXM	190 363	295 736	446 097	416 838	415 475	413 083	441 600	529 528	500 838	499 143	327 569
Loans	KQXN	790 368	774 490	682 767	..	..	..	..	..	..	..	..
Government grants	KQXO	44 974	25 134	..	..	..	..	..	..	..	..	..
Revenue contributions to capital	KQXP	18 665	14 391	11 812	48 997	73 896	111 710	163 228	134 156	197 606	119 641	149 423
Transfer from special funds	KMHZ	6 184	1 904	1 513	2 196	1 218	1 751	7 902	10 679	9 035	2 652	37 929
Other receipts	KMGV	26 004	38 715	23 156	24 649	23 110	42 631	32 554	38 736	29 571	45 067	32 118
Total income	KMGW	6 922 524	7 438 282	..	..	..	..	..	..	..	..	..

1 Excluding government grants towards rate rebates and domestic element of rate support grant (RSG). Including domestic water rate receipts.

2 Until 1982/83, certificated rate rebates paid in respect of supplementary benefit recipients were included with rate income. From 1983/84 onwards, they are included as part of rate rebate grant.

3 From 1989/90 rates refer to non-domestic rates only.

4 From 1989/90 onwards, RSG is not broken down into the three elements.

5 Payment to local authorities in respect of the £140 reduction in community charge awarded.

6 Until 1986/87, government grants not specific to service, eg grants towards job creation, were included under other income. From 1987/88, these are included under other grants and subsidies.

7 Prior to 1984/85 rents are treated as other income. From 1984/85 onwards rents are included with fees and charges.

Source: Scottish Executive, Economic Advice & Statistics: 0131 244 7033

17.20 Income of local authorities from government grants[1]
Scotland

Year ended 31 March

£ thousands

		1987 /88	1988 /89	1989 /90	1990 /91	1991 /92	1992 /93	1993 /94	1994 /95	1995 /96	1996 /97	1997 /98
General fund services[2]	KQYA	278 888	274 707	295 169	339 619	368 717	403 437	428 927	450 056	468 660	487 734	555 034
Education	KQYB	13 323	15 801	11 510	17 809	17 699	20 439	21 791	17 452	17 186	18 324	61 979
Libraries, museums and galleries	KQYC	233	664	57	80	106	154	108	247	123	137	326
Social work	KQYD	3 228	4 434	8 311	10 561	24 817	32 626	36 734	48 091	50 230	57 576	60 046
Law, order and protective services	KQYE	173 228	185 772	211 304	235 477	253 548	274 149	288 835	295 600	312 812	330 767	359 811
Roads	KQYF	30 178	12 641	4 127	4 580	3 325	3 540	1 993	4 315	4 788	403	237
Environmental services	KQYG	5 615	4 491	150	50	56	50	84	82	42	119	159
Planning	KQYH	4 555	770	4 144	971	947	785	1 310	867	3 030	3 337	4 885
Leisure and recreation	KQYI	2 960	2 606	1 522	1 739	1 503	1 396	1 476	1 609	1 830	1 509	1 856
Central administration[3]	KQYJ	357	1 344	867	–	–	–	–	–	–	–	–
Other services	KQYK	45 221	46 184	53 177	68 352	66 716	70 298	76 596	81 793	78 619	75 562	65 735
Housing	KQYL	506 342	537 496	510 201	538 042	577 321	640 127	684 519	723 604	762 172	856 435	921 374
Trading services	KQYM	10 485	9 720	3 165	4 024	4 739	5 405	4 954	4 009	4 557	–	–
Water supply	KQYN	8 001	7 363	3 132	3 998	4 139	4 255	4 823	4 009	4 459	–	–
Ferries	KQYO	213	213	–	–	1	46	47	–	–	–	–
Other trading services	KQYP	2 271	2 144	33	26	599	1 104	84	–	98	–	–
Capital grants[4,5]	KQYQ	–	–	..	..	..	..	..	..	..	..	..
General fund services	KQYR	–	–	..	..	..	..	..	..	..	..	..
Other services	KMGX	–	–	..	..	..	..	..	..	..	..	..
Grants not allocated to specific services[6]	KMGY	2 044 445	2 187 332	2 557 627	2 719 159	3 293 942	3 749 442	3 789 914	3 781 426	3 721 023	3 650 190	3 520 461
Total	KMGZ	2 840 160	3 009 255	..	3 600 844	4 244 719	4 798 411	4 908 314	4 959 095	4 956 412	4 994 362	4 996 869

1 Including grants for capital works.

2 Until 1986/87, excludes government grants not specific to services, eg grants towards job creation. From 1987/88 includes such grants.

3 From 1990/91 Central Administration income is included in the relevant service income.

4 From 1989/90 service breakdown no longer includes capital grants.

5 From 1990-91, the total income from government grants does not include capital grants.

6 Revenue support grant, community charge grant and rate and community charge rebate grants.

Source: Scottish Executive, Economic Advice & Statistics: 0131 244 7033

17.21 Expenditure of local authorities
Northern Ireland
Years ended 31 March

£ thousand

		Out of revenue and special funds										
		1987 /88	1988 /89	1989 /90	1990 /91	1991 /92	1992 /93	1993 /94	1994 /95	1995 /96	1996 /97	1997 /98
Total	KQVA	166 911	167 303	179 404	193 450	203 650	214 107	230 666	251 737	252 794	294 563	304 305
Libraries, museums and art galleries	KQVB	678	683	3 606	3 881	4 030	4 644	5 647	7 214	8 481	10 956	13 928
Environmental health services:												
Refuse collection and disposal	KQVC	25 599	26 418	26 366	29 363	32 131	33 582	39 952	42 109	41 284	52 267	56 246
Public baths	KQVD	3 145	3 105	1 457	1 571	1 808	1 505	1 562	1 648	1 703	1 838	2 585
Parks, recreation grounds, etc	KQVE	56 478	59 871	63 284	68 593	74 439	80 601	91 258	101 319	100 418	111 884	115 302
Other sanitary services	KQVF	19 758	21 543	24 990	24 237	26 251	27 593	32 074	34 582	35 706	39 545	39 682
Housing (grants and small dwellings acquisition)[1]	KQVG	1 265	1 092	1 008	1 130	860	792	873	553	472	489	545
Trading services:												
Gas supply	KQVH	15 529	6 631	7 997	3 472	–	–	–	–	–	..	..
Cemeteries	KQVI	3 674	3 846	4 335	4 492	4 908	5 044	5 352	5 984	5 489	5 120	5 626
Other trading services (including markets, fairs and harbours)	KQVJ	5 242	5 348	6 491	7 340	7 778	7 335	7 123	6 587	4 254	8 672	7 016
Miscellaneous	KQVK	35 543	38 768	39 870	49 371	51 445	53 011	46 825	51 741	54 987	63 792	63 375
Total loan charges	KQVL	12 583	13 588	18 154	19 430	20 226	21 693	19 194	20 797	21 122	24 363	34 823
Loan charges included in terms of expenditure above:												
Allocated to rate fund services	KQVM	10 001	11 221	13 216	15 155	16 777	18 785	..	..	..	..	..
Allocated to trading services	KQVN	1 820	1 969	4 513	3 904	3 102	2 689	..	..	..	..	..
Not allocated	KQVO	762	398	425	371	347	219	..	..	..	..	..

		On capital works[1]										
		1987 /88	1988 /89	1989 /90	1990 /91	1991 /92	1992 /93	1993 /94	1994 /95	1995 /96	1996 /97	1997 /98
Total	KQVP	28 958	33 943	34 955	37 800	42 005	52 908	63 870	56 031	60 629	72 910	69 159
Libraries, museums and art galleries	KQVQ	129	343	683	997	919	2 008	3 833	1 225	1 466	890	1 158
Environmental health services:												
Refuse collection and disposal	KQVR	3 551	5 107	3 917	7 254	6 363	4 629	7 013	8 907	8 772	10 401	9 136
Public baths	KQVS	75	271	73	189	225	156	35	143	213	1 000	2 669
Parks, recreation grounds, etc	KQVT	8 986	8 951	14 343	18 654	19 239	29 205	26 736	20 914	23 732	26 749	28 884
Other sanitary services	KQVU	891	1 333	1 666	1 844	2 053	2 094	1 657	1 883	2 233	1 894	2 739
Housing (including small dwellings acquisition)[1]	KQVV	438	104	84	254	441	164	–	18	–	–	192
Trading services:												
Gas supply	KQVW	112	86	22	–	–	–	–	–	–	–	–
Cemeteries	KQVX	645	1 161	597	795	690	673	267	198	714	322	167
Other trading services (including markets, fairs and harbours)	KQVY	2 520	4 125	4 493	2 762	4 738	3 043	4 231	2 954	357	5 456	6 145
Miscellaneous	KQVZ	11 835	12 462	9 077	5 051	7 337	10 936	20 098	19 789	23 142	26 198	18 069

1 Expenditure met out of loans, government grants for capital works, sales of property and other capital receipts.

Source: Department of the Environment for Northern Ireland: 028 9054 0707

17.22 Income of local authorities
Northern Ireland

Classified according to source Years ended 31 March £ thousand

		1987 /88	1988 /89	1989 /90	1990 /91	1991 /92	1992 /93	1993 /94	1994 /95	1995 /96	1996 /97	1997 /98
Total income	KQWA	203 851	201 189	216 779	234 037	224 082	284 923	..	..	..	..	..
Capital receipts: total	KQWB	27 308	26 421	30 127	33 860	36 022	53 523	63 151	44 567	61 502	67 724	63 242
Loans	KQWC	8 815	11 012	15 557	20 417	19 166	25 736	..	..	..	..	..
Government grants	KQWD	7 534	4 366	5 638	5 607	8 945	16 140	..	..	..	..	..
Other sources	KQWE	10 959	11 043	8 932	7 836	7 911	11 647	..	..	..	..	..
Other income: total	KQWF	130 110	130 646	135 587	145 168	163 500	178 834	..	..	..	..	..
Rates	KQWG	96 475	97 564	101 883	108 829	124 106	136 460	145 306	150 991	163 405	178 897	191 141
Government grants	KQWH	32 950	32 541	33 311	36 005	39 095	42 002	..	..	..	..	..
Housing[1]												
Grants and small dwellings acquisition	KQWI	685	541	393	334	299	372	..	..	..	..	..
Miscellaneous income	KQWJ	21 871	26 454	30 515	36 903	37 619	40 360	..	..	..	..	..
Trading services												
Total (capital and revenue)	KQWK	24 562	17 668	20 550	18 106	16 967	12 206	..	..	..	..	..
Gas: total	KQWL	16 564	7 954	7 599	6 362	–	–	..	..	..	..	..
Loans	KQWM	9	–	–	–	–	–	..	..	..	..	..
Government grants	KQWN	9 954	2 216	5 742	4 883	–	–	..	..	..	..	..
Miscellaneous	KQWO	6 601	5 738	1 857	1 479	–	–	..	..	..	..	..
Cemeteries: total	KQWP	1 507	1 833	2 122	2 427	2 507	2 796	..	..	..	..	..
Loans	KQWQ	497	904	866	1 216	1 063	1 241	..	..	..	..	..
Government grants	KQWR	16	11	99	116	123	243	..	..	..	..	..
Miscellaneous	KQWS	994	918	1 157	1 095	1 321	1 312	..	..	..	..	..
Other trading: total	KQWT	6 491	7 881	10 829	9 317	14 460	9 410	..	..	..	..	..
Loans	KQWU	674	2 739	4 662	4 670	6 365	2 751	..	..	..	..	..
Government grants	KQWV	916	1 330	1 857	23	1 288	1 420	..	..	..	..	..
Miscellaneous	KQWW	4 901	3 812	4 310	4 624	6 807	5 239	..	..	..	..	..

1 Including annuity repayments made by borrowers under the Small Dwellings Acquisition scheme.

Source: Department of the Environment for Northern Ireland: 028 9054 0707

17.23 Income of local authorities from government grants[1]
Northern Ireland

Classified according to services Years ended 31 March £ thousand

		1987 /88	1988 /89	1989 /90	1990 /91	1991 /92	1992 /93	1993 /94	1994 /95	1995 /96	1996 /97	1997 /98
Total	KQXV	51 372	40 464	46 648	46 634	49 450	59 807	..	..	..	..	..
Allocated to specific services:												
Environmental health services	KQXW	6 537	6 500	7 596	8 355	11 179	18 945	..	..	..	..	..
Housing	KQXX	–	–	–	–	–	–	–	..	..	..	..
Other services	KQXY	16 015	4 958	9 017	6 012	4 300	4 960	..	..	..	..	..
Not allocated to specific services[2]	KQXZ	28 820	29 006	30 035	32 267	33 974	35 902	36 139	40 203	41 295	41 267	36 485

1 Including grants for capital works.
2 Assistance to local authorities under various Local Government Acts.

Source: Department of the Environment for Northern Ireland: 028 9054 0707

External trade and investment

18

External trade and investment

External trade *(Table 18.1 and 18.3 to 18.6)*

The statistics in Table 18.1 are on a Balance of Payments (BoP) basis; all other statistics in this section are on an Overseas Trade Statistics (OTS) basis, compiled from information provided to H M Customs and Excise by importers and exporters, which values exports 'f.o.b.' (free on board) and imports 'c.i.f.' (including insurance and freight). In addition to deducting these freight costs and insurance premiums from the OTS figures, coverage adjustments are made to convert the OTS data to a BoP basis. Adjustments are also made to the level of all exports and EU imports to take account of estimated under-recording. The adjustments are set out and described in the annual ONS *'Pink Book' (United Kingdom Balance of Payments)*. These adjustments are made to conform to the definitions in the 5th edition of the IMF Balance of Payments Manual.

Aggregate estimates of trade in goods, seasonally adjusted and on a BoP basis are published monthly in the ONS First Release UK Trade. More detailed figures are available from the ONS Databank and are also published in the Monthly Review of External Trade Statistics (Business Monitor MM24). Detailed figures for EU and non-EU trade on an OTS basis are published by The Stationery Office in Overseas Trade Statistics of the United Kingdom.

A fuller description of how trade statistics are compiled can be found in Statistics on Trade in Goods (Government Statistical Service Methodological Series).

Import penetration and export sales ratios *(Table 18.2)*

The ratios were first introduced in the August 1977 edition of *Economic Trends* in an article 'The Home and Export Performance of United Kingdom Industries'. The article described the conceptual and methodological problems involved in measuring such variables as import penetration.

The industries are grouped according to the 1992 Standard Industrial Classification. The four different ratios are defined as follows:

Ratio 1: percentage ratio of imports to home demand
Ratio 2: percentage ratio of imports to (home demand plus exports)
Ratio 3: percentage ratio of exports to total manufacturers sales
Ratio 4: percentage ratio of exports to (total manufacturers sales plus imports)

Home demand is defined as total manufacturers sales plus imports minus exports.

Ratio 1 is commonly used to describe the import penetration of the home market. Allowance is made for the extent of a domestic industry's involvement in export markets by using Ratio 2; this reduces as exports increase.

Similarly Ratio 3 is the measure normally used to relate exports to total sales by UK producers and Ratio 4 makes an allowance for the extent that imports of the same product are coming into the UK.

Overseas trade in services *(Tables 18.7 and 18.8)*

These data relate to overseas trade in services and cover both production and non-production industries (excluding the Public Sector). In terms of types of services traded this equates to trade in royalties, various forms of consultancy, computing and telecommunications services, advertising and market research and other business services. A separate inquiry covers the Film and Television industries. The surveys cover receipts from the provision of services to residents of other countries (exports) and payments to residents of other countries for services rendered (imports). "Residents of other countries" is defined as companies, governments and individuals.

Sources of data

The OTIS surveys (which consist of a quarterly component addressed to the largest businesses and an annual component for the remainder) are based on a sample of companies derived from the Interdepartmental Business register. The companies are asked to show the amounts for their imports and exports against the geographical area to which they were paid or from which they were received - irrespective of where they were first earned.

The purpose of the OTIS survey is to record international transactions which impact on the UK's Balance of Payments, hence companies are asked to exclude from their earnings trade expenses such as the cost of services purchased abroad. Trade in service exports or imports which are included in invoices for the export or import of goods are excluded as they are already counted in the estimates for trade in goods. However, earnings from third country trade i.e. from arranging the sale of goods between two countries other than the UK and where the goods never physically enter the UK (known as merchanting), are included. Earnings from commodity trading are also included. Together these two comprise "Trade Related Services".

"Royalties" are the largest part of the total trade in services collected in the OTIS survey: these cover transactions for items such as printed matter, sound recordings, performing rights, patents, licences, trade marks, designs, copyrights, manufacturing rights, the use of technical "know-how" and technical assistance.

Balance of payments *(Tables 18.9 to 18.12)*

Tables 18.9 to 18.12 in this section are derived from *United Kingdom Balance of Payments 1999 edition - the ONS Pink Book*. The following general notes to the tables provide brief definitions and explanations of the figures and terms used. Further notes are included in the *Pink Book*.

Summary of Balance of Payments

The Balance of Payments consists of the current account, the capital account, the financial account and the International Investment Position. The **current account** consists of trade in goods and services, income and current transfers. **Income** consists of investment income and a new category compensation of employees. The **capital account** mainly consists of capital transfers and the **financial account** covers financial transactions. The **International Investment Position** covers balance sheet levels of UK external assets and liabilities. Every credit entry in the balance of payments accounts should, in theory, be matched by a corresponding debit entry so that total current, capital and financial account credits should be equal to, and therefore offset by, total debits. In practice there is a discrepancy termed **net errors and omissions.**

The Current Account

Trade in goods

The goods account covers exports and imports of goods. Imports of motor cars from Japan, for example, are recorded as debits in the trade in goods account whereas exports of vehicles manufactured in the UK are recorded as credits. Trade in goods forms a component of the expenditure measure of Gross Domestic Product (GDP).

Trade in services

The services account covers exports and imports of services (eg, civil aviation). Passenger tickets for travel on UK aircraft sold abroad, for example, are recorded as credits in the services account whereas the purchases of airline tickets from foreign airlines by UK passengers are recorded as debits. Trade in services, along with trade in goods, forms a component of the expenditure measure of Gross Domestic Product (GDP).

Income

The income account consists of compensation of employees and investment income and is dominated by the latter. Compensation of employees covers employment income from cross-border and seasonal workers which is less significant in the UK than in other countries. Investment income covers earnings (eg, profits, dividends and interest payments and receipts) arising from foreign investment and financial assets and liabilities. For example, earnings on foreign bonds and shares held by financial institutions based in the UK are recorded as credits in the investment income account, whereas earnings on UK company securities held abroad are recorded as investment income debits. Investment income forms a component of Gross National Income (GNI) but not Gross Domestic Product (GDP).

Current transfers

Current transfers are composed of central government transfers (eg, taxes and payments to and receipts from the European Union) and other transfers (eg, gifts in cash or kind received by private individuals from abroad or receipts from the EU where the UK government acts as an agent for the ultimate beneficiary of the transfer). Current transfers do not form a component either of Gross Domestic Product (GDP) or of Gross National Income (GNI). For example payments to the UK farming industry under the EU Agricultural Guarantee Fund are recorded as credits in the current transfers account while payments of EU agricultural levies by the UK farming industry are recorded as debits in the current transfers account.

Capital Account

Capital account transactions involve transfers of ownership of fixed assets, transfers of funds associated with acquisition or disposal of fixed assets, and cancellation of liabilities by creditors without any counterparts being received in return. The main components are migrants transfers, EU transfers relating to fixed capital formation (regional development fund and agricultural guidance fund) and debt forgiveness. Funds brought into the UK by new immigrants would, for example, be recorded as credits in the capital account, while funds sent abroad by UK residents emigrating to other countries would be recorded as debits in the capital account. The magnitude of capital account transactions are quite minor compared with the current and financial accounts.

Financial Account

While investment income covers earnings arising from overseas investments and financial assets and liabilities, the financial account of the balance of payments covers the flows of such investments. While earnings on foreign bonds and shares held by financial institutions based in the UK are, for example, recorded as credits in the investment income account, the acquisition of such foreign securities by UK based financial institutions are recorded as net debits in the financial account as portfolio investment abroad. Similarly the acquisitions of UK company securities held by foreign residents are recorded in the financial account as net credits as portfolio investment in the UK.

International Investment Position

While the financial account covers the flows of foreign investments and financial assets and liabilities, the International Investment Position records the levels of external assets and liabilities. While the acquisition of foreign securities by UK based financial institutions are recorded in the financial account, as net debits, the total holdings of foreign securities by UK based financial institutions are recorded as levels of UK external assets. Similarly the holdings of UK company securities held by foreign residents are recorded as levels of UK liabilities.

Overseas direct investment
(Tables 18.13 to 18.18)

Direct investment refers to investment that adds to, deducts from or acquires a lasting interest in an enterprise operating in an economy other than that of the investor, the investor's purpose being to have an effective voice in the management of the enterprise. (For the purposes of the statistical inquiry, an effective voice is taken as equivalent to a holding of 10 per cent or more in the foreign enterprise.) Other investments in which the investor does not have an effective voice in the management of the enterprise are mainly portfolio investments and these are not covered here. Cross-border investment by public corporations or in property (which is regarded as direct investment in the national accounts) is not covered here, but is shown in the balance of payments. Similarly direct investment earnings data are shown net of tax in Tables 18.15 and 18.18 but are gross of tax in the balance of payments.

Direct investment is a financial concept and is not the same as capital expenditure on fixed assets. It covers only the money invested in a related concern by the parent company and the concern will then decide how to use the money. A related concern may also raise money locally without reference to the parent company.

The investment figures are published on a net basis, that is, they consist of investments net of disinvestments by a company into its foreign subsidiaries, associate companies and branches.

Definitional changes for 1997

The new European System of Accounts (ESA(95)) definitions were introduced with the 1997 estimates. The changes are as follows:

i) Previously for the measurement of direct investment, an effective voice in the management of an enterprise was taken as the equivalent of a 20 per cent shareholding. This is now 10 per cent.

ii) The Channels Islands (Jersey, Guernsey etc.) and the Isle of Man have been excluded from the definition of the economic territory of the UK. Previously these islands have been considered to be part of the United Kingdom.

iii) Interest received or paid has been replaced by interest accrued in the figures on earnings from direct investment. There is deemed to be little or no impact arising from this definitional change on the estimates.

The definitional changes have been introduced from 1997 only. The data prior to 1997 have not been reworked in Tables 18.13 to 18.18. For clarity, the Offshore Islands are identified separately on the tables. The breaks in the series for the other definitional changes are not quantified but are relatively small. More detailed information on the effect of these changes was published in the business monitor, MA4 – Overseas Direct Investment in February 1999.

Sources of data

The figures in Tables 18.13 to 18.18 are based on annual inquiries into overseas direct investment for 1997. These were sample surveys which involved sending around 900 forms to UK businesses investing abroad and 1700 forms to UK businesses in which foreign parents and associates had invested. The tables also contain some revisions to 1996 as a result of new information coming to light in the course of conducting the 1997 annual inquiries. Further details from the 1997 annual inquiries, including analyses by industry and by components of direct investment, were published in 1999 in MA4, Overseas Direct Investment, one of the ONS Business Monitor series of publications. Updated figures for 1997 and 1998 were published in a First Release in December 1999.

Country allocation

The analysis of inward investment is based on the country of ownership of the immediate parent company. Thus, inward investment in a UK company may be attributed to the country of the intervening overseas subsidiary, rather than the country of the ultimate parent. Similarly, the country analysis of outward investment is based on the country of ownership of the immediate subsidiary. As an example, to the extent that overseas investment in the UK is channelled through holding companies in the Netherlands, the underlying flow of investment from this country is overstated and the inflow from originating countries is understated.

Further information

More detailed statistics on overseas direct investment are available on request from Simon Harrington, Office for National Statistics, Overseas and Financial Division, Room 1.075, Government Buildings, Cardiff Road, Newport, Gwent NP10 8XG. (Telephone 01633 812406, fax 01633 813306, e-mail simon.harrington@ons.gov.uk).

18.1 Trade in goods United Kingdom On a balance of payments basis[1]

		1988	1989	1990	1991	1992	1993	1994	1995	1996	1997	1998
Value(£ millions)												
Exports of goods	BOKG	80 711	92 611	102 313	103 939	107 863	122 039	135 260	153 725	167 403	171 783†	164 132
Imports of goods	BOKH	102 264	117 335	121 020	114 162	120 913	135 358	146 351	165 449	180 489	183 693†	184 897
Balance on trade in goods	BOKI	-21 553	-24 724	-18 707	-10 223	-13 050	-13 319	-11 091	-11 724	-13 086	-11 910†	-20 765
Price index numbers 1995 = 100												
Exports of goods	BQKR	75.5	78.9	81.7	82.8	84.5	95.0	96.9	100.0	100.8	95.2	90.4
Imports of goods	BQKS	76.0	79.3	81.2	82.2	82.9	91.2	94.3	100.0	100.1	93.8	88.0
Terms of trade[2]	BQKT	99.3	99.5	100.6	100.7	101.9	104.2	102.8	100.0	100.7	101.5	102.7
Volume index numbers 1995 = 100												
Exports of goods	BQKU	68.6	72.6	77.2	78.0	79.9	82.8	91.3	100.0	107.7	116.5	118.0
Imports of goods	BQKV	79.9	86.5	86.5	81.9	87.3	90.6	94.6	100.0	109.1	119.0	129.1

1 Statistics of trade in goods on a balance of payments basis are obtained by making certain adjustments in respect of valuation and coverage to the statistics recorded in the *Overseas Trade Statistics*. These adjustments are described in detail in *The Pink Book 1999*.

2 Export price index as a percentage of the import price index.

Source: Office for National Statistics: 020 7533 6064

18.2 Import penetration and export sales ratios for products of manufacturing industry[1,2,3]

United Kingdom: SIC 92

		SIC Division	1994	1995	1996
Ratio 1:Imports/Home Demand					
Description					
Total of divisions below	BAZI		50	52	56
Textiles[2]	BAZJ	*17*	46	47	47
Wearing Apparel: Dressing and dyeing of fur	BAZK	*18*	61	60	65
Chemicals and Chemical products	BAZL	*24*	48	52	57
Ratio 2:Imports/Home Demand plus Exports					
Description					
Total of divisions below	BAZM		34	36	37
Textiles[2]	BAZN	*17*	36	36	36
Wearing Apparel: Dressing and dyeing of fur	BAZO	*18*	47	45	49
Chemicals and Chemical products	BAZP	*24*	31	34	35
Ratio 3:Exports/Sales					
Description					
Total of divisions below	BAZQ		47	49	53
Textiles[2]	BAZR	*17*	34	37	36
Wearing Apparel: Dressing and dyeing of fur	BAZS	*18*	45	46	49
Chemicals and Chemical products	BAZT	*24*	51	53	59
Ratio 4:Exports/Sales plus Imports					
Description					
Total of division below	BAZU		31	32	34
Textiles[2]	BAZV	*17*	22	23	23
Wearing Apparel:Dressing and dyeing of fur	BAZW	*18*	24	25	25
Chemicals and Chemical products	BAZX	*24*	35	35	39

1 The ratios were re-instated in the July 1995 edition of *Monthly Digest of Statistics* after a gap of seven years. They contained figures collected under the PRODCOM Inquiry which was introduced in 1993 and the data was regarded as provisional. Please note that the current tables have been calculated using revised data for 1994.

2 Previously included duplicate trade data for certain headings.

3 Division 23 (Coke, refined petroleum products and nuclear fuel) and SIC 24610 (manufacture of explosives, are excluded from the analysis), SIC 27100 (Basic Iron and Steel and of ferro-alloys) is not incorporated in PRODCOM and therefore also does not form part of the analysis.

4 Division 13 (Mining of Metal Ores) not published in 1995.

5 Division 16 data has a large revision as Excise Duty is now excluded from PRODCOM data.

Source: Office for National Statistics: 01633 813065

18.2 Import penetration and export sales ratios for products of manufacturing industry [1,2,3]

continued

United Kingdom: SIC 92

		SIC Division	1994	1995	1996
Ratio 1 Imports/Home Demand					
Description		*SIC Division*			
Total of divisions below	BAZY		44	45	46
Mining of Metal Ores[4]	BAZZ	13	101	..	..
Other Mining and Quarrying	BBAM	14	106	135	169
Food products and beverages	BBAN	15	21	22	23
Tobacco products	BBAO	16	9	14	14
Textiles	BAZJ	17	46	47	47
Wearing Apparel: Dressing and dyeing of fur	BAZK	18	61	60	65
Tanning and dressing of leather:luggage, handbags, saddlery, harness and footwear	BBAP	19	69	73	74
Wood products of wood and cork (except furniture) articles of straw and plaiting materials	BBAQ	20	42	37	34
Pulp, paper and paper products	BBAR	21	37	39	36
Publishing, printing and reproduction of recorded media[2]	BBAS	22	7	7	7
Chemicals and chemical products	BAZL	24	48	52	57
Rubber and plastic products[2]	BBAT	25	29	29	27
Other non metallic mineral products	BBAU	26	20	20	20
Basic metals [2]	BBAV	27	61	64	66
Fabricated metal products (except machinery and equipment)[2]	BBAW	28	19	20	19
Machinery and equipment not elsewhere classified	BBAX	29	60	61	60
Office machinery and computers	BBAY	30	113	128	102
Electrical machinery not elsewhere classified	BBAZ	31	53	56	58
Radio, television and communication equipment and apparatus	BBBA	32	94	92	98
Medical, precision and optical instruments, watches and clocks	BBBB	33	72	76	79
Motor vehicles, trailers and semi-trailers	BBBC	34	55	53	59
Other transport equipment	BBBD	35	56	54	47
Furniture and manufacturing not elsewhere classified	BBBE	36	48	48	48
Ratio 2 Imports/Home Demand plus Exports					
Description		*SIC Division*			
Total of divisions below	BBBF		32	32	33
Mining of Metal Ores[4]	BBBG	13	98	..	..
Other Mining and Quarrying	BBBH	14	48	54	59
Food products and beverages	BBBI	15	18	19	20
Tobacco products	BBBJ	16	6	8	8
Textiles	BAZN	17	36	36	36
Wearing Apparel: Dressing and dyeing of fur	BAZO	18	47	45	49
Tanning and dressing of leather:luggage, handbags, saddlery, harness and footwear	BBBK	19	54	55	56
Wood products of wood and cork (except furniture) articles of straw and plaiting materials	BBBL	20	40	36	33
Pulp, paper and paper products	BBBM	21	32	35	31
Publishing, printing and reproduction of recorded media[2]	BBBN	22	6	7	6
Chemicals and chemical products	BAZP	24	31	34	35
Rubber and plastic products[2]	BBBO	25	24	24	22
Other non metallic mineral products	BBBP	26	17	16	16
Basic metals[2]	BBBQ	27	44	44	46
Fabricated metal products (except machinery and equipment)[2]	BBBR	28	16	16	16
Machinery and equipment not elsewhere classified	BBBS	29	36	36	35
Office machinery and computers	BBBT	30	56	58	53
Electrical machinery not elsewhere classified	BBBU	31	36	38	38
Radio, television and communication equipment and apparatus	BBBV	32	48	47	52
Medical, precision and optical instruments, watches and clocks	BBBW	33	41	42	44
Motor vehicles, trailers and semi-trailers	BBBX	34	41	39	41
Other transport equipment	BBBY	35	33	31	29
Furniture and manufacturing not elsewhere classified	BBBZ	36	36	35	36

See footnotes on the first part of this table.

Source: Office for National Statistics: 01633 813065

18.2 Import penetration and export sales ratios for products of manufacturing industry[1,2,3]

continued

United Kingdom: SIC 92

		SIC Division	1994	1995	1996
Ratio 3 Exports/Sales					
Description					
Total of divisions below	BBCK		40	42	43
Mining of Metal Ores[4]	BBCL	13	151	..	..
Other Mining and Quarrying	BBCM	14	105	131	158
Food products and beverages	BBCN	15	16	17	16
Tobacco products	BBCO	16	32	46	46
Textiles	BAZR	17	34	37	36
Wearing Apparel:Dressing and dyeing of fur	BAZS	18	45	46	49
Tanning and dressing of leather:luggage, handbags, saddlery, harness and footwear	BBCP	19	50	55	55
Wood products of wood and cork (except furniture) articles of straw and plaiting materials	BBCQ	20	6	6	6
Pulp, paper and paper products	BBCR	21	18	19	19
Publishing, printing and reproduction of recorded media[2]	BBCS	22	10	11	9
Chemicals and chemical products	BAZT	24	51	53	59
Rubber and plastic products[2]	BBCT	25	26	26	25
Other non metallic mineral products	BBCU	26	20	21	22
Basic metals[2]	BBCV	27	50	55	56
Fabricated metal products (except machinery and equipment)[2]	BBCW	28	18	19	18
Machinery and equipment not elsewhere classified	BBCX	29	63	64	64
Office machinery and computers	BBCY	30	115	130	102
Electrical machinery not elsewhere classified	BBDK	31	50	53	56
Radio, television and communication equipment and apparatus	BBDL	32	94	93	98
Medical, precision and optical instruments, watches and clocks	BBDM	33	73	77	80
Motor vehicles, trailers and semi-trailers	BBDN	34	44	44	51
Other transport equipment	BBDO	35	61	62	55
Furniture and manufacturing not elsewhere classified	BBDP	36	39	40	39
Ratio 4 Exports/Sales plus Imports					
Description					
Total of divisions below	BBDQ		27	29	29
Mining of Metal Ores[4]	BBDR	13	2	..	..
Other Mining and Quarrying	BBDS	14	55	60	65
Food products and beverages	BBDT	15	13	14	13
Tobacco products	BBDU	16	30	43	43
Textiles	BAZV	17	22	23	23
Wearing Apparel: Dressing and dyeing of fur	BAZW	18	24	25	25
Tanning and dressing of leather: luggage, handbags, saddlery, harness and footwear	BBDV	19	23	25	24
Wood products of wood and cork (except furniture) articles and straw and plaiting materials	BBDW	20	4	4	4
Pulp, paper and paper products	BBDX	21	13	12	13
Publishing, printing and reproduction of recorded media[2]	BBDY	22	10	10	9
Chemicals and chemical products	BAZX	24	35	35	39
Rubber and plastic products[2]	BBDZ	25	20	20	19
Other non-metallic mineral products	BBEA	26	17	18	19
Basic metals	BBEB	27	28	30	30
Fabricated metal products (except machinery and equipment)[2]	BBEC	28	15	16	15
Machinery and equipment not elsewhere classified	BBED	29	40	41	42
Office machinery and computers	BBEE	30	50	55	48
Electrical machinery not elsewhere classified	BBEF	31	32	33	35
Radio, television and communication equipment and apparatus	BBEG	32	49	49	47
Medical, precision and optical instruments, watches and clocks	BBEH	33	43	45	45
Motor vehicles, trailers and semi-trailers	BBEI	34	26	27	30
Other transport equipment	BBEJ	35	40	43	39
Furniture and manufacturing not elsewhere classified	BBEK	36	25	26	25

See footnotes on the first part of this table.

Source: Office for National Statistics: 01633 813065

18.3 United Kingdom exports of goods, by commodity [1]

£ millions OTS basis seasonally adjusted

		1989	1990	1991	1992	1993	1994	1995	1996	1997	1998
0. Food and live animals	BAFI	4 230	4 344	4 718	5 291	5 720	6 352	7 165	7 117	6 614†	6 322
Of which:											
01. Meat and meat preparations	EPAB	701	613	673	829	965	1 237	1 484	1 099	930†	748
02. Dairy products and eggs	EPAC	503	461	453	535	640	697	831	745	747†	744
04 & 08. Cereals and animal feeding stuffs	EPAD	1 182	1 302	1 406	1 543	1 562	1 528	1 725	2 044	1 813†	1 721
05. Vegetables and fruit	EPAE	283	265	299	333	307	399	462	480	457	410
1. Beverages and tobacco	BAFJ	2 328	2 774	3 032	3 417	3 271	3 745	4 160	4 399	4 562	3 980
11. Beverages	BAFK	1 803	2 116	2 251	2 448	2 623	2 876	3 019	3 183	3 332†	2 910
12. Tobacco	EPAF	526	656	783	971	649	871	1 138	1 213	1 228	1 071
2. Crude materials	BAFL	2 266	2 163	1 918	1 880	2 108	2 405	2 749	2 626	2 507†	2 293
Of which:											
24. Wood, lumber and cork	EPAG	29	27	25	25	22	45	46	52	52	56
25. Pulp and waste paper	EPAH	52	52	38	39	22	50	66	56	64	48
26. Textile fibres	BAFM	498	496	468	489	496	574	642	620	571†	497
28. Metal ores	EPAI	713	634	531	459	618	666	779	668	645†	559
3. Fuels	BOCD	6 176	7 867	7 147	6 970	8 238	8 936	9 438	11 085	10 429	6 990
33. Petroleum and petroleum products	EPBF	5 919	7 544	6 795	6 665	7 801	8 520	8 874	10 424	9 650†	6 494
32, 34 and 35. Coal, gas and electricity	EPBG	256	326	353	308	440	416	565	661	779	494
4. Animal and vegetable oils and fats	BAFN	83	88	94	87	117	171	224	209	265	244
5. Chemicals	BOCH	12 352	13 183	13 783	14 977	17 348	18 859	21 224	22 504	22 059†	22 311
Of which:											
51. Organic chemicals	BAFO	3 374	3 353	3 468	3 701	4 378	4 749	4 981	5 234	5 008†	4 960
52. Inorganic chemicals	BAFP	1 022	954	997	1 185	1 220	1 125	1 248	1 210	1 193†	1 170
54. Medicinal products	TRAE	2 016	2 258	2 558	2 992	3 708	4 017	4 980	5 412	5 459	5 917
57 & 58 Plastics	BOCO	1 934	2 123	2 123	2 139	2 464	2 811	3 310	3 348	3 185†	3 222
6. Manufactures classified chiefly by material	BAFQ	14 519	15 824	15 576	15 485	17 363	19 635	22 698	23 495	22 822†	21 397
Of which:											
63. Wood and cork manufactures	EPAJ	90	116	117	131	131	172	191	229	250	247
64. Paper and paperboard manufactures	EPAK	1 246	1 539	1 625	1 734	1 835	2 056	2 376	2 381	2 327†	2 209
65. Textile manufactures	BAFR	2 207	2 448	2 351	2 457	2 597	2 966	3 319	3 527	3 446†	3 288
67. Iron and steel	BAFS	2 895	3 035	3 013	3 006	3 151	3 676	4 393	4 130	3 667†	3 355
68. Non-ferrous metals	BAFT	1 966	2 192	1 972	1 751	1 913	2 235	2 894	2 723	2 789†	2 456
69. Metal manufactures	BAFU	1 788	2 117	2 182	2 212	2 284	2 654	3 132	3 415	3 403†	3 630
7. Machinery and transport equipment[2]	BAFV	37 750	41 852	43 602	44 422	48 487	56 176	66 337	74 824	79 982†	79 049
71 - 716, 72, 73 & 74. Mechanical machinery	BOCS	12 867	14 423	13 782	14 263	15 070	16 446	18 728	21 005	22 798†	23 267
716, 75, 76 & 77. Electrical machinery	BOCT	13 821	15 196	15 799	16 429	20 741	25 151	31 229	33 725	34 750†	34 990
78. Road vehicles	BAFW	6 071	7 300	8 557	8 895	8 339	9 736	11 852	14 609	15 019†	14 788
79. Other transport equipment	EPBH	4 996	4 935	5 471	4 837	4 342	4 849	4 525	5 492	7 421	6 007
8. Miscellaneous manufactures[2]	BAFX	11 743	13 313	13 104	13 925	14 892	17 463	19 231	21 380	21 908†	20 943
Of which:											
84 & 85. Clothing and footwear	BOCU	1 673	1 973	2 234	2 426	2 726	3 219	3 619	4 029	3 902†	3 560
87 & 88. Scientific and photographic	BOCV	3 927	4 112	4 263	4 459	4 770	5 402	6 014	6 749	7 112†	6 837
9. Other commodities and transactions	BONF	2 287	2 299	1 851	2 070	1 609	1 460	1 758	1 940	1 942†	1 883
TOTAL UK EXPORTS	CGKI	93 724	103 691	104 816	108 509	119 145	135 189	154 976	169 569	173 081†	165 387

1 The numbers on the left hand side of the table refer to the code numbers of the *Standard International Trade Classification*, Revision 3, which was introduced in January 1988. More commodity detail, *not* seasonally adjusted, is available from the monthly *Overseas Trade Statistics of the United Kingdom*, published by The Stationery Office and from the Marketing Agents listed therein.

2 Sections 7 and 8 are shown by broad economic category in Table B4 of the *Monthly Review of External Trade Statistics*.

Source: Office for National Statistics: 020 7533 6064

18.4 United Kingdom imports of goods, by commodity [1]

£ millions OTS basis seasonally adjusted

		1989	1990	1991	1992	1993	1994	1995	1996	1997	1998
0. Food and live animals	BAFY	9 760	10 411	10 389	11 402	11 622	12 385	13 629	14 835	14 019†	13 924
Of which:											
01. Meat and meat preparations	BAFZ	1 828	1 890	1 844	2 030	1 957	1 976	2 339	2 663	2 333†	2 092
02. Dairy products and eggs	BAGA	787	916	872	1 113	1 025	1 144	1 123	1 246	1 142†	1 163
04 & 08. Cereals and animal feeding stuffs	BAGB	1 305	1 411	1 439	1 722	1 820	1 777	1 868	2 020	2 087†	1 893
05. Vegetables and fruit	BAGC	2 726	2 966	3 005	3 118	3 135	3 484	4 019	4 408	4 097†	4 236
1. Beverages and tobacco	BAGD	1 667	1 908	1 936	2 027	2 119	2 262	2 525	2 945	3 015†	3 156
11. Beverages	BAGE	1 323	1 531	1 465	1 563	1 689	1 882	2 045	2 316	2 443†	2 671
12. Tobacco	BAGF	345	377	469	462	433	382	479	626	573†	485
2. Crude materials	BAGG	6 106	5 721	4 681	4 667	4 958	5 547	6 393	6 357	6 186	5 619
Of which:											
24. Wood, lumber and cork	BAGH	1 430	1 408	1 044	1 026	1 111	1 406	1 228	1 288	1 336†	1 202
25. Pulp and waste paper	BAGI	897	779	609	630	545	642	1 069	713	628	535
26. Textile fibres	BAGJ	682	549	452	480	482	633	691	691	646†	494
28. Metal ores	BAGK	1 575	1 478	1 234	1 164	1 165	1 165	1 499	1 619	1 622†	1 485
3. Fuels	BODD	6 430	7 866	7 580	7 018	7 304	6 154	5 896	7 040	6 529†	4 715
33. Petroleum and petroleum products	EPBI	4 867	6 286	5 845	5 325	5 755	4 843	4 684	5 730	5 206	3 532
32, 34 and 35. Coal, gas and electricity	EPBJ	1 565	1 581	1 741	1 689	1 555	1 311	1 209	1 310	1 323†	1 182
4. Animal and vegetable oils and fats	BAGL	385	377	388	421	480	539	604	731	647	591
5. Chemicals	BODH	10 439	10 833	10 980	11 618	13 030	14 623	17 974	18 746	18 021†	18 145
Of which:											
51. Organic chemicals	BAGM	2 682	2 630	2 622	2 794	3 122	3 544	4 810	4 890	4 629†	4 724
52. Inorganic chemicals	BAGN	957	964	1 034	952	997	1 105	1 131	1 288	1 159†	1 074
54. Medicinal products	TRAH	1 064	1 161	1 374	1 663	2 034	2 331	2 862	3 175	3 208	3 457
57 & 58. Plastics	BOEX	3 056	3 228	3 030	3 038	3 302	3 775	4 708	4 478	4 293†	4 051
6. Manufactures classified chiefly by material	BAGO	21 740	21 903	20 522	20 671	21 805	24 551	28 477	29 611	28 865†	28 748
Of which:											
63. Wood and cork manufactures	BAFG	967	951	822	847	890	1 039	1 075	1 121	1 128†	1 138
64. Paper and paperboard manufactures	BAGP	4 015	4 017	3 870	3 804	3 743	4 287	5 317	5 113	4 681†	4 692
65. Textile manufactures	BAGQ	3 771	3 938	3 738	3 943	4 016	4 551	4 948	5 255	5 184†	5 071
67. Iron and steel	BAGR	2 802	2 688	2 621	2 515	2 556	2 975	3 708	3 716	3 436†	3 335
68. Non-ferrous metals	BAGS	3 070	3 004	2 556	2 592	2 917	3 032	3 865	3 860	3 756	3 851
69. Metal manufactures	BAGT	2 480	2 592	2 525	2 579	2 568	2 925	3 340	3 625	3 701†	3 908
7. Machinery and transport equipment[2]	BAGU	46 734	47 164	43 124	47 319	53 425	60 931	69 536	78 426	82 716†	85 681
71 - 716, 72, 73 & 74. Mechanical machinery	BOMY	12 133	12 068	11 073	11 665	11 875	13 432	15 570	17 116	17 630†	17 882
716, 75, 76 & 77. Electrical machinery	BOMZ	18 252	18 451	18 361	20 129	23 890	26 942	32 558	36 741	37 204†	38 354
78. Road vehicles	BAGV	13 003	12 597	10 219	12 120	14 418	16 287	18 247	20 851	22 226†	23 210
79. Other transport equipment	EPBK	3 347	4 052	3 477	3 410	3 244	4 279	3 162	3 723	5 661†	6 240
8. Miscellaneous manufactures[2]	BAGW	17 006	18 171	17 488	19 051	20 184	21 915	23 230	25 864	27 488†	28 969
Of which:											
84 & 85. Clothing and footwear	BOKM	4 516	5 074	5 300	5 629	5 984	6 368	6 893	8 097	8 822†	9 190
87 & 88. Scientific and photographic	BOKN	3 992	4 073	4 090	4 258	4 689	4 989	5 576	6 240	6 306†	6 386
9. Other commodities and transactions	BOMD	1 567	1 740	1 793	1 679	1 256	991	1 358	1 613	1 627†	1 742
TOTAL UK IMPORTS	CGHM	121 827	126 090	118 872	125 867	136 177	149 890	169 611	186 151	189 111†	191 278

1 The numbers on the left hand side of the table refer to the code numbers of the *Standard International Trade Classification*, Revision 3, which was introduced in January 1988. More commodity detail, *not* seasonally adjusted, is available from the monthly *Overseas Trade Statistics of the United Kingdom*, published by The Stationery Office and from the Marketing Agents listed therein.

2 Sections 7 and 8 are shown by broad economic category in Table C4 of the *Monthly Review of External Trade Statistics.*

Source: Office for National Statistics: 020 7533 6064

18.5 United Kingdom exports of goods, by area[1]

£ millions OTS basis seasonally adjusted

		1989	1990	1991	1992	1993	1994	1995	1996	1997	1998
European Union:[2]	BOGB	52 123	59 937	63 969	65 610	66 707	77 297	90 417	97 445	96 120†	95 646
Germany	EQBK	11 367	13 272	14 713	15 337	15 472	17 441	20 440	21 080	20 698†	20 589
France	BAGX	9 662	10 988	11 898	11 608	11 742	13 728	15 411	17 397	16 614†	16 450
Italy	BAHB	4 674	5 637	6 193	6 196	5 859	6 873	7 962	8 169	8 216†	8 611
Netherlands	BAGZ	6 738	7 548	8 278	8 573	7 804	9 653	12 454	13 718	13 934†	12 993
Belgium & Luxembourg	BAGY	4 936	5 673	5 898	5 761	6 876	7 397	8 380	8 668	8 457†	8 417
Denmark	BAHD	1 226	1 419	1 413	1 573	1 554	1 830	2 131	2 254	2 095†	2 046
Irish Republic	BAHC	4 763	5 332	5 303	5 788	6 123	7 204	7 867	8 816	9 367†	9 600
Greece	BAHE	577	684	671	777	887	938	1 059	1 166	1 054†	1 046
Portugal	BAHF	925	1 038	1 087	1 176	1 322	1 269	1 484	1 707	1 751†	1 718
Spain	BAHG	3 259	3 860	4 411	4 552	4 322	5 161	6 160	6 842	6 749†	7 164
Sweden	BAHI	2 455	2 724	2 476	2 459	2 787	3 423	4 197	4 496	4 456†	4 393
Finland	BAHJ	937	1 054	855	1 007	1 078	1 323	1 734	1 845	1 570†	1 433
Austria	BAHL	604	708	773	803	881	1 057	1 138	1 287	1 159†	1 186
Other Western Europe:	BOGC	4 028	4 580	4 529	4 318	5 311	5 727	6 451	7 383	7 994†	7 879
Of which:											
Norway	BAHH	1 057	1 291	1 323	1 421	1 502	2 035	2 007	2 066	2 659†	2 757
Switzerland	BAHK	2 247	2 359	2 105	1 845	2 276	2 460	2 748	3 206	3 007†	2 984
Turkey	BQPE	435	609	741	691	1 048	815	1 157	1 566	1 766†	1 657
Iceland	BOLB	71	89	95	93	147	110	137	155	159	164
North America:	BOGD	14 641	15 235	13 409	14 262	17 720	19 525	20 506	22 436	23 947†	25 061
Of which:											
USA	BOGA	12 185	12 999	11 347	12 231	15 353	16 909	17 948	19 831	20 968†	21 954
Canada	BAHM	2 168	1 902	1 701	1 584	1 844	1 917	1 812	1 975	2 155	2 204
Mexico	EQBR	204	262	274	292	336	390	278	319	431	550
Other OECD countries:	BOGE	5 357	5 815	5 262	5 493	6 611	7 664	9 312	10 922	10 769†	8 997
Of which:											
Japan	BAHN	2 308	2 632	2 257	2 233	2 655	3 000	3 785	4 265	4 177†	3 238
Australia	BAHO	1 712	1 647	1 356	1 377	1 601	1 928	2 121	2 464	2 456	2 276
South Korea	EPBP	494	621	787	660	797	972	1 154	1 303	1 222	692
Poland	BQCA	196	222	350	606	728	701	946	1 353	1 356†	1 220
New Zealand	BAHP	400	440	260	263	333	416	436	473	410	353
Czech Rep	BQCC	..	..	..	..	289	374	567	715	709	714
Oil exporting countries:	BOGF	5 831	5 574	5 718	6 014	6 527	5 738	6 375	8 095	9 786†	7 531
Of which:											
Saudi Arabia	BAHR	2 433	2 013	2 228	1 966	1 826	1 519	1 643	2 484	3 802†	2 694
Kuwait	EPBL	229	181	178	263	311	314	553	580	504	336
Indonesia	EWXH	183	195	198	314	332	366	528	830	702	380
Iran	BAHS	257	385	516	583	497	290	334	398	398†	333
Nigeria	BAHQ	389	501	545	622	635	459	434	433	426	469
Rest of the World	EQBI	11 605	12 516	11 895	12 534	16 130	18 701	21 399	22 660	23 896†	19 964
Of which:											
Eastern Europe	EPBR	1 041	1 018	658	747	1 104	1 485	1 951	2 272	2 679†	2 393
Other America	BPCY	933	1 001	1 415	1 469	1 857	2 378	2 506	2 677	3 194†	3 129
China	BQPC	419	471	322	428	740	846	829	739	922	870
Hong Kong	EPBQ	1 114	1 237	1 387	1 612	2 134	2 297	2 654	2 926	3 213†	2 703
India	ELBV	1 383	1 264	1 017	946	1 131	1 312	1 684	1 706	1 576	1 259
Israel	BAHT	503	568	529	587	878	1 034	1 108	1 267	1 179	1 091
Malaysia	EPBN	441	602	591	636	966	1 311	1 190	1 160	1 206†	684
Pakistan	BQPG	233	253	273	312	330	354	340	344	269	239
Philippines	EWXI	137	158	148	204	308	356	433	397	601	307
Singapore	EPBO	776	1 041	1 020	1 147	1 431	1 769	2 068	2 144	2 041	1 620
South Africa	BAFH	1 039	1 112	1 024	1 079	1 126	1 415	1 832	1 880	1 635†	1 546
Taiwan	EQBQ	410	431	518	559	668	736	962	941	1 034	880
Thailand	EPBM	429	417	466	475	664	749	835	974	862	394

1 More country detail, *not* seasonally adjusted, is available from *Overseas Trade Statistics of the United Kingdom,* published by The Stationery Office and from the Marketing Agents listed therein.

2 In *Overseas Trade Statistics of the United Kingdom* estimates of non-response and below threshold trade with EU member states are not allocated to individual countries. For the purposes of this table, data for the individual EU countries have been adjusted to include these estimates so that comparisons with non-EU countries are more meaningful.

Source: Office for National Statistics: 020 7533 6064

18.6 United Kingdom imports of goods, by area[1]

£ millions OTS basis seasonally adjusted

		1989	1990	1991	1992	1993	1994	1995	1996	1997	1998
European Union:[2]	BOGJ	70 763	72 895	67 562	72 022	72 812	83 627	94 561	101 846	101 041†	102 897
Germany	EQBL	20 257	20 151	17 871	19 153	19 454	22 194	26 429	27 660	25 920†	25 516
France	BAHV	10 899	11 851	11 120	12 305	13 130	15 223	16 487	18 135	17 949†	17 913
Italy	BAHZ	6 734	6 783	6 438	6 808	6 514	7 609	8 328	8 964	9 656†	9 900
Netherlands	BAHX	9 661	10 559	10 115	9 969	8 778	10 214	11 603	12 628	12 466†	13 634
Belgium & Luxembourg	BAHW	5 723	5 773	5 524	5 771	6 534	7 193	8 069	8 761	9 156†	9 573
Denmark	BAIB	2 244	2 294	2 246	2 397	2 087	2 203	2 214	2 397	2 344†	2 188
Irish Republic	BAIA	4 298	4 531	4 450	5 095	5 338	5 985	7 103	7 368	7 473†	7 920
Greece	BAIC	398	403	382	372	312	358	432	399	407†	370
Portugal	BAID	1 047	1 184	1 051	1 177	1 205	1 305	1 479	1 690	1 780†	1 812
Spain	BAIE	2 900	2 992	2 732	3 039	3 188	3 735	4 390	5 131	5 159†	5 836
Sweden	BAIG	3 766	3 620	3 172	3 300	3 495	4 260	4 573	4 852	4 746†	4 437
Finland	BAIH	1 899	1 788	1 536	1 687	1 839	2 306	2 519	2 688	2 575†	2 365
Austria	BAIJ	937	966	925	949	938	1 042	935	1 173	1 410†	1 433
Other Western Europe:	BOGK	8 579	9 388	8 726	8 602	9 798	9 657	10 703	11 829	11 343†	10 309
Of which:											
Norway	BAIF	3 637	4 235	4 254	3 885	4 159	3 823	4 327	4 982	4 927†	3 622
Switzerland	BAII	4 134	4 254	3 754	3 919	4 738	4 818	5 153	5 403	4 896	5 042
Turkey	BQPF	536	551	403	456	533	630	794	933	1 045†	1 165
Iceland	BPFZ	197	258	237	241	252	240	253	269	244	264
North America:	BOGL	16 094	16 924	15 887	15 872	18 462	19 937	23 038	25 743	28 066†	28 782
Of which:											
USA	BOGI	13 521	14 357	13 705	13 715	16 328	17 729	20 268	22 809	25 029	25 656
Canada	BAIK	2 285	2 258	1 919	1 898	1 855	1 883	2 379	2 485	2 554†	2 598
Mexico	EQBS	166	173	149	154	166	241	299	334	383	379
Other OECD countries:	BOGM	10 179	9 871	9 475	10 442	11 937	12 593	14 219	14 355	15 170†	15 644
Of which:											
Japan	BAIM	7 104	6 760	6 754	7 444	8 520	8 832	9 635	8 992	9 409†	9 549
Australia	BAIN	866	1 036	867	1 014	997	1 064	1 112	1 296	1 370	1 428
South Korea	EQBN	1 165	963	925	932	1 077	1 097	1 562	2 039	2 239†	2 308
Poland	BQCB	333	357	313	356	450	546	640	601	620	686
New Zealand	BAIO	437	485	391	429	496	542	578	632	578†	543
Czech Rep	BQCD	..	..	..	..	244	278	325	374	465	579
Oil exporting countries:	BOGN	2 313	2 974	2 786	3 078	3 801	3 267	3 255	3 750	3 971	3 603
Of which:											
Saudi Arabia	BAIQ	502	795	955	964	1 275	812	720	753	997	893
Kuwait	EPBS	151	108	36	128	215	239	150	181	201	186
Indonesia	EWXJ	273	329	415	538	701	783	904	981	1 030†	982
Iran	BAIR	251	279	158	163	244	134	128	119	37	37
Nigeria	BAIP	129	298	248	166	114	124	180	296	123	141
Rest of the World	EQBJ	13 688	13 997	14 394	15 550	19 068	20 385	23 448	28 213	29 014†	29 193
Of which:											
Eastern Europe	EPBV	1 207	1 219	1 159	1 023	1 442	1 767	2 043	2 535	2 776	2 786
Other America	BPGY	1 599	1 584	1 993	2 228	2 428	2 470	2 825	3 167	3 068†	2 909
China	BQPD	529	585	708	953	1 329	1 642	1 941	2 202	2 496†	2 962
Hong Kong	EPBU	2 049	1 970	2 148	2 398	2 998	3 083	3 541	4 074	4 349†	4 591
India	EQAR	702	800	777	864	1 090	1 291	1 434	1 612	1 623†	1 456
Israel	BAIS	479	506	457	486	552	572	693	832	879†	920
Malaysia	EQBM	676	776	914	1 104	1 397	1 204	1 489	2 381	2 026†	1 992
Pakistan	BQPH	212	237	321	273	324	360	364	391	380	357
Philippines	EWXK	234	221	230	241	277	245	353	896	761	901
Singapore	EPBT	902	1 024	1 135	1 193	1 618	1 896	2 210	2 573	2 715	2 463
South Africa	BAIL	866	1 079	965	866	999	971	1 115	1 222	1 392	1 422
Taiwan	EQBO	1 353	1 210	1 270	1 395	1 618	1 583	1 727	2 090	2 341†	2 330
Thailand	EQBP	444	484	635	643	772	914	1 040	1 189	1 223†	1 330

1 More country detail, *not* seasonally adjusted, is available from *Overseas Trade Statistics of the United Kingdom*, published by The Stationery Office and from the Marketing Agents listed therein.

2 In *Overseas Trade Statistics of the United Kingdom* estimates of non-response and below-threshold trade with EU member states are not allocated to individual countries. For the purpose of this table, data for the individual EU countries have been adjusted to include these estimates so that comparisons with non-EU countries are more meaningful.

Source: Office for National Statistics: 020 7533 6064

18.7 Services supplied (exports) and purchased (imports): 1997

£ millions

	Exports	Imports	Net
Business services			
Legal	201	177	24
Accounting	263	106	157
Management consulting	880	330	550
Advertising	758	392	365
Market research	201	67	134
Research and development	1 590	666	924
Insurance: premiums	-	61	-61
claims	4	-	4
Financial Services	136	106	30
Property	20	15	4
Other business services	2 364	1 059	1 305
Telecommunications services			
Communications	890	996	-107
Computer	1 063	323	740
Information	124	67	57
Technical services			
Architectural	107	5	103
Engineering (consulting, process etc.)	2 189	883	1 306
Surveying	23	22	1
Construction	231	144	87
Agriculture	14	3	11
Mining	4	4	-
Other technical	763	298	465
Operational leasing	43	49	-6
Cultural services			
TV and radio services	97	12	85
Music services (excluding royalties)	27	8	19
Other cultural	110	75	35
Royalties	3 578	3 000	578
Trade related services			
Merchanting	321	44	277
Commission	462	349	112
Commodity trading	142	109	33
Management services to affiliated companies	888	424	463
All other services	1 379	350	1 029
WORLD TOTAL	18 870	10 144	8 726

1 Due to rounding, the sum of constituent items may not always equal the total shown.
2 Data excludes the following industries: Financial, Film and TV, Travel and Transport, Public Sector (including Education) and Law Society members.

Source: Office for National Statistics

18.8 Overseas trade in services by country: 1997

£ millions

	Exports	Imports	Net
European Union			
Austria	59	33	26
Belgium/Luxembourg	689	279	410
Denmark	165	138	27
Finland	166	29	136
France	801	626	175
Germany	1 565	837	728
Greece	43	24	19
Irish Republic	844	240	604
Italy	354	231	123
Netherlands	1 010	625	385
Portugal	66	74	-8
Spain	263	103	160
Sweden	305	111	193
Total European Union	6 329	3 351	2 978
EFTA			
Iceland	2	2	-
Norway	358	184	174
Switzerland	748	463	284
Total EFTA	1 107	650	458
Other European countries			
Czech Republic	47	12	34
Poland	54	19	35
Russia	68	30	38
Channel Islands	60	16	44
Isle of Man	3	8	-5
Turkey	96	35	61
Rest of Europe	815	545	270
Total Europe	8 579	4 667	3 912
America			
Brazil	50	14	36
Canada	230	89	140
Mexico	30	4	26
USA	4 528	3 592	936
Rest of America	625	204	421
Total America	5 462	3 903	1 559
Asia			
China	38	17	22
Hong Kong	232	114	117
India	189	47	143
Indonesia	75	27	47
Israel	29	47	-17
Japan	699	342	356
Malaysia	129	62	67
Pakistan	70	18	52
Phillippines	27	8	19
Singapore	143	41	102
South Korea	147	22	125
Taiwan	48	11	38
Thailand	123	16	107
Rest of Asia	1 928	496	1 432
Total Asia	3 877	1 266	2 610
Australia and Oceania			
Australia	244	82	162
New Zealand	30	8	22
Rest of Australia and Oceania	54	45	9
Total Australia and Oceania	328	135	193
Africa			
South Africa	171	44	126
Rest of Africa	432	102	330
Total Africa	603	147	456
Total unallocated	22	26	-3
WORLD TOTAL	18 870	10 144	8 726
Economic Zones			
OCED	13 339	8 166	5 173
NAFTA	4 788	3 686	1 102
Central and Eastern Europe	163	53	110
OPEC	1 723	392	1 331
ASEAN	507	158	350
CIS	129	36	92
NICs1	570	188	382
Offshore Financial centres	595	243	352
ACP	313	63	249

1 Due to rounding, the sum of constituent items may not always equal the total shown.
2 Data excludes the following industries: Financial, Film and TV, Travel and Transport, Public Sector (including Education) and Law Society members.

Source: Office for National Statistics

18.9 Summary of balance of payments in 1998
United Kingdom

£ millions

	Credits	Debits
1. Current account		
A. Goods and services	224 202	232 714
1. Goods	164 132	184 897
2. Services	60 070	47 817
2.1. Transportation	11 505	13 649
2.2. Travel	14 503	20 126
2.3. Communications	1 210	1 454
2.4. Construction	285	108
2.5. Insurance	3 194	570
2.6. Financial	6 318	171
2.7. Computer and information	1 510	436
2.8. Royalties and licence fees	4 061	3 696
2.9. Other business	15 673	5 651
2.10. Personal, cultural and recreational	682	432
2.11. Government	1 129	1 524
B. Income	111 365	96 191
1. Compensation of employees	777	701
2. Investment income	110 588	95 490
2.1 Direct investment	32 426	14 407
2.2 Portfolio investment	27 161	26 477
2.3 Other investment (including earnings on reserve assets)	51 001	54 606
C. Current transfers	15 596	22 122
1. Central government	6 467	6 585
2. Other sectors	9 129	15 537
Total current account	**351 163**	**351 027**
2. Capital and financial accounts		
A. Capital account	1 269	848
1. Capital transfers	1 183	744
2. Acquisition/disposal of non-produced, non-financial assets	86	104
B. Financial account	105 304	114 329
1. Direct investment	40 792	64 077
Abroad		64 077
1.1. Equity capital		46 125
1.2. Reinvested earnings		15 002
1.3. Other capital[1]		2 950
In United Kingdom	40 792	
1.1. Equity capital	29 521	
1.2. Reinvested earnings	5 106	
1.3. Other capital[2]	6 165	
2. Portfolio investment	20 206	34 380
Assets		34 380
2.1. Equity securities		2 453
2.2. Debt securities		31 927
Liabilities	20 206	
2.1. Equity securities	36 946	
2.2. Debt securities	–16 740	
3. Other investment	44 306	16 037
Assets		16 037
3.1 Trade credits		–642
3.2 Loans		465
3.3 Currency and deposits		15 921
3.4 Other assets		293
Liabilities	44 306	
3.1. Trade credits	–	
3.2. Loans	–4 158	
3.3. Currency and deposits	48 208	
3.4. Other liabilities	256	
4. Reserve assets		–166
4.1. Monetary gold		931
4.2. Special drawing rights		–17
4.3. Reserve position in the IMF		752
4.4. Foreign exchange		–1 829
Total capital and financial accounts	**106 573**	**115 177**
Total current, capital and financial accounts	**457 736**	**466 204**
Net errors and omissions	8 468	

1 Other capital transaction on direct investment abroad represents claims on affiliated enterprises less liabilities to affiliated enterprises
2 Other capital transactions on direct investment in the United Kingdom represents liabilities to direct investors less claims on direct investors

Source: Office for National Statistics

18.10 Summary of balance of payments Balances (credits less debits)

United Kingdom

£ millions

	Current account										
	Trade in goods	Trade in services	**Total goods and services**	Compensation of employees	Investment income	**Total income**	**Current transfers**	**Current balance**	Capital account	Financial account	Net errors & omissions
	LQCT	KTMS	KTMY	KTMP	HMBM	HMBP	KTNF	HBOG	FKMJ	HBNT	HHDH
1946	−101	−274	−375	−20	88	68	166	−141	−21	181	−19
1947	−358	−197	−555	−19	159	140	123	−292	−21	552	−239
1948	−152	−64	−216	−20	252	232	96	112	−17	−58	−37
1949	−137	−43	−180	−20	234	214	29	63	−12	−103	52
1950	−54	−4	−58	−21	426	405	39	386	−10	−447	71
1951	−692	32	−660	−21	365	344	29	−287	−15	426	−124
1952	−272	123	−149	−22	265	243	169	263	−15	−229	−19
1953	−244	123	−121	−25	238	213	143	235	−13	−177	−45
1954	−210	115	−95	−27	261	234	55	194	−13	−174	−7
1955	−315	42	−273	−27	175	148	43	−82	−15	34	63
1956	50	26	76	−30	237	207	2	285	−13	−250	−22
1957	−29	121	92	−32	257	225	−5	312	−13	−313	14
1958	34	119	153	−34	303	269	4	426	−10	−411	−5
1959	−116	118	2	−37	269	232	–	234	−5	−68	−161
1960	−404	39	−365	−35	235	200	−6	−171	−6	−7	184
1961	−144	51	−93	−35	259	224	−9	122	−12	23	−133
1962	−104	50	−54	−37	346	309	−14	241	−12	−195	−34
1963	−123	4	−119	−38	416	378	−37	222	−16	−30	−176
1964	−551	−34	−585	−33	420	387	−62	−260	−17	392	−115
1965	−263	−66	−329	−34	467	433	−66	38	−18	49	−69
1966	−111	44	−67	−39	413	374	−89	218	−19	16	−215
1967	−601	157	−444	−39	411	372	−117	−189	−25	172	42
1968	−708	341	−367	−48	357	309	−89	−147	−26	672	−499
1969	−214	392	178	−47	543	496	−120	554	−23	−838	307
1970	−18	455	437	−56	612	556	−82	911	−22	−864	−25
1971	205	590	795	−63	560	497	−51	1 241	−23	−1 441	223
1972	−736	665	−71	−52	518	466	−51	344	−35	462	−771
1973	−2 573	760	−1 813	−68	1 283	1 215	−316	−914	−39	971	−18
1974	−5 241	1 065	−4 176	−92	1 429	1 337	−340	−3 179	−34	3 162	51
1975	−3 245	1 393	−1 852	−102	706	604	−317	−1 565	−36	1 554	47
1976	−3 930	2 465	−1 465	−140	1 385	1 245	−507	−727	−12	420	319
1977	−2 271	3 219	948	−152	−120	−272	−469	207	11	−3 665	3 447
1978	−1 534	3 679	2 145	−140	386	246	−958	1 433	−79	−2 798	1 444
1979	−3 326	3 965	639	−130	540	410	−961	88	−103	−97	112
1980	1 329	3 717	5 046	−82	−1 346	−1 428	−452	3 166	−4	−3 238	76
1981	3 238	3 834	7 072	−66	−369	−435	−88	6 549	−79	−6 612	142
1982	1 879	3 069	4 948	−95	−461	−556	−285	4 107	6	−2 010	−2 103
1983	−1 618	3 941	2 323	−89	1 205	1 116	−140	3 299	75	−4 068	694
1984	−5 409	4 341	−1 068	−94	2 497†	2 403	−126	1 209	107	−8 009†	6 693
1985	−3 416	6 619	3 203	−120	132	12	−988	2 227	185	−2 902	490
1986	−9 617	6 505	−3 112	−156	3 028	2 872	−2 045	−2 285	135	−2 341	4 491
1987	−11 698	6 686	−5 012	−174	1 677	1 503	−2 074	−5 583	333	5 065	185
1988	−21 553	4 330	−17 223	−64	1 355	1 291	−1 605	−17 537	235	15 135	2 167
1989	−24 724	3 917	−20 807	−138	74	−64	−2 620	−23 491	270	19 024	4 197
1990	−18 707	4 010	−14 697	−110	−448	−558	−4 258	−19 513	497	17 529	1 487
1991	−10 223	4 471	−5 752	−63	−1 890	−1 953	−669	−8 374	290	9 990	−1 906
1992	−13 050	5 674	−7 376	−49	2 164	2 115†	−4 821	−10 082†	421	5 716	3 945
1993	−13 319	6 623	−6 696	35	650	685	−4 607	−10 618	309	9 447	862†
1994	−11 091	6 528	−4 563	−170	7 940	7 770	−4 665	−1 458	33	−6 082	7 507
1995	−11 724	8 915	−2 809	−296	6 272	5 976	−6 912	−3 745	534	937	2 274
1996	−13 086	8 897	−4 189	93	8 018	8 111	−4 522	−600	736	1 781	−1 917
1997	−11 910†	12 414†	504†	83	11 087	11 170	−5 051†	6 623	804†	−13 186	5 759
1998	−20 765	12 253	−8 512	76	15 098	15 174	−6 526	136	421	−9 025	8 468

Source: Office for National Statistics

18.11 Balance of Payments: current account
United Kingdom

£ millions

		1988	1989	1990	1991	1992	1993	1994	1995	1996	1997	1998
Credits												
Exports of goods and services												
Exports of goods	LQAD	80 711	92 611	102 313	103 939	107 863	122 039	135 260	153 725	167 403	171 783†	164 132
Exports of services	KTMQ	26 723	29 272	31 188	31 426	35 428	40 039	43 507	48 687	52 900	57 543†	60 070
Total exports of goods and services	KTMW	107 434	121 883	133 501	135 365	143 291	162 078	178 767	202 412	220 303	229 326†	224 202
Income												
Compensation of employees	KTMN	445	476	543	551	551	595	681	887	911	1 007	777
Investment income	HMBN	61 932†	81 799	88 186	84 419	74 549	78 825	82 639	97 086	101 669	106 898	110 588
Total income	HMBQ	62 377	82 275	88 729	84 970	75 100†	79 420	83 320	97 973	102 580	107 905	111 365
Current transfers												
Central government	FJUM	5 160	4 924	4 294	6 746	4 254	5 306	5 008	4 845	6 754	5 979†	6 467
Other sectors	FJUN	4 514	5 161	5 956	7 493	7 386	7 958	7 801	8 518	14 632	9 326†	9 129
Total current transfers	KTND	9 674	10 085	10 250	14 239	11 640	13 264	12 809	13 363	21 386	15 305†	15 596
Total	HBOE	**179 485**	**214 243**	**232 480**	**234 574**	**230 031†**	**254 762**	**274 896**	**313 748**	**344 269**	**352 536**	**351 163**
Debits												
Imports of goods and services												
Imports of goods	LQBL	102 264	117 335	121 020	114 162	120 913	135 358	146 351	165 449	180 489	183 693†	184 897
Imports of services	KTMR	22 393	25 355	27 178	26 955	29 754	33 416	36 979	39 772	44 003	45 129†	47 817
Total imports of goods and services	KTMX	124 657	142 690	148 198	141 117	150 667	168 774	183 330	205 221	224 492	228 822†	232 714
Income												
Compensation of employees	KTMO	509	614	653	614	600	560	851	1 183	818	924	701
Investment income	HMBO	60 577†	81 725	88 634	86 309	72 385	78 175	74 699	90 814	93 651	95 811	95 490
Total income	HMBR	61 086	82 339	89 287	86 923	72 985†	78 735	75 550	91 997	94 469	96 735	96 191
Current transfers												
Central government	FJUO	2 226	2 055	1 995	3 218	3 506	4 156	4 795	4 795	5 056	5 087†	6 585
Other sectors	FJUP	9 053	10 650	12 513	11 690	12 955	13 715	12 679	15 480	20 852	15 269†	15 537
Total current transfers	KTNE	11 279	12 705	14 508	14 908	16 461	17 871	17 474	20 275	25 908	20 356†	22 122
Total	HBOF	**197 022**	**237 734**	**251 993**	**242 948**	**240 113†**	**265 380**	**276 354**	**317 493**	**344 869**	**345 913**	**351 027**
Balances												
Trade in goods and services												
Trade in goods	LQCT	–21 553	–24 724	–18 707	–10 223	–13 050	–13 319	–11 091	–11 724	–13 086	–11 910†	–20 765
Trade in services	KTMS	4 330	3 917	4 010	4 471	5 674	6 623	6 528	8 915	8 897	12 414†	12 253
Total trade in goods and services	KTMY	–17 223	–20 807	–14 697	–5 752	–7 376	–6 696	–4 563	–2 809	–4 189	504†	–8 512
Income												
Compensation of employees	KTMP	–64	–138	–110	–63	–49	35	–170	–296	93	83	76
Investment income	HMBM	1 355†	74	–448	–1 890	2 164	650	7 940	6 272	8 018	11 087	15 098
Total income	HMBP	1 291	–64	–558	–1 953	2 115†	685	7 770	5 976	8 111	11 170	15 174
Current transfers												
Central government	FJUQ	2 934	2 869	2 299	3 528	748	1 150	213	50	1 698	892†	–118
Other sectors	FJUR	–4 539	–5 489	–6 557	–4 197	–5 569	–5 757	–4 878	–6 962	–6 220	–5 943†	–6 408
Total current transfers	KTNF	–1 605	–2 620	–4 258	–669	–4 821	–4 607	–4 665	–6 912	–4 522	–5 051†	–6 526
Total (Current balance)	HBOG	**–17 537**	**–23 491**	**–19 513**	**–8 374**	**–10 082†**	**–10 618**	**–1 458**	**–3 745**	**–600**	**6 623**	**136**

Source: Office for National Statistics

18.12 Balance of payments: summary of international investment position, financial account and investment income

United Kingdom

£ billions

		1988	1989	1990	1991	1992	1993	1994	1995	1996	1997	1998
Investment abroad												
International investment position												
Direct investment	HBWD	104.3	123.2	120.7	126.2	149.7	170.3	173.9	200.9	207.8	226.5	295.0
Portfolio investment	HHZZ	153.7	227.3	197.3	253.0	316.9	456.4	414.4	487.2	535.1	626.7	726.9
Other investment	HLXV	475.6	566.0	555.9	533.6	650.0	694.7	719.0	819.5	864.2	1 066.2	1 105.1
Reserve assets	APDD	28.7	26.3	22.5	26.0	28.3	29.7	30.7	31.8	27.3	22.8	23.3
Total	HBQA	**762.4**	**942.8**	**896.4**	**938.8**	**1 144.9**	**1 351.2**	**1 337.9**	**1 539.4**	**1 634.4**	**1 942.2**	**2 150.4**
Financial account transactions												
Direct investment	-HJYP	20.9	21.5	10.6	9.1	11.1	17.9	22.2	28.2	22.5	38.9	64.1
Portfolio investment	-HHZC	11.8	38.9	15.9	32.4	28.7	89.6	–21.8	39.3	59.8	51.9	34.4
Other investment	-XBMM	21.9	35.5	52.5	–20.5	30.7	45.7	28.9	48.1	137.3	168.4	16.0
Reserve assets	-AIPA	2.8	–5.4	0.1	2.7	–1.4	0.7	1.0	–0.2	–0.5	–2.4	–0.2
Total	-HBNR	**57.3**	**90.4**	**79.0**	**23.6**	**69.0**	**153.8**	**30.4**	**115.3**	**219.2**	**256.8**	**114.3**
Investment income												
Direct investment	HJYW	15.6	18.7	17.5	14.5	15.2	19.1	23.8	26.9	30.9	32.3	32.4
Portfolio investment	HLYX	9.7	12.8	13.9	15.1	17.5	20.4	21.9	25.3	25.3	28.8	27.2
Other investment	AIOP	35.3	48.4	55.1	53.2	40.4	37.8	35.3	43.2	43.9	44.4	49.9
Reserve assets	HHCB	1.4	1.9	1.7	1.7	1.5	1.5	1.6	1.7	1.6	1.4	1.1
Total	HMBN	**61.9†**	**81.8**	**88.2**	**84.4**	**74.5**	**78.8**	**82.6**	**97.1**	**101.7**	**106.9**	**110.6**
Investment in the UK												
International investment position												
Direct investment	HBWI	77.0	100.1	113.5	121.0	123.1	128.2	122.6	137.9	147.6	167.1	207.5
Portfolio investment	HLXW	158.9	194.1	189.2	208.8	247.0	306.9	320.0	386.4	429.6	588.4	679.3
Other investment	HLYD	475.5	597.5	602.0	605.9	752.2	879.7	873.1	1 007.5	1 059.1	1 269.3	1 331.1
Total	HBQB	**711.4**	**891.8**	**904.6**	**935.7**	**1 122.3**	**1 314.8**	**1 315.6**	**1 531.8**	**1 636.3**	**2 024.8**	**2 217.9**
Financial account transactions												
Direct investment	HJYU	12.0	18.6	18.6	9.1	9.2	10.3	6.1	12.9	16.6	22.6	40.8
Portfolio investment	HHZF	30.3	18.2	13.4	11.0	9.4	28.4	33.2	37.2	44.2	27.2	20.2
Other investment	XBMN	30.2	72.6	64.6	13.5	56.1	124.6	–15.0	66.1	160.2	193.9	44.3
Total	HBNS	**72.5**	**109.4**	**96.6**	**33.6**	**74.7**	**163.3**	**24.3**	**116.2**	**220.9**	**243.7**	**105.3**
Investment income												
Direct investment	HJYX	11.9	12.7	9.3	6.4	7.3	12.9	12.3	15.6	18.7	17.3	14.4
Portfolio investment	HLZC	10.8	14.9	15.1	14.3	13.2	14.0	16.5	20.3	20.7	24.9	26.5
Other investment	HLZN	37.9	54.2	64.3	65.6	51.9	51.3	46.0	54.8	54.2	53.6	54.6
Total	HMBO	**60.6†**	**81.7**	**88.6**	**86.3**	**72.4**	**78.2**	**74.7**	**90.8**	**93.7**	**95.8**	**95.5**
Net investment												
International investment position												
Direct investment	HBWQ	27.3	23.1	7.2	5.2	26.6	42.1	51.3	63.0	60.2	59.4†	87.5
Portfolio investment	CGNH	–5.2	33.2	8.1	44.2	69.9	149.5	94.4	100.8	105.5	38.3†	47.6
Other investment	CGNG	0.1	–31.5	–46.0	–72.3	–102.2	–184.9	–154.1	–188.0	–194.9	–203.1†	–226.0
Reserve assets	APDD	28.7	26.3	22.5	26.0	28.3	29.7	30.7	31.8	27.3	22.8	23.3
Net investment position	HBQC	**51.0†**	**51.0**	**–8.2**	**3.1**	**22.5**	**36.4**	**22.2**	**7.6**	**–1.9**	**–82.6**	**–67.5**
Financial account transactions												
Direct investment	HJYV	–8.9	–2.9	8.0	–	–1.8	–7.6	–16.1	–15.2	–6.0	–16.3†	–23.3
Portfolio investment	HHZD	18.5	–20.7†	–2.5	–21.4	–19.3	–61.2	55.0	–2.1	–15.6	–24.8	–14.2
Other investment	HHYR	8.3†	37.1	12.1	34.0	25.5	78.9	–44.0	18.1	22.9	25.5	28.3
Reserve assets	AIPA	–2.8	5.4	–0.1	–2.7	1.4	–0.7	–1.0	0.2	0.5	2.4	0.2
Net transactions	HBNT	**15.1†**	**19.0**	**17.5**	**10.0**	**5.7**	**9.4**	**–6.1**	**0.9**	**1.8**	**–13.2**	**–9.0**
Investment income												
Direct investment	HJYE	3.7	6.0	8.2	8.1	7.9	6.1	11.5	11.2	12.2	15.0	18.0
Portfolio investment	HLZX	–1.1	–2.1	–1.2	0.7	4.3	6.4	5.5	5.0	4.5	3.9	0.7
Other investment	CGNA	–2.6	–5.7	–9.2	–12.3	–11.5	–13.5	–10.6	–11.7	–10.2	–9.3	–4.7
Reserve assets	HHCB	1.4	1.9	1.7	1.7	1.5	1.5	1.6	1.7	1.6	1.4	1.1
Net earnings	HMBM	**1.4†**	**0.1**	**–0.4**	**–1.9**	**2.2**	**0.6**	**7.9**	**6.3**	**8.0**	**11.1**	**15.1**

Source: Office for National Statistics

18.13 Net outward direct investment by UK companies analysed by area and main country[1,2]

£ millions

		1987	1988	1989	1990	1991	1992	1993	1994	1995	1996	1997
Europe	GQBX	2 937	5 773	5 620	5 844	3 978	4 911	6 171	9 241	9 184	13 321	19 735
EU	CAUU	2 250	5 360	5 941	5 103	3 919	4 613	6 146	8 278	9 457	13 432†	15 489
Austria	CBJD	66	41	32	58	45	16	13	102	90	102†	10
Belgium & Luxembourg	CAUV	69	17	301	–211	316	–191	160	132	438	991†	1 521
Denmark	CAUW	–29	50	105	43	66	47	237	64	416	–176†	118
Finland	CBJE	10	6	13	4	–1	7	10	26	112	28†	–5
France	CAUX	249	1 821	1 484	1 158	486	628	471	423	1 515	2 375†	1 520
Germany	CAUY	205	505	797	187	155	536	1 333	1 261	1 478	1 184†	1 273
Greece	CAUZ	7	4	52	42	37	167	44	84	163	106†	303
Irish Republic	CAVA	174	56	299	144	388	895	1 082	100	776	755†	398
Italy	CAVB	222	298	358	548	258	222	282	298	406	421†	442
Netherlands	CAVC	931	2 011	1 644	2 258	985	1 585	2 436	4 615	2 953	6 577†	8 544
Portugal	CAVD	89	109	202	159	61	237	25	169	159	56†	86
Spain	CAVE	269	471	573	699	982	217	–31	460	431	735†	911
Sweden	CBJG	–10	–29	82	15	141	249	84	546	522	277†	370
EFTA	CAVG	539	–5	–163	497	–42	163	–84	645	–594	–12†	2 320
of which												
Norway	CBJF	389	–235	170	32	35	156	92	662	–255	96†	1 999
Switzerland	CBJH	150	232	–333	465	–74	–1	–177	–16	–338	–110†	320
Other European Countries	GQBY	149	417	–159	244	100	135	110	317	322	–97	1 927
of which												
Russia[3]	GLAA	..	..	..	2	–2	..	11	115	39	132	456
UK offshore islands	GLAC	–	–	–	–	–	–	–	–	–	–	–864
America	GQBZ	14 988	12 129	13 000	2 181	2 628	2 657	8 673	8 464	13 460	3 277	14 711
of which												
Bermuda	CBKZ	247	621	25	276	–438	506	586	349	291	142†	–44
Brazil	CBLA	302	260	269	211	89	121	38	291	473	692†	355
Canada	CAVK	1 057	535	542	894	318	–106	5	–4	244	–159†	779
Chile	GQCA	36	62	49	19	80	–22	101	76	220	89	156
Colombia	GQCB	–18	–	48	20	–264	..	–245	204	123	100	242
Mexico	GLAD	114	57	–136	27	39	114	44	42	79	110	762
Panama	GLAE	–33	198	–14	–13	100	211	–100	94	75	103	599
USA	CAVJ	12 591	10 472	11 676	47	2 235	1 321	7 975	6 549	11 840	1 837†	11 288
Asia	GQCI	441	578	798	888	853	959	1 596	2 049	1 657	2 823	2 490
Near and Middle East Countries	CBKF	–142	–81	–7	244	489	44	–239†	253	154	28	373
of which												
Gulf Arabian countries[4]	GQCC	–135	–71	–47	239	476	27	–236	250	116	30	280
Other Asian Countries	GQCD	584	660	806	644	364	915	1 834	1 796	1 503	2 794	2 117
of which												
Hong Kong	CAVN	254	119	–161	–250	–245	–18	456	128	734	730†	–348
India	GLAF	44	14	38	42	2	–27	139	87	61	110	182
Indonesia	GLAG	–6	24	32	–46	74	–59	69	92	–28	155	80
Japan	CAVM	–23	102	230	235	–4	13	–49	245	169	378†	384
Malaysia	CBKN	87	94	–32	159	140	272	363	286	28	184†	755
Singapore	CBKQ	176	191	509	344	232	569	528	590	–48	535†	329
South Korea	GLAH	3	3	27	4	20	37	44	27	47	32	16
Thailand	GLAI	17	25	67	40	24	51	103	177	245	194	106
Australasia & Oceania	GQCE	733	2 362	783	999	1 239	1 280	658	959	2 596	1 843	1 336
of which												
Australia	CBJO	563	1 789	654	913	1 089	989	655	625	2 258	1 472†	1 240
New Zealand	CBJP	127	488	76	–24	66	110	71	264	67	244†	257
Africa	GQCF	60	74	1 290	196	606	300	262	327	707	561	684
of which												
Kenya	GLAJ	53	19	30	47	22	23	33	9	67	24	61
Nigeria	CBJY	–34	–12	..	–12	227	138	347	–197	–271	–94	234
South Africa	CAVO	–38	–8	414	22	151	72	314	170	466	–25†	401
Zimbabwe	CBKD	72	59	51	52	72	59	36	28	16	25	7
World Total	CDQD	19 159	20 916	21 491	10 108	9 304	10 107	17 358	21 040	27 604	21 823†	38 957
OECD	GQCG	17 342	19 194	18 868	7 731	7 673	7 343	14 864	16 880	23 686	17 545	32 743
Central and Eastern Europe[5]	GQCH	–	–	5	2	18	37	41	168	194	201	1 695

1 Net investment includes unremitted profits.
2 Minus sign indicates net disinvestment overseas.
3 Prior to 1995 Russia covers other former USSR countries, the Baltic States and Albania.
4 Includes Abu Dhabi, Bahrain, Dubai, Iraq, Kuwait, Oman, Other Gulf States, Qatar, Saudi Arabia and Yemen.
5 Includes Albania, Belarus, Bulgaria, Croatia, Czech Republic, Estonia, Hungary, Latvia, Lithuania, Poland, Romania, Serbia and Montenegro, Slovakia and Slovenia.
6 .. Data are disclosive.
7 - Indicates nil returns.

Sources: ONS Overseas Direct Investment Inquiries: 01633 813314; Bank of England

18.14 Level of UK outward direct investment - book value of net assets analysed by area and main country at year end

£ millions

		1987	1988	1989	1990	1991	1992	1993	1994	1995	1996	1997
Europe	GQCJ	23 018	26 175	30 626	34 450	37 272	44 213	57 853	66 705	76 434	87 582	102 604
EU	CDLN	20 452	23 163	27 826	31 610	34 312	40 620	53 914	61 674	72 808	83 902†	95 790
Austria	CDLZ	153	169	244	336	384	371	379	454	734	578†	468
Belgium & Luxembourg	CDLO	1 561	1 675	1 814	1 590	2 505	3 804	2 687	3 189	3 172	5 326†	7 332
Denmark	CDLP	550	564	778	757	739	863	1 301	1 538	2 803	2 273†	2 183
Finland	CDMA	63	83	101	102	80	122	104	117	271	215†	215
France	CDLQ	2 985	4 698	6 173	6 898	7 290	8 043	8 619	10 357	12 913	13 128†	11 676
Germany	CDLR	3 939	3 963	4 303	4 217	3 998	3 804	5 974	8 442	9 215	8 943†	9 752
Greece	CDLS	79	81	189	159	156	288	294	203	500	465	366
Irish Republic	CDLT	1 235	1 297	1 659	2 048	2 743	3 333	4 169	4 436	4 587	6 282†	6 039
Italy	CDLU	1 222	1 462	1 941	2 093	2 400	2 427	2 287	2 475	2 698	3 196†	3 042
Netherlands	CDLV	6 304	6 265	6 442	8 287	8 369	11 980	23 190	25 228	29 906	37 676†	48 902
Portugal	CDLW	444	516	739	865	1 032	1 171	1 195	932	1 210	1 190†	1 007
Spain	CDLX	1 422	1 828	2 737	3 498	3 796	3 471	2 820	3 157	3 399	3 477†	3 749
Sweden	CDMD	495	565	707	761	821	942	894	1 146	1 401	1 154†	1 060
EFTA	CDLY	2 098	2 130	2 495	2 511	2 545	2 968	3 303	4 113	2 522	2 161†	1 640
of which												
Norway	CDMC	488	385	549	522	542	736	803†	1 513	1 244	1 171	1 105
Switzerland	CDME	1 609	1 745	1 943	1 989	2 003	2 232	2 500	2 594	1 279	988†	534
Other European Countries	GQCK	468	882	306	329	415	626	636	918	1 103	1 520	5 174
of which												
Russia[1]	GQAA	..	..	10	9	10	20	7	26	120	238	381
UK offshore islands	GQAB	-	-	-	-	-	-	-	-	-	-	2 687
America	GQCU	41 463	55 898	66 515	62 381	62 939	74 868	77 003	75 390	83 275	67 592	82 528
of which												
Bermuda	CDOA	1 782	2 252	2 589	4 630	4 628	5 382	4 869	5 403	5 504	5 295†	4 940
Brazil	CDOB	1 202	1 289	1 452	1 250	1 345	1 880	1 963	2 059	2 323	2 421†	2 222
Canada	CDML	4 938	5 877	6 516	6 219	6 253	6 766	7 162	4 919	5 395	4 563†	5 467
Chile	GQCT	137	190	252	207	363	416	481	439	666	670	964
Colombia	GQCS	..	..	..	..	594	1 109	839	1 071	1 207	1 274	1 199
Mexico	GQAC	273	324	314	405	428	504	419	334	350	553	1 331
Panama	GQAD	51	204	169	100	232	630	435	507	467	744	997
USA	CDMM	29 836	42 613	50 933	45 532	45 367	54 688	57 380	55 174	62 159	49 170†	62 749
Asia	GQCL	6 352	6 927	7 996	8 285	9 633	12 230	14 118	16 095	18 038	19 916	19 331
Near & Middle East Countries	CDNH	347	345	373	314	749	748	698	754	788	676†	923
of which												
Gulf Arabian countries[2]	GQCM	323	315	294	229	657	592	617	660	704	586	664
Other Asian Countries	GQCR	6 004	6 582	7 623	7 970	8 884	11 482	13 419	15 341	17 250	19 239	18 408
of which												
Hong Kong	CDNN	1 947	2 142	2 059	1 654	1 895	2 753	3 569	3 373	4 033	4 636†	4 419
India	GQAE	353	301	312	341	302	245	490	601	498	532	691
Indonesia	GQAF	171	213	281	264	316	222	272	295	418	391	337
Japan	CDMP	980	1 205	1 398	1 591	1 674	1 998	1 936	2 613	2 397	2 437†	1 599
Malaysia	CDNQ	769	709	690	706	845	1 285	1 750	2 119	1 813	2 164†	2 294
Singapore	CDNT	1 317	1 469	2 134	2 449	2 691	3 516	3 706	4 445	5 287	5 822	5 123
South Korea	GQAG	22	26	45	57	69	114	155	183	250	238	171
Thailand	GQAH	73	92	138	205	253	359	404	511	920	1 053	1 008
Australasia & Oceania	GQCN	7 119	9 777	11 284	9 951	10 210	11 222	12 287	13 517	13 985	14 636	14 039
of which												
Australia	CDMO	6 096	8 544	9 979	8 924	9 057	9 605	10 299	11 153	11 365	12 213†	11 147
New Zealand	CDMQ	849	889	1 069	696	700	930	1 443	1 689	1 670	1 640†	1 670
Africa	GQCQ	3 593	3 438	4 526	3 868	4 039	4 080	4 570	5 409	4 955	4 876	5 902
of which												
Kenya	GQAI	289	216	246	267	233	249	165	673	276	237	247
Nigeria	CDNA	153	60	639	325	419	490	754	681	335	321	1 061
South Africa	CDMR	1 860	1 911	2 296	2 012	2 239	2 311	2 622	2 202	2 827	2 429†	2 521
Zimbabwe	CDNF	408	424	435	402	324	411	248	331	262	200†	187
World Total	CDOO	81 545	102 215	120 948	118 935	124 093	146 613	165 831	177 116	196 687	194 601†	224 404
OECD	GQCO	65 854	85 419	100 632	97 637	100 517	118 430	136 247	142 332	159 476	157 652	182 482
Central & Eastern Europe[3]	GQCP	1	1	3	9	36	123	106	460	427	622	750

1 Prior to 1995 Russia covers other former USSR countries, the Baltic States and Albania.
2 Includes Abu Dhabi, Bahrain, Dubai, Iraq, Kuwait, Oman, Other Gulf States, Qatar, Saudi Arabia and Yemen.
3 Includes Albania, Belarus, Bulgaria, Croatia, Czech Republic, Estonia, Hungary, Latvia, Lithuania, Poland, Romania, Serbia and Montenegro, Slovakia and Slovenia.
4 .. Data are disclosive.
5 - Indicates nil returns.

Sources: ONS Overseas Direct Investment Inquiries: 01633 813314; Bank of England

18.15 Net earnings from direct investment overseas by UK companies analysed by area and main country[1,2]

£ millions

		1987	1988	1989	1990	1991	1992	1993	1994	1995	1996	1997
Europe	GQCV	3 021	3 483	4 548	4 727	4 182	3 315	5 182	7 033	8 289	9 700	10 638
EU	CAWG	2 657	3 085	4 045	4 198	3 721	2 499	4 450	6 138	7 251	8 557†	9 463
Austria	CBLQ	51	53	36	49	43	54	35	52	73	82†	68
Belgium & Luxembourg	CAWH	121	263	299	165	259	404	412	387	308	358†	300
Denmark	CAWI	64	64	86	91	71	115	207	115	286	208	191
Finland	CBLR	12	17	17	9	–	24	22	30	43	–	37
France	CAWJ	252	514	805	688	598	279	50	607	619	999†	1 186
Germany	CAWK	438	546	595	623	617	260	381	834	1 032	951†	1 076
Greece	CAWL	3	16	34	22	12	50	58†	–202	–106	172	121
Irish Republic	CAWM	201	161	171	265	414	393	669	516	639	847	1 076
Italy	CAWN	165	182	173	220	175	75	113	219	272	327†	276
Netherlands	CAWO	1 025	964	1 449	1 674	1 141	758	2 459	2 986	3 248	3 748†	4 262
Portugal	CAWP	80	87	135	124	93	83	53	162	188	184†	104
Spain	CAWQ	177	190	178	219	197	–5	39	234	338	493†	562
Sweden	CBLT	68	27	68	50	103	12	–48	200	310	188†	202
EFTA	CAWS	248	194	451	476	433	740	639	792	899	963	991
of which												
Norway	CBLS	75	–14	114	242	214	281	119	267	224	311†	241
Switzerland	CBLU	173	208	337	233	219	459	520	524	675	652	747
Other European Countries	GQCW	116	204	52	53	28	75	95	103	139	180	184
of which												
Russia[3]	GQAJ	..	..	..	1	..	..	2	3	12	24	–22
UK offshore islands	GQAK	–	–	–	–	–	–	–	–	–	–	223
America	GQCX	4 931	6 840	7 525	6 962	4 624	5 530	6 126	8 007	10 254	10 698	11 970
of which												
Bermuda	CBNK	248	251	376	269	449	562	428	446	509	506	70
Brazil	CBNL	288	365	366	227	87	182	193	348	432	650†	386
Canada	CAWW	542	531	650	464	189	175	283	357	374	571†	891
Chile	GQCY	28	38	44	35	88	118	139	195	263	265	230
Colombia	GQCZ	63	32	41	64	21	–40	16	–22	28	23	53
Mexico	GQAL	44	46	8	12	–2	41	43	48	52	74	143
Panama	GQAM	4	14	11	54	17	–27	69	51	45	67	151
USA	CAWV	3 471	4 964	5 499	5 402	3 255	4 065	4 621	5 894	7 803	8 034†	9 465
Asia	GQDA	1 374	1 501	1 873	2 114	2 355	2 636	2 792	3 449	2 501	3 776	3 223
Near & Middle East Countries	CBMS	233	180	221	235	303	342	283	297†	262	241	478
of which												
Gulf Arabian countries[4]	GQDB	231	179	234	249	300	317	273	293	260	212	424
Other Asian Countries	GQDC	1 141	1 321	1 652	1 879	2 053	2 293	2 509	3 152	2 239	3 536	2 744
of which												
Hong Kong	CAYB	487	632	485	439	440	846	972	1 427	1 362	1 463	617
India	GQAN	50	50	63	88	77	–2	–190	121	100	81	133
Indonesia	GQAO	17	1	29	25	71	18	60	48	71	87	67
Japan	CAWY	108	88	125	72	115	182	142	199	212	281†	267
Malaysia	CBNA	135	139	152	182	197	229	248	221	248	332†	319
Singapore	CBND	247	322	611	967	1 010	839	977	781	–144	883†	785
South Korea	GQAP	10	16	14	7	10	5	8	29	40	34	23
Thailand	GQAQ	9	8	30	34	17	28	78	84	75	101	94
Australasia & Oceania	GQDD	839	1 215	1 627	854	541	960	1 233	2 131	2 044	2 269	1 961
of which												
Australia	CBMB	681	1 082	1 590	925	535	809	991	1 776	1 587	1 747†	1 507
New Zealand	CBMC	139	116	15	–98	5	73	213	302	322	319†	279
Africa	GQDE	727	769	1 035	869	986	844	1 465	737	807	1 055	974
of which												
Kenya	GQAR	36	33	47	47	36	41	34	41	46	35	72
Nigeria	CBML	94	116	197	159	273	280	498	–27	–35	61	46
South Africa	CAWZ	403	424	595	410	425	280	488	366	438	503†	521
Zimbabwe	CBMQ	80	75	80	73	81	77	54	58	58	50†	61
World Total	GLAB	10 892	13 808	16 608	15 526	12 688	13 285	16 796	21 355	23 894	27 498†	28 765
OECD	GQDF	7 996	10 294	12 425	11 486	8 279	8 624	11 454	15 578	18 567	20 621	23 092
Central & Eastern Europe[5]	GQDG	–	–	1	2	1	7	11	–24	16	15	–196

1 A minus sign indicates net losses.
2 Net earnings equal profits of overseas branches plus UK companies' receipts of interest and their share of profits of overseas subsidiaries and associates. Earnings are after deducting provisions for depreciation and overseas tax on profits, dividends and interest.
3 Prior to 1995 Russia covers other former USSR countries, the Baltic States and Albania.
4 Includes Abu Dhabi, Bahrain, Dubai, Iraq, Kuwait, Oman, Other Gulf States, Qatar, Saudi Arabia and Yemen.
5 Includes Albania, Belarus, Bulgaria, Croatia, Czech Republic, Estonia, Hungary, Latvia, Lithuania, Poland, Romania, Serbia & Montenegro, Slovakia and Slovenia.
6 .. Indicates data are disclosive.
7 - Indicates nil returns.

Sources: ONS Overseas Direct Investment Inquiries: 01633 813314; Bank of England

18.16 Net direct investment in the UK analysed by area and main country[1 2]

£ millions

		1987	1988	1989	1990	1991	1992	1993	1994	1995	1996	1997
Europe	GQDH	3 530	7 279	7 971	8 042	4 910	4 044	2 202	3 065	4 626	6 931	8 612
EU	CAYO	3 607	5 282	5 617	7 628	4 578	3 433	1 589†	3 367	3 555	4 673	7 330
Austria	CBOB	1	8	1	62	11	–45	13	60	21	20	18
Belgium & Luxembourg	CAYP	400	307	853	296	201	65	–427	357	520	6†	769
Denmark	CAYQ	30	121	198	154	122	91	97	76	68	151†	229
Finland	CBOC	8	26	56	36	8	54	–42	32	–36	2	154
France	CAYR	993	868	1 870	1 551	1 333	802	–37	310	1 004	1 321†	2 693
Germany	CAYS	177	278	460	1 610	396	1 261	656	71	2 090	835†	1 170
Greece	CAYT	..	..	..	..	..	..	..	..	..	..	..
Irish Republic	CAYU	60	121	230	229	137	–32	49	224	–35	221†	1 138
Italy	CAYV	32	83	26	–42	45	–37	80	177	328	–184†	–32
Netherlands	CAYW	1 556	3 142	1 613	2 085	2 335	1 135	1 244	1 915	–633	2 610†	992
Portugal	CAYX	..	..	..	..	..	..	..	..	..	..	..
Spain	CAYY	2	49	130	3	–11	3	14	21	17	66	71
Sweden	CBOE	349	285	185	1 658	5	129	–56	119	184	–379†	129
EFTA	CAZB	–80	2 021	2 310	431	327	674	550†	–176	1 036	2 211	1 258
of which												
Norway	CBOD	141	218	–52	201	–71	–34	49	–131	124	1 060	–71
Switzerland	CBOF	–221	1 803	2 363	229	399	710	501	–44	912	1 151†	1 321
Other European Countries	GQDI	3	–24	44	–17	5	–62	64	–125	36	47	24
of which												
Russia[3]	GQAS	2	–24	28	–26	–11	..	49	–	19	61	19
UK offshore islands	GQAT	–	–	–	–	–	–	–	–	–	–	–26
America	GQDJ	2 902	1 669	7 019	6 266	1 898	3 720	5 993	2 344	8 750	7 210	11 338
of which												
Canada	CAZF	977	168	610	–264	265	–45	33	–246	–438	444†	475
USA	CAZE	1 593	1 288	6 415	5 095	1 879	3 748	5 142	2 138	9 293	6 742†	10 708
Asia	GQDK	1 390	1 092	1 378	1 931	255	630	570	186	–189	412	430
Near & Middle East Countries	GQAU	–6	–27	114	–54	137	251	115	78	58	64	103
Other Asian Countries	GQAV	1 397	1 120	1 264	1 986	118	380	456	110	–248	348	327
of which												
Hong Kong	GQAW	..	109	10	10	53	280	106	45	–124	10	6
Japan	CAZH	857	1 130	1 238	2 091	48	–21	277	4	–379	209†	350
Singapore	GQAX	2	8	4	9	7	9	14	2	40	1	17
South Korea	GQAY	..	..	..	..	..	..	43	2	85	–8	–78
Australasia & Oceania	GQDL	1 140	1 530	990	852	1 328	323	995	387	–647	992	1 211
of which												
Australia	CBOJ	919	1 575	846	957	937	340	995	260	–708	1 096†	1 219
New Zealand	CBOK	..	–45	..	–105	391	–17	–	127	60	–104	–7
Africa	GQAZ	24	–8	47	64	27	99	111	63	117	118	161
of which												
South Africa	CAZJ	..	–30	..	38	30	85	58	50	125	109	149
World Total	CBDH	8 986	11 562	17 405	17 155	8 418	8 816	9 871	6 046	12 654	15 662†	21 751
OECD	GQBA	8 100	11 432	17 188	15 727	8 428	8 221	8 636	5 346	12 511	15 235	21 250
Central & Eastern Europe[4]	GQBB	..	–	4	–	–	–7	–	–141	15	–19	13

1 Net investment includes unremitted profits.
2 A minus sign indicates net disinvestment in the UK.
3 Prior to 1995 Russia covers other former USSR countries, the Baltic States and Albania.
4 Includes Albania, Belarus, Bulgaria, Croatia, Czech Republic, Estonia, Hungary, Latvia, Lithuania, Poland, Romania, Serbia & Montenegro, Slovakia and Slovenia.
5 .. Indicates data are disclosive.
6 - Indicates nil returns.

Sources: ONS Overseas Direct Investment Inquiries: 01633 813314; Bank of England

18.17 Level of UK inward direct investment - book value of net liabilities analysed by area and main country at year end

£ millions

		1987	1988	1989	1990	1991	1992	1993	1994	1995	1996	1997
Europe	GQDM	21 438	28 990	37 610	40 606	45 205	46 928	47 770	47 281	52 144	55 856	62 452
EU	CDOT	16 364	23 593	28 711	33 707	38 036	39 547	40 266†	39 998	43 493	43 774	49 553
Austria	CDPF	36	37	40	104	103	–30	93	428	412	280	81
Belgium & Luxembourg	CDOU	1 076	1 679	1 357	1 634	1 812	1 817	1 801	2 221	2 170	1 233†	2 451
Denmark	CDOV	449	561	696	685	701	877	1 339	1 072	1 127	850†	930
Finland	CDPG	167	192	291	423	364	437	300	356	352	334†	454
France	CDOW	2 407	3 923	6 148	7 427	9 000	8 749	7 880	7 728	8 289	9 147†	14 066
Germany	CDOX	1 643	1 973	2 543	4 239	4 597	5 627	5 921	5 589	8 854	9 508†	10 329
Greece	CDOY	..	..	..	..	..	..	..	..	..	..	..
Irish Republic	CDOZ	58	276	491	1 069	840	865	681	956	686	703†	1 851
Italy	CDPA	293	444	623	597	649	845	875	839	1 223	988†	824
Netherlands	CDPB	8 984	12 871	14 571	14 182	16 547	16 715	18 477	18 511	17 173	18 692†	15 745
Portugal	CDPC	..	..	..	..	..	..	..	..	..	..	..
Spain	CDPD	103	143	194	318	329	356	346	178	164	78	247
Sweden	CDPI	1 135	1 482	1 732	3 025	3 058	3 274	2 542	2 105	2 908	1 871†	2 503
EFTA	CDPE	4 639	4 882	8 248	6 794	7 040	7 308	7 225	7 111	8 188	11 735†	12 127
of which												
Norway	CDPH	538	705	725	748	661	647	701	506	665	1 571	1 611
Switzerland	CDPJ	4 100	4 176	7 521	6 044	6 375	6 657	6 519	6 605	7 523	10 164†	10 424
Other European Countries	GQDN	435	515	651	105	129	73	280	172	463	348	772
of which												
Russia[1]	GQBC	266	306	374	18	18	–51	43	24	80	212	226
UK offshore islands	GQBD	–	–	–	–	–	–	–	–	–	–	384
America	GQDU	30 557	33 333	43 591	49 721	49 682	50 910	55 209	55 918	59 420	62 253	76 784
of which												
Canada	CDPM	3 139	3 762	4 935	4 104	4 427	4 065	4 101	4 593	2 652	3 517†	4 105
USA	CDPN	25 838	28 257	37 093	43 784	44 434	46 001	49 537	49 829	55 129	55 956†	69 751
Asia	GQDO	2 954	3 941	5 729	7 184	7 378	6 935	7 655	7 782	7 934	8 072	8 779
Near & Middle East Countries	GQBE	614	573	1 119	1 110	1 372	1 099	1 427	2 027	1 324	1 274	1 450
Other Asian Countries	GQBF	2 340	3 368	4 609	6 073	6 006	5 836	6 228	5 755	6 609	6 797	7 329
of which												
Hong Kong	GQBG	82	93	402	406	200	422	262	233	19	31	47
Japan	CDPQ	2 039	3 082	3 965	5 648	5 346	4 929	5 427	5 105	5 542	5 888†	6 499
Singapore	GQBH	23	26	29	47	..	..	415	421	500	503	446
South Korea	GQBI	..	..	..	..	–43	–28	–3	–203	12	–206	–305
Australasia & Oceania	GQDP	3 173	5 104	6 297	7 937	8 715	9 135	9 560	9 644	8 527	7 707	8 000
of which												
Australia	CDPP	3 024	5 052	6 081	6 799	7 655	8 083	8 598	7 968	7 021	6 169†	6 501
New Zealand	CDPR	149	52	..	912	832	824	844	1 676	1 506	1 538	1 500
Africa	GQBJ	308	284	357	313	393	501	811	712	861	766	953
of which												
South Africa	CDPS	276	241	..	263	342	438	618	549	665	578	743
World Total	CDPZ	58 430	71 652	93 554	105 760	111 373	114 409	121 005	121 336	128 885	134 654†	156 969
OECD	GQBK	55 217	68 716	89 287	101 702	107 814	110 816	116 082	116 171	123 761	128 475	149 751
Central & Eastern Europe[2]	GQBL	..	..	238	29	29	24	21	10	258	14	22

1 Prior to 1995 Russia covers other former USSR countries, the Baltic States and Albania.
2 Includes Albania, Belarus, Bulgaria, Croatia, Czech Rebuplic, Estonia, Hungary, Latvia, Lithuania, Poland, Romania, Serbia & Montenegro, Slovakia and Slovenia.
3 .. Indicates data are disclosive.
4 - Indicates nil returns.

Sources: ONS Overseas Direct Investment Inquiries: 01633 813314; Bank of England

18.18 Net earnings from direct investment in the UK analysed by area and main country [1,2]

£ millions

		1987	1988	1989	1990	1991	1992	1993	1994	1995	1996	1997
Europe	GQDQ	1 898	2 305	2 734	2 007	1 243	2 482	4 833	3 661	5 432	6 401	5 802
EU	CBDJ	1 695	2 052	2 256	1 689	1 025	1 932	3 556†	3 659	4 657	5 631	5 248
Austria	CBOR	–3	1	–3	–8	..	..	–5	94	30	55†	13
Belgium & Luxembourg	CBDK	39	151	72	–3	–53	–1	224	272	249	209†	186
Denmark	CBDL	37	72	117	88	42	–17	92	18	152	137	171
Finland	CBOS	–6	–13	–10	–43	–71	–36	–52	2	88	74	80
France	CBDM	232	287	297	420	377	340	633	227	836	705†	1 236
Germany	CBDN	158	182	117	169	22	221	576	507	700	706†	569
Greece	CBDO	..	..	..	..	..	..	..	..	..	..	..
Irish Republic	CBDP	79	165	78	–4	–27	–12	74	92	155	141	221
Italy	CBDQ	21	54	30	–73	–55	–24	97	246	196	318	430
Netherlands	CBDR	1 009	1 010	1 443	1 148	927	1 731	1 820	2 076	2 070	3 053†	2 004
Portugal	CBDS	..	..	..	..	..	..	..	..	..	..	..
Spain	CBDT	17	25	28	–	–29	–5	55	17	42	69	106
Sweden	CBOU	114	120	112	13	–56	–177	46	107	127	146	218
EFTA	CBDW	192	233	464	316	146	846	1 188†	–33	685	709	408
of which												
Norway	CBOT	–21	51	42	17	–14	–35	27	12	11	–58	121
Switzerland	CBOV	215	183	423	301	338	885	1 161	–45	675	768†	277
Other European Countries	GQDR	11	20	13	2	–104	–296	88	36	89	60	147
of which												
Russia[3]	GQBM	8	..	9	2	..	..	55	1	15	31	..
UK offshore islands	GQBN	–	–	–	–	–	–	–	–	–	–	72
America	GQDV	4 520	5 210	6 072	5 131	2 948	2 391	4 917	4 992	5 824	7 589	6 805
of which												
Canada	CBEA	60	235	448	273	199	47	234	506	228	281†	65
USA	CBDZ	4 336	4 727	5 492	4 602	2 951	2 284	4 441	4 372	5 606	6 986†	6 515
Asia	GQDS	456	513	481	102	193	107	106	172	459	300	–348
Near & Middle East Countries	GQBO	138	10	29	–119	–89	–31	16	–8	112	89	96
Other Asian Countries	GQBP	319	504	453	222	283	139	90	181	347	211	–444
of which												
Hong Kong	GQBQ	11	85	–27	–23	47	207	52	4	–2	28	21
Japan	CBEC	296	405	456	261	246	–41	45	171	334	151†	–489
Singapore	GQBS	2	2	2	–12	3	–2	4	1	5	–	14
South Korea	GQBT	..	..	..	..	..	..	7	15	–6	–6	–11
Australasia & Oceania	GQDT	119	540	–185	–302	74	211	418	327	472	594	847
of which												
Australia	CBOZ	110	522	–194	–168	92	222	410	330	479	566†	830
New Zealand	CBPA	9	18	9	–134	–18	..	8	–1	–6	28	17
Africa	GQBU	20	44	56	–6	3	..	19	25	49	44	55
of which												
South Africa	CBED	34	35	65	–6	–1	6	9	21	38	35	46
World Total	CBEV	7 013	8 612	9 158	6 932	4 461	5 197	10 293	9 176	12 235	14 928†	13 160
OECD	GQBV	6 709	8 204	8 844	6 831	4 878	5 371	9 998	9 118	12 081	14 388	12 602
Central & Eastern Europe[4]	GQBW	..	..	–	–	–	2	1	24	60	13	18

1 A minus sign indicates net losses.
2 Net earnings equal profits of UK branches plus overseas investors' receipts of interest and their share of the profits of UK subsidiaries and associates. Earnings are after deducting provisions for depreciation and UK tax on profits and interest.
3 Prior to 1995 Russia covers other former USSR countries, the Baltic States and Albania.
4 Includes Albania, Belarus, Bulgaria, Croatia, Czech Republic, Estonia, Hungary, Latvia, Lithuania, Poland, Romania, Serbia & Montenegro, Slovakia and Slovenia.
5 .. Indicates data are disclosive.
6 - Indicates nil returns.

Sources: ONS Overseas Direct Investment Inquiries: 01633 813314; Bank of England

Research and development

19

Research and development
(Tables 19.1 to 19.5)

Research and experimental development (R&D) is defined for statistical purposes as 'creative work undertaken on a systematic basis in order to increase the stock of knowledge, including knowledge of man, culture and society, and the use of this stock of knowledge to devise new applications'.

R&D is financed and carried out mainly by businesses, the Government, and institutions of higher education. A small amount is performed by non-profit-making bodies. Gross Expenditure on R&D (GERD) is an indicator of the total amount of R&D performed within the UK: it has been approximately 2 per cent of GDP in recent years. Detailed figures are reported each year in a First Release published in March ONS (99)107 and the August edition of the ONS's *Economic Trends*. Table 19.1 shows the main components of GERD.

The ONS conducts an annual survey of expenditure and employment on R&D performed by Government, and of Government funding of R&D. The survey collects data on outturn and planning years. Until 1993 the detailed results were reported in the *Annual Review of Government Funded R&D* produced by the Office of Science and Technology (OST). From 1997 the results have appeared in OST's *Science, Engineering and Technology Statistics* publication. Table 19.2 gives some broad totals for gross expenditure by Government (expenditure before deducting funds received by Government for R&D). Table 19.3 gives a breakdown of net expenditure (receipts are deducted).

The ONS conducts an annual survey of R&D in business. Tables 19.4 and 19.5 give a summary of the main trends up to 1997. The latest set of results from the survey are available in a First Release dated 19 November 1999 ONS (99)403 and a Business Monitor (MA14) to be published in January 2000.

Statistics on expenditure and employment on R&D in Higher Education Institutions (HEIs) are based on information collected by Higher Education Funding Councils and HESA (Higher Education Statistics Agency). In 1994 a new methodology was introduced to estimate expenditure on R&D in HEIs. This is based on the allocation of various Funding Council Grants. Full details of the new methodology are contained in SET 1999 available on the Office of Science and Technology Web Site at http://www.dti.gov.uk/ost/

The most comprehensive international comparisons of resources devoted to R&D appear in Main Science and Technology Indicators published by the Organisation for Economic Co-operation and Development (OECD). The Statistical Office of the European Union and the United Nations, also compile R&D statistics based on figures supplied by member states.

To make international comparisons more reliable the OECD have published a series of manuals giving guidance on how to measure various components of R&D inputs and outputs. The most important of these is the Frascati Manual which defines R&D and recommends how resources for R&D should be measured. The UK follows the Frascati Manual as far as possible.

For information on available aggregated data on Research and Development please contact Jane Morgan on 01633 813109, (e-mail jane.morgan@ons.gov.uk).

19.1 Cost of research and development: analysis by sector

	Work performed within each sector											
	1992		1993		1994		1995		1996		1997	
	£m	%	£m	%	£m	%	£m	%	£m	%	£m	%
Sector carrying out the work (£m cash terms)												
Government	1 846	*15*	1 928	*14*	2 051	*15*	1 462	*10*	1 495	*10*	1 427	*10*
Research Councils	-	-	-	-	-	-	581	*4*	575	*4*	590	*4*
Business enterprise	8 489	*67*	9 069	*67*	9 204	*66*	9 254	*65*	9 431	*65*	9 657	*65*
Higher education	2 129	*17*	2 312	*17*	2 623	*19*	2 696	*19*	2 792	*19*	2 896	*20*
Private non-profit	224	*2*	232	*2*	168	*1*	177	*1*	177	*1*	191	*1*
Total	12 689	*100*	13 541	*100*	14 046	*100*	14 172	*100*	14 470	*100*	14 761	*100*

	Finance provided by each sector											
	1992		1993		1994		1995		1996		1997	
	£m	%	£m	%	£m	%	£m	%	£m	%	£m	%
Sector providing the funds (£m cash terms)												
Government	4 239	*33*	4 400	*32*	4 657	*33*	2 611	*18*	2 494	*17*	2 421	*16*
Research Councils	-	-	-	-	-	-	1 078	*8*	1 092	*8*	1 135	*8*
Higher Education Funding Councils	-	-	-	-	-	-	1 018	*7*	1 027	*7*	1 033	*7*
Higher education	99	*1*	103	*1*	116	*1*	119	*1*	120	*1*	123	*1*
Business enterprise[1]	6 461	*51*	6 974	*52*	7 025	*50*	6 796	*48*	6 846	*47*	7 324	*50*
Private non-profit	435	*3*	451	*3*	495	*4*	511	*4*	546	*4*	579	*4*
Abroad	1 455	*11*	1 613	*12*	1 753	*12*	2 039	*14*	2 345	*16*	2 147	*15*
Total	12 689	*100*	13 541	*100*	14 046	*100*	14 172	*100*	14 470	*100*	14 761	*100*

1 Including research associations and public corporations.

Source: Office for National Statistics: 01633 813109

19.2 Gross Central Government expenditure on research and development

£ millions

	1992/93		1993/94		1994/95		1995/96		1996/97		1997/98	
	Intra-mural	Extra-mural[1]	Intra-mural	Extra-mural[1]	Intra-mural	Extra-mural[1]	Intra-mural	Extra-mural[1]	Intra-mural	Extra-mural[1]	Intra-mural	Extra-mural[1]
Defence	681	1478	731	1 680	707	1451	692	1562	804	1429	774	1 639
Research councils	557	603	584	700	589	783	599	823	592	851	608	852
Higher Education Institutes	-	963	-	968	-	1 017	-	1 018	-	1 027	-	1 033
Other programmes	402	641	425	626	414	552	374	572	290	628	253	681
Total (ex NHS)	1 640	3 685	1 739	3 974	1 710	3 803	1 665	3 975	1 686	3 935	1 635	4 205

1 Including work performed overseas and excluding monies spent with other government departments.

Source: Office for National Statistics: 01633 813109

19.3 Net central government expenditure on research and development, using European Union objectives for R&D expenditure

£ millions

		1987 /88	1988 /89	1989 /90	1990 /91	1991 /92	1992 /93	1993 /94	1994 /95	1995 /96	1996 /97	1997 /98
Exploration and exploitation of the earth	KDVP	85.3	99.5	123.7	145.2	143.9	120.5	98.8	106.8	105.2	95.4	81.3
Infrastructure and general planning of land-use	KDVQ	66.8	66.8	71.7	74.1	64.2	85.6	96.7	98.1	94.1	98.8	98.9
Control of environmental pollution	KDVR	56.2	58.4	53.0	70.3	71.9	69.5	108.7	117.2	131.8	128.7	136.2
Protection and promotion of human health (ex NHS)	KDVS	204.3	221.7	264.1	292.0	298.4	341.5	383.1	397.2	416.0	427.4	445.1
Production, distribution and rational utilization of energy	KDVT	170.4	177.3	157.7	141.7	133.0	120.0	96.8	55.5	52.3	43.2	41.0
Agricultural production and technology	KDVU	197.6	209.4	197.7	198.6	216.2	261.1	284.6	263.4	281.9	257.0	268.9
Industrial production and technology	KDVV	414.9	399.2	455.2	478.8	399.0	394.1	458.7	184.4	165.8	144.6	102.8
Social structures and relationships	KDVW	70.5	95.7	101.6	110.8	115.5	141.9	149.1	141.9	137.1	120.7	114.7
Exploration and exploitation of space	KDVX	130.8	148.0	146.3	155.2	134.2	149.1	187.4	161.5	153.0	164.1	164.4
Research financed from General University Funds	KDVY	760.0	800.0	799.0	835.0	919.1	963.3	968.4	1 017.9	1 018.6	1 027.5	1 033.3
Non-oriented research	KDVZ	247.1	249.5	257.3	287.6	287.6	337.1	267.3	612.5	653.5	680.5	671.0
Other civil research	KDWA	13.6	13.8	11.9	11.0	35.5	23.8	34.0	22.2	24.7	20.5	21.6
Defence	KDWB	1 990.3	1 957.2	2 132.7	2 154.7	2 208.8	2 071.1	2 268.6	2 021.9	2 098.9	2 143.5	2 312.1
Total (ex NHS)	KDWC	4 407.8	4 496.5	4 771.8	4 955.1	5 027.4	5 078.6	5 402.3	5 200.4	5 332.9	5 351.8	5 491.2

Source: Office for National Statistics: 01633 813109

19.4 Intramural expenditure on Business Enterprise R&D At Current Prices and Constant Prices

£ millions

(i) Current Prices

		Total				Civil				Defence		
		1995	1996	1997		1995	1996	1997		1995	1996	1997
Total	KDWD	9 254	9 431†	9 657	KDWN	7 863	8 071†	8 214	KDWX	1 391	1 360†	1 443
Manufacturing Total	KDWE	6 917	7 035†	7 360	KDWO	5 626	5 767†	6 055	KDWY	1 291	1 268†	1 304
Chemicals	KDWF	2 515	2 479	2 831	KDWP	2 511	2 477	2 829	KDWZ	3	2	2
Mechanical engineering	KDWG	683	668†	709	KDWQ	418	395	407	KDXA	266	273†	302
Electrical machinery	KJRT	1 245	1 313	1 181	KJTC	823	896	803	KJUL	423	417	377
Aerospace	KDWJ	886	812	893	KDWT	413	359	412	KDXD	473	453	481
Transport equipment	KDWK	833	977†	966	KDWU	823	967†	955	KDXE	10	10	11
Other manufacturing	KDWL	755	787	779	KDWV	639	673	648	KDXF	116	113	131
Services	KDWM	2 337	2 396†	2 297	KDWW	2 237	2 304†	2 158	KDXG	99	92	139

(ii) 1995 Prices

		Total				Civil				Defence		
		1995	1996	1997		1995	1996	1997		1995	1996	1997
Total	KDXH	9 254	9 137†	9 102	KDXR	7 863	7 819†	7 742	KDYB	1 391	1 318†	1 360
Manufacturing Total	KDXI	6 917	6 815†	6 937	KDXS	5 626	5 587†	5 707	KDYC	1 291	1 228†	1 229
Chemicals	KDXJ	2 514	2 402†	2 668	KDXT	2 511	2 400†	2 667	KDYD	3	2	2
Mechanical engineering	KDXK	683	647†	668	KDXU	418	383	384	KDYE	266	264†	285
Electrical machinery	KKKJ	1 246	1 272†	1 113	KKKK	823	868†	757	KKKX	423	404†	355
Aerospace	KDXN	886	787†	842	KDXX	413	348	388	KDYH	473	439	453
Transport equipment	KDXO	833	947†	911	KDXY	823	937†	900	KDYI	10	10	10
Other manufacturing	KDXP	755	762†	734	KDXZ	639	653	611	KDYJ	116	109†	123
Services	KDXQ	2 338	2 321†	2 165	KDYA	2 237	2 232†	2 034	KDYK	99	89	131

Source: Office for National Statistics: 01633 813109

19.5 Sources of funds for R&D within Business Enterprises in the United Kingdom

		Total				Civil				Defence		
		1995	1996	1997		1995	1996	1997		1995	1996	1997
Funds for Business Enterprise R&D £m cash terms												
Total	KDYL	9 254	9 431†	9 657	KDYT	7 863	8 071†	8 214	KDZB	1 391	1 360†	1 443
Government funds	KDYM	1 050	934†	1 005	KDYU	321	242†	288	KDZC	729	693†	717
Overseas funds	KDYN	1 748	2 031†	1 811	KDYV	1 419	1 728†	1 486	KDZD	329	303†	325
Mainly own funds	KDYO	6 457	6 466†	6 841	KDYW	6 124	6 102†	6 439	KDZE	333	364†	401
Funds for Business Enterprise R&D Percentage												
Total	KDYP	*100*	*100*	*100*	KDYX	*100*	*100*	*100*	KDZF	*100*	*100*	*100*
Government funds	KDYQ	*11*	*10*	*10*	KDYY	*4*	*3*	*4*	KDZG	*52*	*51†*	*50*
Overseas funds	KDYR	*19*	*22*	*19*	KDYZ	*18*	*21*	*18*	KDZH	*24*	*22†*	*23*
Mainly own funds	KDYS	*70*	*69*	*71*	KDZA	*78*	*76†*	*78*	KDZI	*24*	*27†*	*28*

Source: Office for National Statistics: 01633 813109

Agriculture, fisheries and food

20

Agriculture, fisheries and food

Agricultural censuses and surveys *(Tables 20.3, 20.5 and 20.13)*

The coverage for holdings detailed in Tables 20.3, 20.5 and 20.13 includes all main and minor holdings for each country. Scotland now includes minor holdings and Northern Ireland data is now based on all active farm business, therefore data here will differ from previous editions.

Estimated quantity of crops and grass harvested *(Table 20.4)*

The estimated yields of sugar beet and hops (Table 20.4) are obtained from production figures supplied by British Sugar plc, and English Hops plc. In Great Britain potato yields are estimated in consultation with the British Potato Council.

Forestry *(Table 20.6)*

Statistics for state forestry are from the Annual Reports and Accounts of Forest Enterprise in Great Britain and the Forest Service in Northern Ireland.

For private forestry in Great Britain, statistics on new planting and restocking are based on records of the Woodland Grant Scheme, and timber removals are estimated from a survey of the largest timber harvesting companies. Productive woodland is based on data obtained from censuses of woodlands and adjusted to reflect subsequent changes. In calculating the GB forest area, figures for Scotland and Wales are based on the 1995 National Inventory of Woodlands; figures for England are based on the 1980 Census of Woodlands.

Average weekly earnings and hours of agricultural and horticultural workers *(Tables 20.7 and 20.8)*

Since 1998, data on the Earnings and Hours of Agricultural and Horticultural workers has been collected via an annual telephone survey. This annual survey collects information for a snapshot in time, relating to the month of September. The survey covers seven main categories of workers and in 1998 data was collected on about 1 200 workers. Prior to 1998, data was collected from a monthly postal survey, which mainly covered male full time workers.

The survey provides data which is used by the Agricultural Wages Board when considering wage claims and by the Ministry of Agriculture, Fisheries and Food in considering the cost of labour in agriculture and horticulture.

Fisheries *(Tables 20.14 and 20.15)*

Data relating to the weight and value of landings of fish in the United Kingdom (Table 20.14) is generally obtained from sales notes completed at fish market auctions.

Fishing fleet information (Table 20.15) is obtained from vessel registers maintained by the Ministry of Agriculture, Fisheries and Food in England and Wales and the Scottish Office Agriculture and Fisheries Department.

Estimated average household food consumption in Great Britain - national food survey *(Table 20.16)*

The sample
The National Food Survey results are derived from the responses of a random sample of private households throughout Great Britain. Although the survey was extended to include Northern Ireland in 1996, the results given are for Great Britain in order to preserve continuity. The person most responsible for the catering arrangements of the household is designated as the main diary keeper; he or she records details of all items of food for human consumption brought into the home during the course of a week. Until 1997 about 8 000 households in Great Britain participated in the Survey. From 1997 a more efficient sample design was introduced and the sample for Great Britain was reduced to around 6 000 households. A description of the sampling methods is given in the annual reports; the latest is *National Food Survey* 1998.

Household
A group of persons living in the same dwelling and sharing common catering arrangements. The size of the household is defined in terms of the number of persons who spend at least four nights in the household during the week of the survey and also have at least one meal a day from the household food supply on at least four days. The head of the household and the main diary keeper are regarded as 'persons' in all cases.

Food obtained for consumption
Purchases of selected foods for home consumption during the Survey week (including purchases in bulk and, exceptionally, school milk) plus any garden or allotment produce, etc, consumed during the week. For a few minor miscellaneous items, expenditure is recorded but not the quantity (eg artificial sweeteners). The survey also includes information on confectionery and soft and alcoholic drinks brought home and, published from 1994, on food and drink consumed outside the home.

Although the quantity of food purchased or obtained free during the survey week will differ from the actual consumption of an individual household, the average quantity of food for a large number of households will equal the average consumption if there is no general change in the level of larder stocks.

20.1 Outputs, inputs and income at current prices

United Kingdom

Calendar years

£ millions

		1988	1989	1990	1991	1992	1993	1994	1995	1996	1997	1998
Outputs[1]												
Cereals:												
wheat	KFKA	1 205†	1 434	1 489	1 771	1 718	1 782	1 723	2 079	2 315	1 854	1 647
barley	KFKB	855†	882	806	749	861	844	847	1 108	1 183	979	777
oats	KFKC	55†	46	54	56	56	73	79	91	84	71	59
rye, mixed corn & triticale	KFKD	3	3	4	3†	4	6	7	9	9	9	9
1. Total production of cereals	KFKF	2 118†	2 365	2 353	2 579	2 638	2 705	2 656	3 287	3 591	2 913	2 493
Other crops:												
oilseed rape	KFKG	240	270	283	349†	268	411	373	379	421	406	399
linseed	KIBT	7	11	15†	39	84	184	39	41	40	51	67
sugar beet	KFKH	239	250	272	278	360	317	307	323	333	300†	274
hops	KFKI	13	15	15	21	18	21	20	17	20	20	14
peas and beans for stockfeed	KFKJ	142	116	117	107	123	186	147	135	129	129†	119
hay and dried grass	KFKK	14	14	22	21	16	17	19	21	21	19†	15
grass and clover seed	KFKL	14	14	14	11	10	10	11	10	16	17	16
straw	LUJX	145	159	213	164	116	219	249	321	211	192	185
unspecified crops[2]	LUJY	46	37	40	44	33	46	55	56	61	56	46
2. Total production of other crops	KFKN	860†	886	991	1 035	1 029	1 411	1 220	1 302	1 250	1 190	1 135
3. Total production of potatoes	KFKO	423†	494	573	521	528	386	712	1 077	637	382	639
Horticulture:												
vegtables	KFKP	891†	929	1 020	978	933	913	992	1 066	1 064	973	1 022
fruit	KFKQ	234†	264	259	282	268	257	231	234	249	180	216
ornamentals	KFKR	417	456	501	517†	526	566	591	617	668	647	658
fruit, vegtables and tomato seeds	KFKS	7	7	6	6	9	7†	8	8	8	8	8
4. Total production of horticulture	KFKT	1 549†	1 656	1 785	1 783	1 736	1 743	1 821	1 925	1 988	1 809	1 903
Livestock												
finished cattle and calves	KFKU	2 001	2 161†	1 975	2 072	2 021	2 392	2 486	2 583	2 551	2 269	1 943
finished sheep and lambs	KFKV	929†	1 009	1 017	1 141	1 164	1 237	1 242	1 338	1 300	1 199	1 142
finished pigs	KFKW	922†	1 061	1 065	1 016	1 141	1 037	1 033	1 175	1 366	1 191	873
poultry	KFKX	960†	1 002	1 079	1 079	1 124	1 271	1 336	1 333	1 501	1 482	1 344
other livestock	KFKY	109†	117	122	127	135	136	135	143	149	156	154
5. Total production of livestock	KFLA	4 920†	5 350	5 257	5 436	5 586	6 074	6 232	6 572	6 867	6 297	5 457
Livestock products:												
milk	KFLB	2 559†	2 727	2 792	2 789	2 924	3 187	3 268	3 508	3 504	3 160	2 729
eggs[3]	KFLC	392†	383	439	410	397	427	432	399	467	428	400
clip wool	KFLD	48	52	49	42	41	30	46	46	40	35†	27
unspecified livestock products	KFLE	17	24	26	21	23	24	22	24	23	25	24
6. Total production of livestock products	KFLF	3 016†	3 186	3 306	3 262	3 386	3 667	3 768	3 977	4 034	3 649	3 180
Capital formation in livestock												
cattle	KUJZ	279	381	321	236	442	651	580	359	229	404	280
sheep	LUKA	195	188	171	125	144	128	154	142	133	213	108
pigs	LUKB	5	13	17	16	20	15	13	11	23	18	2
poultry	LUKC	106	88	106	111	114	120	112	122	130	131	136
7. Total capital formation in livestock	KFLI	585†	670	616	487	719	914	858	634	514	767	525

1 Output is a net of VAT collected on the sale of non-edible products. Figures for total output include subsidies on products, but not other subsidies.
2 Includes turf, other minor crops and arable area payments for fodder maize.
3 Includes the value of duck eggs and exports of eggs for hatching.
4 Gross Output at basic prices includes subsidies (less taxes) on products.
5 Descriptions of terms used in this table are provided in the Annex.
6 Following a revision of the methodology used to calculate the feedbill, it is no longer possible to provide a breakdown of the total for years prior to 1992.
7 Landlords' expenses are included within farm maintenance, miscellaneous expenditure and depreciation of buildings and works.
8 Includes livestock and crop costs, water costs, insurance premia, bank charges, professional fees, rates, vehicle licence costs and other farming costs.
9 Excludes the value of work done by farm labour on own account capital formation in buildings and works.
10 Interest charges on loans for current farming purposes and buildings and less interest on money held on short term deposit.
11 Net rent is the rent paid on tenanted land (including 'conacre' land in Northern Ireland) less landlords' expenses and the benefit value of dwellings on that land.
12 The Farming Income Statistic has been discontinued; reasons are given in the Annex. It would have shown a fall of 46% from 1997 to 1998.

Source: Ministry of Agriculture, Fisheries and Food: 01904 455080

20.1 Outputs, inputs and income at current prices

United Kingdom

continued

Calendar years

£ millions

		1988	1989	1990	1991	1992	1993	1994	1995	1996	1997	1998
Other agricultural activities												
contract work	LUKD	333	360	386	410	439	471	506	562	612	590	574
leasing out milk quota	LUKE	20	30	44	32	42	104	127	136	169	120	108
leasing out ewe premium	LUKF	–	–	–	–	–	4	4	4	3	2	2
leasing out suckler cow premium	LUKG	–	–	–	–	–	1	..	1	1	1	1
8. Total other agricultural activities	LUOS	353	390	430	442	481	578	636	703	784	713	684
9. Total inseparable non-agricultural activities	LUOT	192	211	237	257	278	303	330	337	368	377	399
10. Gross output (at basic prices)[4,5]	KFLT	14 015†	15 207	15 550	15 803	16 381	17 781	18 233	19 815	20 033	18 095	16 414
11. Total subsidies (less taxes) on product	LUOU	399	401	448	660	805	1 763	4 686	2 107	2 782	2 579	2 436
12. Output at market prices (10-11)[5]	LUOV	13 616	14 806	15 102	15 143	15 576	16 018	16 547	17 708	17 251	15 515	13 978
of which												
transactions within the agricultural idustry	LUNP	..	..	..	..	..	..	..	..	..	..	..
feed wheat	LUNQ	61	62	62	95	114	81	75	55	66	62	49
feed barley	LUNR	200	209	228	255	265	233	201	199	206	175	133
feed oats	LUNS	17	18	19	21	18	21	22	16	16	12	10
seed potatoes	LUNT	18	16	29	20	29	14	15	40	31	13	15
straw	LUNU	137	150	203	156	109	207	236	308	200	183	174
contract work	LUNV	333	360	386	410	439	471	506	562	612	590	574
leasing of quota	LUNW	20	30	44	32	42	108	131	142	172	123	110
total capital formation in livestock	LUNX	..	..	..	..	..	..	..	..	..	..	..
Intermediate consumption (formerly known as inputs)												
(Expenditure net of reclaimed VAT)												
Feedingstuffs												
compounds	LUNY	2 565	2 648	2 544	2 505	1 675	1 768	1 693	1 809	1 963	1 777	1 528
straights	LUNZ	–	–	–	–	736	777	828	893	893	760	633
feed purchased from other farms	LUOA	278	288	308	371	398	335	298	270	288	250	192
other costs	LUOB	–	–	–	–	70	72	76	85	80	71	68
13. Total feedingstuffs[6]	KFMB	2 843†	2 936	2 852	2 876	2 878	2 952	2 894	3 057	3 224	2 857	2 422
Seeds												
cereals	KFMC	125	126	121	123	114	114	112	121	124†	114	100
other	KFMD	173†	180	209	192	210	191	210	263	253	228	229
14. Total seeds	KFME	298†	306	330	315	324	305	323	384	377	342	329
15. Total fertilisers and lime	KFMM	768†	897	936	902	839	756	849	944	1 022	999	821
16. Pesticides	KFMN	456†	525	517	545	530	546	560	600	648	679	652
Farm maintenance[7]												
occupier	KCOZ	186	202†	201	202	226	247	282	303	295	297	281
landlord	KCPA	62	61†	59	59	59	57	58	59	62	65	71
17. Total farm maintenance	KCPB	248	262†	260	260	285	304	340	362	357	363	352
Miscellaneous expenditure												
machinery repairs	KFMO	552†	555	577	611	600	637	631	718	740	719	703
machinery fuel and oil	KFMP	283†	291	325	335	328	348	345	360	401	402	382
veterinary expenses and medicines	KCPC	182†	195	211	221	231	252	275	292	296	309	296
power and fuel(mainly electricity)	KCPD	197†	192	213	251	245	245	246	244	252	238	226
imported livestock	LUOC	60	40	40	31	37	35	33	25	19	22	15
straw for bedding	LUOD	137	150	203	156	109	207	236	308	200	183	174
contract work	LUOE	333	360	386	410	439	471	506	562	612	590	574
leasing of quota	LUOF	20	30	44	32	42	108	131	142	172	123	110
other farming costs[7,8]	KCPG	1 084†	1 238	1 302	1 457	1 625	1 672	1 747	1 795	1 945	2 001	2 023
18. Total miscellaneous expenses	KCPH	2 847†	3 052	3 301	3 503	3 655	3 975	4 147	4 447	4 638	4 586	4 503

See footnotes on the first part of this table.

Source: Ministry of Agriculture, Fisheries and Food: 01904 455080

20.1 Outputs, inputs and income at current prices
continued

United Kingdom

Calendar years

£ millions

		1988	1989	1990	1991	1992	1993	1994	1995	1996	1997	1998
19. Total intermediate consumption (formerly known as gross input)[5]	KCPM	7 460†	7 979	8 197	8 400	8 511	8 837	9 114	9 795	10 265	9 827	9 078
20. Gross value added (10-19)[5]	LUOG	6 555	7 229	7 353	7 403	7 870	8 944	9 120	10 020	9 768	8 268	7 336
Consumption of Fixed Capital												
buildings and works	LUOH	582	649	691	636	568	568	582	640	681	671	678
landlord[7]	KCPP	83	91	95	86	75	74	74	80	84†	81	80
other	KCPQ	499	558	596	550	493	494	507	560	598	591†	598
plant, machinery and vehicles	KCPR	944	996	1 037	1 070	1 080	1 112	1 149	1 198	1 258	1 292	..
cattle	LUOI	279	314	283	318	330	577	529	416	315	402	304
sheep	LUOJ	131	132	150	148	120	137	141	174	189	155	80
pigs	LUOK	11	15	14	16	18	14	15	17	18	15	9
poultry	LUOL	101	101	103	109	118	108	108	121	130	123	147
21. Total Consumption of Fixed Capital	KCPS	2 075†	2 233	2 304	2 326	2 265	2 548	2 556	2 601	2 630	2 696	2 548
22. Net value added (at basic prices)(20-21)[5]	KCPT	4 480†	4 996	5 049	5 077	5 605	6 396	6 563	7 419	7 138	5 572	4 787
Other subsidies												
animal disease compensation	LUOM	4	8	6	7	6	11	7	7	6	16	12
set-aside	LUON	–	9	18	25	33	168	222	206	163	90	87
other	LUOO	15	19	19	22	27	37	57	79	84	101	138
23. Total other subsidies	LUOP	19	35	43	54	66	215	286	293	253	206	237
24. Net value added at factor cost(22+23)[5]	LUOQ	4 498	5 031	5 092	5 131	5 671	6 611	6 849	7 712	7 391	5 778	5 025
25. Total compensation of employees[5,9]	LUOR	1 514	1 570	1 701	1 767	1 773	1 775	1 819	1 822	1 863	1 915	1 945
26. Interest[10]	KCPU	682†	930	1 014	843	714	522	528	582	540	613	711
27. Net rent[11]	KCPV	158	144	135	143	153	148	151	149	156	177†	196
28. Total income from farming (22-25-26-27)[12]	KCQB	2 145†	2 387	2 242	2 378	3 032	4 166	4 350	5 160	4 831	3 073	2 174

See footnotes on the first part of this table.

Source: Ministry of Agriculture, Fisheries and Food: 01904 455080

20.2 Output and input volume indices
United Kingdom

Calendar years — 1995=100

		1988	1989	1990	1991	1992	1993	1994	1995	1996	1997	1998
Outputs[1]												
Cereals:												
wheat	LUKH	80.4	96.0	95.5	103.0	98.8	90.4	93.4	100.0	111.4	105.4	109.1
barley	LUKI	120.8	121.0	105.1	96.1	108.8	89.1	88.6	100.0	115.5	112.8	102.6
oats	LUKJ	80.0	67.8	75.1	75.0	71.0	73.1	86.0	100.0	89.5	90.8	91.6
rye, mixed corn & triticale	LUKK	65.6	78.8	95.3	79.4	90.8	86.0	91.1	100.0	96.9	95.7	103.2
1. Total production of cereals	LUKL	93.1	103.0	97.9	99.9	100.8	89.4	91.6	100.0	112.2	107.4	106.6
Other crops:												
oilseed rape	LUKM	84.7	77.6	79.2	105.2	99.4	89.5	100.3	100.0	112.7	122.0	126.5
linseed	LUKN	32.6	42.7	77.3	201.6	279.4	239.0	116.0	100.0	107.3	131.7	177.6
sugar beet	LUKO	96.7	96.2	93.7	103.2	120.4	114.6	103.4	100.0	123.6	131.5	116.3
hops	LUKP	115.5	112.3	110.7	144.0	114.6	126.2	107.0	100.0	127.3	118.1	80.9
peas and beans for stockfeed	LUKQ	164.8	125.0	133.9	115.1	119.0	160.0	127.7	100.0	101.8	123.0	125.8
hay and dried grass	LUKR	118.8	112.7	129.0	129.0	106.9	108.6	104.3	100.0	102.4	97.1	83.6
grass and clover seed	LUKS	146.2	170.4	176.0	135.9	126.1	116.0	132.7	100.0	98.0	107.1	154.9
straw	LUKT	69.6	71.7	86.8	75.5	65.0	74.0	78.5	100.0	85.2	74.4	70.7
unspecified crops[2]	LUKU	102.0	86.0	87.2	85.2	82.0	95.5	122.2	100.0	120.3	105.6	90.4
2. Total production of other crops	LUKV	95.3	89.2	94.5	104.2	105.5	106.4	99.6	100.0	107.5	112.4	110.8
3. Total production of potatoes	LUKW	104.0	97.3	99.8	97.5	121.8	104.6	98.4	100.0	107.5	101.8	92.7
Horticulture:												
vegtables	LUKX	111.0	109.6	104.4	107.0	110.6	107.0	107.5	100.0	105.0	102.5	99.0
fruit	LUKY	103.1	122.4	107.1	112.9	128.4	107.4	104.7	100.0	104.8	73.4	86.7
ornamentals	LUKZ	81.3	84.0	87.7	89.6	88.8	89.9	91.0	100.0	104.0	105.4	106.3
fruit, vegtable and tomato seeds	LULA	117.1	117.7	104.6	105.9	113.9	97.3	97.9	100.0	98.2	100.6	102.2
4. Total production of horticulture	LULB	99.0	102.0	98.4	101.2	105.0	100.9	101.2	100.0	104.6	99.9	100.1
Livestock												
finished cattle and calves	LULC	93.4	100.2	100.5	104.4	96.7	93.0	99.3	100.0	76.2	76.9	77.3
finished sheep and lambs	LULD	87.8	97.6	100.5	103.4	100.0	98.9	99.9	100.0	95.3	92.4	97.9
finished pigs	LULE	102.4	94.7	96.4	100.6	100.2	102.4	105.6	100.0	100.4	108.9	110.3
poultry	LULF	80.7	77.8	81.0	85.7	87.5	92.5	98.1	100.0	103.6	107.1	107.6
other livestock	LULG	94.5	95.9	95.8	94.8	102.4	99.4	97.7	100.0	100.6	103.0	104.4
5. Total production of livestock	LULH	90.4	93.2	94.8	98.7	95.7	95.7	100.1	100.0	90.5	92.5	94.1
Livestock products:												
milk	LULI	102.6	101.3	103.4	100.2	99.1	99.8	101.3	100.0	99.6	100.6	99.1
eggs[3]	LULJ	110.7	97.5	100.9	104.6	102.3	101.3	100.7	100.0	100.3	103.0	104.8
clip wool	LULK	101.3	110.4	110.4	105.5	105.4	99.7	96.7	100.0	95.1	96.9	100.8
unspecified	LULL	83.0	119.4	131.2	101.9	111.2	110.8	91.9	100.0	84.1	96.0	94.9
6. Total production of livestock products	LULM	103.2	100.9	103.2	100.7	99.6	100.0	101.1	100.0	99.5	100.8	99.7
Capital formation in livestock:												
cattle	LULN	88.2	111.4	101.9	78.4	126.2	121.1	119.3	100.0	94.2	105.2	96.0
sheep	LULO	123.5	134.7	115.3	77.6	119.2	99.0	135.4	100.0	83.4	131.2	121.7
pigs	LULP	74.3	132.2	173.2	168.9	183.6	167.6	143.5	100.0	182.1	191.4	31.1
poultry	LULQ	103.3	86.1	95.5	92.9	89.4	101.9	101.0	100.0	101.5	108.5	100.2
7. Total capital formation in livestock	LULR	101.6	114.9	107.1	83.9	119.6	115.0	119.3	100.0	94.7	115.6	104.2
Other agricultural activities												
contract work	LULS	96.2	93.7	92.4	89.9	89.9	91.6	94.4	100.0	104.6	107.3	111.0
leasing out milk quota	LULT	33.1	44.6	56.6	65.4	68.4	79.1	90.9	100.0	102.8	111.7	119.8
leasing out ewe premium	LULU	91.2	91.2	91.2	91.2	91.2	91.2	88.9	100.0	65.0	48.9	43.2
leasing out suckler cow premium	LULV	54.8	54.8	54.8	54.8	54.8	54.8	42.8	100.0	68.8	52.5	52.7
8. Total other non-agricultural activities	LULW	86.5	86.1	86.8	86.1	86.4	89.0	93.5	100.0	104.0	107.9	112.3
9. Total inseparable non-agricultural activities	LULX	79.8	81.3	83.7	85.4	89.1	95.3	101.4	100.0	106.6	106.3	102.8

1 Output is net of VAT collected on the sale of non-edible products. Figures for total output include subsidies on products, but not other subsidies.
2 Includes turf, other minor crops and arable area payments for fodder maize.
3 Includes the value of duck eggs and exports of eggs for hatching.
4 Gross Output at basic prices includes subsidies (less taxes) on products.
5 Descriptions of terms used in this table are provided in the Annex.
6 Landlords' expenses are included within farm maintenance, miscellaneous expenditure and depreciation of buildings and works.
7 Includes livestock and crop costs, water costs, insurance premia, bank charges, professional fees, rates, vehicle licence costs and other farming costs.

Source: Ministry of Agriculture, Fisheries and Food: 01904 455080

20.2 Output and input volume indices

United Kingdom

continued

Calendar years

1995=100

		1988	1989	1990	1991	1992	1993	1994	1995	1996	1997	1998
10. **Gross output** (at basic prices)[4,5]	LULY	94.8	97.1	97.3	98.5	100.1	97.5	99.5	100.0	100.2	100.8	100.5
of which												
transactions within the agricultural industry												
feed wheat	LULZ	108.2	110.2	112.2	161.0	190.4	141.2	145.7	100.0	119.8	139.3	130.4
feed barley	LUMA	100.6	103.4	108.8	120.9	124.3	107.3	105.0	100.0	105.2	113.1	100.0
feed oats	LUMB	101.9	100.2	99.6	114.1	92.4	97.7	124.6	100.0	92.8	88.6	98.3
seed potatoes	LUMC	131.8	138.5	139.9	125.0	175.1	134.7	100.2	100.0	112.2	123.2	101.9
straw	LUMD	100.0	100.0	100.0	100.0	100.0	100.0	100.0	100.0	100.0	100.0	100.0
contract work	LUME	68.4	70.5	86.3	74.4	63.5	72.9	77.6	100.0	84.5	73.3	69.5
leasing of quota	LUMF	96.2	93.7	92.4	89.9	89.9	91.6	94.4	100.0	104.6	107.3	111.0
total capital formation in livestock	LUMG	30.6	41.2	52.3	60.4	63.2	79.3	90.5	100.0	101.5	109.6	117.1
Intermediate Consumption (formerly known as inputs)[5] (Expenditure net of reclaimed VAT)												
Feedingstuffs:												
compounds	LUMH	153.8	150.6	150.3	148.3	95.0	97.8	100.0	100.0	101.1	96.9	102.0
straights	LUMI	88.1	88.1	88.1	88.1	88.1	93.0	99.1	100.0	90.3	88.3	83.1
feed purchased from other farms	LUMJ	102.2	104.4	108.8	128.2	135.2	113.2	114.3	100.0	107.5	117.0	106.2
other costs	LUMK	94.0	94.0	94.0	94.0	94.0	97.4	100.4	100.0	99.0	98.3	100.4
13. Total feedingstuffs	LUML	100.8	99.1	99.4	100.1	96.9	97.9	101.1	100.0	99.2	96.1	97.0
Seeds:												
cereals	LUMM	125.3	119.3	111.4	107.0	92.8	95.9	95.1	100.0	104.5	104.6	96.9
other	LUMN	91.9	94.2	106.9	102.0	107.7	103.4	100.4	100.0	100.3	99.6	99.1
14. Total seeds	LUMO	103.8	103.3	108.6	103.9	102.1	100.6	98.5	100.0	101.6	101.2	98.4
15. Total fertilisers and lime	LUMP	91.1	98.2	101.1	100.8	99.4	95.1	100.7	100.0	97.7	98.5	100.5
16. Pesticides	LUMQ	101.0	110.8	99.9	97.1	93.1	94.5	92.1	100.0	101.4	105.5	108.5
Farm maintenance:[6]												
occupier	LUMR	83.8	83.2	79.1	77.0	85.7	90.9	99.0	100.0	96.2	95.2	88.9
landlord	LUMS	110.8	108.4	106.5	106.5	105.6	104.7	102.7	100.0	98.0	96.2	95.4
17. Total farm maintenance	LUMT	88.4	87.4	83.7	82.1	89.0	93.2	99.6	100.0	96.5	95.4	90.0
Miscellaneous expenditure:												
machinery repairs	LUMU	111.7	107.0	102.9	102.4	93.8	94.8	91.1	100.0	97.5	94.0	92.4
machinery fuel and oil	LUMV	123.4	109.8	108.7	108.7	105.7	105.1	104.1	100.0	96.6	95.4	93.8
veterinary expenses and medicines	LUMW	70.3	74.9	78.4	79.1	81.1	87.7	95.0	100.0	99.9	102.2	97.6
power and fuel (mainly electricity)	LUMX	102.6	91.5	92.2	104.3	96.7	97.8	100.1	100.0	99.8	99.1	100.7
imported livestock	LUMY	370.7	260.9	240.1	181.4	209.3	125.3	124.2	100.0	85.3	108.4	76.4
straw for bedding	LUMZ	68.4	70.5	86.3	74.4	63.5	72.9	77.6	100.0	84.5	73.3	69.5
contract work	LUNA	96.2	93.7	92.4	89.9	89.9	91.6	94.4	100.0	104.6	107.3	111.0
leasing of quota	LUNB	30.6	41.2	52.3	60.4	63.2	79.3	90.5	100.0	101.5	109.6	117.1
other farming costs[6,7]	LUNC	87.9	92.4	89.6	92.4	95.4	98.5	101.5	100.0	99.3	101.7	102.4
18. Total miscellaneous expenses	LUND	93.7	93.0	92.4	93.2	92.0	94.5	96.7	100.0	98.5	99.1	99.0
19. Total intermediate consumption (formerly known as Gross input)[5]	LUNE	96.2	96.5	96.2	96.3	94.5	95.8	98.3	100.0	98.9	98.5	98.8
20. **Gross Value Added** (10-19)[5]	LUNF	93.2	97.8	98.5	100.9	106.5	99.3	100.8	100.0	101.5	103.3	102.1
Consumption of fixed capital:												
buildings and works	LUNG	98.2	98.7	99.2	99.6	99.7	99.8	100.0	100.0	99.5	98.8	97.8
landlord[6]	LUNH	112.0	110.5	109.1	107.5	105.7	103.7	102.0	100.0	97.5	94.8	91.9
other												
plant, machinery and vehicles	LUNI	106.9	105.7	103.9	102.0	100.1	99.1	99.3	100.0	101.5	101.9	100.3
cattle	LUNJ	79.9	83.4	82.4	95.6	79.4	87.7	87.4	100.0	104.7	93.7	93.2
sheep	LUNK	68.8	78.4	82.8	76.7	81.8	86.0	102.4	100.0	96.8	79.5	79.6
pigs	LUNL	105.1	91.9	91.5	103.2	104.0	101.0	109.8	100.0	92.1	101.1	111.1
poultry	LUNM	99.2	101.6	94.6	93.9	95.6	92.1	98.3	100.0	102.7	102.9	109.8
21. Total consumption of Fixed Capital	LUNN	96.0	97.0	96.3	97.1	94.5	95.5	96.9	100.0	101.2	98.7	98.0
22. **Net value added** (at basic prices)[5]	LUNO	91.8	97.9	99.2	102.4	111.9	100.8	102.4	100.0	101.6	105.0	103.6

See footnotes on the first part of this table.

Source: Ministry of Agriculture, Fisheries and Food: 01904 455080

20.3 Agriculture-land use
United Kingdom

Area at the June census[1]

Thousand hectares

		1988	1989	1990	1991	1992	1993	1994	1995	1996	1997	1998
Total Agricultural area	KFEU	18 992	18 929	18 884	18 854	18 849	18 890	18 850	18 746	18 753	18 653	18 593
Crops	KSJA	5 256	5 139	5 015	4 957	4 982	4 520	4 470	4 544	4 722	4 990	4 972
Bare fallow	KIJQ	62	69	68	67	53	50	46	43	37	29	34
Total tillage	KIJR	5 318	5 208	5 082	5 025	5 035	4 569	4 516	4 586	4 759	5 020	5 005
All grass under 5 years old	KFEM	1 636	1 561	1 606	1 608	1 584	1 582	1 456	1 407	1 395	1 405	1 303
Total arable land	KFEN	6 954	6 770	6 689	6 633	6 618	6 152	5 972	5 993	6 154	6 425	6 308
All grasses 5 years old and over	KFEO	5 222	5 303	5 316	5 327	5 279	5 274	5 388	5 375	5 354	5 282	5 350
Total tillage and grass	KFEP	12 176	12 073	12 005	11 960	11 897	11 426	11 360	11 368	11 507	11 706	11 658
Sole right rough grazing	KFEQ	5 011	4 997	4 965	4 950	4 943	4 879	4 825	4 785	4 760	4 657	4 624
Set aside[2]	KSJB	..	..	72	97	160	677	728	633	509	306	314
All other land on agricultural holdings including woodland	KSJC	569	623	607	614	618	680	712	734	751	763	777
Total land on agricultural holdings	KSJD	17 756	17 693	17 648	17 620	17 619	17 661	17 626	17 520	17 527	17 432	17 372
Common rough grazing (estimated)	KFER	1 236	1 236	1 236	1 233	1 230	1 229	1 224	1 226	1 226	1 221	1 221
Crops	KSJE	5 256	5 139	5 015	4 957	4 982	4 520	4 470	4 544	4 722	4 990	4 972
Cereals	KSJF	3 901	3 876	3 660	3 502	3 489	3 033	3 043	3 182	3 359	3 514	3 420
Wheat	KFDA	1 886	2 083	2 014	1 981	2 067	1 759	1 811	1 859	1 976	2 036	2 045
Barley	KFDB	1 881	1 654	1 518	1 395	1 299	1 166	1 108	1 193	1 269	1 359	1 255
Oats	KFDC	121	119	107	104	100	92	109	112	96	100	98
Mixed corn	KFDD	5	5	4	4	4	3	3	3	3	2	2
Rye[3]	KFDE	7	7	8	9	8	6	7	8	8	9	10
Triticale[4]	KIHX	..	8	9	11	11	7	6	7	7	8	10
Other arable crops (excluding potatoes)	KSJG	965	881	970	1 074	1 115	1 128	1 073	1 003	996	1 126	1 193
Oilseed rape	KFEH	347	321	390	440	421	377	404	354	356	445	506
Sugar beet not for stock feeding[5]	KFEG	201	197	194	196	197	197	195	196	199	196	189
Hops[6]	KIHZ	4	4	4	4	4	3	3	3	3	3	3
Peas for harvesting dry and field beans	KSJH	260	215	216	203	208	244	228	195	178	197	213
Linseed[7]	KSJI	..	17	34	92	144	150	58	54	49	73	99
Other crops	KSJJ	153	127	132	141	141	157	185	201	211	211	211
Potatoes	KGRS	181	175	177	177	181	171	164	172	178	166	164
Horticultural	KSJK	209	208	208	204	198	188	189	187	189	185	177
Vegetables grown in the open	KSJL	142	141	142	139	135	126	127	130	132	126	123
Orchard fruit	KSJM	37	36	34	34	33	32	32	28	28	30	28
Soft fruit	KSJN	15	15	15	15	14	13	13	12	12	11	10
Ornamentals[8]	KSJO	13	14	14	14	14	14	14	15	14	14	13
Glasshouse crops	KFEF	2	2	2	2	2	2	2	2	2	2	2

1 Includes estimates for minor holdings for all countries. In previous editions minor holdings were excluded for Scotland and therefore data here will differ from previous editions. Also, Northern Ireland data are now based on all active farm business.
2 Figures are for England only in 1990 and 1991 and Great Britain only in 1992.
3 Figures are for England and Wales only.
4 Figures are for Great Britain only.
5 Figures are for England and Wales only.
6 Figures are for England and Wales only for 1989 and England only from 1990.
7 Figures are for England and Wales only from 1989 to 1991 and 1992 Great Britain only.
8 Includes non-commercial orchards.

Source: Agricultural Departments: 01904 455332

20.4 Estimated quantity of crops and grass harvested[1,2]
United Kingdom

Thousand tonnes

		1988	1989	1990	1991	1992	1993	1994	1995	1996	1997	1998
Cereals												
Wheat	KFQA	11 751	14 033	14 035	14 362	14 095	12 890	13 320†	14 310	16 100	15 020	15 470
Barley	KFQB	8 778	8 073	7 911†	7 632	7 368	6 050	5 950	6 840	7 790	7 830	6 630
Oats	KFQC	548	529	520†	522	503	480	600	615	590	575	585
Mixed corn for threshing	KFQD	22	18	17	15	16	14†	13	16	12	11	6
Rye for threshing	KFQE	34	36	40	49	37	30	43	43	51	53	42
Maize for threshing[3]	KFQF	875	940	1 035	1 695	1 975	2 730	..	..	..	..	..
Potatoes												
Early crop[4]	KFQG	420	366	437	359	411	426	337	373	325	386	337
Main crop[4]	KFQH	6 522	5 956	6 106	6 000	7 402	6 647	6 203	6 033	6 900	6 739	6 168
Fodder crops												
Beans for stockfeeding	KFQI	596	450	474	423	445	612	454	319	318	377†	423
Turnips, swedes[5]	KFQJ	3 325	905	920	835	760	620	..	..	..	..	..
Fodder beet and mangolds[5]	KFQK	735	600	640	715	710	655	..	..	..	..	..
Maize for threshing or stock-feeding[3]	KFQL	875	940	..	..	..	..	..	..	..	..	..
Kale, cabbage, savoys and kohl rabi	KFQN	735	565	535	610	515	485	..	..	..	..	..
Peas harvested dry for stock-feeding	KFQP	360	277	320	224	209	266	246	229	240	297	311
Other crops												
Sugar beet[6,7]	KFQQ	8 150	8 115	7 902	8 701	10 148	9 666	8 720	8 431	10 420	11 084†	9 802
Rape grown for oilseed	KFQR	1 040	950	1 258	1 278	1 166	1 100	1 243†	1 224	1 412	1 527	1 569
Hops	KFQS	5	5	5	6	5	6	5	4	6	5	4
Horticultural crops												
Vegetables grown in the open												
Brussels sprouts	KFQW	146	124	100	97	115	98	105	79	80	79	73
Cabbage (including savoys and spring greens)	KFQX	465	460	393	388	402	381	403	344†	363	312	284
Cauliflowers	KFQY	344	308	306	312	341	309	289	245†	238	196	192
Carrots	KFQZ	494	503	486	544	608	592	633	517	617	621†	618
Parsnips	KFRA	55	47	47	49	60	64	70	63	75	92†	95
Turnips and swedes	KFRB	152	140	128	156	158	141	137	148	123	126†	125
Beetroot	KFRC	91	93	89	87	74	93	87	73	72	74	70
Onions, Dry bulb	KFRD	249	256	217	213	235	274	274	225†	285	336	342
Onions, Salad	KFRE	27	30	20	23	28	24	25	24	28	26	25
Leeks	KFRF	64	66	64	68	72	60	61	63†	56	47	53
Broad beans	KFRG	12	12	13	20	17	11	12	11	9	10	13
Runner beans including French	KFRH	51	44	28	37	39	46	36	36†	35	35	30
Peas, Green for market	KFRI	12	8	8	8	6	6	6	8	7	8	7
Peas, Green for processing	KFRJ	191	206	238	223	211	209	181	191	217	167†	153
Celery	KFRK	52	56	46	49	44	41	43	39	38	39	35
Lettuce	KFRL	215	206	197	194	206	145	178	193†	188	159	153
Rhubarb	KFRM	28	30	27	28	20	20	21	20	21	20†	18
Protected crops												
Tomatoes	KFRN	128	130	134	133	122	110	109†	113	117	115	109
Cucumbers	KFRO	96	96	101	104	90	104	83	88	86	82	84
Lettuce	KFRP	47	48	50	47	44	37	33	30	26	24	21
Fruit crops												
Total Dessert Apples	KFRQ	132.5	202.4	179.0	148.3	164.7	167.1	175.9	138.5	105.4	96.0	97.8
Total Culinary Apples	KFRR	110.7	197.6	125.6	157.1	161.7	158.6	130.1	134.9	118.3	91.0†	84.5
Pears	KFRS	44.1	36.9	34.1	36.0	30.8	30.7	32.5	29.7	35.8	33.0	26.3
Plums	KFRT	20.6	10.7	7.2	21.5	21.3	11.9	11.1	14.5	19.7	12.2	6.5
Cherries	KFRU	1.3	3.2	1.5	1.0	3.8	2.4	1.3	3.5	3.6	0.6	0.9
Soft fruit	KFRV	91.5	86.3	87.2	88.6	93.9	93.8	75.7	73.7	81.1†	57.9	61.2

1 For vegetables; output marketed for the calendar year; for horticultural crops output marketed for the crop year.
2 Except for sugar beet and hops, the production area for England and Wales is the area returned at the June census together with estimates for very small holdings (known as minor holdings). In Scotland and Northern Ireland the area returned at June is also the production area except that estimates for minor holdings are included in Scotland for potatoes and in Nothern Ireland for barley, oats and potatoes.
3 From 1979 maize for threshing is included with maize for stockfeeding.
4 Revised basis of calculation adopted which prevents direct comparison with post - 1980 figures and earlier years.
5 Before 1977 fodder beet was included with turnips and swedes for stockfeeding. In 1977, as a result of changes in the census categories, fodder beet was included with mangolds in England and Wales but continued to be included with turnips and swedes for stockfeeding in Scotland and Northern Ireland. Scotland collected fodder beet separately for the first time in 1986.
6 The production area for sugar beet is provided by British Sugar plc.
7 From 1991 figures are adjusted to constant 16% sugar content.

Source: Agricultural Departments: 01904 455332

20.5 Cattle, sheep, pigs and poultry on agricultural holdings[1]
United Kingdom

At June each year

Thousands

		1988	1989	1990	1991	1992	1993	1994	1995	1996	1997	1998
Total cattle and calves	KFSB	12 008	12 101	12 192	12 003	11 924	11 851	11 954	11 857	12 040	116 637†	11 519
of which:												
dairy cows	KFCT	2 913	2 866	2 848	2 771	2 683	2 668	2 716	2 603	2 587	2 478	2 439
beef cows	KFCU	1 404	1 525	1 632	1 700	1 731	1 784	1 809	1 840	1 864	1 862	1 947
heifers in calf	KFCV	840	799	763	738	768	803	775	775	818	848	787
Total sheep and lambs	KFSG	41 495	43 588	44 469	44 166	44 540	44 436	43 813	43 304	42 086	42 823	44 471
of which:												
ewes and shearlings	KSJP	19 373	20 368	20 787	20 665	20 732	20 881	20 861	20 830	20 550	20 696	21 260
lambs under one year old	KFSF	20 834	21 897	22 380	22 213	22 607	22 394	21 758	21 350	20 443	21 032	22 138
Total pigs[2]	KFSQ	8 024†	7 606	7 548	7 695	7 707	7 853	7 892	7 627	7 590	8 072	8 146
of which:												
sows in pig and other sows for breeding	KSJQ	715†	671	671	690	683	698	691	654	649	683	675
gilts in pig	KSJR	103†	98	110	109	110	117	106	101	107	116	103
Total fowls	KFSV	131 699	121 029	125 357	128 025	124 842	131 093	126 653	127 035	..	..	..
of which:												
table fowls including broilers	KFSU	75 785	70 509	73 944	76 111	73 748	79 940	75 696	77 177	..	106 937	98 244
laying fowls[2]	KFSS	37 583	34 101	33 624	33 416	33 348	32 965	32 682	31 837	..	34 286	29 483
growing pullets	KFSR	11 321	9 487	10 530	11 102	10 849	10 750	10 388	10 210	..	11 510	9 860

1 Includes estimates for minor holdings for all countries. In previous editions minor holdings were excluded for Scotland and therefore data here will differ from previous editions. Also Northern Ireland data are now based on all active farm business.

2 Figures for Scotland and Northern Ireland have been revised and therefore data here will differ from previous editions.

3 Excludes fowls laying eggs for hatching.

Source: Agricultural Departments: 01904 455332

20.6 Forestry

Years ending 31 March

Area - thousand hectares, Volume - million cubic metres

		1988 /89	1989 /90	1990 /91	1991 /92	1992 /93	1993 /94	1994 /95	1995 /96	1996 /97	1997 /98	1998 /99
Forest area[1]												
United Kingdom	KUAA	2 380	2 400	2 411	2 427	2 438	2 455	2 469	2 485	2 504	2 521	2 659
Great Britain	KUAB	2 307	2 326	2 336	2 350	2 361	2 377	2 390	2 405	2 423	2 440	2 577
Northern Ireland	KUAC	73	74	75	77	77	78	79	80	81	81	82
Forestry Commission (Great Britain)												
Productive woodland	KUAD	888	864[5]	859	855	845	827	815	804	795	785	780
Net disposals of forestry land[2]	KUAE	2.5	5.1	2.2	4.7	6.5	14.1	9.6	8.2	9.6	10.3	4.1
New planting[3]	KUAF	4.1	4.1	3.5	3.0	2.4	1.4	0.9	0.4	0.5	0.1	0.1
Restocking[3]	KUAG	8.5	7.9	7.6	8.3	8.5	7.9	7.9	7.5	7.9	7.8	7.9
Volume of timber removed[4]	KUAH	3.6	3.5	3.4	3.9	4.1	4.3	4.3	4.2	4.7	4.7	5.3
Total estates	KUAI	1 143	1 140	1 133	1 128	1 115	1 100	1 089	1 080	1 073	1 062	1 057
Private forestry (Great Britain)												
Productive woodland	KUAJ	1 231	1 266	1 281	1 298	1 317	1 344	1 367	1 392	1 417	1 444	1 542
New planting[3]	KUAK	25.4	15.6	15.6†	14.3	15.5	16.1	18.5	15.3	16.4	16.2	16.2
Restocking[3]	KUAL	4.9	6.3	7.2	7.9	8.1	8.5	6.2	5.8	6.6	5.6	5.5
Volume of timber removed[4]	KUAM	3.2	3.6	3.3	3.1	2.9	3.0	3.8	4.2	4.0	4.4	3.9
State afforestation in Northern Ireland												
Land under plantation	KUAN	58.3	58.5	58.8	60.6	60.8	60.8	61.0	61.0	60.9	61.0	61.0
Plantable land acquired during year	KUAO	0.4	0.6	0.4	0.3	0.4	0.4	0.1	0.1	0.1	0.1	0.3
Total area planted during year	KUAP	1.0	1.0	1.0	0.9	0.9	0.8	0.8	0.8	0.6	0.7	0.7
Total estates	KUAQ	74.3	74.6	74.8	75.1	75.2	75.5	75.5	75.3	75.5	75.5	75.7

1 Includes unproductive woodland.
2 Disposals less acquisitions, for plantations and plantable land.
3 New planting is planting on ground not previously carrying forest; Restocking is replacing trees after felling.
4 Standing volume overbark; private forestry figures are for calendar years ending previous December.
5 The apparent decrease in 1989/90 is mainly the result of re-classification of certain woodland types within the Forestry Commission.

Sources: Department of Agriculture (Northern Ireland); Forestry Commission: 0131 334 0303

20.7 Average weekly earnings and hours of male agricultural workers[1,2,3]
England and Wales

For the month of September 1994 to 1998

		1994	1995	1996	1997	1998
Average weekly earnings (£)	LQML	248.33	275.30	289.19	285.12	303.94
95% confidence interval		(+/-£8.48)	(+/-£10.03)	(+/-£13.34)	(+/-£11.43)	(+/-£12.92)
Average weekly hours worked	LQMM	50.2	50.6	52.5	50.8	51.0
95% confidence interval		(+/-1.1)	(+/-1.3)	(+/-1.8)	(+/-1.3)	(+/-1.5)
Average earnings/hours (£)	LQMN	4.95	5.44	5.51	5.61	5.96
95% confidence interval		(+/-£0.10)	(+/-£0.11)	(+/-£0.12)	(+/-£0.12)	(+/-£0.18)
Number of workers in the sample		268	246	176	251	292

1 Total earnings of regular full time male workers, including all payments-in-kind, valued where applicable in accordance with the Agricultural Wages Order.
2 Aged 20 and over in September 1994 and 19 and over from September 1995.
3 Results exclude earnings and hours of hire farm managers.

Source: Agricultural Departments: 01904 455332

20.8 Average weekly earnings and hours of different types of agricultural workers[1,2]
England and Wales

For the month of September 1998

	Full time		Part time[3]		Casual[4]		
	Male	Female	Male	Female	Male	Female	Managers
Average weekly earnings (£)	303.94	210.28	97.75	109.48	154.58	113.95	401.09
95% confidence interval	(+/-£12.92)	(+/-£15.29)	(+/-£10.79)	(+/-£7.63)	(+/-£15.47)	(+/-£10.67)	(+/-£25.12)
Average weekly hours worked	51.0	42.7	21.5	23.8	33.4	28.3	..
95% confidence interval	(+/-£1.5)	(+/-£1.6)	(+/-£2.0)	(+/-£1.3)	(+/-£2.7)	(+/-£2.4)	
Average earnings/hour (£)	5.96	4.92	4.55	4.60	4.63	4.03	..
95% confidence interval	(+/-£0.18)	(+/-£0.26)	(+/-£0.26)	(+/-£0.17)	(+/-£0.18)	(+/-£0.14)	
Number of workers in the sample	292	59	124	209	191	151	138

1 Total earnings of workers, including all payments-in-kind, valued where applicable in accordance with the Agricultural Wages Order.
2 Aged 19 and over.
3 Part time workers are defined as working less than 39 basic hours per week.
4 Casual workers are employed on a temporary basis.

Source: Agricultural Departments: 01904 455332

20.9 Sales for food of agricultural produce and livestock
United Kingdom

		Unit	1988	1989	1990	1991	1992	1993	1994	1995	1996	1997	1998
Cereals:													
Wheat[1,2]	KCQK	Thousand tonnes	3 395	4 228	4 347	4 269	4 212	3 865	4 234	4 674	4 845	4 749	4 670
Barley	KCQL	"	4 560	5 026	3 791	3 849	3 599	2 762	3 190	3 776	3 692	3 411	3 483
Oats[3]	KCQM	"	196	239	222	215	222	208	215	216	235	259	273
Potatoes[4]	KCQN	"	6 018	6 085	5 911	5 799	5 994	6 357	6 491	5 923†	6 123	6 275	5 967
Milk[5]:													
Milk utilised for liquid consumption	KCQO	Million litres	6 858	6 859	6 892	6 893	6 966	7 026	6 926	6 922	6 838	6 755†	6 743
Milk utilised for manufacture	KCQP	Million litres	7 481	7 258	7 489	7 022	6 876	6 880	7 134	6 918	6 934†	7 061	6 875
Total milk available for domestic use[6]	KCQQ	Million litres	14 709	14 485	14 751	14 282	14 189†	14 228	14 420	14 255	14 193	14 266	..
Hen eggs in shell	KCQR	Million dozens	884	763	804	818	806	791	787	774	775	794	793
Animals slaughtered:													
Cattle and calves:													
Cattle	KCQS	Thousands	3 345	3 414	3 479	3 568	3 355	2 944	3 089	3 266	2 291	2 264	2 297
Calves	KCQT	"	35	28	46	49	33	19	22	26	24	20	32
Total	KCQU	"	3 380	3 442	3 524	3 617	3 387	2 963	3 112	3 292	2 315	2 284	2 329
Sheep and lambs	KCQV	"	17 128	19 618	20 012	20 918	19 428	18 864	18 962	19 311	18 049	16 660	18 688
Pigs:													
For bacon:													
Used wholly[7]	KCQW	"	2 034	1 966	1 903	14 091	14 143	14 265	14 681	14 021	13 897	15 132†	15 872
Used in part	KCQX	"	3 494	3 449	3 281	..	..	..	..	..	..	..	..
Other uses	KCQY	"	9 872	8 767	8 694	..	..	..	..	..	..	..	..
Sows and boars	KCQZ	"	384	331	325	366	373	356	389	355	324	363	415
Total	KCRA	"	15 784	14 514	14 203	14 457	14 515	14 620	15 069	14 376	14 221	15 496	16 286
Poultry:[8] slaughtered	KCRB	Millions	686	644	670	693	688	715	754†	774	814	842	846

TIME SERIES. The figures for cereals and for animals slaughtered relate to periods of 52 weeks (53 weeks in 1987 and 1992).

1 Flour millers' receipts of home-grown wheat.
2 Following the receipt of additional data, figures have been revised back to 1992 and are not strictly compatible with those for earlier years.
3 Oatmeal millers' receipts of home-grown oats.
4 Total sales for human consumption in the UK.
5 Data to 1994 sourced from the Milk Marketing Boards. Data from 1995 sourced from surveys run by the agricultural departments. 1994 includes two months of data sourced from the surveys run by the agricultural departments.
6 The totals of liquid consumption and milk used for manufacture may not add up to the total available for domestic use because of adjustments for dairy wastage, stock changes and other uses, such as farmhouse consumption, milk fed to stock and on farm waste.
7 From 1991, data series quotes the number of clean pigs slaughtered for all pig meat.
8 Total fowls, ducks, geese and turkeys.

Source: Ministry of Agriculture, Fisheries and Food: 01904 455332

20.10 Stocks of food and feedingstuffs[1]
United Kingdom

At end-December in each year

Thousand tonnes

		1988	1989	1990	1991	1992	1993	1994	1995	1996	1997	1998
Wheat and flour (as wheat)	KCRC	918	918	860	1 461	1 809†	1 226	1 205	1 031	986	951	1 033
Barley (GB only)	KCRD	1 127	1 043	1 120	1 489	2 015†	1 391	1 365	1 529	1 488	1 430	1 317
Maize	KCRE	64	70	41	36	108†	21	31	32	24	44	66
Oilcake and meal[2]	KCRF	190	148	153	101	103	86	104	97	85	79	..
Oilseeds and nuts (crude oil equivalent)	KCRG	26	24	28	22	12	20	30	35	28	19	..
Vegetable oil (as crude oil)	KCRI	86	85	109	85	94	91	110	93	101	101	..
Marine oil (as crude oil)	KCRJ	12	11	9	9	8	8	18	17	10	4	..
Butter[3]	KCRK	82	55	74	65	63	34	19	11	14	7	11
Meat and offal[4]	KCRL	109.9	85.0	150.8	212.0	218.2	198.7	102.5	76.9	129.5	161.9	156.8
Raw coffee[5]	KCRM	8	7	11	10	10	12	13	11	8	7	8
Tea[6]	KCRN	50	51	48	43	40	44	43	38	39	37	38
Sugar	KCRO	978	860	824	824	888	1 069	1 016	766	807	1 003	928
Chocolate and sugar confectionery[7]	KCRP	..	..	..	..	..	..	..	..	..	..	..

1 Recorded stocks, including stocks in bond or held by the main processors.
2 Excluding castor meal, cocoa cake and meal.
3 In addition to stocks in public cold stores surveyed by MAFF, closing stocks include all intervention stocks in private cold stores.
4 Stocks of imported and home-produced meat and offal held in public cold stores, excluding poultrymeat, bacon and ham.
5 Including manufacturers' stocks and additional public warehouses.
6 Covering stocks held by primary wholesalers and held in public/private warehouses.
7 Manufacturers' stocks only.

Source: Ministry of Agriculture, Fisheries and Food: 01904 455332

20.11 Processed food and animal feedingstuffs: production
United Kingdom

Thousand tonnes

		1988	1989	1990	1991	1992	1993	1994	1995	1996	1997	1998
Flour milling:												
Wheat milling: total[8]	KFTA	5 112	5 032	4 875	4 796	5 022	5 236	5 290	5 338	5 501	5 535	5 714
Home produced[8]	KFTB	3 356	4 161	4 273	4 185	4 192	3 781	4 106	4 598	4 772	4 667	4 590
Imported	KFTC	1 756	871	605	611	830	1 455	1 184	740	729	868	1 125
Flour produced[8]	KFTD	3 973	3 928	3 879	3 851	3 975	4 120	4 232	4 292	4 454	4 439	4 532
Offals produced[8]	KFTE	1 157	1 076	999	968	1 045	1 123	1 078	1 079	1 111	1 112	1 223
Oat milling:												
Oats milled by oatmeal millers	KFTF	183	221	231	216	225	207	209	217	250	259	272
Products of oat milling	KFTG	109	129	132	124	121	101	98	105	119	126	150
Seed crushing:												
Oilseeds and nuts processed	KFTH	1 870	1 725	1 877	1 710	1 807	1 903	2 035	2 122	2 466	2 587	..
Crude oil produced, including production of maize oil	KFTI	659	589	630	583	636	658	739	779	862	909	..
Oilcake and meal produced, excluding castor meal, cocoa cake and meal	KFTJ	1 162	1 077	1 193	1 081	1 125	1 187	1 242	1 305	1 541	1 614	..
Production of home-killed meat: total including meat subsequently canned	KFTK	2 322†	2 330	2 376	2 455	2 392	2 262	2 358	2 393	2 094	2 162	2 270
Beef	KFTL	944	976	1 000	1 018	970	858	915	974†	702	696	697
Veal	KFTM	2	1	2	2	1	1	1	1	1	1	1
Mutton and lamb	KFTN	322	366	370	385	362	349	352	364	345	322	356
Pork	KFTO	788	718	735	772	786	783	807	761	778	875†	937
Offal[1]	KFTP	265†	268	270	278	274	270	284	293	268	269	278
Production of poultry meat[2]	KFTQ	1 148	1 116	1 148	1 211	1 226	1 289	1 344	1 397	1 461	1 502	..
Production of bacon and ham, including meat subsequently canned	KFTR	206	204	191	193	189	216	233	245	241	239	236
Production of milk products:												
Butter[3]	KFTS	147	140	151	132	127	141	148	132	130	139	137
Cheese (including farmhouse)	KFTT	303†	288	320	309	332	338	341	362	377	377	367
Condensed milk: includes skim concentrate and condensed milk used in manufacture of chocolate crumb	KFTU	183	207	204	198	206	191	196	181	195	198	167
Milk powder: excluding buttermilk and whey powder												
Full cream	KFTV	104	95	70	79	84	71	83	90	83	96	97
Skimmed	KFTW	136	133	166	143	106	132	142	117	108	109	107
Cream, fresh and sterilised; including farm cream[3]	KFTX	180†	208	230	254	240	255	274	282	281	268	266
Sugar: production from home-grown sugar-beet (as refined sugar)	KFTY	1 304	1 267	1 241	1 220	1 472	1 433	1 263	1 469	1 324	1 524	1 557
Production of compound fats:												
Margarine and other table spreads[4]	KFTZ	469	489	486	481	477	493	487	490	491	461	446
Solid cooking fats	KFUA	104	121	122	117	116	129	109	110	123	109	131
Production of other processed foods:												
Jam and marmalade	KFUB	174	181	–	..	..	..	..	..	..	..	..
Syrup and treacle[6]	KFUC	53	52	51	51	49	52	50	48	–	–	..
Canned vegetables[5]	KFUD	701	732[3]	–	–	..	..	..	..	..	..	..
Canned and bottled fruit[5]	KFUE	36	37	..	..	..	..	..	..	..	..	..
Chocolate confectionery	KSJS	485	498	539	534	532	563	573	562	576†	569	569
Sugar confectionery	KSJT	314	309	320	303	300	302	322	322	351	345†	338
Cocoa beans excluding re-exports	KFVX	101	115	125	148	159	171	163	160	189	174	172
Soups, canned and powdered	KFUF	314	–	..	..	..	..	..	..	..	..	..
Canned meat	KFUG	86	118	91	147	..	..	..	..	..	..	..
Canned fish	KFUH	8	9	..	..	..	..	..	..	..	..	..
Biscuits, total disposals of home produced	KFUI	724	–	–	..	..	..	..	..	..	..	..
Breakfast cereals, other than oatmeal and oatmeal flakes	KFUJ	268	278	286	290	303	302	325	335	343	359	349
Glucose	KFUM	499	542	546	557	568	580	593	607	642	637	644
Production of soft drinks (million litres):												
Concentrated	KFUN	566	555	557	..	..	..	..	..	..	..	..
Unconcentrated	KFUO	3 628	4 112	4 316	..	..	..	..	..	..	..	..
Compound feedingstuffs: total[7]	KFUP	10 729	11 115	11 163	11 175	11 564	11 795	12 093	12 352	12 607	12 203	
Cattle food[7]	KFUQ	3 780	3 947	3 909	3 798	4 098	4 271	4 347	4 476†	4 430	3 926†	3 844
Calf food[7]	KFUR	345	347	324	280	242	261	286	307	277	227†	198
Pig food[7]	KFUS	2 180	2 158	2 276	2 361	2 489	2 561	2 547	2 453†	2 566	2 759	2 740
Poultry food[9]	KFUT	3 476†	3 216	3 212	3 062	2 737	3 123	3 179	3 243	3 286	3 324	3 213
Other compounds[7]	KFUU	753	1 045	843	873	948	925	1 043	1 130	1 249	1 168†	1 210

TIME SERIES. The figures relate to periods of 52 weeks (53 weeks in 1987 and 1992) with the following exceptions which are on a calendar year basis: butter, cheese, cream, canned meat, soft drinks, condensed milk and milk powder, canned vegetables, canned and bottled fruit, jam and marmalade, and soups.

Please note production of Poultry Feed by Integrated Poultry Units has been;

1992 = 2 555 ('000) tonnes
1993 = 2 679 ('000) tonnes
1994 = 2 767 ('000) tonnes
1995 = 2 892 ('000) tonnes
1996 = 3 012 ('000) tonnes
1997 = 3 072 ('000) tonnes
1998= 3 105 ('000) tonnes

These figures have been revised following the receipt of more up-to-date information.

1 Including poultry offal.
2 Total of fowl, ducks, geese and turkeys (carcass weight).
3 Includes cream from the residual elements of low fat milk production.
4 Table spreads are only included from 1986 onwards.
5 From 1981 the method of collecting these figures has changed. They are therefore not comparable with previous years.
6 This survey ceased at the end of 1995.
7 Figures from 1992 onwards have been revised to include data on blended cattle and sheep feeds and are therefore not strictly compatible with earlier years. In 1996, mainly as a result of the introduction of the Beef Assurance Scheme, the Ministry became aware of a number of feed mills that it had not been surveying. Data from these mills has now been included in these figures back to 1992. Total production from these mills has been in the range of 110 000 to 135 000 tonnes each year.
8 Following the receipt of additional data, figures have been revised back to 1992 and are not strictly compatible with those for earlier years.
9 It was found that data for Poultry food had previously contained some mills whose production or part of their production should have been recorded against the Integrated Poultyr Unit (IPU) series. Production for these sites has now been moved from the Poultry food series to the IPU series. This data was available from 1992 onwards. Earlier years have been estimated in order to make the Poultry food series comparable.

Source: Ministry of Agriculture, Fisheries and Food: 01904 455332

20.12 Food and animal feedingstuffs: disposals

Thousand tonnes

		1988	1989	1990	1991	1992	1993	1994	1995	1996	1997	1998
Flour[8]	KFPU	3 866	3 858	3 830	3 798	3 932	4 076†	4 200	4 290	4 457	4 434	4 534
Sugar (as refined sugar): total disposals	KFPV	2 316	2 354	2 332	2 297	2 319	2 154	2 196	2 177	2 200	2 040	2 143
For food in the United Kingdom[1]	KFPW	2 301	2 336	2 320	2 280	2 292	2 123	2 165	2 157	2 180	2 007	2 106
Syrup and treacle[5]	KFPX	53	52	51	51	49	52	50	48	–	–	..
Meat and fish:												
Fresh and frozen meat and offal, including usage for canning:												
Beef and veal	KFPY	1 235	1 219	1 104	1 130	1 145	1 094	1 228†	1 206	804	865	878
Mutton and lamb	KFPZ	465	506	517	511	497	470	474	504	501	469†	495
Pork	KFVA	850	811	814	847	877	895	922	926	963	1 039†	1 100
Offal[6]	KFVB	351	345	329†	324	335	316	327	353	311	319	314
Poultry-meat[3,4]	KFVC	1 167	1 160	1 217	1 256	1 336	1 373	1 448†	1 486	1 567	1 548	1 595
Bacon and ham, including usage for canning	KFVD	462	463	451	446	428	453	457	473†	502	477	469
Dairy products:												
Butter	KFVI	411	283	245	245	258†	279	284	254	240	256	238
Cheese	KFVJ	487†	476	515	509	560	523	559	577	617	614	631
Condensed milk[2]	KFVK	192	212	218	211	218	202	208	195	206	213	180
Milk powder, excluding buttermilk and whey powder:												
Full cream	KFVM	103	99	76	81	91	75	91	97	94	105	107
Skimmed	KFVN	143	149	170	157	136	144	149	138	95	104	91
Hen eggs in shell	KFVO	901	781	868	853	831	821†	837	818	833	838	821
Oils (as crude oil):												
Vegetable oil	KFVP	1 555	1 501	1 465	1 515	1 542	1 589	1 600	1 619	1 934	1 982	..
Marine oil for the manufacture of margarine and compound fat	KFVQ	129	141	133	120	100	100	110	105	55	33	..
Potatoes	KFVR	6 018	6 085	5 911	5 799	5 994	6 357	6 491	5 923†	6 123	6 275	5 967
Other foods:												
Chocolate confectionery	KFVT	564†	578	603	621	640	667	696	674	698	672	692
Sugar confectionery, excluding medicated	KFVU	366	359	366	362	367	366†	382	380	410	419	413
Tea excluding re-exports	KFVV	164	162	145	152	147	157	149	141	148	152†	145
Raw coffee	KFVW	103	98	104	101	108	116	119	105	116	118	122
Barley:[9]												
For brewing and distilling and for food	KFVY	4 554	5 095	3 858	3 917	3 436	2 682	3 092	3 574	3 579	3 439	3 502
Maize (including maize meal): total disposals	KFVZ	1 330	1 405	1 659	1 510	1 349†	1 344	1 312	1 345	1 307	1 364	1 273
Animal feed	KCRT	202	176	292	145	64†	81	104	90	92	108	23
Oilcake and meal	KCRQ	3 163	3 174	3 569	3 470	3 646	3 891	4 130	4 338	4 280	4 207	..
Wheat milling offals	KCRR	1 292	1 169	1 133	1 064	1 086	1 170	1 121	1 131	1 156	1 159	1 275
Fish and meat meal for animal feed,[7] figures relate to sales	KCRS	300	439	394	383	430	382†	385	378	257	233	216

TIME SERIES. The figures relate to periods of 52 weeks (53 weeks in 1987 and 1992) with the following exceptions which are on a calendar year basis: fish and potatoes; condensed milk; milk powder; butter and sugar.

1 Including sugar used in the manufacture of other foods subsequently exported. Excluding sugar in imported manufactured foods.

2 Includes skim concentrate and condensed milk used in the manufacture of chocolate crumb.

3 Provisional.

4 Total of fowls, ducks, geese and turkeys.

5 This survey ceased at the end of 1995.

6 Including poultry offal.

7 The use of mammalian meat and bone meal was banned from 29 March 1996. The use of poultry meat and bone meal was not affected by the ban.

8 Following the receipt of additional data, figures have been revised back to 1992 and are not strictly compatible with those for previous years.

9 From 1992 figures exclude sales of barley and barley screenings and are not strictly compatible with those from earlier years.

Source: Ministry of Agriculture, Fisheries and Food: 01904 455332

20.13 Number of workers employed in agriculture [1,2]
United Kingdom

At June in each year

Thousands

	Regular workers					Seasonal or casual workers			All workers			Salaried managers[3]
		Whole - time		Part - time								
	Total	Male	Female	Male	Female	Total	Male	Female	Total	Male	Female	
	KAXV	KAXW	KAXX	KAXY	KAXZ	KAYA	KAYB	KAYC	KAYD	KAYE	KAYF	KAYG
1988	194.2	120.7	14.6	29.9	28.9	92.9	56.2	36.7	287.1	206.8	80.3	7.9
1989	186.6	114.7	15.2	29.2	27.6	88.3	54.0	34.3	274.9	197.8	77.1	7.8
1990	183.3	109.7	15.7	29.9	28.0	90.5	55.6	35.0	273.8	195.2	78.6	8.1
1991	176.9	104.7	15.0	29.7	27.4	86.6	53.8	32.8	263.5	188.2	75.3	7.9
1992	169.9	99.9	14.8	29.1	26.1	86.2	54.4	31.9	256.2	183.3	72.8	7.8
1993	165.3	96.5	13.7	29.8	25.3	85.4	55.0	30.4	250.7	181.3	69.4	7.6
1994	161.0	93.6	13.2	30.0	24.2	82.2	53.9	28.4	243.2	177.5	65.7	7.8
1995	157.4	90.4	13.0	30.0	24.1	83.7	56.5	27.2	241.2	176.8	64.3	7.7
1996	156.4	89.2	12.6	31.2	23.4	81.5	55.6	25.8	237.9	176.0	61.9	7.8
1997	154.2†	87.5	12.5†	31.2	22.9†	80.9	55.4†	25.6†	235.0†	174.1†	61.0†	7.8
1998	155.6	88.0	13.1	29.7	24.7	79.5	55.0	24.4	235.0	172.8	62.2	12.3

See chapter text.

1 Includes estimates for minor holdings for all countries. In previous editions minor holdings were excluded for Scotland and therefore data here will differ from previous editions. Also, Northern Ireland data are now based on all active business.

2 Figures exclude school children, farmers, partners and directors and their spouses and most trainees.

3 Great Britain only.

Source: Agricultural Departments: 01904 455332

20.14 Landings of fish by United Kingdom vessels: live weight and value into United Kingdom

		Quantity (thousand tonnes)							Value (£ thousand)					
		1993	1994	1995	1996	1997	1998		1993	1994	1995	1996	1997	1998
Total all species	KSJU	678.2	687.6	725.6	635.9	602.7	552.5	KSLN	425 593	454 466	478 209	491 045	467 694	448 005
Total wet fish	KSJV	575.7	576.5	601.0	504.3	467.2	428.3	KSLO	318 431	326 720	333 174	343 045	314 712	323 297
Brill	KSJX	0.4	0.5	0.5	0.5	0.5	0.4	KSLP	1 502	1 927	2 201	2 639	2 244	1 674
Catfish	KSJY	1.7	1.4	1.0	1.1	1.0	0.8	KSLQ	1 867	1 669	1 474	1 449	1 152	1 070
Cod	KSJZ	64.8	65.7	74.4	75.7	71.0	72.7	KSLR	64 915	65 090	65 772	69 752	67 641	79 450
Dogfish	KSKA	10.4	8.2	10.9	9.7	8.6	7.4	KSLS	7 327	6 291	7 684	7 002	5 586	5 535
Haddock	KSKB	87.1	92.9	85.3	89.1	82.6	82.8	KSLT	54 710	61 017	54 734	54 333	44 778	57 131
Hake	KSKC	4.3	3.2	3.1	2.8	2.7	2.5	KSLU	11 535	8 363	7 361	7 346	6 375	5 117
Lemon Soles	KSKD	5.0	4.8	4.6	5.1	5.2	4.7	KSLV	11 073	11 931	10 115	11 897	11 786	10 620
Ling	KSKE	7.7	8.1	9.8	9.2	9.4	10.0	KSLW	5 210	5 832	7 483	6 945	6 568	8 291
Megrims	KSKF	4.3	4.7	5.1	6.0	6.0	5.3	KSLX	8 543	9 210	9 227	10 430	9 467	8 573
Monks or Anglers	KSKG	16.7	17.7	22.2	29.7	25.6	18.5	KSLY	32 271	34 328	39 451	51 468	45 947	38 540
Plaice	KSKH	19.4	17.5	15.5	12.5	12.9	11.5	KSLZ	21 499	19 937	17 573	16 013	15 484	13 185
Pollack (Lythe)	KSKI	3.4	2.9	3.3	2.9	3.0	2.6	KSMA	3 074	2 564	2 935	2 563	2 491	2 423
Saithe	KSKJ	12.2	12.5	12.8	13.3	12.3	10.1	KSMB	4 671	4 875	5 625	5 674	4 894	4 971
Sand Eels	KSKK	2.9	8.7	7.3	9.3	14.5	11.6	KSMC	124	424	398	505	815	742
Skates and Rays	KSKL	7.3	7.3	7.6	8.3	7.2	6.8	KSMD	5 899	6 322	6 543	7 451	6 215	6 403
Soles	KSKM	2.6	2.8	2.8	2.5	2.3	2.0	KSME	12 795	13 454	14 200	14 068	14 800	13 960
Turbot	KSKN	0.9	1.0	0.8	0.8	0.7	0.6	KSMF	4 521	5 238	4 667	5 006	4 263	3 614
Whiting	KSKO	45.6	41.8	39.8	37.3	34.5	26.7	KSMG	20 748	20 135	18 918	18 962	15 939	13 380
Whiting, Blue	KSKP	2.3	1.9	3.4	3.5	12.4	27.8	KSMH	137	90	180	190	693	2 003
Whitches	KSKQ	2.0	2.1	2.3	2.3	2.3	1.9	KSMI	2 305	2 867	2 915	3 068	2 263	1 587
Other Demersal	KSKR	8.9	8.9	12.9	13.3	13.2	12.7	KSMJ	8 730	10 709	17 028	16 614	16 000	14 794
Total Demersal[1]	KSKS	310.1	314.7	325.4	334.8	327.9	319.3	KSMK	283 860	292 276	296 485	313 377	285 402	293 882
Herring[2]	KSKT	87.2	86.1	95.9	72.3	57.4	39.5	KSML	9 716	9 756	11 105	8 742	6 430	4 897
Horse Mackerel	KSKU	3.9	9.5	26.2	22.0	5.5	5.1	KSMM	378	835	2 636	2 966	878	783
Mackerel[2]	KSKV	161.8	155.9	140.4	60.8	63.2	54.4	KSMN	22 055	21 230	20 860	15 645	19 822	21 754
Pilchards	KSKW	4.8	2.1	6.8	6.8	4.7	4.7	KSMO	632	277	680	959	696	765
Sprats	KSKX	7.0	7.6	5.9	7.2	8.3	5.0	KSMP	954	1 535	1 080	1 175	1 386	1 112
Tuna	KSKY	0.3	0.4	0.1	..	..	..	KSMQ	652	707	246	66	46	44
Other Pelagic	KSKZ	0.5	0.1	0.2	0.3	0.1	0.2	KSMR	185	104	82	116	52	61
Total Pelagic	KSLA	265.6	261.7	275.6	169.5	139.3	108.9	KSMS	34 571	34 444	36 689	29 668	29 309	29 416
Cockles	KSLB	21.4	22.3	25.4	24.2	19.5	12.1	KSMT	3 440	3 070	3 544	3 266	3 632	4 162
Crabs	KSLC	15.2	19.2	21.7	20.3	22.5	27.2	KSMU	15 810	20 397	21 346	22 226	24 264	32 290
Lobsters	KSLD	1.0	1.1	1.3	1.3	1.5	1.5	KSMV	9 353	10 688	11 896	11 888	13 060	14 343
Mussels	KSLE	7.8	10.3	9.5	12.3	19.0	12.7	KSMW	1 431	1 986	2 976	4 574	3 102	3 502
Nephrops	KSLF	28.4	29.8	31.1	29.0	31.1	28.6	KSMX	48 883	56 366	60 704	57 164	63 480	56 810
Periwinkles	KSLG	1.9	2.3	2.4	2.4	2.9	2.0	KSMY	1 150	1 554	1 954	2 046	2 735	1 892
Queens	KSLH	7.5	3.0	2.9	2.3	5.6	8.1	KSMZ	2 879	1 789	1 929	1 225	1 954	2 887
Scallops	KSLI	9.3	14.0	15.7	17.1	18.5	20.1	KSNA	13 787	21 209	24 279	27 036	27 576	30 156
Shrimps/Prawns	KSLJ	1.5	1.4	2.7	2.7	1.0	2.2	KSNB	1 858	1 981	3 517	3 257	960	2 751
Squid	KSLK	1.7	1.3	1.6	1.4	1.6	2.2	KSNC	2 482	2 824	3 783	2 987	4 044	4 362
Other shellfish	KSLL	6.8	6.2	10.4	18.7	12.2	7.6	KSND	6 494	5 881	9 106	12 332	8 176	7 553
Total shellfish	KSLM	102.5	111.1	124.6	131.6	135.5	124.2	KSNE	107 567	127 746	145 035	148 000	152 983	160 708

1 Includes fish roes.
2 Includes transshipments ie caught by UK boats but not actually landing at UK ports. These quantities are transshipped to foreign vessels in coastal waters and are later recorded as exports.

Source: Agricultural Departments: 01904 455332

20.15 United Kingdom fishing fleet

At 31 December each year[1]

Number

		1992	1993	1994	1995	1996	1997	1998
By size								
10m & under	KSNF	7 376	7 666	7 194	6 319	5 606	5 474	5 487
10.01 - 12.19m	KSNG	1 450	1 361	1 167	1 016	800	732	628
12.20 - 17.00m	KSNH	787	751	680	622	540	523	491
17.01 - 18.29m	KSNI	232	220	193	187	164	162	154
18.30 - 24.38m	KSNJ	697	657	610	574	509	471	443
24.39 - 30.48m	KSNK	200	210	211	212	223	227	226
30.49 - 36.58m	KSNL	120	124	126	127	114	104	89
over 36.58m	KSNM	117	119	116	117	117	119	121
Total over 10m	KSNN	3 603	3 442	3 103	2 855	2 467	2 338	2 152
Total UK fleet (excluding Islands)	KSNO	10 979	11 108	10 297	9 174	8 073	7 812	7 639
By segment								
Pelagic gears	KSNP	76	69	68	67	58	49	50
Beam trawl	KSNQ	227	240	212	220	215	153	123
Demersal trawls & seines	KXET	..	..	..	..	1 040	..	..
Demersal trawls	KSIX	1 039	988	854	856	..	..	..
Nephrop trawls	KSIY	566	560	593	528	411	..	..
Seines	KSIZ	255	203	197	165	..	..	..
Demersal, Seines & Nephrops	JZCI	..	..	..	..	..	1 428	1 318
Lines & Nets	KSNR	339	329	300	267	224	214	187
Shellfish: mobile	KSNS	214	181	206	194	265	227	241
Shellfish: fixed	KSNT	306	312	305	283	339	352	311
Distant water	KSNU	16	14	13	12	15	13	14
Under 10m	KSNV	7 831	8 128	7 607	6 757	6 091	6 022	6 027
Non-active/non-TAC	KSNW	692	668	472	371	–	–	–
Other: Mussel Dredgers	JZCJ	..	..	..	..	..	3	2
Total UK fleet[2]	KSNX	11 561	11 692	10 827	9 720	8 658	8 461	8 271

1 Prior to 1990 the figures referred to vessels that were active; after 1990 the figures refer to vessels in the registered fleet for which the data are currently under review.
2 The UK figures here include Channel Islands and Isle of Man.

Source: Ministry of Agriculture, Fisheries and Food: 01904 455332

20.16 Estimated household food consumption by all households in Great Britain

Grammes per person per week

		1988	1989	1990	1991	1992	1993	1994	1995	1996	1997	1998
Liquid wholemilk[1] (ml)	KPQM	1 513	1 377	1 232	1 104	995	898	870	812	776	712	693
Fully skimmed (ml)	KZBH	177	190	193	198	213	217	207	204	137	158	164
Semi skimmed (ml)	KZBI	350	432	516	579	752	814	863	899	935	978	945
Other milk and cream (ml)	KZBJ	240	237	228	247	262	249	252	255	259	248	243
Cheese	KPQO	117	115	113	117	114	109	106	108	111	109	104
Butter	KPQP	57	50	46	44	41	40	39	36	39	38	39
Margarine	KPQQ	108	98	91	89	79	70	43	41	36	26	26
Low and reduced fat spreads	KZBK	38	46	45	47	51	52	74	72	79	77	69
All other oils and fats (ml for oils)	KPQR	69	66	67	60	74	69	70	69	71	62	62
Eggs (number)	KPQS	2.67	2.29	2.20	2.25	2.08	1.92	1.86	1.85	1.87	1.78	1.74
Preserves and honey	KPQT	52	50	48	51	45	42	43	39	41	41	38
Sugar	KPQU	196	183	171	167	156	151	144	136	144	128	119
Beef and veal	KPQV	180	171	149	152	141	133	131	121	101	110	109
Mutton and lamb	KPQW	78	85	83	86	71	66	54	54	66	56	59
Pork	KPQX	94	89	84	82	72	80	77	71	73	75	76
Bacon and ham, uncooked	KPQY	99	95	86	85	77	77	77	76	77	72	76
Bacon and ham, cooked (including canned)	KPQZ	32	35	32	33	33	35	38	39	33	41	40
Poultry uncooked	JZCH	213	207	211	202	216	222	209	215	233	221	218
Cooked poultry (not purchased in cans)	KYBP	16	13	15	14	15	16	20	22	23	33	33
Other cooked and canned meats	KPRB	74	70	62	60	68	60	63	63	62	52	49
Offals	KPRC	18	16	14	13	12	11	9	9	7	7	5
Sausages, uncooked	KPRD	70	72	68	62	61	60	61	63	63	63	60
Other meat products	KPRE	161	165	163	173	183	194	203	211	207	209	216
Fish, fresh and processed (including shellfish)	KPRF	74	74	69	65	67	71	71	68	72	70	70
Canned fish	KPRG	25	27	29	30	32	30	30	29	31	31	29
Fish and fish products, frozen	KPRH	44	47	47	43	43	44	44	46	50	46	46
Potatoes (excluding processed)	KPRI	1 033	1 009	996	959	901	875	812	803	805	745	715
Fresh green vegetables	KPRJ	293	290	277	259	250	240	245	225	233	251	246
Other fresh vegetables	KPRK	475	486	459	461	475	477	464	470	489	497	486
Frozen potato products	KYBQ	76	80	73	82	92	98	103	99	113	106	111
Other frozen vegetables	KPRL	109	110	112	117	106	105	107	101	94	94	88
Potato products not frozen	JZCF	62	62	62	59	75	80	82	89	92	90	89
Canned beans	KPRM	132	126	124	123	120	112	111	117	125	122	118
Other canned vegetables (excl. potatoes)	KPRN	132†	122	117	120	124	114	103	110	113	104	99
Other processed vegetables (excl. potatoes)	LQZH	41	42	41	43	50	52	55	48	55	52	54
Apples	KPRO	204	206	201	190	187	179	180	183	175	179	181
Bananas	KPRP	101	113	125	129	144	151	162	176	185	195	198
Oranges	KPRQ	90	84	81	76	72	62	65	66	63	62	63
All other fresh fruit	KPRR	200	204	198	216	216	224	238	247	263	276	274
Canned fruit	KPRS	61	61	52	52	51	48	46	45	43	44	37
Dried fruit, nuts and fruit and nut products	KPRT	38	36	36	39	38	39	36	34	36	35	34
Fruit juices (ml)	KPRU	211	214	202	250	222	236	240	244	258	277	304
Flour	KPRV	103	93	91	81	81	82	62	57	70	54	55
Bread	KPRW	859	834	797	752	755	757	758	756	752	746	742
Buns, scones and teacakes	KPRX	32	33	34	39	40	39	38	36	47	43	41
Cakes and pastries	KPRY	73	69	70	79	76	79	85	85	87	93	88
Biscuits	KPRZ	149	149	149	147	148	142	138	135	150	138	137
Breakfast cereals	KPSA	126	126	127	134	132	129	134	135	140	135	136
Oatmeal and oat products	KPSB	18	16	15	19	15	14	11	11	13	16	11
Other cereals and cereal products	JZCG	174	192	187	204	217	218	218	251	304	293	270
Tea	KPSC	47	46	43	42	39	36	38	39	38	36	35
Instant coffee	KPSD	15	14	14	15	14	13	13	12	13	11	12
Canned soups	KPSE	79	77	68	69	70	66	68	64	72	70	71
Pickles and sauces	KPSF	65	66	67	69	72	77	77	80	84	92	96

Note: Chapter text contains a description of the National Food Survey.

1 Including also school and welfare milk.

Sources: Ministry of Agriculture, Fisheries and Food (National Food Survey); 020 7270 8563

Production

21

Production

Production survey *(Table 21.1)*

The Annual Sample Survey of Production Industries provides information about the structure of industry in the United Kingdom. The results meet a wide range of needs for government, economic analysts and the business community at large. In official statistics the surveys are an important source for the national accounts and input-output tables, but they also provide weights for the indices of production and producer prices. Survey results also enable the United Kingdom to meet statistical requirements of the European Union.

Contributors for the survey were selected under a stratified random sampling scheme. The strata used were industry and employment size group. The sampling fractions varied by industry but, in general, smaller enterprises had a lower chance of selection. All the largest enterprises in each industry were selected.

Results are available on the *PACSTAT* CD ROM and in a summary volume.

Manufacturers sales by industry *(Table 21.2)*

This table shows the total manufacturers sales for products classified to the 1992 Standard Industrial Classification and collected under the PRODCOM (Products of the European Community) Inquiry since its introduction in 1993. Some data is not available for confidentiality reasons or where data has not been published for a given period.

Number of local units in manufacturing industries in 1999 *(Table 21.3)*

This table shows the number of local units (sites) in manufacturing by employment sizebands. The classification breakdown is at division level (2 digit) as classified to the 1992 Standard Industrial Classification held on the Inter Departmental Business Register (IDBR). This register became fully operational in 1995 and combines information on VAT traders and PAYE employers in a statistical register comprising 2 million enterprises (businesses), representing nearly 99% of economic activity. Business Monitor PA1003 - Size Analysis of United Kingdom Businesses provides further details and contains detailed information on enterprises in the UK including size, classification and location. Additionally, this information is available for manufacturing local units.

Total inland energy consumption *(Table 21.4)*

This table shows energy consumption by fuel and final energy consumption by fuel and class of consumer. Primary energy consumption covers consumption of all primary fuels for energy purposes. Primary fuels are coal, gas, oil and primary electricity (i.e. electricity generated by hydro and nuclear stations and also electricity imported from France through the interconnector). This measure of energy consumption includes energy that is lost by converting primary fuels into secondary fuels, ie the energy lost burning coal to generate electricity or the energy used by refineries to separate crude oil into fractions, in addition to losses in distribution. The other common way of measuring energy consumption is to measure the energy content of the fuels supplied to consumers. This is called final energy consumption. It is net of fuel used by the energy industries, conversion, transmission and distribution losses. The figures are presented on a common basis, measured as energy supplied and expressed in million tonnes of oil equivalent. So far as practicable the user categories have been grouped on the basis of the SIC92 although the methods used by each of the supply industries to identify end users are slightly different. Chapter 1 of the *Digest of UK Energy Statistics* (published by The Stationery Office) gives more information on these figures.

Coal *(Table 21.5*

Since 1995, aggregate data on coal production has been obtained from the Coal Authority. In addition main coal producers provide data in response to an annual DTI inquiry which covers production (deepmined and opencast), trade, stocks and disposals. HM Customs and Excise also provide trade data for solid fuels. The DTI collects information on the use of coal from Iron and Steel Statistics Bureau and consumption of coal for electricity generation is covered by data provided by the electricity generators.

Gas *(Table 21.6)*

Production figures, covering the production of gas from the UKCS offshore and onshore gas fields and gas obtained during the production of oil, are obtained from returns made under the DTI's Petroleum Production Reporting System. Additional information is used on imports and exports of gas and details from the operators of gas terminals in the UK to complete the picture.

DTI carry out an annual survey of gas suppliers to obtain details of gas sales to the various categories of consumer. Estimates are included for the suppliers with the smallest market share since the DTI inquiry covers only the largest suppliers (i.e. those more than about 0.5% per cent share of the UK market).

Electricity *(Tables 21.7 - 21.9)*

The electricity Tables 21.7 to 21.9 cover all generators and suppliers of electricity in the United Kingdom.

The relationship between generation, supply, availability and consumption is as follows:

Electricity generated
less electricity used on works
equals electricity supplied (gross)
less electricity used in pumping at pumped storage stations
equals electricity supplied (net)
plus imports (net of exports) of electricity
equals electricity available
less losses and statistical differences
equals electricity consumed.

In Table 21.9 all fuels are converted to the common unit of million tonnes of oil equivalent, ie the amounts of oil which would be needed to produce the output of electricity generated from those fuels.

More detailed statistics on energy are given in the *Digest of United Kingdom Energy Statistics 1999.* Readers may wish to note that following consultation made about the presentation of figures in the 1998 edition, the production and consumption of fuels are now presented using commodity balances. A commodity balance shows the flows of an individual fuel through from production to final consumption, showing its use in transformation and energy industry own use.

Oil and oil products *(Tables 21.10 - 21.12)*

The data on the production of crude oil, condensates and natural gases given in Table 21.10 are collected by the DTI direct from the operators of production facilities and terminals situated on UK territory, either on land or on the UK continental shelf. Data is also collected from the companies on their trade in oil and oil products. This data is used in preference to the foreign trade as recorded by HM Customs & Excise in the Overseas Trade Statistics.

Data on the internal UK oil industry (ie on the supply, refining and distribution of oil and oil products in the UK) is collected by the UK Petroleum Industry Association. This data, reported by individual refining companies and wholesalers and supplemented where necessary by data from other sources, provides the contents of Tables 21.11 and 21.12. The data is presented in terms of deliveries to the inland UK market. This is regarded as an acceptable proxy for actual consumption of products. The main shortcoming is that whilst changes in stocks held by companies in central storage areas are taken into account, changes in the levels of stocks further down the retail ladder (such as stocks held on petrol station forecourts) are not. This is not thought to result in a significant degree of difference in the data.

Iron and steel *(Tables 21.13 - 21.15)*

Iron and steel industry
The general definition of the UK iron and steel industry is based on groups 271 "ECSC iron and steel", 272 "Tubes", and 273 "Primary Transformation" of the UK Standard Industrial Classification (1992), except those parts of groups 272 and 273 which cover cast iron pipes, drawn wire, cold formed sections and Ferro alloys.

The definition excludes certain products which may be made by works within the industry such as refined iron, finished steel castings, steel tyres, wheels, axles and rolled rings, open and closed die forgings, colliery arches and springs. Iron foundries and steel stockholders are also considered to be outside of the industry.

Statistics
The statistics for the UK iron and steel industry are compiled by ISSB Ltd. from data collected from UK steel producing companies with the exception of trade data which is based on HM Customs data.

Crude steel is the total of usable ingots, usable continuously cast semi finished products and liquid steel for castings.

Production of finished products is the total production at the mill of that product after deduction of any material which is immediately scrapped.

Deliveries are based on invoiced tonnages and will include deliveries made to steel stockholders and service centres by the UK steel industry.

For more detailed information on definitions etc please contact ISSB Ltd. on 020 7343 3900.

Construction *(Tables 21.21 - 21.22)*

Table 21.21 shows the value of contractors' output in the construction industry in Great Britain. Contractors' output is defined as the amount chargeable to customers for building and civil engineering work done in the relevant period. The data comes from surveys run by the Department of the Environment, Transport and the Regions (DETR). As well as being an important input to the National Accounts, it is used by the government and the construction industry in their efforts to fully understand the industry, and also by Eurostat.

Table 21.22 shows the value of new orders in the construction industry; this is also collected by DETR. This information relates to contracts for new construction work awarded to main contractors by clients in both the public and private sectors; it also includes speculative work, undertaken on the initiative of the firm, where no contract is awarded. New orders are used as a good indicator of future output.

Production

Motor vehicle production *(Table 21.25)*

The figures represent the output of United Kingdom based manufacturers classified to Class 34.10 (motor vehicles) of the Standard Industrial Classification 1992. They are derived from the Motor Vehicle Production Inquiry (MVPI).

These figures include vehicles produced in the form of kits for assembly. The value of the kit must be 50% or more of the value of a corresponding complete vehicle.

Drink and tobacco *(Tables 21.26 and 21.27)*

Data for these tables are derived by Customs and Excise from the systems for collecting excise duties. Alcoholic drinks and tobacco products become liable to duty when released for consumption in the UK. Figures for releases include both home-produced products and commercial imports. Production figures are also available for potable spirits distilled and beer brewed in the UK.

Alcoholic drink *(Table 21.26)*
The figures for Imported and other spirits released for home consumption include gin and other UK produced spirits, for which a breakdown is not available.

Since June 1993 beer duty has been charged when the beer leaves the brewery or other registered premises. Previously duty was chargeable at an earlier stage (the worts stage) in the brewing process, and an allowance was made for wastage. Figures for years prior to 1994 include adjustments to bring them into line with current data. The change in June 1993 also led to the availability of data on the strength; a series in hectolitres of pure alcohol is shown from 1994.

Made wine with alcoholic strength from 1.2% to 5.5% is termed 'coolers'. Included in coolers are alcoholic lemonade and similar products of appropriate strength. Prior to 1989 coolers were included under the heading of made wine other than coolers.

Tobacco Products *(Table 21.27)*
Releases of cigarettes and other tobacco products tend to be higher in the period before a Budget. Products may then be stocked, duty paid, before being sold.

21.1 Production survey: summary table
United Kingdom

	Estimates for all firms							
	(£ million)							
			Stocks and work in progress					
	Gross output (Production)[1]	Gross value added	At end of year	Change during year	Capital expenditure *less* disposals	Wages and salaries	Average number of persons employed[2] (000s)	Gross value added per person employed (£)
Standard Industrial Classification: Revised 1992								
Production and construction								
Sections C-F								
	KSCC	KSCD	KSCE	KSCF	KSCG	KSCH	KSCI	KSCJ
1993	484 871.8	153 044.2	..	..	19 814.6	81 196.8	5 701.0	26 845
1994	506 477.7	163 110.6	..	..	20 809.9	83 607.6	5 507.8	29 615
1995	..	176 754.7	72 058.7	..	23 798.6	76 335.9	5 667.2	31 189
1996	576 309.5	185 760.7	68 398.2	430.3	23 672.2	88 606.3	5 470.0	33 960
1997	591 462.2	193 254.7	68 974.9	614.4	27 185.4	94 429.7	5 502.6	35 121
Production industries (Revised definitions)								
Sections C-E								
	KSCK	KSCL	KSCM	KSCN	KSCO	KSCP	KSCQ	KSCR
1993	405 928.1	133 440.3	51 958.9	387.0	18 697.0	68 824.0	4 598.6	29 018
1994	435 895.6	143 203.5	52 940.7	3 170.4	19 560.0	70 847.0	4 608.6	31 073
1995	473 200.5	152 601.4	58 212.2	4 888.6	22 499.1	72 291.5	4 655.7	32 777
1996	494 056.1	159 731.8	55 475.4	–5.6	22 485.7	74 137.1	4 454.1	35 862
1997	503 731.0	163 793.2	55 902.4	–541.1	25 429.2	77 758.4	4 483.6	36 532
Mining and quarrying								
Section C								
	KSCS	KSCT	KSCU	KSCV	KSCW	KSCX	KSCY	KSCZ
1993	6 591.4	2 319.2	1 092.1	–68.7	346.6	1 777.3	79.0	29 360
1994	5 147.9	1 811.8	562.6	–83.9	271.8	843.8	45.3	40 003
1995	5 643.0	2 254.4	566.2	–84.1	298.0	895.4	47.8	47 120
1996	5 977.2	2 586.8	618.8	11.9	473.7	1 018.4	54.9	47 110
1997	5 461.5	2 450.1	583.8	–11.0	350.7	947.7	48.2	50 877
Mining and quarrying of energy producing materials								
Subsection CA								
	KSDA	KSDB	KSDC	KSDD	KSDE	KSDF	KSDG	KSDH
1993	3 597.5	1 227.6	..	..	219.5	1 293.9	48.1	25 523
1994	2 006.0	752.5	372.1	–78.3	72.9	339.0	14.1	53 408
1995	2 447.6	888.2	364.8	–101.1	92.4	401.3	17.0	52 305
1996	2 250.8	949.6	369.4	–12.3	157.1	443.0	18.8	50 460
1997	1 997.7	891.7	355.2	–6.0	138.0	414.1	16.2	55 171
Mining and quarrying except energy producing materials								
Subsection CB								
	KSDI	KSDJ	KSDK	KSDL	KSDM	KSDN	KSDO	KSDP
1993	2 993.9	1 091.6	..	..	127.2	483.3	30.9	35 334
1994	3 142.0	1 059.3	190.5	–5.6	198.9	504.8	31.2	33 949
1995	3 195.3	1 366.2	201.4	16.9	205.6	494.1	30.9	44 268
1996	3 726.4	1 637.2	249.4	24.1	316.6	575.4	36.1	45 363
1997	3 463.8	1 558.4	228.6	–5.0	212.8	533.6	32.0	48 708
Manufacturing (Revised definition)								
Section D								
	KSDQ	KSDR	KSDS	KSDT	KSDU	KSDV	KSDW	KSDX
1993	357 294.0	115 177.3	48 462.8	718.4	13 047.6	62 725.7	4 310.7	26 719
1994	388 551.0	125 741.1	50 640.6	3 915.7	14 011.3	65 899.7	4 366.5	28 797
1995	422 186.5	135 386.1	55 955.3	5 074.5	17 405.7	67 518.0	4 433.6	30 536
1996	442 799.0	140 898.9	53 512.6	90.2	17 836.7	69 783.1	4 256.1	33 106
1997	452 009.6	144 518.1	53 919.4	–503.1	20 031.1	73 474.2	4 299.3	33 614

See footnotes on the last part of this table.

Source: Office for National Statistics: 01633 812929

21.1 Production survey: summary table
United Kingdom

continued

	Estimates for all firms							
	(£ million)							
			Stocks and work in progress					
	Gross output (Production)[1]	Gross value added	At end of year	Change during year	Capital expenditure *less* disposals	Wages and salaries	Average number of persons employed[2] (000s)	Gross value added per person employed (£)
Standard Industrial Classification: Revised 1992								
Manufacture of food; beverages and tobacco Subsection DA								
	KSDY	KSDZ	KSEA	KSEB	KSEC	KSED	KSEE	KSEF
1993	66 239.8	16 739.6	6 242.6	-36.9	2 321.6	7 262.0	571.8	29 277
1994	68 474.0	17 608.9	6 253.9	269.3	2 290.3	7 509.1	562.9	31 283
1995	72 609.9	17 640.0	6 765.6	377.2	2 444.2	7 366.7	533.1	33 092
1996	75 880.6	19 345.2	6 984.5	501.9	2 346.5	7 529.3	498.0	38 845
1997	76 355.6	19 616.4	7 449.8	259.2	2 561.7	8 110.5	494.3	39 686
Manufacture of textile and textile products Subsection DB								
	KSEG	KSEH	KSEI	KSEJ	KSEK	KSEL	KSEM	KSEN
1993	14 823.3	5 820.3	2 308.9	119.6	398.1	3 584.5	392.4	14 834
1994	15 912.9	6 163.8	2 914.6	424.8	470.5	3 638.6	370.9	16 616
1995	16 982.8	6 374.2	2 869.8	197.4	467.0	3 720.5	380.1	16 768
1996	17 306.4	6 803.5	2 786.1	15.3	575.3	3 872.0	356.6	19 077
1997	18 307.3	7 128.6	3 087.8	44.6	482.0	4 103.0	364.4	19 560
Manufacture of leather and leather products Subsection DC								
	KSEO	KSEP	KSEQ	KSER	KSES	KSET	KSEU	KSEV
1993	2 555.0	970.4	384.2	18.7	52.9	622.2	58.2	16 661
1994	2 747.4	976.0	473.9	51.9	65.6	592.8	59.0	16 547
1995	2 292.3	788.5	358.3	24.9	60.5	568.6	54.7	14 425
1996	2 355.8	886.6	382.3	3.4	48.9	425.5	39.6	22 407
1997	2 144.6	771.0	403.1	47.6	48.0	420.3	36.0	21 407
Manufacture of wood and wood products Subsection DD								
	KSEW	KSEX	KSEY	KSEZ	KSFA	KSFB	KSFC	KSFD
1993	4 598.2	1 435.6	575.0	36.9	132.0	888.7	78.1	18 383
1994	5 533.2	1 598.5	682.4	82.2	130.6	1 041.5	83.2	19 205
1995	5 777.1	1 814.1	713.5	14.1	158.3	1 094.9	87.4	20 758
1996	5 601.0	1 799.9	629.6	-6.7	170.1	1 080.1	87.0	20 696
1997	5 761.2	2 034.0	670.4	24.6	235.2	1 110.4	91.1	22 331
Manufacture of pulp, paper and paper products; publishing and printing Subsection DE								
	KSFE	KSFF	KSFG	KSFH	KSFI	KSFJ	KSFK	KSFL
1993	32 112.5	13 549.0	2 329.1	66.4	1 644.0	7 193.2	453.6	29 873
1994	35 611.0	14 802.5	2 735.5	332.5	1 761.4	8 033.7	471.9	31 366
1995	39 119.0	16 236.2	3 150.9	437.3	2 121.0	8 056.6	481.8	33 697
1996	40 630.4	16 678.0	2 828.3	–194.1	1 784.9	8 690.4	467.8	35 649
1997	40 201.6	16 494.4	2 860.8	–68.8	1 997.2	8 973.0	470.6	35 047
Manufacture of coke, refined petroleum products and nuclear fuel Subsection DF								
	KSFM	KSFN	KSFO	KSFP	KSFQ	KSFR	KSFS	KSFT
1993	22 144.8	2 347.1	1 378.6	–96.5	721.6	670.0	29.7	79 007
1994	21 864.2	2 270.2	1 092.6	25.7	580.1	675.3	28.0	81 202
1995	22 422.6	2 912.2	1 220.0	227.3	752.2	670.3	27.2	107 044
1996	25 984.9	2 558.5	1 128.3	–121.3	716.8	714.3	27.9	91 756
1997	23 831.4	2 089.6	1 178.8	–240.7	700.4	678.1	25.6	81 770

See footnotes on the last part of this table.

Source: Office for National Statistics: 01633 812929

21.1 Production survey: summary table
United Kingdom

continued

	Estimates for all firms							
	(£ million)							
			Stocks and work in progress					
	Gross output (Production)[1]	Gross value added	At end of year	Change during year	Capital expenditure *less* disposals	Wages and salaries	Average number of persons employed[2] (000s)	Gross value added per person employed (£)
Standard Industrial Classification: Revised 1992								
Manufacture of chemicals, chemical products and man-made fibres								
Subsection DG	KSFU	KSFV	KSFW	KSFX	KSFY	KSFZ	KSGA	KSGB
1993	38 009.5	12 614.7	4 916.9	142.2	1 882.4	5 252.5	274.3	45 993
1994	40 575.8	13 510.8	5 111.0	259.1	1 888.9	5 339.5	271.1	49 843
1995	44 721.5	15 411.9	5 808.8	447.2	2 285.9	5 735.7	281.7	54 707
1996	45 588.4	15 440.3	5 717.1	264.8	2 758.2	5 702.1	267.1	57 810
1997	46 359.1	14 802.0	5 540.7	–46.6	2 830.2	5 996.0	269.2	54 990
Manufacture of rubber and plastic products								
Subsection DH	KSGC	KSGD	KSGE	KSGF	KSGG	KSGH	KSGI	KSGJ
1993	14 952.7	5 820.0	1 522.1	52.4	767.8	3 257.1	236.5	24 606
1994	16 681.9	6 382.2	1 670.5	167.9	764.8	3 444.6	239.4	26 660
1995	18 102.5	6 553.9	1 820.2	171.2	952.7	3 613.0	248.0	26 427
1996	19 175.5	7 122.8	1 891.7	27.5	983.8	3 836.8	250.0	28 496
1997	19 989.7	7 701.9	1 863.8	23.8	1 124.5	4 075.3	265.7	28 990
Manufacture of other non-metallic mineral products								
Subsection DI	KSGK	KSGL	KSGM	KSGN	KSGO	KSGP	KSGQ	KSGR
1993	9 084.7	3 664.6	1 202.3	–81.4	412.8	2 114.9	150.6	24 329
1994	10 699.1	4 533.4	1 343.5	97.9	423.2	2 280.6	159.8	28 366
1995	11 855.6	4 976.9	1 462.0	121.4	607.6	2 437.5	167.1	29 776
1996	11 349.0	4 976.5	1 437.3	94.3	671.9	2 374.3	156.8	31 732
1997	11 531.6	4 908.4	1 474.3	9.2	576.6	2 441.7	154.7	31 725
Manufacture of basic iron and of ferro-alloys								
Subsection DJ	KSGS	KSGT	KSGU	KSGV	KSGW	KSGX	KSGY	KSGZ
1993	33 965.9	12 219.6	3 966.0	61.2	962.7	7 816.5	529.7	23 067
1994	38 171.4	13 612.0	4 204.0	257.2	1 128.8	8 131.3	541.9	25 118
1995	43 489.1	15 889.6	4 711.8	515.9	1 369.6	8 647.9	582.8	27 263
1996	43 725.9	15 344.2	4 452.6	7.7	1 572.2	8 860.9	551.8	27 806
1997	41 867.9	15 580.0	4 246.0	–21.9	1 653.0	8 822.1	538.8	28 915
Manufacture of machinery and equipment not elsewhere specified								
Subsection DK	KSHA	KSHB	KSHC	KSHD	KSHE	KSHF	KSHG	KSHH
1993	26 198.2	9 575.3	4 678.4	–70.3	714.0	6 177.5	393.5	24 334
1994	28 808.0	10 553.3	4 940.5	161.4	873.1	6 570.6	403.4	26 163
1995	33 839.5	12 014.6	5 661.6	376.1	1 080.1	6 993.0	423.1	28 397
1996	34 700.4	12 478.2	5 198.5	69.0	1 177.8	6 993.3	400.2	31 177
1997	36 531.9	13 541.9	5 510.1	137.9	1 141.7	7 476.5	404.7	33 465
Manufacture of electrical and optical equipment								
Subsection DL	KSHI	KSHJ	KSHK	KSHL	KSHM	KSHN	KSHO	KSHP
1993	42 414.7	13 784.5	6 925.9	334.0	1 493.1	8 402.2	532.4	25 891
1994	45 490.6	15 745.2	6 958.1	575.3	1 793.3	8 623.2	544.3	28 926
1995	50 427.2	16 717.5	7 373.7	733.2	2 370.1	8 472.2	537.5	31 102
1996	54 777.8	17 480.6	6 975.1	–213.2	2 193.6	8 977.3	533.0	32 797
1997	58 459.6	18 048.2	6 956.5	–50.8	2 810.1	9 536.3	544.1	33 174

See footnotes on the last part of this table.

Source: Office for National Statistics: 01633 812929

21.1 Production survey: summary table
United Kingdom

continued

	Estimates for all firms							
	(£ million)							
			Stocks and work in progress					
	Gross output (Production)[1]	Gross value added	At end of year	Change during year	Capital expenditure *less* disposals	Wages and salaries	Average number of persons employed[2] (000s)	Gross value added per person employed (£)
Standard Industrial Classification: Revised 1992								
Manufacture of transport equipment Subsection DM								
	KSHQ	KSHR	KSHS	KSHT	KSHU	KSHV	KSHW	KSHX
1993	40 436.6	12 982.0	10 790.2	97.0	1 301.3	7 211.9	419.1	30 977
1994	45 485.8	13 617.6	10 716.8	1 082.0	1 512.2	7 240.4	408.2	33 360
1995	47 033.2	13 425.3	12 335.0	1 305.9	2 303.0	7 418.3	408.3	32 885
1996	51 840.5	14 833.4	11 493.3	–432.7	2 414.9	7 867.0	406.3	36 512
1997	55 880.4	16 080.5	10 808.1	–682.2	3 302.9	8 600.2	414.6	38 786
Manufacture not elsewhere classified Subsection DN								
	KSHY	KSHZ	KSIA	KSIB	KSIC	KSID	KSIE	KSIF
1993	9 758.1	3 654.6	1 242.6	75.0	243.5	2 272.6	190.9	19 148
1994	12 495.7	4 366.8	1 543.4	128.4	328.4	2 778.7	222.4	19 634
1995	13 514.3	4 631.2	1 704.2	125.5	433.5	2 722.9	220.8	20 976
1996	13 882.5	5 151.1	1 607.9	74.3	421.8	2 859.9	214.0	24 074
1997	14 787.8	5 721.1	1 869.2	60.8	567.6	3 130.8	225.6	25 363
Electricity, gas and water supply Section E								
	KSIG	KSIH	KSII	KSIJ	KSIK	KSIL	KSIM	KSIN
1993	42 042.6	15 943.8	2 404.0	–262.7	5 302.8	4 321.0	208.9	76 328
1994	42 196.6	15 650.6	1 737.5	–661.4	5 276.9	4 103.4	196.8	79 517
1995	45 371.0	14 960.9	1 690.7	–101.8	4 795.5	3 878.1	174.3	85 841
1996	45 279.9	16 246.1	1 344.0	–107.7	4 175.3	3 335.5	143.2	113 483
1997	46 259.9	16 824.9	1 399.2	–27.0	5 047.4	3 336.6	136.1	123 597
Construction Section F								
	KSIO	KSIP	KSIQ	KSIR	KSIS	KSIT	KSIU	KSIV
1993	78 943.7	19 603.9	..	..	1 117.6	12 372.8	1 102.4	17 782
1994	70 582.1	19 907.0	..	..	1 249.9	12 760.6	899.2	22 139
1995	..	24 153.3	13 846.6	..	1 299.5	4 044.4	1 011.5	23 879
1996	82 253.4	26 028.8	12 922.8	435.9	1 186.5	14 469.3	1 015.9	25 621
1997	87 731.2	29 461.6	13 072.5	1 155.5	1 756.2	16 671.3	1 019.0	28 912

See footnotes on the last part of this table.

Source: Office for National Statistics: 01633 812929

21.1 Production survey: summary table
United Kingdom

continued

	Estimates for all firms							
	(£ million)							
			Stocks and work in progress					
	Gross output (Production)[1]	Gross value added	At end of year	Change during year	Capital expenditure *less* disposals	Wages and salaries	Average number of persons employed[2] (000s)	Gross value added per person employed (£)
Standard Industrial Classification: Revised 1980								
Production and construction Divisions 1 - 5[3]								
	KABW	KABX	KAFE	KAFF	KAFG	KAFH	KAFI	KAFJ
1986	311 772	107 141	..	−224	13 669	59 705	6 468	16 565
1987	345 812	117 216	..	..	14 797	..	6 448	18 177
1988	386 112	131 584	..	..	18 026	..	6 487	20 283
1989	421 674	144 135	..	..	21 000	..	6 494	22 195
1990	454 031	149 788	..	..	21 666	..	6 364	21 530
1991	440 770	142 849	..	..	20 736	..	5 919	24 134
1992	447 431	145 012	..	..	20 248	..	5 698	25 448
Production industries (Revised definition) Divisions 1 - 4[3]								
	KAFK	KAFL	KAFM	KAFO	KAGP	KAIA	KAIB	KAIC
1986	268 313	93 958	47 028	225	13 032	48 854	5 346	17 576
1987	297 798	102 748	49 501	2 113	14 019	52 040	5 309	19 354
1988	326 814	114 523	52 676	..	16 789	56 202	5 341	21 443
1989	355 406	123 709	55 391	..	19 578	61 594	5 334	23 194
1990	380 790	129 106	55 699	..	20 591	65 782	5 202	24 821
1991	374 221	124 543	52 984	..	20 115	67 159	4 884	25 501
1992	385 042	127 369	54 423	..	19 616	67 934	4 683	27 199
Manufacturing (Revised definition) Divisions 2 - 4								
	KAID	KAIE	KAIF	KAIG	KAIH	KAII	KAJF	KAJG
1986	232 499	80 955	42 979	515	8 705	43 426	4 878	16 595
1987	254 683	89 745	44 797	2 111	9 754	46 429	4 874	18 412
1988	283 434	99 934	48 338	3 459	12 170	50 664	4 932	20 261
1989	309 020	108 291	51 510	3 530	14 499	55 492	4 953	21 863
1990	319 295	111 051	51 546	376	14 308	59 712	4 840	22 945
1991	308 909	105 606	48 722	−2 202	13 100	60 793	4 538	23 274
1992	319 515	108 144	50 167	819	12 094	61 727	4 369	24 752
Energy and water supply industries Division 1[3,4]								
	KAJH	KAJI	KAJJ	KAJK	KAJL	KAJM	KAJN	KAJO
1986	35 814	13 003	4 049	−740	4 327	5 428	468	27 799
1987	43 116	13 003	4 704	2	4 269	5 611	435	29 916
1988	43 380	14 589	4 338	281	4 620	5 539	409	35 704
1989	46 386	15 418	3 882	−132	5 080	6 102	380	40 508
1990	61 495	18 055	4 152	100	6 282	6 070	362	49 920
1991	65 312	18 937	4 262	−40	7 016	6 366	346	54 674
1992	65 526	19 226	4 256	−56	7 522	6 207	314	61 262
Coal extraction, coke ovens and manufacture of solid fuels Classes 11 and 12								
	KAJP	KAJQ	KAJR	KAJS	KAJT	KAJU	KAJV	KAJW
1986	4 007	2 513	546	8	507	1 828	160	15 691
1987	3 859	2 266	468	−72	507	1 878	132	17 166
1988	3 924	2 375	557	89	401	1 639	112	21 207
1989	3 700	2 030	533	−24	347	1 724	90	22 556
1990	3 466	1 853	566	29	297	1 406	76	24 382
1991	3 550	1 881	751	109	215	1 237	66	28 415
1992	3 029	1 401	602	−152	123	1 066	50	28 020
Extraction of mineral oil and natural gas Class 13[5]								
	KAMY	KANE	KAPF	KAQL	KARG	KARY	KATL	KAYU
1986	11 861	9 000	..	−89	2 579	..	23	391 322
1987	12 664	10 088	..	54	2 042	..	28	360 275
1988	10 062	7 625	..	−56	2 214	..	29	262 941
1989	10 411	7 457	..	28	2 706	..	31	240 532
1990	11 924	7 923	..	−90	3 568	..	37	214 143
1991	12 168	7 510	..	..	5 126	..	33	227 561
1992	12 132	7 866	..	..	5 420	..	30	262 193

See footnotes on the last part of this table.

Source: Office for National Statistics: 01633 812929

21.1 Production survey: summary table
United Kingdom

continued

	Estimates for all firms							
	(£ million)							
			Stocks and work in progress					
	Gross output (Production)[1]	Gross value added	At end of year	Change during year	Capital expenditure *less* disposals	Wages and salaries	Average number of persons employed[2] (000s)	Gross value added per person employed (£)
Standard Industrial Classification: Revised 1980								
Mineral oil processing								
Class 14	KAYV	KAYW	KAYX	KAYY	KAYZ	KAZA	KAZB	KAZC
1986	9 080	1 337	994	–541	301	199	14	93 139
1987	16 328	1 349	996	–20	223	226	15	91 198
1988	15 173	1 609	738	–176	334	254	16	102 685
1989	16 528	1 960	791	222	303	268	15	131 910
1990	19 016	2 492	1 029	192	343	279	14	174 213
1991	19 010	1 484	728	–279	416	292	13	110 581
1992	18 606	1 434	788	53	420	310	13	107 281
Other energy and water supply								
Classes 15 to 17[4]	KAZD	KAZE	KAZF	KAZG	KAZH	KAZK	KAZN	KAZO
1986	22 727	9 153	2 509	–208	3 518	3 401	293	31 215
1987	22 929	9 388	3 240	94	3 535	3 508	287	32 711
1988	24 282	10 605	3 043	368	3 885	3 646	282	37 607
1989	26 157	11 428	2 558	–330	4 430	4 109	276	41 390
1990	39 013	13 710	2 556	–121	5 642	4 384	270	50 778
1991	42 752	15 572	2 783	..	6 384	4 836	267	58 366
1992	43 893	16 391	2 866	..	6 979	4 831	250	65 564
Extraction of minerals and ores other than fuels; manufacture of metals, mineral products and chemicals								
Division 2	KAZP	KAZQ	KAZR	KAZS	KAZT	KAZU	KBAB	KBAC
1986	46 618	15 700	7 068	–216	2 249	6 669	655	23 961
1987	51 645	18 427	7 505	470	2 680	7 226	653	28 223
1988	58 159	21 202	8 058	610	3 216	7 946	662	32 036
1989	63 288	22 068	8 679	563	3 907	8 789	676	32 641
1990	62 684	20 839	8 666	1	4 008	9 385	656	31 782
1991	59 157	19 453	8 263	–494	3 365	9 562	614	31 685
1992	59 953	19 796	8 159	–77	3 000	9 680	585	33 835
Extraction and preparation of metalliferous ores								
Class 21	KBAD	KBAG	KBAH	KBAI	KBAJ	KBAL	KBAM	KBAN
1986	25	4	6	–	6	12	1	4 486
1987	19	–1	12	6	11	11	1	1 925
1988	3	2	1	–	–	2	–	10 454
1989	3	1	–	–	–	2	–	4 447
1990	–	–	–	–	–	–	–	–6 773
1991	–	–	–	–	–	–	–	–
1992	–	–	–	–	–	–	–	–
Metal manufacturing								
Class 22	KBAO	KBAQ	KBAR	KBAS	KBAT	KBBB	KBBC	KBBD
1986	11 134	2 956	2 008	–25	417	1 593	153	19 278
1987	12 368	3 539	2 055	120	461	1 622	142	24 848
1988	14 549	4 247	2 298	219	580	1 766	141	30 124
1989	15 600	4 236	2 302	–13	688	1 878	141	30 002
1990	14 685	3 638	2 086	–183	692	1 980	136	26 578
1991	12 700	3 063	1 844	–246	504	1 913	124	24 745
1992	12 319	2 943	1 723	–121	402	1 810	112	26 262
Extraction of minerals nes								
Class 23	KBBE	KBBG	KBBH	KBBI	KBBJ	KBBL	KBBM	KBBN
1986	659	343	41	–3	30	98	10	34 578
1987	670	350	43	3	43	102	9	37 191
1988	743	370	41	–2	70	111	10	39 028
1989	717	371	41	5	63	116	9	39 648
1990	717	385	52	12	66	126	9	42 803
1991	698	368	48	–1	35	123	8	43 679
1992	637	309	42	–	34	120	7	40 772

See footnotes on the last part of this table.

Source: Office for National Statistics: 01633 812929

21.1 Production survey: summary table
United Kingdom

continued

	Estimates for all firms							
	(£ million)							
			Stocks and work in progress					
	Gross output (Production)[1]	Gross value added	At end of year	Change during year	Capital expenditure *less* disposals	Wages and salaries	Average number of persons employed[2] (000s)	Gross value added per person employed (£)
Standard Industrial Classification: Revised 1980								
Manufacture of non-metallic mineral products Class 24								
	KBBO	KBBQ	KBBR	KBBS	KBBT	KBBV	KBBW	KBBX
1986	9 021	3 848	1 170	–13	431	1 780	200	19 274
1987	10 013	4 364	1 185	10	601	1 964	202	21 583
1988	11 786	5 308	1 257	54	822	2 251	214	24 816
1989	12 936	5 650	1 455	190	1 028	2 526	220	25 719
1990	12 631	5 256	1 567	114	891	2 616	210	25 063
1991	11 196	4 510	1 502	–25	480	2 517	190	23 732
1992	10 865	4 237	1 496	–26	407	2 473	180	23 599
Chemical industry Class 25								
	KBBY	KBDB	KBDC	KBDD	KBDE	KBDG	KBDH	KBDI
1986	24 889	8 221	3 738	–171	1 318	3 061	281	29 298
1987	27 620	9 814	4 099	324	1 504	3 392	287	34 166
1988	30 153	10 869	4 357	344	1 712	3 681	287	37 862
1989	33 014	11 394	4 776	380	2 048	4 124	296	38 449
1990	33 601	11 120	4 838	42	2 271	4 507	290	38 265
1991	33 553	11 100	4 755	–209	2 235	4 855	283	39 270
1992	35 050	11 894	4 779	84	2 113	5 143	278	42 725
Production of man-made fibres Class 26								
	KBDJ	KBDL	KBDM	KBDN	KBDO	KBDQ	KBDR	KBDS
1986	889	329	105	–4	47	125	11	30 454
1987	956	362	112	6	59	135	11	33 113
1988	925	405	104	–6	32	136	10	39 752
1989	1 018	418	106	2	81	143	10	42 824
1990	1 048	439	123	17	88	155	9	46 606
1991	1 010	411	114	–13	111	153	9	45 542
1992	1 081	414	118	–15	43	133	7	54 948
Metal goods, engineering and vehicle industries Division 3								
	KBDT	KBDV	KBDW	KBDX	KBDY	KBEB	KBEC	KBED
1986	87 583	34 104	23 855	879	3 183	20 242	2 156	15 821
1987	95 497	36 453	24 394	716	3 182	21 286	2 116	17 231
1988	108 319	41 058	26 144	1 775	4 092	23 199	2 131	19 269
1989	120 878	45 674	28 134	2 259	5 149	25 400	2 130	21 440
1990	126 974	47 571	27 960	–38	5 300	27 730	2 111	22 535
1991	120 192	43 669	26 121	–1 456	4 834	28 132	1 964	22 233
1992	124 847	43 259	27 552	568	4 259	28 078	1 863	23 222
Manufacture of metal goods nes Class 31								
	KBEE	KBEG	KBEH	KBEI	KBEJ	KBEL	KBEM	KBEN
1986	10 611	4 298	1 581	27	350	2 610	322	13 339
1987	11 331	4 560	1 653	100	407	2 750	316	14 428
1988	12 954	5 229	1 872	176	536	3 088	325	16 090
1989	14 340	5 776	1 962	139	644	3 450	335	17 265
1990	15 364	6 272	2 014	–45	629	3 816	338	18 524
1991	14 400	5 729	1 828	–180	490	3 700	308	18 596
1992	14 054	5 694	1 746	–55	454	3 705	293	19 423
Mechanical engineering Class 32								
	KBEO	KBEQ	KBER	KBES	KBET	KBEV	KBEW	KBEX
1986	23 541	9 651	5 886	–62	711	5 958	615	15 681
1987	24 351	9 966	5 710	51	696	5 973	586	16 995
1988	27 463	11 079	5 905	356	801	6 491	586	18 901
1989	31 186	12 622	6 277	456	1 029	7 170	590	21 401
1990	33 296	13 300	6 184	–100	1 136	7 902	588	22 596
1991	31 270	12 568	5 541	–467	930	8 001	547	22 963
1992	31 223	12 491	5 485	–386	819	8 069	521	23 965

See footnotes on the last part of this table.

Source: Office for National Statistics: 01633 812929

21.1 Production survey: summary table
United Kingdom

continued

	Estimates for all firms							
	(£ million)							
			Stocks and work in progress					
	Gross output (Production)[1]	Gross value added	At end of year	Change during year	Capital expenditure *less* disposals	Wages and salaries	Average number of persons employed[2] (000s)	Gross value added per person employed (£)
Standard Industrial Classification: Revised 1980								
Electrical and electronic engineering and manufacturing of office machinery and data processing equipment								
Classes 33 and 34	KBEY	KBFR	KBFV	KBFW	KBFX	KBGG	KBGN	KBHO
1986	24 729	9 914	5 676	135	950	5 389	590	16 795
1987	27 401	10 861	5 868	261	999	5 796	590	18 408
1988	31 379	12 434	6 411	369	1 288	6 372	599	20 758
1989	34 104	12 650	6 772	450	1 444	6 878	590	21 440
1990	35 254	12 750	6 810	51	1 300	7 236	566	22 526
1991	33 996	12 008	6 196	–412	1 162	7 578	534	22 504
1992	34 785	11 389	6 152	–332	1 115	7 520	504	22 597
Manufacture of motor vehicles and parts thereof								
Class 35	KBHP	KBHQ	KBHR	KBHS	KBHT	KBHU	KBHV	KBHW
1986	15 410	4 657	2 907	173	677	2 688	261	17 844
1987	17 629	5 112	2 925	82	612	2 804	258	19 814
1988	20 882	5 564	3 363	409	925	3 180	265	20 967
1989	23 399	6 686	3 637	301	1 339	3 519	267	25 067
1990	23 660	6 331	3 880	180	1 522	3 826	268	23 663
1991	21 053	4 768	3 061	–837	1 778	3 807	246	19 399
1992	23 526	5 956	3 183	88	1 385	3 842	234	25 463
Manufacture of other transport equipment								
Class 36	KBHX	KBHY	KBHZ	KBJU	KBJV	KBJW	KBJX	KBJY
1986	10 889	4 401	7 055	599	387	2 871	284	15 485
1987	11 854	4 695	7 461	366	354	3 183	282	16 640
1988	12 245	5 218	7 688	347	402	3 192	268	19 449
1989	14 163	6 417	8 540	883	526	3 414	262	24 497
1990	15 552	7 291	8 143	–69	570	3 911	265	27 487
1991	15 545	6 806	8 584	410	324	4 012	257	27 090
1992	17 040	5 876	9 988	1 100	334	3 881	234	25 148
Instrument engineering								
Class 37	KBJZ	KBLI	KBNN	KBNO	KBNP	KBNQ	KBNR	KBNS
1986	2 674	1 183	750	7	109	726	82	14 350
1987	2 932	1 259	776	20	114	779	83	15 128
1988	3 396	1 533	906	118	140	875	87	17 685
1989	3 685	1 522	947	30	167	968	88	17 393
1990	3 847	1 627	928	–56	144	1 039	85	19 056
1991	3 928	1 791	911	30	150	1 033	78	22 909
1992	4 217	1 852	998	27	153	1 061	77	24 053
Other manufacturing industries								
Division 4	KBNT	KBNU	KBNV	KBNW	KBNX	KBNY	KBNZ	KBRO
1986	98 298	31 151	12 056	–147	3 273	16 513	2 067	15 067
1987	107 540	34 865	12 897	926	3 892	17 917	2 106	16 556
1988	116 957	37 674	14 135	1 074	4 862	19 519	2 140	17 607
1989	124 854	40 548	14 696	708	5 442	21 305	2 147	18 889
1990	129 638	42 641	14 920	413	4 999	22 596	2 073	20 568
1991	129 560	42 485	14 338	–251	4 901	23 100	1 960	21 682
1992	134 715	45 088	14 456	327	4 835	23 970	1 921	23 469
Food, drink and tobacco manufacturing								
Class 41/42	KBRP	KBRQ	KBRR	KBRS	KBRT	KBRU	KBRV	KBRW
1986	47 023	10 801	4 808	–384	1 275	4 735	582	18 572
1987	49 272	11 696	4 844	66	1 453	5 082	594	19 692
1988	52 004	12 134	5 190	267	1 660	5 406	592	20 483
1989	55 029	13 268	5 482	314	1 909	5 882	598	22 191
1990	58 298	14 390	5 867	407	2 076	6 364	591	24 347
1991	60 237	14 875	5 914	139	2 132	6 774	577	25 796
1992	61 826	15 616	5 845	295	2 102	6 980	564	27 693

See footnotes on the last part of this table.

Source: Office for National Statistics: 01633 812929

21.1 Production survey: summary table
United Kingdom

continued

	Estimates for all firms							
	(£ million)							
			Stocks and work in progress					
	Gross output (Production)[1]	Gross value added	At end of year	Change during year	Capital expenditure *less* disposals	Wages and salaries	Average number of persons employed[2] (000s)	Gross value added per person employed (£)
Standard Industrial Classification: Revised 1980								
Textile industry Class 43								
	KBRX	KBRY	KBUE	KBUF	KBUG	KBUH	KBUI	KBUJ
1986	6 788	2 503	1 207	–4	265	1 504	229	10 929
1987	7 571	2 873	1 338	125	271	1 623	228	12 596
1988	7 930	2 980	1 448	96	353	1 709	225	13 234
1989	7 942	2 962	1 484	10	296	1 791	216	13 745
1990	7 648	2 941	1 338	–82	251	1 785	196	15 035
1991	7 384	2 842	1 236	–101	234	1 795	182	15 625
1992	7 544	2 956	1 253	8	281	1 843	172	17 130
Footwear and clothing industries and manufacture of leather and leather goods Classes 44 and 45								
	KBUK	KBUL	KBUM	KBUN	KBUO	KBUP	KBUR	KBUS
1986	7 143	2 864	1 342	45	174	1 798	341	8 399
1987	7 800	3 104	1 509	198	194	1 904	341	9 103
1988	8 217	3 229	1 618	125	212	2 042	341	9 470
1989	8 321	3 288	1 565	–4	188	2 066	320	10 277
1990	8 499	3 448	1 620	82	192	2 104	298	11 570
1991	7 965	3 219	1 436	–56	156	2 060	270	11 944
1992	8 305	3 405	1 462	19	144	2 123	258	13 198
Timber and wooden furniture industries Class 46								
	KBUT	KBUV	KBUW	KBUX	KBUY	KBUZ	KBVA	KBVB
1986	7 112	2 393	1 119	51	197	1 513	193	12 428
1987	8 335	2 884	1 230	155	271	1 647	198	14 544
1988	9 783	3 339	1 405	137	351	1 925	213	15 705
1989	10 138	3 478	1 418	82	397	2 082	215	16 207
1990	10 295	3 459	1 383	1	260	2 186	207	16 715
1991	9 464	3 165	1 232	–120	225	2 110	189	16 755
1992	9 448	3 137	1 205	–41	259	2 130	186	16 883
Manufacture of paper and paper products; printing and publishing Class 47								
	KBVC	KBVE	KBVF	KBVG	KBVH	KBVI	KBVJ	KBVK
1986	19 396	8 364	1 982	91	870	4 653	439	19 065
1987	21 914	9 327	2 156	169	1 070	5 051	446	20 908
1988	24 824	10 464	2 451	213	1 521	5 517	456	22 956
1989	27 389	11 469	2 589	199	1 771	6 142	469	24 454
1990	28 304	12 004	2 534	–46	1 398	6 582	462	25 983
1991	28 357	12 102	2 489	–23	1 474	6 719	442	27 377
1992	30 168	13 023	2 559	39	1 300	7 006	438	29 759
Processing of rubber and plastics Class 48								
	KBVL	KBVM	KBVN	KBVO	KBVP	KBVR	KBVS	KBVT
1986	8 379	3 274	1 125	44	419	1 777	204	16 084
1987	9 869	3 876	1 280	135	549	2 011	214	18 075
1988	11 028	4 214	1 362	124	651	2 250	225	18 729
1989	12 542	4 704	1 494	64	763	2 578	237	19 867
1990	13 085	5 026	1 555	46	710	2 794	234	21 451
1991	12 886	4 976	1 465	–104	581	2 872	224	22 257
1992	13 923	5 510	1 525	9	644	3 067	224	24 579
Other manufacturing industries Class 49								
	KBVU	KBVV	KBVW	KBVX	KBVY	KBVZ	KBWA	KBWB
1986	2 458	952	475	11	73	534	81	11 747
1987	2 778	1 106	540	78	86	597	84	13 223
1988	3 171	1 314	662	112	115	670	88	14 914
1989	3 492	1 378	666	43	119	763	93	14 906
1990	3 507	1 372	622	6	111	786	85	16 128
1991	3 269	1 305	568	14	99	770	77	16 975
1992	3 501	1 441	607	–2	104	821	79	18 204

See footnotes on the last part of this table.

Source: Office for National Statistics: 01633 812929

21.1 Production survey: summary table
United Kingdom

continued

	Estimates for all firms							
	(£ million)							
			Stocks and work in progress					
	Gross output (production)[1]	Gross value added	At end of year	Change during year	Capital expenditure *less* disposals	Wages and salaries	Average number of persons employed[2] (000s)	Gross value added per person employed (£)
Standard Industrial Classification: Revised 1980								
Construction Division 5								
	KBWC	KBWE	KBWF	KBWG	KBWH	KBWI	KBWJ	KBWK
1986	43 459	13 183	..	..	637	10 851	1 122	11 751
1987	48 013	14 168	..	..	779	11 948	1 140	12 696
1988	59 298	17 061	..	..	1 237	13 261	1 146	14 880
1989	66 268	20 426	..	..	1 422	13 018	1 160	17 605
1990	73 240	20 681	..	..	1 076	16 311	1 163	17 782
1991	66 549	18 306	..	..	621	15 442	1 035	17 683
1992	62 390	17 642	..	..	631	16 000	1 016	17 373

1 Figures for gross output include a substantial amount of duplication represented by the total value of partly manufactured goods sold by one industrial establishment to another. The extent of duplication varies from one census industry to another.

2 The figures include working proprietors but exclude outworkers.

3 Figures for mineral oil and natural gas not included.

4 Figures for stocks and work in progress exclude water undertakings and work in progress in the gas industry.

5 Figures for stocks and work in progress exclude goods on hand for sale.

Source: Office for National Statistics: 01633 812929

21.2 Manufacturers sales by industry[1]

£ millions

Industry		SIC (92)	1995	1996	1997	1998
Other mining and quarrying						
Quarrying of stone for construction	KSPF	14110	..	..	..	..
Quarrying of limestone, gypsum and chalk	KSPG	14120	..	..	..	..
Quarrying of slate	KSPH	14130	..	..	..	..
Operation of gravel and sand pits	KSPJ	14210	..	..	..	..
Mining of clays and kaolin	KSPK	14220	..	..	..	..
Mining of chemical and fertilizer minerals	KSPL	14300	..	..	..	..
Production of salt	KSPM	14400	..	..	..	..
Other mining and quarrying n.e.c.	KSPN	14500	..	..	..	..
Manufacture of food products and beverages						
Production and preserving of meat	KSPO	15110	..	..	..	..
Production and preserving of poultry meat	KSPP	15120	..	..	..	..
Bacon and ham production	KSPQ	15131	..	..	..	..
Other meat and poultry meat processing	KSPR	15139[2]	..	..	..	..
Processing and preserving of fish and fish products	KSPS	15200	..	..	..	..
Processing and preserving of potatoes	KSPT	15310	..	..	1 148	1 151
Fruit and vegetable juice	KSPU	15320	488	488	..	405
Processing and preserving of fruit and vegetables n.e.c.	KSPV	15330	1 894	2 024	1 966	1 935
Crude oils and fats	KSPW	15410	..	..	..	..
Refined oils and fats	KSPX	15420	..	..	..	..
Margarine and similar edible fats	KSPY	15430	..	..	..	..
Operation of dairies	KTEH	15510	6 013	..	5 790	5 750
Ice cream	KSPZ	15520	..	..	..	..
Grain mill products	KSQA	15610	2 706	2 942	2 815	2 620
Starches and starch products	KSQB	15620	488	494	451	394
Prepared feeds for farm animals	KSPI	15710	2 251	2 558	2 673	2 285
Prepared pet foods	KSQC	15720	1 476	..	1 459	1 401
Bread; fresh pastry goods and cakes	KSQD	15810	..	..	..	..
Rusks and biscuits; preserved pastry goods and cakes	KSQE	15820	3 109	3 281	..	3 103
Sugar	KSQF	15830	..	..	..	..
Cocoa; chocolate and sugar confectionery	KSQG	15840	2 967	3 060	3 264	3 241
Macaroni, noodles, couscous and similar farinaceous products	KSQH	15850	..	..	..	..
Processing of tea and coffee	KSQI	15860	1 499	1 348	1 379	1 485
Condiments and seasonings	KSQJ	15870	..	..	..	..
Homogenised food preparations and dietetic foods	KSQK	15880	..	..	..	..
Manufacture of other food products n.e.c.	KSQL	15890	..	..	1 491	1 677
Distilled potable alcoholic beverages	KSQM	15910	2 080	2 263	2 030	2 087
Production of ethyl alcohol from fermented materials	KSQN	15920	..	..	..	..
Wines	KSQO	15930	..	..	..	..
Cider and other fruit wines	KSQP	15940	..	..	..	..
Other non-distilled fermented beverages	KSQQ	15950	..	..	..	..
Beer	KSQR	15960	..	..	..	..
Malt	KSQS	15970	..	..	..	..
Mineral waters and soft drinks	KSQT	15980	..	..	..	..
Manufacture of tobacco products						
Tobacco products	KSQU	16000	2 375	2 616	2 622	2 484
Manufacture of textiles						
Preparation and spinning of textile fibres	KSQV	17100	1 370	1 358	840	1 009
Textile weaving	KSQW	17200	..	..	..	..
Finishing of textiles	KSQX	17300	837	896	778	717
Soft furnishings	KSQY	17401	..	..	..	..
Canvas goods, sacks etc	KSQZ	17402	151	160	145	123
Household textiles	KSRA	17403	..	..	..	..
Carpets and rugs	KSRB	17510	..	..	..	..
Cordage, rope, twine and netting	KSRC	17520	79	79	100	82

1 The data are collected under the PRODCOM inquiry which was introduced in 1993. The inquiry replaced the previous QSI/ASI inquiries.
2 Previously 15132.

Source: Office for National Statistics: 01633 813065

21.2 Manufacturers sales by industry[1]

continued

£ millions

Industry		SIC (92)	1995	1996	1997	1998
Manufacture of textiles continued						
Nonwovens and articles made from nonwovens, except apparel	KSRD	17530	..	..	..	..
Lace	KSRE	17541	..	..	..	..
Narrow fabrics	KSRF	17542	..	..	..	..
Other textiles n.e.c.	KSRG	17549[2]	..	..	..	..
Knitted and crocheted fabrics	KSRH	17600	605	..	608	..
Knitted and crocheted hosiery	KSRI	17710	..	..	495	452
Knitted and crocheted pullovers, cardigans and similar	KSRJ	17720	917	920	1 007	759
Manufacture of wearing apparel; dressing and dyeing of fur						
Leather clothes	KSRK	18100	..	..	..	..
Workwear	KSRL	18210	235	253	..	290
Men's outerwear	KSRM	18221	..	..	..	..
Other women's outerwear	KSRN	18222	2 135	1 498	1 460	1 297
Men's underwear	KSRO	18231	..	..	..	..
Women's underwear	KSRP	18232	1 019	998	902	911
Hats	KSRQ	18241	..	..	..	..
Other wearing apparel and accessories	KSRR	18249[3]	..	..	..	..
Dressing/dyeing of fur; articles of fur	KSRS	18300	..	..	..	..
Tanning and dressing of leather; manufacture of luggage, handbags, saddlery, harness and footwear						
Tanning and dressing of leather	KSRT	19100	416	443	374	356
Luggage, handbags and the like, saddlery and harness	KSRU	19200	229	247	256	243
Footwear	KSRV	19300	..	..	..	..
Manufacture of wood and of products of wood and cork, except furniture; manufacture of articles of straw and plaiting materials						
Sawmilling and planing of wood, impregnation of wood	KSRW	20100	..	..	..	..
Veneer sheets	KSRX	20200	..	..	..	..
Builders' carpentry and joinery	KSRY	20300	..	..	..	..
Wooden containers	KSRZ	20400	498	568	515	504
Other products of wood	KSSA	20510	434	425	323	385
Articles of cork, straw and plaiting materials	KSSB	20520	33	34	30	27
Manufacture of pulp, paper and paper products						
Paper and paperboard	KSSC	21120	–	..	..	..
Corrugated paper and paperboard, sacks and bags	KSSD	21211	4 075	3 880	3 498	3 309
Cartons, boxes, cases and other containers	KSSE	21219[4]	797	746	..	..
Household and sanitary goods and toilet requisites	KSSF	21220	3 525	3 688	3 575	3 498
Paper stationery	KSSG	21230	881	928	813	731
Wallpaper	KSSH	21240	..	..	..	301
Other articles of paper and paperboard n.e.c.	KSSI	21250	784	819	793	779
Publishing, printing and reproduction of recorded media						
Publishing of books	KSSJ	22110	2 610	2 724	2 729	2 849
Publishing of newspapers	KSSK	22120	..	3 131	3 453	3 560
Publishing of journals and periodicals	KSSL	22130	5 115	5 650	5 757	6 039
Publishing of sound recordings	KSSM	22140	123	132	..	..
Other publishinig	KSSN	22150	410	423	413	444
Printing of newspapers	KSSO	22210	265	238	207	..
Printing n.e.c.	KSSP	22220	8 518	9 372	10 409	9 968
Bookbinding and finishing	KSSQ	22230	436	416	378	359
Composition and plate-making	KSSR	22240	776	723	618	566
Other activities related to printing	KSSS	22250	338	410	519	671
Reproduction of sound recording	KSST	22310	310	372	..	336
Reproduction of video recording	KSSU	22320	..	..	..	194
Reproduction of computer media	KSSV	22330	..	119	92	122
Manufacture of chemicals and chemical products						
Industrial gases	KSSW	24110	449	479	535	625
Dyes and pigments	KSSX	24120	..	..	..	..
Other inorganic basic chemicals	KSSY	24130	..	..	..	..
Other organic basic chemicals	KSSZ	24140	..	..	..	..
Fertilizers and nitrogen compounds	KSTA	24150	..	..	..	..

1 The data are collected under the PRODCOM inquiry which was introduced in 1993. The inquiry replaced the previous QSI/ASI inquiries.
2 Previously 17543.
3 Previously 18242.
4 Previously 21212.

Source: Office for National Statistics: 01633 813065

21.2 Manufacturers sales by industry[1]

continued

£ millions

Industry		SIC (92)	1995	1996	1997	1998
Manufacture of chemicals and chemical products continued						
Plastics in primary forms	KSTB	24160	..	..	..	..
Synthetic rubber in primary forms	KSTC	24170	..	..	..	..
Pesticides and other agro-chemical products	KSTD	24200	1 487	1 597	1 159	..
Paints, varnishes and similar coatings, printing ink and mastic	KSTE	24300	..	..	2 229	2 468
Basic pharmaceutical products	KSTF	24410	650	564	633	527
Pharmaceutical preparations	KSTG	24420	5 738	5 966	5 969	6 030
Soap and detergents, cleaning and polishing preparations	KSTH	24510	2 175	2 311	1 889	1 827
Perfumes and toilet preparations	KSTI	24520	..	..	..	..
Explosives	KSTJ	24610	103	..	122	123
Glues and gelatines	KSTK	24620	..	..	..	..
Essential oils	KSTL	24630	..	..	..	..
Photographic chemical material	KSTM	24640	..	..	1 158	1 204
Prepared unrecorded media	KSTN	24650	273	297	..	134
Other chemical products n.e.c.	KSTO	24660	2 231	2 381	2 193	2 206
Man-made fibres	KSTP	24700	..	..	..	..
Manufacture of rubber and plastic products						
Rubber tyres and tubes	KSTQ	25110	1 117	1 118	1 006	1 014
Retreading and rebuilding of rubber tyres	KSTR	25120	165	180	..	..
Other rubber products	KSTS	25130	1 580	1 695	1 745	1 811
Plastic plates, sheets, tubes and profiles	KSTT	25210	..	..	..	..
Plastic packing goods	KSTU	25220	2 627	2 827	2 652	2 543
Builders' ware of plastic	KSTV	25230	2 688	3 123	3 184	3 300
Other plastic products	KSTW	25240	3 580	3 828	3 863	3 796
Manufacture of other non-metallic mineral products						
Flat glass	KSTX	26110	..	..	..	..
Shaping and processing of flat glass	KSTY	26120	..	..	.	..
Hollow glass	KSTZ	26130	694	732	742	714
Glass fibres	KSUA	26140	372	343	320	315
Manufacturing and processing of other glass incl technical glassware	KSUB	26150	214	208	..	307
Ceramic household and ornamental articles	KSUC	26210	742	765	737	651
Ceramic sanitary fixtures	KSUD	26220	173	172	..	..
Ceramic insulators and insulating fittings	KSUE	26230	..	..	..	23
Other technical ceramic products	KSUF	26240	22	26	..	..
Other ceramic products	KSUG	26250	..	..	14	13
Refractory ceramic products	KSUH	26260	..	..	495	..
Ceramic tiles and flags	KSUI	26300	..	93	118	103
Bricks, tiles and construction products in baked clay	KSUJ	26400	..	..	601	582
Cement	KSUK	26510	..	647	721	741
Lime	KSUL	26520	..	..	..	..
Plaster	KSUM	26530	..	..	..	..
Concrete products for construction purposes	KSUN	26610	1 496	1 442	1 623	1 545
Plaster products for construction purposes	KSUO	26620	245	..	262	284
Ready mixed concrete	KSUP	26630	..	981	..	994
Mortars	KSUQ	26640	..	111	..	..
Fibre cement	KSUR	26650	94	82	..	..
Other articles of concrete, plaster and cement	KSUS	26660	133	135	126	..
Cutting, shaping and finishing of stone	KSUT	26700	..	..	..	..
Abrasive products	KSUU	26810	..	..	..	..
Other non-metallic mineral products n.e.c.	KSUV	26820	..	..	..	..
Manufacture of basic metals						
Cast iron tubes	KSUW	27210	..	..	..	..
Steel tubes	KSUX	27220	..	..	..	..
Cold drawing	KSUY	27310	..	..	..	..

1 The data are collected under the PRODCOM inquiry which was introduced in 1993. The inquiry replaced the previous QSI/ASI inquiries.

Source: Office for National Statistics: 01633 813065

21.2 Manufacturers sales by industry[1]

continued

£ millions

Industry		SIC (92)	1995	1996	1997	1998
Manufacture of basic metals continued						
Cold rolling of narrow strip	KSUZ	27320	..	..	..	..
Cold forming or folding	KSVA	27330	..	..	..	..
Wire drawing	KSVB	27340	..	..	..	..
Other first processing of iron and steel n.e.c.	KSVC	27350	..	..	..	..
Precious metals production	KSVD	27410	..	..	..	..
Aluminium production	KSVE	27420	..	..	..	..
Lead, zinc and tin production	KSVF	27430	..	..	..	..
Copper production	KSVG	27440	..	..	..	..
Other non-ferrous metal production	KSVH	27450	..	..	..	..
Casting of iron	KSVI	27510	759	851	813	760
Casting of steel	KSVJ	27520	202	228	222	183
Casting of light metals	KSVK	27530	..	253	272	279
Casting of other non-ferrous metals	KSVL	27540	570	508	513	477
Manufacture of fabricated metal products, except machinery and equipment						
Metal structures and parts of structures	KSVM	28110	..	..	..	..
Builders' carpentry and joinery of metal	KSVN	28120	..	..	..	..
Tanks, reservoirs and containers of metal	KSVO	28210	..	..	..	..
Central heating radiators and boilers	KSVP	28220	..	..	..	..
Steam generators, except central heating hot water boilers	KSVQ	28300	..	..	..	..
Forging, pressing, stamping and roll forming of metal	KSVR	28400	..	..	..	..
Treatment and coating of metals	KSVS	28510	942	1 052	1 075	1 150
General mechanical engineering	KSVT	28520	2 921	2 813	2 826	2 808
Cutlery	KSVU	28610	..	240	207	194
Tools	KSVV	28620	1 141	1 115	1 107	1 061
Locks and hinges	KSVW	28630	..	..	..	..
Steel drums and similar containers	KSVX	28710	..	..	..	..
Light metal packaging	KSVY	28720	1 263	1 417	1 298	1 250
Wire products	KSVZ	28730	..	..	..	..
Fasteners, screw machine products, chain and spring	KSWA	28740	..	..	..	..
Other fabricated metal products n.e.c.	KSWB	28750	..	..	..	..
Manufacture of machinery and equipment not elsewhere classified						
Engines and turbines, except aircraft, vehicles and cycle engines	KSWC	29110	..	..	..	..
Pumps	KSWD	29121	..	..	..	..
Compressors	KSWE	29122	1 044	1 100	1 191	1 216
Taps and valves	KSWF	29130	..	1 362	1 296	1 385
Bearings, gears, gearing and driving elements	KSWG	29140	1 146	1 282	1 249	1 177
Furnaces and furnace burners	KSWH	29210	379	374	..	547
Lifting and handling equipment	KSWI	29220	..	..	..	..
Non-domestic cooling and ventilation equipment	KSWJ	29230	2 483	2 282	2 385	2 594
Other general purpose machinery n.e.c.	KSWK	29240	..	..	..	..
Agricultural tractors	KSWL	29310	945	..	1 049	1 060
Other agricultural and forestry machinery	KSWM	29320	662	576	541	572
Machine tools	KSWN	29400	1 828	1 945	2 115	2 019
Machinery for metallurgy	KSWO	29510	125	149	113	99
Machinery for mining	KSWP	29521	..	..	..	..
Earth-moving equipment	KSWQ	29522	..	..	..	..
Equipment for concrete crushing and screening and roadworks	KSWR	29523	..	..	..	..
Machinery for food, beverage and tobacco processing	KSWS	29530	..	..	..	..
Machinery for textile, apparel and leather production	KSWT	29540	437	391	393	297
Machinery for paper and paperboard production	KSWU	29550	304	378	287	317
Other special purpose machinery n.e.c.	KSWV	29560	2 054	2 232	2 254	2 097
Weapons and ammunition	KSWW	29600	..	..	..	..

1 The data are collected under the PRODCOM inquiry which was introduced in 1993. The inquiry replaced the previous QSI/ASI inquiries.

Source: Office for National Statistics: 01633 813065

21.2 Manufacturers sales by industry[1]

continued

£ millions

Industry		SIC (92)	1995	1996	1997	1998
Manufacture of machinery and equipment not elsewhere classified continued						
Electric domestic appliances	KSYR	29710	..	..	..	..
Non-electric domestic appliances	KSWX	29720	..	..	..	..
Manufacture of office machinery and computers						
Office machinery	KSWY	30010	..	..	..	..
Computers and other information processing equipment	KSWZ	30020	..	..	..	..
Manufacture of electrical machinery and apparatus not elsewhere classified						
Electric motors, generators and transformers	KSXA	31100	..	..	..	..
Electricity, distribution and control apparatus	KSXB	31200	..	..	..	..
Insulated wire and cable	KSXC	31300	..	..	..	..
Accumulators, primary cells and batteries	KSXD	31400	489	533	521	507
Lighting equipment and electric lamps	KSXE	31500	1 123	1 248	1 241	1 199
Electrical equipment for engines and vehicles n.e.c.	KSXF	31610	..	..	..	..
Other electrical equipment n.e.c.	KSXG	31620	..	..	..	..
Manufacture of radio, television and communication equipment and apparatus						
Electronic valves and tubes and other electronic components	KSXH	32100	..	..	..	..
Telegraph and telephone apparatus and equipment	KSXI	32201	..	..	..	..
Radio and electronic capital goods	KSXJ	32202	..	..	..	..
Television and radio receivers, sound or video recording etc	KSXK	32300	..	..	..	..
Manufacture of medical, precision and optical instruments, watches and clocks						
Medical and surgical equipment and orthopaedic appliances	KSXL	33100	..	..	..	..
Instruments and appliances for measuring, checking, testing etc	KSXM	33200	..	..	..	..
Industrial process control equipment	KSXN	33300	..	..	..	..
Optical instruments and photographic equipment	KSXO	33400	..	..	..	..
Watches and clocks	KSXP	33500	77	85	91	88
Manufacture of motor vehicles, trailers and semi-trailers						
Motor vehicles	KSXQ	34100	..	..	..	..
Bodies (coachwork) for motor vehicles (exc caravans)	KSXR	34201	..	..	..	..
Trailers and semi-trailers	KSXS	34202	1 081	1 049	951	..
Caravans	KSXT	34203	..	..	..	..
Parts and accessories for motor vehicles and their engines	KSXU	34300	8 409	8 198	7 906	8 580
Manufacture of other transport equipment						
Building and repairing of ships	KSXV	35110	..	..	..	1 585
Building and repairing of pleasure and sporting boats	KSXW	35120	..	..	..	423
Railway and tramway locomotives and rolling stock	KSXX	35200	..	..	..	..
Aircraft and spacecraft	KSXY	35300	..	..	..	..
Motorcycles	KSXZ	35410	..	..	..	..
Bicycles	KSYA	35420	149	148	157	115
Invalid carriages	KSYB	35430	..	..	..	..
Other transport equipment n.e.c.	KSYC	35500	..	..	..	..
Manufacture of furniture; manufacturing not elsewhere classified						
Chairs and seats	KSYD	36110	..	..	..	..
Other office and shop furniture	KSYE	36120	..	..	..	..
Other kitchen furniture	KSYF	36130	..	..	..	..
Other furniture	KSYG	36140	..	..	..	..
Mattresses	KSYH	36150	..	..	..	..
Striking of coins and medals	KSYI	36210	..	..	..	..
Jewellery and related articles n.e.c.	KSYJ	36220	..	..	..	..
Musical instruments	KSYK	36300	47	60	62	60
Sports goods	KSYL	36400	293	289	259	263
Games and toys	KSYM	36500	..	..	..	..
Imitation jewellery	KSYN	36610	..	..	..	..
Brooms and brushes	KSYO	36620	..	..	..	..
Miscellaneous stationers' goods	KSYP	36631	272	285	278	264
Other manufacturing n.e.c.	KSYQ	36639[2]	..	..	..	..

1 The data are collected under the PRODCOM inquiry which was introduced in 1993. The inquiry replaced the previous QSI/ASI inquiries.
2 Previously 36632.

Source: Office for National Statistics: 01633 813065

21.3 Number of local units in manufacturing industries in 1999

United Kingdom

Standard Industrial Classification 1992 Division by Employment Sizeband

	Employment size								
	1 - 9	10 - 19	20 - 49	50 - 99	100 - 199	200 - 499	500 - 999	1 000+	Total
Number of local units									
Division									
15/16 Food products; beverages & tobacco	6 345	1 790	1 200	570	500	435	135	40	11 015
17 Textiles & textile products	3 965	820	720	370	300	150	20	-	6 345
18 Wearing apparel; dressing & dyeing of fur	5 060	980	665	260	185	125	20	-	7 295
19 Leather & leather products	740	170	155	75	45	30	-	-	1 220
20 Wood & wood products	6 960	1 020	550	140	70	20	5	-	8 765
21 Pulp, paper & paper products	1 690	395	470	225	190	115	15	-	3 095
22 Publishing, printing & reproduction of recorded media	24 255	3 135	1 665	605	325	180	35	15	30 205
23 Coke, refined petroleum products & nuclear fuel	205	40	45	25	15	10	10	5	345
24 Chemicals, chemical products & man-made fibres	2 540	570	585	365	280	200	65	30	4 635
25 Rubber & plastic products	4 425	1 290	1 155	505	335	175	35	5	7 935
26 Other non-metallic mineral products	4 965	795	560	330	185	105	20	5	6 975
27 Basic metals	1 540	365	365	190	155	110	20	10	2 760
28 Fabricated metal products, except machinery & equipment	21 935	4 230	2 575	870	380	170	20	-	30 185
29 Machinery & equipment not elsewhere classified	9 420	2 225	1 760	700	435	265	60	20	14 885
30 Office machinery & computers	1 070	150	125	40	40	40	10	10	1 485
31 Electrical machinery & apparatus not elsewhere classified	3 995	710	650	285	230	160	40	10	6 070
32 Radio, television & communication equipment & apparatus	1 925	305	270	125	125	80	30	20	2 880
33 Medical, precision & optical instruments, watches & clocks	3 385	650	575	265	175	90	25	5	5 170
34 Motor vehicles, trailers & semi-trailers	1 520	355	370	175	165	150	50	30	2 810
35 Other transport equipment	1 835	270	225	100	85	75	35	40	2 655
36/37 Manufacturing not elsewhere classified	16 470	1 775	950	370	225	120	20	-	19 930
Total manufacturing (15/37)	124 235	22 030	15 640	6 590	4 440	2 800	670	245	176 655

1 The analysis follows the Division in the UK Standard Industrial Classification of Economic Activities 1992 - revised edition.
2 The data in this table is taken from the ONS publication Business Monitor *PA1003 - Size Analysis of United Kingdom Businesses* 1999 edition. The count of units refers to local units i.e. individual sites, rather than whole businesses.
3 All counts have been rounded to avoid disclosure.

Source: Office for National Statistics: 01633 813269

21.4 Total inland energy consumption

Heat supplied basis

Million tonnes of oil equivalent[1]

		1988	1989	1990	1991	1992	1993	1994	1995	1996	1997	1998
Inland energy consumption of primary fuels and equivalents[2]	KLWA	214.4	212.8	214.9	220.2	217.7†	221.0	218.5	218.7	230.3	225.6	228.9
Coal[3]	KLWB	70.0†	67.0	66.9	67.1	63.0	55.0	51.3	48.9	47.5	42.4	42.9
Petroleum[4]	KLWC	74.7	76.3	78.3	77.8	78.3	78.5	77.6	75.7†	77.9	75.3	74.8
Primary electricity[5]	KLWD	18.1	19.2	17.7	19.2	20.4	23.4	23.2†	23.1	23.6	25.1	25.1
Natural gas[6]	KLWE	51.3†	49.5	51.2	55.3	55.1	63.0	64.8	69.2	81.3	82.8	85.9
less Energy used by fuel producers and losses in conversion and distribution	KLWF	65.8	66.5	67.6	68.3	66.8	68.2	66.2	68.6	70.9	71.0†	72.5
Total consumption by final users[2]	KLWG	148.6	145.0	146.4	150.4	150.0	153.0	153.2	152.2	162.0	154.5†	156.2
Final energy consumption by type of fuel												
Coal (direct use)[7]	KLWH	9.7	8.9	8.1	8.6	8.1	7.6	6.9	5.5	4.9†	4.4	3.8
Coke and breeze	KLWI	5.6	4.6	4.3	4.0	3.9	3.9	1.2†	1.2	1.0	0.9	0.9
Other solid fuel[7]	KLWJ	0.9†	0.8	0.8	0.8	0.7	0.8	0.8	0.7	0.8	0.7	0.6
Coke oven gas	KLWK	0.6†	0.6	0.6	0.6	0.5	0.6	0.8	0.7	1.3	1.3	1.2
Natural gas (direct use)[6]	KLWL	46.3†	44.8	46.1	49.7	48.4	49.3	49.9	51.8	57.4	54.4	55.9
Electricity	KLWM	22.8	23.3	23.6	24.2	24.2	24.6	24.4	25.2†	26.2	26.5	27.1
Petroleum (direct use)[8]	KLWN	62.0	62.7	63.3	63.5	64.6	65.4	65.2†	63.8	66.1	65.4	65.8
Final energy consumption by class of consumer												
Agriculture	KLWP	1.4	1.3	1.3	1.4	1.4	1.4	1.4	1.3	1.4	1.3	1.4
Iron and steel industry	KLWQ	8.2	8.0	6.9	6.6	6.5	7.0	7.7	7.7	4.7†	4.6	4.4
Other industries	KLWR	32.7	30.2	30.9	30.2	29.1	29.9	30.7	30.7	31.2†	30.1	30.5
Railways[9]	KLWS	0.8†	0.7	0.7	0.7	0.7	0.7	0.7	0.7	1.3	1.2	0.5
Road transport	KLWT	36.2	37.8	38.8	38.5	39.4	39.5	39.7	39.3	40.7†	41.2	40.1
Water transport	KLWU	1.2	1.4	1.4	1.4	1.4	1.4†	1.3	1.2	1.3	1.3	1.2
Air transport	KLWV	6.9	7.3	7.3	6.9	7.4	7.9	8.1	8.5	8.9	9.3	10.2
Domestic	KLWW	42.3	40.2	40.8	44.8	44.0	45.5	44.0	42.7	48.1	44.9†	46.0
Public administration	KLWX	8.3	7.7	7.7	8.5	9.1	8.1	8.3	8.5	8.9†	8.6	8.2
Commercial and other services[10]	KLWY	10.3	10.1	10.3	11.0	10.5	11.2	11.0	11.5	12.0†	12.1	12.1

1 Estimates of the gross calorific values used for converting the statistics for the various fuels to them are given in the *Digest of United Kingdom Energy Statistics 1997.*
2 Including, from 1988, small amounts of primary heat sources (solar, geothermal, etc). Up to 1988 includes natural gas used for non-energy purposes (eg petrochemicals).
3 Including net trade and stock change in other solid fuels. Including, from 1988, solid renewable sources (wood, waste, etc).
4 Refinery throughput of crude oil, *plus* net foreign trade and stock change in petroleum products. Petroleum products not used as fuels (chemical feedstock, industrial and white spirits, lubricants, bitumen and wax) are excluded.
5 Primary electricity comprises nuclear, natural flow hydro, net imports of electricity, and from 1988 generation of wind stations.
6 Natural gas includes colliery methane, non-energy use of natural gas up to 1988, and from 1988, landfill gas and sewage gas.
7 Includes from 1988 solid renewable sources.
8 Includes briquettes, ovoids, Phurnacite, Coalite, etc, and from 1988 wood, waste etc used for heat generation.
9 Including fuel used at transport premises from 1990.
10 Including fuel used at transport premises prior to 1990.

Source: Department of Trade and Industry: 020 7215 5187

21.5 Coal: supply and demand

Million tonnes

			1988	1989	1990	1991	1992	1993	1994	1995	1996	1997	1998
Supply													
Production of deep-mined coal	KLXA		83.8	79.6	72.9	73.4	65.8	50.5	31.9	35.2	32.2	30.3	25.0
Production of opencast coal	KLXB		17.9	18.6	18.1	18.6	18.2	17.0	16.8	16.4	16.3	16.7	15.0
Total	KLXC		101.7	98.2	91.0	92.0	84.0	67.5	48.7	51.6	48.5	47.0	40.0
Recovered slurry, fines, etc	KLXD		2.4	1.5	1.7	2.2	0.5	0.7	0.1†	1.5	1.7	1.5	1.4
Imports	KLXE		11.7	12.1	14.8	19.6	20.3	18.4	15.1	15.9	17.8	19.8	21.2
Total	KLXF		115.8	111.8	107.5	113.8	104.8	86.6	64.9†	69.0	68.0	68.3	62.6
Change in colliery stocks	KLXG	KSOL	0.7	1.3	–0.9	2.8	2.2	1.8	–4.2	–4.2	–3.0	0.6	–0.3
Change in stocks at opencast sites	KLXH		0.5	1.5	–0.1	–0.8	0.6	0.5	–0.5				
Total supply	KLXI		114.6	109.0	108.5	111.8	102.0	84.5	69.6†	73.1	71.0	67.7	62.9
Home consumption													
Electricity supply industry[1]	KLXJ		84.3	82.1	84.0	83.6	78.5	66.1	62.4	59.6	54.9	47.0†	48.3
Coke ovens	KLXK		10.9	10.8	10.9	10.0	9.0	8.5	8.6	8.7	8.6	8.8	8.7
Low temperature carbonization plants	KLXL	KYAM	0.8	0.8	0.8	0.8	0.6	1.3	1.2	1.0	0.9	0.9	0.6
Manufactured fuel plants	KLXM		1.2	0.9	0.8	0.7	0.7						
Collieries	KLXO		0.2	0.1	0.1	0.1	0.1	–	–	–	–	–	–
Industry[2,3]	KLXQ		7.2	6.8	6.3	6.4	6.6	5.3	4.9	4.5	3.6	3.0†	2.7
Domestic:													
House coal[3,4]	KLXR	LUQM	3.7	3.1	2.5	2.7	2.9	2.7	1.9	1.2	1.3	1.3	2.2
Anthracite and dry steam coal[4,5,6]	KLXS		1.3	1.3	1.2	1.6	1.3	1.9	2.0	1.5	1.4	1.3	
Miners' coal	KLXT		0.8	0.7	0.6	0.5	0.4	–	–	–	–	–	–
Public services	KLXU		0.9	0.9	0.9	0.8	0.7	0.6	0.5	0.3	0.4	0.5	0.3
Miscellaneous	KLXV		0.2	0.2	0.3	0.3	0.2	0.2	0.2	0.2	0.2	0.2	0.1
Total home consumption	KLXW		111.5	107.6	108.3	107.5	100.6	86.6	81.7	77.0	71.3	63.3	62.9
Overseas shipments and bunkers	KLXX		1.8	2.0	2.5	1.7	1.0	1.1	1.2	0.9	1.0	1.1	0.9
Total consumption and shipments	KLXY		113.3	109.6	110.8	109.2	101.6	87.7	82.9	77.9	72.3	64.1†	63.8
Change in distributed stocks[7]	KLXZ		1.7	0.4	–0.4	3.6	1.2	–3.6	–13.9	–4.4	–0.8	3.3	–1.0
Balance[8]	KLYA		–0.4	1.0	1.9	–1.0	–0.8	0.4	–0.6†	–0.4	–0.5	0.3	0.1
Stocks at end of year													
Distributed[7]	KLYB		28.8	29.2	28.7	32.3	33.5	29.9	16.0	11.6	10.8	14.1	13.1
At collieries	KLYC	KSOM	5.6	6.9	6.0	8.8	10.9	12.7	8.5	7.1	4.2	4.8	4.6
At opencast sites	KLYD		1.7	3.2	3.0	2.2	2.8	3.3	2.8				
Total stocks	KLYE		36.2	39.2	37.8	43.3	47.2	45.9	27.3	18.7	14.9†	18.9	17.7

TIME SERIES. Figures relate to periods of 52 weeks.

For 1990, figures relate to 52 weeks estimate for period ended 29 December 1990.

1 Includes quantities used in the production of steam for sale.
2 Colliery and opencast disposals to industry.
3 Includes estimated proportions of steam coal imports.
4 Colliery and opencast disposals to merchants.
5 Including disposals of imported anthracite.
6 Anthracite is also consumed under other categories, including miners' coal and manufactured fuel plants.
7 Great Britain. Stock change excludes industrial and domestic stocks.
8 This is the balance between supply and consumption, shipments and changes in known distributed stocks.

Source: Department of Trade and Industry: 020 7215 2717

21.6 Fuel input and gas output: gas sales[5]

Public supply

		Units	1988	1989	1990	1991	1992	1993	1994	1995	1996	1997	1998
Fuel input to gas industry													
Petroleum	KLZA	Million tonnes	–	–	–	–	–	–	..	..	..	..	..
Petroleum gases[1]	KIJY	Giga-watt hours	29	88	29	41	51	50	52	..	..	..	..
Total to gas works	KIKB	"	176	88	29	41	51	50	..	..	..	..	..
Natural gas for direct supply	KIKC	"	559 267	549 450	568 037	616 194	619 921	699 050	724 832	777 483	908 072	..	..
Total fuel input	KIKD	"	559 443	549 538	568 066	616 235	619 972	702 041	729 426	..	..	..	..
Fuel input to gas industry		Million tonnes of coal or coal equivalent											
Petroleum	KLZH		–	–	–	–	–	–	..	..	..	..	..
Petroleum gases[1]	KLZI	"	–	–	–	–	–	–	..	..	..	..	..
Total to gas works	KLZL	"	–	–	–	–	–	–	..	..	..	..	..
Natural gas for direct supply	KLZM	"	76.3	75.0	77.5	84.1	84.6	95.8	99.5	..	..	..	..
Total fuel input	KLZN	"	76.3	75.0	77.5	84.1	84.6	95.8	99.5	..	..	..	..
Gas output and sales		Giga-watt hours											
Gas output:													
Town gas	KIKE	"	117	29	29	29	28	32	30	..	..	..	..
Natural gas supplied direct	KIKF	"	559 267	549 450	568 037	616 194	619 921	699 050	724 832	777 483	908 072	..	..
Gross total available	KIKG	"	559 385	549 479	568 066	616 223	619 949	699 082	724 862	..	..	..	..
Own use[2]	KIKH	"	–3 517	–2 960	–2 383	–3 007	–2 651	–2 930	–3 090	–3 311	–4 576	..	..
Statistical difference[3]	KIKI	"	–9 613	–8 763	2 638	362	–26 166	29 667	–15 842	..	..	..	..
Total sales	KIKJ	"	546 255	537 756	553 045	594 854	591 132	666 485	705 930	..	..	..	..
Analysis of gas sales													
Fuel producers													
Power stations[4]	KIKK	"	11 166	6 108	6 404	6 561	17 894	81 778	114 575	145 790	190 691	243 205†	257 828
Coal extraction and manufacture of solid fuels	KIKL	"	410	451	338	630	1 042	415	266	368	344	193	67
Coke ovens	KIKM	"	29	59	91	62	108	191	1	1	–	–	–
Petroleum refineries	KIKN	"	322	293	272	279	1 940	2 449	1 933	2 922	3 118	3 213	3 362
Nuclear fuel production	KIKO	"	440	410	422	496	508	565	550	467	874	923	989
Production and distribution of other energy	KIKP	"	117	193	145	270	447	178	114	352	437	487	549
Total final producers	KIKQ	"	12 485	7 514	7 672	8 298	21 939	85 576	117 439	149 900	195 464	248 021†	262 795
Final users:													
Iron and steel industry	KIKR	"	13 071	13 693	13 594	12 565	13 908	15 577	20 327	20 689	21 460†	20 577	20 574
Other industries	KIKS	"	132 380	141 319	146 122	139 488	132 936	132 719	143 979	149 097	164 520	161 445†	171 007
Domestic	KIKT	"	300 515	290 557	300 410	333 963	330 100	340 162	329 710	326 010	375 841	345 532	355 895
Public administration	KIKU	"	36 370	34 828	35 376	40 030	43 817	38 725	41 119	47 208	52 011	52 645	52 319
Agriculture	KIKV	"	879	938	1 001	1 087	1 286	1 277	1 227	1 210	1 420	1 443	1 344
Miscellaneous	KIKW	"	50 525	48 907	48 844	59 394	54 769	58 917	58 790	61 502	66 504	62 882†	64 242
Total final users	KIKX	"	533 740	530 242	545 347	586 527	576 816	587 377	595 152	605 715	682 255†	644 524	665 381
Total sales	KIKY	"	546 225	537 756	553 019	594 825	598 755	672 952	712 592	755 615	877 719†	892 545	928 176

Note : The breakdown of consumption by industrial users is made according to the 1980 Standard Industrial Classification, though only data from 1984 are available on this basis.

1 Butane, propane, ethane and refinery tail gases.

2 Used in works, offices, showrooms, etc.

3 Supply greater than recorded demand (-).Includes losses in distribution.

4 Pre-1987 auto-production of electricity is included under the relevant industry, post 1987 it is included under power stations. However, 1987 and 1988 are calculated on a different basis to 1989-1995.

5 It is no longer possible to present information on fuels input into the gas industry and gas output and sales in the same format as in the previous versions of this table. As such, users are directed to chapter 4 of the 1999 edition of the *Digest of United Kingdom Energy Statistics* published by the DTI, where more detailed information on gas production and consumption in the UK is available.

Source: Department of Trade and Industry: 020 7215 2717

21.7 Electricity: generation, supply and consumption
United Kingdom

Gigawatt-hours[6]

		1988	1989	1990	1991	1992	1993	1994	1995	1996	1997	1998
Electricity generated												
Major power producers[1]: total	KLUA	288 511	292 896	298 495	301 490	300 177	305 434	306 726	313 958	326 287	324 143	334 972
Conventional steam stations	KLUB	222 887	219 712	230 376	229 190	217 228	187 786	174 943	169 866	160 565	133 132	134 317
Combined cycle gas turbine stations	KJCS	..	..	..	312	2 991	22 811	36 971	48 720	65 880	86 974	93 832
Nuclear stations[7]	KLUC	63 456	71 734	65 749	70 543	76 807	89 353	88 282	88 964	94 671	98 146	100 140
Gas turbines and oil engines	KLUD	464	529	432	356	358	359	244	190	226	459	221
Hydro-electric stations:												
Natural flow	KLUE	4 171	4 002	4 393	3 777	4 591	3 522	4 317	4 096	2 801	3 337	4 240
Pumped storage	KLUF	2 121	1 910	1 982	1 523	1 697	1 437	1 463	1 552	1 556	1 486	1 624
Renewables other than hydro[2]	KLUG	1	2	4	3	45	165	506	570	588	609	590
Other generators[3]: total	KLUH	20 314	21 690	21 244	21 385	20 864	17 669	18 252	20 084†	21 096	21 235	23 276
Conventional steam stations[4]	KLUI	14 294	15 241	15 014	15 230	15 049	14 865	14 866	16 576	17 200	16 606†	16 559
Combined cycle gas turbine stations	KJCT	..	..	292	310	409	607	902	937	1 232	1 500	2 848
Hydro-electric stations (natural flow)	KLUK	762	747	814	847	840	780	777	742	560	790	986
Renewables other than hydro[2]	KILA	659	708	683	784	1 028	1 417	1 707	1 829†	2 104	2 641	3 532
All generating companies: total	KLUL	308 825	314 585	319 739	322 875	321 043	323 102	324 978†	334 042	347 383	345 378	358 248
Conventional steam stations[4]	KLUM	237 181	234 953	245 390	244 420	232 277	202 651	189 809	186 442	177 765	149 738†	150 876
Combined cycle gas turbine stations	KJCU	..	..	292	622	3 400	23 418	37 873	49 657	67 112	88 474	96 680
Nuclear stations	KLUN	63 456	71 734	65 749	70 543	76 807	89 353	88 282	88 964	94 671	98 146	100 140
Gas turbines and oil engines	KLUO	464	529	432	356	358	359	244	190	226	459	221
Hydro-electric stations:												
Natural flow	KLUP	4 933	4 749	5 207	4 624	5 431	4 302	5 094	4 838	3 361	4 127	5 226
Pumped storage	KLUQ	2 121	1 910	1 982	1 523	1 697	1 437	1 463	1 552	1 556	1 486	1 624
Renewables other than hydro[2]	KLUR	670	710	687	787	1 073	1 582	2 213	2 399†	2 692	3 250	4 122
Electricity used on works: Total	KLUS	20 226	20 263	19 611	20 111	20 237	19 287	17 491†	17 411	17 697	16 371	17 777
Major generating companies[1]	KLUT	18 694	18 610	17 891	18 424	18 485	17 391	16 696†	16 510	16 674	15 404	16 471
Other generators[3]	KLUU	1 532	1 653	1 720	1 687	1 752	1 896	795†	901	1 023	967	1 306
Electricity supplied (gross)												
Major power producers[1]: total	KLUV	269 817	274 286	280 604	283 066	281 692	287 264	290 030	297 448	309 612	308 812†	318 501
Conventional steam stations[5]	KLUW	211 502	208 675	218 957	217 947	205 897	178 312	166 883†	162 084	153 170	127 075	127 581
Combined cycle gas turbine stations	KJCV	..	..	..	309	2 964	22 611	36 815	48 525	65 604	86 682†	93 005
Nuclear stations	KLUX	55 642	63 602	58 664	62 761	69 135	80 979	79 962	80 598†	85 820	89 341	91 186
Gas turbines and oil engines[5]	KLUY	430	494	403	310	311	324	233	181	216	436	211
Hydro-electric stations:												
Natural flow	KLUZ	4 160	3 992	4 384	3 767	4 579	3 513	4 265	4 051	2 763	3 299	4 228
Pumped storage	KLVA	2 025	1 812	1 892	1 465	1 635	1 388	1 417	1 502	1 507	1 439	1 569
Renewables other than hydro[2]	KLVB	1	2	3	3	37	136	455	506	533	540	722
Other generators[3]: total	KLVC	18 782	20 037	19 524	19 698	19 112	16 522	17 457	19 183†	20 073	20 268	21 970
Conventional steam stations[4]	KLVD	13 442	14 326	14 082	14 312	14 033	13 836	14 183	15 815	16 340	15 583	..
Combined cycle gas turbine stations	KJCW	..	..	280	298	394	584	866	900	1 180	1 425	2 706
Hydro-electric stations (natural flow)	KLVF	669	654	806	839	832	772	769	730	554	751	880
Renewables other than hydro[2]	KIKZ	643	681	656	753	987	1 360	1 639	1 738†	1 999	2 509	3 123
All generating companies: total	KLVG	288 599	294 323	300 128	302 764	300 804	303 815	307 487	316 630†	329 686	329 080	340 471
Conventional steam stations[4,5]	KLVH	224 944	223 001	233 039	232 259	219 930	192 148	181 066†	177 899	169 510	142 950	143 411
Combined cycle gas turbine stations	KJCX	..	..	280	607	3 358	23 195	37 681	49 425	66 784	88 107†	95 711
Nuclear stations	KLVI	55 642	63 602	58 664	62 761	69 135	80 979	79 962	80 598	85 820	89 341	91 186
Gas turbines and oil engines[5]	KLVJ	430	494	403	310	311	324	233	181	216	436	210
Hydro-electric stations:												
Natural flow	KLVK	4 914	4 730	5 190	4 666	5 411	4 285	5 034	4 781	3 317	4 050	5 108
Pumped storage	KLVL	2 025	1 812	1 892	1 465	1 635	1 388	1 417	1 502	1 507	1 439	1 569
Renewables other than hydro[2]	KLVM	644	684	659	756	1 024	1 496	2 094	2 244†	2 532	3 049	3 845
Electricity used in pumping												
Major power producers[1]	KLVN	2 888	2 572	2 626	2 109	2 257	1 948	2 051	2 282	2 430	2 477	2 594
Electricity supplied (net): Total	KLVO	265 711	291 751	297 502	300 654	298 547	301 868	305 436	314 349†	327 254	326 603	337 877
Major power producers[1]	KLVP	266 929	271 714	277 978	280 956	279 435	285 316	287 979	295 166	307 181	306 335†	315 907
Other generators[3]	KLVQ	18 782	20 037	19 524	19 698	19 112	16 552	17 457	19 183†	20 073	20 268	21 970
Net imports	KGEZ	12 830	12 631	11 990	16 408	16 694	16 716	16 887	16 313	16 755†	16 574	12 468
Electricity available	KGIZ	298 540	304 382	309 408	317 062	315 241	318 584	322 323	330 662†	344 009	343 177	350 345
Losses in transmission etc	KGKW	24 036	24 983	24 988	26 221	23 788	22 838	30 541	28 432†	29 724	25 691	26 076
Electricity consumption: Total	KGKX	274 505	279 399	284 420	290 841	291 453	295 746	291 782	302 230	314 285	317 486	324 269
Fuel industries	KGKY	9 163	9 001	9 986	9 794	9 984	9 615	7 518	8 289	8 629	8 235	7 575
Final users: total	KGKZ	265 342	270 398	274 434	281 048	281 468	286 130	284 264	293 942	305 656	309 251	316 694
Industrial sector	KGLZ	97 143	99 417	100 643	99 570	95 277	96 842	95 067	99 909	103 129	104 743	107 226
Domestic sector	KGMZ	92 362	92 270	93 793	98 098	99 482	100 456	101 407	102 210	107 513	104 455	109 610
Other sectors	KGNZ	75 837	78 711	79 997	83 380	86 711	88 833	87 790	91 823	95 014	100 053	99 858

1 Generating companies corresponding to the old public sector supply system, ie AES Electric Ltd., Anglian Power Generation, Barking Power Ltd., British Nuclear Fuels plc., Coolkeeragh Power Ltd., Corby Power Ltd., Derwent Cogeneration Ltd., Eastern Merchant Generation Ltd., Elm Energy & Recycling (UK) Ltd., Fellside Heat and Power Ltd., Fibrogen Ltd., Fibropower Ltd., First Hydro Ltd., Humber Power Ltd., Hydro-Electric, Indian Queens Power Ltd., Keadby Generation Ltd., Lakeland Power Ltd., Magnox Electric plc., Medway Power Ltd., Midlands Power (UK) Ltd., National Power, NIGEN, Nuclear Electric, Peterborough Power Ltd., PowerGen, Premier Power Ltd., Regional Power Generators Ltd., Rocksavage Power Company Ltd., Scottish Nuclear, Scottish Power, South East London Combined Heat & Power Ltd., South Western Electricity, Teesside Power Ltd., Fibrothetford Ltd.

2 Including wind and biofuels.

3 Larger establishments in the industrial and transport sectors generating 1 Gigawatt-hour or more a year.

4 For other generators, conventional steam stations cover all types of station not separately listed

5 Prior to 1986 gas turbines and oil engines are included with conventional steam stations.

6 1 Gigawatt-hour equals 1 million kilowatt hours.

7 Nuclear generators are now included under "major power producers" only.

Source: Department of Trade and Industry: 020 7215 5190

21.8 Electricity: plant capacity and demand
United Kingdom

Megawatts

		1991 end-March	1992 end-March	1993 end-March	1994 end-March	1995 end-March	1996 end-March	1996 end-December[8]	1997 end-December[8]	1998 end-December[8]
Major power producers[1]:										
Total declared net capability	KGON	69 320	66 336	63 377	66 901	64 923	66 100	69 090†	68 288	68 390
Conventional steam stations	KGOO	51 365	48 309	44 860	41 143	38 453	38 242	38 230	37 395	35 039
Combined cycle gas turbine stations	KJCZ	..	229	1 129	5 463	8 364	9 034	12 052†	12 252	14 680
Nuclear stations[7,9]	KGOP	11 353	11 353	11 353	11 894	12 037	12 762	12 916	12 946	12 956
Gas turbines and oil engines	KGOQ	3 130	2 968	2 539	2 248	1 895	1 890	1 721†	1 526	1 492
Hydro-electric stations:										
Natural flow	KGOR	1 302	1 308	1 314	1 314	1 314	1 314	1 313	1 311	1 327
Pumped storage	KGOS	2 787	2 787	2 787	2 787	2 788	2 788	2 788	2 788	2 788
Renewables other than hydro	KGOT	3	2	15	52	72	71	70	70	108
Other generators[2]:										
Total capacity of own generating plant[3]	KGOU	4 205	4 210	4 122	3 622	3 818	3 910	4 182	4 310†	4 624
Conventional steam stations[4]	KGOV	3 279	3 211	3 037	3 096	3 257	3 311	3 407†	3 317	3 311
Combined cycle gas turbine stations	KJDA	76	102	150	150	153	151	190	323†	569
Hydro-electric stations (natural flow)	KGOX	108	122	110	111	111	114	144	183	145
Renewables other than hydro	KILB	122	155	205	265	297	334	441†	487	599
All generating companies: Total capacity[3]	KGOY	73 525	70 546	67 499	68 523	68 741	70 011	73 272†	72 598	73 014
Conventional steam stations[4]	KGOZ	54 644	51 520	47 897	44 239	41 710	41 553	41 637†	40 712	38 350
Combined cycle gas turbine stations	KJDC	76	331	1 279	5 613	8 517	9 185	12 242†	12 575	15 249
Nuclear stations[7]	KGPM	11 353	11 353	11 353	11 894	12 037	12 762	12 916	12 946	12 956
Gas turbines and oil engines	KGPN	3 130	2 968	2 539	2 248	1 895	1 890	1 721†	1 526	1 492
Hydro-electric stations:										
Natural flow	KGPO	1 410	1 430	1 424	1 425	1 425	1 428	1 457	1 494	1 472
Pumped storage	KGPP	2 787	2 787	2 787	2 787	2 788	2 788	2 788	2 788	2 788
Renewables other than hydro	KGPQ	125	157	220	317	369	405	511†	557	707
Major power producers[1]:										
Simultaneous maximum load met[5]	KGPR	54 068	54 472	51 663	54 848	52 362	55 611	56 815	56 965	56 312
Major power producers[1]:										
System load factor[6] (percentage)	KGQY	*62.2*	*62.9*	*66.6*	*63.8*	*67.3*	*65.4*	*65.5†*	*65.4*	*67.2*

1 See footnote 1 to Table 21.7.
2 See footnote 3 to Table 21.7.
3 Capacity figures for other generators are as at end-December of the previous year.
4 For other generators, conventional steam stations cover all types of stations not separately listed.
5 Maximum load in year to end of March.
6 The average hourly quantity of electricity available during the year ended March expressed as a percentage of the maximum demand.
7 The 1995 figure includes 300 MW of the 1 188 MW capacity of Sizewell B which began to produce electricity in March 1995.
8 From 1996 data are on a calendar year basis.
9 Nuclear generators are now included under "major power producers" only.

Source: Department of Trade and Industry: 020 7215 5190

21.9 Electricity: fuel used in generation
United Kingdom

Million tonnes of oil equivalent

		1988	1989	1990	1991	1992	1993	1994	1995	1996	1997	1998
Major power producers[1]: total all fuels	KGPS	73.20	74.10	72.20	72.90	71.30	72.32	71.65	72.71	74.67†	73.19	74.92
Coal	FTAJ	51.9	50.7	49.0	49.0	46.0	38.3	35.9	35.0	32.4†	27.7	28.9
Oil[2]	FTAK	5.4	5.6	6.8	5.9	5.0	4.4	3.6	3.1	3.0	1.2	0.9
Gas[3]	KGPT	–	–	–	–	1.0	6.3	9.1	11.4	15.2	19.3	20.3
Nuclear	FTAL	15.7	17.7	16.3	17.4	18.5	21.6	21.2	21.3	22.2	23.0	23.3
Hydro (natural flow)	FTAM	0.3	0.3	0.4	0.3	0.4	0.3	0.4	0.4	0.3	0.3	0.4
Other fuels used by UK companies[3]	KGPU	–	–	–	–	–	0.1	0.1	0.1	0.2	0.2	0.1
Net imports	KGPV	1.1	1.1	1.0	1.4	1.4	1.4	1.5	1.4	1.4	1.4	1.1
Other generators[4]: total all fuels	KGPW	5.6	6.1	5.2	5.4	6.7	4.5	3.5	3.8	3.8	4.7†	4.6
Transport undertakings												
Gas	KGPX	0.2	0.2	0.2	0.2	0.2	0.2	0.2	0.2	0.2	0.2	0.3
Undertakings in industrial sector												
Coal	KGPY	1.2	0.9	1.0	1.0	1.0	1.3	1.2	1.1	1.0	1.1†	1.1
Oil	KGPZ	1.2	1.6	1.6	1.7	3.1	1.4	0.5	0.5	0.5	0.7†	0.6
Gas	KGQM	0.7	0.3	0.3	0.3	0.3	0.8	0.6†	0.9	1.0	1.4	1.6
Hydro (natural flow)	KGQO	0.1	0.1	0.1	0.1	0.1	0.1	0.1	0.1	–	–	–
Other fuels	KGQP	1.0	1.7	1.8	1.9	1.1	1.0	1.0	1.0	1.1	1.2	1.0
All generating companies[4]: total fuels	KGQQ	79.4	80.2	77.4	78.3	78.0	76.8	75.2	76.5	78.4†	77.9	79.5
Coal	KGQR	53.1	51.6	50.0	50.0	46.9	39.6	37.1	36.1	33.4†	28.9	30.0
Oil	KGQS	7.1	7.1	8.4	7.6	8.1	5.8	4.1	3.6	3.5	1.9	1.4
Gas[3]	KGQT	1.0	0.5	0.6	0.6	1.5	7.0	9.9	12.5	16.4	20.9	22.2
Nuclear	KGQU	15.7	17.7	16.3	17.4	18.5	21.6	21.2	21.3	22.2	23.0	23.3
Hydro (natural flow)	KGQV	0.4	0.4	0.4	0.4	0.5	0.4	0.4	0.5	0.3	0.4	0.4
Other fuels used by UK companies[3,5]	KGQW	1.0	1.7	1.8	0.9	1.1	1.0	1.1	1.2	1.2	1.4	1.2
Net imports	KGQX	1.1	1.1	1.0	1.4	1.4	1.4	1.5	1.4	1.4	1.4	1.1

1 See footnote 1 to Table 21.7.
2 Includes oil used in gas turbine and diesel plant for lighting up coal fired boilers and Orimulsion.
3 For 1990 and 1991 gas used by major power producers was included with other fuels for reasons of confidentiality.
4 See footnote 3 to Table 21.7. Prior to 1987 'other generators' covers only transport undertakings and industrial hydro and nuclear stations. For years 1987 to 1989 figures for fuel used by other generators are largely estimated.
5 Main fuels included are coke oven gas, blast furnace gas, waste products from chemical processes and sludge gas.
6 Nuclear generators are now included under "major power producers" only.

Source: Department of Trade and Industry: 020 7215 5190

21.10 Indigenous production, refinery receipts, imports and exports of oil[1]

Thousand tonnes

		1988	1989	1990	1991	1992	1993	1994	1995	1996	1997	1998
Total indigenous petroleum production[2]	KMBA	114 459	91 710	91 064	91 260	94 251	100 188	126 812†	129 894	129 741	128 234	132 633
Crude petroleum[3]:												
Refinery receipts total	KMBB	83 925	88 840	89 735	92 523	92 789	97 400	93 771	93 572	96 721	97 032	94 071
Indigenous[4]	KMBC	40 582	39 585	37 754	35 932	35 472	36 680	42 174	44 872	49 449	47 589	46 382
Other[5]	KMBD	730	904	916	772	832	852	427	1 110	997	794	1 255
Net foreign arrivals[6]	KMBE	42 613	48 351	51 065	55 819	56 485	59 868	51 170	47 590	48 275	48 649	46 434
Foreign trade												
Imports[6]	KMBF	44 272	49 500	52 710	57 084	57 683	61 701	53 096	48 749	50 099	49 994	47 957
Exports												
Indigenous	KMBG	69 965	49 130	54 022	52 396	54 441	60 165	77 466	77 926	76 937	74 708	77 765
Other[7]	KMBH	1 967	1 332	1 878	1 406	1 536	2 225	2 359	1 761	2 219	1 805†	2 051
Petroleum products												
Foreign trade												
Imports[6]	KMBI	9 219	9 479	11 005	10 140	10 567	10 064	10 441	9 878	9 316	8 705†	11 327
Exports[6]	KMBJ	17 176	17 873	18 002	20 677	21 899	24 890	24 644	24 418	26 018	29 118	26 895
Net imports[6]	KMBK	–7 957	–8 394	–6 997	–10 537	–11 332	–14 826	–14 203	–14 540	–16 702	20 413†	–15 568
Bunkers[8]	KMBL	1 831	2 396	2 538	2 486	2 546	2 478	2 313	2 465	2 665	2 962	3 080

1 The term 'indigenous' is used in this table for convenience to include oil from the UK Continental Shelf as well as the small amounts produced on the mainland.
2 Crude oil *plus* condensates and petroleum gases derived at onshore treatment plants.
3 Includes process(partly refined) oils.
4 Includes condensate for distillation.
5 Mainly recycled products.
6 Foreign trade as recorded by the petroleum industry and may differ from figures published in *Overseas Trade Statistics*.
7 Re-exports of imported crude which may include some indigenous oil in blend.
8 International marine bunkers.

Source: Department of Trade and Industry: 020 7215 5184

21.11 Throughput of crude and process oils and output of refined products from refineries[1]

Thousand tonnes

		1988	1989	1990	1991	1992	1993	1994	1995	1996	1997	1998
Throughput of crude and process oils	KMAU	85 662	87 699	88 692	92 001	92 334	96 274	93 162	92 743	96 660	97 024	93 797
less: Refinery fuel[2]:	KMAA	5 484	5 816	5 838	6 058	6 080	6 383	6 256	6 481	6 623	6 572	6 468
Losses	KMAB	340	491	568	467	471	308	261	129	152	86	233
Total output of refined products	KMAC	79 838	81 392	82 286	85 476	85 783	89 583	86 645	86 133	89 885	90 366	87 096
Gases:												
Butane and propane	KMAE	1 581	1 568	1 514	1 664	1 583	1 575	1 605	1 815	1 828	1 950	1 978
Other petroleum	KMAF	68	90	106	134	172	162	132	133	144	139	200
Naphtha and other feedstock	KMAG	1 856	2 073	2 139	2 515	3 040	2 696	2 794	2 711	2 824	2 854	2 335
Aviation spirit	KMAH	–	–	–	–	–	–	–	–	–	–	–
Wide-cut gasoline	KMAI	–	–	–	–	–	–	–	–	–	–	–
Motor spirit	KMAJ	26 409	27 237	26 724	27 793	27 980	28 394	27 562	27 254	28 046	28 260	27 392
Industrial and white spirit	KMAK	112	105	121	136	150	159	143	143	136	128	136
Kerosene:												
Aviation turbine fuel	KMAL	6 725	7 092	7 541	7 037	7 681	8 341	7 697	7 837	8 305	8 342	7 942
Burning oil	KMAM	2 344	2 309	2 344	2 446	2 450	2 707	2 967	2 924	3 510	3 336	3 471
Gas/diesel oil	KMAN	23 925	23 292	23 402	26 057	25 650	27 361	27 137	27 169	28 903	28 778	27 859
Fuel oil	KMAO	12 495	13 020	13 805	13 205	12 388	13 183	11 378	10 969	11 479	11 747	11 066
Lubricating oil	KMAP	970	1 050	974	973	1 163	1 264	1 296	1 261	1 111	1 231	1 134
Bitumen	KMAQ	2 296	2 393	2 454	2 302	2 336	2 450	2 569	2 459	2 189	2 258	2 190
Petroleum wax	KMAR	63	54	40	37	62	59	64	46	41	65	59
Petroleum coke	KMAS	541	564	586	555	535	621	679	759	714	598	694
Other products	KMAT	509	510	569	620	593	613	623	653	655	680	630

1 Crude and process oils comprise all feedstocks, other than distillation benzines, for treatment at refinery plants. Refinery production does not cover further treatment of finished products for special grades such as in distillation plant for the preparation of industrial spirits.
2 Comprising 3 126 thousand tonnes gases, 2 300 thousand tonnes fuel oil and 1 196 thousand tonnes other products in 1996.

Source: Department of Trade and Industry: 020 7215 5184

21.12 Deliveries of petroleum products for inland consumption

Thousand tonnes

		1988	1989	1990	1991	1992	1993	1994	1995	1996	1997	1998
Total (including refinery fuel)	KMCA	77 801	78 844	79 781	80 564	81 550	82 173	81 213	80 175	82 013	79 073	78 437
Total (excluding refinery fuel)	KMCB	72 317	73 028	73 943	74 506	75 470	75 970†	74 957	73 694	75 390	72 501	71 969
Gases:												
Butane and propane:												
For gas works	KMCC	35	33	37	42	40	41	47	43	48	45	46
Other uses	KMCD	1 052	1 045	1 009	1 172	1 049	1 094	1 153	1 158	1 272	1 148	1 085
Other gases:												
For gas works	KMCE	–	–	–	–	–	–	–	–	–	–	–
Other uses	KMCF	59	49	48	63	58	48	88	98	101	73	69
Feedstock:												
For petroleum chemical plants	KMCG	5 942	5 816	5 115	5 935	5 970	5 921	6 199	6 227	6 228	6 088	6 321
Naphtha (LDF) for gasworks[1]	KMCH	10	–	–	–	–	–	–	–	–	–	–
Aviation spirit	KMCI	28	29	26	24	27	27	29	29	32	37	36
Wide-cut gasoline	KMCJ	–	–	–	–	–	–	–	–	–	–	–
Dealers:												
4 star	KMCK	20 599	18 325	15 586	13 793	12 481	11 046	9 503	7 973	7 043	6 138	4 595
3 star	KMCL	17	2	–	–	–	–	–	–	–	–	–
2 star	KMCM	1 669	321	–	–	–	–	–	–	–	–	–
Unleaded	KMCN	247	4 543	8 060	9 587	10 942	12 193	12 859	13 529	14 926	15 694	12 696
Commercial consumers:												
4 star	KMCO	548	582	472	360	294	217	178	149	135	112	91
3 star	KMCP	51	14	–	–	–	–	–	–	–	–	–
2 star	KMCQ	109	31	–	–	–	–	–	–	–	–	–
Unleaded	KMCR	11	105	197	282	326	311	303	302	305	307	321
Motor spirit: total	BHOD	23 251	23 923	24 315	24 022	24 043	23 767	22 843	21 953	22 409	22 251	17 703
Industrial and white spirits	KMCS	246	207	171	162	159	164	170	178	184	195	179
Kerosene:												
Aviation turbine fuel	BHOE	6 200	6 564	6 589	6 176	6 666	7 106	7 284	7 660	8 049	8 411	9 241
Burning oil	KMCT	1 992	1 937	2 058	2 383	2 472	2 625	2 655	2 774	3 336	3 343	3 575
Gas/diesel oil:												
Derv fuel	BHOI	9 370	10 118	10 652	10 694	11 132	11 806	12 914	13 457	14 365	14 976	15 143
Other	BHOJ	8 456	8 323	8 046	8 031	7 871	7 782	7 491	7 227	7 631	7 326	7 244
Fuel oil	BHOK	11 865	11 125	11 997	11 948	11 481	10 770	9 275	7 975	6 854	3 936	2 935
Lubricating oils	BHOL	849	839	822	759	786	806	795	895	864	872	813
Bitumen	BHOM	2 342	2 423	2 491	2 514	2 555	2 523	2 595	2 420	2 146	2 015	1 967
Petroleum wax	KMCU	67	77	55	49	47	48	47	44	44	44	18
Petroleum coke	KMCV	136	120	112	154	682	778	911	1 008	1 210	1 095	887
Miscellaneous products	KMCW	419	399	403	380	433	484	461	549	617	644	538

1 Including a small quantity supplied for use as fuel by other consumers.

Source: Department of Trade and Industry: 020 7215 5184

21.13 Iron and steel
Summary of steel supplies, deliveries and stocks
United Kingdom

		Finished product weight - thousand tonnes										
		1988	1989	1990	1991	1992	1993	1994	1995	1996	1997	1998
Supply, disposal and consumption												
UK producers' home deliveries	KLTA	10 780	10 907	9 711	7 951	7 315	7 567	7 827	8 256	8 378	8 612	8 255
Imports excl steelworks receipts	KLTB	4 228	4 424	4 446	4 509	4 418	4 132	5 012	5 384	5 136	5 760	6 411
Total deliveries to home market (a)	KLTC	15 007	15 331	14 157	12 460	11 733	11 698	12 839	13 640	13 514	14 372	14 666
Total exports (producers, consumers, merchants)	KLTD	6 168	6 180	6 550	7 444	7 718	7 621	8 120	8 228	8 906	9 062	7 996
Exports by UK producers	KLTE	5 902	5 973	6 370	7 082	7 587	7 536	7 873	7 828	8 297	8 528	7 876
Derived consumers' and merchants' exports (b)	KLTF	266	207	180	362	131	85	247	400	609	534	120
Net home disposals (a)-(b)	KLTG	14 742	15 124	13 977	12 098	11 602	11 614	12 592	13 240	12 905	13 838	14 546
Consumers' and merchants' stock change[1]	KLTH	30	220	−290	−400	60	60	390	..	..	..	..
Estimated home consumption	KLTI	14 712	14 904	14 267	12 498	11 542	11 554	12 202	13 240	12 905	13 838	14 546
Stocks												
Producers												
- ingots & semis	KLTJ	1 311	1 182	1 245	1 035	933	1 005	946	1 068	767	946	714
- finished steel	KLTK	1 633	1 692	1 563	1 719	1 573	1 425	1 389	1 274	1 515	1 358	1 503
Consumers	KLTL	1 480	1 700	1 590	1 400	1 410	1 300	1 470	..	..	..	..
Merchants	KLTM	1 230	1 230	1 050	840	890	1 060	1 280	..	..	..	..

		Crude steel equivalent - million tonnes										
		1988	1989	1990	1991	1992	1993	1994	1995	1996	1997	1998
Estimated home consumption												
Crude steel production[2]	KLTN	18.95	18.74	17.84	16.47	16.21	16.62	17.28	17.60	17.99	18.50	17.32
Producers' stock change[1]	KLTO	0.06	−0.09	−0.08	−0.07	−0.30	−0.09	−0.12	0.01	−0.07	0.03	−0.11
Re-usable material	KLTP	0.08	0.10	0.10	0.07	0.06	0.08	0.09	0.08	0.07	0.06	0.02
Total supply from home sources	KLTQ	18.97	18.93	18.02	16.61	16.57	16.79	17.49	17.67	18.13	18.53	17.45
Total imports[3]	KLTR	5.95	6.25	5.96	6.21	5.99	5.44	6.58	7.05	7.01	7.49	8.38
Total exports[3]	KLTS	7.39	7.49	7.66	8.72	9.07	8.95	9.55	9.63	10.26	10.43	9.25
Net home disposals	KLTT	17.53	17.69	16.32	14.10	13.49	13.28	14.52	15.09	14.88	15.59	16.58
Consumers' and merchants' stock change[1]	KLTU	0.04	0.29	−0.37	−0.50	0.07	0.07	0.48	..	..	..	..
Estimated home consumption	KLTV	17.49	17.40	16.69	14.60	13.42	13.21	14.04	15.09	14.88	15.59	16.58

TIME SERIES. The figures relate to periods of 52 weeks (53 weeks in 1992).

1 Increases in stock are shown as + and decreases in stock (ie deliveries from stock) as -.
2 Includes liquid steel for castings.
3 Based on HM Customs Statistics, reflecting total trade rather than producers' trade.

Source: Iron and Steel Statistics Bureau: 020 8686 9050 ext 126

21.14 Iron and steel
Iron ore, manganese ore, pig iron and iron and steel scrap
United Kingdom

Thousand tonnes

		1988	1989	1990	1991	1992	1993	1994	1995	1996	1997	1998
Iron ore												
Jurassic	KLOA	222	32	53	57	29	–	–	–	–	–	–
Hematite	KLOB	2	–	–	–	–	–	–	–	–	–	–
Production: total	KLOC	224	32	53	57	29	–	–	–	–	–	–
Home produced	KLOD	237	37	53	57	29	–	–	–	–	–	–
Imported	KLOE	19 233	18 563	17 935	17 833	17 235	17 507	18 161	18 670	19 720	20 820	19 532
Consumption: total	KLOF	19 470	18 600	17 988	17 890	17 264	17 507	18 161	18 670	19 720	20 820	19 532
Manganese ore												
Consumption	KLOG	307	339	340	383	308	152	64	32	48	37	22
Pig iron (and blast furnace ferro-alloys)												
Average number of furnaces in blast during period	KLOH	*11*	*11*	*10*	*8*	*8*	*8*	*8*	*8*	*9*	*9*	*9*
Production												
Steelmaking iron	KLOI	12 943	12 551	12 218	11 834	11 469	11 534	11 943	12 236	12 830	13 054	12 746
Foundry iron	KLOJ	113	87	102	50	73	–	–	–	–	–	–
Speigeleisen and ferro-manganese	KLOK	107	143	143	178	137	45	–	–	–	–	–
In blast furnaces: total	KLOL	13 163	12 781	12 463	12 062	11 679	11 579	11 943	12 236	12 830	13 054	12 746
In steel works and steel foundries	KLOM	13 022	12 771	12 358	11 836	11 463	11 554	11 889	12 121	12 753	13 044	12 746
In iron foundries	KLON	146	146	195	181	214	..	..	..	..	..	..
Consumption of pig iron: total	KLOO	13 168	12 917	12 553	12 017	11 677	11 554	11 889	12 121	12 753	13 044	12 746
Pig iron stocks (end of period)	KLOP	95	85	..	..	..	..	..	..	..	..	..
Iron and steel scrap												
Steelworks and steel foundries												
Circulating scrap	KLOQ	2 846	2 817	2 489	2 332	2 449	2 303	2 326	2 390	2 639	2 459	2 380
Purchased receipts	KLOR	4 654	5 172	4 730	3 694	3 739	4 149	4 533	4 688	4 130	5 418	4 045
Consumption	KLOS	7 700	7 904	7 251	6 085	6 190	6 550	6 874	7 012	6 828	7 207	6 408
Stocks (end of period)	KLOT	382	467	430	365	365	267	253	319	260	236	253
Iron foundries												
Arisings	KLOU	418	453	693	582	587	..	..	..	..	..	..
Consumption	KLOV	1 166	1 137	1 609	1 457	1 556	..	..	..	..	..	..
Stocks (end of period)	KLOW	65	64	..	..	..	..	..	..	..	..	..

TIME SERIES. The figures relate to periods of 52 weeks (53 weeks in 1992).

Source: Iron and Steel Statistics Bureau: 020 8686 9050 ext 126

21.15 Iron and steel
Number of furnaces and production of steel

United Kingdom

		Unit	1988	1989	1990	1991	1992	1993	1994	1995	1996	1997	1998
Steel furnaces (in existence at end of period)[1]		Number											
Total	KLPA	"	259	229	214	209	206	202	202	192	192	192	190
Open hearth	KLPB	"	–	–	–	–	–	–	–	–	–	–	–
Oxygen converters	KLPC	"	14	14	14	14	11	11	11	11	11	11	11
Electric	KLPD	"	248	215	200	195	195	191	191	181	181	181	179
Stock and tropenas	KLPE	"	–	–	–	–	–	–	–	–	–	–	–
Production of crude steel (thousand tonnes)													
Total	KLPF	"	18 950	18 740	17 841	16 474	16 212	16 625	17 286	17 604	17 992	18 499	17 315
by process													
Open hearth	KLPG	"	–	–	–	–	–	–	–	–	–	–	–
Oxygen converters	KLPH	"	14 008	13 627	13 169	12 540	12 092	12 330	12 909	13 082	13 758	13 986	13 426
Electric	KLPI	"	4 942	5 113	4 672	3 934	4 120	4 295	4 377	4 522	4 234	4 513	3 889
Stock and tropenas	KLPJ	"	–	–	–	–	–	–	–	–	–	–	–
by cast method													
Cast to ingot	KLPK	"	5 361	3 469	2 692	2 201	2 083	2 140	2 033	2 174	1 892	1 660	784
Continuously cast	KLPL	"	13 356	15 031	14 909	14 085	13 958	14 319	15 079	15 250	15 912	16 653	16 346
Steel for castings	KLPM	"	233	240	240	189	171	166	174	180	188	186	185
by quality													
Non alloy steel	KLPN	"	17 610	17 371	16 641	15 496	15 195	15 558	16 062	16 243	16 708	17 193	16 145
Stainless and other alloy steel	KLPO	"	1 340	1 369	1 200	978	1 017	1 067	1 224	1 361	1 284	1 306	1 170
Production of finished steel products (all quantities)[2]		Thousand tonnes											
Rods and bars for reinforcement (in coil and lengths)	KLPP	"	1 168	1 219	1 241	1 151	1 130	1 229	1 269	1 154	1 182	1 118	1 012
Wire rods and other rods and bars in coil	KLPQ	"	1 349	1 417	1 407	1 272	1 327	1 427	1 524	1 642	1 536	1 565	1 588
Hot rolled bars in lengths	KLPR	"	1 067	1 243	1 336	1 192	1 178	1 140	1 275	1 311	1 499	1 716	1 529
Bright steel bars[3]	KLPS	"	342	340	318	257	281	295	363	424	357	385	336
Light sections other than rails	KLPT	"	335	326	325	324	349	294	306	286	298	302	224
Heavy and light rails and accessories	KLPU	"	395										
Other heavy sections	KLPV	"	2 100										
	KGQZ			2 540	2 583	2 297	2 361	2 408	2 412	2 549	2 557	2 397	2 280
Hot rolled plates, sheets and strip in coil and lengths	KLPW	"	8 051	8 076	8 089	7 270	7 145	7 230	7 715	8 077	8 512	8 956	8 535
Cold rolled plates and sheets in coil and lengths	KLPX	"	3 952	3 976	3 749	3 592	3 532	3 635	3 835	4 100	4 221	4 437	4 288
Cold rolled strip[3]	KLPZ	"	394	388	357	286	222	229	243	267	246	255	259
Tinplate	KLQW	"	1 018	960	855	742	805	829	767	791	739	754	769
Other coated sheet	KLQX	"	1 603	1 568	1 657	1 647	1 795	1 865	2 021	2 306	2 366	2 534	2 584
Tubes and pipes[3]	KLQY	"	1 482	1 519	1 465	1 260	1 187	1 155	1 136	1 183	1 317	1 310	1 276
Forged bars[3]	KLQZ	"	17	16	16	7	4	2	2	3	3	3	3

TIME SERIES. The figures relate to periods of 52 weeks (53 weeks in 1992).

1 Includes steel furnaces at steel foundries.
2 Includes material for conversion into other products listed in the table.
3 Based on producers' deliveries.

Source: Iron and Steel Statistics Bureau: 020 8686 9050 ext 126

21.16 Non-ferrous metals

Thousand tonnes

		1987	1988	1989	1990	1991	1992	1993	1994	1995	1996	1997
Copper												
Production of refined copper:												
Primary	KLAA	54.0	49.3	48.6	47.0	16.6	10.4	10.7	11.1	12.0	13.0	9.1
Secondary	KLAB	68.3	74.7	70.4	74.6	53.5	31.7	35.9	35.6	43.0	43.6	51.3
Home consumption:												
Refined	KLAC	327.7	327.7	324.7	317.2	269.4	308.3	325.0	377.3	397.9	396.0	408.5
Scrap (metal content)	KLAD	137.7	132.1	129.7	126.3	118.5	83.2	77.9	88.0	81.0	81.0	69.0
Stocks (end of period)[1,2]	KLAE	11.7	12.2	14.5	11.7	9.3	9.7	9.3	8.1	7.5	6.6	12.8
Analysis of home consumption (refined and scrap):[3] total	KLAF	465.4	459.8	454.4	443.5	387.9	391.5	403.0	468.0	493.2	477.3	488.0
Wire[4]	KLAG	220.3	217.8	221.7	220.2	191.9	247.0	253.9	306.2	321.4	309.4	312.5
Rods, bars and sections	KLAH	62.5	57.3	55.5	54.4	52.7	51.5	53.2	54.9	59.0	58.3	58.3
Sheet, strip and plate	KLAI	62.5	66.3	59.2	54.4	37.1	31.0	30.7	33.0	37.1	34.0	36.5
Tubes	KLAJ	79.1	77.5	77.7	73.0	65.5	62.0	65.2	73.9	75.7	75.6	70.1
Castings and miscellaneous	KLAK	41.0	40.9	40.4	41.5	40.7	..	..	..	..	..	..
Zinc												
Slab zinc:												
Production	KLAL	81.4	76.0	79.8	93.3	100.7	96.8	102.4	101.3	106.0	96.9	107.7
Home consumption	KLAM	188.1	192.8	194.5	193.0	183.7	190.1	195.9	196.5	198.4	195.7	194.8
Stocks (end of period)	KLAN	14.0	13.2	13.9	12.2	11.2	11.4	11.4	10.5	9.8	10.5	10.1
Other zinc (metal content):												
Consumption	KLAO	52.6	51.7	49.6	52.4	49.5	46.7	45.4	45.0	46.8	41.3	41.5
Analysis of home consumption (slab and scrap): total	KLAP	240.7	244.4	244.1	245.4	233.2	236.8	241.3	241.5	245.2	237.1	236.3
Brass	KLAQ	55.8	54.1	51.1	51.4	47.2	40.9	41.6	42.6	45.2	39.1	41.6
Galvanized products	KLAR	95.8	101.4	104.3	104.7	96.6	103.4	105.1	107.4	110.7	110.3	108.4
Zinc sheet and strip	KLAS	4.0	4.1	4.0	3.6	3.6	4.1	4.0	4.6	3.0	3.0	3.3
Zinc alloy die castings	KLAT	44.0	43.5	43.1	44.4	45.4	45.2	46.5	46.5	46.5	46.5	46.5
Zinc oxide	KLAU	22.8	23.2	21.7	21.1	20.3	22.5	20.7	21.6	21.6	20.7	20.6
Other products	KLAV	18.3	18.2	19.8	20.2	20.2	20.8	23.4	18.8	18.2	17.5	16.1
Refined lead												
Production[5,6]	KLAW	347.0	373.8	350.0	329.4	311.0	346.8	363.8	352.5	320.7	351.4	384.1
Home consumption[6,7]												
Refined lead	KLAX	287.5	302.5	301.3	301.6	263.8	263.6	263.6	267.6	285.4	272.8	270.4
Scrap and remelted lead[6]	KLAY	36.6	37.0	35.0	32.5	33.6	38.7	35.2	38.5	41.6	43.4	39.1
Stocks (end of period)[8]												
Lead bullion	KLAZ	26.1	18.6	17.0	18.0	22.8	36.3	20.7	10.2	9.5	32.9	15.5
Refined soft lead at consumers	KLBA	24.4	26.7	25.7	22.3	21.8	24.5	25.0	23.5	24.9	28.8	29.1
In LME Warehouses (UK)	KLBB	2.8	–	11.0	12.1	8.0	9.0	9.5	6.1	0.4	3.0	2.4
Analysis of home consumption (refined and scrap): total	KLBC	324.1	339.4	336.4	334.0	297.4	302.3	298.8	306.1	327.0	316.2	309.5
Cables	KLBD	10.0	11.0	12.6	10.4	8.6	9.3	8.9	9.3	9.8	9.8	9.7
Batteries (excluding oxides)	KLBE	48.0	49.2	50.5	51.0	51.9	53.4	48.7	52.5	52.7	52.3	54.7
Oxides and compounds:												
Batteries	KLBF	49.0	51.4	51.7	52.8	54.2	53.1	53.9	55.2	56.2	54.9	56.1
Other uses[9]	KLBG	82.7	76.8	72.6	73.7	55.0	55.6	56.3	53.9	53.8	56.1	54.5
Sheets and pipes	KLBH	86.1	102.4	98.1	96.8	79.8	82.8	82.7	84.6	101.2	94.1	91.1
White lead	KLBI	..	..	..	..	..	..	..	..	..	..	..
Solder	KLBJ	7.4	9.0	9.1	8.0	7.7	7.4	7.4	7.4	7.4	7.4	7.4
Alloys	KLBK	13.6	13.5	13.4	14.0	12.2	14.7	14.1	15.3	15.9	12.1	9.4
Other uses	KLBL	27.3	26.1	28.4	27.5	28.0	26.2	26.8	27.9	30.0	29.5	26.6

See footnotes on the second part of this table.

Source: World Bureau of Metal Statistics: 01920 461274

21.16 Non-ferrous metals

continued

Thousand tonnes

		1988	1989	1990	1991	1992	1993	1994	1995	1996	1997	1998
Tin												
Tin ore (metal content):												
Production	KLBM	3.4	4.0	3.4	2.3	2.0	2.2	1.9	2.1	2.3	0.4	..
Tin metal[11]:												
Production[12]	KLBO	16.8	10.8	12.0	5.2	–	–	–	–	–	–	..
Home consumption[12]	KLBT	10.2	10.2	10.4	10.3	10.4	10.4	10.4	10.5	10.5	10.1	..
Exports and re-exports[13]	KLBQ	14.0	5.4	5.7	2.9	0.2	0.3	1.2	2.7	0.6	0.3	..
Stocks (end of period):												
Consumers	KLBR	1.0	1.0	1.0	1.0	1.0	1.0	1.0	1.0	1.0	1.0	..
Merchants and others	KLBS	..	..	..	..	..	..	..	..	..	..	..
Analysis of home consumption (excluding scrap): total	KLBT	10.2	10.2	10.4	10.3	10.4	10.4	10.4	10.5	10.5	10.1	..
Tinplate	KLBU	3.4	3.6	3.6	3.6	3.6	3.6	3.6	3.6	3.6	3.6	..
Alloys	KLBV	3.4	3.4	3.4	3.3	3.4	3.4	3.3	3.4	3.5	3.6	..
Solder	KLBW	1.0	1.0	1.0	1.1	1.1	1.1	1.1	1.1	1.1	1.1	..
Other uses	KLBX	2.3	2.2	2.4	2.3	2.3	2.3	2.4	2.4	2.3	2.3	..
Primary aluminium[14]												
Production	KLBY	300.2	297.3	289.8	293.5[13]	244.2	239.1	231.2	237.9	240.0	247.7	258.4
Despatches to consumers	KLBZ	541.3	494.2	520.2	485.4	587.2	815.3	610.6	459.4	593.7	456.4	648.5
Secondary aluminium												
Production	KLCA	200.0	220.0	201.4	195.1	197.3	236.2	224.3	229.7	261.0	242.7	274.8
Exports	KLCC	45.0	61.2	59.7	64.8	83.5	98.3	124.8	145.8	152.2	153.3	156.6
Fabricated aluminium												
Total despatches[15]	KLCD	544.4	544.8	547.7	518.0	598.2	592.0	643.1	648.3	633.5	663.2	668.5
Rolled products[16]	KLCE	225.7	236.2	261.0	269.4	328.9	309.3	346.4	359.2	327.9	350.4	352.5
Extrusions and tubes[17]	KLCF	175.8	163.1	157.9	129.1	132.8	136.2	142.7	142.1	149.6	160.8	168.0
Castings	KLCH	122.0	121.0	111.3	97.3	116.3	133.0	143.4	147.0	156.0	152.0	148.0
Refined nickel												
Production (including ferro-nickel)	KLCM	28.0	26.1	26.5	28.6	28.0	28.0	28.4	35.1	42.0	36.6	..

1 Unwrought copper (electrolytic, fire refined and blister).
2 Reported stocks of refined copper held by consumers and those held in London Metal Exchange (LME) warehouses in the United Kingdom.
3 Copper content.
4 Consumption for high-conductivity copper and cadmium copper wire represented by consumption of wire rods, production of which for export is also included.
5 Lead reclaimed from secondary and scrap material and lead refined from bullion and domestic ores.
6 From 1975, figures for production and consumption of refined lead include antimonial lead, and for scrap and remelted lead, exclude secondary antimonial lead.
7 Including toll transactions involving fabrication.
8 Excluding goverment stocks.
9 Includes values for White Lead wef 1986.
10 Separate data no longer available, see footnote 9.
11 Including production from imported scrap and residues refined on toll.
12 Primary and secondary metal.
13 Including re-exports on toll transactions.
14 Including primary alloys. Despatches to consumers are calculated as the despatches by UK industry plus imports, minus exports but deliveries to LME warehouses are not included.
15 Includes wrought, cast products, and excludes foil products.
16 Includes foil stock and excludes foil products.
17 Excluding forging bars, wirebars, and almost two-thirds of despatches of hot rolled rod.

Sources: World Bureau of Metal Statistics: 01920 461274; Aluminium Federation: 0121 456 1103

21.17 Cotton, man-made fibres and wool

		Unit	1988	1989	1990	1991	1992	1993	1994	1995	1996	1997	1998
Raw cotton[1]													
Imports[9]	KLKC	Thousand	48	43	32	23	24	24	..	..	..	..	..
Home consumption: total	BKCA	tonnes	42	37	30	20	14	6	..	..	..	..	..
Stocks (end of period)	BKCB	"	4	3	1	1	–	1	..	..	..	..	..
Cotton waste													
Imports	KLKD	"	28.4	29.2	36.1	31.3	35.2	25.3	..	..	..	..	..
Cotton linters													
Imports	KLKE	"	13.6	18.0	13.7	16.1	19.0	5.2	..	..	..	..	..
Man-made fibres[2]													
Production: total	KLKF	"	280.1	272.5	273.2	267.4	261.9	240.6	206.9	195.3	..	..	..
Continuous filament yarn	KLKG	"	105.3	108.5	101.7	92.6	93.8	86.7	67.1	68.1	68.2	..	..
Staple fibre	KLKH	"	174.9	164.0	171.5	174.8	168.1	153.9	139.8	127.2	115.5	..	..
Cotton and man-made fibre yarn (including waste spinning)													
Production of single yarn[3]:													
100% Cotton	KJEK	"	33.2	25.9	20.4	11.2	6.4	7.3	5.3	5.5	5.8	4.8	3.9
Predominantly Cotton	KJEL	"	2.2	2.4	2.2	1.4	2.3	7.3	8.0	7.1	7.0	5.2	4.9
100% man-made fibre	KJEM	"	17.1	13.1	11.0	8.1	7.6	10.2	10.4	9.1	7.8	6.8	5.1
Mixtures including 50/50 blends	KJEN	"	29.8	26.6	24.4	21.3	19.3	16.3	16.4	15.9	15.5	14.1	13.3
Spindle activity (includes waste spinning)	KJEO	Millions	1.00	0.83	0.52	0.42	0.33	0.28	0.27	0.26	0.23	0.19	0.15
Yarn consumption:													
Cotton and predominantly Cotton yarns	KJEP	Thousand	31.5	30.6	28.0	24.1	23.6	20.7	19.1	17.9	19.0	11.8	10.8
100% man-made fibre spun yarns	KJEQ	tonnes	2.9	2.7	2.8	2.5	2.1	2.6	2.6	2.5	2.3	..	..
Spun mixture yarns	KJER	"	26.2	24.2	25.5	24.7	23.9	21.7	20.0	34.1	35.8	..	..
Continuous filament													
Man-made fibre yarn	KJES	"	34.6	35.2	34.4	31.8	33.0	33.0	32.0	30.2	33.0	..	..
Other	KJET	"	3.2	3.2	3.0	1.5	0.4	0.2	0.2	0.2	–	..	..
Other	LQYW	"	..	..	..	..	..	..	..	..	..	22.0	18.5
Cotton and man-made fibre weaving													
Production of woven cloth:													
Cotton	KLKV	Million linear metres	219	206	169	154	142	109	86	80	83	78	74
Man-made fibres	KLKW	"	212.5	217.4	223.4	197.1	195.8	191.3	187.8	179.6	184.6	161.5	152.4
Cotton/man-made fibre mixtures	KLKX	"	44.6	35.7	36.5	35.1	34.3	30.7	29.6	28.9	26.9	22.7	19.4
Loom activity													
Average number of looms running on cotton and man-made fibres	KLKY	Thousands	11.6	10.6	8.9	7.9	7.6	6.8	5.9	5.9	5.6	5.0	4.6
Wool													
Virgin wool (clean weight):													
Production[4]	KLKZ	Million kilo-grammes	44	47	48	47	47	44	..	..	..	..	..
Imports[9]	KLLA	"	89	77	88	93	102	67	..	..	..	..	..
Exports[5]	KLLB	"	34	33	23	40	32	19	..	..	..	..	..
Stocks at 31 August	KLLC	"	21	21	20	28	26	19	..	..	..	..	..
Consumption:													
Wool	KLLD	"	104.5	97.2	93.6	87.9	92.3	89.6	86.5	84.4	79.9	..	..
Hair	KLLE	"	7.0	6.5	5.9	5.6	5.0	3.5	2.3	0.9	1.1	..	..
Man-made fibres	KLLF	"	53.9	45.9	40.6	37.7	36.4	38.8	39.6	31.7	37.3	..	..
Other fibres[6]	KLLG	"	10.0	8.9	6.4	5.6	5.7	5.9	5.4	7.0	6.7	..	..
Tops													
Production: total	KLLH	"	80.0	69.1	60.5	61.9	64.3	61.5	63.4	52.8	52.3	..	..
Wool and hair	KLLI	"	42.8	38.7	34.2	36.0	38.6	36.4	38.2	32.9	29.2	..	..
Man-made fibres	KLLJ	"	37.2	30.4	26.3	25.9	25.7	25.1	25.2	19.9	23.1	..	..
Worsted yarns[9]:													
Production	KLLK	"	60.2	52.4	47.2	44.8	47.0	44.4	43.0	39.1	40.3	..	..
Semi-worsted yarns[7]:													
Production	KLLL		13.5	12.7	11.4	7.6	7.0	6.8	7.1	5.5	5.0	..	..
Woollen yarn:													
Production	KLLM	"	78.3	74.9	72.7	66.1	68.2	67.0	68.1	63.7	62.6	..	..
Woven woollen and worsted fabrics[8]:													
Deliveries	KLLN	Million square metres	89.0	85.4	79.0	72.1	71.8	70.4	75.9	68.3	68.6	..	..
Blankets:													
Deliveries	KLLO	"	7.2	7.0	7.3	6.2	5.4	5.7	4.7	5.4	6.0	..	..

TIME SERIES. Figures for consumption of raw cotton, and production and consumption of cotton and man-made fibre yarn are for periods of 52 weeks (53 weeks in 1986).

1 From 1978 figures for consumption of raw cotton are for calendar months.
2 Figures are based on returns from producers (excluding waste) and include all man-made fibres in commercial production.
3 Includes waste spinning.
4 Estimated.
5 Including imported wool and wool from imported skins, scoured, etc in the United Kingdom.
6 Including noils, broken tops, wastes, mungo and shoddy.
7 Including all yarn spun on the worsted system.
8 Includes mixture and man-made fibre fabrics classified as wool or worsted, but excludes blankets.
9 From 1993 trade data does not include UK trade with other EU Member States.

Sources: British Man-made Fibres Federation; Textile Statistics Bureau: 0161 620 7272

21.18 Fertilisers

Deliveries to UK agriculture by Members of the Fertiliser Manufacturers' Association

Years ending 30 June[1]

Thousand tonnes

		1989	1990	1991	1992	1993	1994	1995	1996	1997	1998	1999
Nutrient Content												
N (nitrogen):												
Straight	KGRM	676	707	628	587	537	553	602	685	615	664	618
Compounds	KGRN	487	483	462	378	377	413	449	444	420	406	430
P_2O_5 (phosphate)	KGRO	337	318	278	227	218	236	272	284	287	296	291
K_2O (potash)	KGRP	431	417	364	287	284	299	343	353	356	360	357
Compounds - total product	KGRQ	3 049	2 967	2 742	2 254	2 216	2 378	2 656	2 688	2 664	2 620	2 727

1 Pre 1990 the year ended is 31 May.

Source: Fertiliser Manufacturers' Association: 01780 720422

21.19 Minerals: production

Thousand tonnes

			Great Britain										
			1988	1989	1990	1991	1992	1993	1994	1995	1996	1997	1998
Limestone	KLEA		103 410	107 908	9 775	91 999	86 000	90 069	102 844	90 933	82 442	84 252	85 382
Sandstone	KLEB		16 031	16 748	14 952	12 928	11 586	12 100	13 494	15 017	12 581	12 457	13 545
Igneous rock	KLEC		44 636	46 809	49 542	46 008	48 630	49 209	50 014	49 641	43 731	42 370	39 838
Clay/shale	KLED		18 534	19 011	15 864	13 038	12 155	10 891	12 464	13 930	11 804	11 322	12 230
Industrial sand	KLEE		4 340	4 380	4 132	4 201	3 615	3 587	4 038	4 344	4 861	4 704	4 662
Chalk	KLEF		14 516	13 877	13 129	10 317	9 171	9 076	10 236	9 949	9 239	9 550	9 934
Fireclay	KLEG		1 057	1 052	892	867	572	479	679	708	536	338	577
Barium sulphate	KLEH		76.2	70.0	67.6	85.5	76.7	32.6	34.0	73.7	93.0	57.0	64.0
Calcium fluoride	KLEI		103.7	122.1	118.5	77.9	76.1	70.3	49.9	45.9	..	58.0	52.0
Copper	KLEJ		0.7	0.5	1.0	0.2	–	–	–	–	–	–	–
Lead	KLEK		1.2	2.2	1.4	0.2	..	..	1.8	..	..	..	1.0
Tin	KLEL		3.4	3.8	3.4	2.3	2.0	2.2	1.9	2.0	2.1	2.0	–
Zinc	KLEM		5.5	5.8	6.7	1.0	..	..	..	..	..	–	–
Iron ore: crude	KLEN		227	4	4	..	4	..	2	2	1	2	2
Iron ore: iron content	KLEO		50	2	2	..	2	1	1	1	1	1	1
Calcspar	KLEP		23	22	34	8	4	3	..	..	..	13	15
China clay	KLEQ	KILC	4 352	4 190	4 042	3 744	2 732	2 852	2 977	3 076	2 654	2 798	2 866
Ball clay		KIMS					..	..	913	..	..	..	..
Chert and flint	KLER		11	..	14	5	..	..	..	..	..	..	..
Fuller's earth	KLES		277	265	228	202	203	153	193	150	183	162	111
Lignite	KLET		18	4	5	3	3	3	2	–	–	–	–
Rock salt	KLEU		877	594	815	2 088	..	..	..	..	..	..	..
Salt from brine	KLEV		1 426	..	..	1 319	..	..	..	..	..	..	..
Salt in brine	KLEW		3 827	..	..	..	3 401	4 076	4 009	3 548	3 512	3 561	..
Anhydrite	KLEX		..	..	..	..	..	..	..	..	–	–	–
Dolomite	KLEY		19 861	21 271	20 674	19 454	18 539	17 985	17 616	17 966	16 555	17 282	15 632
Gypsum	KLEZ		..	..	..	..	..	..	..	..	..	..	..
Slate[1]	KLFA		708	590	359	360	326	462	402	275	408	347	425
Soapstone and talc	KLFB		14	15	15	11	5	5	5	4	5	5	5
Sand and gravel (land-won)	KLFC		110 516	110 504	98 993	85 479	78 341	79 380	86 341	78 031	70 489	74 362	73 016
Sand and gravel (marine dredged)	KLFD		19 638	20 728	17 179	12 439	10 557	10 090	11 331	11 625	11 508	12 004	12 952

1 Includes waste used for constructional fill, and powder and granules used in manufacturing.

		Northern Ireland										
		1988	1989	1990	1991	1992	1993	1994	1995	1996	1997	1998
Clay and shale	KLFF	..	..	..	..	..	..	..	..	–	–	–
Sand and gravel	KLFG	3 871	4 554	4 030	3 832	3 697	4 318	5 109	5 262	7 684	5 138	5 300
Basalt and igneous rock (other than granite)	KLFH	7 324	7 463	7 691	7 787	9 024	8 557	6 480	7 564	6 974	6 286	6 107
Limestone	KLFI	2 409	3 485	2 866	2 861	3 398	3 236	..	3 703	4 122	3 500	3 892
Sandstone[2]	KLFJ	2 870	2 845	3 090	3 679	3 304	3 959	5 480	4 779	4 941	6 042	6 584
Granite	KLFL	..	..	..	..	..	..	..	..	–	–	–
Others[3]	KLFN	..	..	..	..	..	647	896	812	1 392	625	473

2 Prior to 1993 the 'Sandstone' heading was called 'Grit and conglomerate'. The new heading is all encompassing and was confirmed as correct with the Geological Survey in Northern Ireland.

3 Rock salt, Chalk, Diatomite and Fireclay.

Source: Office for National Statistics: 01633 812082

21.20 Building materials and components: production
Great Britain

		Unit	1988	1989	1990	1991	1992	1993	1994	1995	1996	1997	1998
Building bricks (excluding refractory and glazed)	KLGA	Millions	4 682	4 654	3 802	3 212	3 000	2 639	3 114	3 256	3 046	2 997	3 000
Cement[1]	KLGB	Thousand tonnes	16 506	16 849	14 740	12 297	11 006	11 039	12 307	11 805	12 214	12 638	12 409
Building sand[2]	KLGC	"	23 415	23 290	20 948	18 079	16 769	17 406	18 534	17 389	14 655	15 337	13 810
Concreting sand	KLGD	"	39 174	41 223	37 213	31 239	28 573	28 021	30 977	29 390	28 659	30 130	30 244
Gravel[3]	KLGE	"	67 566	66 718	58 010	48 598	43 557	44 043	48 162	42 877	38 683	40 899	41 914
Crushed rock aggregates:													
used as roadstone (coated)	KLGF	"	28 860	23 733	26 430	26 387	26 647	27 238	28 512	28 972	26 270	23 906	23 131
roadstone (uncoated)	KLGG	"	54 187	66 015	61 742	60 748	53 471	54 412	51 121	49 307	40 893	40 186	36 816
fill and ballast	KLGH	"	59 989	59 689	54 640	45 669	48 919	52 141	65 779	56 140	50 982	51 396	51 623
concrete aggregate	KLGI	"	17 978	19 356	18 804	15 203	14 929	15 786	16 345	16 419	14 748	18 300	20 146
Ready mixed concrete[1]	KLGJ	Million m^3	28.8	29.6	26.8	22.5	20.8	20.8	22.9	21.7	20.9	22.3	23.0
Fibre cement products	KLGK	Thousand tonnes	251.0	220.7	234.7	133.7	121.1	128.7	154.1	160.5	146.2	163.5	160.9
Clay roofing tiles[1]	KLGL	Thousand m^2	3 459	3 756	2 912	2 191	1 495	..	..	..	..	..	..
Concrete roofing tiles	KLGM	"	38 818	35 787	31 510	26 359	21 490	24 574	28 149	26 118	24 651	24 958	24 981
Concrete building blocks:[1]													
dense aggregate	KLGN	"	44 404	45 564	39 297	32 456	29 732	30 116	36 997	36 933	34 996	37 250	39 439
lightweight aggregate	KLGO	"	33 530	31 041	23 768	18 581	17 479	19 235	22 048	18 147	16 316	17 783	19 110
aerated concrete	KLGP	"	32 095	31 394	28 089	23 594	20 984	24 936	28 503	23 207	24 554	27 505	26 113
Concrete pipes[1]	KLGQ	Thousand tonnes	700	717	737	723	564	..	..	..	..	..	..
Roofing & architectural slates	KLGR	"	40.9	46.6	44.1	41.0	41.6	37.3	44.9	42.0	48.5	44.6	46.2
Gypsum (excluding anhydrite)	KLGS	"	..	..	..	..	..	..	..	..	..	..	..
Plaster	KLGT	"	..	..	..	..	..	..	..	..	..	..	..
Plasterboard	KLGU	Thousand m^2	..	..	..	..	..	..	..	..	..	..	..
Unglazed floorquarries[1]	KLGV	"	2 681	1 180	1 070	889	889	..	..	..	..	..	..
Unglazed tiles[1]	KLGW	"	–	1 521	1 349	934	493	..	..	..	..	..	..

1 United Kingdom.
2 Including sand and gravel used for coating.
3 Figures include hoggin, concrete aggregate, other purposes (excluding fill) and fill.

Source: Department of the Environment, Transport and the Regions: 020 7890 5593

21.21 Construction: Value of output in Great Britain[1]

£ millions

		1988	1989	1990	1991	1992	1993	1994	1995	1996	1997	1998[2]
All work: total	FGAY	44 853	52 150	55 307	51 115	47 472	46 323	49 439	52 643	55 243	58 352	62 060
New work: total	BLAB	24 763	29 320	30 762	27 726	24 814	23 556	25 086	26 672	27 926	29 928	32 491
New housing: total	KLQA	8 655	8 083	6 680	5 796	6 084	6 628	7 417	7 135	7 013	7 983	8 430
For public sector	BLAC	880	966	934	793	1 243	1 415	1 671	1 660	1 421	1 232	1 069
For private sector	BLAD	7 775	7 117	5 746	5 003	4 841	5 213	5 746	5 475	5 592	6 751	7 361
Infrastructure: total	KIAM	3 367	4 017	4 965	6 062	5 716	5 544	5 149	5 660	6 338	6 311	6 182
Other new work: total (excluding infrastructure)	KLQB	12 741	17 220	19 118	15 867	13 015	11 384	12 521	13 877	14 575	15 635	17 878
For public sector	BLAE	3 075	3 903	4 414	4 142	4 181	4 045	4 384	4 661	4 441	3 756	4 151
For private sector	KLQC	9 666	13 317	14 704	11 725	8 834	7 339	8 137	9 217	10 134	11 879	13 727
Private Industrial	BLAF	2 763	3 425	3 394	2 622	2 234	2 208	2 489	3 008	3 119	3 491	3 810
Private Commercial	BLAG	6 903	9 892	11 310	9 103	6 600	5 131	5 648	6 209	7 015	8 388	9 917
Repair and maintenance: total	BLAH	20 090	22 830	24 544	23 389	22 658	22 767	24 353	25 971	27 317	28 423	29 569
Housing: total[2]	KLQD	11 530	13 092	13 839	13 001	12 586	12 809	13 767	14 595	15 035	15 755	16 202
For public sector	BLBK	4 485	4 943	5 384	4 938	4 991	5 439	5 963	6 465	6 637	6 629	6 506
For private sector	BLBL	7 045	8 149	8 455	8 063	7 595	7 370	7 804	8 130	8 398	9 126	9 696
Public other work	BLAJ	4 488	4 982	5 488	5 291	5 087	4 916	5 211	5 398	5 252	5 079	5 220
Private other work	BLAK	4 073	4 755	5 218	5 098	4 985	5 042	5 375	5 978	7 030	7 590	8 147

1 Output by contractors, including unrecorded estimates by small firms and self-employed workers, and output by public sector direct labour departments - classified to construction in the *1980 Standard Industrial Classification.*

2 Provisional.

Source: Department of the Environment, Transport and the Regions: 020 7890 5583

21.22 Construction: Value of new orders obtained by contractors[1]
Great Britain

£ millions

		1988	1989	1990	1991	1992	1993	1994	1995	1996	1997	1998
New work: total	FHAA	26 298	27 142	22 491	19 455	17 493	19 965	21 285	22 065	22 834	24 806	27 477
Public housing	BLBC	882	872	683	875	1 246	1 668	1 386	1 182	1 073	995	933
Private housing[2]	BLBD	7 893	6 497	4 855	4 552	4 016	4 874	5 721	4 905	5 416	6 253	5 997
New housing: total	FGAU	8 775	7 370	5 538	5 427	5 263	6 542	7 106	6 087	6 487	7 248	6 930
Infrastructure:												
Water	KIBV	157	251	321	509	669	421	412	500	640	733	957
Sewerage	KIBW	332	332	491	429	469	447	389	394	481	656	737
Electricity	KIBX	154	271	187	301	281	211	170	218	294	382	359
Roads	KIBY	1 101	1 226	1 425	1 415	1 322	1 435	1 356	1 531	1 710	928	821
Gas, communications, air	KIBZ	230	339	391	441	554	642	494	904	745	693	745
Railways[3]	KIDP	128	362	160	125	200	623	412	351	524	416	573
Harbours	KIDQ	206	180	216	203	252	220	218	273	270	182	287
Total	KIDR	2 307	2 690	3 190	3 423	3 746	3 998	3 451	4 170	4 664	3 991	4 479
of which												
- Public	KIDS	1 908	2 369	2 029	2 090	1 690	2 472	2 211	2 327	1 671	1 352	1 505
- Private	KIDT	399	592	1 161	1 333	2 056	1 525	1 240	1 843	2 993	2 639	2 974
Other public non-housing:												
Factories	KIDU	237	188	149	89	128	111	111	94	91	72	84
Warehouses	KIDV	21	48	14	14	54	23	38	29	14	27	20
Oil, steel, coal	KIDW	180	115	81	13	47	30	12	13	4	4	2
Schools and colleges	KIDX	435	471	527	584	584	655	658	710	707	749	770
Universities	KIDY	66	139	146	150	203	353	376	373	355	273	405
Health	KIDZ	711	824	663	578	644	697	752	717	681	491	769
Offices	KIFP	414	519	570	492	499	684	469	393	379	391	292
Entertainment	KIFQ	341	382	283	250	189	281	308	285	259	342	432
Garages	KIFR	88	75	56	44	28	42	49	51	28	34	19
Shops	KIFS	38	26	24	25	23	26	14	21	12	35	35
Agriculture	KIFT	18	6	5	17	14	44	22	12	8	33	17
Miscellaneous	KIFU	659	1 043	601	421	351	450	844	508	419	441	660
Total	KIFV	3 209	3 836	3 117	2 677	2 763	3 397	3 654	3 206	2 956	2 894	3 504
Private industrial:[2]												
Factories	KIFW	2 099	2 269	2 094	1 830	1 006	1 221	1 451	2 055	1 603	2 184	1 878
Warehouses	KIFX	786	706	648	438	426	429	498	594	663	901	1 014
Oil, steel, coal	KIFY	56	73	108	79	72	27	51	76	71	64	79
Total	KIFZ	2 941	3 049	2 850	2 347	1 503	1 677	1 999	2 725	2 337	3 149	2 971
Private commercial:[2]												
Schools, universities	KIHP	191	179	169	175	121	134	115	105	156	189	351
Health	KIHQ	191	295	271	369	193	179	255	288	277	356	651
Offices	KIHR	4 585	5 271	4 215	2 215	1 691	1 471	1 777	2 123	2 169	2 506	3 472
Entertainment	KIHS	1 356	1 418	1 195	1 111	747	751	928	940	1 407	1 847	2 244
Garages	KIHT	375	405	364	261	234	308	300	301	265	344	315
Shops	KIHU	2 048	2 086	1 345	1 222	1 034	1 278	1 453	1 871	1 795	1 937	2 154
Agriculture	KIBN	157	127	127	105	94	108	120	124	123	148	146
Miscellaneous	KIBO	163	149	109	123	103	122	127	126	198	198	259
Total	KIBP	9 066	9 927	7 796	5 581	4 218	4 351	5 075	5 877	6 390	7 525	9 593

1 Classified to construction in the *1980 Standard Industrial Classification.* Total figures only are available in this series prior to 1985.

2 Figures for private sector include work to be carried out by contractors on their own initiative for sale.

3 Infrastructure orders for 1987 include The Channel Tunnel Project.

Source: Department of the Environment, Transport and the Regions: 020 7890 5583

21.23 Total engineering
Total turnover of UK based manufacturers[1,2,3]
Standard Industrial Classification 1992

£ millions

Activity heading	Product group		1992	1993	1994	1995	1996	1997	1998
	Class 29: **Manufacture of machinery and equipment not elsewhere classified**								
2911	Manufacture of engines and turbines except aircraft, vehicle & cycle engines	BBEM	1 515	1 653	1 649†	1 764	1 616	1 417	1 492
2912	Manufacture of pumps and compressors	BBEN	2 211	2 368	2 466	2 780	2 739†	2 698	2 897
2913	Manufacture of taps and valves	BBEO	1 342	1 441†	1 504	1 589	1 569	1 536	1 648
2914	Manufacture of bearings, gears, gearing and driving elements	BBEP	1 356	1 381	1 505	1 616	1 720	1 574	1 562
2922	Manufacture of lifting and handling equipment	BBEQ	3 069	2 775†	3 286	3 453	3 265	2 937	2 990
2923	Manufacture of non-domestic cooling and ventilation equipment	BBER	2 219	2 888	2 812	2 925	3 058	3 268	3 350
2924	Manufacture of other general purpose machinery not elsewhere classified	BBES	2 347	2 473	2 571	2 474	2 669	2 877	2 820
2940	Manufacture of machine tools	BBET	1 906	1 881	2 097	2 301	2 413	2 508	2 420
2952	Manufacture of machinery for mining, quarrying and construction	BBEU	1 207	1 864	2 230	2 231	2 203	2 496	2 575
2953	Manufacture of machinery for food, beverage and tobacco processing	BBEV	808	1 107	1 133	1 343	1 247	1 209	1 117
2954	Manufacture of machinery for textile, apparel and leather production	BBEW	469	504	443†	495	460	419	379
2956	Manufacture of other special purpose machinery not elsewhere classified	BBEX	1 972	2 155	2 444†	2 163	2 298	2 097	2 004
2971	Manufacture of electric domestic appliances	BBEY	1 914	2 122	2 314†	2 233	2 230	2 266	2 247
	Total	BBEZ	22 335	24 612†	26 454	27 367	27 487	27 302	27 501
	Class 30 : **Manufacture of electrical and optical equipment**								
3001	Manufacture of office machinery	BBFA	936	1 065	1 189	1 232	1 449	1 279	1 354
3002	Manufacture of computers and other information processing equipment	BBFB	9 122	9 218	10 769†	12 105	12 166	13 518	14 068
	Total	BBFC	10 058	11 958†	11 958	13 337	13 615	14 797	15 422
	Class 31 : **Manufacture of electrical machinery and apparatus not elsewhere classified**								
3110	Manufacture of electric motors, generators and transformers	BBFD	1 925	1 888†	2 020	2 057	2 280	2 305	2 275
3120	Manufacture of electricity distribution and control apparatus	BBFE	2 626	2 999	3 155	3 428	3 784	3 582	3 488
3130	Manufacture of insulated wire and cable	BBFF	1 365	1 625	2 156	2 384	2 581	2 152	1 873
3140	Manufacture of accumulators, primary cells and primary batteries	BBFG	602	689	721	613	664	655	606
3150	Manufacture of lighting equipment and electric lamps	BBFH	1 224	1 365	1 496	1 589	1 732	1 670	1 575
3161	Manufacture of other electrical equipment for engines and vehicles not otherwise classified	BBFI	1 224	1 270	1 339	1 247	1 226	1 148	1 065
3162	Manufacture of other electrical equipment not elsewhere classified	BBFJ	1 955	2 532	2 677	2 808	2 764	2 705†	2 924
	Total	BBFK	10 941	12 368†	13 564	14 126	15 031	14 217	13 806
	Class 32 : **Manufacture of radio, television and communication equipment and apparatus**								
3210	Manufacture of electronic valves and tubes and other electronic components	BBFL	3 133	4 734	5 923	6 362	6 270	6 102	6 108
3220	Manufacture of television and radio transmitters and apparatus for line telephony and line telegraphy	BBFM	3 836	4 684†	5 902	6 616	6 815	7 029	7 786
3230	Manufacture of television and radio receivers, sound or video recording or reproducing apparatus and associated goods	BBFN	2 586	3 241	3 858†	4 257	4 978	4 707	4 425
	Total	BBFO	9 555	12 659†	15 683	17 235	18 063	17 838	18 319
	Class 33 : **Manufacture of medical, precision and optical instruments, watches and clocks**								
3310	Manufacture of medical and surgical equipment and orthopaedic appliances	BBFP	1 935	1 986	1 883†	2 015	1 967	1 772	1 716
3320	Manufacture of instruments and appliances for measuring, checking, testing, navigating and other purposes, except industrial process control equipment	BBFQ	5 087	6 307	6 030†	6 066	6 693	6 440	6 147
3340	Manufacture of optical instruments and photographic equipment	BBFR	853	1 089	1 205	1 233	1 249	1 283	1 356
	Total	BBFS	7 875	9 382	9 118†	9 314	9 909	9 495	9 219

1 The figures shown represent the output of UK based manufacturers classified to Subsections DK and DL of the Standard Industrial Classification 1992. The figures shown are derived from the monthly production inquiry (MPI) and include estimates for non-responders and for establishments which are not sampled.

2 Orders on hand figures are given for the end of the period to which they relate.

3 The data on this table are not seasonally adjusted.

Source: Office for National Statistics: 01633 812786

21.24 Volume index numbers of turnover and orders for the engineering industries
United Kingdom

	Total			Home			Export		
	Orders on hand end of period (1995 average = 100)	1995 average monthly sales = 100: New orders[1]	1995 average monthly sales = 100: Turnover	Orders on hand end of period (1995 average = 100)	1995 average monthly sales = 100: New orders[1]	1995 average monthly sales = 100: Turnover	Orders on hand end of period (1995 average = 100)	1995 average monthly sales = 100: New orders[1]	1995 average monthly sales = 100: Turnover
Total Engineering industries *SIC 1992 Class 29, 30, 31,32 and 33*									
	FGWA	FGWB	FGVT	FGVU	FGVV	FGVW	FGVX	FGVY	FGVZ
1991	99.3	85.2	86.8	96.7	85.4	88.1	103.4	84.8	84.9
1992	95.2	84.7	85.9	90.2	85.5	87.5	103.0	83.6	83.7
1993	95.8	87.4	87.2	91.8	92.4	91.9	101.8	80.5	80.8
1994	104.9	98.0	95.3	105.4	103.0	98.8	104.1	91.1	90.5
1995	98.2	98.0	100.0	100.0	98.4	100.0	95.6	97.6	100.0
1996	94.2	102.5	103.7	94.3	98.0	99.7	94.0	108.7	109.1
1997	96.9	107.3	106.5	97.4	101.9	101.0	96.2	114.7	114.1
1998	93.6	111.0	111.9	88.8	101.6	104.2	101.1	124.0	122.6
Percentage change 1998 on 1997[2]	*-3.4*	*3.4*	*5.0*	*-8.8*	*-0.3*	*3.2*	*5.1*	*8.1*	*7.4*
Manufacture of Machinery and Equipment *SIC 1992 Class 29*									
	FGVB	FGVD	FGVF	FGVH	FGVJ	FGVL	FGVN	FGVP	FGVR
1991	102.5	99.4	101.0	94.8	98.5	101.7	114.9	101.1	99.6
1992	94.6	94.6	97.4	82.4	94.8	99.1	114.2	94.2	94.4
1993	95.0	96.3	96.1	84.5	100.6	99.8	111.8	89.0	89.8
1994	106.7	103.9	99.8	107.4	110.7	102.8	105.5	92.4	94.7
1995	97.6	96.8	100.0	97.9	96.7	100.0	97.2	96.9	100.0
1996	96.2	98.8	99.3	99.7	97.0	96.3	90.4	101.9	104.4
1997	100.2	99.5	98.1	105.4	97.3	95.4	92.0	103.3	102.7
1998	84.2	92.1	97.8	84.9	86.3	93.3	82.9	102.1	105.4
Percentage change 1998 on 1997[2]	*-16.0*	*-7.4*	*-0.3*	*-19.4*	*-11.3*	*-2.2*	*-9.9*	*-1.2*	*2.6*
Manufacture of Electrical and Optical Equipment *SIC 1992 Class 30, 31,32 and 33*									
	FGVC	FGVE	FGVG	FGVI	FGVK	FGVM	FGVO	FGVQ	FGVS
1991	96.7	76.5	78.2	98.3	76.4	78.7	94.4	76.6	77.5
1992	95.7	78.7	79.0	96.8	79.1	79.5	94.3	78.3	78.3
1993	96.4	82.0	81.8	98.0	86.7	86.4	94.0	76.2	76.2
1994	103.4	94.5	92.6	103.6	97.7	96.1	103.0	90.5	88.3
1995	98.8	98.8	100.0	101.8	99.5	100.0	94.4	97.9	100.0
1996	92.6	104.7	106.3	89.7	98.7	102.1	96.8	112.1	111.5
1997	94.2	112.0	111.6	90.6	105.1	104.8	99.4	120.5	119.9
1998	101.4	122.4	120.5	92.1	112.2	111.8	115.1	135.1	131.3
Percentage change 1998 on 1997[2]	*7.6*	*9.3*	*8.0*	*1.7*	*6.6*	*6.7*	*15.8*	*12.1*	*9.5*

1 Net of cancellations.
2 The percentage changes have been calculated on unrounded values and then rounded to the nearest half percentage point.

Source: Office for National Statistics: 01633 812786

21.25 Motor vehicle production
United Kingdom

Number

		1988	1989	1990	1991	1992	1993	1994	1995	1996	1997	1998
Motor vehicles												
SIC 1992, Class 34-10												
Passenger cars: total	GKAJ	1 226 835	1 299 082	1 295 610	1 236 900	1 291 880	1 375 524	1 466 823	1 532 084	1 686 134	1 698 001†	1 748 258
1 000 c.c. and under	GKAB	129 446	133 135	93 039	15 918	22 037	98 034	98 178	95 198	108 645	119 894	112 044†
Over 1 000 c.c. but not over 1 600c.c.	GKAD	764 289	716 784	809 219	496 822	793 307	709 615	729 397	814 873	845 084	829 086	814 595†
Over 1 600 c.c. but not over 2 800 c.c.	GKAF	260 231	375 309	325 116	193 972	437 951	515 487	573 357	528 444	635 861	653 154	720 556†
Over 2 800 c.c.	GKAH	72 869	73 854	68 236	26 001	38 585	52 388	65 891	93 569	96 544	95 881	101 063†
Commercial vehicles: total	GKDF	317 343	326 590	270 346	217 141	248 453	193 467	227 815	233 001	238 314	237 703	227 379†
Of which:												
Light commercial vehicles	GKDH	250 053	267 135	230 510	105 633	216 477	171 141	197 285	199 346	205 372	210 942	203 629†
Trucks:												
Under 7.5 tonnes	GKDJ	19 732	17 687	10 515	5 379	9 558	4 755	8 154	9 523	9 812	6 254†	5 006
Over 7.5 tonnes	GKDL	24 887	21 083	13 674	7 673	11 113	8 269	10 016	11 717	9 229	7 930†	7 002
Motive units for articulated vehicles	GKCV	6 171	5 827	3 327	1 444	2 788	2 283	2 794	3 476	208	2 573†	2 492
Buses, coaches and mini buses	GKDN	16 500	14 858	12 320	5 593	8 517	7 019	9 566	8 939	9 254	10 004†	9 250

TIME SERIES. Figures for motor vehicles relate to periods of 52 weeks (53 weeks in 1983, 1988 and 1993).

Source: Office for National Statistics: 01633 812963

21.26 Alcoholic drink
United Kingdom

			1988	1989	1990	1991	1992	1993	1994	1995	1996	1997	1998
Spirits[1]		Thousand hectolitres of alcohol											
Production	KMEA	"	3 568	4 211	4 676	4 476	4 219	3 974	4 114	4 507	4 868	5 297†	5 145
Released for home consumption													
Home produced whisky	KMEE	"	452	430	413	383	356	374	383	310	321	312	289
Imported and other	KMEG	"	583	570	567	529	505	504	531	481	495	533	505
Total	KMEH	"	1 035	1 000	980	912	861	878	914	791	815	845	794
Beer[2]		Thousand hectolitres											
Production	KMEI	"	61 885	62 132	61 783	59 552	57 616	56 746	58 333	56 800	58 072	59 139	56 652
Released for home consumption	KMEL	"	65 023	65 303	65 184	63 038	60 973	59 177	60 575	59 129	59 894	61 114	58 835
		Thousand hectolitres of pure alcohol											
Production	JYXJ		..	..	..	..	..	..	2 345	2 298	2 360	2 406	2 333
Released for home consumption	JYXK		..	..	..	..	..	..	2 453	2 410	2 448	2 504	2 439
Wine of fresh grapes													
Released for home consumption		Thousand hectolitres											
Fortified	KMEM		451	327	378	357	345	335	329	358†	359	324	370
Still table	KMEN	"	5 680	5 871	5 893	5 910	6 166	6 471	6 759	6 577	6 995	7 653	7 985
Sparkling	KMEO	"	352	392	365	316	292	296	297	315	358	382	416
Total	KMEP	"	6 483	6 669	6 636	6 583	6 803	7 102	7 385	7 250†	7 712	8 358	8 772
Made-wine													
Released for home consumption													
Other than coolers	KMEQ	"	593	512	521	436	439	505	470	516	513	481†	485
Coolers[3]	KJDD	"	..	79	185	263	433	485	549	903	1 781	1 151†	1 243
Cider and perry													
Released for home consumption	KMER	"	3 103	3 263	3 666	3 713	4 398	4 496	4 811	5 575	5 656	5 513	5 548

1 Potable spirits distilled.
2 A new system was introduced for beer duty in June 1993. The figures in this table include adjustments to data prior to this date to bring them into line with current data.
3 Made wine with alcoholic strength 1.2% to 5.5%. Includes alcoholic lemonade of appropriate strength and similar product. Coolers were included with other made wine prior to 1989.

Source: HM Customs and Excise: 020 7865 5249

21.27 Tobacco products: released for home consumption
United Kingdom

			1988	1989	1990	1991	1992	1993[1]	1994	1995	1996	1997	1998
Cigarettes:		Thousand million											
Home produced	KMFA	"	87.9	87.5	87.4	85.3	76.4	86.4	79.4	70.8	73.8	71.1	67.8
Imported	KMFB	"	9.4	10.3	10.3	10.3	9.3	8.8	13.2	29.5	9.5	9.9	7.5
Total	KMFC	"	97.3	95.9	97.8	95.6	85.7	95.2	92.6	80.3	83.3	81.0	75.3
Cigars:		Million kg.											
Home produced	KMFD	"	2.4	2.3	2.1	1.9	1.8	1.7	1.6	1.5	1.4	1.3	1.2
Imported	KMFE	"	0.2	0.2	0.2	0.1	0.1	0.1	0.1	0.1	0.1	0.1	0.1
Total	KMFF	"	2.6	2.4	2.3	2.0	1.9	1.8	1.7	1.6	1.5	1.4	1.3
Hand-rolling tobacco:													
Home produced	KMFG	"	4.5	4.3	4.1	4.1	3.7	3.5	3.0	2.4	2.1	1.8	1.7
Imported	KMFH	"	–	–	–	0.1	0.1	0.1	0.1	0.1	0.1	0.1	0.1
Total	KMFI	"	4.5	4.3	4.1	4.1	3.7	3.6	3.2	2.6	2.3	1.9	1.8
Other smoking and chewing tobacco:													
Home produced	KMFJ	"	2.5	2.2	2.1	2.0	1.8	1.8	1.5	1.3	1.2	1.1	1.0
Imported	KMFK	"	–	0.1	0.1	0.1	0.2	0.1	0.1	0.1	0.1	0.1	0.1
Total	KMFL	"	2.5	2.3	2.2	2.1	2.0	1.9	1.6	1.4	1.3	1.2	1.1

1 1993 contained two Budgets, in March and November.

Source: HM Customs and Excise: 020 7865 5249

Banking, insurance, etc

22

22.1 Bank of England

£ millions

		December										
		1988	1989	1990	1991	1992	1993	1994	1995	1996	1997	1998
Issue Department												
Liabilities:												
Notes in circulation	AEFA	16 071	16 849	17 283	17 466	17 542	18 218	20 055	21 262	22 407	23 715	24 573
Notes in Banking Department	AEFB	9	11	7	4	8	12	5	7	12	5	7
Assets:												
Government securities[1]	AEFC	10 339	13 946	14 672	11 791	7 808	6 816	11 468	14 552	16 524	16 416	15 826
Other securities[2]	AEFD	5 741	2 914	2 618	5 679	9 742	11 414	8 592	6 717	5 896	7 304	8 754
Banking Department												
Liabilities:												
Total[3]	AEFE	3 203	5 398	8 613	5 825	5 623	11 095	6 192	7 114	6 229	7 221	6 802
Public deposits[4]	AEFF	94	69	44	104	97	6 205	938	1 159	1 001	1 192	237
Bankers' deposits[5]	AEFH	1 310	1 750	1 842	1 813	1 553	1 700	1 855	2 001	2 021	2 800	1 388
Reserves and other accounts	AEFI	1 784	3 565	6 713	3 894	3 959	3 175	3 385	3 941	3 193	3 214	5 163
Assets:												
Total	KCYT	3 203	5 398	8 615	5 825	5 623	11 095	6 192	7 114	6 229	7 221	6 802
Government securities	AEFJ	882	1 354	1 432	1 346	1 237	1 174	1 050	1 090	1 232	1 373	1 352
Advances and other accounts	AEFK	661	726	2 146	2 443	3 935	9 411	4 696	5 499	2 339	5 388	3 302
Premises, equipment and other securities	AEFL	1 651	3 307	5 030	2 031	443	498	441	518	2 646	455	2 141
Notes and coin	AEFM	9	11	7	5	8	12	5	7	12	5	7

1 Including the historic liability of the Treasury of £11 million until 1993.
2 Including gilt and Treasury bill repurchase agreements (from 1994 - previously in "Government securities")
3 The only liability not shown separately is the Bank's capital (held by the Treasury) which has been constant at £14.6 million.
4 Excluding local government and public corporations' deposits which are included under Reserves and other accounts.
5 Up to 19 August 1981 these constituted the current accounts held at the Bank by the banks and discount houses. From the introduction of new arrangements for monetary control on 20 August, they consist of operational deposits held mainly by the clearing banks and non-operational cash ratio deposits for which institutions authorised under the Banking Act, deposit - taking UK branches of "European Authorised Institutions" and (from 1998) building societies are liable.

Source: Bank of England: 020 7601 3236

22.2 Value of inter-bank clearings[1]

£ billions

		1988	1989	1990	1991	1992	1993	1994	1995	1996	1997	1998
Bulk paper clearings												
Cheque (formerly general)	KCYY	953	1 030	1 089	1 084	1 052	1 065	1 075	1 097	1 161	1 196	1 210
Credit (formerly credit clearing)	KCYZ	103	107	109	104	101	97	93	92	94	94	93
High-value clearings												
Town	KCZA	7 693	6 754	4 776	2 228	1 387	1 069	681	59	-	-	-
CHAPS	KCZB	11 289	14 733	18 880	19 050	20 929	23 545	25 052	26 719	28 881	36 032	41 501
Electronic clearing	KCZC	423	526	668	772	803	836	941	1 055	1 251	1 432	1 602

1 Excludes inter-branch clearings and clearings in Scotland and Northern Ireland.

Source: Association of Payment Clearing Services (APACS): 020 7601 5948

22.3 Other banks' balance sheet[1]

£ millions

		1989	1990	1991	1992	1993	1994	1995	1996	1997	1998
Sterling liabilities											
Notes outstanding & cash loaded cards	TBFA	1 560†	1 678	1 840	2 084	2 186	2 456	2 576	2 717	2 832	2 929
Sight deposits											
UK banks	TBFB	12 928†	12 893	10 229	14 243	14 768	16 222	16 983	20 138	44 573	37 839
UK building societies	TBFC	..	..	..	..	..	..	..	..	950	1 277
UK public sector	TBFD	3 049†	3 697	2 953	2 952	3 335	3 301	3 824	3 641	3 781	3 003
Other UK residents	TBFE	127 698†	138 310	144 516	148 011	161 907	164 451	192 297	211 179	271 232	295 069
Non-residents	TBFF	13 113†	14 760	14 466	15 094	15 075	16 425	18 690	17 707	37 730	43 528
Time deposits											
UK banks	TBFG	76 888†	84 736	72 344	78 640	68 531	75 629	90 044	95 798	99 783	111 970
UK building sicieties	TBFH	–	–	–	–	–	–	–	–	5 682†	4 362
UK public sector	TBFI	6 671†	4 699	2 885	3 193	4 223	4 908	7 896	6 087	9 060	9 749
Other UK residents	TBFJ	130 547†	148 934	154 223	157 979	162 053	166 505	190 707	210 045	284 630	295 925
of which TESSAs	TBFK	–	–	2 453†	4 331	5 896	7 517	10 314	9 389	20 393	21 568
of which SAYE	TBFL	–	–	–	–	–	–	–	–	2 255†	2 604
Non-residents	TBFM	56 464†	67 349	57 718	58 550	58 583	62 905	68 274	63 995	91 040	97 954
Acceptances granted	TBFN	–	–	–	–	–	–	–	–	19 952†	16 568
Liabilities under sale and repurchase agreements											
of which British govt. securities	TBFU	–	–	–	–	–	–	–	22 668†	47 297	55 561
UK banks	TBFP	–	–	–	–	–	–	–	13 718†	29 089	43 314
UK building societies	TBFQ	..	..	..	..	..	..	..	..	20	32
UK public sector	TBFR	–	–	–	–	–	–	–	2 279†	6 044	–
Other UK residents	TBFS	–	–	–	–	–	–	–	7 703†	18 114	20 918
Non-residents	TBFT	–	–	–	–	–	–	–	683†	5 664	5 469
CDs and other short-term paper issued	TBFV	42 767†	53 225	52 307	50 912	52 783	63 880	70 277	96 390	119 266	138 248
Total sterling deposits	TBFW	470 124†	528 604	511 641	529 575	541 260	574 224	658 993	749 364	1 046 610	1 125 223
Sterling items in suspense and transmission	TBFX	9 178†	11 920	11 349	10 317	11 011	8 412	10 423	12 120	16 055	17 372
Net derivatives	TBFY	..	..	..	..	..	..	..	..	8 186	11 242
Accrued amounts payable	TBFZ	..	..	..	..	..	..	..	..	20 713	24 632
Sterling capital and other internal funds	TBGA	56 331†	62 020	65 908	68 640	71 833	78 775	82 160	90 464	103 463	100 967
Total sterling liabilities	TBGB	537 193†	604 222	590 738	610 616	626 290	663 868	754 152	854 666	1 197 858	1 282 366
Foreign currency liabilities											
Sight and time deposits											
UK banks	TBGC	79 675†	75 833	76 901	89 532	95 921	105 813	104 769	91 314	90 858	77 128
UK building societies	TBGD	..	..	..	..	..	..	..	..	1 027	639
UK public sector	TBGE	..	..	..	..	..	..	..	..	226	149
Other UK residents	TBGF	..	..	..	..	..	..	..	..	64 188	60 513
Non-residents	TBGG	457 618†	440 307	432 000	543 883	580 951	618 769	696 380	605 439	716 573	766 934
Acceptances granted	TBGH	..	..	..	..	..	..	..	..	743	730
Sale and repurchase agreements											
UK banks	TBGJ	–	–	–	–	–	–	–	17 114†	21 311	30 669
UK building societies	TBGK	..	..	..	..	..	..	..	..	–	–
UK public sector	TBGL	..	..	..	..	..	..	..	..	22	–
Other UK residents	TBGM	..	..	..	..	..	..	..	..	25 716	26 742
Non-residents	TBGN	–	–	–	–	–	–	–	61 307†	100 935	118 909
CDs and other short-term paper issued	TBGO	83 911†	70 973	65 648	73 814	61 017	67 869	89 032	101 005	131 620	124 151
Total foreign currency deposits	TBGP	666 716†	637 509	608 685	754 851	790 361	851 856	972 832	975 619	1 153 220	1 206 562
Items in suspense and transmission	TBGQ	5 864†	4 979	5 637	11 261	20 714	14 779	13 881	21 272	35 713	25 027
Net derivatives	TBGR	..	..	..	..	..	..	..	..	8 653	4 857
Accrued amounts payable	TBGS	..	..	..	..	..	..	..	..	21 996	25 184
Capital and other internal funds	TBGT	23 893†	19 721	18 697	20 833	21 581	21 081	26 430	25 535	31 676	44 752
Total foreign currency liabilities	TBGU	696 473†	662 209	633 019	786 945	832 656	887 716	1 013 143	1 022 426	1 251 258	1 306 381
Total liabilities	TBGV	1 233 666†	1 266 430	1 223 757	1 397 561	1 458 946	1 551 584	1 767 295	1 877 092	2 449 116	2 588 747

1 The implementation of the review of banking statistics at end-September 1997 has resulted in several changes to this table:
a) The table now includes the business of all monthly and quarterly reporting banks in the UK; it formerly covered only the business of monthly reporting institutions.
b) The Channel Islands and Isle of Man are no longer treated as part of the UK for statistical purposes. Banking institutions in the Channel Islands and Isle of Man no longer have the option of being within the UK banking sector and their business, along with the business of offshore island branches of UK mainland banks, is now excluded from the figures within this table. Additionally, the business of the UK banking sector with offshore island residents and entities has been reclassified from UK residents to non-residents.
c) The table now contains more comprehensive detail of business with building societies. This business was previously included indistinguishably within the UK private sector elements of the table.
d) The aggregate balance sheet of the banking sector has been inflated because it is now reported on an accruals basis rather than a cash basis (accrued amounts payable/receivable are shown under both liabilities and assets respectively). Additionally, acceptances have been brought onto the balance sheet and are shown under both Liabilities and assets.

2 With effect from 1998, the balance sheet of the Banking Department of the Bank of England is excluded from this table, and other banks' business with the Issue Department is reclassified from "UK public sector" to "UK banks".

Source: Bank of England: 020 7601 3236

22.3 Other banks' balance sheet[1]

continued

£ millions

		1989	1990	1991	1992	1993	1994	1995	1996	1997	1998
Sterling assets											
Notes and coins	TBGW	3 890†	3 956	3 920	4 454	4 711	4 983	5 357	4 812	5 225	6 698
With UK central bank											
Cash ratio deposits	TBGX	1 477†	1 690	1 576	1 402	1 415	1 490	1 682	1 888	2 566	1 068
Other	TBGY	164†	3	27	–42	183	103	113	533	217	383
Market loans											
UK banks	TBGZ	83 018†	91 047	77 568	87 691	79 281	88 582	103 006	109 500	139 996	148 139
UK bank CDs	TBHB	14 423†	20 271	19 661	20 620	19 028	23 632	26 767	36 922	62 584	65 510
UK bank commercial paper	TBHC	..	..	..	..	..	..	..	..	29	130
UK building societies CDs etc and deposits	TBHD	2 114†	3 357	3 673	4 433	5 360	5 358	4 387	4 891	4 241	4 505
Non-residents	TBHE	25 575†	28 707	23 978	33 734	43 571	44 582	47 470	50 585	79 367	85 820
Acceptances granted											
UK building societies	TBHF	..	..	..	..	..	..	..	..	–	–
UK public sector	TBHG	..	..	..	..	..	..	..	..	–	–
Other UK residents	TBHH	..	..	..	..	..	..	..	..	18 573	15 241
Non-residents	TBHI	..	..	..	..	..	..	..	..	1 379	1 327
Bills											
Treasury bills	TBHJ	1 898†	3 242	4 045	1 952	1 561	4 166	10 404	1 652	554	779
UK bank bills	TBHA	..	..	..	..	..	..	..	..	18 221	14 110
UK building societies	TBHK	..	..	..	..	..	..	..	..	–	–
Other UK	TBHL	..	..	..	..	..	..	..	..	1 117	1 220
Non-residents	TBHM	..	..	..	..	..	..	..	..	309	207
Claims under sale and repurchase agreements											
of which British govt. securities	TBHT	–	–	–	–	–	–	–	26 030†	47 158	56 639
UK banks	TBHO	–	–	–	–	–	–	–	16 853†	27 611	41 969
UK building societies	TBHP	..	..	..	..	..	..	..	..	345	134
UK public sector	TBHR	..	..	..	..	..	..	..	..	21 283	23 803
Other UK residents	TBHR	..	..	..	..	..	..	..	..	21 283	23 803
Non-residents	TBHS	..	..	..	..	..	..	..	..	6 873	5 907
Advances											
UK public sector	TBHU	1 761†	1 701	1 788	2 397	3 309	3 526	3 549	2 912	3 872	3 402
Other UK residents	TBHV	318 294†	359 586	366 459	366 096	371 086	380 615	426 936	460 107	636 162	672 812
Non-residents	TBHW	14 952†	14 863	13 825	12 336	12 268	12 163	13 403	16 759	21 102	21 040
Banking dept. lending to central govt. (net)	TBNU	1 321†	1 671	1 526	1 277	–5 407	–947	–1 801	948	–2 741	..
Investments											
British government stocks	TBHX	5 190†	4 201	3 873	4 848	13 753	14 366	17 135	19 910	23 078	14 714
Other public sector	TBHY	262†	306	334	236	204	303	409	303	282	215
UK banks	TBHZ	..	..	..	..	..	..	..	..	11 922	13 415
UK building societies	TBIA	2 531†	3 105	3 966	4 767	5 492	4 769	4 639	4 898	2 875	2 222
Other UK residents	TBIB	..	..	..	..	..	..	..	..	45 133	48 780
Non-residents	TBIC	..	..	..	..	..	..	..	..	9 342	11 833
Items in suspense and collection	TBID	15 383	18 070	17 356	16 426	17 135	16 477	17 613	19 830	23 526	23 888
Accrued amounts receivable	TBIE	..	..	..	..	..	..	..	..	15 003	17 351
Other assets	TBIF	10 506	12 045	12 346	11 724	11 146	10 242	10 525	10 267	12 218	12 593
Total sterling assets	TBIG	541 933	609 785	597 389	617 158	630 527	668 960	755 937	852 075	1 192 263	1 259 219
Foreign currency assets											
Market loans and advances											
UK banks	TBIH	75 596	72 229	72 961	87 596	96 129	99 885	98 305	83 920	90 367	72 262
UK banks' CDs etc.	TBII	8 220	10 161	11 081	11 239	8 642	7 226	9 345	7 793	13 634	11 064
UK building societies CDs etc. and deposits	TBIJ	..	..	..	..	..	..	..	..	83	259
UK public sector	TBIK	50	39	29	4 061	3 421	28	38	36	25	45
Other UK residents	TBIL	77 923	70 419	59 592	65 836	71 326	76 538	102 061	72 213	76 356	83 968
Non-residents	TBIM	473 219	446 041	412 552	504 567	489 439	534 589	604 783	505 953	598 541	616 832
Claims under sale and repurchase agreement											
UK banks	TBIO	–	–	–	–	–	–	–	21 189	24 184	31 900
UK building societies	TBIP	..	..	..	..	..	..	..	..	–	–
UK public sector	TBIQ	..	..	..	..	..	..	..	..	22	–
Other UK residents	TBIR	..	..	..	..	..	..	..	..	55 945	39 764
Non-residents	TBIS	–	–	–	–	–	–	–	75 376	121 101	147 562
Acceptances granted	TBIT	..	..	..	..	..	..	..	..	743	729
Bills	TBIU	4 556	5 562	7 835	11 784	15 398	10 490	14 429	9 660	12 728	15 240
Investments											
British government stocks	TBIV	..	..	..	..	..	..	..	..	3 453	4 755
Other public sector	TBIW	..	..	..	..	..	..	..	..	–	–
UK banks	TBIX	..	..	..	..	..	..	..	..	2 850	4 310
UK building societies	TBIY	109	141	335	701	889	997	728	701	414	526
Other UK residents	TBIZ	..	..	..	..	..	..	..	..	4 055	4 584
Non-residents	TBJA	42 281†	44 562	53 940	79 547	114 904	128 507	154 398	161 045	186 289	234 564
Items in suspense and collection	TBJB	5 049†	3 753	4 557	9 981	19 371	15 572	17 598	21 664	40 175	30 228
Accrued amounts receivable	TBJC	..	..	..	..	..	..	..	..	23 672	27 820
Other assets	TBJD	2 539†	1 510	1 370	1 586	2 791	1 868	1 817	1 809	2 214	3 111
Total foreign currency assets	TBJE	691 733†	656 645	626 368	780 403	828 419	882 624	1 011 358	1 025 017	1 256 850	1 329 525
Total assets	TBJF	2 467 333†	2 532 860	2 447 513	2 795 122	2 917 892	3 103 168	3 534 590	3 754 183	2 449 113	2 588 743
Holdings of own sterling acceptances	TBJG	20 643†	20 286	19 081	21 988	21 814	18 274	18 721	21 521	1 823	2 137
Holdings of own FC acceptances	TBJH	4 529†	4 824	5 474	1 439	914	640	1 153	1 031	291	168
Eligible banks' total sterling acceptances	TBJI	20 260†	19 929	18 870	21 812	21 726	18 168	18 476	21 220	21 411	18 663
Eligible liabilities	TBJJ	356 611†	400 584	401 451	406 885	425 353	440 032	507 279	564 648	768 082	808 798

See footnote on first part of this table.

Source: Bank of England: 020 7601 3236

22.4 Industrial analysis of bank lending to UK residents[1]

£ millions

	UK residents: Total to	UK residents: of which sterling	Agriculture, hunting and forestry	Fishing	Mining & quarrying	Manufacturing: Total	Manufacturing: Food, beverages & tobacco	Manufacturing: Textiles & leather	Manufacturing: Pulp, paper, publishing & printing
Amounts outstanding (sterling & other currencies)									
Loans & advances (including under repo & sterling commercial paper)									
	TBOA	TBOB	TBOC	TBOD	TBOE	TBOF	TBOG	TBOH	TBOI
1997	793 907	661 560	6 476	333	4 997	45 073	8 579	2 056	5 471
1998	824 053	700 275	7 013	336	5 299	52 531	9 819	2 155	7 346
Acceptances									
	TBQA	TBQB	TBQC	TBQD	TBQE	TBQF	TBQG	TBQH	TBQI
1997	19 216†	18 573	36	–	494	5 901	2 857	118	218
1998	15 920	15 241	3	–	351	5 889	2 993	88	276
Total									
	TBSA		TBSC	TBSD	TBSE	TBSF	TBSG	TBSH	TBSI
1997	812 802		6 512	333	5 491	50 974	11 436	2 174	5 689
1998	839 633		7 015	336	5 650	58 421	12 812	2 244	7 621
of which in sterling									
	TBUA		TBUC	TBUD	TBUE	TBUF	TBUG	TBUH	TBUI
1997	680 133		6 443	326	3 584	39 923	10 258	1 799	4 537
1998	715 516		6 942	329	3 399	41 508	10 086	1 822	5 048
Facilities granted									
	TCAA		TCAC	TCAD	TCAE	TCAF	TCAG	TCAH	TCAI
1997	1 066 387		18 143	398	12 164	110 274	27 233	4 039	11 650
1998	1 060 474		9 444	427	10 105	113 673	26 178	3 888	14 979
of which in sterling									
	TCCA		TCCC	TCCD	TCCE	TCCF	TCCG	TCCH	TCCI
1997	833 932		8 811	385	5 285	73 636	18 228	3 230	8 115
1998	866 812		9 321	420	4 788	69 933	15 450	3 193	8 169

	Manufacturing: Chemicals, man-made fibres, rubber & plastics	Manufacturing: Non-metallic mineral products & metals	Manufacturing: Machinery, equipment & transport equipment	Manufacturing: Electrical, medical & optical equipment	Manufacturing: Other manufacturing	Electricity, gas and water supply: Electricity, gas & heated water	Electricity, gas and water supply: Cold water purification & supply	Construction
Amounts outstanding (sterling & other currencies)								
Loans & advances (including under repo & sterling commercial paper)								
	TBOJ	TBOK	TBOL	TBOM	TBON	TBOO	TBOP	TBOQ
1997	5 844	5 812	6 733	5 515	5 063	6 232	1 017	8 080
1998	7 794	6 130	8 036	6 062	5 190	6 046	990	8 667
Acceptances								
	TBQJ	TBQK	TBQL	TBQM	TBQN	TBQO	TBQP	TBQQ
1997	491	368	850	621	378	1 054	–	47
1998	695	253	765	445	372	541	1	101
Total								
	TBSJ	TBSK	TBSL	TBSM	TBSN	TBSO	TBSP	TBSQ
1997	6 335	6 181	7 583	6 136	5 441	7 286	1 017	8 126
1998	8 489	6 385	8 800	6 507	5 562	6 587	991	8 768
of which in sterling								
	TBUJ	TBUK	TBUL	TBUM	TBUN	TBUO	TBUP	TBUQ
1997	4 635	4 343	5 662	4 458	4 232	7 100	975	6 926
1998	5 510	4 408	6 367	4 153	4 114	6 213	956	7 580
Facilities granted								
	TCAJ	TCAK	TCAL	TCAM	TCAN	TCAO	TCAP	TCAQ
1997	17 174	12 432	14 369	12 864	10 513	17 821	4 657	14 336
1998	17 038	11 915	16 080	14 478	9 118	17 883	4 189	14 768
of which in sterling								
	TCCJ	TCCK	TCCL	TCCM	TCCN	TCCO	TCCP	TCCQ
1997	9 107	8 275	9 935	9 216	7 531	14 457	3 813	11 980
1998	9 751	8 206	11 004	7 448	6 662	13 735	3 325	12 749

1 This is a series of statistics based on the Standard Industrial Classification 1992 and comprises loans, advances (including under reverse repos), finance leasing, acceptances, facilities and holdings of sterling commercial paper. It includes lending under the Department of Trade and Industry special scheme for domestic shipbuilding. Holdings of investments and bills and adjustments for transit items are not included. Figures are supplied by monthly reporting banks and grossed to cover quarterly reporters. They exclude lending to building societies and to residents of the Channel Islands and Isle of Man.

Source: Bank of England: 020 7601 3236

22.4 Industrial analysis of bank lending to UK residents[1]

continued

£ millions

	Wholesale and retail trade: Total	Sale & repair of motor vehicles & fuel	Other wholesale trade	Other retail trade & repair	Hotels and restaurants	Transport, storage & communication	Real estate, renting, computer and other business activities: Total	Development, buying, selling, renting of real estate	Renting of machinery & equipment
Amounts outstanding (sterling & other currencies)									
Loans & advances (including under repo & sterling commercial paper)									
	TBOR	TBOS	TBOT	TBOU	TBOV	TBOW	TBOX	TBOY	TBPA
1997	31 934	8 247	12 729	10 958	11 551	15 172	51 489	34 227	3 014
1998	34 697	8 665	12 714	13 318	11 826	15 810	56 670	39 289	3 881
Acceptances									
	TBQR	TBQS	TBQT	TBQU	TBQV	TBQW	TBQX	TBQY	TBRA
1997	3 558	379	2 460	719	168	198	402	41	120
1998	2 169	306	1 153	710	210	176	346	19	83
Total									
	TBSR	TBSS	TBST	TBSU	TBSV	TBSW	TBSX	TBSY	TBTA
1997	35 492	8 626	15 190	11 677	11 719	15 370	51 891	34 269	3 134
1998	36 866	8 972	13 866	14 028	12 037	15 986	57 016	39 307	3 963
of which in sterling									
	TBUR	TBUS	TBUT	TBUU	TBUV	TBUW	TBUX	TBUY	TBXA
1997	29 478	8 342	10 075	11 061	10 992	13 260	48 659	33 352	-7†
1998	29 930	8 490	8 893	12 547	10 779	13 556	53 712	38 120	334
Facilities granted									
	TCAR	TCAS	TCAT	TCAU	TCAV	TCAW	TCAX	TCAY	TCBA
1997	58 589	12 293	24 720	21 576	15 726	36 838	74 102	45 359	4 127
1998	58 135	12 991	21 129	24 015	15 750	28 789	79 842	51 886	5 046
of which in sterling									
	TCCR	TCCS	TCCT	TCCU	TCCV	TCCW	TCCX	TCCY	TCDA
1997	45 504	11 055	16 056	18 392	14 059	26 119	67 399	43 167	3 737
1998	45 954	11 712	14 251	19 991	13 660	21 786	73 533	49 610	4 462

	Real estate, renting, computer and other business activities: Computer & related activities	Legal, accountancy, consultancy & other business activities	Public administration & defence	Education	Health & social work	Recreational, personal & community service activities: Recreational, cultural & sporting activities	Personal & community services activities	Financial intermediation (excl. insurance & pension funds): Total	Financial leasing corporations
Amounts outstanding (sterling & other currencies)									
Loans & advances (including under repo & sterling commercial paper)									
	TBPB	TBPC	TBPD	TBPE	TBPF	TBPH	TBPG	TBPI	TBPJ
1997	1 405	12 842	3 687	2 178	6 401	4 224	3 629	210 185	34 989
1998	1 574	11 927	2 985	2 434	6 498	4 937	3 386	197 647	37 721
Acceptances									
	TBRB	TBRC	TBRD	TBRE	TBRF	TBRH	TBRG	TBRI	TBRJ
1997	49	191	3	–	–	18	40	6 725	1 629
1998	30	214	–	–	1	35	12	5 377	1 518
Total									
	TBTB	TBTC	TBTD	TBTE	TBTF	TBTH	TBTG	TBTI	TBTJ
1997	1 455	13 033	3 690	2 178	6 402	4 242	3 670	216 910	36 617
1998	1 602	12 141	2 985	2 434	6 499	4 972	3 398	203 024	39 239
of which in sterling									
	TBVB	TBVC	TBVD	TBVE	TBVF	TBVH	TBVG	TBVI	TBVJ
1997	980	11 352	3 666	2 160	6 323	3 858	3 349	114 846	35 829
1998	1 221	10 814	2 974	2 411	6 406	4 413	3 043	117 321	38 379
Facilities granted									
	TCBB	TCBC	TCBD	TCBE	TCBF	TCBH	TCBG	TCBI	TCBJ
1997	2 625	21 991	5 453	2 778	7 578	7 362	5 307	271 437	40 762
1998	3 152	19 758	4 830	3 305	8 095	8 177	4 790	236 651	44 601
of which in sterling									
	TCDB	TCDC	TCDD	TCDE	TCDF	TCDH	TCDG	TCDI	TCDJ
1997	1 770	18 725	5 396	2 752	7 427	5 909	4 622	130 307	39 266
1998	2 336	17 125	4 780	3 270	7 876	6 398	4 283	134 804	43 358

See footnote on first part of this table.

Source: Bank of England: 020 7601 3236

22.4 Industrial analysis of bank lending to UK residents[1]

continued

£ millions

	Financial intermediation (excl. insurance & pension funds)								
	Non-bank credit grantors, excl. credit unions	Credit unions	Factoring corporations	Mortgage & housing credit corporations	Investment & unit trusts excl. money market mutual funds	Money market mutual funds	Bank holding companies	Securities dealers (f)	Other financial intermediaries
Amounts outstanding (sterling & other currencies)									
Loans & advances (including under repo & sterling commercial paper)									
	TBPK	TBPL	TBPM	TBPN	TBPO	TBPP	TBPQ	TBPR	TBPS
1997	14 113	37	2 202	9 964	5 585	22	7 737	100 822	34 713
1998	13 665	29	2 746	10 217	4 187	15	10 985	81 761	36 322
Acceptances									
	TBRK	TBRL	TBRM	TBRN	TBRO	TBRP	TBRQ	TBRR	TBRS
1997	1 906†	–	63	44	177	–	167†	163	2 576
1998	1 638	–	66	22	59	–	46	151	1 878
Total									
	TBTK	TBTL	TBTM	TBTN	TBTO	TBTP	TBTQ	TBTR	TBTS
1997	16 018†	37	2 264	10 008	5 762	22	7 904†	100 985	37 290
1998	15 303	29	2 812	10 239	4 246	15	11 030	81 912	38 200
of which in sterling									
	TBVK	TBVL	TBVM	TBVN	TBVO	TBVP	TBVQ	TBVR	TBVS
1997	14 292†	37	2 226	9 940	3 046	22	5 295†	25 119	19 041
1998	13 774	6	2 671	10 239	2 669	12	8 760	21 448	19 483
Facilities granted									
	TCBK	TCBL	TCBM	TCBN	TCBO	TCBP	TCBQ	TCBR	TCBS
1997	20 468†	53	2 545	10 466	10 135	30	9 016†	125 040	52 922
1998	18 214	31	3 169	11 341	7 199	15	11 463	92 352	48 265
of which in sterling									
	TCDK	TCDL	TCDM	TCDN	TCDO	TCDP	TCDQ	TCDR	TCDS
1997	16 214†	53	2 474	10 373	5 160	30	5 555†	27 605	23 578
1998	15 424	6	2 955	11 335	5 034	13	9 038	23 344	24 296

		Activities auxiliary to financial intermediation		Individuals & individual trusts		
	Insurance companies & pension funds	Fund management activities	Other	Total	Lending secured on dwellings inc. bridging finance	Other loans & advances
Amounts outstanding (sterling & other currencies)						
Loans & advances (including under repo & sterling commercial paper)						
	TBPT	TBPU	TBPV	TBPW	TBPX	TBPY
1997	12 020	1 681	2 736	364 814	300 491	64 323
1998	15 621	1 681	2 412	387 217	314 592	72 625
Acceptances						
	TBRT	TBRU	TBRV			
1997	197	15	37			
1998	320	7	58			
Total						
	TBTT	TBTU	TBTV	TBTW	TBTX	TBTY
1997	12 216	1 696	2 773	364 814	300 491	64 323
1998	15 941	1 038	2 470	387 217	314 592	72 625
of which in sterling						
	TBVT	TBVU	TBVV	TBVW	TBVX	TBVY
1997	11 293	968	1 415	364 589	300 444	64 145
1998	14 950	410	1 680	387 022	314 558	72 464
Facilities granted						
	TCBT	TCBU	TCBV	TCBW	TCBX	TCBY
1997	19 024	3 015	3 998	386 605	307 466	79 139
1998	23 715	1 751	3 559	412 486	322 671	89 815
of which in sterling						
	TCDT	TCDU	TCDV	TCDW	TCDX	TCDY
1997	16 211	1 532	2 025	386 305	307 405	78 900
1998	20 507	935	2 536	412 246	322 635	89 611

See footnote on first part of this table.

Source: Bank of England: 020 7601 3236

22.5 Industrial analysis of bank deposits from UK residents[1]

£ millions

	Total from UK residents	Agriculture, hunting and forestry	Fishing	Mining & quarrying	Manufacturing: Total	Manufacturing: Food, beverages & tobacco	Manufacturing: Textiles & leather	Manufacturing: Pulp, paper, publishing & printing
Amounts outstanding (sterling & other currencies)								
Deposit liabilities (including under repos)	TDAA	TDAB	TDAC	TDAD	TDAE	TDAF	TDAG	TDAH
1997	683 013	3 082†	102	3 707†	35 778†	4 332†	1 319†	3 764†
1998	712 068	3 065	111	3 669	34 681	3 618	1 350	3 119
of which in sterling	TDCA	TDCB	TDCC	TDCD	TDCE	TDCF	TDCG	TDCH
1997	592 861	3 049†	97†	2 990†	28 245†	3 676†	1 151†	2 774†
1998	624 664	3 016	101	2 677	28 163	2 871	1 107	2 675

	Manufacturing: Chemicals, man-made fibres, rubber & plastics	Manufacturing: Non-metallic mineral products & metals	Manufacturing: Machinery, equipment & transport equipment	Manufacturing: Electrical, medical & optical equipment	Manufacturing: Other manufacturing	Electricity, gas and water supply: Electricity, gas & heated water	Electricity, gas and water supply: Cold water purification & supply	Construction
Amounts outstanding (sterling & other currencies)								
Deposit liabilities (including under repos)	TDAI	TDAJ	TDAK	TDAL	TDAM	TDAN	TDAO	TDAP
1997	5 382†	4 965†	6 599†	5 900†	3 517†	4 135†	550†	7 273†
1998	4 945	4 789	6 781	5 869	4 211	4 102	756	7 382
of which in sterling	TDCI	TDCJ	TDCK	TDCL	TDCM	TDCN	TDCO	TDCP
1997	3 801†	4 429†	5 330†	4 280†	2 803†	3 955†	491†	7 054†
1998	3 710	4 377	5 393	4 440	3 590	3 842	730	7 132

	Wholesale and retail trade: Total	Wholesale and retail trade: Sale & repair of motor vehicles & fuel	Wholesale and retail trade: Other wholesale trade	Wholesale and retail trade: Other retail trade & repair	Hotels and restaurants	Transport, storage & communication	Real estate, renting, computer and other business activities: Total	Real estate, renting, computer and other business activities: Development, buying, selling, renting of real estate	Real estate, renting, computer and other business activities: Renting of machinery & equipment
Amounts outstanding (sterling & other currencies)									
Deposit liabilities (including under repos)	TDAQ	TDAR	TDAS	TDAT	TDAU	TDAV	TDAW	TDAX	TDAY
1997	20 946†	2 324†	10 663†	7 960†	2 502†	10 657†	41 282†	10 916†	734†
1998	20 874	2 321	9 896	8 658	2 407	10 678	46 229	13 435	872
of which in sterling	TDCQ	TDCR	TDCS	TDCT	TDCU	TDCV	TDCW	TDCX	TDCY
1997	17 114†	2 220†	7 233†	7 661†	2 350†	8 850†	38 490†	10 794†	592†
1998	17 785	2 217	7 314	8 254	2 370	8 692	43 214	13 077	770

1 This is a series of statistics based on the Standard Industrial Classification 1992 and includes borrowing under sale and repurchase agreements (repo). Adjustments for transit items are not included. Figures are supplied by monthly reporting banks and grossed to cover quarterly reporters. They exclude deposits from building societies and residents of the Channel Islands and Isle of Man.

Source: Bank of England: 020 7601 3236

22.5 Industrial analysis of bank deposits from UK residents[1]

continued

£ millions

	Real estate, renting, computer and other business activities					Recreational, personal & community service activities		Financial intermediation (excl. insurance & pension funds)	
	Computer & related activities	Legal, accountancy, consultancy & other business activities	Public administration & defence	Education	Health & social work	Recreational, cultural & sporting activities	Personal & community services activities	Total	Financial leasing corporations
Amounts outstanding (sterling & other currencies)									
Deposit liabilities (including under repos)	TDAZ	TDBA	TDBB	TDBC	TDBD	TDBF	TDBE	TDBG	TDBH
1997	3 098†	26 535†	10 601†	3 579†	6 066†	4 501†	9 054†	122 515†	2 886†
1998	4 007	27 914	10 270	3 854	6 664	4 551	9 173	123 382	3 871
of which in sterling	TDCZ	TDDA	TDDB	TDDC	TDDD	TDDF	TDDE	TDDG	TCDJ
1997	2 452†	24 653†	16 503†	3 462†	6 005†	4 033†	8 586†	67 258†	39 266
1998	3 329	26 038	10 145	3 758	6 579	4 058	8 803	69 681	43 358

	Financial intermediation (excl. insurance & pension funds)								
	Non-bank credit grantors, excl. credit unions	Credit unions	Factoring corporations	Mortgage & housing credit corporations	Investment & unit trusts excl. money market mutual funds	Money market mutual funds	Bank holding companies	Securities dealers	Other financial intermediaries
Amounts outstanding (sterling & other currencies)									
Deposit liabilities (including under repos)	TDBI	TDBJ	TDBK	TDBL	TDBM	TDBN	TDBO	TDBP	TDBQ
1997	5 722†	183†	268†	632†	19 343†	546†	3 285†	56 076†	33 574†
1998	2 937	103	301	567	15 268	563	2 872	55 440	41 462
of which in sterling	TDDI	TDDJ	TDDK	TDDL	TDDM	TDDN	TDDO	TDDP	TDDQ
1997	4 194†	183†	262†	615†	13 938†	509†	3 242	21 622	20 392†
1998	2 455	103	285	562	11 891	530	2 451	22 927	24 926

	Insurance companies & pension funds	Activities auxiliary to financial intermediation		Individuals & individual trusts
		Placed by fund managers	Other	
Amounts outstanding (sterling & other currencies)				
Deposit liabilities (including under repos)	TDBR	TDBS	TDBT	TDBU
1997	48 989†	20 064†	8 685†	312 943†
1998	56 183	27 781	9 122	327 134
of which in sterling	TDDR	TDDS	TDDT	TDDU
1997	41 538†	15 052†	6 143†	311 595†
1998	49 191	22 425	6 732	325 568

See footnote on first part of this table.

Source: Bank of England: 020 7601 3236

22.6 Public sector net cash requirement and other counterparts to changes in money stock during the year

£ millions, not seasonally adjusted

		1988	1989	1990	1991	1992	1993	1994	1995	1996	1997	1998
Public sector net cash requirement (surplus-)	ABEN	−12 132†	−9 282	−1 687	6 961	28 439	43 374	39 344	35 446	24 688	11 857	−6 532
Sales(-) of public sector debt to M4 private sector	KHGZ	3 809	12 736	−767†	−5 685	−20 268	−30 799	−23 426	−21 688	−19 241	−14 126	1 550
M4 lending[1]	AVBS	82 362†	89 132	70 597	37 050	25 536	22 575	31 605	57 743	59 129	68 311	63 378
External and foreign currency finance of the public sector	KHJP	3 462	−3 385	2 971†	−3 102	−12 703	−10 829	361	−3 688	−10 921	−4 553	−4 760
Other external and foreign currency flows[2]	AVBW	−11 957	−13 657	−12 259	1 216	7 649	14 682	−7 497†	−3 401	17 917	24 922	12 965
Net non-deposit liabilities (increase-)	AVBX	−13 036†	−10 789	−8 484	−8 469	−10 794	−15 204	−15 099	−8 305	−12 137	−6 241	−7 842
Money stock (M4)	AUZI	52 514	64 755	50 371	27 971	17 902	23 798†	25 292	56 107	59 433	80 169	58 758

1 Bank and building society lending, plus holdings of commercial bills by the Issue Department of the Bank of England.
2 Including sterling lending to non-residents sector.

Source: Bank of England: 020 7601 5468

22.7 Money stock and liquidity

£ millions

		1988	1989	1990	1991	1992	1993	1994	1995	1996	1997	1998
Amounts outstanding at end-year												
Notes and coin in circulation with the M4 private sector[1]	VQKT	14 446	15 355	15 253	15 690	16 770	17 795	18 749	20 007	20 843	22 477	23 318
UK private sector sterling non-interest bearing sight deposits[2,3,6,7]	AUYA	34 829	31 240	29 282	29 388	28 659	34 515	33 487	35 435	38 433	38 936	38 429
Money stock (M2)[5,6,7]	VQXV	256 317	279 953	309 810	335 928	373 243	394 510	410 469	437 102	460 294	484 828	515 786
Money stock M4[6,7,8]	AUYM	358 233	426 322	477 128†	504 114	517 786	543 860	566 847	623 444	682 941	722 199	781 882
Changes during the year[4]												
Notes and coin in circulation with the M4 private sector[1]	VQLU	1 043	898	–102	437	1 087	1 025	954	1 256	836	1 646	1 138
UK private sector sterling non-interest bearing sight deposits[2,3,6,7]	AUZA	4 062	–3 928	–2 056	–780	–318	4 201	1 084	1 783	3 527	5 366†	536
Money stock (M2)[5,6,7]	AUZE	30 977†	21 064	18 332	22 988	9 811	18 851	17 614	26 570	24 358	36 013	31 471
Money stock M4[6,7,8]	AUZI	52 514	64 755	50 371	27 971	17 902	23 798†	25 292	56 107	59 433	80 169	58 758

1 The estimates of levels of coin in circulation include allowance for wastage, hoarding, etc.
2 Non-interest bearing deposits are confined to those with institutions included in the United Kingdom banks sector (See Table 22.3).
3 Banks' data from 1986 onwards are adjusted by a proportion of sterling items in transit and suspense.
4 As far as possible the changes exclude the effect of changes in the number of contributors to the series, and also of the introduction of new statistical returns. Changes are not seasonally adjusted.
5 With effect from the flow for December 1992, M2 comprises the UK non-monetary financial institutions and non-public sector, i.e. M4 private sector's holdings of notes and coin together with its sterling denominated retail deposits with UK monetary financial institutions.
6 Revised rules on netting of customers' credit balances against their borrowing increased the UK private sector's outstanding balances of deposits and borrowing by £2.5 bn at end-December 1993. Within retail deposits, £1.7 bn of the increase was in NIB bank deposits. Re-netting during 1994 amounted to £1.7 bn. Changes data have been adjusted to exclude these effects.
7 Building societies' data from 1992 onwards are affected by the revised treatment of building society transit items within M4.
8 From end-1986, the discrepancy in banks' reporting of their domestic sterling interbank business has been allocated mainly to the wholesale component of M4, rather than to net non-deposit sterling liabilities.

Source: Bank of England: 020 7601 5468

22.8 Selected retail banks' base rate[1]

Percentage rates operative between dates shown

Rate

Date of change	New rate
1986 Jan 9	12.50
Mar 19	11.50
Apr 8	11.00-11.50
Apr 9	11.00
Apr 21	10.50
May 23	10.00-10.50
May 27	10.00
Oct 14	10.00-11.00
Oct 15	11.00
1987 Mar 10	10.50
Mar 18	10.00-10.50
Mar 19	10.00
Apr 28	9.50-10.00
Apr 29	9.50
May 11	9.00
Aug 6	9.00-10.00
Aug 7	10.00
Oct 23	9.50-10.00
Oct 29	9.50
Nov 4	9.00-9.50
Nov 5	9.00
Dec 4	8.50
1988 Feb 2	9.00
Mar 17	8.50-9.00
Mar 18	8.50
Apr 11	8.00
May 17	7.50-8.00
May 18	7.50
Jun 2	7.50-8.00
Jun 3	8.00
Jun 6	8.00-8.50
Jun 7	8.50
Jun 22	8.50-9.00

Date of change	New rate
Jun 23	9.00
Jun 28	9.00-9.50
Jun 29	9.50
Jul 4	9.50-10.00
Jul 5	10.00
Jul 18	10.00-10.50
Jul 19	10.50
Aug 8	10.50-11.00
Aug 9	11.00
Aug 25	11.00-12.00
Aug 26	12.00
Nov 25	13.00
1989 May 24	14.00
Oct 5	15.00
1990 Oct 8	14.00
1991 Feb 13	13.50
Feb 27	13.00
Mar 22	12.50
Apr 12	12.00
May 24	11.50
Jul 12	11.00
Sep 4	10.50
1992 May 5	10.00
Sep 16[1]	12.00
Sep 17[1]	10.00-12.00
Sep 18	10.00
Sep 22	9.00
Oct 16	8.00-9.00
Oct 19	8.00
Nov 13	7.00

Date of change	New rate
1993 Jan 26	6.00
Nov 23	5.50
1994 Feb 8	5.25
Sep 12	5.75
Dec 7	6.25
1995 Feb 2[1]	6.25-6.75
Feb 3	6.75
Dec 13	6.50
1996 Jan 18	6.25
Mar 8	6.00
Jun 6	5.75
Oct 30	5.75-6.00
Oct 31	6.00
1997 May 6	6.25
Jun 6	6.25-6.50
Jun 9	6.50
Jul 10	6.75
Aug 7	7.00
Nov 6	7.25
1998 Jun 4	7.50
Oct 8	7.25
Nov 5	6.75
Dec 10	6.25
1999 Jan 7	6.00
Feb 4	5.50
Apr 8	5.25
Jun 10	5.00
Sep 8	5.25

1 Data obtained from Barclays Bank, Lloyds TSB Bank, HSBC, National Westminster Bank whose rates are used to compile this series. Where all the rates did not change on the same day a spread is shown.

Source: Bank of England: 020 7601 4342

22.9 Average three month sterling money market rates[4]

Rate per cent

	1988	1989	1990	1991	1992	1993	1994	1995	1996	1997	1998
Treasury bills:[1] KDMM											
January	8.37	12.45	14.50	13.00	9.97	6.05	4.88	5.93	6.08	6.01	6.84
February	8.79	12.39	14.45	12.39	9.80	5.37	4.76	6.16	5.96	5.81	6.88
March	8.27	12.41	14.57	11.64	10.10	5.38	4.83	6.09	5.81	5.92	6.95
April	7.74	12.47	14.59	11.25	9.97	5.33	4.88	6.30	5.80	6.09	7.02
May	7.54	12.54	14.50	10.84	9.42	5.30	4.81	6.20	5.82	6.15	6.99
June	8.88	13.59	14.38	10.72	9.42	5.19	4.88	6.37	5.58	6.37	7.29
July	10.05	13.29	14.32	10.52	9.43	5.13	5.09	6.62	5.49	6.60	7.22
August	11.13	13.32	14.31	10.20	9.65	5.06	5.34	6.59	5.54	6.81	7.19
September	11.53	13.44	14.26	9.66	9.16	5.16	5.39	6.52	5.54	6.88	6.94
October	11.54	14.46	13.37	9.86	7.47	5.15	5.44	6.57	5.55	6.94	6.54
November	12.07	14.45	12.92	9.98	6.49	4.95	5.63	6.44	6.02	7.09	6.31
December	12.54	14.50	12.96	10.10	6.39	4.87	5.87	6.21	6.08	7.04	5.72
Eligible bill:[2] KDMY											
January	8.42	12.58	14.55	13.34	10.08	6.43	5.07	6.41	6.18	6.15	7.28
February	8.88	12.54	14.53	12.55	9.87	5.59	4.95	6.54	6.00	5.99	7.24
March	8.51	12.56	14.67	11.75	10.14	5.62	4.90	6.40	5.87	6.01	7.25
April	7.90	12.66	14.63	11.44	10.18	5.66	4.93	6.44	5.83	6.26	7.24
May	7.63	12.70	14.54	11.05	9.56	5.67	4.92	6.54	5.85	6.31	7.20
June	8.63	13.65	14.40	10.78	9.52	5.62	4.94	6.44	5.66	6.50	7.41
July	10.15	13.41	14.33	10.61	9.59	5.63	5.01	6.66	5.57	6.80	7.49
August	10.92	13.36	14.34	10.39	9.80	5.59	5.44	6.64	5.60	6.95	7.40
September	11.70	13.54	14.31	9.85	9.43	5.58	5.54	6.55	5.63	7.02	7.20
October	11.61	14.41	13.40	10.02	7.78	5.50	5.74	6.58	5.76	7.10	6.91
November	11.81	14.54	13.10	10.09	6.70	5.31	5.89	6.49	6.10	7.27	6.52
December	12.63	14.37	13.14	10.21	6.69	5.04	6.12	6.29	6.18	7.31	6.05
Interbank rate:[3] AMIJ											
January	8.93	13.12	15.16	13.97	10.65	6.94	5.39	6.56	6.36	6.32	7.48
February	9.27	13.02	15.11	13.25	10.37	6.16	5.22	6.75	6.16	6.19	7.46
March	8.89	13.03	15.29	12.40	10.62	5.98	5.16	6.66	6.05	6.20	7.48
April	8.30	13.13	15.21	11.95	10.62	5.98	5.21	6.67	6.00	6.38	7.44
May	8.03	13.14	15.15	11.53	10.06	5.97	5.17	6.72	6.02	6.45	7.41
June	8.93	14.15	14.97	11.24	9.98	5.89	5.13	6.64	5.85	6.66	7.63
July	10.55	13.90	14.95	11.09	10.15	5.95	5.20	6.80	5.73	6.96	7.71
August	11.40	13.85	15.00	10.89	10.35	5.84	5.53	6.79	5.75	7.15	7.66
September	12.15	14.02	14.91	10.29	9.99	5.91	5.67	6.72	5.77	7.21	7.38
October	12.02	15.02	14.03	10.40	8.32	5.76	5.91	6.73	5.94	7.26	7.14
November	12.33	15.11	13.64	10.48	7.21	5.57	6.06	6.64	6.30	7.54	6.89
December	13.15	15.12	13.81	10.78	7.15	5.33	6.37	6.49	6.35	7.62	6.38
Certificate of deposits:[3] KOSA											
January	8.82	13.05	15.10	13.86	10.53	6.82	5.29	6.48	6.31	6.27	7.44
February	9.10	12.96	15.05	13.18	10.27	6.00	5.14	6.67	6.11	6.14	7.42
March	8.77	12.96	15.20	12.33	10.51	5.87	5.09	6.59	6.01	6.15	7.43
April	8.16	13.03	15.15	11.86	10.50	5.84	5.12	6.62	5.96	6.33	7.40
May	7.88	13.03	15.09	11.44	9.95	5.86	5.10	6.66	5.98	6.39	7.37
June	8.78	14.04	14.90	11.13	9.89	5.77	5.06	6.56	5.79	6.62	7.59
July	10.35	13.85	14.90	10.99	10.04	5.81	5.11	6.72	5.69	6.92	7.66
August	11.13	13.80	14.94	10.79	10.21	5.71	5.45	6.73	5.71	7.12	7.61
September	11.99	13.98	14.87	10.21	9.83	5.79	5.58	6.66	5.74	7.17	7.34
October	11.90	14.94	13.97	10.30	8.17	5.65	5.81	6.68	5.89	7.22	7.09
November	12.17	15.04	13.53	10.37	7.04	5.46	5.97	6.58	6.25	7.50	6.82
December	13.04	15.08	13.68	10.67	7.02	5.23	6.25	6.44	6.29	7.57	6.32
Local authority deposits:[3] KDPX											
January	8.88	13.04	15.02	13.94	10.61	6.84	5.30	6.52	6.31	6.27	7.43
February	9.25	13.01	15.02	13.24	10.34	6.09	5.12	6.69	6.13	6.15	7.40
March	8.86	12.96	15.20	12.45	10.57	5.93	5.08	6.59	6.02	6.14	7.40
April	8.22	13.09	15.19	11.89	10.59	5.95	5.14	6.60	5.98	6.33	7.38
May	7.98	13.10	15.12	11.50	10.02	5.94	5.12	6.68	6.00	6.38	7.34
June	8.86	14.04	14.94	11.23	9.94	5.86	5.08	6.58	5.80	6.57	7.56
July	10.44	13.86	14.91	11.05	10.10	5.90	5.18	6.76	5.69	6.90	7.64
August	11.28	13.77	14.96	10.88	10.24	5.80	5.44	6.74	5.71	7.11	7.55
September	12.06	13.94	14.92	10.32	9.82	5.85	5.62	6.65	5.72	7.19	7.35
October	11.95	14.93	14.03	10.45	8.28	5.70	5.84	6.68	5.86	7.21	7.08
November	12.24	15.00	13.60	10.44	7.17	5.51	5.99	6.60	6.24	7.49	6.85
December	13.05	15.01	13.78	10.75	7.08	5.26	6.32	6.43	6.30	7.56	6.35

1 Average rate of discount at weekly (Friday) tender.
2 Working day average of the mean of the bid and offer discount rate.
3 Working day average of the mean bid and offer yield.
4 A full definition of these series is given in Section 7 of the ONS Financial Statistics Explanatory Handbook.

Source: Bank of England: 020 7601 4342

22.10 Average foreign exchange rates[1,2]

	1988	1989	1990	1991	1992	1993	1994	1995	1996	1997	1998
Sterling exchange rate index (1990 = 100) AJHX											
January	102.0920†	108.1170	96.5021	102.9858	99.5979	88.8336	92.0038	88.5324	83.2346	95.8913	104.6623
February	101.6192†	107.5687	98.3450	102.9625	99.7257	84.7457	90.4067	87.3905	83.8176	97.4245	104.6965
March	104.9260†	106.0021	95.4663	101.8699	98.9134	86.3847	89.7636	85.6399	83.5274	97.3640	106.7993
April	107.0049†	105.4098	95.4921	101.5699	100.2161	88.8721	89.2014	84.4638	83.7533	99.4789	107.0961
May	107.4803†	104.2779	96.5078	100.9390	101.7402	88.9631	88.8886	84.3257	84.5903	99.0484	103.3721
June	105.1148†	100.9669	99.0978	99.4506	101.7323	88.3278	89.1179	84.0808	85.9614	100.4005	105.4381
July	104.7294†	101.9619	102.3685	99.6321	101.1578	90.4918	87.9294	83.5918	85.7003	104.5099	105.3140
August	106.3019†	101.3353	104.2089	99.8946	100.5146	90.2605	87.7557	84.4026	84.7153	102.5315	104.5648
September	104.8220†	100.9536	102.7583	100.1457	96.5358	89.7669	88.1372	84.8425	86.1091	100.3838	103.3355
October	105.7744†	99.0453	103.7342	99.6039	88.7454	89.3793	89.0762	84.2733	88.3683	101.1132	100.6723
November	106.5869†	96.8825	102.9682	99.9458	86.2431	90.2148	89.0754	83.2937	91.9568	103.7891	100.5664
December	107.8140†	95.1479	101.9799	100.0058	88.1800	90.9959	89.0820	82.8901	93.8311	104.3532	100.3812
Sterling/US Dollar AJFA											
January	1.7981	1.7748	1.6523	1.9348	1.8127	1.5325	1.4940	1.5747	1.5306	1.6587	1.6353
February	1.7578	1.7541	1.6962	1.9655	1.7781	1.4386	1.4799	1.5720	1.5364	1.6246	1.6407
March	1.8324	1.7146	1.6246	1.8265	1.7238	1.4625	1.4917	1.6005	1.5271	1.6063	1.6620
April	1.8765	1.7022	1.6374	1.7502	1.7576	1.5472	1.4837	1.6074	1.5145	1.6295	1.6733
May	1.8694	1.6295	1.6778	1.7252	1.8109	1.5481	1.5029	1.5868	1.5152	1.6334	1.6366
June	1.7790	1.5532	1.7094	1.6499	1.8556	1.5099	1.5252	1.5949	1.5418	1.6446	1.6507
July	1.7042	1.6244	1.8081	1.6503	1.9186	1.4963	1.5463	1.5953	1.5539	1.6702	1.6437
August	1.6983	1.5955	1.8998	1.6841	1.9412	1.4911	1.5427	1.5681	1.5502	1.6034	1.6320
September	1.6833	1.5715	1.8786	1.7249	1.8559	1.5261	1.5651	1.5584	1.5597	1.6015	1.6822
October	1.7382	1.5882	1.9454	1.7226	1.6577	1.5037	1.6057	1.5779	1.5862	1.6329	1.6952
November	1.8090	1.5724	1.9647	1.7787	1.5275	1.4806	1.5886	1.5623	1.6626	1.6890	1.6620
December	1.8282	1.5957	1.9257	1.8258	1.5536	1.4904	1.5595	1.5398	1.6647	1.6597	1.6705
Sterling/Deutsche Mark AJFH											
January	2.9770	3.2558†	2.7934	2.9195	2.8564	2.4748	2.6039	2.4089	2.2364	2.6603	2.9709
February	2.9825†	3.2451	2.8427	2.9083	2.8772	2.3618	2.5666	2.3599	2.2523	2.7218	2.9745
March	3.0725†	3.1994	2.7688	2.9315	2.8645	2.4067	2.5232	2.2490	2.2562	2.7248	3.0364
April	3.1375†	3.1808	2.7615	2.9788	2.8948	2.4679	2.5198	2.2176	2.2819	2.7882	3.0317
May	3.1658†	3.1719	2.7896	2.9634	2.9359	2.4884	2.4913	2.2390	2.3226	2.7820	2.9037
June	3.1235†	3.0721	2.8788	2.9404	2.9171	2.4957	2.4837	2.2324	2.3558	2.8401	2.9584
July	3.1441†	3.0738	2.9616	2.9471	2.8598	2.5660	2.4256	2.2145	2.3362	2.9917	2.9544
August	3.2070	3.0736†	2.9829	2.9362	2.8156	2.5288	2.4128	2.2642	2.2980	2.9497	2.9189
September	3.1425†	3.0650	2.9499	2.9251	2.6826	2.4719	2.4245	2.2764	2.3490	2.8630	2.8599
October	3.1596†	2.9628	2.9654	2.9111	2.4551	2.4629	2.4414	2.2320	2.4236	2.8678	2.7780
November	3.1624†	2.8768	2.9174	2.8840	2.4240	2.5174	2.4458	2.2146	2.5132	2.9257	2.7951
December	3.2066†	2.7764	2.8791	2.8557	2.4551	2.5494	2.4505	2.2176	2.5834	2.9519	2.7876

1 Working day average
2 A full definition of these series is given in Section 7 of the ONS Explanatory Handbook.

Source: Bank of England: 020 7601 4342

22.11 Average FTSE Actuaries share indices[1,2]

	1988	1989	1990	1991	1992	1993	1994	1995	1996	1997	1998
FTSE 100 AJNO											
January	1 767.30	1 891.73	2 367.84	2 106.64	2 520.19	2 790.29	3 431.29	3 028.27	3 715.78	4 166.48	5 242.10
February	1 746.52	2 045.16	2 296.99	2 278.36	2 543.31	2 840.17	3 396.40	3 051.68	3 738.08	4 316.57	5 657.73
March	1 807.48	2 074.01	2 248.53	2 449.81	2 495.24	2 897.07	3 206.10	3 078.24	3 697.50	4 349.78	5 861.80
April	1 786.53	2 065.76	2 191.33	2 514.14	2 549.44	2 837.45	3 130.94	3 198.38	3 792.32	4 312.51	5 974.51
May	1 782.53	2 136.55	2 230.57	2 493.32	2 702.34	2 830.13	3 089.23	3 288.31	3 758.41	4 622.62	5 936.72
June	1 847.21	2 147.33	2 376.98	2 495.14	2 604.44	2 874.70	2 980.32	3 351.55	3 734.02	4 649.40	5 847.04
July	1 856.07	2 246.82	2 361.90	2 530.10	2 443.25	2 850.73	3 036.57	3 426.50	3 707.21	4 843.16	5 987.19
August	1 830.25	2 353.95	2 195.05	2 600.50	2 344.11	3 019.28	3 178.50	3 486.95	3 841.75	4 945.56	5 555.77
September	1 773.85	2 376.46	2 081.02	2 622.16	2 452.41	3 028.12	3 098.35	3 534.27	3 927.10	5 010.35	5 171.92
October	1 842.89	2 201.81	2 091.89	2 580.86	2 587.09	3 125.15	3 046.77	3 531.80	4 020.99	5 145.13	5 063.84
November	1 818.03	2 205.54	2 089.68	2 505.09	2 712.92	3 111.59	3 086.85	3 580.31	3 969.54	4 846.24	5 595.58
December	1 769.68	2 358.54	2 164.10	2 410.50	2 777.08	3 313.68	3 026.62	3 650.05	4 038.89	5 087.53	5 695.61
FT Non-Financials AJMG											
January	992.40	1 064.79	1 291.51	1 115.00	1 353.77	1 496.94	1 818.98	1 631.29	1 919.36	2 104.75	2 455.76
February	986.84	1 158.47	1 249.22	1 204.72	1 374.45	1 522.74	1 818.96	1 631.03	1 941.86	2 150.64	2 586.40
March	1 021.06	1 182.26	1 219.10	1 308.69	1 354.06	1 552.97	1 745.46	1 631.47	1 953.49	2 178.24	2 708.91
April	1 011.11	1 174.75	1 186.16	1 342.30	1 387.84	1 522.32	1 708.89	1 690.67	2 021.61	2 149.68	2 776.91
May	1 009.62	1 214.28	1 204.65	1 331.16	1 476.08	1 522.69	1 690.44	1 738.74	2 013.00	2 219.45	2 847.14
June	1 042.22	1 222.95	1 285.06	1 330.25	1 419.66	1 539.22	1 618.42	1 767.52	1 999.60	2 231.81	2 833.04
July	1 054.34	1 273.48	1 275.24	1 342.46	1 319.06	1 521.63	1 642.28	1 806.86	1 958.89	2 284.21	2 859.68
August	1 044.04	1 323.05	1 184.05	1 379.55	1 258.02	1 614.52	1 726.77	1 846.24	2 002.27	2 326.86	2 658.60
September	1 002.78	1 329.32	1 116.53	1 404.22	1 301.02	1 623.43	1 680.38	1 868.17	2 043.06	2 369.90	2 480.53
October	1 044.29	1 223.70	1 119.39	1 387.17	1 364.42	1 653.67	1 639.72	1 851.30	2 069.12	2 439.08	2 413.68
November	1 034.07	1 212.80	1 113.82	1 353.28	1 425.57	1 643.34	1 654.73	1 854.75	2 040.44	2 332.02	2 630.55
December	994.40	1 280.70	1 145.07	1 299.58	1 473.45	1 733.10	1 622.39	1 879.54	2 056.24	2 401.00	2 651.00
Financials AMAA											
January	1 256.21	1 350.05	1 621.71	1 333.92	1 385.89	1 718.78	2 619.71	2 084.15	2 916.24	3 674.44	4 968.04
February	1 247.33	1 443.61	1 583.14	1 464.83	1 384.60	1 822.76	2 607.79	2 125.71	2 942.72	3 919.61	5 542.60
March	1 293.07	1 439.29	1 520.56	1 591.19	1 346.68	1 888.45	2 324.02	2 170.51	2 837.38	3 876.31	5 732.81
April	1 274.13	1 404.13	1 500.51	1 594.89	1 353.29	1 901.37	2 240.03	2 264.66	2 866.93	3 825.30	5 801.80
May	1 294.51	1 431.67	1 476.32	1 548.06	1 505.57	1 926.91	2 164.18	2 350.00	2 898.78	4 331.40	5 497.77
June	1 359.75	1 401.36	1 556.43	1 507.79	1 452.21	2 011.94	2 111.15	2 411.52	2 867.16	4 281.17	5 274.89
July	1 366.21	1 463.87	1 543.97	1 513.55	1 352.32	2 065.80	2 129.92	2 466.78	2 886.69	4 468.02	5 352.53
August	1 338.44	1 534.84	1 406.57	1 584.38	1 267.47	2 175.74	2 197.58	2 526.20	3 078.43	4 647.61	4 914.36
September	1 288.20	1 560.19	1 284.25	1 605.07	1 348.16	2 174.42	2 181.71	2 599.72	3 149.04	4 634.33	4 410.37
October	1 338.13	1 451.59	1 312.97	1 531.22	1 483.95	2 310.58	2 137.44	2 691.46	3 292.82	4 787.52	4 319.56
November	1 322.25	1 485.83	1 337.08	1 455.47	1 622.84	2 289.99	2 188.27	2 805.33	3 300.68	4 407.61	4 877.76
December	1 288.36	1 607.03	1 386.04	1 348.81	1 645.58	2 502.99	2 134.40	2 879.80	3 418.82	4 756.91	4 955.84
Investments Trusts DEPG											
January	1 271.38	1 505.58	1 949.83	1 497.25	1 822.17	2 064.63	3 058.15	2 609.12	3 079.31	3 198.05	3 436.17
February	1 307.36	1 645.71	1 850.43	1 642.43	1 825.74	2 217.35	3 056.69	2 604.17	3 142.90	3 328.88	3 635.46
March	1 347.03	1 682.71	1 783.47	1 835.51	1 808.41	2 244.24	2 901.35	2 583.43	3 106.91	3 340.35	3 855.17
April	1 331.57	1 686.87	1 757.29	1 863.97	1 826.01	2 225.74	2 844.45	2 648.57	3 216.41	3 234.90	3 935.37
May	1 327.85	1 771.28	1 800.58	1 864.72	1 945.40	2 256.79	2 812.57	2 762.00	3 237.58	3 377.50	4 036.89
June	1 378.24	1 766.20	1 882.72	1 859.45	1 837.53	2 330.08	2 709.20	2 791.09	3 170.26	3 422.71	3 976.40
July	1 419.22	1 836.10	1 878.07	1 854.11	1 726.09	2 372.09	2 725.64	2 874.20	3 075.76	3 453.98	3 960.55
August	1 403.78	1 943.11	1 688.83	1 885.60	1 626.22	2 563.16	2 882.04	2 938.91	3 147.91	3 512.81	3 606.48
September	1 381.18	1 963.65	1 581.00	1 926.31	1 688.05	2 549.64	2 831.12	2 969.59	3 182.58	3 493.30	3 166.36
October	1 429.26	1 847.86	1 537.79	1 905.34	1 801.19	2 663.51	2 732.79	2 924.52	3 204.46	3 534.96	3 118.92
November	1 433.84	1 868.74	1 540.42	1 834.04	1 933.38	2 669.09	2 732.46	2 932.60	3 137.47	3 275.77	3 478.60
December	1 401.17	1 979.77	1 561.66	1 747.99	2 000.31	2 880.07	2 686.29	2 998.07	3 129.25	3 370.41	3 480.98
FTSE All Share AJMA											
January	901.60	976.16	1 184.51	1 010.79	1 203.08	1 351.60	1 710.35	1 501.87	1 818.72	2 041.99	2 455.05
February	896.46	1 059.28	1 146.53	1 095.30	1 218.70	1 384.99	1 709.09	1 506.17	1 839.78	2 105.82	2 624.19
March	927.96	1 076.81	1 115.82	1 190.90	1 199.04	1 415.40	1 619.80	1 511.04	1 836.04	2 121.89	2 740.51
April	918.65	1 066.84	1 087.87	1 217.27	1 225.16	1 393.70	1 582.06	1 566.98	1 892.47	2 092.74	2 799.69
May	920.66	1 099.89	1 100.35	1 203.75	1 310.73	1 397.10	1 559.27	1 614.69	1 889.96	2 203.40	2 814.73
June	954.74	1 101.70	1 170.76	1 198.76	1 260.28	1 419.94	1 497.44	1 643.59	1 875.05	2 207.24	2 775.66
July	965.18	1 148.56	1 162.07	1 208.52	1 171.62	1 413.38	1 517.70	1 680.80	1 844.67	2 269.94	2 804.19
August	954.36	1 195.36	1 075.21	1 244.49	1 114.82	1 498.55	1 591.47	1 718.13	1 899.46	2 324.27	2 597.71
September	919.36	1 204.37	1 008.65	1 265.42	1 156.91	1 505.06	1 554.01	1 743.26	1 938.37	2 357.80	2 397.79
October	957.55	1 110.88	1 009.41	1 245.25	1 221.44	1 544.80	1 516.65	1 739.84	1 973.94	2 422.75	2 338.00
November	948.92	1 108.38	1 007.00	1 210.28	1 284.96	1 535.11	1 533.46	1 755.17	1 951.91	2 293.45	2 571.19
December	914.63	1 175.26	1 040.25	1 156.90	1 324.08	1 628.88	1 502.42	1 783.30	1 976.41	2 388.17	2 595.88

1 Working day average
2 A full definition of these series is given in Section 7 of the ONS Explanatory Handbook.

Source: Bank of England: 020 7601 4342

22.12 Average Zero Coupon Yields[1,2]

	1988	1989	1990	1991	1992	1993	1994	1995	1996	1997	1998
Nominal Five Year Yield KORU											
January	9.27	10.12	11.17	10.46	9.37	7.14	5.84	8.68	6.85	7.14	6.17
February	9.19	9.95	11.60	9.93	9.14	6.78	6.19	8.57	7.12	6.82	6.10
March	8.85	10.10	12.44	10.00	9.56	6.66	6.93	8.49	7.47	7.07	6.10
April	8.71	10.50	12.71	9.98	9.32	6.97	7.58	8.30	7.53	7.27	5.94
May	8.93	10.51	12.34	10.11	8.91	7.23	8.00	7.98	7.54	6.93	5.95
June	9.19	10.80	11.71	10.32	8.91	7.15	8.47	7.81	7.47	6.96	6.05
July	9.74	10.29	11.71	10.09	8.92	6.83	8.31	7.92	7.31	7.00	6.13
August	9.94	10.06	11.89	9.85	9.31	6.44	8.52	7.73	7.20	6.97	5.81
September	10.15	10.43	11.86	9.45	9.00	6.43	8.75	7.50	7.19	6.72	5.30
October	9.74	10.67	11.30	9.51	7.71	6.30	8.69	7.58	6.98	6.52	4.94
November	10.04	10.77	10.95	9.59	7.08	6.28	8.53	7.23	7.18	6.70	4.90
December	10.42	10.81	10.70	9.58	7.35	5.86	8.54	6.90	7.20	6.45	4.51
Nominal Ten Year Yield AJTT											
January	9.56	9.52	10.27†	10.17	9.21	8.36	6.35	8.60	7.48	7.51	5.96
February	9.27†	9.31	10.66	9.81	9.09	8.12	6.79	8.52	7.82	7.15	5.90
March	9.01†	9.49	11.47	10.02	9.40	7.80	7.47	8.47	8.10	7.40	5.85
April	9.05	9.71	11.76†	9.93	9.15	7.93	7.88	8.33	8.05	7.57	5.68
May	9.24	9.70†	11.50	10.12	8.76	8.20	8.31	8.08	8.09	7.07	5.71
June	9.27†	10.05	11.04	10.31	8.84	8.00	8.72	8.05	8.07	7.03	5.58
July	9.52†	9.72	11.09	10.03	8.74	7.63	8.53	8.23	7.95	6.91	5.64
August	9.58	9.50†	11.45	9.79	9.04	7.12	8.58	8.10	7.86	6.95	5.39
September	9.72†	9.75	11.52	9.43	8.95	7.01	8.79	7.95	7.86	6.69	5.02
October	9.38†	9.92	11.22	9.54	8.68	6.93	8.67	8.12	7.56	6.36	4.91
November	9.48†	10.02	10.99	9.58	8.25	6.93	8.53	7.80	7.55	6.46	4.82
December	9.70†	10.01	10.05	9.15	8.31	6.33	8.23	7.53	7.51	6.22	4.43
Nominal Twenty Year Yield KOSW											
January	9.22	8.51	9.16	9.89	9.30	9.46	6.76	8.41	8.00	7.77	5.99
February	9.15	8.30	9.54	9.72	9.18	9.18	7.11	8.35	8.31	7.42	5.91
March	8.85	8.19	10.13	9.87	9.47	8.83	7.70	8.33	8.50	7.64	5.76
April	8.84	8.41	10.37	9.79	9.31	8.78	7.92	8.25	8.34	7.79	5.62
May	8.99	8.46	10.06	9.90	9.00	8.92	8.25	8.08	8.40	7.23	5.70
June	8.94	8.68	9.92	10.11	9.10	8.75	8.40	8.15	8.42	7.15	5.43
July	9.01	8.57	10.06	9.87	8.80	8.31	8.25	8.38	8.33	6.88	5.44
August	8.78	8.48	10.50	9.68	8.95	7.70	8.37	8.30	8.26	6.94	5.33
September	8.84	8.66	10.52	9.50	9.12	7.49	8.51	8.21	8.26	6.72	4.94
October	8.59	8.78	10.51	9.57	9.91	7.36	8.43	8.45	7.97	6.45	4.95
November	8.52	9.03	10.42	9.61	9.49	7.30	8.37	8.14	7.84	6.42	4.79
December	8.52	9.00	10.04	9.50	9.36	6.77	8.27	7.95	7.72	6.21	4.47
Real Ten Year Yield KOSX											
January	3.76	3.62	3.61	3.99	4.21	3.62	2.76	3.90	3.45	3.51	3.10
February	3.60	3.47	3.82	4.00	4.15	3.34	2.89	3.87	3.57	3.35	3.05
March	3.57	3.39	4.00	4.01	4.24	3.17	3.13	3.87	3.70	3.49	2.99
April	3.42	3.53	4.06	3.97	4.40	3.28	3.34	3.78	3.66	3.59	2.90
May	3.47	3.54	4.12	4.03	4.31	3.37	3.59	3.60	3.73	3.57	2.92
June	3.48	3.71	4.11	4.21	4.25	3.34	3.82	3.61	3.78	3.68	2.84
July	3.53	3.47	4.25	4.29	4.34	3.26	3.87	3.65	3.72	3.62	2.77
August	3.61	3.26	4.16	4.29	4.53	3.21	3.80	3.51	3.60	3.61	2.66
September	3.77	3.34	4.20	4.12	4.52	3.12	3.84	3.48	3.59	3.53	2.60
October	3.57	3.58	4.16	4.09	3.79	3.05	3.84	3.68	3.44	3.24	2.65
November	3.47	3.52	3.97	4.11	3.47	2.98	3.84	3.57	3.47	3.25	2.39
December	3.61	3.53	3.96	4.25	3.72	2.81	3.86	3.49	3.48	3.11	2.15
Real Twenty Year Yield KOSY											
January	4.10	3.67	3.64	4.10	4.30	3.85	3.01	3.85	3.54	3.62	3.04
February	3.90	3.51	3.81	4.10	4.29	3.65	3.16	3.87	3.65	3.44	3.04
March	3.89	3.46	4.01	4.16	4.48	3.51	3.37	3.87	3.75	3.53	2.96
April	3.78	3.54	4.09	4.08	4.55	3.50	3.44	3.80	3.72	3.64	2.83
May	3.80	3.57	4.05	4.11	4.39	3.60	3.62	3.62	3.80	3.63	2.81
June	3.88	3.74	4.00	4.21	4.30	3.57	3.84	3.63	3.84	3.66	2.61
July	3.94	3.58	4.15	4.27	4.37	3.50	3.87	3.66	3.77	3.55	2.56
August	3.90	3.50	4.14	4.31	4.57	3.35	3.79	3.57	3.71	3.55	2.50
September	3.94	3.55	4.22	4.20	4.47	3.25	3.81	3.57	3.70	3.46	2.45
October	3.77	3.60	4.24	4.22	3.93	3.19	3.80	3.71	3.57	3.19	2.54
November	3.64	3.58	4.20	4.25	3.78	3.20	3.79	3.61	3.58	3.17	2.31
December	3.67	3.57	4.07	4.35	3.96	3.06	3.81	3.55	3.58	3.04	2.05

1 Working day average.
2 Data revised following technical changes to the yield curve estimation model.

Source: Bank of England: 020 7601 4342

22.13 Average Rates on Representative British Government Stocks[1]

	1988	1989	1990	1991	1992	1993	1994	1995	1996	1997	1998
5 Year Conventional Rate KORP											
January	9.48	10.23	11.51	10.66	9.57	6.88	5.76	8.61	6.78	7.19	6.33
February	9.44	10.07	11.90	10.16	9.34	6.72	6.05	8.52	7.02	6.86	6.24
March	9.12	10.23	12.78	10.22	9.75	6.62	6.72	8.44	7.56	7.08	6.26
April	8.89	10.61	13.01	10.23	9.54	6.91	7.33	8.26	7.43	7.30	6.11
May	9.08	10.62	12.67	10.27	9.09	7.10	7.74	7.96	7.61	6.98	6.14
June	9.37	10.96	12.01	10.41	9.12	7.01	8.22	7.79	7.52	7.01	6.31
July	9.88	10.49	12.03	10.25	9.13	6.70	8.06	7.90	7.35	7.09	6.14
August	10.13	10.27	12.06	10.04	9.50	6.35	8.31	7.69	7.21	7.02	5.84
September	10.32	10.70	12.01	9.67	9.13	6.34	8.61	7.45	7.20	6.78	5.34
October	9.92	11.03	11.48	9.74	7.80	6.17	8.57	7.54	7.01	6.59	4.88
November	10.22	11.14	11.12	9.79	7.15	6.09	8.44	7.16	7.22	6.79	4.86
December	10.57	11.23	10.99	9.83	7.38	5.66	8.49	6.83	7.26	6.60	4.45
10 year Conventional Rate KORQ											
January	9.79	10.14	10.73	10.42	9.47	8.22	6.23	8.66	7.42	7.53	6.07
February	9.62	9.91	11.16	10.11	9.32	7.91	6.61	8.59	7.75	7.17	6.02
March	9.32	10.08	12.06	10.25	9.67	7.66	7.29	8.53	8.05	7.41	5.97
April	9.29	10.37	12.33	10.17	9.43	7.82	7.68	8.39	8.05	7.60	5.81
May	9.49	10.40	12.06	10.32	9.06	8.06	8.13	8.12	8.08	7.13	5.85
June	9.60	10.75	11.46	10.53	9.14	7.87	8.54	8.08	8.04	7.10	5.77
July	9.92	10.38	11.53	10.24	9.08	7.49	8.37	8.23	7.91	7.01	5.67
August	10.00	10.13	11.84	10.00	9.35	6.98	8.52	8.10	7.81	7.05	5.56
September	10.17	10.45	11.80	9.65	9.17	6.90	8.80	7.92	7.80	6.77	5.10
October	9.78	10.70	11.46	9.72	8.68	6.81	8.70	8.08	7.51	6.47	4.93
November	9.96	10.80	11.13	9.76	8.26	6.77	8.57	7.75	7.56	6.59	4.87
December	10.27	10.83	10.66	9.69	8.38	6.29	8.53	7.45	7.54	6.34	4.49
20 Year Conventional Rate KORR											
January	9.39	9.17	10.08	10.06	9.23	8.74	6.53	8.45	7.73	7.71	6.04
February	9.21	8.98	10.49	9.75	9.12	8.44	6.88	8.43	8.04	7.35	5.98
March	8.95	9.07	11.30	9.89	9.44	8.19	7.49	8.40	8.28	7.58	5.90
April	8.93	9.33	11.60	9.81	9.25	8.42	7.81	8.30	8.26	7.74	5.73
May	9.08	9.33	11.33	10.00	8.93	8.58	8.18	8.09	8.31	7.21	5.79
June	9.13	9.68	10.90	10.20	8.99	8.36	8.48	8.08	8.31	7.15	5.59
July	9.28	9.35	10.92	9.98	8.85	7.98	8.35	8.30	8.21	6.93	5.63
August	9.25	9.18	11.27	9.78	9.06	7.46	8.46	8.19	8.12	6.98	5.43
September	9.39	9.44	11.16	9.44	9.08	7.31	8.65	8.06	8.11	6.74	5.02
October	9.03	9.63	10.97	9.50	9.15	7.18	8.56	8.26	7.84	6.45	4.92
November	9.08	9.80	10.80	9.57	8.78	7.12	8.46	7.93	7.77	6.50	4.79
December	9.26	9.78	10.25	9.46	8.78	6.57	8.39	7.70	7.67	6.32	4.49
10 Year Index-Linked Rate KORS											
January	3.85	3.77	3.78	4.19	4.29	3.60	2.70	3.89	3.42	3.44	3.01
February	3.71	3.59	4.01	4.19	4.18	3.23	2.81	3.87	3.57	3.23	2.94
March	3.54	3.51	4.27	4.19	4.29	3.07	3.07	3.86	3.70	3.41	2.89
April	3.20	3.66	4.38	4.09	4.47	3.11	3.25	3.79	3.66	3.55	2.80
May	3.23	3.67	4.41	4.14	4.41	3.30	3.51	3.58	3.74	3.52	2.83
June	3.23	3.88	4.38	4.28	4.40	3.31	3.78	3.58	3.80	3.62	2.81
July	3.24	3.63	4.56	4.34	4.51	3.22	3.85	3.61	3.82	3.68	2.67
August	3.37	3.47	4.47	4.34	4.76	3.16	3.82	3.52	3.59	3.59	2.55
September	3.59	3.55	4.50	4.18	4.68	3.09	3.85	3.46	3.57	3.47	2.59
October	3.40	3.71	4.49	4.16	3.82	3.03	3.84	3.65	3.41	3.17	2.66
November	3.43	3.69	4.23	4.19	3.40	2.96	3.84	3.54	3.42	3.23	2.39
December	3.69	3.68	4.14	4.34	3.61	2.75	3.85	3.45	3.41	3.01	2.11
20 Year Index-Linked rate KORT											
January	4.13	3.78	3.76	4.23	4.34	3.84	2.96	3.91	3.58	3.62	3.01
February	3.94	3.60	3.94	4.23	4.32	3.64	3.11	3.89	3.70	3.43	3.01
March	3.92	3.53	4.14	4.25	4.49	3.50	3.35	3.89	3.82	3.55	2.92
April	3.81	3.65	4.23	4.17	4.59	3.51	3.45	3.81	3.77	3.65	2.80
May	3.85	3.68	4.24	4.23	4.42	3.62	3.64	3.64	3.84	3.61	2.79
June	3.92	3.84	4.20	4.35	4.35	3.57	3.88	3.67	3.88	3.65	2.61
July	3.97	3.66	4.34	4.40	4.42	3.48	3.90	3.71	3.72	3.68	2.56
August	3.94	3.54	4.32	4.40	4.61	3.32	3.83	3.62	3.75	3.54	2.51
September	3.99	3.60	4.40	4.25	4.55	3.22	3.87	3.60	3.74	3.43	2.51
October	3.80	3.73	4.43	4.25	4.00	3.15	3.86	3.74	3.60	3.17	2.58
November	3.69	3.72	4.29	4.27	3.77	3.14	3.85	3.62	3.59	3.16	2.35
December	3.80	3.70	4.20	4.39	3.94	2.98	3.86	3.55	3.58	3.02	2.12

1 Working day average.

Source: Bank of England: 020 7601 4342

22.14 Securities quoted on the Stock Exchange[1]

At last working day in March

£ millions

		1986	1987	1988	1989	1990	1991	1992	1993	1994	1995	1996
Total of all securities at market values	KDRV	1 247 580	1 547 845	1 427 397	1 861 091	2 106 147	2 195 213	2 054 927	2 605 093	3 225 921	3 342 381	4 077 109
British government and government guaranteed stocks[2]												
Nominal values	KDRW	128 850	138 777	142 857	133 780	117 857	118 702	130 143	150 803	200 282	217 283	262 246
Market values	KDRX	138 417	147 550	151 207	137 195	109 431	121 183	132 806	172 114	221 960	228 242	292 057
Other securities at market values:												
Irish government stocks	KDRY	8 398	9 727	11 026	10 730	11 834	11 878	13 359	14 003	15 366	14 445	17 320
Corporation stocks, public boards, etc	KDRZ	1 149	912	644	763	309	276	246	213	423	434	671
Dominion and foreign government and corporation stocks	KDSA	83 656	104 460	102 703	142 501	123 575	138 957	138 977	205 555	282 018	335 613	355 169
Company securities												
Total	KDSB	1 015 960	1 285 196	1 161 817	1 569 902	1 860 998	1 922 919	1 769 539	2 213 208	2 706 154	2 763 647	3 429 210
Loan capital	KDSC	9 004	12 013	11 820	13 621	16 788	14 129	15 164	16 407	22 216	21 758	24 872
Preference and preferred capital	KDSD	8 259	22 438	26 887	35 459	39 859	32 918	26 822	15 226	13 124	11 390	14 673
Ordinary and deferred capital[3]	KDSE	998 697	1 250 745	1 123 110	1 520 822	1 804 351	1 875 872	1 727 553	2 181 575	2 670 814	2 730 499	3 389 664

1 The stock exchanges of the United Kingdom and the Republic of Ireland form one exchange (The Stock Exchange).

2 Excluding marketable unlisted securities; including all outstanding amounts of 4 per cent Victory Bonds and 4 per cent Funding Loan, 1960-90 (that is, amounts for death duties and held by the National Debt Commissioners are included).

3 From 1978 shares of no par value are no longer distinguished separately.

Source: Council of The Stock Exchange

22.15 Capital issues and redemptions[1,2]

£ millions

		1989	1990	1991	1992	1993	1994	1995	1996	1997	1998
Total issues and redemptions											
Gross issues	KDSF	17 275	13 742	20 049†	12 712	29 561	28 678	23 149	28 215	26 466	21 625
Gross redemptions	KDSG	5 762†	5 359	3 765	7 011	11 400	11 866	10 467	14 123	14 579	11 607
Issues *less* redemptions: total	KDSH	11 513†	8 383	16 284	5 701	18 161	16 812	12 682	14 092	11 887	10 018
Loan capital	KDSI	13 344†	13 001	11 498	12 744	22 561	24 640	20 888	29 493	29 960	27 915
Preference shares	DEDO	858	781	1 245	589	1 700†	603	2 875	710	–881	–330
Ordinary shares	DEDF	3 457	3 449	10 849	6 224	16 672	14 617	9 777	10 273	8 614†	4 637
UK borrowers: total	DEDX	18 620	13 462	22 970†	13 638	29 216	25 241	15 462	24 231	23 638	20 684
Local government	DEEA	–	–	1	1	–1	–	–	–	–	–
Listed public companies	KDSN	18 620†	13 462	22 969	13 637	29 217	25 241	15 462	24 231	23 638	20 684
Non-resident borrowers: total	KDSO	5 274	2 878	5 132†	1 829	9 289	3 285	2 153	9 452	10 790	3 935
Central government[3] State, local government[3]	KIMI (KDSP, KDSQ)	2 636	1 142	2 956†	2 001	5 525	749	652	1 929	2 724	2 889
Companies[4]	KHDR	2 638	1 736	2 176†	–172	3 764	2 536	1 501	7 523	8 066	1 046

1 See footnotes on the second part of this table.

Source: Bank of England: 020 7601 3149

22.15 Capital issues and redemptions[1,2]

continued

£ millions

		1989	1990	1991	1992	1993	1994	1995	1996	1997	1998
UK listed public companies											
All companies: total	KDSN	18 620†	13 462	22 969	13 637	29 217	25 241	15 462	24 231	23 638	20 684
Loan capital	KDSS	14 305†	9 232	10 875	6 824	10 845	10 021	2 810	13 248	15 905	16 377
Preference shares	KDST	858	781	1 245	589†	1 700	603	2 875	710	–881	–330
Ordinary shares	KDSU	3 457	3 449	10 849†	6 224	16 672	14 617	9 777	10 273	8 614	4 637
Financial corporations: total	DEEC	10 384	6 737	7 815†	5 427	12 556	11 774	–441	11 030	6 711	6 576
Loan capital	KDSW	8 448	5 868	6 027†	3 898	7 488	6 972	–4 105	7 508	4 142	5 936
Preference shares	KDSX	264	214	660	459	1 221	138	1 538	868	–536	175
Ordinary shares	DELC	1 672	655	1 128	1 070	3 847	4 664	2 126	2 654	3 105	465
Other companies: total	DEEB	8 236	6 725	15 154	8 210	16 661†	13 467	15 903	13 201	16 927	14 108
Loan capital	KDTA	5 857†	3 364	4 848	2 926	3 357	3 049	6 915	5 740	11 763	10 441
Preference shares	KDTB	594	567	585	130†	479	465	1 337	–158	–345	–505
Ordinary shares	DEEY	1 785	2 794	9 721	5 154	12 825	9 953	7 651	7 619	5 509†	4 172
UK listed public companies											
Financial companies: total	KDTD	10 383	6 735	7 817†	5 426	12 553	11 774	–442	11 032	8 170	6 828
Monetary financial institutions	KDTE	6 789	4 614	2 856†	1 972	4 764	3 763	–748	9 618	22	2 161
Insurance corporations	DELR	183	–157	63	394	1 694	501†	578	1 007	1 805	617
Investment trust companies	DELS	938	439	841	709	3 027	4 118	1 415	1 537	927	427
Other[5]	KDTH	2 473	1 839	4 057†	2 351	3 068	3 392	–1 687	–1 130	5 416	3 623
Other companies: total	DEEB	8 236	6 725	15 154	8 210	16 661†	13 467	15 903	13 201	16 927	14 108
Manufacturing industries: total	KDTJ	3 782†	2 522	5 611	4 460	7 100	4 811	9 536	1 888	1 849	–1 019
Minerals and metal manufacture	DEKN	610	209	1 170	1 116	448	76	534	–947	–549	–325
Chemicals and allied industries	DEKO	–59	628	338	51	2 043	283†	1 759	1 124	854	–295
Metal goods, engineering and vehicles	DEKP	597	612	1 165	1 197	1 441	1 535	1 097	365	676	94
Electrical and electronic engines	DEKQ	196	104	219	339	119†	376	868	447	451	145
Food, drink and tobacco	DEKR	1 589	40	1 289	561	1 102†	1 242	3 730	–112	–279	–1 080
Other manufacturing	DEKS	849	929	1 430	1 196	1 947	1 299	1 548	1 011	696†	442
Energy	DEKT	–1 414	704	817	539	1 413	1 173	948	1 821	1 781	1 829
Water	DEKU	–6	87	724	168	212	251	648	354	509	570
Construction	DEKV	131	533	852	238	662	1 284	53	623	703	54
Distribution, hotels and repairs	DEKW	1 544	1 048	3 121	1 064	2 191†	1 137	901	854	1 481	1 242
Transport and communications	DEKX	691	499	843	418	872	912	1 134	3 744	4 769	6 946
Property companies	DEKY	1 774	338	1 402	212	2 007	1 873	1 370	1 907	2 798†	1 196
Services	DEKZ	1 741	993	1 782	1 118	2 207†	2 025	1 310	2 011	3 039	3 295

1 The estimates relate to new money raised on the main stock market by issues of ordinary, preference and loan capital (public issues, offers for sale, issues by tender, placings and issues to shareholders and employees) by listed public companies and local government in the United Kingdom; and by non-resident borrowers split between central governent, state and local governments and companies. The estimates include UK local government negotiable bonds (of not less than one year) issued to or through the agency of banks, discount houses, issuing houses or brokers. Mortgages, bank advances and any other loans redeemable in less than twelve months are excluded; so also are loans from UK government funds (including the former Industrial Reorganisation Corporation and the National Enterprise Board) but not government subscriptions to company issues made *pari passu* with the market. Issues to shareholders are included only if the sole or principal share register is maintained in the United Kingdom. Estimates of issues are based on the prices at which securities are offered to the market. Subscriptions are recorded under the periods in which they are due to be paid. Redemptions exclude share buy-backs. They relate to fixed interest securities of the kinds included as issues; conversion issues in lieu of cash repayment are included in the gross figures of both issues and redemptions. These figures include issues of debentures and loan stock carrying the right of conversion into, or subscription to, equity capital. Estimates for these issues *less* redemptions are: 1977 -5; 1978 -21; 1979 23; 1980 178; 1981 194; 1982 8; 1983 47; 1984 101; 1985 320.
The division between United Kingdom and non-resident company borrowers is determined by the location of the registered office. The industrial classification of companies is according to the primary occupation of the borrowing company or group and is based up to 1982 on the former *Standard Industrial Classification 1968* and from 1983 on the *Revised Standard Industrial Classification 1980.*

2 The estimates exclude issues on the unlisted securities market which was launched by the Stock Exchange in Novembere 1980. These issues are mainly of ordinary shares by industrial and commercial companies. Estimates of new money raised are: 1980 88; 1981 54; 1982 87; 1983 164; 1984 159; 1985 181; 1986 320; 1987 967; 1988 632; 1989 767; 1990 364; 1991 261.

3 Series now combined.

4 Non-resident companies including public corporations.

5 'Other' includes special finance agencies (listed public companies engaged in the provision of medium and long-term finance to industry eg ICFC)
and those finance houses and other consumer credit grantors not included under banks and building societies.

Source: Bank of England: 020 7601 3149

22.16 Building societies[1]
United Kingdom

Number and balance sheets

			1989	1990	1991	1992	1993	1994	1995[6]	1996[6]	1997[6]	1998
Societies on register	KRNA	number	126	117	110	105	101	96	94	88	82	74
Share investors	KRNB	thousands	36 805	36 948	37 925	37 533	37 809	38 150	38 998	37 768	19 234	21 195
Depositors	KRNC	"	4 490	4 299	4 698	3 879	3 686	5 369	6 143	6 718	882	820
Borrowers	KRND	"	6 699	6 724	6 998	7 055	7 140	7 222	6 906	6 586	2 703	2 934
Liabilities:		£ million										
Shares	KRNE	"	143 359.3	160 538.2	177 519.4	187 108.4	194 975.1	201 812.2	200 682.0	196 546.4	90 092.8	103 289.8
Deposits and loans	KRNF	"	30 532.5	40 695.5	49 516.6	57 067.5	62 301.2	69 925.2	67 513.8	73 919.1	31 033.7	33 311.2
Taxation and other	KRNG	"	3 071.6	3 768.8	3 093.9	2 559.5	2 565.7	2 939.2	3 306.2	3 727.4	1 338.8	1 586.4
General reserves	KRNH	"	8 680.7	10 206.1	11 430.4	12 634.4	14 269.5	16 312.3	17 218.3	17 940.3	7 331.2	7 926.4
Other Capital	KRNI	"	1 368.0	1 639.7	2 419.4	3 144.7	3 900.9	4 125.7	3 498.0	4 762.3	1 643.9	1 550.7
Assets: total	KRNJ	"	187 012.1	216 848.3	243 979.7	262 514.5	278 012.4	295 114.6	292 218.3	296 895.5	131 440.4	147 664.5
Mortgages	KRNK	"	151 491.7	175 745.4	196 945.6	210 994.5	221 237.6	240 297.2	236 841.0	241 472.9	107 531.5	118 288.4
Investments KRNL Cash KRNM	KHVZ	" "	30 932.2	35 050.9	39 513.6	42 909.1	47 174.2	50 786.7	50 894.1	51 016.7	21 869.8	27 102.0
Other	KRNN	"	4 588.2	6 052.0	7 520.5	8 610.1	9 600.6	4 030.7	4 483.2	4 405.9	2 039.1	2 274.1

Current transactions

			1989	1990[5]	1991[5]	1992[5]	1993[5]	1994[5]	1995[6]	1996[6]	1997[6]	1998
Shares:		£ million										
Received	KRNO	"	97 386.2	..	..	..	..	..	..	..	..	..
Interest thereon	KRNP	"	11 775.6	..	..	..	..	..	..	..	..	..
Withdrawn (including interest)	KRNQ	"	87 744.1	..	..	..	..	..	..	..	..	..
Deposits:												
Received	KRNR	"	62 271.1	..	..	..	..	..	..	..	..	..
Interest thereon	KRNS	"	2 887.8	..	..	..	..	..	..	..	..	..
Withdrawn (including interest)	KRNT	"	54 149.8	..	..	..	..	..	..	..	..	..
Mortgages:												
Advances	KRNU	"	42 032.2	43 081.0	42 948.0	34 989.0	32 259.0	34 829.0	34 673.0	38 488.0	28 771.7	21 988.3
Repayments of principal	KRNV	"	37 862.3	..	..	..	..	..	..	..	..	..
Interest[3]	KRNW	"	18 395.5	..	..	..	..	..	..	..	..	..
Management expenses	KRNX	"	2 093.2	2 363.0	2 591.0	2 723.7	2 952.5	3 136.7	3 352.6	3 555.3	2 270.5	1 501.7
Percentage rate of interest[4]												
Paid on shares	KRNY		*8.03*	..	..	..	..	..	..	..	..	..
Paid on deposits	KRNZ		*10.12*	..	..	..	..	..	..	..	..	..
Received on mortgage advances	KDUL		*12.21*	..	..	..	..	..	..	..	..	..

1 The figures for each year relate to accounting years ending on dates between 1 February of that year and 31 January of the following year.

2 1988 and subsequent years include Northern Ireland societies, responsibility for which was acquired under the Building Societies Act 1986.

3 Includes amounts recoverable from HM Government under Option Mortgage Scheme and MIRAS (Mortgage interest relief at source).

4 Based on the mean of the amounts outstanding at the end of the previous and the current year.

5 Apart from Mortgage Advances and Management expenses no new data is available for 1990 onwards. This is due to procedural changes.

6 The societies which have converted to the banking sector, namely Cheltenham & Gloucester (August 1995), National & Provincial (August 1996), Alliance & Leicester (April 1997), Halifax (June 1997), Woolwich (July 1997), Bristol & West (July 1997), and Northern Rock (October 1997) have been included in flow figures (using flows up to the date of conversion), but have been excluded from end year balances.

Source: Building Societies Commission: 020 7676 1000

22.17 Consumer credit

£ millions

		1988	1989	1990	1991	1992	1993	1994	1995	1996	1997	1998
Total amount outstanding	AILA	42 544	48 406	52 578	53 555	52 609	52 330	..	..	..	..	..
	VZRD	43 075[9]	49 036[9]	53 471[9]	54 416[9]	53 486[9]	53 295[9]	58 051[9]	68 205[9]	77 494	88 100†	102 270
Total net lending[1]	-AIKL	6 727	6 521	4 439	2 245	284	2 423	..	..	..	..	..
	VZQC	6 860†	6 773	4 616	2 318	477	2 650	5 743	8 234	11 215	12 013	14 538
Retailers[2]	AAPP	189	–1	50	37	25	69	83	–133	75	208†	55
Building societies' class 3 loans [3]	ALPY	214	303	223	–45	–65	46	89	238	383	120	–22
Banks	AIKN	5 954	5 713	3 765	2 523	785	840	3 704	5 606	7 682	9 027†	11 479
of which												
Credit cards	VTFY	675	540	1 761	774	126	700	..	..	..	..	..
	VZQS	675	540	1 761	774	138	719	1 483	2 103	3 029	3 507	4 854
Loans on personal accounts [4]	VTGA	5 226	5 007	1 981	1 805	615	156	..	..	..	..	..
Other	VZQT	6 185†	6 233	2 855	1 544	339	1 931	4 261	6 132	8 186	8 505	9 684
Insurance companies [5]	-AIKQ	70	59	156	68	29	165	..	..	..	..	..
	RSBK	71†	59	157	69	29	170	–178	39	–81	4	–16
Non-bank credit companies	-AGSJ	352	607	254	–417	–483	1 332	1 856	2 222	2 805	2 287	2 409
Other specialist lenders	VZQQ	432	699	421	–266	–297	1 525	2 045	2 485	3 156	2 654	2 810
Total gross lending	VZQG	11 327†	12 376	13 299	12 530	11 913	54 094	75 079	89 115	103 217	116 136	133 930
Narrower coverage[6]												
Total amount outstanding	RLWE	24 226	27 057	30 255	30 624	30 120	32 204	37 027	..	..	..	..
Total net lending[7]	RLWF	3 913	3 165	3 683	1 060	367	3 042	5 515	..	..	..	..
Total new credit advanced[8]	RLBY	36 306	41 521	46 336	47 098	48 865	54 859	62 050	..	..	..	..

1 Figures for net lending refer to changes in amounts outstanding adjusted to remove distortions caused by revaluations of debt outstanding, such as write-offs.
2 Self-financed credit advanced by food retailers, clothing retailers, household goods retailers, mixed business retailers (other than co-operative societies) and general mail order houses.
3 Class 3 loans advanced under the terms of the Building Societies' Act 1986.
4 Excludes loans for house purchase and bridging finance.
5 Prior to 1985 includes only policy loans.
6 Data relating to the narrower coverage covers finance houses and other specialist credit grantors, bank credit cards (operated under the VISA and Mastercard systems), and unsecured loans by building societies (since the end of 1986). Data no longer available from 1995.
7 Before 1987, net lending equals changes in amounts outstanding.
8 A high proportion of credit advanced in certain types of agreement, notably on credit cards, is repaid within a month. This reflects use of such agreements as a method of payment rather than a way of obtaining credit.

Source: Office for National Statistics: 01633 812789

22.18 End-year assets and liabilities of investment trust companies, unit trusts[1] and property unit trusts[2]

£ millions

		1988	1989	1990	1991	1992	1993	1994	1995	1996	1997	1998
Investment trust companies												
Short-term assets and liabilities (net):	CBPL	383	864	1 070	730	390	623	273	627	1 076	1 157	2 283
Cash and UK bank deposits	AHAG	279	462	642	397	340	387	443	1 009	1 087	1 577	2 728
Other short-term assets	CBPN	402	742	657	711	645	1 030	772	738	794	1 445	1 678
Short-term liabilities	-CBPS	–298	–340	–229	–378	–595	–794	–942	–1 120	–805	–1 865	–2 123
Medium and long-term liabilities and capital:	-CBPO	–19 893	–24 496	–20 382	–23 417	–28 895	–36 140	–40 180	–43 882	–50 911	–54 117	–50 006
Issued share and loan capital	-CBPQ	–4 335	–4 692	–5 193	–5 076	–4 827	–8 286	–10 978	–13 250	–8 330	–8 625	–8 837
Foreign currency borrowing	-CBPR	–122	–70	–119	–135	–407	–473	–354	–1 061	–638	–658	–615
Other borrowing	-CBQA	–670	–698	–596	–733	–802	–849	–1 354	–622	–823	–1 296	–1 736
Reserves and provisions, etc	-AHBC	–14 766	–19 036	–14 474	–17 473	–22 859	–26 532	–27 494	–28 949	–41 120	–43 538	–38 818
Investments:	CBPM	19 298	23 251	19 108	22 392	28 586	35 300	39 586	43 410	50 034	53 076	47 682
British government securities	AHBF	360	177	326	415	996	1 013	2 490	1 194	1 422	1 052	819
UK company securities:												
Loan capital and preference shares	CBGZ	878	903	687	720	814	854	1 000	846	832	1 287	1 359
Ordinary and deferred shares	CBGY	8 817	11 121	9 878	10 637	12 825	14 892	15 926	19 384	25 046	29 082	24 729
Overseas company securities:												
Loan capital and preference shares	CBHA	581	454	257	355	514	533	896	740	279	977	773
Ordinary and deferred shares	AHCC	8 116	10 259	7 392	9 105	11 943	16 886	17 873	19 485	21 047	17 810	17 843
Other investments	CBPT	546	337	568	1 160	1 494	1 122	1 401	1 761	1 408	2 868	2 159
Unit trusts												
Short-term assets and liabilities:	CBPU	1 522	2 026	2 315	1 698	2 099	2 673	3 266	3 116	3 822	5 048	6 886
Cash and UK bank deposits	AGYE	1 261	1 683	1 864	1 318	1 793	2 579	3 102	3 326	3 895	4 731	6 020
Other short-term assets	CBPW	409	488	710	628	920	1 013	1 364	986	1 201	869	1 346
Short-term liabilities	-CBPX	–148	–145	–259	–248	–614	–919	–1 200	–1 196	–1 274	–552	–480
Foreign currency borrowing	-AGYK	–9	–63	–31	–47	–8	–39	–21	–1	–	–	–
Investments:	CBPZ	37 613	52 994	41 608	50 459	58 785	88 479	83 495	104 069	125 841	144 038	163 050
British government securities	CBHT	457	374	411	523	664	959	1 414	1 774	2 716	3 202	3 771
UK company securities:												
Loan capital and preference shares	CBHU	1 529	1 690	1 337	1 337	1 664	2 906	2 970	3 298	5 029	6 627	9 290
Ordinary and deferred shares	RLIB	21 967	28 808	25 529	29 546	33 356	49 657	43 335	59 122	67 509	86 864	93 410
Overseas company securities:												
Loan capital and preference shares	CBHV	509	453	347	516	570	864	1 001	2 145	1 288	1 896	1 801
Ordinary and deferred shares	RLIC	12 958	21 401	13 791	18 124	21 862	32 904	33 473	36 062	47 346	42 931	51 119
Other assets	CBQE	193	268	193	413	669	1 189	1 302	1 668	1 953	2 518	3 659
Property unit trusts												
Short-term assets and liabilities (net)	AGVC	71	78	136	86	103	237	119	280	255	328	176
Property	CBQG	857	1 020	948	1 241	1 373	1 492	2 197	1 807	2 582	3 895	2 813
Other assets	AGVL	196	26	21	–	34	60	11	11	11	168	202
Long-term borrowing	-AGVM	–	–6	–6	–8	–42	–42	–	–131	–45	–247	–106

Note: Assets are shown as positive: liabilities as negative.
1 Including open ended investment companies (OEICs).
2 Investments are at market value.

Source: Office for National Statistics: 01633 812789

22.19 Self-administered pension funds: market value of assets
United Kingdom

£ millions, end-year

		1992	1993	1994	1995	1996	1997	1998
Total pension funds[1]								
Total net assets	AHVA	381 997†	480 547	443 467	508 600	543 883	656 873	699 191
Short-term assets	RYIQ	18 492	20 279†	22 617	26 114	31 521	35 368	39 005
British government securities	AHVK	25 189†	34 280	41 855	52 659	57 784	80 534	91 084
UK local authority long-term debt	AHVO	34	81	250	81	89	156†	183
Overseas government securities	AHVT	10 529	11 044	11 092	11 721	11 800	13 079†	15 493
UK company securities								
Ordinary shares	AHVP	202 311	251 099	219 189	256 625	276 001	339 687	334 648
Other	AHVQ	5 906	5 759	3 935	7 064	6 181	5 617	8 168
Overseas company securities								
Ordinary shares	AHVR	63 276	84 118	74 813	82 164	84 163	104 187	108 884
Other	AHVS	1 787	2 103	3 045	1 184	4 909	3 851	3 842
UK loans and mortgages	RLDQ	232	260	44	34	83	160	22
UK land, property and ground rent	AHWA	19 914	21 931†	24 353	21 317	21 638	24 176	24 355
Authorised unit trust units	AHVU	8 569	13 188	13 345	15 212	21 767	21 979	30 596
Property unit trusts	AHVW	1 745	1 905	2 463	2 485	2 666	3 219†	3 211
Other assets	RKPL	28 794†	38 761	31 318	36 352	30 628	32 978	47 136
Total liabilities	GQFX	4 781	4 261	4 852	4 412	5 347	8 118	7 436

1 These figures cover funded schemes only and therefore exclude the main superannuation arrangements in the central government sector.

Source: Office for National Statistics: 01633 812729

22.20 Insurance companies: balance sheet Market values

£ millions, end year

		1990	1991	1992	1993	1994	1995	1996	1997	1998
Long-term insurance companies										
Assets										
Total current assets (gross)	RYEW	15 749	15 303	18 101	16 925	16 690	24 171	31 699	42 795†	46 165
Agents' and reinsurance balances (net)	AHNY	892	788	799	457	−209	−157	−232	155	1 383
Other debtors[1]	RKPN	3 105	4 070	4 198	4 368	4 562	7 565	12 982	15 708	18 210
British government securities	AHNJ	31 137	36 584	50 970	72 575	64 921	80 268	90 996	107 847†	127 903
UK local authority securities etc	AHNN	820	577	667	772	815	1 322	1 088	914†	1 722
UK company securities[2]	RKPO	114 482	145 590	172 204	237 020	217 034	272 554	304 587	386 734	438 666
Overseas company securities	RKPP	20 067	28 120	32 101	49 087	48 195	59 950	62 378	73 428	82 122
Overseas government securities	AHNS	4 935	5 996	8 793	8 874	6 871	8 793	7 554	8 471†	17 515
Loans and mortgages	RKPQ	7 303	7 771	8 345	7 885	6 833	7 305	6 653	8 271	11 027
UK land, property and ground rent	AHNX	34 828	32 185	30 074	33 939	35 914	35 596	36 209	42 275†	45 903
Overseas land, property and ground rent	RGCP	236	160	124	144	151	118	114	98†	252
Other investments	RKPR	844	1 447	2 663	3 096	4 644	2 162	3 886	3 416	5 654
Total	RFXN	234 398	278 591	329 039	435 142	406 421	499 647	557 914	690 112	796 522
Net value of direct investment in:										
Non-insurance subsidiaries and associate companies in the United Kingdom	RYET	3 175	3 044	2 569	2 288	2 547	2 773	3 033	3 426	3 035
UK associate and subsidiary insurance companies and insurance holding companies	RYEU	85	345	639	1 186	504	701	575	−239	148
Overseas subsidiaries and associates	RYEV	350	471	773	1 016	1 034	987	986	1 104	1 087
Total assets	RKBI	238 008	282 451	333 020	439 632	410 506	504 108	562 508	694 403†	800 792
Liabilities										
Borrowing:										
Borrowing from UK banks	RGDF	2 467	1 701	1 162	1 234	1 570	1 907	2 234	3 027†	3 252
Other UK borrowing	RGDE	954	1 372	1 007	553	982	796	1 349	786†	1 040
Borrowing from overseas	RGDD	163	292	603	381	176	79	90	104†	148
Long-term business:										
Funds	RKDC	197 686	238 145	280 276	354 711	357 263	424 866	470 893	581 009†	669 301
Claims admitted but not paid	RKBM	841	927	951	1 035	1 085	1 419	1 441	1 436	1 712
Provision for taxation net of amounts receivable:										
UK authorities	RYPI	−585	−907	−1 119	−141	−470	502	2 568	4 207	5 443
Overseas authorities	RYPJ	−24	−13	−14	−20	29	−11	9	25	67
Provision for recommended dividends	RYPK	116	145	46	83	76	195	276	368	359
Other creditors and liabilities	RYPL	2 603	3 642	3 763	4 020	4 399	5 979	6 303	8 083	12 509
Excess of assets over above liabilities:										
Excess of value of assets over liabilities in respect of long-term funds	RKBR	32 978	36 032	43 511	74 160	42 608	63 255	71 817	89 790†	96 456
Minority interests in UK subsidiary companies	RKTI	–	–	30	30	–	3	–	2	–
Shareholders' capital and reserves in respect of general business	RKBS	266	414	1 952	890	1 143	2 050	2 576	3 862	6 299
Other reserves including profit and loss account balances	RKBT	543	701	852	2 696	1 645	3 068	2 952	1 704†	4 206
Total liabilities	RKBI	238 008	282 451	333 020	439 632	410 506	504 108	562 508	694 403†	800 792

See footnotes on the second part of this table.

Source: Office for National Statistics: 01633 812729

22.20 Insurance companies: balance sheet
Market values

continued

£ millions, end year

		1990	1991	1992	1993	1994	1995	1996	1997	1998
Other than long-term insurance companies										
Assets										
Total current assets (gross)	RYME	5 866	5 774	6 523	6 467	7 426	8 318	11 559	12 628†	8 524
Agents' and reinsurance balances (net)	AHMX	4 869	5 796	6 380	5 887	6 123	7 494	11 569	9 405	10 528
Other debtors[1]	RKPS	2 217	2 833	1 765	1 847	2 118	3 403	6 097	5 998	6 277
British government securities	AHMJ	5 464	6 359	8 378	11 474	12 320	14 363	16 893	15 666†	16 409
UK local authority securities etc	AHMN	22	33	49	59	50	56	42	16†	14
UK company securities[2]	RKPT	10 065	9 877	10 480	14 533	14 312	17 425	17 825	18 845	18 440
Overseas company securities	RKPU	3 233	3 776	3 747	4 163	3 578	4 422	5 072	6 594	8 676
Overseas government securities	AHMS	2 712	3 534	4 660	5 324	5 064	6 511	9 546	8 215†	10 459
Loans and mortgages	RKPV	1 870	1 166	1 147	1 234	1 321	1 337	1 593	1 385	1 335
UK land, property and ground rent	AHMW	3 288	3 034	2 398	2 375	2 121	2 100	2 077	2 842†	1 146
Overseas land, property and ground rent	RYNK	79	91	185	80	89	128	120	149†	107
Other investments	RKPW	676	673	633	418	536	665	716	2 465	2 366
Total	RKAL	40 361	42 946	46 345	53 861	55 058	66 222	83 106	84 208	84 281
Net value of direct investment in:										
Non-insurance subsidiaries and associate companies in the United Kingdom	RYNR	1 418	1 724	2 617	2 214	2 474	2 449	3 195	6 950	5 553
UK associate and subsidiary insurance companies and insurance holding companies	RYNS	1 153	682	1 087	1 835	1 738	1 642	7 170	4 204	6 424
Overseas subsidiaries and associates	RYNT	6 823	8 534	9 908	12 275	11 854	15 485	14 859	16 402	14 239
Total assets	RKBY	49 755	53 886	59 957	70 185	71 124	85 798	108 330	111 764	110 497
Liabilities										
Borrowing:										
Borrowing from UK banks	RYMB	1 054	910	434	721	1 382	1 584	1 524	3 029†	1 825
Other UK borrowing	RYMC	852	1 208	1 677	1 989	2 354	2 370	2 536	2 996†	1 551
Borrowing from overseas	RYMD	1 086	1 202	1 296	1 103	1 626	1 876	1 976	1 202†	1 600
General business technical reserves	RKCT	29 139	33 832	38 005	39 746	42 374	47 493	58 618	59 527†	60 775
Long-term business:										
Funds	RKTF	–	–	–	–	–	–	–	–	–
Claims admitted but not paid	RKTK	195	–	–	–	–	–	–	–	–
Provision for taxation net of amounts receivable:										
UK authorities	RYPO	–14	–148	–295	235	397	841	807	1 253	1 197
Overseas authorities	RYPP	12	18	24	14	22	16	22	7	11
Provision for recommended dividends	RYPQ	951	635	650	794	874	1 098	1 407	2 048	1 318
Other creditors and liabilities	RYPR	1 598	2 180	1 921	1 987	2 551	2 955	3 886	3 873	3 793
Excess of assets over above liabilities:										
Excess of value of assets over liabilities in respect of long-term funds	RKCG	61	–	–	–	–	–	–	–	–
Minority interests in UK subsidiary companies	RKCH	4	17	33	80	52	22	24	60	68
Shareholders' capital and reserves in respect of general business	RKCI	13 349†	13 035	14 552	21 355	17 628	25 545	35 069	35 172	34 144
Other reserves including profit and loss account balances	RKCJ	1 471	997	1 660	2 161	1 864	1 998	2 461	2 597	4 215
Total liabilities	RKBY	49 755	53 886	59 957	70 185	71 124	85 798	108 330	111 764	110 497

1 Including outstanding interest, dividends and rents (net).
2 Including authorised unit trust units.

Source: Office for National Statistics: 01633 812729

22.21 Industrial and provident societies[1]
Great Britain

		1987	1988	1989	1990	1991	1992	1993	1994	1995	1996	1997
Number of societies	KRFQ	10 486	10 871	11 243	11 329	11 388	11 348	11 004	10 738	10 656	10 601	10 584
Number of members (thousands)	KRFR	11 232	12 006	12 131	10 830	10 198	10 402	9 989	10 362	9 652	9 128	10 382
Assets (£ millions)	KRFS	18 660†	19 812	22 677	27 007	31 876	35 671	42 023	35 158	38 479	41 528	47 031

1 The annual returns from which these figures are derived are mainly made up to dates varying between September of the year shown and January of the following year.

Source: Registry of Friendly Societies: 020 7663 5222

22.22 Co-operative trading societies[1]
Great Britain

		1987	1988	1989	1990	1991	1992	1993	1994	1995	1996	1997
Number of societies												
General trading societies:												
Retail societies	KREA	203	200	210	195	184	173	148	133	127	127	131
Principal wholesale societies[2]	KREB	2	2	2	2	2	2	2	2	1	1	1
Other wholesale and productive societies	KREC	243	244	219	200	198	200	176	157	138	126	121
Agricultural and fishing trading societies	KRED	641	646	626	703	1 200	1 160	1 123	1 099	1 063	1 041	1 021
Number of members (thousands)												
General trading societies:												
Retail societies	KREE	7 429	7 646	7 588	7 750	6 985	7 220	6 471	6 820	5 872	5 965	6 083
Principal wholesale societies[2]	KREF	0.2	0.2	0.2	0.2	0.2	0.2	0.2	0.2	0.1	0.1	0.1
Other wholesale and productive societies	KREG	53	53	54	58	183	182	181	48	45	45	45
Agricultural and fishing trading societies	KREH	295	286	277	308	312	297	302	312	266	255	254

1 These societies are registered under the Industrial and Provident Societies Acts and are included in Table 22.23. See also footnote 1 to that table.
2 Co-operative Wholesale Society Ltd.

Source: Registry of Friendly Societies: 020 7663 5222

22.23 Individual insolvencies

Number

		1988	1989	1990	1991	1992	1993	1994	1995	1996	1997	1998
England and Wales												
Bankruptcies[1,2]	AIHW	7 717	8 138	12 058	22 632	32 106	31 016	25 634	21 933	21 803	19 892	19 647
Individual voluntary arrangements[3]	AIHI	779	1 224	1 927	3 002	4 686	5 679	5 103	4 384	4 466	4 545	4 901
Deeds of arrangement	AIHO	11	3	2	6	2	8	2	2	2	4	1
Total	AIHK	8 507	9 365	13 987	25 640	36 794	36 703	30 739	26 319	26 271	24 441	24 549
Scotland												
Sequestrations[4]	KRHA	1 401	2 301	4 350	7 665	10 845	6 828	2 182	2 188	2 503	2 502	3 016
Northern Ireland												
Bankruptcies[5,6]	KRHB	164	238	286	367	406	474	438	399	415	393	394
Individual voluntary arrangements[7]	KJRK	..	..	..	2	42	67	84	64	101	84	121
Deeds of arrangement	KRHC	–	–	–	–	–	–	–	–	–	–	–
Total	KRHD	164	238	286	369	448	541	522	463	516	477	515

1 Comprises receiving and administration orders under the Bankruptcy Act 1914 and bankruptcy orders under the Insolvency Act 1986.
2 Orders later consolidated or rescinded are included in these figures.
3 Introduced under the Insolvency Act 1986.
4 Sequestrations awarded but not brought into operation are included in these figures.
5 Comprises bankruptcy adjudication orders, arrangement protection orders and orders for the administration of estates of deceased insolvents.
6 Orders later set aside or dismissed are included in these figures.
7 Introduced under the Insolvency Northern Ireland order 1989.

Source: Department of Trade and Industry: 020 7215 3291/3305

22.24 Company insolvencies

Number

		1988	1989	1990	1991	1992	1993	1994	1995	1996	1997	1998
England and Wales												
Compulsory liquidations	AIHR	3 667	4 020	5 977	8 368	9 734	8 244	6 597	5 519	5 080	4 735	5 216
Creditors' voluntary liquidations	AIHS	5 760	6 436	9 074	13 459	14 691	12 464	10 131	9 017	8 381	7 875	7 987
Total	AIHQ	9 427	10 456	15 051	21 827	24 425	20 708	16 728	14 536	13 461	12 610	13 203
Scotland												
Compulsory liquidations	KRGA	228	229	251	304	310	286	242	252	266	254	338
Creditors' voluntary liquidations	KRGB	168	199	219	312	360	265	202	189	175	223	228
Total	KRGC	396	428	470	616	670	551	444	441	441	477	566
Northern Ireland												
Compulsory liquidations	KRGD	62	69	73	112	79	73	69	72	68	60	53
Creditors' voluntary liquidations	KRGE	54	75	55	71	77	85	52	37	54	53	46
Total	KRGF	116	144	128	183	156	158	121	109	122	113	99

Source: Department of Trade and Industry: 020 7215 3291/3305

22.25 Industry analysis: bankruptcies and deeds of arrangement[1]
England and Wales

Number

		1988	1989	1990	1991	1992	1993	1994	1995	1996	1997	1998
Industry												
Self-employed												
Agriculture and horticulture	KRFY	162	142	198	266	313	277	231	218	168	155	157
Manufacturing:												
Food, drink and tobacco	KRFZ	25	9	31	48	56	34	33	30	31	18	21
Chemicals	KRLA	3	3	4	12	9	7	23	8	5	5	7
Metals and engineering	KRLB	158	180	240	424	634	612	523	396	411	413	378
Textiles and clothing	KRLC	73	77	63	152	174	160	95	114	91	76	81
Timber and furniture	KRLD	94	62	116	233	317	207	176	158	118	98	96
Paper, printing and publishing	KRLE	60	39	80	143	205	161	142	142	117	104	101
Other	KRLF	30	38	67	120	125	169	133	146	117	116	110
Total	KRLG	443	408	601	1 132	1 520	1 350	1 125	994	890	830	794
Construction and transport:												
Construction	KRLH	1 590	1 652	2 348	3 812	4 692	4 361	3 362	2 783	2 713	2 182	1 919
Transport and communication	KRLI	527	601	953	1 620	2 038	1 754	1 402	1 138	1 227	1 162	1 060
Total	KHGP	3 462	2 253	3 301	5 432	6 730	6 115	4 764	3 921	3 940	3 344	2 979
Wholesaling:												
Food, drink and tobacco	KRLJ	53	41	57	68	114	114	94	103	77	62	53
Motor vehicles	KRLK	6	4	8	21	48	21	28	33	36	28	20
Other	KRLL	69	67	81	122	220	191	160	122	101	78	92
Total	KHGQ	128	112	146	211	382	326	282	258	214	168	165
Retailing:												
Food, drink and tobacco	KRLM	447	401	595	895	1 001	1 107	981	782	662	546	514
Motor vehicles and filling stations	KRLN	163	131	155	362	399	412	343	316	327	276	238
Other	KRLO	459	490	807	1 442	2 159	2 087	1 615	1 566	1 268	1 048	971
Total	KHGR	1 069	1 022	1 557	2 699	3 559	3 606	2 939	2 664	2 257	1 870	1 723
Services:												
Financial institutions	KRLP	86	95	143	247	266	292	241	185	125	105	79
Business services	KRLQ	325	386	662	1 284	1 859	1 843	1 537	1 354	1 176	1 117	1 057
Hotels and catering	KRLR	625	718	867	1 481	2 366	2 437	2 102	1 956	1 736	1 603	1 309
Total	KHGS	1 036	1 199	1 672	3 012	4 491	4 572	3 880	3 495	3 037	2 825	2 445
Other	KHGT	646	724	1 014	1 857	2 530	2 315	1 893	1 732	2 161	2 077	2 157
Total: self-employed	KRLT	5 601	5 860	8 489	14 609	19 525	18 561	15 114	13 282	12 667	11 269	10 420
Other individuals												
Employees	KRLU	686	856	1 172	1 639	2 588	2 507	2 279	1 981	2 471	2 625	3 141
No occupation and unemployed	KRLV	652	698	1 107	2 811	4 325	4 816	3 696	2 859	3 294	3 051	3 384
Directors and promoters of companies	KRLW	345	305	427	667	965	862	628	484	368	310	272
Occupation unknown	KRLX	444	419	865	2 906	4 703	4 270	3 917	3 327	3 003	2 637	2 430
Total: other individuals	KRLY	2 127	2 278	3 571	8 023	12 581	12 455	10 520	8 651	9 136	8 623	9 227
Total bankruptcies and deeds of arrangements[1]	KRLZ	7 728	8 138	12 060	22 632	32 106	31 016	25 634	21 933	21 803	19 892	19 647

1 From January 1991 Industrial Analysis excludes Deeds of Arrangement.

Source: Department of Trade and Industry: 020 7215 3291/3305

22.26 Industry analysis: company insolvencies[1]
England and Wales

Number

		1988	1989	1990	1991	1992	1993	1994	1995	1996	1997	1998
Industry												
Agriculture and horticulture	KRMA	73	78	111	135	191	157	166	99	89	51	65
Manufacturing:												
Food, drink and tobacco	KRMB	88	105	109	171	215	213	142	130	163	93	89
Chemicals	KRMC	75	85	97	134	141	91	108	69	65	31	57
Metals and engineering	KRMD	708	697	972	1 344	1 621	1 381	932	681	658	591	594
Textiles and clothing	KRME	811	959	921	1 052	1 120	917	736	567	568	596	526
Timber and furniture	KRMF	242	302	391	527	508	333	252	267	249	181	149
Paper, printing and publishing	KRMG	326	425	552	856	830	777	579	452	438	364	426
Other	KRMH	480	468	792	939	1 014	878	859	681	599	613	652
Total	KRMI	2 730	3 041	3 834	5 023	5 449	4 590	3 608	2 847	2 740	2 469	2 493
Construction and transport:												
Construction	KRMJ	1 471	1 638	2 445	3 373	3 830	3 189	2 401	1 844	1 610	1 419	1 325
Transport and communication	KRMK	548	589	932	1 246	1 261	1 082	774	706	682	540	504
Total	KHGU	2 019	2 227	3 377	4 619	5 091	4 271	3 175	2 550	2 292	1 959	1 829
Wholesaling:												
Food, drink and tobacco	KRML	125	162	235	287	388	231	244	205	183	158	139
Motor vehicles	KRMM	91	69	107	152	186	142	112	83	95	41	60
Other	KRMN	487	428	724	841	672	639	638	678	429	340	364
Total	KHGV	703	659	1 066	1 280	1 246	1 012	994	966	707	539	563
Retailing:												
Food, drink and tobacco	KRMO	170	165	244	291	406	388	299	246	236	219	186
Motor vehicles and filling stations	KRMP	121	136	174	245	339	229	226	195	227	132	120
Other	KRMQ	795	738	1 181	1 578	1 732	1 388	1 186	1 127	956	891	847
Total	KHGW	1 086	1 039	1 599	2 114	2 477	2 005	1 711	1 568	1 419	1 242	1 153
Services:												
Financial institutions	KRMR	159	167	303	394	563	421	259	198	222	111	101
Business services	KRMS	843	952	1 558	2 396	2 788	2 415	1 807	1 525	1 500	1 528	1 617
Hotels and catering	KRMT	359	371	489	748	1 010	912	777	692	708	609	626
Total	KJRS	1 361	1 490	2 350	3 538	4 361	3 748	2 843	2 415	2 430	2 248	2 344
Other	KHGX	1 455	1 922	2 714	5 118	5 610	4 925	4 231	4 091	3 784	4 102	4 756
Total company insolvencies	KHGY	9 427	10 456	15 051	21 827	24 425	20 708	16 728	14 536	13 461	12 610	13 203

1 Including partnerships.

Source: Department of Trade and Industry: 020 7215 3291/3305

Service industry

23

Service industry

Annual retailing inquiries *(Table 23.1)*

The 1996 retail inquiry covers mainly businesses which operate in and are engaged in retail distribution. It is undertaken on a stratified random sample of 11 000 businesses classified to the retail division of the Standard Industrial Classification (SIC 1992). The results were, however supplemented by information on commodity sales collected from legal units classified outside retailing which were believed to undertake retail activity. The inquiry is conducted primarily for national accounts purposes, provides a benchmark for the various series of short term statistics and analyses the structure of retail trade.

The inquiry covers the whole of the United Kingdom, that is England, Scotland, Wales and (for the first time), Northern Ireland. The Annual retail inquiry forms part of a family of annual distribution and service inquiries conducted by the Office for National Statistics (ONS); these include the motor trades, catering, wholesaling, property and service trades. The inquiry collects data on total turnover, sales by non-store activity, outlets, questions relating to trading in second-hand goods, or from market stalls and road side pitches. A detailed breakdown of commodity sales, stocks, purchases and capital expenditure is also collected. From 1996, it is also planned to collect data on detailed purchases and taxes and levies.

The register used for the 1996 annual retail inquiry was the Inter-Departmental Business Register (IDBR) which consists of companies, partnerships, sole proprietorships, public authorities, central government departments, local authorities and non-profit making bodies. The main administrative sources of the IDBR are HM Customs and Excise for VAT details and Inland Revenue for PAYE information. A point to note is that the register used to compile the retail results shown in the previous publications *(Annual Abstract of Statistics: 1998 Edition and Business Monitor SDA 25: Results of the 1994 Retailing Survey)* was based on information available from HM Customs and Excise on VAT registrations. The register now used supplements this with information obtained from PAYE sources. This source provides information about many additional, though small, businesses which contribute to the economy even though they are not registered for VAT.

A second major change was the method of classifying businesses to individual industries. The previous method made use of a commodity based flow chart scheme (shown in Business Monitor SDA 25). Classification is now obtained directly from the register which collects information via ONS employment surveys. In most cases, the classification obtained would be the same but there are instances where the classification is different. The main advantage of the employment based classification is that it produces a better correlation between employment and output data, enabling more consistent estimates.

The method of classification is that each legal unit (i.e. an individual company, partnership, sole proprietor, etc.) is classified to a single activity, whether it is wholly or mainly engaged in that activity. The nature of a business can naturally change with time, possibly as a result of another business being absorbed in a take-over. Some businesses have a very significant secondary activity, perhaps completely different from their main activity, and a small change of direction can lead to a new main activity and to the reclassification of the business. Other changes may arise from improvements to classification data held on the IDBR as a result of new information received about individual businesses.

Sector Review - Retailing - Data for 1996

Final results for the 1996 Annual Retailing Inquiry were published in the above Sector Review towards the end of September 1998. The Sector Review replaces the previous publication, Business Monitor SDA 25 and incorporates some substantial changes from previously published data.

Annual Retailing - Data for 1995

Data for the 1995 Annual Retailing Inquiry was not released in the form of a Business Monitor but has been available since the early part of 1998. The data is available on request from the Office for National Statistics *(Enquiries, Tel: 01633 812921/2476)* if required.

Retail trade: index numbers of value and volume *(Table 23.2)*

The monthly retail sales estimates cover the retail trades (excluding the motor trades) in Great Britain. Until the end of 1991, the statistics were based on returns from a voluntary panel of 3 500 retailers. However, as a part of a package of measures to improve economic statistics, the inquiry was made compulsory from January 1992 and the sample size was increased to approximately 5 000. The new inquiry provides more soundly-based estimates as it covers all large retailers including those who did not contribute to the old voluntary inquiry and a random sample of smaller retailers. The use of statutory powers means that it has been possible to improve the sample design at the detailed level and reduce the sampling error associated with the results.

During 1998, the retail sales index was rebased using detailed information from the 1995 annual retailing inquiry. The reference year is now set at 1995=100. Details of the work, together with revised figures for January 1990 to December 1997, were published in *ONS News Release* (98) 349 on 21 October 1998.

The latest summary statistics are published each month by *First Release.* More disaggregated value indices (not seasonally adjusted) are published each month in *Business Monitor SDM28*, available from ONS Direct, price £110 a year.

Motor trades and catering inquiries *(Tables 23.3 and 23.4)*

The motor trades and catering inquiries are conducted on a sample basis from the ONS's Inter Departmental Business Register (IDBR). In general the inquiries are addressed solely to businesses classified to these trades, in that businesses who may be engaged in the catering or motor trades as a subsidiary activity are excluded. (An exception is that the catering inquiry includes certain large groups of managed public houses owned by breweries.) The results generally relate to the total activity of all businesses registered for VAT, including those with a VAT turnover beneath the mandatory registration threshold, and or businesses obtained from PAYE source.

23.1 Retail businesses[1], 1996

By Standard Industrial Classification (SIC 92)

United Kingdom

	Businesses	Outlets	Total turnover[2,3]	Value added tax in total turnover	Retail turnover[2,3]	Purchases[4]
	Number	Number	£ million	£ million	£ million	£ million
By class level of SIC 92						
Total Retail Businesses	**206 964**	**320 622**	**193 236**	**18 681**	**181 538**	**119 546**
Non-specialised stores (NSS)						
Retail Sales in non-specialised stores	**37 559**	**59 508**	**89 656**	**7 201**	**80 903**	**62 664**
NSS mainly food, drink or tobacco	22 157	37 556	71 968	4 815	63 612	52 074
NSS other sale	15 402	21 953	17 687	2 386	17 291	10 590
Specialised stores (SS)						
Retail sales of food, drink or tobacco	**58 321**	**74 134**	**14 565**	**1 069**	**14 072**	**9 826**
Fruit and vegetables	6 841	8 979	1 279	21	1 236	881
Meat and meat products	10 662	12 459	2 464	7	2 360	1 671
Fish	1 997	2 551	245	1	238	167
Bread, cakes, flour or sugar confectionery	4 931	9 506	1 463	55	1 344	406
Alcoholic and other drinks	6 313	11 172	3 773	534	3 636	2 689
Tobacco products	17 127	18 537	3 940	389	3 896	3 000
Other sales of food, drinks or tobacco	10 450	10 931	1 400	61	1 362	1 012
Retail sales of pharmaceutical and medical goods, cosmetics and toilet articles	**7 321**	**13 789**	**7 375**	**551**	**7 299**	**5 125**
Dispensing chemists	6 643	11 537	6 043	373	6 005	4 403
Medical and orthopaedic goods	194	514	128	12	108	45
Cosmetic and toilet articles	484	1 738	1 203	167	1 186	677
Other sales of new goods in specialised stores	**90 291**	**153 795**	**70 488**	**8 530**	**68 746**	**36 558**
Textiles	1 592	3 019	847	115	820	450
Clothing	15 274	33 133	24 398	2 598	23 908	11 671
Footwear and leather goods	6 265	13 865	3 783	379	3 723	1 676
Furniture, lighting and household articles	9 941	14 616	6 764	1 003	6 615	3 532
Electrical household appliances, radio and television goods	7 447	13 245	8 280	1 204	7 982	5 081
Hardware, paints and glass	6 124	8 231	6 345	936	6 251	3 409
Books, newspapers and stationery	9 590	16 650	4 542	310	4 382	2 855
SS other sales	34 057	51 035	15 529	1 985	15 065	7 885
Floor coverings	1 988	3 016	1 463	218	1 449	750
Photographic, optical and precision equipment, office supplies & equipment (including computers, etc)	5 330	8 513	3 047	367	2 982	1 307
SS other sales (nes)	26 739	39 505	11 019	1 400	10 634	5 828
Retail sales of second-hand goods	**5 146**	**7 013**	**1 511**	**87**	**1 451**	**968**
Retail sales not in stores	**6 321**	**9 734**	**9 204**	**1 181**	**8 682**	**4 319**
Mail order houses	2 102	3 053	7 199	963	6 824	3 272
Stalls and markets	1 031	2 961	301	14	261	205
Other non-store retail sales	3 188	3 720	1 704	204	1 597	841
Repair of personal and household goods	**2 005**	**2 649**	**438**	**61**	**385**	**86**

1 The figures in this table include retail outlets and retail turnover of sales by mail order, party-plan, automatic vending machines, market stalls, roadside pitches or door-to-door.
2 Inclusive of VAT.
3 Includes retail hire from non-hire retail businesses.
4 Exclusive of VAT.

Source: Office for National Statistics: 01633 812921/2476

23.1 Retail businesses[1], 1996

continued

By Standard Industrial Classification (SIC 92)

United Kingdom

	Stocks[4]		Total turnover[4] divided by end-year stocks	Gross margin[4]		Value added
	Beginning of year	End of year		Amount	As a percentage of total turnover	
	£ million	£ million	Ratio	£ million	%	£ million
By class level of SIC 92						
Total Retail Businesses	**16 243**	**17 347**	**10.1**	**56 114**	**32.1**	**36 618**
Non-specialised stores (NSS)						
Retail Sales in non-specialised stores	**4 903**	**5 198**	**15.9**	**20 085**	**24.4**	**14 249**
NSS mainly food, drink or tobacco	3 104	3 234	20.8	15 209	22.6	10 868
NSS other sale	1 799	1 964	7.8	4 875	31.9	3 381
Specialised stores (SS)						
Retail sales of food, drink or tobacco	**661**	**691**	**19.5**	**3 701**	**27.4**	**2 397**
Fruit and vegetables	16	17	72.9	379	30.1	258
Meat and meat products	35	37	67.3	788	32.1	569
Fish	3	3	71.4	77	31.7	57
Bread, cakes, flour or sugar confectionery	39	41	34.8	1 003	71.3	619
Alcoholic and other drinks	280	282	11.5	552	17.0	332
Tobacco products	239	259	13.7	570	16.1	341
Other sales of food, drinks or tobacco	48	52	25.9	331	24.7	221
Retail sales of pharmaceutical and medical goods, cosmetics and toilet articles	**679**	**716**	**9.5**	**1 735**	**25.4**	**1 241**
Dispensing chemists	557	574	9.9	1 285	22.7	992
Medical and orthopaedic goods	10	11	10.4	72	62.0	45
Cosmetic and toilet articles	112	130	8.0	378	36.4	205
Other sales of new goods in specialised stores	**8 577**	**9 166**	**6.8**	**25 990**	**41.9**	**16 173**
Textiles	150	159	4.6	291	39.7	182
Clothing	2 184	2 432	9.0	10 378	47.6	7 034
Footwear and leather goods	625	635	5.4	1 738	51.1	1 159
Furniture, lighting and household articles	730	816	7.1	2 316	40.2	1 285
Electrical household appliances, radio and television goods	1 081	1 109	6.4	2 024	28.6	1 093
Hardware, paints and glass	1 049	1 097	4.9	2 047	37.8	1 110
Books, newspapers and stationery	589	568	7.4	1 356	32.0	752
SS other sales	2 168	2 351	5.8	5 841	43.1	3 559
Floor coverings	127	136	9.1	504	40.5	300
Photographic, optical and precision equipment, office supplies & equipment (including computers, etc)	213	240	11.2	1 399	52.2	878
SS other sales (nes)	1 828	1 975	4.9	3 938	40.9	2 382
Retail sales of second-hand goods	**582**	**625**	**2.3**	**498**	**35.0**	**283**
Retail sales not in stores	**811**	**914**	**8.8**	**3 807**	**47.5**	**2 076**
Mail order houses	657	759	8.2	3 065	49.2	1 636
Stalls and markets	15	17	17.2	84	29.2	56
Other non-store retail sales	139	138	10.9	658	43.9	384
Repair of personal and household goods	**30**	**37**	**10.1**	**298**	**79.2**	**199**

See footnotes on the first part of this table.

Source: Office for National Statistics: 01633 812921/2476

23.1 Retail businesses[1], 1996

continued

By Standard Industrial Classification (SIC 92)

United Kingdom

	Businesses	Outlets	Total turnover[2,3]	Value added tax in total turnover	Retail turnover[2,3]	Purchases[4]
	Number	Number	£ million	£ million	£ million	£ million
By group level of SIC 92 and number of outlets						
Total Retail Businesses	**206 964**	**320 622**	**193 236**	**18 681**	**181 538**	**119 546**
Businesses having:						
1 outlet	181 880	181 880	40 270	3 689	39 228	25 353
2-9 outlets	23 906	64 487	24 334	2 689	23 081	13 078
10-99 outlets	1 021	25 054	23 774	2 745	22 590	13 643
100 or more outlets	156	49 203	104 858	9 558	96 639	67 472
Retail sales in non-specialised stores	**37 559**	**59 508**	**89 656**	**7 201**	**80 903**	**62 664**
Businesses having:						
1 outlet	34 910	34 910	7 363	574	7 272	5 339
2-9 outlets	2 436	6 479	3 222	327	3 067	1 975
10-99 outlets	174	3 706	10 451	1 054	9 682	6 894
100 or more outlets	39	14 414	68 619	5 246	60 882	48 456
Specialised stores						
Retail sales of food, drink or tobacco	**58 321**	**74 134**	**14 565**	**1 069**	**14 072**	**9 826**
Businesses having:						
1 outlet	54 139	54 139	8 887	570	8 695	6 373
2-9 outlets	4 080	10 797	1 997	77	1 841	1 246
10-99 outlets	89	2 144	813	69	737	518
100 or more outlets	13	7 054	2 869	352	2 799	1 690
Retail sales of pharmaceutical and medical goods, cosmetics and toilet articles	**7 321**	**13 789**	**7 375**	**551**	**7 299**	**5 125**
Businesses having:						
1 outlet	6 139	6 139	2 963	158	2 944	2 177
2-9 outlets	1 135	3 324	1 671	115	1 632	1 163
10-99 outlets	43	1 115	623	62	613	396
100 or more outlets	5	3 211	2 117	216	2 111	1 388
Other retail sales of new goods in specialised stores	**90 291**	**153 795**	**70 488**	**8 530**	**68 746**	**36 558**
Businesses having:						
1 outlet	75 309	75 309	17 325	2 009	16 791	9 343
2-9 outlets	14 241	38 159	11 513	1 394	10 977	6 002
10-99 outlets	650	16 935	10 904	1 408	10 587	5 354
100 or more outlets	92	23 393	30 747	3 718	30 392	15 860
Retail sales of second-hand goods in stores	**5 146**	**7 013**	**1 511**	**87**	**1 451**	**968**
Businesses having:						
1 outlet	4 234	4 234	1 109	60	1 059	745
2-9 outlets	878[					
10-99 outlets	[34][	2 780	402	26	392	222
100 or more outlets	[ ][					
Retail sales not in stores	**6 321**	**9 734**	**9 204**	**1 181**	**8 682**	**4 319**
Businesses having:						
1 outlet	5 372	5 372	2 534	290	2 355	1 328
2-9 outlets	915	3 180	5 556	733[	]	2 492
10-99 outlets	[35][	1 183	1 114	158][	6 327][	499
100 or more outlets	[ ][			][	][	
Repair of personal and household goods	**2 005**	**2 649**	**438**	**61**	**385**	**86**
Businesses having:						
1 outlet	1 778	1 778	205	27	201	48
2-9 outlets	222[					
10-99 outlets	[4][	871	233	35	185	38
100 or more outlets	[ ][					

See footnotes on the first part of this table.

Source: Office for National Statistics: 01633 812921/2476

23.1 Retail businesses[1], 1996

continued

By Standard Industrial Classification (SIC 92)

United Kingdom

	Stocks[4]		Total turnover[4] divided by end-year stocks	Gross margin[4]		Value added
	Beginning of year	End of year		Amount	As a percentage of total turnover	
	£ million	£ million	Ratio	£ million	%	£ million
By group level of SIC 92 and number of outlets						
Total Retail Businesses	**16 243**	**17 347**	**10.1**	**56 114**	**32.1**	**36 618**
Businesses having:						
1 outlet	4 251	4 492	8.1	11 470	31.4	7 703
2-9 outlets	2 771	3 047	7.1	8 842	40.9	5 858
10-99 outlets	2 284	2 507	8.4	7 610	36.2	4 902
100 or more outlets	6 937	7 301	13.1	28 191	29.6	18 155
Retail sales in non-specialised stores	**4 903**	**5 198**	**15.9**	**20 085**	**24.4**	**14 249**
Businesses having:						
1 outlet	508	549	12.4	1 491	22.0	1 019
2-9 outlets	271	291	10.0	940	32.5	712
10-99 outlets	748	797	11.8	2 552	27.2	1 961
100 or more outlets	3 376	3 561	17.8	15 102	23.8	10 557
Specialised stores						
Retail sales of food, drink or tobacco	**661**	**691**	**19.5**	**3 701**	**27.4**	**2 397**
Businesses having:						
1 outlet	350	362	23.0	1 955	23.5	1 311
2-9 outlets	64	71	27.1	680	35.4	446
10-99 outlets	41	44	16.9	230	30.8	128
100 or more outlets	206	214	11.7	835	33.2	513
Retail sales of pharmaceutical and medical goods, cosmetics and toilet articles	**679**	**716**	**9.5**	**1 735**	**25.4**	**1 241**
Businesses having:						
1 outlet	254	264	10.6	637	22.7	497
2-9 outlets	146	150	10.3	398	25.5	288
10-99 outlets	61	66	8.5	170	30.2	119
100 or more outlets	219	236	8.0	530	27.9	337
Other retail sales of new goods in specialised stores	**8 577**	**9 166**	**6.8**	**25 990**	**41.9**	**16 173**
Businesses having:						
1 outlet	2 442	2 555	6.0	6 087	39.7	4 323
2-9 outlets	1 675	1 836	5.5	4 276	42.3	2 998
10-99 outlets	1 342	1 502	6.3	4 302	45.3	2 451
100 or more outlets	3 118	3 274	8.3	11 325	41.9	6 401
Retail sales of second-hand goods in stores	**582**	**625**	**2.3**	**498**	**35.0**	**283**
Businesses having:						
1 outlet	474	515	2.0	344	32.8	195
2-9 outlets						]
10-99 outlets	108	110	3.4	154	41.0	89]
100 or more outlets						]
Retail sales not in stores	**811**	**914**	**8.8**	**3 807**	**47.5**	**2 076**
Businesses having:						
1 outlet	209	231	9.7	940	41.9	441
2-9 outlets	506	581	8.3	2 405	49.9	1 333
10-99 outlets	96	102	9.4	462	48.4	301]
100 or more outlets						]
Repair of personal and household goods	**30**	**37**	**10.1**	**298**	**79.2**	**199**
Businesses having:						
1 outlet	15	17	10.8	132	74.0	94
2-9 outlets						]
10-99 outlets	14	21	9.7	166	83.9	105]
100 or more outlets						]

See footnotes on the first part of this table.

Source: Office for National Statistics: 01633 812921/2476

23.2 Retail trade: Index numbers of value and volume of sales[1]

Great Britain

Weekly average 1995 = 100, not seasonally adjusted

Value		Sales in 1995 £m	1988	1989	1990	1991	1992	1993	1994	1995	1996	1997	1998
All retailing:total	EAFY	166 681	70.4	75.3	80.5	84.1	87.3	92.0	96.3	100.0	105.4	112.0	116.4
All retailing:large	EAFZ	119 937	65	69	74	78	83	89	94	100	107	115	121
All retailing:small	EAGA	46 743	85	92	99	100	100	101	102	100	102	103	104
Predominantly food stores:total	EAFS	74 914	62.9	67.9	74.1	80.2	85.7	90.1	94.4	100.0	105.3	110.7	116.3
Predominantly food stores:large	EAGK	57 713	58	62	68	75	82	87	93	100	107	113	120
Predominantly food stores:small	EAGL	17 201	87	93	100	102	102	102	100	100	100	102	103
Non specialised food stores:total	EAGB	60 602	57	62	68	75	81	87	93	100	107	113	119
Non specialised food stores:large	EAGC	54 878	56	60	66	73	80	86	92	100	107	114	121
Non specialised food stores:small	EAGD	5 724	80	84	91	95	96	97	98	100	102	104	104
Specialised food stores	EAPP	14 311	90	96	101	103	105	105	102	100	100	101	103
Retail sale of fruit and vegetables	EAOZ	1 283	101	110	117	115	108	102	98	100	100	102	112
Retail sale of meat and meat products	EAPA	2 515	124	132	132	129	122	116	107	100	97	92	89
Retail sale of fish, crustaceans & molluscs	EAPB	266	132	144	157	147	128	123	109	100	95	91	78
Retail sale of bread cakes and confectionery	EAPC	1 796	79	84	91	95	99	102	102	100	100	103	109
Retail sale of alcohol and other beverages	EAPD	3 208	79	83	89	90	95	95	96	100	101	104	105
Retail sale of tobacco products	EAPE	3 944	86	90	97	102	108	109	106	100	100	103	106
Other specialised food stores	EAPF	1 296	96	101	107	108	105	110	105	100	97	100	102
Predominantly non-food stores:total	EAFT	83 184	74.7	79.6	84.2	85.8	87.2	92.4	97.0	100.0	105.8	113.6	116.8
Predominantly non-food stores:large	EAGM	55 351	70	74	78	81	83	89	95	100	107	118	122
Predominantly non-food stores:small	EAGN	27 832	83	90	97	97	96	100	102	100	104	105	107
Non specialised predominantly non-food stores:total	EAGE	15 035	71.9	76.1	80.7	84.5	88.1	93.4	96.4	100.0	108.2	115.5	116.9
Non specialised predominantly non-food stores:large	EAGF	12 070	73	77	80	84	87	92	95	100	108	115	117
Non specialised predominantly non-food stores:small	EAGG	2 964	66	74	83	91	98	106	103	100	109	116	115
Pharmaceutical,medical,cosmetic and toilet goods	EAPQ	3 165	62	69	74	81	89	97	97	100	102	109	113
Textiles,clothing,footwear and leather:total	EAFU	26 100	74.9	79.6	84.7	85.8	87.2	91.2	96.8	100.0	104.2	111.8	112.4
Textiles clothing footwear and leather:large	EAGO	20 352	69	73	77	80	84	89	95	100	106	116	118
Textiles clothing footwear and leather:small	EAGP	5 746	100	108	118	111	101	102	103	100	97	96	93
Retail sale of textiles	EAPG	699	83	94	105	105	106	127	117	100	99	105	106
Retail sale of clothing:total	EAGH	21 535	74	78	83	85	87	90	96	100	106	113	115
Retail sale of clothing:large	EAGI	17 606	68	73	77	80	84	89	95	100	108	118	120
Retail sale of clothing:small	EAGJ	3 928	97	103	110	106	99	99	101	100	95	92	92
Retail sale of footwear and leather goods	EAPH	3 864	82	88	96	91	87	92	100	100	97	105	98
Household goods stores:total	EAFV	19 770	78.8	82.0	84.1	85.9	86.9	92.9	97.0	100.0	107.9	118.3	124.8
Household goods stores:large	EAGQ	13 273	67	69	70	75	81	87	94	100	109	122	130
Household goods stores:small	EAGR	6 496	102	109	115	108	101	106	104	100	106	112	114
Retail sale of furniture, lighting etc	EAPI	6 400	92	95	96	94	94	101	103	100	108	122	128

1 Please see notes in the chapter text.

Source: Office for National Statistics: 01633 812609

23.2 Retail trade: Index numbers of value and volume of sales[1]
Great Britain

continued

Weekly average 1995 = 100, not seasonally adjusted

		Sales in 1995 £m	1988	1989	1990	1991	1992	1993	1994	1995	1996	1997	1998
Retail sale of electrical household appliances	EAPJ	7 727	76	77	79	80	79	86	91	100	108	119	124
Retail sale of hardware paint and glass	EAPK	5 642	72	77	81	87	91	96	100	100	107	114	122
Other specialised non-food stores: total	EAFW	22 280	72.4	79.5	85.4	86.3	87.1	92.5	97.3	100.0	104.1	110.0	114.8
Other specialised non-food stores: large	EAGS	9 655	73	81	89	85	81	89	94	100	103	119	124
Other specialised non-food stores: small	EAGT	12 624	70	77	83	87	92	95	100	100	105	103	108
Retail sale of books newspapers and periodicals	EAPL	4 078	69	76	84	88	92	97	99	100	104	109	117
Retail sale of floor covering	EAPM	1 045	92	94	93	91	92	98	99	100	107	116	99
Photo, optical and precision equipment office supplies and equipment	EAWH	2 676	78	98	113	112	111	97	97	100	104	103	104
Other retail sale in specialised stores nes	EAWK	10 201	75	83	89	87	83	90	95	100	105	112	120
Second-hand goods stores	EAQA	1 112	65	68	68	67	68	74	105	100	94	104	102
Other retail sale (non-store) and repair: total	EAFX	8 583	90.7	94.2	97.4	99.0	99.3	102.1	104.3	100.0	102.8	107.7	113.9
Other retail sale (non-store) and repair: large	EAGU	6 872	84	87	89	90	91	97	102	100	104	112	122
Other retail sale (non-store) and repair: small	EAGV	1 710	101	109	116	119	118	114	110	100	96	90	82
Retail sale via mail order houses	EAPN	6 629	79	82	87	89	92	98	103	100	103	109	117
Non-store retail exc mail order	EAPO	1 544	108	111	113	113	110	107	106	100	100	102	101
Repair of personal and household goods	EAPR	408	80	80	84	99	106	115	106	100	104	110	110
Volume													
All retailing	EAHC	166 681	90.5	92.4	93.0	91.8	92.4	95.3	98.8	100.0	103.1	108.5	111.7
Predominantly food stores	EAGW	74 914	86.0	88.0	89.4	90.4	93.1	95.0	98.1	100.0	101.8	105.9	108.8
Predominantly non-food stores	EAGX	83 184	91.5	93.5	93.9	91.0	90.3	94.2	98.4	100.0	104.4	111.2	114.3
Non specialised predominantly non-food stores	EAHI	15 035	93.0	94.0	94.3	92.9	93.6	97.0	99.0	100.0	106.0	111.5	111.5
Textiles,clothing, footwear and leather	EAGY	26 100	87.5	88.5	89.9	88.0	89.1	92.3	97.5	100.0	104.3	111.2	112.0
Household goods stores	EAGZ	19 770	90.1	91.3	89.7	87.0	86.5	92.2	97.3	100.0	106.8	117.2	125.2
Other specialised non-food stores	EAHA	22 280	96.4	100.9	102.5	97.1	93.8	96.7	100.1	100.0	101.4	105.7	109.2
Other retail sale (non-store) and repair	EAHB	8 583	113.3	111.8	109.4	105.7	103.8	105.0	106.1	100.0	101.4	105.7	111.6

1 Please see notes in the chapter text.

Source: Office for National Statistics: 01633 812609

23.3 Motor trades[1]
United Kingdom

	Number of businesses	Total turnover	Motor trades turnover	Retail sales: New cars	Retail sales: Other new motor vehicles and motorcycles	Sales to other dealers: New cars	Sales to other dealers: Other new motor vehicles and motorcycles	Gross sales of used motor vehicles and motorcycles	Turnover from sales of petrol, diesel, oil and other petroleum products	Other motor trades sales and receipts (including parts and accessories, workshop receipts)	Non-motor trades turnover
	Number	£ million	£ million	£ million	£ million	£ million	£ million	£ million	£ million	£ million	£ million
Sale, maintenance and repair of motor vehicles and motorcycles; retail sale of automotive fuel (SIC 92 50.00)											
	MKEQ	CMRH	CMRI	CMRJ	CMRK	CMRL	CMRM	CMRN	CMRO	CMRP	CMRQ
1993[2]	71 679	80 748	78 408†	14 490	1 945	11 367	2 257	16 695	10 985	20 669	2 340
1994[2]	69 148	89 360	86 562	16 012	2 774	13 025	2 888	18 821	11 050	21 991	2 798
1995	70 278	99 171	96 055	17 889	3 126	13 902	2 407	22 739	11 661	24 332	3 116
1996	71 119	107 104	103 839	19 577	3 195	15 834	2 291	25 158	11 516	26 268	3 265
1997	71 024	119 303	115 712	23 284	3 465	15 739	2 481	28 580	12 751	29 412	3 591
Sale of motor vehicles (SIC 92 50.10)											
	MKER	EWRI	FDFZ	FDGA	FDGB	FDGC	FDGD	FDGE	FDGF	FDGG	FDHJ
1993[2]	39 957	52 879	52 305	..	..	..	..	..	1 341	15 388	574
1994[2]	38 693	59 220	58 509	..	..	..	..	..	1 350	16 388	711
1995	39 473	68 096	67 356	16 259	2 697	13 856	2 315	20 513	1 307	10 408	740
1996	36 383	75 717	74 820	18 582	2 833	15 802	2 190	23 421	1 251	10 741	897
1997	33 873	84 988	84 137	22 709	3 013	15 714	2 351	27 121	1 174	12 055	851
Maintenance and repair of motor vehicles (SIC 92 50.20)											
	MKES	FDHK	FDHL	FDHM	FDHN	FDHO	FDHP	FDHQ	FDHR	FDHS	FDHT
1993[2]	17 844	8 501	8 405	..	..	..	..	..	212	2 450	96
1994[2]	17 280	9 520	9 401	..	..	..	..	..	214	2 609	119
1995	17 363	8 790	8 562	1 219	166	31	1	1 444	729	4 972	228
1996	19 709	8 134	8 020	784	127	19	1	1 005	541	5 543	114
1997	21 509	7 962	7 796	411	80	20	5	661	229	6 390	166
Sale of motor vehicle parts and accessories (SIC 92 50.30)											
	MKET	FDIW	FDIX	FDIY	FDIZ	FDJA	FDJB	FDJC	FDJD	FDJE	FDJF
1993[2]	4 844	7 631	7 549	..	..	..	..	..	216	2 227	82
1994[2]	4 690	8 547	8 446	..	..	..	..	..	218	2 372	101
1995	4 801	9 291	8 960	220	97	10	–	255	128	8 249	331
1996	5 829	9 908	9 652	69	24	9	–	116	77	9 358	256
1997	6 497	10 458	10 284	9	19	–	–	56	76	10 124	174
Sale, maintenance and repair of motorcycles and related parts and accessories (SIC 92 50.40)											
	MKEU	FDKI	FDKJ	FDKK	FDKL	FDKM	FDKN	FDKO	FDKP	FDKQ	FDKR
1993[2]	1 249	899	889	..	..	..	..	..	23	263	10
1994[2]	1 209	1 007	995	..	..	..	..	..	23	280	12
1995	1 276	1 038	1 008	94	165	4	90	286	10	359	30
1996	1 231	1 004	981	15	211	–	100	337	3	315	23
1997	1 517	1 282	1 261	–	353	–	125	392	2	389	21
Retail sale of automotive fuel (SIC 92 50.50)											
	MKEV	FDLV	FDLW	FDLX	FDLY	FDLZ	FDMA	FDMB	FDMC	FDMD	FDME
1993[2]	7 785	10 838	9 260	..	..	..	..	..	9 193	341	1 578
1994[2]	7 276	11 067	9 212	..	..	..	..	..	9 246	342	1 855
1995	7 365	11 957	10 172	96	1	2	–	241	9 487	344	1 785
1996	7 967	12 341	10 366	128	1	4	–	280	9 643	311	1 975
1997	7 628	14 613	12 234	155	–	5	–	350	11 270	454	2 379

1 Figures are exclusive of VAT.
2 Data for 1993 and 1994 was collected on a different basis from that for 1995 onwards.

Source: Office for National Statistics: 01633 812264

23.3 Motor trades[1]
United Kingdom

continued

	Purchases of goods and services											
	Total purchases	Energy, water and materials	Used motor vehicles and motorcycles	Parts used solely in repair and servicing activities	Goods for resale	Hiring, leasing and renting of plant, machinery and vehicles	Commercial insurance premiums	Road transport services	Telecommunication services	Computer and related services	Advertising and marketing services	Other services
	£ million	£ million	£ million	£ million	£ million	£ million	£ million	£ million	£ million	£ million	£ million	£ million
Sale, maintenance and repair of motor vehicles and motorcycles; retail sale of automotive fuel (SIC 92 50.00)												
	CMNR	CMRS	COBU	CMRT	CMRU	CMRV	CMRW	CMRX	CMRY	CMRZ	CMSA	CMSB
1993[2]	71 609	..	15 145	..	48 215	..	..	..	..	..	..	..
1994[2]	78 514	..	17 036	..	52 560	..	..	..	..	..	..	..
1995	86 533	..	20 426	..	56 554	..	..	..	..	..	..	..
1996	93 277†	497	22 640	5 650†	58 667†	182	324	296	223	130	1 298	3 371†
1997	102 979	554	25 537	5 643	65 125	234	335	345	233	130	1 305	3 538
Sale of motor vehicles (SIC 92 50.10)												
	FDGH	FDGI	FDGJ	FDGK	FDGL	FDGM	FDGN	FDGO	FDGP	FDGQ	FDGR	FDGS
1993[2]	46 717	..	..	..	29 608	..	..	..	..	..	..	..
1994[2]	51 932	..	..	..	32 789	..	..	..	..	..	..	..
1995	60 577	..	18 440	..	36 861	..	..	..	..	..	..	..
1996	67 582	244	21 147	3 000	39 246	85	160	199	121	87	1 093	2 200
1997	75 235	247	24 241	2 848	43 621	96	158	216	124	91	1 113	2 480
Maintenance and repair of motor vehicles (SIC 92 50.20)												
	FDHU	FDHV	FDHW	FDHX	FDHY	FDHZ	FDIA	FDIB	FDIC	FDID	FDIE	FDIF
1993[2]	7 497	..	..	..	4 835	..	..	..	..	..	..	..
1994[2]	8 334	..	..	..	5 355	..	..	..	..	..	..	..
1995	6 519	..	1 303	..	2 536	..	..	..	..	..	..	..
1996	5 615	113	886	2 231	1 558	58	88	21	44	17	88	511
1997	5 057	162	603	2 305	1 170	95	102	33	51	14	62	460
Sale of motor vehicle parts and accessories (SIC 92 50.30)												
	FDJG	FDJH	FDJI	FDJJ	FDJK	FDJL	FDJM	FDJN	FDJO	FDJP	FDJQ	FDJR
1993[2]	6 740	..	..	..	4 299	..	..	..	..	..	..	..
1994[2]	7 492	..	..	..	4 761	..	..	..	..	..	..	..
1995	7 703	..	222	..	6 560	..	..	..	..	..	..	..
1996	7 980†	68	99	286†	6 934†	31	41	68	39	21	81	312†
1997	8 207	63	55	292	7 147	30	40	89	41	22	92	336
Sale, maintenance and repair of motorcycles and related parts and accessories (SIC 92 50.40)												
	FDKT	FDKU	FDKV	FDKW	FDKX	FDKY	FDKZ	FDLA	FDLB	FDLC	FDLD	FDLE
1993[2]	794	..	..	..	506	..	..	..	..	..	..	..
1994[2]	883	..	..	..	560	..	..	..	..	..	..	..
1995	871	..	243	..	535	..	..	..	..	..	..	..
1996	843	3	267	29	481	1	5	4	2	2	14	35
1997	1 101	12	334	46	625	1	7	4	4	1	18	49
Retail sale of automotive fuel (SIC 92 50.50)												
	FDMF	FDMG	FDMH	FDMI	FDMJ	FDMK	FDML	COBV	COBW	COBX	COBY	COBZ
1993[2]	9 860	..	..	..	8 967	..	..	..	..	..	..	..
1994[2]	9 874	..	..	..	9 096	..	..	..	..	..	..	..
1995	10 863	..	218	..	10 062	..	..	..	..	..	..	..
1996	11 254	69	242	103	10 448	6	29	3	16	4	21	313
1997	13 379	70	304	152	12 562	12	28	3	13	2	20	213

1 Figures are exclusive of VAT.
2 Data for 1993 and 1994 was collected on a different basis from that for 1995 onwards.

Source: Office for National Statistics: 01633 812264

23.3 Motor trades[1] United Kingdom

continued

	Taxes and levies			Capital Expenditure			
	Total taxes	Local authority rates	Other taxes	Total acquisitions	Total disposals	Net capital expenditure	Work of a capital nature carried out by your own staff (included in acquisitions)
	£ million	£ million	£ million	£ million	£ million	£ million	£ million
Sale, maintenance and repair of motor vehicles and motorcycles; retail sale of automotive fuel (SIC 92 50.00)							
	CMSC	CMSD	CMSE	CMSF	CMSG	CMSH	CMSI
1993[2]	..	..	..	1 107	400	707	..
1994[2]	..	..	..	1 092	381	711	..
1995	..	..	..	851	349	502	..
1996	588	459	129	853	365	488	7
1997	677	464	213	1 537	614	922	14
Sale of motor vehicles (SIC 92 50.10)							
	FDGT	FDGU	FDGV	FDGW	FDGX	FDGY	FDGZ
1993[2]	..	..	..	748	278	470	..
1994[2]	..	..	..	706	259	447	..
1995	..	..	..	584	284	300	..
1996	325	213	112	637	321	316	5
1997	367	245	122	941	423	518	7
Maintenance and repair of motor vehicles (SIC 92 50.20)							
	FDIG	FDIH	FDII	FDIJ	FDIK	FDIL	FDIM
1993[2]	..	..	..	122	55	67	..
1994[2]	..	..	..	115	51	64	..
1995	..	..	..	72	14	58	..
1996	104	101	3	74	14	60	1
1997	141	88	53	250	46	204	4
Sale of motor vehicle parts and accessories (SIC 92 50.30)							
	FDJS	FDJT	FDJU	FDJV	FDJW	FDJX	FDJY
1993[2]	..	..	..	108	40	68	..
1994[2]	..	..	..	102	37	65	..
1995	..	..	..	99	18	81	..
1996	109	102	7	95	16	79	1
1997	99	73	26	236	81	154	3
Sale, maintenance and repair of motorcycles and related parts and accessories (SIC 92 50.40)							
	FDLF	FDLG	FDLH	FDLI	FDLJ	FDLK	FDLL
1993[2]	..	..	..	13	5	8	..
1994[2]	..	..	..	12	4	8	..
1995	..	..	..	11	2	9	..
1996	8	6	2	11	2	9	–
1997	12	11	1	26	6	21	–
Retail sale of automotive fuel (SIC 92 50.50)							
	COCA	COCB	COCC	COCD	COCE	COCF	COCG
1993[2]	..	..	..	116	22	94	..
1994[2]	..	..	..	157	30	127	..
1995	..	..	..	85	31	54	..
1996	42	37	5	36	12	24	–
1997	57	47	11	84	58	25	–

1 Figures are exclusive of VAT.
2 Data for 1993 and 1994 was collected on a different basis from that for 1995 onwards.

Source: Office for National Statistics: 01633 812264

23.3 Motor trades[1]
United Kingdom

continued

	Stocks			Employment costs			Gross margin		
	Increase during the year	Value at end of year	Total turnover divided by end-year stocks	Total employment costs	Wages and salaries	National insurance and pension contributions	Amount	As a percentage of total turnover	Value added at basic prices
	£ million	£ million	Quotient	£ million	£ million	£ million	£ million	Percentage	£ million
Sale, maintenance and repair of motor vehicles and motorcycles; retail sale of automotive fuel (SIC 92 50.00)									
	CMSJ	CMSK	CMSL	CMSM	COBP	COBQ	COBR	COBS	COBT
1993[2]	1 124	7 656†	10.5	5 052	4 519	533	18 512	22.9	10 263
1994[2]	1 384	9 518	9.4	5 248	4 692	556	21 148	23.7	12 230
1995	950	10 256	9.7	5 846	5 228	618	23 141	23.3	13 588
1996	885	11 085	9.7	6 384	5 724	660	21 032	20.7	14 590†
1997	1 375	12 030	9.9	7 179	6 413	766	24 373	21.4	17 500
Sale of motor vehicles (SIC 92 50.10)									
	FDHA	FDHB	FDHC	FDHD	FDHE	FDHF	FDHG	FDHH	FDHI
1993[2]	839	5 505	9.6	3 518	3 141	377	..	..	7 001
1994[2]	1 032	6 925	8.6	3 681	3 287	394	..	..	8 320
1995	726	7 973	8.5	3 156	2 806	350	13 521	19.9	8 245
1996	745	8 753	8.7	3 478	3 111	367	13 069	18.0	8 772
1997	1 212	9 533	8.9	3 844	3 420	424	15 490	18.9	10 850
Maintenance and repair of motor vehicles (SIC 92 50.20)									
	FDIN	FDIO	FDIP	FDIQ	FDIR	FDIS	FDIT	FDIU	FDIV
1993[2]	141	916	9.3	558	498	60	..	..	1 145
1994[2]	174	1 152	8.3	583	521	62	..	..	1 360
1995	57	643	13.7	1 256	1 132	124	5 008	57.0	2 328
1996	30	556	14.6	1 405	1 265	140	3 489	59.1	2 548
1997	46	542	14.7	1 550	1 380	170	3 930	69.5	2 902
Sale of motor vehicle parts and accessories (SIC 92 50.30)									
	FDJZ	FDKA	FDKB	FDKC	FDKD	FDKE	FDKF	FDKG	FDKH
1993[2]	121	799	9.6	507	453	54	..	..	1 012
1994[2]	149	1 005	8.5	530	474	56	..	..	1 204
1995	129	1 132	8.2	966	866	100	2 638	28.4	1 717
1996	81	1 226	8.1	1 047	937	110	2 670	27.7†	2 003†
1997	51	1 289	8.1	1 224	1 102	122	3 015	29.7	2 279
Sale, maintenance and repair of motorcycles and related parts and accessories (SIC 92 50.40)									
	FDLM	FDLN	FDLO	FDLP	FDLQ	FDLR	FDLS	FDLT	FDLU
1993[2]	14	94	9.6	60	54	6	..	..	119
1994[2]	17	118	8.5	63	56	7	..	..	141
1995	15	186	5.6	87	78	9	275	26.5	182
1996	16	215	4.7	71	63	8	243	24.9	174
1997	28	266	4.8	93	82	11	305	24.7	208
Retail sale of automotive fuel (SIC 92 50.50)									
	COCH	COCI	COCJ	COCK	COCL	COCM	COCN	CMQN	CMQO
1993[2]	9	342	31.7	409	373	36	..	..	987
1994[2]	12	318	34.8	391	354	37	..	..	1 205
1995	23	322	37.1	381	346	35	1 700	14.2	1 117
1996	13	335	36.8	383	348	35	1 561	12.8	1 094
1997	38	400	36.5	470	429	41	1 633	11.3	1 262

1 Figures are exclusive of VAT.
2 Data for 1993 and 1994 was collected on a different basis from that for 1995 onwards.

Source: Office for National Statistics: 01633 812264

23.4 Catering and allied trades United Kingdom

	Number of businesses	Total turnover[2]	Taxes and levies[3] Total taxes	Local authority rates	Other taxes	Capital expenditure[3] Capital acquisitions	Capital disposals	Net capital expenditure	Work of a capital nature carried out by your own staff (included in acquisitions)
	Number	£ million	£ million	£ million	£ million	£ million	£ million	£ million	£ million
Total catering and allied trades (SIC 92 55.00)									
	MKEK	CMKX	CMLM	CMLJ	CMLL	CMLP	CMLQ	CMLK	CMLR
1993[1]	110 471	36 404	..	..	..	1 960	371	1 588	..
1994[1]	108 459	38 349	..	..	..	2 293	506	1 788	..
1995	111 330	40 480	..	..	..	2 954	576	2 378	..
1996	109 532	43 183	1 016	816	200	3 131	466	2 665	13
1997	109 197	47 688	1 187	903	284	3 961	499	3 462	13
Hotels and motels (SIC 92 55.11 and 55.12)									
	MKEL	CMLW	CMML	CMMI	CMMK	CMMO	CMMP	CMMJ	CMMQ
1993[1]	11 878	6 713	..	..	..	466	62	404	..
1994[1]	12 167	7 263	..	..	..	508	72	436	..
1995	12 444	8 526	..	..	..	641	131	510	..
1996	12 047	9 258	253	234	19	819	204	615	3
1997	11 382	9 841	261	231	29	1 100	103	997	8
Camping sites and other provision of short-stay accommodation (SIC 92 55.21 to 55.23)									
	MKEM	CMMV	CMNK	CMNH	CMNJ	CMNN	CMNO	CMNI	CMNP
1993[1]	2 787	1 655	..	..	..	183	20	163	..
1994[1]	3 105	1 350	..	..	..	125	19	106	..
1995	3 036	1 492	..	..	..	203	23	180	..
1996	3 065	1 679	45	45	..	166	16	150	1
1997	3 264	1 946	55	51	4	258	37	221	2
Restaurants or cafes, take-away food shops (SIC 92 55.30)									
	MKEN	CMNU	CMOJ	CMOG	CMOI	CMOM	CMON	CMOH	CMOO
1993[1]	46 287	9 032	..	..	..	346	61	285	..
1994[1]	44 726	10 081	..	..	..	443	92	351	..
1995	46 132	11 474	..	..	..	681	87	594	..
1996	45 938	12 067	255	233	22	654	69	585	1
1997	46 791	13 642	293	267	26	774	68	706	1
Licensed clubs with entertainment, independent, tenanted, managed public houses or wine bars (SIC 92 55.40)[6]									
	MKEO	CMOT	CMPI	CMPF	CMPH	CMPL	CMPM	CMPG	CMPN
1993[1]	45 814	16 174	..	..	..	907	213	694	..
1994[1]	44 754	16 743	..	..	..	1 160	315	845	..
1995	45 954	15 410	..	..	..	1 361	324	1 037	..
1996	44 705	16 048	439	280	159	1 428	172	1 256	7
1997	43 939	17 272	532	317	215	1 732	288	1 444	2
Canteen operator, catering contractor (SIC 92 55.51 and 55.52)									
	MKEP	CMPS	CMQH	CMQE	CMQG	CMQK	CMQL	CMQF	CMQM
1993[1]	3 705	2 830	..	..	..	57	15	42	..
1994[1]	3 707	2 912	..	..	..	57	7	50	..
1995	3 764	3 578	..	..	..	68	11	57	..
1996	3 777	4 131	24	24	..	64	5	59	1
1997	3 821	4 988	47	37	9	97	3	94	–

1 Data for 1993 and 1994 was collected on a different basis from that for 1995 onwards.
2 Inclusive of VAT.
3 Exclusive of VAT.
4 The 1995, 1996 and 1997 total turnover figure used to calculate these data excludes VAT.
5 Figures for 1995, 1996 and 1997 are Great Britain only.
6 Includes figures for managed public houses owned by breweries.

Source: Office for National Statistics: 01633 812264

23.4 Catering and allied trades
United Kingdom

continued

	Stocks[3]		Purchases of goods and services[3]									
	Increase during year	Value at end of year	Total purchases	Energy, water and materials	Goods for resale	Hiring, leasing of plant, machinery etc.	Commercial insurance premiums	Road transport services	Telecommunication services	Computer and related services	Advertising and marketing services	Other services
	£ million	£ million	£ million	£ million	£ million	£ million	£ million	£ million	£ million	£ million	£ million	£ million
Total catering and allied trades (SIC 92 55.00)												
	CMLN	CMLO	CMLI	CMKZ	CMLA	CMLB	CMLC	CMLD	CMLE	CMLF	CMLG	CMLH
1993[1]	-7	804	20 429	..	11 926	..	..	..	..	..	..	..
1994[1]	29	898	20 615	..	8 875	..	..	..	..	..	..	..
1995	37	790	20 679	..	8 786	..	..	..	..	..	..	..
1996	39	811	20 697†	6 331	8 261†	215	292	64	158	37	497	4 842†
1997	51	867	22 185	8 311	7 583	249	299	104	190	47	573	4 829
Hotels and motels (SIC 92 55.11 and 55.12)												
	CMMM	CMMN	CMMH	CMLY	CMLZ	CMMA	CMMB	CMMC	CMMD	CMME	CMMF	CMMG
1993[1]	–1	118	3 115	..	1 175	..	..	..	..	..	..	..
1994[1]	–34	120	3 266	..	706	..	..	..	..	..	..	..
1995	2	118	3 593	..	732	..	..	..	..	..	..	..
1996	–10	134	3 538	1 255	568	68	86	14	66	16	179	1 284
1997	–	141	3 614	1 485	523	68	74	28	62	17	167	1 192
Camping sites and other provision of short-stay accommodation (SIC 92 55.21 to 55.23)												
	CMNL	CMNM	CMNG	CMMX	CMMY	CMMZ	CMNA	CMNB	CMNC	CMND	CMNE	CMNF
1993[1]	–3	63	777	..	431	..	..	..	..	..	..	..
1994[1]	4	60	625	..	274	..	..	..	..	..	..	..
1995	3	58	731	..	275	..	..	..	..	..	..	..
1996	4	67	789	171	261	8	28	6	10	3	49	252
1997	8	88	875	211	291	11	23	6	10	3	50	270
Restaurants or cafes, take-away food shops (SIC 92 55.30)												
	CMOK	CMOL	CMOF	CMNW	CMNX	CMNY	CMNZ	CMOA	CMOB	CMOC	CMOD	CMOE
1993[1]	13	155	5 055	..	2 825	..	..	..	..	..	..	..
1994[1]	24	187	5 238	..	1 146	..	..	..	..	..	..	..
1995	11	184	5 876	..	1 848	..	..	..	..	..	..	..
1996	6	191	5 934†	2 293	1 740	22	68	16	31	5	123	1 636†
1997	17	217	6 635	3 167	1 582	34	85	26	47	7	178	1 510
Licensed clubs with entertainment, independent, tenanted, managed public houses or wine bars (SIC 92 55.40)[6]												
	CMPJ	CMPK	CMPE	CMOV	CMOW	CMOX	CMOY	CMOZ	CMPA	CMPB	CMPC	CMPD
1993[1]	–17	428	10 108	..	6 467	..	..	..	..	..	..	..
1994[1]	29	482	10 126	..	6 520	..	..	..	..	..	..	..
1995	18	370	8 748	..	5 563	..	..	..	..	..	..	..
1996	30	348	8 501	1 345	5 330†	89	96	18	42	4	130	1 447†
1997	14	348	8 877	2 007	4 955	103	103	19	60	12	166	1 451
Canteen operator, catering contractor (SIC 92 55.51 and 55.52)												
	CMQI	CMQJ	CMQD	CMPU	CMPV	CMPW	CMPX	CMPY	CMPZ	CMQA	CMQB	CMQC
1993[1]	1	40	1 374	..	1 028	..	..	..	..	..	..	..
1994[1]	6	49	1 360	..	229	..	..	..	..	..	..	..
1995	3	60	1 731	..	368	..	..	..	..	..	..	..
1996	9	71	1 934	1 267	362	28	12	10	9	9	13	221
1997	12	73	2 183	1 441	232	33	15	24	11	9	12	406

See footnotes on the first page of this table.

Source: Office for National Statistics: 01633 812264

23.4 Catering and allied trades
United Kingdom

continued

	Employment costs[3]			Gross margin[4]			Accommodation	
	Total employment costs	Wages and salaries	National insurance and pension contributions	Amount	As a percentage of turnover	Value added at basic prices[4]	Number of establishments	Letting bedplaces
	£ million	£ million	£ million	£ million	Percentage	£ million	Number	Number
Total catering and allied trades (SIC 92 55.00)								
	CMKY	CMKV	CMKW	CMQP	CMQQ	CMQR	CMLS	CMLT
1993[1]	7 150	6 513	638	24 472	67.2	15 968	..	..
1994[1]	7 939	7 296	642	29 504	76.9	17 764	..	..
1995	7 850	7 236	614	25 856	74.7	13 963	..	..
1996	8 332	7 679	653	28 825†	77.8†	16 201†	..	..
1997	9 331	8 611	720	33 590	81.7	18 717	22 417	1 652 556
Hotels and motels (SIC 92 55.11 and 55.12)								
	CMLX	CMLU	CMLV	CMQS	CMQT	CMQU	CMMR	CMMS
1993[1]	1 667	1 514	153	5 537	82.5	3 598	..	..
1994[1]	1 936	1 770	166	6 523	89.8	3 964	..	..
1995	2 088	1 914	174	6 585	90.0	3 723	13 337	716 434
1996	2 211	2 027	184	7 319	92.7	4 333	12 780†	696 006†
1997	2 325	2 130	195	7 883	93.8	4 770	13 317	740 750
Camping sites and other provision of short-stay accommodation (SIC 92 55.21 to 55.23)								
	CMMW	CMMT	CMMU	CMQV	CMQW	CMQX	CMNQ	CMRR
1993[1]	281	239	42	1 221	73.8	875	..	..
1994[1]	233	209	24	1 080	80.0	729	..	..
1995	239	220	20	1 036	79.2	580	3 506	610 894
1996	258	238	20	1 223	82.6	695	3 572	760 625†
1997	318	291	27	1 444	83.6	858	3 671	815 599
Restaurants or cafes, take-away food shops (SIC 92 55.30)								
	CMNV	CMNS	CMNT	CMQY	CMQZ	CMRA	CMOP	CMOQ
1993[1]	1 702	1 560	142	6 220	68.9	3 990	..	..
1994[1]	1 895	1 748	147	8 959	88.9	4 867	..	..
1995	2 029	1 877	152	7 991	81.3	3 963	..	..
1996	2 125	1 974	151	8 639	83.3	4 420†	..	..
1997	2 498	2 317	181	10 175	86.7	5 097	759	29 243
Licensed clubs with entertainment, independent, tenanted, managed public houses or wine bars (SIC 92 55.40)[6]								
	CMOU	CMOR	CMOS	CMRB	CMRC	CMRD	CMPO	CMPP
1993[1]	2 476	2 264	212	9 690	59.9	6 048	..	..
1994[1]	2 793	2 574	219	10 252	61.2	6 646	..	..
1995	2 338	2 162	176	7 575	57.7	4 390	5 303	56 898
1996	2 453	2 259	194	8 377†	61.2†	5 054†	5 580	63 957†
1997	2 664	2 464	200	9 860	66.6	5 725	..	66 964
Canteen operator, catering contractor (SIC 92 55.51 and 55.52)								
	CMPT	CMPQ	CMPR	CMRE	CMRF	CMRG		
1993[1]	1 024	936	88	1 803	63.7	1 457		
1994[1]	1 082	995	87	2 689	92.3	1 558		
1995	1 157	1 063	94	2 670	88.0	1 307		
1996	1 285	1 181	104	3 265	90.2	1 695		
1997	1 526	1 409	117	4 229	95.1	2 268		

See footnotes on the first page of this table.

Source: Office for National Statistics: 01633 812264

SOURCES

This index of sources gives the titles of official publications or other sources containing statistics allied to those in the tables of this *Annual Abstract*. These publications provide more detailed analyses than are shown in the *Annual Abstract*. This index includes publications to which reference should be made for short-term (monthly or quarterly) series. No entry is made in this index for items where the data have been obtained from departmental records. Further advice on published statistical sources is available from the National Statistics Public Enquiry Service on the numbers provided on page ii.

Subject	Table number in *Abstract*	Government department or other organisation	Official publication or other source
1. Area			
	1.1	Ordnance Survey Ordnance Survey of Northern Ireland Office for National Statistics	Regional Trends (annual, The Stationery Office (TSO))
2. Parliamentary elections			
Elections	2.1	Home Office	Electoral statistics (series EL) Return of election expenses Vachers Parliamentary Companion
By-elections	2.2	Home Office	Social Trends (annual, TSO)
3. Overseas aid			
	3.1, 3.2	Department for International Development	British Aid Statistics (annual)
4. Defence			
	4.1 - 4.13	Ministry of Defence/ DASA	UK Defence Statistics 1999 (TSO)
5. Population and vital statistics			
Population	5.1 - 5.3, 5.5, 5.6	Office for National Statistics	*England and Wales*: Census reports 1911, 1921, 1931, 1951, 1961, 1971, 1981 and 1991 Census 1991, Key Population and Vital Statistics; Great Britain Digest of Welsh Statistics (annual, National Assembly for Wales)
		General Register Office (Scotland)	*Scotland*: Census reports 1951, 1961, 1971, 1981 and 1991 Census 1991, Key statistics for urban areas: Scotland
		General Register Office (Northern Ireland)	*Northern Ireland*: Census of population 1951, 1961, 1966 and 1971, 1981 and 1991
		Office for National Statistics	*England and Wales*: Series FM (Family statistics), DH (Deaths), MB (Morbidity), PP (Population estimates and projections), MN (Migration) and VS (Key population and vital statistics) Series PP1, Population estimates: The Registrar General's estimates of the population of regions and local government areas of England and Wales Population Trends (quarterly TSO) Health Statistics Quarterly (TSO)
		General Register Office (Scotland)	*Scotland*: Annual report of the Registrar General for Scotland Annual estimate of the population of Scotland
		General Register Office (Northern Ireland)	*Northern Ireland*: Annual report of the Registrar General
Projections	5.1 - 5.3	Government Actuary's Department Office for National Statistics	Series PP2, Population projections - national figures
Migration	5.7, 5.8	Office for National Statistics	Series MN (International migration) Population Trends (quarterly, TSO)
	5.9, 5.10	Home Office	Control of immigration statistics United Kingdom (annual)

Subject	Table number in *Abstract*	Government department or other organisation	Official publication or other source
Vital statistics	5.4, 5.11 - 5.21	Office for National Statistics	*England and Wales*: Series FM (Births, marriages and divorce statistics), DH (Deaths), MB (Morbidity), PP (Population estimates and projections), MN (International migration) and VS (Key population and vital statistics) Population Trends (quarterly, TSO)
		General Register Office (Scotland)	*Scotland*: Annual report of the Registrar General for Scotland Quarterly return of births, deaths and marriages
		General Register Office (Northern Ireland)	*Northern Ireland*: Annual report of the Registrar General Quarterly return of births, deaths and marriages
	5.13	Northern Ireland Court Service	Northern Ireland Judicial Statistics (annual)
	5.17	Scottish Executive	
	5.22	Government Actuary's Department	*England and Wales*: Interim Life Table *Scotland*: Interim Life Table *Northern Ireland*: Annual Report of the Registrar General
6. Education			
	6.1 - 6.9	Education Departments	Scottish Educational Statistics (annual, TSO) Northern Ireland Education Statistics (annual) Compendium of Education Statistics for the United Kingdom (annual) Northern Ireland Annual Abstract of Statistics Statistics of Education and Training in Wales (annual, Welsh Office)
	6.10 - 6.14	Higher Education Statistics Agency Higher Education Funding Council for England	Students in Higher Education Institutions 1997/98 Resources of Higher Education Institutions 1997/98
7. Labour market			
Labour Force Survey	7.1 - 7.3, 7.6, 7.9, 7.10, 7.12, 7.15- 7.17	Office for National Statistics	Labour Market Trends (monthly, TSO)
	7.4, 7.5	Office for National Statistics	
	7.7	Cabinet Office	Civil Service Statistics (annual) Monthly Digest of Statistics (TSO)
	7.8	Department of the Environment Transport and the Regions (DETR) Scottish Executive National Assembly for Wales	Labour Market Trends (monthly, TSO) Annual Review of the Engineering Industry Training Board
Claimant count	7.11, 7.13 7.14, 7.27	Office for National Statistics	Labour Market Trends (monthly, TSO)
	7.18	Office for National Statistics	Labour Market Trends (monthly TSO) Monthly Digest of Statistics (TSO)
New Earnings Survey	7.19, 7.20, 7.23, 7.25	Office for National Statistics	New Earnings Survey (annual, ONS)
Average Earnings Index	7.21, 7.22	Office for National Statistics	Labour Market Trends (monthly, TSO) Monthly Digest of Statistics (TSO)
	7.24	Department of Economic Development (Northern Ireland)	New Earnings Survey Northern Ireland (annual) (some details in Northern Ireland Annual Abstract of Statistics)
	7.26	Department of Trade and Industry	Labour Market Trends (monthly, TSO)
	7.28	Training and Employment Agency (Northern Ireland)	

Subject	Table number in *Abstract*	Government department or other organisation	Official publication or other source
8. Personal income, expenditure and wealth			
	8.1	Office for National Statistics	Economic Trends, March 1997 (monthly, TSO)
	8.2	Board of Inland Revenue	Inland Revenue Statistics (annual, TSO) Economic Trends (monthly, TSO)
	8.3 - 8.5	Office for National Statistics	Family Expenditure Survey, (annual) (1990 onwards edition-Family Spending) (annual, TSO)
9. Health			
National health service	9.1	NHS Executive	Appropriation Accounts (annual) Health and Personal Social Services Statistics for England (annual)
		National Assembly for Wales	Health and Personal Social Services Statistics for Wales (annual)
	9.2	Scottish Health Service, Common Services Agency	
	9.3	Central Services Agency (Northern Ireland) Department of Health and Social Services (Northern Ireland)	Summary of Health and Personal Social Services Accounts (annual) Hospital Statistics (annual)
	9.4	NHS Executive Scottish Health Service, Common Services Agency National Assembly for Wales	Health and Personal Social Services Statistics for England (annual)
Public health	9.5	Communicable Disease Surveillance Centre	Communicable Disease Statistics Series MB2 (annual)
		Scottish Health Service, Common Services Agency	Scottish Health Statistics (annual)
		Department of Health and Social Services (Northern Ireland)	Annual report of the Registrar General Northern Ireland Quarterly return of births, deaths and marriages
	9.6 - 9.8	Health and Safety Executive Department of Social Security	Health and Safety Statistics
10. Social protection			
Social security pensions, benefits and allowances	10.1, 10.2, 10.4, 10.5	Department of Social Security Department of Health and Social Services (Northern Ireland)	National Insurance Fund Account (annual) Northern Ireland National Insurance Fund Account (annual)
	10.3 10.6 - 10.16	Department of Social Security Department of Health and Social Services (Northern Ireland)	Social Security Statistics (annual, TSO) Health and Personal Social Services Statistics for England (annual) Welsh Office: Health and Personal Social Services Statistics for Wales (annual)
Social services	10.17 - 10.21	Office for National Statistics Department for Education and Employment	Appropriation Accounts (annual) Northern Ireland Annual Abstract of Statistics
Housing	10.22	Office for National Statistics	
11. Crime and justice			
	11.1	Home Office	Monthly Digest of Statistics (TSO)
	11.2	Home Office	*England and Wales*: Report of Her Majesty's Chief Inspector of Constabulary (annual)
		Scottish Executive Justice Department	*Scotland*: Report of Her Majesty's Chief Inspector of Constabulary for Scotland (annual)

Subject	Table number in *Abstract*	Government department or other organisation	Official publication or other source
	11.2 continued	Royal Ulster Constabulary	*Northern Ireland*: Chief Constable's Report
	11.3 - 11.11	Home Office	Criminal Statistics, England and Wales (annual) Prison statistics, England and Wales (annual) Digest of Welsh Statistics (annual, Welsh Office) Offences relating to motor vehicles, England and Wales (annual bulletin)
	11.12 - 11.17	Home Office	HM Prison Service Annual Report and Accounts April 1998 - March 1999
		Scottish Executive Justice Department	Recorded Crime in Scotland, 1998
	11.18, 11.19	Scottish Executive Justice Department	Prison Statistics Scotland, 1998 (annual, TSO) Scottish Prison Service Annual Report 1998-99
	11.20 - 11.23	Northern Ireland Office	Chief Constable's Report A Commentary on Northern Ireland Crime Statistics
12. Lifestyles			
	12.1	Department for Culture, Media and Sport	DCMS Annual Report
	12.2	Office for National Statistics	Labour Market Trends (monthly, TSO)
	12.3	Office for National Statistics	GB Cinema exhibitor statistic First Release Monthly Digest of Statistics (TSO)
	12.4	Department for Culture, Media and Sport	British Film Institute Film and Television Handbook
	12.5	Department for Culture, Media and Sport	Digest of Tourist Statistics (annual) Sightseeing in the UK (annual)
		Office for National Statistics	Travel Trends (annual, TSO) Overseas Travel and Tourism First Release Monthly Digest of Statistics (TSO)
	12.6	Office for National Statistics	General Household Survey Report
	12.7	Gaming Board for Great Britain Office for National Statistics	Social Trends (annual, TSO)
Households	12.8	Office for National Statistics General Register Office (Scotland)	Census 1991 National Report Great Britain Part II 1991 Census. Household & Family Composition (10%) G.B.
13. Environment, water and housing			
Environment	13.1 - 13.17	DETR	Digest of Environmental Statistics (annual, TSO)
	13.8, 13.9	Scottish Executive	
Housing	13.18 - 13.21	Office for National Statistics DETR	Housing and Construction Statistics (quarterly) Housing return for Scotland (quarterly) Housing in England (annual, TSO)
		National Assembly for Wales	Welsh Housing Statistics (annual, National Assembly for Wales)
		Scottish Executive	Housing return for Scotland (quarterly)
		Department of the Environment for Northern Ireland	Northern Ireland Housing Statistics (annual)
	13.22	Office of Water Services (OFWAT)	Digest of environmental protection and water statistics (annual)

Subject	Table number in *Abstract*	Government department or other organisation	Official publication or other source
14. Transport and communications			
Road transport	14.1 - 14.7 14.10, 14.11 14.14	DETR	Transport Statistics Great Britain (annual, TSO) Vehicle Licensing Statistics (annual, TSO) Monthly Digest of Statistics (TSO)
	14.12, 14.13	DETR	Road accidents in Great Britain (annual, TSO) Monthly Digest of Statistics (TSO) Road accidents Wales (annual, National Assembly for Wales)
	14.8, 14.9	Department of the Environment for Northern Ireland	Transport statistics NI
Rail transport	14.15 - 14.18	DETR	Transport Statistics Great Britain (annual, TSO) Health and Safety Executive: Industry and Services (annual) Bulletin of Rail Statistics (quarterly)
	14.19, 14.20	Department of the Environment for Northern Ireland	Northern Ireland Annual Abstract of Statistics
Air transport	14.21 - 14.25	Civil Aviation Authority	Monthly Digest of Statistics (TSO) Civil Aviation Authority; Annual and Monthly Statistics Accidents to aircraft on the British Register (annual)
Sea transport	14.26, 14.27	DETR	Maritime Statistics (annual, TSO) Monthly Digest of Statistics (TSO)
Passenger movement	14.28	DETR Civil Aviation Authority	Monthly Digest of Statistics (TSO)
Communications	14.29	Royal Mail Parcel Force Subscription Services Ltd Post Office Counters Ltd	Monthly Digest of Statistics (TSO) Post Office report and accounts (annual)
15. National accounts			
	15.1 - 15.22	Office for National Statistics	United Kingdom National Accounts (annual, TSO) Monthly Digest of Statistics (TSO) Economic Trends (monthly, TSO) UK Economic Accounts (quarterly, TSO) Consumer Trends (quarterly, TSO)
16. Prices			
Producer prices	16.1, 16.2	Office for National Statistics	Producer Price Index Press Notice (monthly) Business Monitor MM22, Producer Price Indices (monthly, TSO) Monthly Digest of Statistics (TSO)
Consumer prices	16.3 - 16.5	Office for National Statistics	Monthly Digest of Statistics (TSO) Labour Market Trends (monthly, TSO) Business Monitor MM23, Consumer Price Indices (monthly, TSO)
	16.6, 16.7	Ministry of Agriculture, Fisheries and Food	Agricultural Statistics, United Kingdom (annual) MAFF Statistical Notice Agricultural Price Indices (Monthly) Monthly Digest of Statistics (TSO)
	16.8	Ministry of Agriculture, Fisheries and Food	Annual Review of Agriculture (annual)
17. Government finance			
Central government	17.1 - 17.3	Office for National Statistics	Financial Statistics (monthly, TSO)
	17.4 - 17.7	HM Treasury Office for National Statistics	Consolidated Fund and National Loans Fund Accounts Financial Statistics (monthly, TSO)
	17.8	Bank of England	

Subject	Table number in *Abstract*	Government department or other organisation	Official publication or other source
Saving	17.9	Department for National Savings	Accounts of National Savings Bank Investment Deposit Accounts (annual) Ordinary Deposit Accounts (annual)
Central government	17.10, 17.11	Board of Inland Revenue	Inland Revenue Statistics (annual, TSO)
Rateable values	17.12	Board of Inland Revenue	Rates and Rateable values in England and Wales (annual) (Up to 1989)
Local authorities	17.13, 17.14	Department of the Environment, Transport and the Regions	Local government financial statistics (England)(annual)
		National Assembly for Wales	Welsh local government financial statistics (annual)
		Public Works Loan Board	Annual report of the Public Works Loan Board
		Scottish Executive, Economic Advice and Statistics	Local financial returns (Scotland) (annual)
		Department of the Environment for Northern Ireland	
		Chartered Institute of Public Finance and Accountancy	
	17.15, 17.16	Department of the Environment, Transport and the Regions	Local government financial statistics (England) (annual)
		National Assembly for Wales	Welsh local government financial statistics (annual)
	17.17	Department of the Environment, Transport and the Regions	Local government financial statistics (England) (annual)
	17.18 - 17.20	Scottish Executive, Economic Advice and Statistics	Local financial returns (Scotland) (annual)
	17.21 - 17.23	Department of the Environment, Transport and the Regions Northern Ireland	District Council - Summary of Statement of Accounts (annual)
18. External trade and investment			
	18.1 - 18.8	Office for National Statistics	Business Monitor MM24, Monthly Review of External Trade Statistics (monthly, TSO) Overseas Trade Analysed in Terms of Industries MQ10 (quarterly, TSO) Monthly Digest of Statistics (monthly, TSO) HM Customs & Excise: Overseas Trade Statistics of the UK (monthly, quarterly and annual)
	18.9 - 18.18	Office for National Statistics Bank of England	United Kingdom Balance of Payments (annual, TSO) Quarterly figures: UK Economic Accounts Financial Statistics (monthly, TSO)
19. Research and development			
	19.1 - 19.5	Office for National Statistics	Business Monitor MA14, Research and Development in UK Business (annual, ONS)
20. Agriculture, fisheries and food			
Agriculture	20.1, 20.2	Agricultural Departments	Agriculture in the United Kingdom 1998 (annual)
	20.3 - 20.5	Agricultural Departments	Agricultural Statistics; United Kingdom (annual) Scottish Agricultural Economics (annual) Welsh Agricultural Statistics (annual, National Assembly for Wales)

Subject	Table number in *Abstract*	Government department or other organisation	Official publication or other source
	20.6	Forestry Commission	Great Britain: Annual Report and Accounts of the Forestry Commission
		Department of Agriculture (Northern Ireland)	Northern Ireland Annual Abstract of Statistics
	20.7, 20.8	Agricultural Departments	
Food	20.9 - 20.12	Ministry of Agriculture, Fisheries and Food	Monthly Digest of Statistics (TSO)
	20.13	Agriculture Departments	Agricultural Statistics, United Kingdom (annual) As for notes 20.3 - 20.5
Fisheries	20.14, 20.15	Ministry of Agriculture, Fisheries and Food; Scottish Executive Agricultural Departments	*England and Wales*: Sea fisheries statistical tables (annual) *Scotland*: Fisheries of Scotland report (annual) Scottish Sea fisheries statistics (annual, TSO)
Food consumption	20.16	Ministry of Agriculture, Fisheries and Food	National Food Survey
21. Production			
Production survey	21.1	Office for National Statistics	Manufacturing Summary Volume (Business Monitor PA 1002) (annual, ONS) PACSTAT (CD-ROM) (annual, ONS)
Manufacturers sales	21.2	Office for National Statistics	Product Sales and Trade PRA1 to 92 and PRQ1 to 33 (ONS)
	21.3	Office for National Statistics	Size analysis of United Kingdom Businesses (Business Monitor PA1003) (annual, ONS)
Energy	21.4 - 21.12	Department of Trade and Industry (Energy Policy, Technology Analysis and Coal Unit)	Digest of United Kingdom Energy Statistics (annual) Energy Trends (monthly) PACSTAT (CD-ROM) (annual, ONS)
Iron and steel	21.13 - 21.15	Iron and Steel Statistics Bureau	Iron and steel industry: annual statistics published by the Iron and Steel Statistics Bureau Corporation Regional Trends (annual, TSO)
Industrial materials	21.16	World Bureau of Metal Statistics Aluminium Federation	World Metal Statistics (monthly) PACSTAT (CD-ROM) (annual, ONS)
	21.17, 21.18	British man-made fibres Federation Textile Statistics Bureau Fertiliser Manufacturers' Association	Monthly Digest of Statistics (TSO) PACSTAT (CD-ROM) (Annual, ONS)
	21.19	Office for National Statistics	Minerals (Business Monitor PA 1007) (annual, ONS) Natural Environment Research Council: United Kingdom Minerals Yearbook
		Department of Economic Development (Northern Ireland)	Northern Ireland Annual Abstract of Statistics
Building and construction	21.20	Office for National Statistics DETR	Minerals (Business Monitor PA 1007) (annual, ONS) Monthly Digest of Statistics (TSO) Housing and Construction Statistics (quarterly and annual)
	21.21, 21.22	DETR	Housing and Construction Statistics (quarterly and annual)
Engineering	21.23, 21.24	Office for National Statistics	PACSTAT (CD-ROM) (annual, ONS)
Motor vehicle production	21.25	Office for National Statistics	Business Monitor PM 34.10, (monthly, ONS) PACSTAT (CD-ROM) (annual, ONS) Sector Review- Motor Trades (formerly Business Monitor SDA27) (annual, TSO)

Subject	Table number in *Abstract*	Government department or other organisation	Official publication or other source
Drink and tobacco	21.26, 21.27	HM Customs and Excise	Annual report of the Commissioners of HM Customs and Excise
		Office for National Statistics	Monthly Digest of Statistics (TSO)
22. Banking, insurance, etc			
Banking	22.1 22.3 - 22.5	Bank of England	Bank of England Annual Report and Accounts Bank of England Quarterly Bulletin
	22.2	Association of Payment Clearing Services	Annual report of the Bankers' Clearing House
	22.6	Bank of England	Financial Statistics (monthly, TSO)
	22.7	Bank of England	Bank of England Quarterly Bulletin
	22.8	Bank of England	Bank of England Quarterly Bulletin
	22.9 - 22.13	Bank of England	Monthly Digest of Statistics (TSO) Financial Statistics (monthly, TSO)
	22.14	Council of The Stock Exchange	Stock Exchange Fact Book (quarterly)
	22.15	Bank of England	Financial Statistics (monthly, TSO) Registry of Friendly Societies
Other financial institutions	22.16	Building Societies Commission	Report of the Chief Registrar incorporating the Report of the Industrial Assurance Commissioner (annual)
	22.17	Office for National Statistics	Business Monitor SDQ7, Assets and Liabilities of Finance Houses and Other Credit Companies (quarterly, ONS)
	22.18	Office for National Statistics	Financial Statistics (monthly, TSO) Monthly Digest of Statistics (TSO) Business Monitor MQ5, Insurance Companies; Pension Funds and Trusts Investments (quarterly, ONS) First Release
	22.19, 22.20	Office for National Statistics	Financial Statistics (monthly, TSO) Business Monitor MQ5, Insurance Companies; Pension Funds and Trusts Investments (quarterly, ONS)
	22.21 - 22.22	Registry of Friendly Societies	Annual Report of the Registry of Friendly Societies
		Friendly Society Commission	Annual Report of the Friendly Societies Commission
Insolvency	22.23 - 22.26	Department of Trade and Industry	Insolvency Annual Report (DTI) Companies (DTI) Financial Statistics (monthly, TSO)
23. Service industry			
Retail trades	23.1	Office for National Statistics	Sector Review - Retailing (formerly Business Monitor SDA25) (TSO)
	23.2	Office for National Statistics	Business Monitor SDM 28, Retail Sales (monthly, ONS) Monthly Digest of Statistics (TSO)
Motor trades	23.3	Office for National Statistics	Sector Review- Motor Trades (formerly Business Monitor SDA 27) (TSO)
Catering	23.4	Office for National Statistics	Sector Review- Catering and Allied Trades (formerly Business Monitor SDA 28 (TSO)

INDEX

Figures indicate table numbers

INDEX

Figures indicate table numbers

C

Figures indicate table numbers

INDEX

Figures indicate table numbers

Figures indicate table numbers

INDEX

Figures indicate table numbers

Figures indicate table numbers

O

P

Q

INDEX

Figures indicate table numbers

Figures indicate table numbers

INDEX

Figures indicate table numbers

W

Y

Z

UNITS OF MEASUREMENT

Length

1 millimetre (mm)		= 0.039 370 1 inch
1 centimetre (cm)	= 10 millimetres	= 0.393 701 inch
1 metre (m)	= 1 000 millimetres	= 1.093 61 yards
1 kilometre (km)	= 1 000 metres	= 0.621 371 mile
1 inch (in.)		= 25.4 millimetres or 2.54 centimetres
1 foot (ft.)	= 12 inches	= 0.304 8 metre
1 yard (yd.)	= 3 feet	= 0.914 4 metre
1 mile	= 1 760 yards	= 1.609 34 kilometres

Area

1 square millimetre (mm^2)		= 0.001 55 square inch
1 square metre (m^2)	= one million square millimetres	= 1.195 99 square yards
1 hectare (ha)	= 10 000 square metres	= 2.471 05 acres
1 square kilometre (km^2)	= one million square metres	= 247.105 acres
1 square inch (sq. in.)		= 645.16 square millimetres or 6.451 6 square centimetres
1 square foot (sq. ft.)	= 144 square inches	= 0.092 903 square metre or 929.03 square centimetres
1 square yard (sq. yd.)	= 9 square feet	= 0.836 127 square metres
1 acre	= 4 840 square yards	= 4 046.86 square metres or 0.404 686 hectare
1 square mile (sq. mile)	= 640 acres	= 2.589 99 square kilometres or 258.999 hectares

Volume

1 cubic centimetre (cm^3)		= 0.061 023 7 cubic inch
1 cubic decimetre (dm^3)	= 1 000 cubic centimetres	= 0.035 314 7 cubic foot
1 cubic metre (m^3)	= one million cubic centimetres	= 1.307 95 cubic yards
1 cubic foot (cu. ft.)		= 0.028 316 8 cubic metre or 28.316 8 cubic decimetres
1 cubic yard (cu. yd.)	= 27 cubic feet	= 0.764 555 cubic metre

Capacity

1 litre (l)	= 1 cubic decimetre	= 0.220 gallon
1 hectolitre (hl)	= 100 litres	= 22.0 gallons
1 pint		= 0.568 litre
2 pints	= 1 quart	= 1.137 litres
8 pints	= 1 gallon	= 4.546 09 cubic decimetres or 4.546 litres
36 gallons (gal.)	= 1 bulk barrel	= 1.636 56 hectolitres

Weight

1 gram (g)		= 0.035 274 0 ounce
1 hectogram (hg)	= 100 grams	= 3.527 4 ounces or 0.220 462 pound
1 kilogram (kg)	= 1 000 grams or 10 hectograms	= 2.204 62 pounds
1 tonne (t)	= 1 000 kilograms	= 1.102 31 short tons or 0.984 2 long ton
1 ounce avoirdupois (oz.)		= 28.349 5 grams
1 pound avoirdupois (lb.)	= 16 ounces	= 0.453 592 37 kilogram
1 hundredweight (cwt.)	= 112 pounds	= 50.802 3 kilograms
1 short ton	= 2 000 pounds	= 907.184 74 kilograms or 0.907 184 74 tonne
1 long ton (referred to as ton)	= 2 240 pounds	= 1 016.05 kilograms or 1.016 05 tonnes
1 ounce troy	= 480 grains	= 31.103 5 grams

Energy	British thermal unit (Btu)	= 0.252 kilocalorie (kcal) = 1.05 506 kilojoule (kj)
	Therm	= 100 000 British thermal units = 25 200 kcal = 105 506 kj
	Megawatt (Mw)	= 10^6 watts
	Gigawatt hour (GWh)	= 10^6 kilowatt hours = 34 121 therms

Food and drink	Butter	23 310 litres milk	= 1 tonne butter (average)
	Cheese	10 070 litres milk	= 1 tonne cheese
	Condensed milk	2 550 litres milk	= 1 tonne full cream condensed milk
		2 953 litres skimmed milk	= 1 tonne skimmed condensed milk
	Milk	1 million litres	= 1 030 tonnes
	Milk powder	8 054 litres milk	= 1 tonne full cream milk powder
		10 740 litres skimmed milk	= 1 tonne skimmed milk powder
	Eggs	17 126 eggs	= 1 tonne (approximate)
	Sugar	100 tonnes raw sugar	= 95 tonnes refined sugar
	Beer	1 bulk barrel	= 36 gallons irrespective of gravity

Shipping	Gross tonnage	= The total volume of all the enclosed spaces of a vessel, the unit of measurement being a ton of 100 cubic feet.
	Deadweight tonnage	= Deadweight tonnage is the total weight in tons of 2 240 lb. that a ship can legally carry, that is the total weight of cargo, bunkers, stores and crew.

Printed in the United Kingdom for The Stationery Office
J101278 C45 01/00 473620 19585